Ecological Genomics of Algae

Senjie Lin
Department of Marine Sciences
University of Connecticut
Groton, CT 06340, USA

CRC Press
Taylor & Francis Group
Boca Raton London New York

CRC Press is an imprint of the
Taylor & Francis Group, an **informa** business
A SCIENCE PUBLISHERS BOOK

Cover credit:
The ocean view photo is from an open source.
Other images are from the author's own laboratory.

First edition published 2026
by CRC Press
2385 NW Executive Center Drive, Suite 320, Boca Raton FL 33431

and by CRC Press
4 Park Square, Milton Park, Abingdon, Oxon, OX14 4RN

CRC Press is an imprint of Taylor & Francis Group, LLC

Library of Congress Cataloging-in-Publication Data (applied for)

ISBN: 978-1-032-38225-8 (hbk)
ISBN: 978-1-032-38227-2 (pbk)
ISBN: 978-1-003-34405-6 (ebk)

DOI: 10.1201/9781003344056

Typeset in Times New Roman
by Prime Publishing Services

Preface

The intricate world of algae, with its astounding diversity and ecological significance, has captivated the scientific community for decades. The advent of genomic technologies has ushered in a new era of exploration, allowing us to delve into the molecular underpinnings of algal biology with unprecedented depth and precision. This book, "The Ecological Genomics of Algae", aims to serve as a comprehensive resource for graduate students, postdoctoral researchers, professors, and professional researchers who are dedicated to studying algae from a genomic perspective.

The journey through the 21 chapters of this book begins with foundational concepts of ecological genomics and the array of techniques, aimed at helping beginners. It then traverses the evolutionary landscape of algae and the process of photosynthesis, setting the stage for an in-depth exploration of genome evolution and adaptation. An examination of the genomes of various algal groups ensues, covering from cyanobacteria, green and red eukaryotes, to various groups of chromalveolates. Additionally, the genomes of seaweeds are treated in a separate chapter to offer insights into the complexity of multicellularity.

Understanding how algae perceive and respond to their environment is crucial. Therefore, several chapters are dedicated to the sensing of environmental stimuli and signal transduction, stress response and adaptation, and the genomic bases of responses to temperature changes. Furthermore, responses to carbon dioxide variations and ocean acidification are explored, shedding light on how algae navigate the challenges posed by climate change.

The chapters on energy and nutrient acquisition provide a molecular perspective on how algae sustain their growth and metabolism in diverse environments. The genomic basis of defense mechanisms against grazing and microbial insults highlights the survival strategies that enable algae to thrive in competitive ecosystems. In addition, an investigation of the genomic bases of sexual reproduction and the cell cycle reveals their intricacies. The last two chapters address the fascinating realms of algal symbioses and algae-microbes interactions, exposing the multidimensional interconnectedness of algae and other life forms in aquatic ecosystems.

Numerous individuals generously assisted throughout this work. Special thanks are due to the editors of CRC Taylor and Francis, for their patience and support throughout this endeavor. Dr. Minglei Ma, Ms. Xiaoyu Wang, and Mr. Shuaishuai Wu helped with phylogenetic tree construction and functional domain searches of various genes. Dr. Hanhua Hu provided summaries on diatom C_4 pathway and CO_2 adaptation. Mss. Yifan Gu, Yulin Huang, Linli Jia, Shutao Wu, Dr. Tangcheng Li, and Mr. Yujie Wang helped prepare figures and tables. Ms. Jiashun Li and Dr. Kaidian Zhang gathered information and prepared the CRISPR/Cas workflow. Jingtian Wang prepared the genomics workflow figure. Richard Antosca assisted with literature search and permission requests for reusing published illustrations. I am indebted to many researchers and alga enthusiasts or artists who shared images of various algae, particularly to Dr. Brian E. Lapointe for *Sargassum* bloom photographs, Dr. Hao Luo for *Noctiluca* bioluminescence scene, and Dr. Sallie Chisholm for microscopic images of *Prorochlorococcus*. Yifan Gu's artistic flair beautifully enhanced the cover design.

Special gratitude is extended to my Ph.D. advisor, Professor Edward J. Carpenter, for his unwavering support and encouragement of my pursuit of molecular work on phytoplankton from my early days as a novice. I also owe my career in plankton ecology to Dr. Darcy Lonsdale from Stony Brook University, who initially took me on as a graduate student to study the molecular basis of thermal adaptation in copepods. Although that plan was aborted, her confidence in me early on encouraged me till today.

I thank my wife Ingrid and children (Daniel, Nina, and Matthew) for their love, which sustains me through my career and this adventure.

I dedicate this book to my late parents for their selfless love and care.

It has been a special privilege and blessing to have the opportunity and strength to explore the intricacies of algal life. The insights presented in this book, even just a scratch on the surface, are breathtaking and awe-inspiring. My ultimate gratitude goes to the Most High!

Senjie Lin
June 4, 2024

Contents

Preface iii

1. Concepts and Techniques 1
2. Evolution of Algae and Photosynthesis 27
3. Genome Evolution, Plasticity, and Adaptation 46
4. Genomes of Cyanobacteria 60
5. Genomes of Green Algae 74
6. Genomes of Rhodophytes 89
7. Genomes of Diatoms 98
8. Genomes of Dinoflagellates 116
9. Genomes of Haptophytes 150
10. Genomes of Cryptophytes and Other Algal Phyla 162
11. Genomes of Seaweeds and Multicellularity 172
12. Sensing of Environmental Stimuli and Signal Transduction 188
13. Stress Response and Adaptation 216
14. Temperature and Global Warming 242
15. CO_2 and Ocean Acidification 261
16. Energy Acquisition—Photosynthesis and Rhodopsin Proton Pump 272
17. Nutrient Acquisition and Regulation 282
18. Defense Against Grazing and Microbial Insults 312
19. Life Cycle, Sex, and the Cell Cycle 329
20. Genomics of Algal Symbioses 355
21. Genomics of Algae-Microbes Interactions 377

References 385

Index 455

1

Concepts and Techniques

Ecological Genomics

Ecological genomics is a scientific discipline that studies the structure and function of genomes to understand the relationship between organisms and their biotic and abiotic environments (van Straalen and Roelofs 2012). In essence, ecological genomics aims to uncover the genetic mechanisms underlying organisms' responses to their natural environments (Ungerer et al. 2007). This is achieved by applying functional genomic approaches to identify and characterize genes of ecological and evolutionary relevance.

Functional genomics seeks to describe the functions and interactions of genes and proteins using genome-wide approaches, contrasting with the gene-by-gene approach of classical molecular biology. It integrates data from various processes related to DNA sequence, gene expression, and protein function, such as coding and noncoding transcription, protein translation, and interactions between proteins and nucleic acids. These data are used to model interactive and dynamic networks that regulate gene expression, cell differentiation, and cell cycle progression (Bunnik and Le Roch 2013).

Practically, unraveling the genes and pathways underlying a physiological trait or ecological condition (e.g., stress response) can lead to the identification of indicator genes that diagnose the expression of the trait or the status of the condition. This inverse approach in ecological genomics has significant potential for ecosystem monitoring, conservation, and restoration.

Core Questions of Ecology

Ecologists generally address two major questions: (1) Who is present in the environment, and how abundant are they? (2) What are they doing there? For the first question, ecologists analyze a community to document the composition and relative abundances of species. This can be done using morphological examination or, more desirably due to greater speed and higher accuracy, DNA barcoding (this often involves metabarcoding or environmental DNA), or a combination of both methods. Accurate species resolution and identification is essential but often challenging. Microalgal species are difficult to distinguish due to their small size and lack of distinct features, and macroalgae present challenges due to their morphological plasticity, as a result of environmental adaptation. Molecular markers (Table 1.1) such as SSU, LSU, or ITS of rRNA genes have been extremely helpful. Additionally, as elaborated in the next chapter, phylogenomics can provide more robust resolution of different lineages of algae. Evolutionary ecologists or biologists may also ask how organisms came into being, an extended "Who" question that requires molecular phylogenetics or phylogenomics to address.

Table 1.1. Genes that have been explored as DNA barcode in major groups of algae*.

	ITS	28S	18S	rbcL	mtCOI	mtCOB	tufA	UPA	Cp16S
Algae									
Bacillariophyta	√	√		√	√	√			
Dinophyta	√	√			√	√			
Chlorophyta	√			√	√		√		
Rhodophyta		√		√	√			√	
Phaeophyta	√			√	√		√		
Chrysophyta	√		√	√	√	√			√
Raphidophyta	√			√	√				
Haptophyta	√								√
Cryptophyta		√		√	√				√

* ITS, 18S, 28S: nuclear rRNA cistron; rbcL: Rubisco; mtCOI, mtCOB: mitochondrial cytochrome C subunit 1 and cytochrome B; tufA: chloroplast elongation factor Tu; UPA: universal plastid amplicon (23S); CP16S: plastid 16S.

For the second question, functional genomic analysis is performed to uncover what genes an organism possesses, what functions these genes serve, how their expression is regulated by environmental conditions, and how they modulate metabolic processes and physiological performance as a result. Functional genomics, in a broad sense, includes transcriptomics and proteomics. From an ecological perspective, these multi-omics are integral, and inevitably, transcriptomics and proteomics will come into discussion in this book, even though the focus is on genomics.

Why Genomics?

A phenotype of an organism is determined by its genotype (including epigenetic elements) and environmental factors. Physiology of a cell is determined by the cell's genome and its interaction with the environment, and the interaction is often mediated by epigenetic mechanisms as well as genetically wired signaling pathway (Fig. 1.1). Therefore, to understand the genetic underpinning of a particular phenotype, one must identify and understand the function of associated genes, including those that sense the environmental cues and transducing signals to the nucleus. For instance, the phenotype of using a specific type of nutrient is determined by the genes responsible for the uptake and assimilation of that nutrient. The expression of these genes is triggered by the availability of the nutrient and regulated by epigenetic and/or genetic mechanisms (Fig. 1.2).

By the same token, to gain a comprehensive understanding of an organism's total physiology, it is crucial to explore the underlying regulatory machinery, including the genes it possesses and how they are regulated. By studying the structure and expression profile of an organism's genome, valuable insights into its physiology can be inferred. Furthermore, analyzing genome structure and function often leads to the discovery of previously unknown physiologies or traits. For instance, the identification of proton pump rhodopsin, a photosystem-independent photoenergy harvesting mechanism, exemplifies the power of the genomic approach. This approach proves particularly valuable in studying microalgal populations within natural microbial assemblages, where direct physiological measurements for individual species are impractical. Additionally, the genomic approach enables the analysis of ancient samples, such as those embedded deep in sea sediment, to reconstruct taxon identity and gain insights into paleoclimate and paleoenvironmental conditions.

Common Assumptions in Ecological Genomics

The assumptions commonly used in genomics play a crucial role in understanding genetic functions and physiological traits. Firstly, the assumption that a gene with significant similarity to a functionally

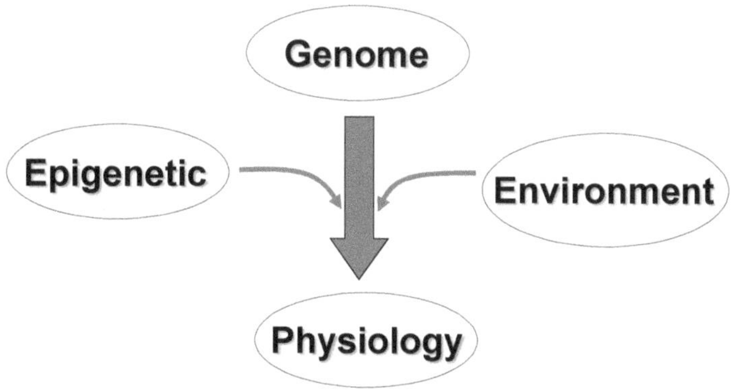

Figure 1.1. The physiology of an organism is determined by the organism's genome and its interaction with the environment. The interaction is often mediated through epigenetic mechanisms as well as genetic signaling pathways.

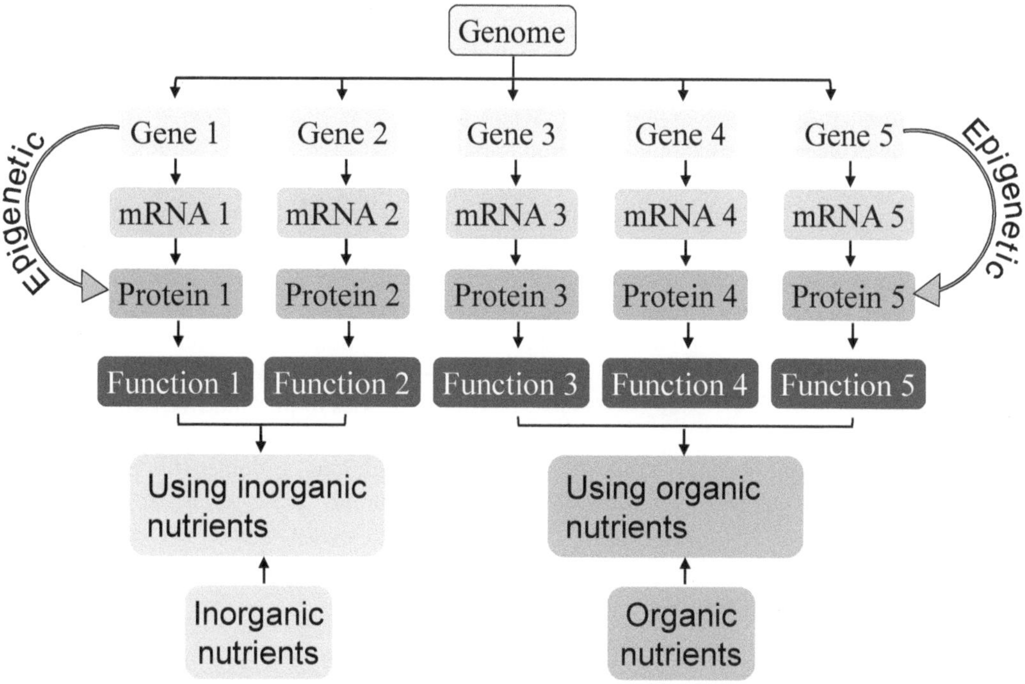

Figure 1.2. Schematic of the classical gene-to-mRNA-to-protein dogma and the more recently recognized importance of epigenetic regulation. The phenotype of using a type of nutrient is determined by the genes responsible for the uptake and assimilation of that nutrient, the expression of which is triggered by the availability of the nutrient and regulated by an epigenetic mechanism and/or a genetic mechanism

characterized gene shares the same function forms the basis of functional annotation using the Basic Local Alignment Search Tool (BLAST) analysis. In BLAST analyses, the threshold similarity is minimally set at the E value of -10^{-3}. However, more stringent values, such as 10^{-50} for reasonably conserved genes, are often required to confidently identify genes with similar functions.

Secondly, the assumption that increased gene expression corresponds to elevated physiology suggests that highly demanded functions in an organism are achieved through high expression levels of relevant genes. This can be evolutionarily fixed through gene duplication or gene family expansion, strengthening the encoded function. For example, the presence of multiple copies of the ribulose-1,5 bisphosphate carboxylase/oxygenase (Rubisco) gene in dinoflagellates, driven by a high demand for this enzyme (Zhang and Lin 2003), exemplifies this assumption. This is known as *superfunctionalization.* However, some of the gene copies (called *paralogs*) may have been diverged from the parental gene to the extent that it has lost its function (*nonfunctionalization*) or assumed a new function (*neofunctionalization*).

Superfunctionalization is often assumed in the ecological context. The premise is that when a gene exists in multiple copies in a species, its transcript level will be higher than if it exists in single copy, and its protein level will be higher as well. It is important to note, however, that the activity of a protein or enzyme is not only determined by the primary sequence, but also by the secondary (alpha-helix, beta-sheet, and loop), tertiary (interactions of different regions within a peptide), and quaternary (interactions of multiple peptides) structures, which can be influenced by post-translational modifications.

Genomics in Brief

Genome Structure

A genome encompasses all the DNA within a cell or organelle. Prokaryotes, such as cyanobacteria in the realm of algae, lack organelles and are thus considered to have only one genome. Approximately 90% of bacteria possess a single circular chromosome, while about 10% of species have more than one chromosome (typically fewer than three), which can be either linear or circular. All sequenced cyanobacteria to date contain a single circular chromosome.

However, bacteria often harbor plasmids—double-stranded circular DNA molecules that are physically separate from chromosomes (hence termed extrachromosomal DNA) and replicate independently. Plasmids frequently carry genes that provide a selective advantage, such as antibiotic resistance. Some cyanobacteria, including *Synechocystis* and *Synechococcus*, have been reported to contain plasmid DNA (Shintani et al. 2015). Plasmids are also found in archaea and eukaryotes. Eukaryotic plasmids are widespread across algae, land plants, fungi, yeast, and other eukaryotes (Lee et al. 2016). Plasmids have been identified in red algae since the 1990s (Goff and Coleman 1990, La Claire II et al. 1997, 2000), with more recent discoveries of plasmid DNA integrated into host genomes (Lee et al. 2016).

Besides the nuclear genome, eukaryotic cells contain additional genomes: the mitochondrial genome and, in photosynthetic organisms, the chloroplast genome. These organellar genomes are, to varying degrees, under the control of the nuclear genome. Unless specified, the term "genome" typically refers to the nuclear genome. Most eukaryotic cells have a single nucleus, but multinucleate species exist in some lineages. For instance, coenocytic (also known as siphonous) algae, which are single-celled but with pseudo-branches, contain numerous nuclei (for more information, see Chapter 5).

Genomic DNA consists of both coding and non-coding regions. The coding regions encompass genes that encode proteins or functional RNA, while non-coding DNA includes intergenic sequences, repetitive elements, and transposons. Some non-coding DNA is transcribed and plays a role in gene regulation. In eukaryotes, genes account for only a small fraction of the genome, and this fraction decreases steeply as genome size increases (ranging from 81.6 to 1.2%). This is in stark contrast to prokaryotes and organelles, where the majority of DNA is genic (47–97%). Generally, the larger the genome, the smaller the fraction of coding sequence (Hou et al. 2009). Consequently, genome size may be determined by selection pressure for optimal fitness that constrains a scaling ratio of cell size to genome size (Gregory 2001, Lin 2006).

DNA possesses a complex three-dimensional structure, with dynamics and functions that extend far beyond what its primary nucleotide sequence suggests. The advent of chromosome conformation capture (Hi-C) technology has begun to unveil the intricate interactions within and between chromosomes. In a DNA-DNA interaction plot, where both the row and column represent the same chromosome (Fig. 1.3), the strongest interaction—defined by the technique as a region's interaction with itself—appears on the diagonal line. Signals near this diagonal indicate interactions

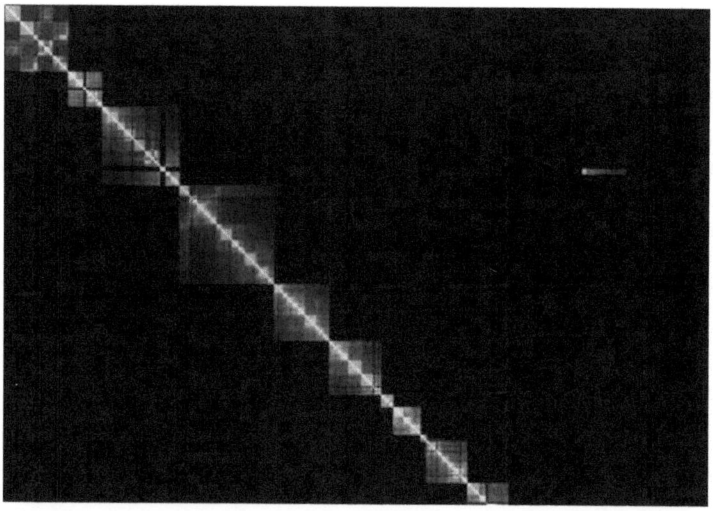

Figure 1.3. Three-dimensional DNA interactions detected by Hi-C based assembly of the longest scaffold (> 120 Mbp) of the dinoflagellate *Fagacium kawagutii* genome. The square scatter of contact points around the diagonal line indicates numerous interacting points along the chromosome, whereas the two strong signals far below and above the diagonal line indicate large loops formed with downstream and upstream regions. Based on data from Li et al. 2020.

with upstream (above the diagonal) or downstream (below the diagonal) regions of the chromosome. Any breaks in the diagonal line suggest gaps in the chromosome assembly. When the row and column represent different chromosomes, the signals in the plot reveal inter-chromosomal interactions.

This Hi-C technology provides a powerful tool for visualizing and analyzing the spatial organization of chromosomes, offering insights into genome structure and function that were previously inaccessible through traditional sequencing methods.

Core Genome and Flexible Genome

The core genome refers to the constitutively expressed portion of an organism's genetic material. This set of genes maintains relatively stable expression levels, either at the transcriptomic or proteomic level, across varied environmental conditions. By definition, housekeeping genes are integral components of the core genome. In contrast, the flexible genome (or proteome), sometimes termed the accessory genome (or proteome), responds significantly to environmental changes. This portion of the genome exhibits variable expression levels at the transcriptomic or proteomic level. It represents genes or their encoded proteins involved in physiological responses to changing environmental conditions.

When comparing different species or populations that have adapted to their respective environments (e.g., different latitudes), the flexible genome may contain unique elements specific to each species or population. This variability in the flexible genome reflects the diverse adaptive strategies employed by organisms in response to their particular environmental challenges. Understanding the interplay between the core and flexible genomes provides valuable insights into an organism's ability to maintain essential functions while adapting to environmental fluctuations.

Genes and Gene Expression Regulation

Genes, Introns, Splicing, and Alternative Splicing

In prokaryotes, genes consist of contiguous DNA sequences. In contrast, eukaryotic genes are often interrupted by non-coding sequences called introns, with the dispersed coding sequences termed exons. When genes containing introns are transcribed into RNA (pre-mRNA), the introns are removed and the exons are joined together through a process called splicing. This generates mature mRNA, which also acquires a 5' G cap and a 3' poly-A tail.

Typically, one gene produces one mRNA. However, alternative splicing can generate multiple different mRNAs encoding distinct proteins from a single gene by skipping certain exons or splicing at different positions. Additionally, alternative transcription can produce multiple transcripts by initiating or terminating transcription at various sites.

While a gene sequence usually corresponds to a single protein sequence, *trans*-splicing can generate hybrid proteins from two partial transcripts. *Trans*-splicing joins parts of different RNA molecules. This occurs in various organisms, including algae (such as euglenophytes and dinoflagellates), nematodes, *Drosophila*, and humans. In dinoflagellates and euglenophytes, however, *trans*-splicing primarily affects the untranslated regions (UTRs), potentially influencing gene expression regulation without altering the protein's amino acid sequence.

These diverse mechanisms of gene expression and RNA processing highlight the complexity of genetic information flow and its regulation in different organisms, particularly in eukaryotes.

Polycistronic, Monocistronic, and Polyprotein Genes

A polycistronic gene is an array of multiple protein-coding sequences (open reading frames) under the control of a single regulatory system (promoter, terminator), which is transcribed as one transcript. That multi-gene transcript is translated into individual proteins. In contrast, monocistronic genes are individual genes with their own promoter and terminator, which are transcribed and translated separately. In prokaryotes and organelles, most genes are polycistronic, with genes in the same pathway organized in a cluster (operon; e.g., the *nif* gene operon). Polycistronic genes are uncommon in eukaryotes but do exist in some lineages such as Kinetoplastida and *Caenorhabditis elegans* (Blumenthal 2004, Michaeli 2011). In algae, polycistronic genes have been demonstrated in various chlorophyte lineages (Gallaher et al. 2021).

Strictly speaking, the operon-like genes in Kinetoplastida is not polycistronic but intermediate between mono- and poly-cistronic. This is because their primary transcripts that contain a tandem repeat array of genes are *trans*-spliced into individual mature mRNAs, each of which is translated into a protein. Therefore, their gene structure and transcription resemble polycistronic genes, but their RNA processing and translation are in line with monocistronic genes. This intermediate type may occur in dinoflagellates as well, because they have many tandemly repeated genes and operate *trans*-splicing (Zhang et al. 2007, Lin 2011).

Another type of gene organization and expression involve polyprotein genes. The structure and transcription of polyprotein genes also resemble those of polycistronic genes. However, their polyprotein transcripts are translated into a long multi-protein polypeptide, which is then cleaved proteolytically to produce mature individual proteins. Polyprotein expression is less common in eukaryotes than viruses, but there are notable examples of eukaryotic organisms that utilize polyproteins, such as the TY1 retrotransposon in yeast, certain plant storage proteins such as those in seeds, and prohormones in mammals. Among algae, dinoflagellate form II Rubisco genes are an outstanding example of polyprotein genes. After translation, the polyprotein is presumably guided to the chloroplast by the signal peptide at the N-terminus, where it is first cleaved into individual proteins with a transit peptide, which is cleaved off after the proteins have entered the chloroplast (Zhang and Lin 2003, Lin 2011).

mRNA Editing

RNA editing refers to the process in which single or multiple nucleotides are altered in RNA during or after transcription, leading to differences in the nucleotide sequence between a gene and its corresponding mRNA. This phenomenon is observed not only in mRNA but also in tRNA and rRNA, with a higher occurrence in mitochondrial and plastid genes, as well as nuclear genes, depending on the organism (Gray 2012).

From a phylogenetic perspective, RNA editing is sporadic and occurs independently in unrelated lineages. Among algal phyla, dinoflagellates have been extensively studied in terms of RNA editing. It was initially discovered in mitochondrial mRNA and subsequently observed in chloroplast mRNA and rRNA in dinoflagellates, involving single nucleotide changes (Lin et al. 2002, Dang and Green 2009, Mungpakdee et al. 2014). The extent and types of RNA editing appear to increase as dinoflagellates evolve (Lin et al. 2008). It is worth noting that RNA editing occurs not only in typical dinoflagellate plastids with peridinin as the major accessory pigment, but also tertiary replacement plastids, which contain fucoxanthin as the major accessory pigment, or taxa with chlorophyll *b*-containing plastids arising from serial secondary endosymbiosis (Klinger et al. 2018, Matsuo et al. 2022). These observations indicate the adaptation of acquired organelles to the cellular environment of the host.

Apart from single-nucleotide editing, RNA editing can also involve the replacement of long-tract nucleotides, as found in Kinetoplastida and potentially in euglenophyte algae. More recent studies utilizing high-throughput sequencing have reported mRNA editing of nuclear genes, although these findings still require experimental verification. The functional significance of mRNA editing is still not fully understood. In dinoflagellate mitochondria, mRNA editing appears to increase the GC content of affected genes, potentially improving translation efficiency since dinoflagellate mitochondria likely import cytoplasmic tRNA, while mitochondrial genes are generally more AT-rich than nuclear genes (Lin et al. 2002). It can be deduced, however, that mRNA editing is inherently mutagenic, allowing mutated genes to persist, which would otherwise be eliminated through evolution (Gray 2012).

Multi-Tiered Gene Expression Regulation

Phenotypic differences can arise from variations in gene structure or gene regulation. Interestingly, evolutionary developmental biology places a strong emphasis on regulatory changes as the primary mechanism driving evolutionary changes in body plans. In contrast, physiologists tend to emphasize structural changes in proteins as the main mechanism of adaptation (van Straalen and Roelofs 2012).

Gene expression involves several steps. Initially, the coding DNA is transcribed into RNA through the action of DNA-dependent RNA polymerase (Pol). Messenger RNA (mRNA), which codes for proteins, is synthesized by Pol II. Additionally, Pol II transcribes microRNA (miRNA) and small nuclear RNA (snRNA). Ribosomal RNA subunits (excluding 5S rRNA), crucial components of ribosomes, are transcribed by Pol I. Pol III is responsible for synthesizing transfer RNA (tRNA) and 5S rRNA. Following transcription, mRNA is recruited by ribosomes, which assemble amino acids into peptides based on the triplet codons carried in the mRNA.

Cells need to swiftly respond to changes in the external environment, such as temperature or nutrient availability, in order to adapt and survive in new conditions. This rapid response is initiated by the activation of gene transcription, known as inducible gene expression. While many genes are induced by environmental stimuli, there are also genes that are constantly expressed regardless of the environmental conditions. These constitutively expressed genes, also referred to as housekeeping genes, play fundamental roles in cellular functions.

Transcriptional Regulation. When changes in mRNA abundance directly determine the changes in the abundance of the encoded protein, this gene is regulated transcriptionally. This is the most common mode of gene regulation. Rapid activation of gene expression in response to environmental stimuli primarily occurs through the regulation of RNA Pol II-dependent transcription. This process involves the regulated recruitment of the transcription machinery to the promoter of the gene being expressed, chromatin remodeling, and the release of paused polymerase. In addition, components of signal transduction cascades also play an integral part. For in-depth review of steps in transcriptional regulation, the readers are referred to Weake and Workman (2010). In organisms where transcriptional regulation is the major mode of regulation, transcriptomic profiling can characterize responses to environmental changes.

Post-Transcriptional Regulation. mRNA abundance not only depends on transcriptional activity (transcriptionally regulated), but also on the stability and half-life of mRNA. One of the mechanisms regulating mRNA stability is the action of miRNA, which will be elaborated later.

Translational Regulation. When mRNA abundance of a gene does not change in response to environmental changes, but protein abundance does, this gene is regulated translationally. Translation efficiency can be modulated by miRNA or mRNA methylation, depending on the species and type of genes.

Post-Translational Regulation. When neither transcript nor protein abundance changes in response to environmental changes, but the activity of the encoded protein (often an enzyme) does, the regulation lies at the post-translational level. There are many types of Post-Translational protein Modifications (PTMs), mainly including protein phosphorylation (including polyphosphorylation), methylation, glycosylation, acetylation, ubiquitination, and SUMoylation.

Phosphorylation and dephosphorylation are common reversible PTMs, which primarily occur on serine and threonine residues. In eukaryotes, cell cycle proteins (e.g., Cyclin-Dependent protein Kinases [CDKs]) and signaling pathways (e.g., Mitogen Activated Protein Kinase [MAPK] system) are commonly regulated by phosphorylation and dephosphorylation, mediated by protein kinases and phosphatases. For example, phosphorylation of cyclin-dependent kinase has been detected in the dinoflagellate *Karenia brevis*. Phosphorylation is implicated in circadian clock control in dinoflagellates (Jadhav et al. 2022). Phosphorylation of MAPK-like proteins has been reported in intertidal macroalgae (*Ulva*, *Ellisolandia*, and *Cystoseira*) under stress conditions (Parages et al. 2014). In the chlorophyte model *Chlamydomonas reinhardtii*, environmentally induced photosynthetic state transitions involve phosphorylation of photosystem II (PSII) core and its light-harvesting antennae (LHCII) proteins (Turkina et al. 2006).

Protein methylation, the addition of methyl groups to proteins, occurs on arginine and lysine residues and at the N- and C-termini of various proteins. Protein methylation regulates diverse cellular functions, including signaling, protein trafficking, gene expression, DNA damage repair, and protein-protein interactions. It plays a key role in cellular signal transduction pathways, acting both proximal to receptor-mediated responses and as an epigenetic regulator (Cheng et al. 2012). In *C. reinhardtii*, seven protein arginine N-methyltransferases (PRMTs) have been identified, and arginine methylation regulates the activity of the cytosolic translation repressor protein NAB1 (Blifernez et al. 2011). Arginine methylation is also important in *Chlamydomonas* flagella for flagellar growth resorption (Schneider et al. 2008, Mizuno and Sloboda 2017).

Glycosylation is a ubiquitous mode of PTM with two major types: N-glycosylation and O-glycosylation. O-glycosylation occurs on serine or threonine residues of proteins, while N-glycosylation involves the transfer of an oligosaccharide to asparagine residues in the sequence Asn-X-Ser/Thr within nascent polypeptides (Schoberer et al. 2018). Research on protein glycosylation has focused more on N-glycosylation (Mathieu-Rivet et al. 2020).

Protein acetylation involves the transfer of an acetyl group to lysine residues, catalyzed by lysine acetyltransferases (KATs). This modification plays a critical role in regulating gene expression, DNA repair, silencing, and cell cycle progression (Galdieri et al. 2014). In *C. reinhardtii*, lysine acetylome dynamics were observed under different trophic modes, with high occupancy levels on key proteins under heterotrophic conditions (Füßl et al. 2021).

Ubiquitination involves the attachment of ubiquitin to lysine residues on substrate proteins, resulting in protein monoubiquitination or polyubiquitination. Ubiquitination plays roles in protein degradation, subcellular localization, and kinase activation. Monoubiquitination signals nonproteolytic events such as endocytosis and DNA repair (Swatek and Komander 2016). Ubiquitin can also be acetylated or phosphorylated, altering the signaling outcome. Ubiquitin is expressed as polyubiquitin or fused with ribosomal proteins, acting as a chaperone for ribosome biogenesis before its cleavage (van de Heuvel et al. 2021).

SUMOylation involves the covalent attachment of a small ubiquitin-like modifier (SUMO) protein to lysine residues in target proteins via an enzymatic cascade (Wilkinson and Heley 2010). SUMOylation modulates protein stability, solubility, localization, and interaction profiles. In *C. reinhardtii*, SUMO protease SMT7 modulates ribosomal protein L30 (RPL30), regulating cell-size checkpoint function (Lin et al. 2020).

Transcription Factor

Transcription Factors (TFs) are proteins involved in regulating the transcription of genes (DNA) into mRNA. They contain DNA-binding domains, which bind to specific sequences of DNA located upstream of coding sequences. Some TFs bind to a promoter sequence near the transcription start site and facilitate the formation of the transcription initiation complex. Other TFs bind to regulatory sequences such as enhancer sequences and can either stimulate or repress transcription of the cognate gene. These regulatory sequences can be thousands of base pairs upstream or downstream of the gene being transcribed.

TF genes are an important source of the diversity and change that underlie the evolution of organisms (Carroll 2000). TFs show lineage-specific patterns and are related to organismal complexity (Petroll et al. 2021). For instance, the basic Helix-Loop-Helix (bHLH) is a highly conserved TF, and its copy number seems to scale with the complexity of the organism. Only one bHLH gene was found in the unicellular rhodophyte *Cyanidioschyzon merolae*, several in chlorophyte algae, but over 50 in higher plants and over 100 in animals. Based on this trend of expansion in evolution, algae of secondary and tertiary endosymbiosis-dependent lineages might possess a high number of bHLH genes. However, it is surprising that only seven and five were found in *P. tricornutum* and *T. pseudonana*, respectively, and none was found in *E. huxleyi* (Rayko et al. 2010).

The survey of TFs in stramenopiles by Rayko et al. (2010) revealed lineage-specific diversity and that Heat Shock Factor (HSF) was overamplified and constituted the most abundant class of TFs in the two diatom models, *P. tricornutum* and *T. pseudonana.* In these two diatoms, 212 and 258 TFs were identified, whereas 156, 283, and 419 TFs were found in the genomes of *Aureococcus anophagefferens*, *Ectocarpus siliculosus*, and *Emiliania huxleyi*, respectively. Furthermore, Myb and C2H2-type zinc finger TFs were the most abundant TF families in all other stramenopile genomes examined. The study also detected the presence of diatom and lineage-specific gene fusions, in particular a class of putative photoreceptors with light-sensitive Per-Arnt-Sim (PAS) and DNA-binding (basic-leucine zipper, bZIP) domains and an HSF-AP2 domain fusion. An analysis of expression data indicated that many of the TFs were transcribed and may be involved in specific responses to environmental stimuli.

The mechanisms by which TFs regulate the dynamics of gene transcription in algae are not yet well understood. Conventional chromatin immunoprecipitation (ChIP) studies typically measure transcription factor occupancy, which indicates the overall transcriptional activity in the genome. Although a single TF can typically bind to numerous sites throughout the genome, *cis*-regulatory activity at enhancers requires the concerted action of multiple binding events. TFs bind to enhancers in a combinatorial manner, facilitated through direct and indirect cooperative mechanisms. The combinatorial nature of enhancer occupancy allows genes to be regulated in complex patterns in both space and time. DNA and proteins can act together as a scaffold to cooperatively recruit TFs to enhancers. Enhancers undergo changes, in which their occupancy by TFs and the position of nucleosomes with or without post-translational modification reflect their inactive, poised, or active state. Multiple elements regulating the same gene seem to assemble in a three-dimensional structure with the promoter, having synergistic or repressive influences on each other (Spitz and Furlong 2012).

Conventional ChIP results give no indication of kinetics and are poor predictors of transcription factor function at a given locus. A study investigating genome-wide protein-DNA binding dynamics in the budding yeast *Saccharomyces cerevisiae* using competition ChIP with a sequence-specific *S. cerevisiae* TF, Rap1, revealed how a TF regulates transcriptional dynamics (Lickwar et al. 2012). The results indicated that long Rap1 residence was coupled to transcriptional activation, whereas fast binding turnover, referred to as 'treadmilling', was linked to low transcriptional output. This finding led to the proposition that transcription factor binding turnover is a major point of regulation in determining the functional consequences of transcription factor binding, mediated mainly by control of competition between transcription factors and nucleosomes. This model predicts a clutch-like mechanism that rapidly engages a treadmilling transcription factor into a stable binding state, or vice versa, to modulate transcription factor function.

Evolution of Cis- and Trans- Regulation

Cis regulatory factors bind to 5' sites close to the Transcription Start Site (TSS) of a gene, also known as the proximal promoter. In contrast, *trans* regulatory factors bind to 5' sites distant from the TSS, referred to as the distal promoter. Typically, the proximal promoter is about 200 to 1500 bp upstream of the TSS, whereas the distal promoter can be many kilobases upstream. However, there is no well-defined distance that distinguishes the two types of promoters. Evolutionary changes in *cis*-regulatory gene expression involve substitutions in Transcription Factor Binding Sites (TFBSs), insertions and deletions altering the spacing of TFBSs, and recombinatory events creating new combinations. Evolutionary changes in *trans*-regulation concern the structure of a regulatory factor or the expression of such factors (van Straalen and Roelofs 2012).

There is functional differentiation between *cis*- and *trans*-acting factors based on studies of model organisms. In *Drosophila*, studies on *D. melanogaster* and *D. simulans* suggest that genetic variation in gene expression between individuals of the same species is mostly due to divergence in *trans*-acting factors, whereas differences between two species tend to stem from differences in *cis*-regulation. Furthermore, evolutionary changes in *cis*-regulation tend to cause large and variable effects on gene expression. In the plant model *Arabidopsis*, loci regulated by *trans*-acting factors tend to elicit small and additive phenotypic effects (van Straalen and Roelofs 2012).

Promoter sequences are expected to evolve faster than coding sequences because they are not constrained by an open reading frame. Studies have shown that promoter polymorphisms can have evolutionary consequences. Research on *Drosophila* has demonstrated that polymorphisms in promoters can affect differential transcriptional responses and lead to differential phenotypes with microevolutionary consequences (Wray et al. 2003). Furthermore, it appears that divergence between species is more often associated with differences in *cis*-regulation of genes, while variation across intraspecific strains is often attributable to divergent *trans*-acting influences on gene expression.

Regulatory RNA

Cells express many RNAs that do not encode proteins and are not ribosomal or transfer RNAs. Increasingly, these non-coding RNAs are found to have important regulatory functions. One class of regulatory RNAs is microRNA (miRNA), which is typically 22 nucleotides long. Most miRNAs are transcribed from DNA sequences into primary miRNA (pri-miRNA), processed into precursor miRNAs (pre-miRNAs), and finally into mature miRNAs. Typically, a miRNA interacts with the 3'-UTR of target transcripts to suppress translation (e.g., in most cases of animal miRNA-mRNA interactions). However, depending on the species, miRNAs can interact with the 5'-UTR, coding region, and even gene promoters (e.g., in most plant miRNA-mRNA interactions). Furthermore, miRNAs can activate gene expression under certain conditions. Additionally, miRNAs may be shuttled between different subcellular compartments to control translation and even transcription. For a deeper understanding of the mechanisms underlying miRNA regulation of gene expression, readers are referred to a review by Dexheimer and Cochella (2020).

It is noteworthy that miRNAs can regulate not only endogenous genes but also genes in interacting organisms, even across kingdoms. For instance, miRNAs from food species or pathogens may regulate genes in consumer or host organisms (Zhang et al. 2012, Weiberg et al. 2013). Based on *in silico* prediction of target mRNAs, Lin et al. (2015) found the potential of miRNAs from the endosymbiotic algae (Symbiodiniaceae) to regulate genes of their coral host to facilitate the establishment and maintenance of their mutualistic relationships. Similar findings were reported later, indicating the potential of Symbiodiniaceae miRNAs to repress immunity in host corals (Zhu et al. 2022).

Another class of regulatory RNA is long non-coding RNA (lncRNA). lncRNAs are RNAs longer than 200 nucleotides that are not translated into functional proteins (Statello et al. 2021). By this definition, lncRNAs encompass a large and highly diverse collection of transcripts that differ in their biogenesis and genomic origin. It has been suggested that the human genome harbors more than 16,000 lncRNA genes, but the number can exceed 100,000. These lncRNAs are transcribed mainly by RNA polymerase II but can also be transcribed by other RNA polymerases. They can come from intergenic regions as well as sense or antisense transcripts that overlap with other genes. The transcribed lncRNAs are often capped by 7-methyl guanosine (m7G) at their 5' ends, polyadenylated at their 3' ends, and spliced similarly to mRNAs. lncRNAs are known to regulate gene expression by interacting with DNA, RNA, and proteins (Statello et al. 2021). In eukaryotes, lncRNAs serve as molecular scaffolds that link transcription factors to target genes, thereby altering their transcription (Moison et al. 2021).

Epigenetics

Epigenetic regulation of gene expression plays important roles in the organism's response to environmental changes. Epigenetic mechanisms confer phenotypic plasticity and can also have *trans*-generation effects (van Straalen and Roelofs 2012). Therefore, epigenetic regulation of gene expression is particularly relevant to ecological genomics. The major epigenetic mechanisms include DNA methylation, histone modifications, mRNA modification, and the processes mediated by the non-coding RNAs (Peschansky and Wahlestedt 2014, Gilbert et al. 2016).

DNA Methylation

DNA methylation is a heritable epigenetic mark involving the covalent transfer of a methyl group from *S*-Adenyl Methionine (SAM) to a nucleotide residue (Ferrari et al. 2023). The dynamics of DNA methylation serves as a driver of plastic responses to the environment across cells, organisms, and populations (Bogan and Soojin 2024). Although all nucleotides can be methylated, DNA methylation generally refers to the addition of a methyl group onto the C5-cytosine to form 5-methylcytosine (5-mC), as well as onto the N6-adenine to form 6-methyladenine (6mA). In plants, cytosines are methylated in both symmetrical (CG or CHG) or asymmetrical (CHH, where H is A, T, or C) contexts. DNA methylation is regulated by a family of DNA methyltransferases (Dnmts), including Dnmt1, Dnmt2, Dnmt3A, Dnmt3B, and Dnmt3L. Of these enzymes, Dnmt3A and Dnmt3B can establish a new methylation pattern to unmodified DNA and are thus known as *de novo* Dnmt. Dnmt1 functions during DNA replication to copy the DNA methylation pattern from the parental DNA strand onto the newly synthesized daughter strand. Dnmt3L lacks the catalytic domain present in other Dnmt enzymes, but it associates with the Dnmt3a and Dnmt3b and stimulates their methyltransferase activity. Dnmt2 methylates cytosine-38 in the anticodon loop of aspartic acid transfer RNA instead of DNA.

Histone Modifications

Histone tails undergo a variety of covalent modifications including acetylation, methylation, phosphorylation, ubiquitination, and sumoylation, regulating key cellular process such as gene transcription, DNA replication, and DNA repair (Cedar and Bergman 2009). Trimethylation of histone H3 lysine 9 (H3K9), histone H3 lysine 27 (H3K27), and histone H4 lysine 20 (H4K20) has been suggested to be a prerequisite for subsequent DNA methylation, which appears to be attributable to physical associations between the components of these histone methylation systems and one or more DNMTs.

RNA Methylation

In cells, RNA undergoes over 170 covalent chemical modifications, adding another layer of complexity to the gene regulation machinery (Wiener and Schwartz 2021). The most prevalent is the methylation of adenosine at one of the several distinct N atoms of the nucleobase (N^1, N^2 or N^6) or the 2' oxygen of the ribose (Zaccara et al. 2019, Boulias and Greer 2023). However, RNA methylation can also occur at the fifth position of cytosine (m^5C) or N^7 position of guanosine (m^7G). The presence of N^6-methyladenosine (m^6A) in mRNAs was established in the 1970's and was linked to mRNA instability shortly after. These findings were followed by the identification of the responsible gene methyltransferase-like protein 3 (METTL3) and the establishment of its role in developmental arrest in sporulation and seed development in *Arabidopsis*.

Many studies have linked m6A to translation upregulation via three distinct mechanisms: (1) involving the canonical m6A reader YTHDF1 (a.k.a DF1), which is proposed to bind to the eukaryotic translation initiation factor elF3; (2) involving direct binding of 5' UTR m^6A to elf3; or (3) involving direct translation activation by METTL3. Depending on organisms, m^6A may also play a role in regulating splicing or nuclear export of mRNA. However, m^6A occurs more in rRNAs than in mRNAs, but it also exists in small nuclear RNAs and several species of regulatory RNAs.

In algae, research on RNA methylation has just begun. Recent research detected m^1A in dinoflagellates, in which m1A methylation seemed to regulate mRNA fates by decreasing translation efficiency and may play a role in coping with nutrient stress (Li et al. 2023).

Major Techniques

The genomic approach as applied to physiological and ecological studies involves a similar workflow, typically starting with collection and preservation of samples, DNA or RNA extraction, followed by DNA sequencing, sequence assembly, gene annotation. This is completed with downstream analyses such as gene ontology and pathway mapping, comparative genomics analysis, phylogenomics, and differential gene expression analyses. In this workflow, Polymerase Chain Reaction (PCR) is involved in various steps. On many occasions, the function of a gene needs to be experimentally determined or verified. For this purpose, gene transformation and overexpression, genome editing, or gene silencing are the most common techniques. These major techniques will be briefly introduced below, but for full technical details, readers are referred to Slatko et al. (2018).

PCR: Trouble-Shooting Tips

Polymerase chain reaction (PCR) and DNA sequencing are two major pillars of molecular biology technology with significant positive impacts on marine ecology. PCR, invented by Kary Mullis (Mullis et al. 1986), revolutionized life sciences by enabling rapid *in vitro* DNA replication. Mullis received the Nobel Prize in Chemistry in 1993 for this groundbreaking technology.

PCR analyses should always include a positive control and a negative control. The positive control uses a template known to be amplified by the experiment's primers. Failure to generate a detectable amplicon suggests an issue with the chemical system or the PCR machine. The negative control contains all PCR components except the template and should produce no amplicon; if it does, the system is contaminated with DNA from the study organism or other sources (e.g., human, bacteria, fungi).

If the positive control generates an amplicon correctly and both the negative control and experimental samples do not, the experimental samples might be truly negative, or the template

DNA might be too impure for amplification. Diluting the DNA template with pure water by 10- or 100-fold can reduce inhibitory impurities and often result in more efficient amplification. Usually, amplification failure is due to poor quality, not low quantity, of the DNA template.

DNA Barcoding, eDNA, and Metabarcoding

DNA barcoding is a diagnostic technique for species identification that uses a short, standardized DNA region known as a barcode (Lahaye et al. 2008). To qualify as a barcode, the DNA sequence must exhibit significantly greater divergence between species than within species, be easy to PCR-amplify, sequence, and analyze, and be universal (Lin et al. 2014). Practically, the inter-specific genetic distances based on the barcode should be markedly greater than the intra-specific distances, with a detectable gap. The sequence should be relatively short so one sequencing read can cover it completely, and it should be flanked by highly conserved sequences, allowing the design of a "universal" set of primers to amplify as many taxa as possible. This universal attribute of the primers is highly desirable for analyzing microalgal communities (metabarcoding).

In practice, DNA is extracted from species or environmental samples (eDNA), subjected to PCR amplification of the barcode, and then sequenced. A phylogenetic tree is used to compare the genetic distance between taxa. According to the "easy" requirement of DNA barcoding, the simplest evolutionary model, UPGMA (Unweighted Pair Group Method with Arithmetic Mean), is initially used in phylogenetic analysis. However, this has evolved into using more advanced models to account for uneven mutation rates of different nucleotides.

Currently, no universal barcodes are available for all algae, let alone all eukaryotes. Multiple gene regions have been explored and found to provide different resolving power. Overall, nuclear ITS, the large subunit of rubisco (rbcL), and mitochondrial cytochrome C subunit 1 (COI) are most commonly used, with COI originally proposed as the model barcode (Hebert et al. 2003). The use of dual barcodes has been suggested for macroalgae (Saunders and McDevit 2012), with the 28S D2/3 region barcode serving as the common secondary barcode alongside the more group-specific primary barcode (e.g., ITS for dinoflagellates). For more technical details and precautions about DNA barcoding and metabarcoding, readers are referred to existing reviews (Bartolo et al. 2020, Barki et al. 2021).

Whole-assemblage metabarcoding and metatranscriptome

For field (ecological) genomic studies, two opposing approaches are available: single-cell genomics and whole-assemblage genomics. In single-cell genomics, individual cells of the same species can be isolated using a Fluorescence-Assisted Cell Sorter (FACS) based on cell size, surface features, and pigment characteristics. DNA or RNA can then be extracted from these single cells for genome or transcriptome sequencing, allowing researchers to reveal single-species genomic or transcriptomic characteristics in their natural environment. Technical details of single-cell sequencing will be provided in a separate section later.

The second approach involves sampling assemblages for bulk DNA and RNA extraction. The DNA can be used for metabarcoding using rDNA SSU (e.g., the V4 region) for both eukaryotes (18S rDNA) and prokaryotes (16S rDNA). To quantitatively reconstruct the whole assemblage taxon profile, samples can be spiked with external DNA or cells of known quantity as a "standard" (Tkacz et al. 2018, Barlow et al. 2020). For RNA analyses, instead of using the polyA-based mRNA enrichment commonly employed in eukaryotic transcriptomics, a rRNA-depleting method is used to enrich both prokaryotic and eukaryotic mRNA for sequencing. This Whole-Assemblage Molecular Ecological (WAME) approach allows for the comparison of transcriptomic activities of prokaryotes and eukaryotes within the same assemblage. This approach has been applied to research on harmful algal bloom formation and regulatory mechanisms (Yu et al. 2023) as well as photosynthetic and non-photosynthetic carbon fixation in the ocean (Li et al. 2024).

Microsatellite or Simple Sequence Repeat (SSR)

The necessity for population-level genetic analysis has led to the use of markers that exhibit greater variability than conserved markers such as rRNA genes. Microsatellites, also known as Simple Sequence Repeats (SSRs) in plant genetics, are a potent type of such markers. These short tandem repeats of 1-6 base pair nucleotides are found abundantly in both protein-coding and noncoding regions of the genome. Due to their high diversity, codominance, and stability, SSRs have become widely used in population genetics and evolutionary biology.

In the past, the application of SSRs was limited due to the high cost and labor intensity involved in developing these markers. This process required the identification of repeat sequences and flanking conserved sequences for primer design, which was implemented using PCR, cloning, and sequencing. However, the advent of high-throughput sequencing and the development of analysis software, such as GMATA by Wang and Wang (2016), has made it possible to easily identify a large number of SSRs from extensive sequence datasets. While the initial identification of SSR markers was based on genomic DNA, transcriptomes are increasingly utilized for this purpose.

Genome Sequencing and Bioinformatic Analyses

Although the choice of specific sample processing and data analyses may generally depend on the size and complexity of the genomes, a general workflow (Fig. 1.4) is similar for studying any genomes. The workflow includes genome sequencing, genome assembly, genome annotation, pathway constructing, comparative analyses, and other downstream analysis.

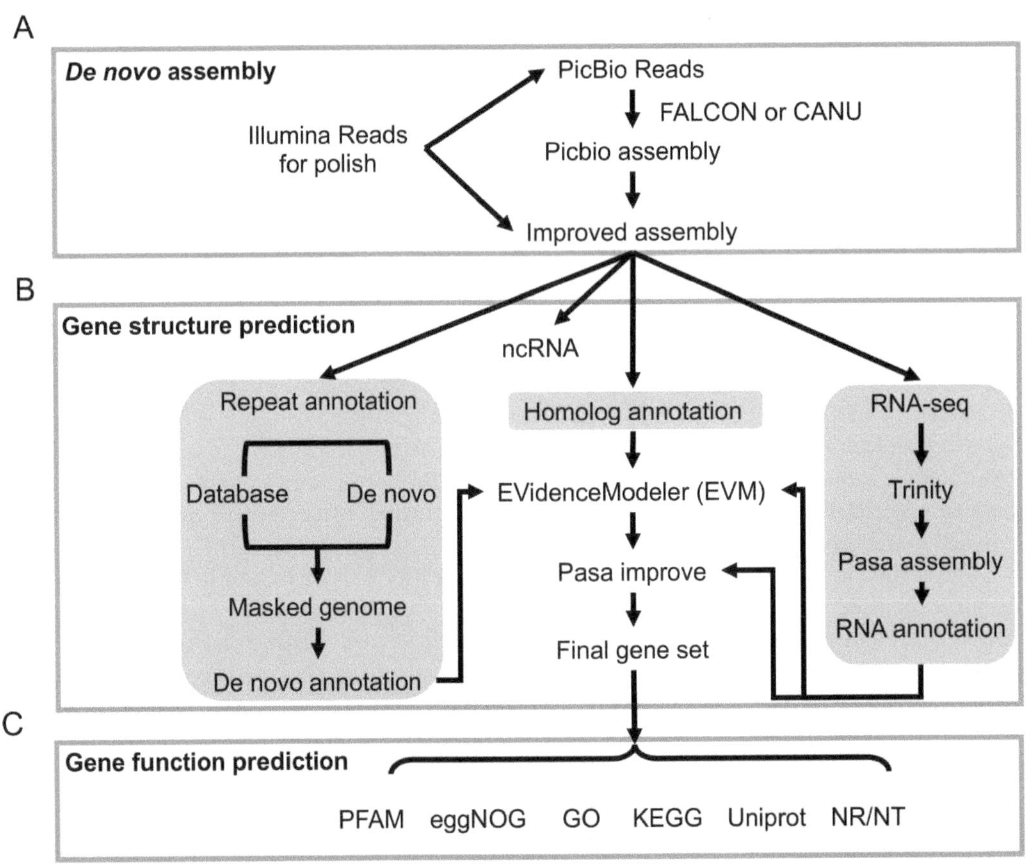

Figure 1.4. Workflow of genomics.

Genome Sequencing

The invention of DNA sequencing revolutionized life sciences by enabling the "reading" of DNA sequences more easily and quickly than ever imagined. Beginning in the 1970s, two main methods emerged: the chemical method developed by Walter Gilbert and Allan Maxam at Harvard University, and the dideoxy termination method by Frederic Sanger at the MRC Laboratory in Cambridge, UK. The Sanger method eventually prevailed, evolving into various faster and more cost-effective techniques. Sanger sequencing, which stops DNA synthesis using dideoxy nucleotides (ddNTPs) and separates synthesized DNA fragments by gel or capillary electrophoresis, was instrumental in creating the draft human genome, albeit at a cost of billions of dollars and a decade of effort. It remains useful for single-gene studies but has been largely supplanted by Next-Generation Sequencing (NGS) and Third-Generation sequencing (single-molecule sequencing) for larger-scale DNA sequencing. Gilbert and Sanger shared half of the 1980 Nobel Prize in Chemistry for their sequencing technology contributions, with the other half awarded to Paul Berg for his work on nucleic acid biochemistry and recombinant DNA.

Several NGS platforms have emerged, including pyrosequencing (454 Life Sciences/Roche Diagnostics), Solexa (Illumina, Inc.), Ion Torrent (Life Technologies), and SOLiD (ABI), all of which deploy massively parallel sequencing to process numerous sequences simultaneously. The 454 platform uses emulsion-PCR to attach single DNA strands to microbeads, each bead in an oil droplet where a new DNA strand is synthesized. The incorporation of nucleotides releases pyrophosphate, causing a fluorescence signal that is recorded. Illumina's technology also involves sequencing-by-synthesis, where nucleotide incorporation emits a signal. Ion Torrent detects nucleotide incorporation by measuring ion conductivity changes due to proton release. The SOLiD technique uses fluorescently labeled dinucleotide probes that hybridize to the DNA template, generating a signal upon a match. Each technology has unique benefits: 454 offers the longest read lengths, SOLiD the largest data output, and Illumina remains dominant due to its cost efficiency, large data output, and reasonable read length (~ 250 bp).

PacBio provides very long reads (> 10 kb) using Single-Molecule Real-Time (SMRT) sequencing, while Oxford Nanopore offers another long-read single-molecule sequencing platform. Despite the evolution of sequencing technologies, each platform's specific attributes make them suitable for different applications in genomics and molecular biology.

Genome Assembly

Genome assembly involves arranging nucleotide sequences into the correct order to reconstruct the DNA sequence of the original organism. This process is necessary due to the relatively short lengths of sequence reads, which are typically much shorter than most genomes or even individual genes. Assembly is an algorithm-driven automated process that can be approached using three methods: (1) greedy, (2) Overlap-Layout-Consensus (OLC) and Hamiltonian path, and (3) de Bruijn graph and Eulerian path.

The greedy algorithm aligns sequence reads with the highest similarity, prioritizing maximum sequence overlap. Early assemblers like Phrap, TIGR assembler, and CAP were based on this algorithm and proved highly useful. The Phred–Phrap–Consed suite of programs has been widely used. The OLC algorithm constructs a directed graph from reads and their overlaps using pairwise comparisons. Assemblers such as Arachne, CABOG (Celera Assembler), Newbler, Minimus, Edena, and MIRA employ this method. The de Bruijn graph method is effective for assembling short reads (< 100 bp) but has also shown success with longer reads. Assemblers using this algorithm include Euler-SR, Oases, Velvet, ALLPATH, ABySS, and SOAPdenovo.

Genome sequence assembly can be done through "mapping and assembly" or "*de novo* assembly". When a reference genome is available, resequenced reads are aligned to the reference genome and then assembled, a method known as "mapping and assembly". Bowtie is an efficient short-read aligner for this purpose, while Bowtie 2 is preferred for longer reads due to its increased speed, sensitivity, and reduced memory usage. In contrast, "*de novo* assembly" is used when no reference genome is available, with paired reads preferred over single reads to aid in generating scaffolds. Genome assembly progresses hierarchically from sequence reads to contigs, contigs to scaffolds (supercontigs), and scaffolds to chromosomes.

Third-generation sequencing technologies, such as PacBio and Nanopore, greatly benefit genome assembly by producing long reads that overcome challenges posed by regions rich in repeats and high GC content. These technologies have become the standard in genome assembly research, offering solutions to the complexities encountered by second-generation sequencing methods. However, third-generation sequencing comes with a trade-off in accuracy compared to second-generation sequencing. The relatively lower cost and superior accuracy of second-generation sequencing make it suitable for enhancing third-generation sequencing data via read correction. Figure 1.4A outlines an assembly approach relying on third-generation sequencing data, with PacBio as a representative example, and using Illumina reads for sequence correction.

After sequence reads are organized into contigs, these contigs are further organized into longer sequences called scaffolds, which can be integrated into chromosomes. Linking contigs into scaffolds and then chromosomes is greatly facilitated by Hi-C linkage analysis (Ghurye et al. 2019). Hi-C is a sequencing-based assay initially developed to investigate the 3D architecture of the genome within a cell nucleus by quantifying the frequency of contact between all pairs of loci in the genome (Lieberman-Aiden et al. 2009). The frequency of contact between two loci is highly associated with their one-dimensional distance. Hi-C data offers linkage details across diverse length scales, covering tens of megabases, and finds utility in genome scaffolding. Shortly after its introduction, Hi-C was applied to produce chromosome-scale scaffolds. SALSA2 is one of the programs developed to provide accurate chromosome-scale assemblies (Ghurye et al. 2019).

Genome Annotation

Genome annotation is the process of identifying functional elements along the sequence of a genome, thereby giving biological meaning to it. It includes deriving the structural elements such as locations, sequences, and structures of genes (Fig. 1.4B) and functional information such as functions of genes (Fig. 1.4C).

Structural Annotation. This process is typically implemented through three main methods: *ab initio* prediction, homology-based annotation, and transcriptome-assisted annotation. In *de novo* gene prediction, repetitive sequences, widespread in eukaryotic genomes, either clustered or dispersed between genes, are identified. Annotating repetitive sequences not only provides insights into the distribution characteristics of specific biological genomes but also reduces unnecessary resource consumption for the structural annotation of genomic genes. By masking the positions annotated to repetitive sequences, gene structures can be predicted *ab initio.*

Homology-based gene prediction leverages known genes from closely related species to align sequences and identify homologous regions. By utilizing gene signals such as splicing signals, start and stop codons, gene structures can be predicted using these homologous sequences as a foundation. Compared to the "needle in a haystack" approach of *ab initio* prediction, homology-based prediction significantly narrows down potential regions in the genome, making gene identification more efficient. Transcriptome-assisted gene prediction further enhances this process by using high-quality transcriptome data to complement incomplete genes obtained through other annotation methods, thereby improving the overall completeness of the final gene structure annotation. By integrating information from these three prediction methods with different weights, the ultimate gene annotation results are formulated, followed by gene function annotation.

Functional Annotation. Based on gene prediction results, deduced protein sequences are extracted from the genome and compared to mainstream databases for functional annotation. The general workflow for functional annotation includes the following steps: (1) selecting and downloading databases, (2) constructing BLASTp indices, (3) using BLASTp to align protein sequences to the databases, and (4) compiling the results. Some software integrates multiple processes of database retrieval and functional annotation, and some even provide web servers for online annotation. Examples include eggNOG-mapper and InterProScan, which integrate eggNOG and InterPro databases, respectively. Prominent databases used in this process include the NCBI non-redundant database, UniProt, eggNOG (a database grouping proteins into orthologous groups at different taxonomic levels), PFAM (a classification system for protein domain annotation), GO (Gene Ontology, a database for gene ontology annotation), and KEGG (Kyoto Encyclopedia of Genes and Genomes, a database for metabolic pathway annotation).

Pathway Mapping

A gene may belong to a Gene Ontology (GO) hierarchy and be part of a metabolic pathway. Organizing functionally annotated genes within GO terms or pathways facilitates the functional characterization of the genome. The most popular resource for GO and pathway analyses is the Kyoto Encyclopedia of Genes and Genomes (KEGG). KEGG (https://www.kegg.jp) is a manually curated database that integrates diverse biological entities, organized into systems, genomic, chemical, and health information categories (Kanehisa et al. 2023). Each entry in KEGG is identified by a unique KEGG identifier (kid), typically composed of a prefix followed by a five-digit number. These entries can be accessed by appending /entry/kid to the URL. Additionally, appending /pathway/kid, /brite/kid, and /genome/kid to the URL allows users to launch the KEGG pathway map viewer, the Brite hierarchy viewer, and the newly released KEGG genome browser, respectively.

Single-Cell Genome Sequencing

Single-cell sequencing was initially developed for studying marine prokaryotes (Zhang et al. 2006) and has since been extended to marine microeukaryotes (Sieracki et al. 2019). The major technical steps in single-cell sequencing involve isolating single cells, typically using Fluorescence-Activated Cell Sorting (FACS), and performing Multiple Displacement Amplification (MDA) with isothermal random priming and extension. This extension relies on a high-processivity, low-error, strand-targeted polymerase (Φ29 polymerase). The development, evolution, and applications of single-cell genomics technology have been systematically reviewed (Gawad et al. 2016).

Single-cell sequencing technology has primarily been employed to study environmental plankton, revealing previously unrecognized diversity and functional genes. For instance, Strassert et al. (2018) used FACS-based single-cell isolation for plankton from the Pacific, obtaining seven distantly related species. The sorted cells underwent single-cell genome sequencing, yielding 0.5–3.1 Mbp per cell, confidently assigned to marine alveolates and encoding between 97 and 917 predicted genes. Despite the low coverage, this effort encourages future single-cell sequencing for complex protists such as dinoflagellates and related alveolates.

The single-cell sequencing approach has also been applied in transcriptomics. For example, sequencing the transcriptome of cells isolated from the field led to the discovery of new dinoflagellate species (Cooney et al. 2020). The partial transcriptome data enabled multi-protein (221 proteins) phylogenetic analysis, revealing that the genus *Abeienium*, previously regarded as a subgroup of *Noctilucales*, is distinct and basal to that family. Additionally, the technology was extended to sequence compartments of the giant single-cell green alga *Acetabularia acetabulum*, revealing distinct subcellular localization of different functional sets of gene transcripts (e.g., photosynthesis gene transcripts at the cap, nuclear activity gene transcripts in the basal end where the nucleus is located) (Andresen et al. 2021).

Analysis of Whole Genome Duplication (WGD) and Gene Family Expansion/Contraction

Whole-Genome Duplication (WGD) is a major force for rapid genome evolution, increasing organism complexity, enhancing adaptation through dosage effects (e.g., gene family expansion), and inducing speciation and biodiversity by creating reproductive isolation from sister populations. Three major methods are available for detecting WGD (Yang et al. 2019). The first method is all-against-all chromosome (or scaffold) synteny analysis within the genome (Fig. 1.5), which requires a well-assembled genome. The other two methods are based on phylogenetics and the distribution of pairwise paralogous synonymous substitutions per synonymous site (dS), which are more suitable for less well-assembled genomic data.

The phylogenetic method estimates WGDs through gene count data, counting the number of gene copies in various gene families across a group of taxa along the phylogeny while considering gene birth and death rates. The dS-based method assumes that each gene family has constant rates of birth or loss, while WGDs violate this assumption and produce peaks in cumulative distributions of pairwise dS between paralogs within a genome (Maere et al. 2005). In the dS-based method, which has become the most widely used, pairwise paralogs are identified among the genomes being compared,

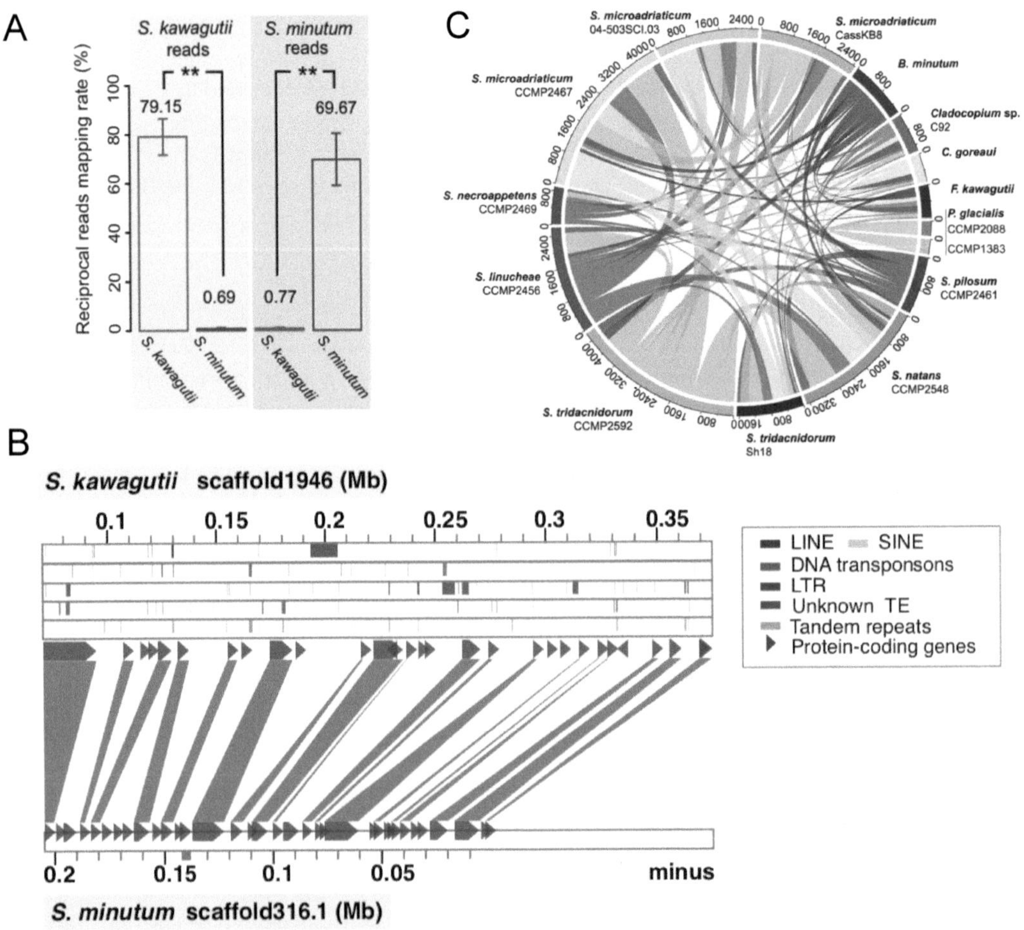

Figure 1.5. Comparative analysis of Symbiodiniaceae genomes showing low whole-genome similarity. (A) Reciprocal sequence reads to genome assembly mapping. (B) Feature distribution and synteny with *Breviolum minutum* (formerly *Symbiodinium minutum*) of scaffold1946 of Fugacium kawaguii (formerly *S. kawagutii*). A & B are adopted from Lin et al. (2015) with permission from Science. C is reproduced from González-Pech et al. 2021 under a Creative Commons CC BY 4.0 license.

and the distribution of fourfold synonymous third-codon transversion rate (4DTV) or dS between paralog pairs is estimated to determine the WGDs. To implement this for lower-quality genome assemblies, an improved computational algorithm named WGDdetector was recently developed (Yang et al. 2019). Additionally, a method developed for detecting copy-number aberrations and WGDs in cancer (Zaccaria and Raphael 2020) has potential applications in algal research, offering the added benefit of identifying gene copy variations occurring in specific populations or associated with specific traits.

Comparative Genomic Analyses

Comparing sequenced genomes from related organisms is powerful for understanding the evolution of the genomes. Whole-genome comparison often reveals striking differences in closely related and morphologically similar taxa. For instance, two species in the family Symbiodiniaceae of dinoflagellates, *Fugacium kawagutii* and *Breviolum minutum*, showed less than 1% of similarity at the whole genome level in the reciprocal comparison (Fig. 1.5A). This can be visualized from the rarity of the synteny between the assembled scaffolds of the genomes (Fig. 1.5B). Low similarity was also found between more species in the family (Fig. 1.5C). However, despite the extreme genome divergence, their protein-coding sequences are much more similar, with 4.4–21.4% shared genes between the symbiodiniacean species (Table 1.2). For *F. kawagutii* and *B. minutum*, > 17% of their proteomes are shared.

Furthermore, comparison of genomes could help with the identification of novel genes and regulatory elements in the genomes being compared. This extends from the functional annotation of genomes, which is based on the comparison of the predicted proteome to the proteome of a previously studied model species or related species in the public databank. Typically, the annotation process yields a percentage of predicted genes that match genes in the databank, and a percentage that match none or hypothetical proteins, but whether these are real genes or artificial (essentially random open reading frames) needs to be verified. A method named *Reading frame conservation test* has been developed (Kellis et al. 2003), in which the nucleotide sequence conservation of a predicted open reading frame is examined between the species in question and a better studied species. In yeast, this facilitates the classification of each ORF in *Sacchromyces cerevisiae* as biological meaningful or meaningless, depending on the proportion of the sequence over which conservation with other species is observed. Each of the other species is considered a 'voter', 'approving' or 'rejecting' the sequence in *S. cerevisiae* (van Straalen and Roelofs 2012).

In addition, comparative genomic analysis can reveal the horizontal gene transfer. When the best hit of BLAST is a gene from a distant organism in a different domain of life, it either has come from horizontal gene transfer or sample contamination. Beyond the genome level analysis, other forms of comparative analyses often lead to surprising findings. For instance, the comparison of full-length transcripts within and between dinoflagellates species resulted in the discovery of spliced leader *trans*-splicing in this phylum (Zhang et al. 2007). Comparison of the genomic and cDNA sequences of the same mitochondrial gene within species unveiled the unusual RNA processing, mRNA editing, in the phylum of dinoflagellates (Lin et al. 2002).

Table 1.2. Common (Comm) and unique (Uniq) genes identified by comparative genomic analysis*.

	Slin	Spil	Snec	Bmin	Snat	Fkaw	Stri	Cgor
Comm	3445	3445	3445	6804	7544	8140	8512	3445
Uniq	4892	5790	6254	7995	8105	11116	12741	43077
% shared	10.7	14.7	9.7	17.0	21.4	18.0	18.7	4.4
Total	32053	23436	35672	40120	35270	45192	45474	79161

* *Slin, Symbiodinium linucheae; Spil, S. pilosum; Snec, S. necroappetens; Bmin, Breviolum minutum; Snat, S. natans; Fkaw, Fugacium kawagutii; Stri, S. tridacnidorum; Cgor, Cladocopium goreaui.*

Phylogenetic and Phylogenomic Analyses

Phylogenetic analysis is used to determine the evolutionary relationships among a group of organisms, typically producing a phylogenetic tree composed of nodes and branches. The nodes represent taxonomic units, with internal nodes denoting ancestors and terminal nodes (or leaves) representing existing entities such as populations, species, or higher taxonomic levels. Branches measure the divergence between two nodes. A phylogenetic tree can be constructed using one gene, multiple genes, or the whole genome, the latter known as phylogenomic analysis. There are alignment-based and alignment-free methods to infer phylogenies.

Alignment-Based Phylogeny

Alignment-based phylogeny begins with the alignment of multiple sequences, looking for correspondence of individual or groups of bases or amino acid residues that are in the same order in those sequences. This approach requires the presence of conserved bases or residues to anchor the alignment and variable positions to provide the phylogenetic signal. Single gene markers such as ribosomal RNA small subunit or multiple gene markers concatenated are appropriate. There are two major categories of methods: distance-based and character-based. The distance-based method (e.g., UPGMA and NJ) assesses the distance between two sequences in pairwise comparisons, while the character-based method (e.g., ML and MB) considers where the differences occur in the aligned sequences and the specific changes from one base or residue to another. These methods are available in commonly used software packages such as Phylip and MEGA, which, along with Mr. Bayes, are freely available for download.

The consideration of different mutational types is the basis of many evolutionary models embedded in phylogenetic analysis programs. For nucleotide sequences, purine-to-purine, or pyrimidine-to-pyrimidine changes (transitional mutations) occur more frequently than purine-to-pyrimidine or pyrimidine-to-purine changes (transversional mutations). The commonly used Kimura 2-parameter (K2P) model addresses the uneven pace of these mutations. For amino acid sequences, mutations at the third codon position are more likely to occur than at the first two positions because the former often do not change the encoded amino acid (synonymous substitution), while the latter typically do (nonsynonymous substitution). Different amino acids have varying likelihoods of being replaced by others, leading to the development of various protein evolution models. The pioneering Dayhoff and JTT protein models have been improved to more recent ones such as WAG and LG.

Given the rapidly growing number of substitution models, selecting the one that best fits the data is important. This can be done independently or as part of the tree inference. For example, the MEGA program incorporates this function, making it convenient. The selected model and parameters are then used to inform the tree inference program. For nucleotide sequences, the General Time Reversible model with gamma distribution of site-specific rates of evolution and invariable sites (GTR+Γ+I) is the most comprehensive and widely applicable, albeit computationally demanding. For protein sequences, WAG and LG are commonly used, sometimes coupled with +Γ+I.

Alignment-based phylogeny faces challenges when genes or proteins have complex structures, sequences are too divergent to produce a reliable alignment, or when the number of sequences being analyzed is so large that it becomes computationally unfeasible. The complexity can also arise from lack of similarity on the primary level due to shuffling and recombination events.

Alignment-free Phylogenetic Approach

The alignment-free phylogenetic approach bypasses multiple sequence alignment and the computationally intensive tree inference step. It is equivalent to the distance method in the alignment-based approach, except that sequence similarity or dissimilarity is quantified without using or producing an alignment at any step of the analysis. As such, alignment-free methods are computationally less expensive and suitable for whole-genome comparisons. They have been shown to produce accurate phylogenetic relationships among thousands of bacterial and archaeal genomes. One approach uses k-mers, which are short sub-sequences of defined length k. In this method, the proportion of shared k-mers between each pair of genomes is calculated to obtain a pairwise distance. These distances are then used to derive the phylogenetic relationships of the genomes based on the neighbor-joining algorithm. By using different similarity thresholds, the distances can create similarity networks and explore different levels of relatedness among the lineages being studied. Alignment-free phylogeny also offers the flexibility to choose different regions of the genome for analysis, such as protein-coding sequences, noncoding sequences, and repeats, to interrogate differential evolutionary trajectories or selection pressures.

Alignment-free phylogeny is a rapidly evolving field. The applicability and intuitive basis of k-mers in alignment-free approaches have been contended, especially for not considering substitution models of molecular evolution. It is important to note that the optimal k length varies among different datasets and needs to be determined empirically in each study.

Inferring phylogenetic trees requires replication. From a fixed set of sequence data, only pseudo-replicates are available. All phylogenetic programs use bootstrapping, i.e., taking different regions of the sequences in each replication, to reiterate the analysis. Hundreds or thousands of trees are generated, and either a consensus tree or the best tree out of them is accepted as the result.

Molecular Clock

A molecular clock (also known as a gene clock or evolutionary clock) is a method based on a constant mutation rate of a gene to measure the distance between organisms in terms of years and date the emergence of a lineage. While phylogenetic analysis provides the distance between organisms, the dates for the emergence of the organisms and the time interval between them need to be constrained by other evidence, such as fossils, for some of the nodes. This methodology requires that the gene used to reconstruct the phylogeny mutates at a constant rate over time and in different lineages of organisms. The notion of a molecular clock originated from the discovery that for hemoglobin, the number of amino acid differences between lineages changes almost linearly with time as estimated from fossil evidence. The concept was later developed into the Neutral Theory of Molecular Evolution by Motoo Kimura (1983).

There are two major challenges in molecular clock dating. First, no perfect molecular clock gene marker exists, and uneven mutation rates became evident as its application expanded. Nevertheless, compromise is often made to use the best available gene for clock analysis while relaxing the constant rate requirement (referred to as a relaxed clock) to yield an approximate estimation of the timing of evolutionary events. The ribosomal RNA gene, such as the small subunit (SSU), is highly conserved across the tree of life and was initially promising as a clock-like marker. However, it is now known that rRNA genes have very different mutation rates in different groups of organisms. One way to improve the accuracy of analysis is to use multiple genes. Second, fossils are hard to come by for eukaryotic algae. Recently, a study using synchrotron-based X-ray fluorescence and absorption found evidence of intracellular bound chlorophylls in 1-billion-year-old fossils of the eukaryotic alga *Arctacellularia tetragonala*, ascertaining the existence of eukaryotic algae one billion years ago.

Population Genomics and Genome Scan

Population genetics profiles allele frequencies and characterize population genetic structure by assessing disequilibrium of alleles at certain loci. Microsatellites, for instance, are commonly used loci. Population genomics expands population genetic to the genome-wide level by utilizing high-throughput genome sequencing and analyzing polymorphic loci (e.g., SNPs) throughout the genome. Fst is used to test significance for the difference between populations for each locus. When Fst is plotted as the function of genomic position locus by locus, a map results. This process is called genome scan.

Functional Genetics

Typically, only about 50% of predicted genes from an algal genome match functional characterized genes in the database. Unraveling the functions of the many uncharacterized genes remains challenging. Even within the 50% of genes that match known genes, their functions in a not well-studied organism still need to be experimentally verified. One way to find or verify the function of these genes is to construct an expression plasmid or phage and transform it into a model organism for expression and phenotypical observation. Another widely used approach is gene knockout, where the gene of interest is disrupted to eliminate its functional presence, and the consequent phenotypic change is documented. Arguably, this is the ultimate way to experimentally determine the function of a gene.

Cell Transformation and Gene Overexpression

Cell transformation for gene overexpression can be performed in the native organism (i.e., homologous expression) or in a foreign species (i.e., heterologous expression). The choice depends on the ease of introducing the gene and ensuring its expression. Model organisms frequently used for gene overexpression include both prokaryotes and eukaryotes. *Escherichia coli* is a commonly used prokaryotic model for expressing genes from other prokaryotes. Transformation of *E. coli* is straightforward with a standard protocol. However, *E. coli* may not be suitable for expressing eukaryotic genes for functional examination because the proteins expressed in *E. coli* may not fold properly and often form inclusion bodies (denatured forms). In such cases, eukaryotic cells are more appropriate.

Yeast is commonly used as an expression system for eukaryotic genes and has been employed in the study of algal genes. The yeast system is advantageous not only because it can express algal genes but also because numerous mutants have been created and are available. Using a mutant strain with a known gene mutation, the function of an algal gene can be verified by transforming the homologous gene into the mutant yeast cell line. Restoration of the eliminated function confirms the function of the algal gene. This method is known as *functional complementation* method.

Homologous expression is ideal but not always feasible for algae. However, although many species are still recalcitrant to genetic manipulation, the list of species that can be transformed with a recombinant gene construct is growing (Faktorová et al. 2020). Several model species are particularly amenable to genetic manipulation and are thus widely used as chassis for transgenic work. *Chlamydomonas reinhardtii* and *Phaeodactylum tricornutum* represent the best-developed models for green algae and diatoms, respectively. In contrast, dinoflagellates are generally recalcitrant to transformation due to their tough cell covering. Recent development of the field includes transformation of a chloroplast genome in a symbiodiniacean species (Nimmo et al. 2019) and the nuclear genomes of *Oxyrrhis marina, Karlodinium veneficum*, and *Breviolum minutum* (Faktorová et al. 2020, Sprecher et al. 2020, Gornik et al. 2022).

One strategy to improve gene expression is to insert the gene of interest between the flanking DNA sequences or downstream of the promoter region of a naturally highly expressed gene in the species. Promoters of Light-Harvesting Complex (LHC) genes, for instance, are often used in expression plasmids.

Genome Editing

Traditionally, gene knockout is achieved by creating a library of mutants using radiation or chemical treatment and screening for phenotypic changes to identify the altered genotype, a forward genetic approach. This technique, initially developed in the early 1900s, has advanced to transposon insertions but remained a method that creates random mutations. While this technique has been instrumental in the discovery of numerous genes underlying a wide range of phenotypes, it has in recent decades been superseded by targeted genomic alteration, i.e., the reverse genetic approach.

For a detailed history of the technical advances, readers are referred to the excellent review by Dana Carroll (2017). Here, a brief introduction is provided to suffice for an adequate understanding of the principles and technical procedures.

The work of targeted gene disruption began in the 1970s and 1980s, but the methods available then were inefficient and laborious. Since the turn of the century, Zinc Finger Nucleases (ZFNs) and Transcription Activator-Like Effector Nucleases (TALENs) were discovered to act as site-specific nucleases, cutting DNA at strictly defined sites and revolutionizing gene knockout technology (Bibikova et al. 2002, Joung and Sander 2013). The ease of experimentation and efficiency of mutagenesis were profoundly improved with the emergence of the CRISPR/Cas9 technique. This technique takes advantage of the bacterial "immune" system, which retains short sequences of invading viruses as an imprint of infection. These sequences are known as Clustered Regularly Interspaced Short Palindromic Repeats CRISPR). When the same viruses infect the bacterium again, the "memory" triggers the CRISPR-associated (Cas) proteins encoded adjacent to the repeat clusters to cleave and inactivate the invading viral genomes, guided by short RNAs (crRNAs) transcribed from the CRISPR arrays (Carroll 2017). Both the processing of the crRNAs and the cleavage of the invading DNA involve a small *trans*-acting RNA (tracrRNA). Assembling the crRNA with tracrRNA and the enzyme needed for cleavage (Cas9) led to the editing toolkit that is now most widely used.

The general workflow (Fig. 1.6) of the CRISPR/Cas9 gene editing starts with the identification of the target site of the gene in question that contains the Protospacer-Adjacent Motif (PAM), 5'-NGG-3', where N represents any base, situated downstream of the target site. The second step is to design the guide RNA (gRNA), which directs the Cas9 enzyme to the target site in the genomic DNA. Next, the Cas enzyme and gRNA are delivered into the cells. This can be done in one of three major forms: plasmids carrying the genes encoding the Cas enzyme and the gRNA, the encoded RNA synthesized *in vitro*, or a protein-RNA complex (Li et al. 2024). Other Cas enzymes besides Cas9 are available, but the design of PAM and gRNA needs to be optimized for different Cas enzymes.

Gene Silencing Using Small Interfering RNA (siRNA)

Small interfering RNA (siRNA) is a class of double-stranded non-coding RNA molecules (dsRNA), typically 20–24 base pairs in length, similar to microRNA (miRNA), and operating within the RNA interference (RNAi) pathway (Dana et al. 2017). siRNA interferes with the expression of specific genes with precisely complementary nucleotide sequences by degrading mRNA after transcription, thereby preventing translation. The mechanism operates in several steps (Fig. 1.7):

1. Cleavage by Dicer: Long dsRNA is cleaved by an endo-ribonuclease named Dicer, which cuts the dsRNA into siRNA.
2. Formation of RISC: The siRNA is recruited into a set of proteins to form the RNA-induced silencing complex (RISC).
3. Unwinding of siRNA: Within the RISC complex, the siRNA is unwound to form single-stranded siRNA.
4. Target mRNA Recognition: The single-stranded siRNA in the RISC complex scans and finds a precisely complementary mRNA.
5. mRNA Cleavage: Once the single-stranded siRNA binds to its target mRNA, it induces mRNA cleavage.

This mechanism shares the molecular machinery with miRNA, but miRNA has broader specificity and does not require a 100% complementary sequence as the target. To exploit the natural siRNA system as a gene silencing tool, either synthesized siRNA is used, or an expression vector is constructed carrying siRNA with an RNA polymerase III promoter (e.g., U6 or H1). The siRNA sequence in the expression gene cassette is modified to introduce a short loop between the two strands, making the transcribed siRNA more stable than the natural siRNA.

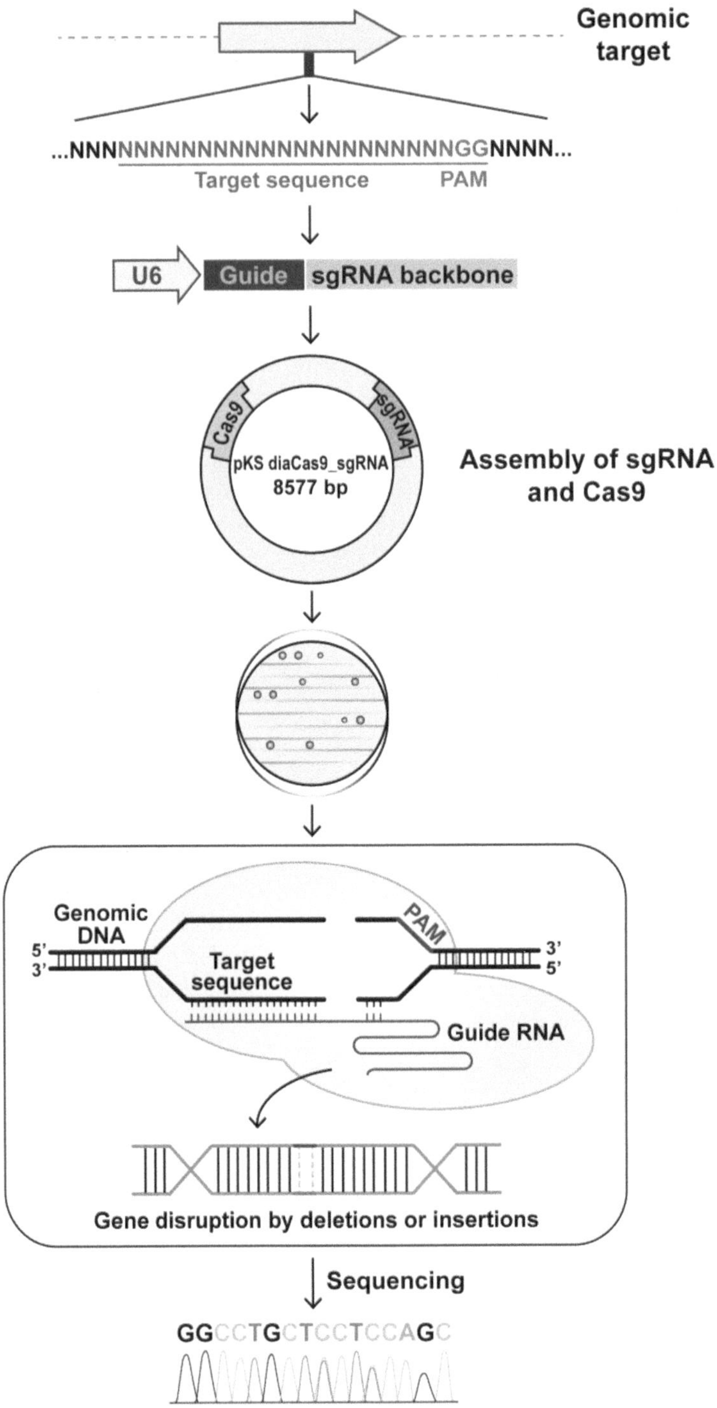

Figure 1.6. Workflow of CRISPR/Cas9 genome editing.

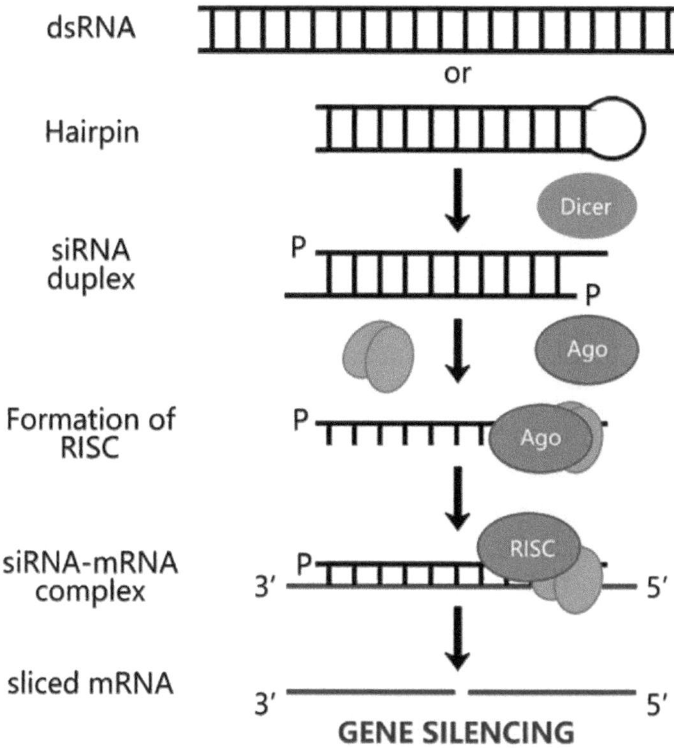

Figure 1.7. Principle of SiRNA silencing gene expression (degrading mRNA).

Interactomics

Complex biological systems function through interaction networks of proteins and other components, which occur at all levels of cellular function, including architecture, metabolism, signaling, and the availability of cellular energy. The recent technological development of high-throughput methods for interactomics research is dramatically increasing our knowledge of protein interaction networks and the regulatory functions of component proteins. The two most frequently used methods include Yeast two-Hybrid (Y2H) screening and affinity purification of complexes followed by mass spectrometry analysis (AP/MS).

Two-Hybrid Screening

Yeast two-hybrid (Y2H) screening represents a sensitive *in vivo* method for identifying and analyzing protein-protein and protein-DNA interactions, which are crucial for the regulatory functions of proteins (Brückner et al. 2009). The premise behind the technique is the activation of downstream reporter gene(s) by the binding of a transcription factor onto an Upstream Activating Sequence (UAS). For two-hybrid screening, the transcription factor is split into two separate fragments, the DNA-Binding Domain (BD) and Activating Domain (AD), respectively. The BD domain is responsible for binding to UAS and the AD domain is responsible for the activation of transcription (Young 1998). In this method, the protein of interest serves as the bait, whereas its binding partner, which is the focus of the study, acts as the prey, such that the "bait-prey" interaction can be detected, facilitating the identification of the prey.

QTL and eQTL

To link genes in the genome to ecologically important traits, the analysis of quantitative trait (QTL) is a useful approach. A QTL is a polymorphic genome segment containing one or more genes that affect the variation in a quantitative phenotypic trait, such as temperature tolerance, cell size, or growth rate. The position of a QTL in the genome is usually established by linkage to a set of polymorphic marker loci, each segregating in a Mendelian fashion, abundant and evenly spread throughout the genome (van Straalen and Roelofs 2012). Traditionally, QTL mapping uses controlled crosses, preferably starting with two inbred parent strains that differ in the trait of interest. When the offspring from two such inbred parents are mated, recombination breaks up the linkage between traits in the parental chromosome.

Thanks to the increasing accessibility of the high throughput genome sequencing, QTL is now often studied using Genome-Wide Association Study (GWAS). Gene polymorphic sites (SNPs) are usually the target of analysis. Like QTL mapping, GWAS uses linkage disequilibrium as a tool to link specific phenotypes to genes; however, it is not based on genotyping experimental populations but populations varying naturally. GWAS takes advantage of natural variations and the patterns of linkage and recombination that have evolved over thousands of generations (van Straalen and Roelofs 2012).

GWA studies have demonstrated that most of sequence variants are found in non-coding regions of the genome and are therefore likely to be involved in gene regulation. The analysis of such variants in the context of gene expression measured in cells or tissues has spawned a new field in genetics, i.e., expression quantitative trait loci (eQTLs). An eQTL is a locus that explains a fraction of the genetic variance of a gene expression phenotype. So an eQTL study is the discovery of genetic variants that explain variation in gene expression levels (Nica and Dermitzakis 2013). eQTL offers not only promises for the characterization of functional sequence variation but also for the understanding of basic processes of gene regulation and interpretation of genome-wide association trends. To capture the complexity of genome-wide gene regulation, a massive number of markers are required to cover complete genome in eQTL analyses.

Knowing the Limitation of the Genomic Approach

It is important to note that, like any other subdiscipline of science, genomics has its limitations. First, genome analysis is reliant on database, and the outcome can only be as accurate and thorough as the database itself. Interestingly, the quality and coverage of the database, in turn, depend on advances of genomics and experimental biology (functional genetics). Second, when the genomic approach is applied to the natural assemblages, there are additional limitation. The database may not cover the species of interest, and it is challenging to reconstruct genomes of the constituent species, particularly low-abundance taxa. This challenge is related to sequencing depth coverage. Finally, and more importantly, information from the genome indicates the genetic potential, not necessarily the activity, of the organism. Similarly, transcriptomic, and proteomic information provides evidence that the organism is expressing the genes and carrying out metabolic functions of these genes. However, whether observed changes, or absence thereof, in transcript and protein abundances reflect corresponding trends of activity depends on which level these genes are regulated at in the organisms. In particular, the lack of transcriptomic or proteomic changes do not necessarily indicate that the function of the encoded proteins does not change, as changes may occur in the posttranscriptional or posttranslational level. It is imperative to be aware of these limitations to accurately interpret genomic (more generally, 'omic) data.

2

Evolution of Algae and Photosynthesis

Emergence Of Photosynthesis, Oxygenic Photosynthesis, and Cyanobacteria

In the Earth's history (4.543 Byr), the long early stage (until 2.45 Bya) was largely or entirely anoxic (Holland 2006) due to the lack of oxygenic photosynthesis (Fig. 2.1). The earliest form of photosynthesis was anoxygenic, carried out by anaerobic photosynthetic bacteria such as *Chlorobium*, *Rhodopseudomonas*, *Chloroflexus* or *Heliobacillus* (Mulkidjanian et al. 2006). Oxygen emerged in the atmosphere about 2.32-2.4 Bya (Bekker et al. 2004, Lyons et al. 2014), about halfway of the Earth's history. A second major oxygenation event, Neoproterozoic Oxidation Event (NOE), occurred about 800–600 million years ago (Mya), significantly increasing atmospheric oxygen concentrations (Scott et al. 2008) to levels similar to those today (Sánchez-Baracaldo and Carbona 2019). Over geological time, atmospheric O_2 accumulated as its production, accompanied by carbon fixation, by oxygenic phototrophs exceeded its consumption by the respiration. The imbalance resulted in the drawdown of carbon dioxide and burial of organic carbon in marine sediments (Holland 2006). NOE has been linked to the origin of animals, major glaciation events, disruptions to the carbon cycle, and the emergence of marine planktonic groups (Brocks et al. 2017).

Oxygen production is a water oxidation process attributed to the emergence of the phylum Cyanobacteria (Soo et al. 2017). Cyanobacteria are believed to have emerged from the fusion of ancestral anoxygenic phototrophs (Hohmann-Marriott and Blankenship 2011), specifically a purple photosynthetic bacterium and a green sulfur photosynthetic bacterium, which contributed photosystem I (PSI) and photosystem II (PSII), respectively. Anoxygenic phototrophs have either PSI- or PSII-like photosystems, while cyanobacteria possess both. Hypotheses regarding the evolution of the two photosystems include the emergence of a 'protocyanobacteria' containing two anoxygenic photosystems from an ancient duplication of a single system (Mulkidjanian et al. 2006, Martin et al. 2018) and the evolution of photosystems via horizontal gene transfers (Raymond et al. 2002). Comparative structural biology and phylogeny now suggest that photosystems uniquely evolved the ability to perform water oxidation from the beginning, further specializing into what we now know as oxygenic photosynthesis (Sánchez-Baracaldo and Cardona 2019).

The crown group Cyanobacteria emerged along with the evolutionary efficiency improvement of the water oxidation enzymes. Besides, the core proteins of cyanobacterial PSII (D1 and D2) originated from an ancient gene duplication event. D1 and D2 are associated with two core antenna subunits named CP43 and CP47, which are believed to also have arisen from a duplication event. Taken the recent information together, it seems that water oxidation predated the two duplication

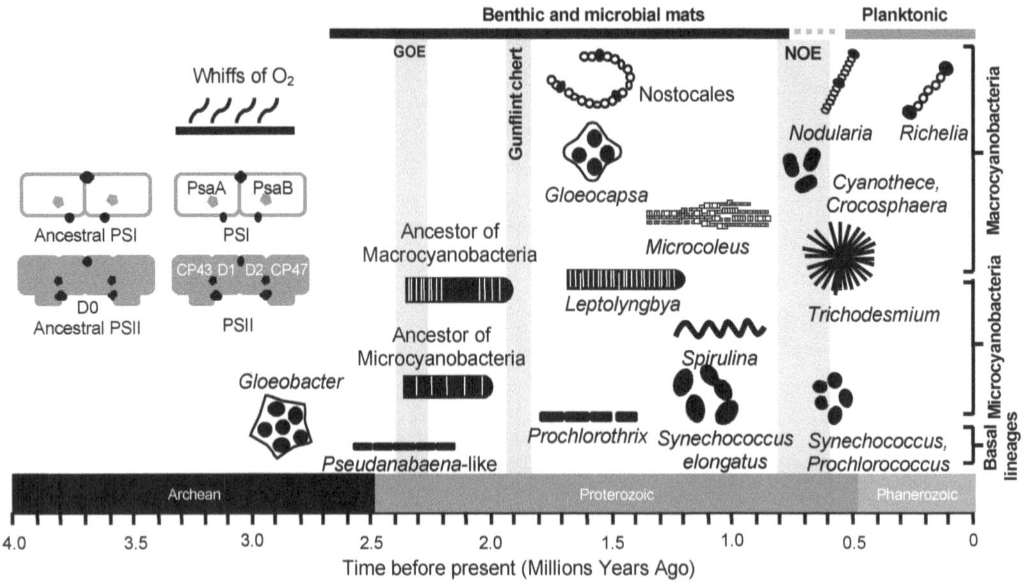

Figure 2.1. Evolution of PSI, PSII and cyanobacterial lineages. Age estimates are shown for PSI, PSII, crown group Cyanobacteria, major clades and taxa. GOE, the Great Oxidation Event. NOE, Neoproterozoic Oxidation Event. Cartoons are not drawn to scale. Ancestral forms of oxygenic photosynthesis powered by homodimeric PSII and PSI emerged in the early Archean or early Paleoproterozoic. D0 denotes an ancestral core subunit before the gene duplication that led to D1 and D2; it is thought to have assembled into a primordial water-splitting photosystem. The most recent common ancestor of Cyanobacteria inherited a heterodimeric photosystem shared by all extant oxygenic phototrophs. Taxa with smaller cell diameter (basal lineages and Microcyanobacteria) are shown at the bottom and larger cell diameter (Macrocyanobacteria) at the top. Major cyanobacterial clades radiated into many diverse forms after the GOE. Marine planktonic Cyanobacteria evolved towards the end of the Precambrian and the Cretaceous periods. Based on information from Sánchez-Baracaldo and Cardona (2019).

events in PSII, which predated the most recent common ancestor of Cyanobacteria (Sánchez-Baracaldo and Cardona 2019).

The core subunits of both PSII and PSI are highly conserved, but the antenna vary substantially between groups of algae as they are adapted to distinct habitat light conditions (Bai et al. 2021). Their core protein compositions in both photosystems are as shown in Fig. 2.2. There are 11–15 subunits in PSI, with ten of them being common to all photosynthetic organisms, and 21–30 in PSII, with 20 shared across all photosynthetic organisms. Haptophytes possess the fewest subunits, having 11 in PSI and 22 in PSII. *Arabidopsis thaliana* features the highest number of subunits in PSI with 16, followed by chlorophytes with 15, whereas cyanobacteria have the most subunits in PSII, totaling 30.

Emergence of Photosynthetic Eukaryotes: Primary Endosymbiosis

The evolution of photosynthetic eukaryotes gave rise to the immense diversity of photoautotrophs including land plants. The advent of plastids represents a major biological innovation, which is mediated by endosymbiosis (Sagan 1967, Margulis 1970). This process involved three major episodes of ecological innovations: primary, secondary endosymbiosis, and tertiary endosymbiosis (Fig. 2.3). Primary endosymbiosis, in which a heterotrophic ancestor engulfed the fateful cyanobacterium and enslaved it as a symbiont, which subsequently evolved into chloroplast (Cavalier-Smith 1998, Rodriguez-Expeleta et al. 2005, Weber et al. 2006), resulted in the emergence of Archaeplastida, a monophyletic group that includes the glaucophytes, rhodophytes, and chlorophytes.

It is generally believed that primary endosymbiosis occurred in a single event, approximately 1.5 Bya (Yoon et al. 2004, Blair et al 2005). The green algae include Chlorophyta and some lineages

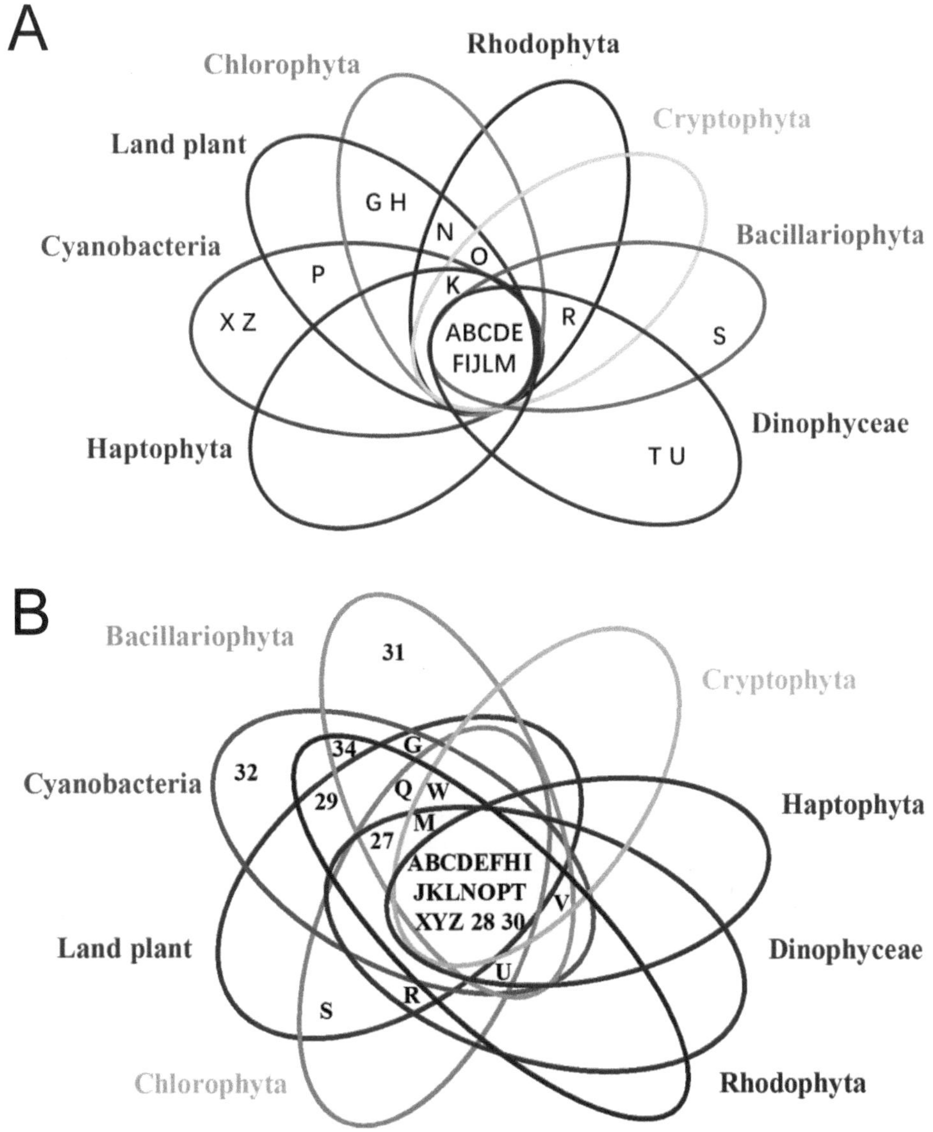

Figure 2.2. Comparison of photosystem (PS) components across lineages. (A) PSI. Reproduced from Lin et al. 2024. B) PSII. Information source: Fromme et al. 2001, Pi et al. 2018, Qin et al. 2019, Gisriel et al. 2022, Su et al. 2019, 2022, Zhao et al. 2023, Li et al. 2024.

in the Streptophyta, the phylum that also contains land plants (embryophytes). Within Streptophyta, Charophyceae had been traditionally considered the closest algal lineage to land plants. However, a recent study suggested that Zygnematophyceae may be closer (Cheng et al. 2019). Characteristic of the "primary" chloroplasts is an envelope consisting of two layers of membrane: the outer layer from the membrane of the host's food vacuole, the inner layer from the cell membrane of the cyanobacterium. Genome analyses indicate that over 4,500 genes encoding plant proteins have cyanobacterial origin (Martin et al. 2002), providing genomic evidence in support of the primary endosymbiosis hypothesis. The taxonomic identity of the cyanobacterial ancestor of chloroplast has remained elusive, but the closest relative of the Archaeplastida documented to date is the genus Gloeomargarita (Fig. 2.4), which emerged about 1.9 Bya (Moreira et al. 2017, Sanchez-Baracaldo et al. 2017).

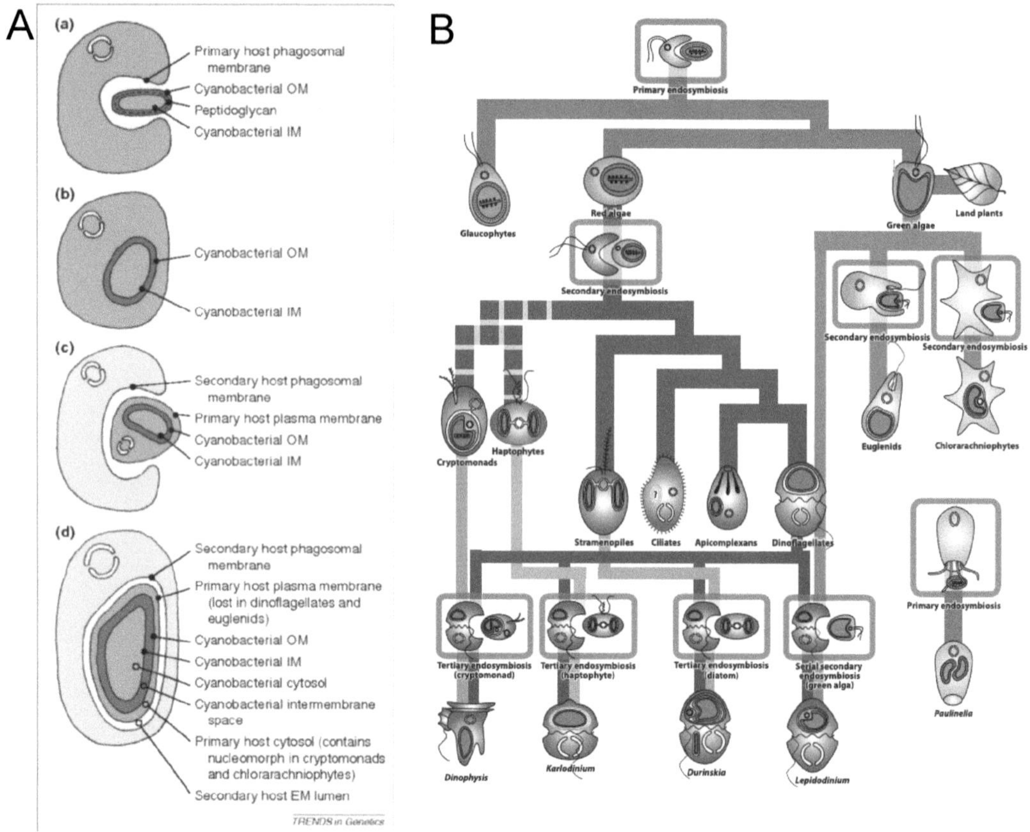

Figure 2.3. Evolution of chloroplasts through primary, secondary, and tertiary endosymbiosis. (A) different plastid envelope structures among the different endosymbiosis pathways. (a,b) In red and green algae that evolved by primary endosymbiosis, the cyanobacterial peptidoglycan wall has been lost. (c,d) Secondary endosymbiosis. Abbreviations: EM, endomembrane; IM, inner membrane; OM, outer membrane. Reproduced with permission from Archibald and Keeling 2002. (B) Schematic depicting major endosymbiotic events in plastid evolution. At the top, primary plastids originated through the endosymbiotic integration of a cyanobacterium, resulting in two-membrane plastids in glaucophytes (which retain the peptidoglycan wall between the two plastid membranes), red algae, green algae, and their close relatives, the land plants. Green algae were integrated in two independent secondary endosymbioses, giving rise to plastid-bearing euglenids and chlorarachniophytes (middle right). The origin of red algal secondary plastids (middle left) is more contentious, in particular regarding whether the cryptomonad and haptophyte plastids (dashed red lines) originated from the same endosymbiosis as those of stramenopiles and alveolates (the latter consisting of ciliates, apicomplexans, and dinoflagellates). An additional round of endosymbiosis occurred in dinoflagellates (bottom). Serial secondary endosymbiosis involved the integration of another primary plastid (a green alga), resulting in *Lepidodinium*. Tertiary endosymbiosis involved the integration of another secondary alga. The greatest degree of integration involves a haptophyte, but other putative tertiary plastids were acquired from cryptomonads and diatoms. Independent of this entire history, another primary endosymbiotic origin of plastids in the euglyphid testate amoeba *Paulinella* appears to be under way involving a cyanobacterium related to *Synechococcus* (bottom right). Reproduced with permission from Keeling (2013).

Exception to the "Single Primary Endosymbiosis Event" (SPEE) is an independent event that occurred very recently (124 Mya) giving rise to the photosynthetic species of *Paulinella* (Gabr et al. 2022). The plastid of *Paulinella*, named chromatophore or cyanelle, originated from a cyanobacterium. It is regarded as in an early evolutionary stage of integration into a mature plastid, and as such, only a low number of symbiont genes have transferred to the nucleus while a surprisingly massive number of chromatophore-targeted genes are ancestral host genes (Singer et al. 2017). Furthermore, retrotransposition led to a large plastid proteome (Singer et al. 2017) and is believed to play an important role in domesticating the endosymbiont towards a permanent plastid (Calatrava et al. 2022).

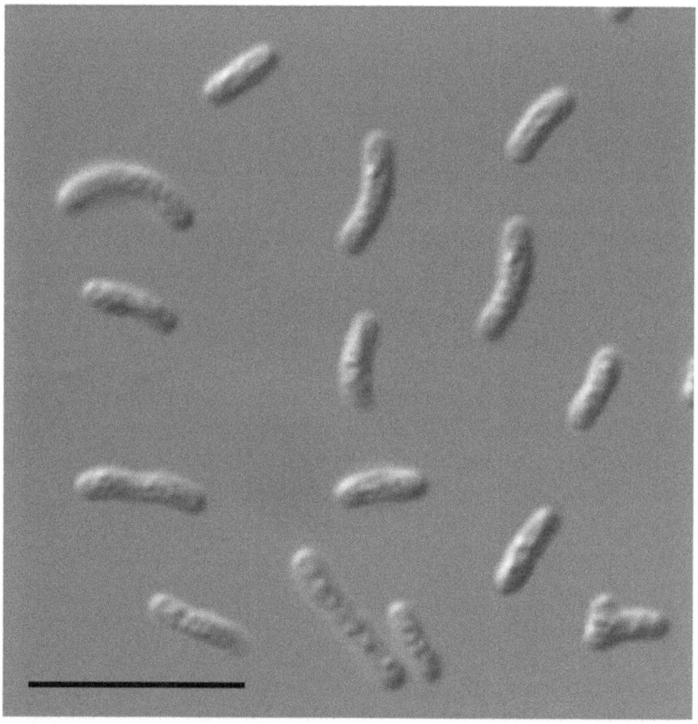

Figure 2.4. Micrograph of *Gloeomargarita* cells, which are the closest relative of the cyanobacterial ancestor that became the chloroplast of Archaeplastida. Reproduced with permission from Moreira et al. (2017).

The only other known exception to the SPEE notion of plastid evolution is the ciliate *Pseudoblepharisma tenue*, which, in addition to a photosynthetic symbiont that is an enslaved green alga, also has a photosynthetic prokaryote as a symbiont, a purple bacterium with a reduced genome, instead of a cyanobacterium (Muñoz-Gómez et al. 2021).

Radiation of Eukaryotic Algae: Secondary Endosymbiosis

Subsequent to the emergence of primary plastids, eukaryotic photosynthetic organisms radiated into a highly diverse flora through secondary and tertiary endosymbiosis. During the secondary endosymbiosis evolution, there was a prominent event involving a bacterial endosymbiont. Based on phylogenetic analyses, genes from *Chlamydia* are present in Viridiplantae (green algae and land plants), glaucophytes, rhodophytes, and diatoms, evidence that *Chlamydia* was once an endosymbiont in the common ancestor of these lineages and horizontal gene transfer occurred early in the evolution of the Plantae (Becker et al. 2008).

Secondary endosymbiosis gave birth to the majority of the algal flora existing today, including ochrophytes, dinoflagellates, haptophytes, and cryptophytes. Ochrophyta, also known as Stramenopiles or Heterokontophyta, is a supergroup with four subdivision that includes major taxonomic phyla such as diatoms, brown algae, raphidophytes, chrysophytes, pelagophytes, and silicoflagellates, among others (Graham and Wilcox 2000). There are two opposing hypotheses to account for the establishment of the "secondary plastids": one Single Secondary Endosymbiosis event (SSE) versus multiple independent events (known as serial eukaryote-eukaryote endosymbiosis or SEE). The SSE hypothesis perfectly explains the evolution of the green secondary plastids in the euglenophytes and chlorarachniophytes, which were derived from the chlorophyte. The secondary endosymbiosis in the red family is much more complex, and that is where the SSE and SEE disagree.

The SSE hypothesis is best represented by the chromalveolate hypothesis (Cavalier-Smith 1999, 2003, Keeling 2009). According to this hypothesis, a heterotrophic protist captured a red alga and retained it as a photosynthetic organelle, giving rise to the supergroup Chromalveolata, which includes cryptophytes, haptophytes, stramenopiles, and dinoflagellates. This hypothesis is well supported by molecular data showing that the plastids in chromalveolate taxa are of red alga origin and many of the plastid proteins are encoded in the host's nucleus (a result of organelle to nucleus gene transfer) (Harper and Keeling 2003, Bhattacharya et al. 2004, Li et al. 2006).

In the SEE hypothesis (Archibald et al. 2009, Bodyl et al. 2009, Baurain et al. 2010, Stiller et al. 2014), at least four secondary endosymbiosis events have been proposed to give birth to the highly diverse photosynthetic protists with red or green symbionts and phaeophytes. These secondary chloroplasts, compared to primary plastids, would have two additional membranes in the plastid envelope contributed by the host's food vacuole membrane and the endosymbiont alga's cell membrane, but during evolution, one of these additional membranes might have been lost. The algae arising from the secondary endosymbiosis include the macroalgae phaeophytes and the microalgal lineages diatoms and other stramenopiles, dinoflagellates, haptophytes, and cryptophytes as the descendants of the red alga ("red line"), and euglenophytes and chlorarachniophytes as the descendants of the green alga ("green line"). The "red-line" lineages possess Chl *c* as the accessory chlorophyll whereas the "green-line" possess Chl *b* instead. The biosynthesis of Chl *c* requires a 2-oxoglutarate (2OG)–dependent dioxygenase, which was only found in 2023 (Jiang et al. 2023).

During integration of the plastid, the nucleus of the algal endosymbiont typically is lost and some of the genes are transferred to the host nucleus, as indicated by the mosaic nature of the nucleus-encoded plastid protein genes (e.g., Nosenko et al. 2006). However, in the rare cases, the endosymbiont's nucleus is retained, albeit in a highly reduced form. Named nucleomorph, this symbiont nucleus vestige is best known in cryptophytes and chlorarachniophytes, holding 487 and 284 genes, respectively. Besides these two lineages, recent studies indicated that some dinoflagellates with green-type plastids also retain a nucleus remnant of their chlorophyte endosymbionts (Sarai et al. 2020). In Sarai et al. (2020), the detection of spliced leader *trans*-splicing signature on the green algal genes indicated that these green algal genes have been transferred into the dinoflagellate nuclei. This interesting result is evidence that the dinoflagellate and the green plastid are integrated, suggesting that the symbiont nucleus in the dinoflagellate plastid is indeed a nucleomorph. These dinoflagellates are thus excellent models for studying organellogenesis as the plastid-to-nucleus gene transfer is ongoing. Despite the intense evolutionary inquiries, what contribution the presence of the nucleomorph can make to the physiology and ecological performance of the algae is unclear and warrants further investigation.

Among the major secondary plastid lineages, cryptophytes, with no fossil available, are estimated to have emerged anywhere from 500 to 200 Ma (Zauner et al. 2000). For dinoflagellates, the fossil record has dated their occurrence at ~ 230 Ma, whereas molecular clock dating suggests its emergence as early as 650 Ma, with the gap between the two estimates potentially filled by presence of unclassified cysts (acritarchs) and the signature sterol dinosteranes in the time interval (Dale 2023). Fossil records have indicated the presence of haptophyte coccoliths at ~ 220 Ma, and molecular clock estimates suggest haptophytes may have an even earlier emergence (Delwiche 2007, De Vargas et al. 2007). Diatoms, with ample fossil records due to their siliceous frustules, are a rather late arrival in the geologic record (~ 190 Ma [Sims et al. 2006]), but with an emergence as early as 240 Ma based on molecular clock data (Kooistra and Medlin 1996, Medlin et al. 1997).

Further Diversification of Eukaryotic Algae: Serial Secondary Endosymbiosis and Tertiary Endosymbiosis

The secondary plastids in some lineages have been replaced by another primary plastid, in a process termed serial secondary endosymbiosis. This occurs in dinoflagellates, which have unparalleled propensity to lose and regain plastids. While the majority of dinoflagellates possess the secondary endosymbiosis plastid of red alga origin (peridinin-type), at least in the genus of *Lepidodinium* the chloroplast is of green alga origin ('green' family). *L. viride* and *L. chlorophorum* have acquired a plastid of prasinophyte origin (Kamikawa et al. 2015 and references therein). The evolutionary pathway is known as serial secondary endosymbiosis because the acquired plastid is a primary plastid (of green algal origin), which is in the same evolutionary status as the original, peridinin-containing plastid (of red alga origin).

The trajectory of algal plastid evolution is remarkably complicated by the numerous events of plastid loss and replacement in dinoflagellates (Dorrell and Howe 2015). Three separate lineages of dinoflagellates harbor secondary plastids that secondarily replaced the original peridinin-containing plastids. These are named tertiary replacement plastids. One of these three groups is represented by the Gymnodiniales genera *Karenia* and *Karlodinium*, which contain plastids of haptophyte origin (Tengs et al. 2000). Another lineage of dinoflagellates that owns a tertiary plastid is *Dinophysis*, a genus that does not possess peridinin plastids but harbors plastids of either haptophyte origin (Koike et al. 2005, Hongo et al. 2019), cryptophyte origin (Hackett et al. 2003), or multiple sources (Qiu et al. 2011, Nishitani et al. 2012). However, as still open to debate, at least some of these plastids are kleptoplastic and transient rather than permanent plastids.

Unintegrated 'Plastids': Symbionts and Kleptoplastids

Dinoflagellate taxa such as *Kryptoperidinium, Durbinska*, and *Galeidinium* contain a diatom endosymbiont. The diatom symbiont does not only contribute its chloroplast, which appears to be the only chloroplast in the dinoflagellate, but also functional mitochondria and nucleus, as diatom mRNAs encoding proliferating cell nuclear antigen (NCBI accession numbers OR268365, OR268366) and mitochondrial proteins (Imanian and Keeling 2007) have been detected in *Kryptoperidinium foliaceum*.

Kleptoplastidy evolved in diverse marine eukaryotes including heterotrophic relatives of microalgae, with dinoflagellates being one of the most common hosts. Kleptoplastidy is regarded as a transient state of evolution from heterotrophy to phototrophy. The source of the Kleptoplastids is taxonomically wide, frequently reported from cryptophytes, haptophytes, chlorophytes, and diatoms. The dinoflagellate *Pfiesteria* spp. use peduncle to funnel prey (cryptophytes) intracellular materials; these dinoflagellates retain chloroplasts while consuming the rest (Lewitus et al. 1999). The kleptoplastids were observed to be active in photosynthesis for a period (Feinstein et al. 2002). Besides, as indicated earlier, *Dinophysis* spp. harbor a plastid of either haptophyte origin (Koike et al. 2005, Hongo et al. 2019) or cryptophyte origin (Hackett et al. 2003), or both in some cases (Qiu et al. 2011). In one report, 14 cells of *D. mitra* were found to harbor > 100 different plastid sequences, suggesting kleptoplastids from many source species (Nishitani et al. 2012). The status of the plastids in *Dinophysis*, whether they are fully integrated and permanent or kleptoplastic and transient, is still under debate (see Chapter 8). It is noteworthy that *Dinophysis* spp. are unable to ingest food directly but rely on *Mesodinium rubrum* to get "second hand" food. This means that the kleptoplastid transfer takes different, more complicated, route than counterparts in other protists. Haptophyte-originated plastids were also found in an Antarctic dinoflagellate related to genera *Karenia* and *Karlodinium* (Gast et al. 2007).

The recently described euglenid *Rapaza viridis* feeds on a specific strain of the green alga *Tetraselmis* sp. and retains the chloroplast (Yamaguchi et al. 2012). Furthermore, the kleptoplastid is broken into many small transcriptionally active units. These units remain active for several days. Proteins encoded by *R. viridis* and targeted into the kleptoplastids are found with a signal peptide and a transit peptide. This might represent an intermediate state of evolution from prey-derived endosymbiont to mature chloroplast (Karnkowski et al. 2023). More discussion on kleptoplastids in various groups of organisms will be given in Chapter 20.

The Change of Tide: From 'Green Dominance' to 'Red Dominance' in the Ocean

The green algal lineage, from which all land plants have descended, is thought to have dominated eukaryotic phytoplankton from ~ 600 Mya until a few hundred million years ago (Knoll et al. 2007), with planktonic red algal species being of secondary importance. In the modern ocean, the green-type lineages, including the Archaeplastida flora and descendant groups (euglenophytes) gave way to the red-type family. The shift coincided with the geochemical transition from a highly reduced, anoxygenic, paleo-atmosphere to the oxygenic modern atmosphere during the Paleozoic to Mesozoic transition (~ 250 Mya) (Quigg et al. 2003, Zhang et al. 2022). Some metal elements such as iron are more soluble in a reduced environment whereas others such as copper are more soluble in an oxygenic environment. The decreasing availability of iron and increasing availability of copper from the ancient anoxic to modern oxic environments might have driven the transition from green-type algal dominance to red-type algal dominance. Recently, genomic analyses revealed striking accordance of the evolutionary trend in metal-binding proteins with that of the algal lineages (Zhang et al. 2022). In essence, the percentages of Fe-, Mn-, and Zn-binding proteins in phytoplankton are positively correlated with the half-saturation coefficients (*Ks*) of these metals, respectively. Furthermore, algal genome data show significantly lower percentage of metalloproteins in the secondary endosymbiosis-derived lineages (SEN) compared to the primary endosymbiosis-derived (PEN) lineages (Archaeplastida), and accordingly, the SEN lineages have lower metal requirements for Fe, Zn, and Cu than the PEN lineages.

Chloroplast Genomes and Proteomes

Chloroplasts are semi-autonomous organelles with several thousand proteins that participate mainly in the photosynthesis process but also in the biosynthesis pathways of many relevant compounds, e.g., fatty acids, amino acids, hormones, vitamins, nucleotides, and secondary metabolites (Dobrogojski et al. 2020). In the course of evolution, many genes of the ancestral chloroplasts have been transferred into the nuclear genome, such that most of the chloroplast proteins are nucleus-encoded. The number of chloroplast-encoded genes ranges from 0 to 315 in plants (Dobrogojski et al. 2020), and from 100 to 250 in most algal species (Lang and Nedelcu 2012), but an exceptionally small number of genes are found in lineages such as peridinin-containing dinoflagellates (16–17 genes) (He et al. 2024). The chloroplast-to-nucleus transferred genes in plants include important genes such as chloroplast translation initiation factor 1 (INFA), ribosomal proteins RPL22 and RPL32, the NAD(P)H dehydrogenase (NDH) complex, which is attached to PSI and mediates cyclic electron transport during the light reaction, and a large group of genes involved in DNA repair and recombination processes.

In plants cpDNA varies from 15,553 bp in *Asarum minus* to 521,168 bp in *Floydiella terrestris* (Dobrogojski et al. 2020). In algae, most chloroplast genomes range from 35–190 Kbp (see Table 3.1 in Chapter 3). These are over 20-fold smaller than the genomes of cyanobacteria (~ 1 to ~ 3 Mbp), indicating a huge genome reduction when the endosymbiont evolved into a chloroplast. Intriguingly, the proteome serving in plastids is also drastically smaller than that in a cyanobacterium (Rockwell et al. 2014). Despite the much smaller genome in the chloroplast, the expression of the chloroplast genes is regulated by much more complex systems than cyanobacteria genes; it is strongly dependent on post-transcriptional regulation, including the polycistronic mRNA processing, intron splicing, and RNA editing. In algae, RNA editing is also known to occur in dinoflagellates and euglenophytes, and transcripts of dinoflagellate plastid-encoded genes contain a poly U tail instead of poly A tail characteristic of eukaryotic mRNA (see Chapter 8).

Due to EGT, the nucleus of an algal species retains traces of plastid history. Phylogenetic analysis of the nuclear-encoded plastid-targeted protein genes can shed light on the historical plastid imprints. With this approach, Frommolt et al. (2008) found that the common ancestor of the chromist acquired carotenoid biosynthesis genes from green algae. This suggests that this chromist ancestor might have had a green plastid before the acquisition of the red plastid existing today. For diatoms, even though the widespread algae possess plastids derived from red algae, their nuclear genomes encompass a significant proportion of genes that seem to originate from green algae. Anywhere between 13 and 70% (Moustafa et al. 2009, Deschamps and Moreira 2012), and probably ~ 50% (Morozov and Galachyants 2019), of EGTs can be traced from cyanobacteria to green algae to diatoms, while the rest are of red algal origin. Similar mosaic sources of nuclear-encoded plastid protein genes have been reported in dinoflagellates (Nosenko et al. 2006, Minge et al. 2010). The mosaicism of plastid proteomes renders algae a "recycle center".

Evolution of Photosystems and Algae with Single Photosystem

PSII and PSI

Green sulfur bacteria and purple bacteria are considered ancestors of cyanobacteria because they possess reaction centers resembling PSI and PSII, respectively (Imhoff et al. 2018). Ancestral forms of oxygenic photosynthesis powered by homodimeric PSII and PSI emerged in the early Archean or early Paleoproterozoic (Fig. 2.1). PSII is functionally upstream of PSI but is so named because it was discovered after PSI. By convention, gene names for photosystems are formed by Psb + subunit letter for PSII and Psa + subunit letter for PSI. The exceptions are D1 and D2 of PSII, with their genes designated as PsbA and PsbD, respectively. PSII is composed of proteins D1 and D2 as the reaction center protein family, CP43 (encoded by PsbC) and CP47 (encoded by PsbB) as the light-harvesting proteins, and a set of coenzymes and cofactors (manganese center, also known as oxygen evolving center or OEC; chlorophyll, carotenes, heme B559, phaeophytin, and plastoquinone). Ancestral PSII operated with the core protein D0, which was subsequently duplicated leading to D1 and D2 and assembled into a primordial water-splitting photosystem (Cardona et al. 2019). How PSII evolved to acquire specific traits that facilitated oxygen evolution is unclear, however. The most recent common ancestor of Cyanobacteria inherited a heterodimeric photosystem (PSI and PSII) shared by all extant oxygenic phototrophs. Photosystem II has maintained remarkable stability over billions of years; however, the continuous duplication of the D1 subunit, responsible for controlling photochemistry and catalysis, has allowed the enzyme to adapt to diverse environmental conditions and evolve new enzymatic functions beyond water oxidation (Oliver et al. 2023).

PSI employs light energy to facilitate the transfer of electrons across the thylakoid membrane from plastocyanin to ferredoxin. In the end, electrons transferred by PSI are utilized to produce the moderate-energy hydrogen carrier NADPH. Additionally, the photon energy absorbed by PSI generates a proton-motive force, which is utilized to produce ATP. PSI consists of over 110 cofactors, a significantly larger number compared to Photosystem II.

Single PS

Although most of the naturally occurring algae operate on the existence of both PSI and PSII, natural or mutated species with only one of these do exist. As will be elaborated in Chapter 19, a geographically widespread but uncultivable unicellular cyanobacterium capable of N_2-fixation, UYCN-A, has no PSII (Zehr et al. 2008). By flow cytometric cell sorting, the authors isolated a thousand cells of UYCN-A among the co-existing cyanobacteria (*Synechococcus* and *Prochlorococcus*) and picoeukaryotes. The genomic DNA was amplified using the Multiple Displacement Amplification (MDA) technique and sequenced using 454 Titanium technology to achieve approximately 10x depth coverage. The data showed that the overall sequencing depth was sufficient as a substantial number of genes from the contaminating cyanobacteria *Prochlorococcus* and *Cyanothece*- or *Crocosphaera*-like species (all in low abundance in the sorted cell population) were retrieved, and the complete nitrogenase gene cluster of UCYN-A was obtained. The absence of any PSII protein genes in the dataset was evidence that this photosystem had been lost in evolution, which is adaptive to the fact that nitrogenase is sensitive to oxygen and that this species tends to form symbiotic relationship with a eukaryotic alga (Thompson et al. 2012). In contrast, for the chlorophyte *Chlamydomonas reinhardtii*, PSI-deficient mutants can stably grow photoautotrophically on CO_2 and produce O_2 and H_2 (Greenbaum et al. 1995). These mutants, created by metronidazole enrichment, are stable both under anaerobic and aerobic conditions at a photon flux of 200 μmol photons m^{-2} s^{-1} and can survive as high light intensity as 2000 μmol photons m^{-2} s^{-1} with air, although they are less resistant to photoinhibition (Lee et al. 1996).

Evolution of Chlorophylls

Chlorophylls (Chls) are crucial for charge separation and electron transfer in the reaction centers of photosynthetic apparatus for oxygenic photosynthesis. The light absorption capability of these photosynthetic organisms is determined by their specific chlorophylls. Anaerobic phototrophic bacteria operate on bacteriochlorophylls (bChls) *a, b, c, d, e, f,* or *g*, all with a phytol tail, and their absorption maxima are in the infra-red region. bChl is widely present in photosynthetically active proteobacteria, purple bacteria and sulfur oxidizing bacteria and absorbs maximally at 805 nm; bChl *c, d* and *e* widely occur in green sulfur bacteria and absorbs light in the 720–755 nm region; bChl *g* absorbs red light of 670 nm or 788 nm (Smith et al. 1982). Similarly, oxygenic photosynthetic organisms (cyanobacteria, eukaryotic algae, and plants) have multiple (five) types of chlorophylls. Chl *a*, Chl *b*, Chl *d*, and Chl *f* are Mg-chlorin-type chlorophylls with a phytol chain esterified at $C17^3$ on ring D. Chl *b*, Chl *d*, and Chl *f* differ from Chl *a* due to formyl substitutions on ring A and ring B of the macrocycle, which increase the hydrophilicity of the chlorophylls, and Chl *c* (c1, c2, or c3) differs from other chlorophylls for its lack of the phytol tail. Chl *a* unites all oxygenic photosynthetic organisms, including cyanobacteria and chloroplasts; Chl *b* is unique to green lineages (chlorophytes, euglenophytes, prasinophytes, streptophytes); Chl *c* occurs widely in non-green (except rhodophyte) chloroplasts; Chl *d* and *f* are present in chlorophytes and cyanobacteria (*d*) or stromatolites (*f*). Chl *e* is from the algae *Tribonema bombycinum* and *Vaucheria bamata*, characterized by absorption bands at 415 and 654 nm in methanol, but neither its chemical nature nor the organism producing it has received necessary investigation (Chen 2019). Whereas Chl *a, b, c,* and *e* absorb blue light, Chl *d* and *f* absorb red and infra-red light (Larkum and Kühl 2005, Chen 2019).

The biosynthesis of bacteriochlorophyll and chlorophyll both proceeds through chlorophyllide *a*, hence sharing a backbone set of steps. Because bacteriochlorophyll is biochemically more complex to synthesize on the modern biosynthetic pathway, it has been proposed that chlorophyll preceded bacteriochlorophyll in its evolution (Ward and Shih 2021 and references therein). However, alternative hypotheses have also been proposed invoking the complete bacteriochlorophyll synthesis pathway in the last common ancestor of extant phototrophs followed by secondary simplification of the pathway in Cyanobacteria (Xiong et al. 2000). Because other molecular components of photosynthesis produce incongruent phylogenetic orders, it also has been posited that bacteriochlorophyll might have emerged indepenedently through horizontal gene tranfer (Ward and Shih 2021). The issue remains to be resolved with more data.

Evolution of Photosynthetic Carbon Fixation and Rubisco

Seven mechanisms have been identified for carbon fixation (Kono et al. 2017), including reductive tricarboxylic acid (rTCA) cycle, reductive acetyl-CoA (or Wood-Ljungdahl [WL]) pathway, 3-hydroxypropionate/4-hydroxybutyrate (3-HP/4-HB) pathway, dicarboxylate/4-hydroybutyrate cycle (DC/4-HB) pathway, Reductive Hexulose-Phosphate (RHP) pathway, and Calvin-Benson-Bassham (CBB) cycle. Among these, CBB is the most dominant carbon fixing mechanism, occurring in all oxygenic photosynthetic organisms (cyanobacteria, algae, and plants) and in some prokaryotic lineages including some alpha-, beta-, and gamma-proteobacteria, sulfobacillus, and Oscillochloridaecea (Hügler and Sievert 2011).

The key enzymes for the Calvin-Benson-Bassham (CBB) cycle are ribulose 1,5-bisphosphate (RuBP) carboxylase/oxygenase (Rubisco), Fructose 1,6-bisphosphatase (FBPase), Sedoheptulose-bisphosphatase (SBPase), and Phosphoribulokinase (PRK). Rubisco catalyzes the carboxylation of RuBP to give two molecules of 3-phosphoglycerate (PGA). This is the rate-limiting step of photosynthetic carbon fixation. SBPase operates at a pivotal junction within the Calvin cycle, where it regulates the pathway's divergence between the regeneration of the acceptor molecule ribulose 1,5-bisphosphate (RuBP) and its conversion to starch biosynthesis. FBPase catalyzes the hydrolysis of fructose 1,6-bisphosphate to fructose 6-phosphate. PRK is responsible for catalyzing the ATP-dependent phosphorylation of ribulose 5-phosphate (RuP) to produce ribulose 1,5-bisphosphate (RuBP), both being intermediates in the CBB.

The Rubisco-coding gene exists in four forms: Forms I, II, III, and IV. The majority of the highly diverse algae use form I Rubisco for carbon fixation. Even though the form I Rubisco gene is generally conserved, substantial variation still exists, which further divides this form into four subgroups: Ia, Ib, Ic, and Id. Dinoflagellates either possess a form II or a form I Rubisco depending on the lineages. Typical peridinin dinoflagellates operate on form II Rubisco, which is believed to be acquired from anaerobic photosynthetic bacteria through horizontal gene transfer (Morse et al. 1995). Other photosynthetic dinoflgellates have plastids of haptophyte (Kareniaceae), cryptophyte, chlorophyte, or diatom origin, which operate with form I Rubisco. It is believed that these lineages lost their original peridinin-containing plastids and acquired their plastids from chlorophyte (named serial secondary plastid) and cryptophytes or haptophytes (both named tertiary plastids). It is worth mentioning that *Gertia stigmatica* is a species in the Kareniaceae family, but its chloroplast is a peridinin-containing type (Takahashi et al. 2019), a notable exception to the general trend.

Evolution of Carbon Concentrating Mechanisms (CCMs)

When oxygenic photosynthesis first emerged (in cyanobacteria) over a billion years ago, the environment was highly reduced and CO_2-rich. In correspondence, the CO_2-fixing enzyme ribulose-1,5-bisphosphate carboxylase/oxygenase (Rubisco) in the CBB was originally best suited to high CO_2 and low O_2 conditions. As such, Rubisco is very inefficient to capture CO_2 converting it to glucose and is sensitive to O_2. The carboxylase activity relative to oxygenase activity of Rubisco is influenced by the ratio of $[CO_2]/O_2]$. The CO_2 half saturation coefficient for Rubisco mostly falls in the range from 10 to 100 μM but can be over 200 μM in some photosynthetic eukaryotes (Flamholz et al. 2019), while typical CO_2 concentration in the contemporary ocean is 10–15 μM (see Chapter 15). Therefore, Rubisco in marine algae is less than half-saturated with CO_2 in the modern ocean (Falkowski 1997), even in the face of Ocean Acidification (OA).

To compensate for the low affinity of Rubisco for CO_2, except for some terrestrial chlorophytes and some freshwater rhodophytes, most algae have evolved the Carbon Concentrating Mechanisms (CCMs) so that cells can take up the abundant bicarbonate in the ocean as the source of CO_2 for the photosynthetic apparatus, and the efficiency of CO_2 fixation is enhanced (Giordano et al. 2005). There are two types of CCMs, one known as biophysical and the other biochemical. The biophysical CCM is composed of bicarbonate transporter to take up HCO_3^- and Carbonic Anhydrase (CA) to facilitate the interconversion between HCO_3^- and CO_2. There are several forms of CA, named alpha, beta, gamma, delta, epsilon, and theta, which have different subcellular localizations (extracellular, cytoplasmic, plastidic). These enzymes typically contain Zn or Co as cofactor but, strikingly, one CA in diatoms uses Cd as the cofactor, which is the only known case in which the toxic heavy metal has biological functions. Cyanobacteria possess a CA localized in the carboxysome; α-carboxysomes contain a β-CA (CsoSCA), so does a subset of β-carboxysomes (CcaA) (Kimber 2014). A recent study showed that CsoSCA in a *Cyanbobium* strain was a hexamer in solution and its activity was regulated by binding to RuBP, the substrate of Rubisco (Pulsford et al. 2024).

Biochemical CCM includes the C4 and CAM pathways. Most photosynthetic organisms directly fix the CO_2 coming from the environment or converted from HCO_3^- imported from the environment to produce two three-carbon compounds, hence named C3 phototrophs. For C3 plants, when the leaves open the stromata (small holes) to allow infusion of CO_2, water vapor leaks out. This water loss is problematic, particularly in dry and hot areas. Convergent evolution has occurred multiple times in different lineages of plants giving rise to the C4 pathway to ameliorate the water loss problem and enhance the efficiency of carbon fixation. In this pathway, which occurs in about 3% of plants, HCO_3^- is first fixed onto phosphoenolpyruvate (PEP) by PEP carboxylase (PEPC) in the mesophyll to produce oxaloacetate, which is further converted to Malic Acid (MA) or aspartic acid (Asp), a four-carbon compound. The four-carbon compounds then diffuse to the bundle sheath cells where they are decarboxylated (e.g., by Malic Enzyme [ME] or PEPCK, for MA or Asp, respectively) to release CO_2 to be fixed by Rubisco and pyruvate to be recycled back in the mesophyll. The C4 pathway has the advantage that PEPC does not fix O_2, unlike Rubisco, thus avoiding photorespiration, and stores CO_2 in the C_4 molecule for use by Rubisco without the need to open stromata, thus avoiding water loss. For algae, there is no water vapor loss issue, but low efficiency of carbon fixation is similar to land plants. Algae were generally known only to use C3 pathway, but C_4 pathway has been documented in various lineages of algae, such as the macrophytes *Udotea flabellum* (Reiskind and Bowes 1991) and *Ulva prolifera* (Liu et al. 2020), the diatom *Conticribra* (=*Thalassiosira*) *weissflogii* (Reinfelder et al. 2000), the euglenophyte *Eutreptiella* sp. (Kuo et al. 2013), and the dinoflagellates *Fugacium kawagutii* (Lin et al. 2015) and *Prorocentrum shikokuense* (Zhang et al. 2021). However, the presence of C4-metabolic enzymes is not definitive evidence that the species performs C_4 photosynthesis. As to be detailed in Chapter 7, model diatoms *T. pseudonana* and *P. tricornutum* seem not to perform C4 photosynthesis even though they possess all the C4 enzymes (Roberts et al. 2007, Ewe et al. 2018).

In some plants like Crassulaceae, which live in arid environments, the PEPC (C4) and Rubisco (C3) reactions are temporally separated, a pathway is called Crassulacean Acid Metabolism (CAM). During the night, CAM plants open their stomata to allow CO_2 to enter and be fixed in the cytoplasm of mesophyll cells by a PEP reaction similar to that of C4 pathway. The resulting organic acids are stored in the vacuoles. During the day, the CO_2-storing organic acids are released from the vacuoles of mesophyll cells and enter the stroma of the chloroplasts, where an enzymatic reaction releases the CO_2 for Rubisco to complete fixation. This process is completed without opening of stromata, and thus the risk of water loss is avoided. Occurrence of CAM in aquatic plants have been implicated. Some studies also have suspected that CAM occurs in some algae to avoid photorespiration, but definitive evidence is still lacking.

Carbon Storage

The Photosynthetically produced Organic Carbon (PhOC) is stored in different chemical forms of polysaccharides in different lineages of algae (Table 2.1), and the storage in algae is distinctive from that in animals and other organisms. Two of the three Archaeplastida lineages, chlorophytes and glaucophytes, store PhOC as starch, as do the chromalveolates, dinoflagellates, apicomplexans, and cryptophytes. The third Archaeplastida lineage, rhodophytes, produces floridean starch. Ochrophytes and haptophytes store chrysolaminarin although some ochrophytes can also store lipids. As demonstrated in *Guillardia theta*, the storage polysaccharide granules in cryptophytes are composed of both amylose and amylopectin fractions with a chain length distribution and crystalline organization similar to those of starch in green algae and land plants (Deschamps et al. 2006). Accordingly, *G. theta* possesses an enzyme resembling the amylose-synthesizing granule-bound starch synthase 1 in green plants, and it is the major polypeptide bound to the polysaccharide matrix. The enzyme contains a nonfunctional transit peptide in its bipartite topogenic signal sequence, which

Table 2.1. Predominant photosynthetic pigments, storage products, and cell wall components for the major algal groups.

	Chlorophylls	Accessory pigments	Storage products	Cell covering
Cyanobacteria (Chloroxybacteria)	chlorophyll *a* chlorophyll *b* (species w/o phycobilins) chlorophyll *d*, *e*, or *f* (rare species)	phycocyanin, allophycocyanin phycoerythrin, β-carotene	cyanophycin granules, cyanophytan starch (glycogen)	peptidoglycan
Glaucophyta	*chlorophyll a*	phycocyanin, allophycocyanin β-carotene, xanthophylls	starch	some cellulosic
Euglenophyta	*chlorophyll a, c*	β-carotene, other carotenes, xanthophylls	paramylon	proteinaceous pellicle beneath plasma membrane
Cryptophyta	*chlorophyll a, c*	phycocyanin, phycoerythrin α, β-carotene, xanthophylls		proteinaceous pellicle beneath plasma membrane
Haptophyta	*chlorophyll a, c*	β-carotene, xanthophylls	chrysolaminaran	$CaCO_3$ scales common
Dinophyta	*chlorophyll a, c*	β-carotene, xanthophylls	starch	cellulosic plates in vesicles beneath plasma membrane
Ochrophyta (diatoms, chrysophytes, brown algae, etc.)	*chlorophyll a, c*	β-carotene, xanthophylls	chrysolaminaran, lipids	some naked; some with silica/ organic scales; cellulose, alginates in some
Eustigmatophytes	*chlorophyll a*	β-carotene, xanthophylls	chrysolaminaran	present, but unknown composition
Rhodophyta	*chlorophyll a*	phycocyanin, phycoerythrin allophycocyanin α, β-carotene, xanthophylls	floridean starch	cellulose, sulfated polysaccharides; some calcified
Chlorophyta	*chlorophyll a, b*	β-carotene, other carotenes, xanthophylls	starch	wall of cellulose/other polymers; scales on some; some naked; some calcified

is cleaved without leading the enzyme to cross the two inner plastid membranes, consistent with the perplastidial locality of starch synthesis in *G. theta*. The enzyme exhibits similar affinities for ADP- and UDP-glucose, but the maximum reaction velocity (Vmax) measured with UDP-glucose is twofold higher, suggesting that *G. theta* mainly utilizes a UDP-glucose-based pathway to synthesize starch.

Cytochrome P450

Cytochrome P450 enzymes (P450s) are a superfamily of heme (iron protoporphyrin) proteins widely found in humans, animals, plants, algae, fungi, bacteria, archaea, and even viruses (Nelson 2018). They are named "P450" due to their characteristic spectroscopic absorption peak at 450 nm when their reduced heme-iron center binds with carbon monoxide. This family of proteins are involved in both primary and secondary metabolisms, ranging from the synthesis of membrane sterols and phytohormones, to signaling molecules (Zheng et al. 2022). One of their major functions is catalyzing common monooxygenation (e.g., hydroxylation and epoxidation) reactions to modify core structure and improve aqueous solubility of metabolites or detoxify pollutants. In algae, P450s are involved in the metabolism of a wide range of metabolites such as xanthophylls, fatty acids, polychlorinated biphenyls, jasmonates, and retinoic acids. Based on the type of redox partners, algal P450 catalytic systems can be classified into two major types (I and II). Blue-green algae possess Type I catalytic systems, which partners with FAD-containing ferredoxin reductase (FdR) and a small iron-sulfur ([Fe-S]) protein ferredoxin (Fdx). In contrast, eukaryotic algae mainly use the Type II P450 catalytic system, which relies on a single FAD/FMN-containing flavoprotein to transfer the reducing equivalents. While the Type I systems are composed of cytosolically soluble proteins, the algal Type II systems are usually endomembrane-associated (Zheng et al. 2022). A high number of P450-encoding genes is present in the genomes of algae as well as other organisms, but they are poorly understood, despite their immense ecological and biotechnological implications.

Flagella

Cells from all three domains of life on Earth employ mobile macromolecular devices that extend from the cell surface to generate forces enabling cells to navigate through fluid media (Khan and Scholey 2018). The motility organelles of the three domains are fundamentally different in their properties and ancestry, despite their common function, indicating convergent evolution (Fig. 2.5). Flagella emerged about 2.75 Byr ago, before the origin of proteobacteria (Cavalier-Smith 2006). Bacterial flagella and archaeal archaella both utilize rotary filaments, composed of distinct subunits originating from unrelated ancestors, to generate torque. However, they employ different energy sources: chemiosmotic ion gradients for bacteria and ATP hydrolysis catalyzed by the FlaI motor for archaea. In contrast, eukaryotic flagella (or cilia) assemble through kinesin-2-driven intraflagellar transport, utilizing microtubules and ATP-hydrolyzing dynein motors to propel movement via a sliding filament mechanism, producing a variety of waveforms (Khan and Scholey 2018).

In bacteria, including cyanobacteria, the flagellum undergoes tip-oriented assembly. This process begins with subunit synthesis in the cytoplasm, followed by secretion into the flagellar filament's central channel via the type III secretory complex and proton-motive force. Subunits are then translocated along the channel via facilitated diffusion. Flagellar rotation is powered by chemiosmotic mechanisms driven by ion gradient dissipation. Gram-negative bacteria possess L and P rings due to their extra outer membrane, absent in Gram-positive bacteria. Polar flagella are anchored in large outer wall disk structures. Stator complexes oscillate between bound and free states. The FliI ATPase, along with accessory proteins FliH and FliJ, aids in ATP hydrolysis to load flagellin subunits into the export machinery's membrane component (Khan and Scholey 2018).

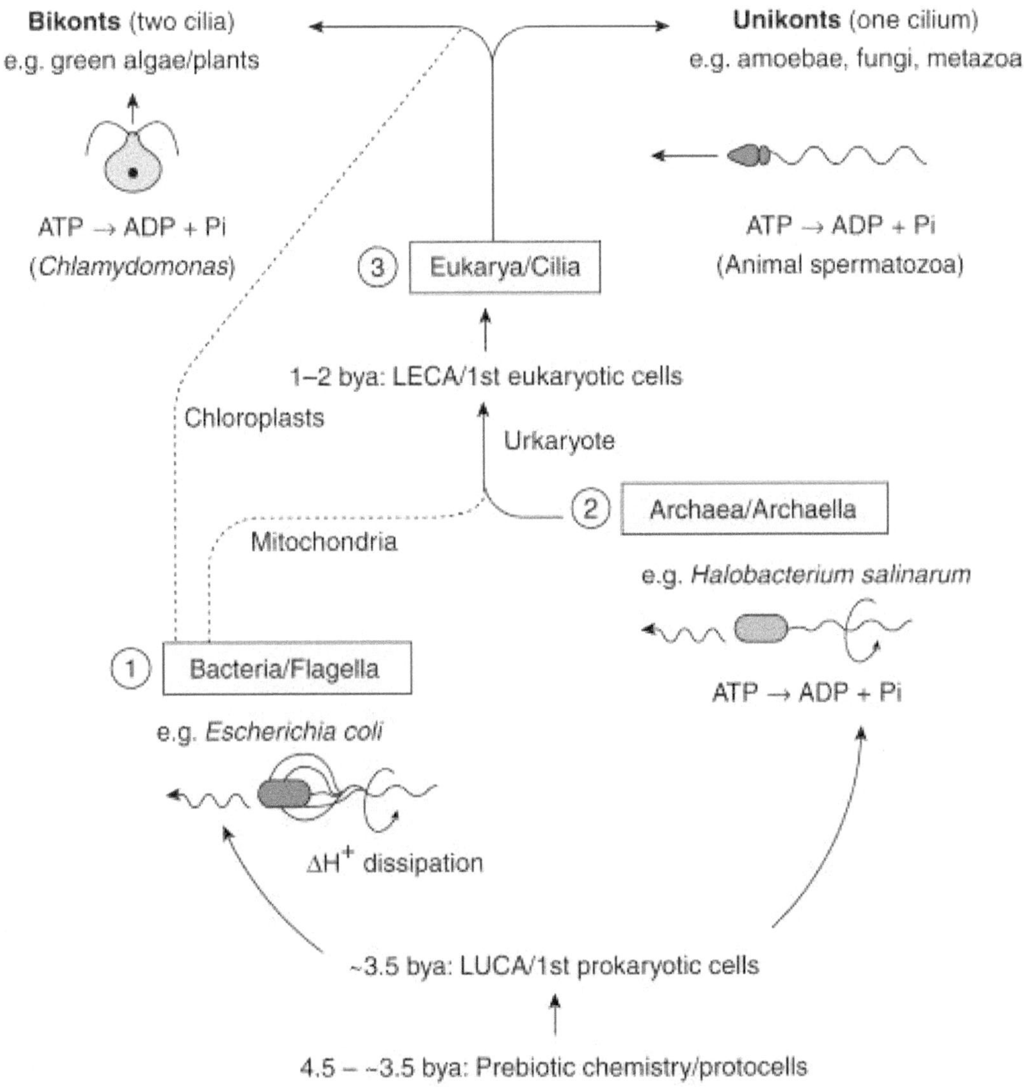

Figure 2.5. Evolution of the three domains of life and their archaella, flagella and cilia. Reproduced with permission from Khan and Scholey 2018.

Cyanobacteria are capable of gliding motility, but lack flagella (Cavalier-Smith 2006), having apparently lost them in evolution.

Eukaryotic cilia/flagella are cellular projections that are ~ 10-fold larger and more complex than bacterial flagella (Fig. 2.6). Comprising an axoneme of nine doublet microtubules made of tubulin projecting from a basal body located at the cell surface, with their plus ends distal, cilia/flagella are surrounded by a cytosolic matrix and a specialized ciliary/flagellar membrane. Cilia/flagella serve various functions, including:

1. Cell motility: they facilitate swimming in low Reynold's number environments and intraflagellar transport (IFT), which aids surface attachment and gliding motility.
2. Sensory reception: Acting as antennae, they detect sensory cues such as chemical gradients or light photons, transmitting signals to the cytoplasm or nucleus, thus regulating gene expression and cell function.

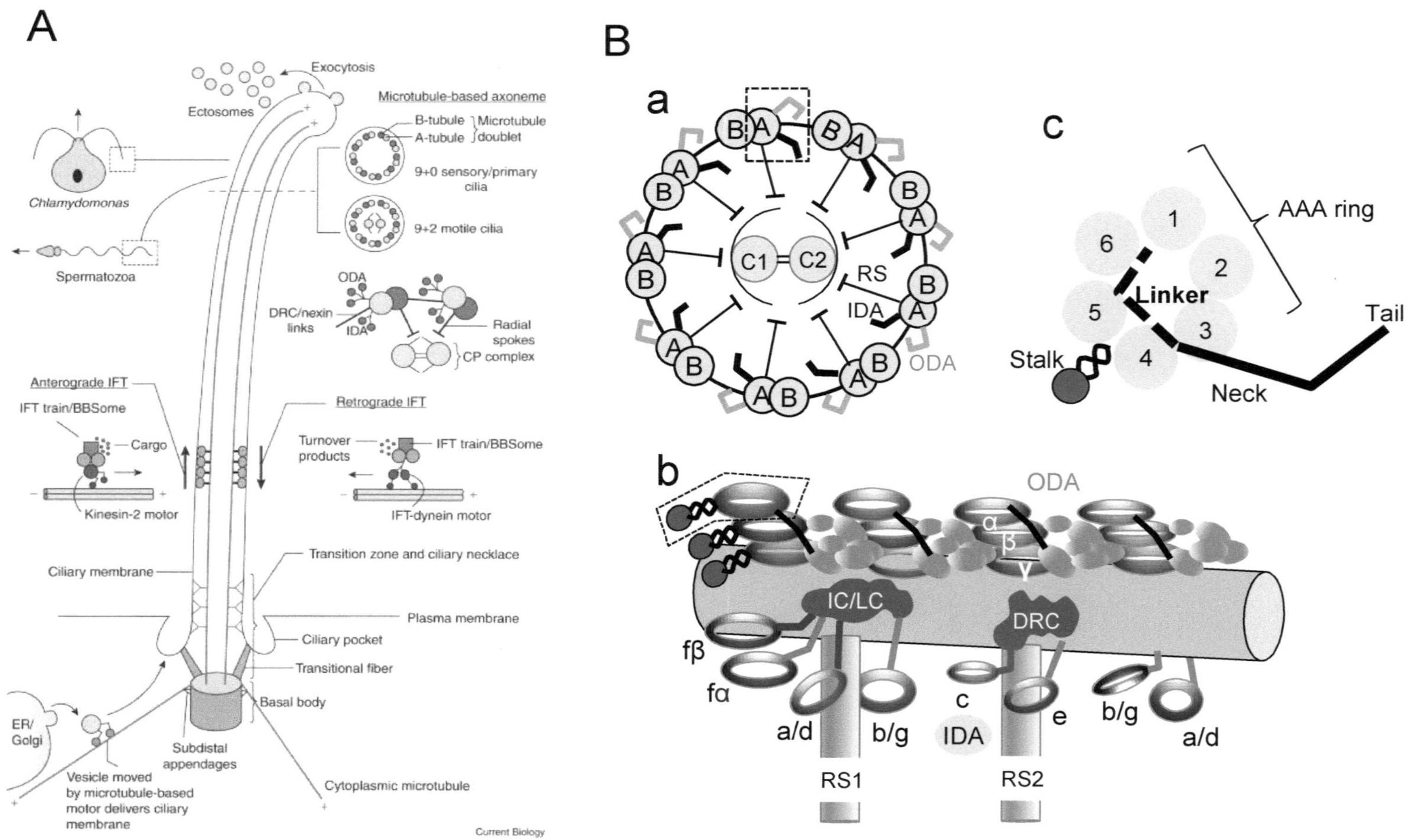

Figure 2.6. Schematic architecture of a eukaryotic flagellum based on *Chlamydomonas reinhardtii*. (A) Assembly and function of a flagellum. (B) Tubule arrangement and dynein arms. (a) ('9 + 2') axoneme structure of *C. reinhardtii* flagella. ODA, outer dynein arm; IDA, inner dynein arm; RS, radial spoke; C1&C2, central pairs; A&B, A- and B-microtubules. (b) Closer-up of IDAs and ODAs on one A-microtubule (the region enclosed by a square in (a)). IC/LC of dynein f and the dynein regulatory complex (DRC) are indicated. Possible locations of IC/LC of ODA is shown by grey bulbs. The stalk of dynein is drawn only in one ODA. (c) Closer-up of one dynein heavy chain (extracted from (b) enclosed by dotted line). The six AAA domains, the tail, the linker, the neck and the stalk are indicated. (A) is reproduced with permission from Khan and Scholey 2018 and (B) is based on Movassagh et al. 2010.

3. Signaling-dependent control: They regulate processes such as autophagy-mediated degradation and recycling of cellular components.
4. Intercellular transmission and reception: Secreting bioactive extracellular vesicles in an IFT-dependent manner from their tips, they facilitate the transmission and reception of signals between cells (Khan and Scholey 2018).

Most algae have flagella throughout or at some point of the life history. For instance, vegetative cells of diatoms do not have flagella but their gametes do. Functionally, flagella can be motile or nonmotile. Structurally, some flagella are smooth and others with hairs. The group of stramenopile has two flagella, one smooth and the other with tripartite hairs (hence also known as heterokonts). These hairs are glycoproteins produced in the Peripheral Endoplasmic Reticulum (PER) and are moved out of the cell through exocytosis. Flagella-enabled motility facilitates exploration of nutrient sources to depths and photoenergy to surface through chemotactic and phototactic clues, some of the ecologically important functions.

The chlorophyte *C. reinhardtii* has served as a model for the anatomy of flagella. As shown in Fig. 2.6, its flagellar axoneme consists of nine outer doublet microtubules with a central pair (9+2), characteristic of motile cilia (Merchant et al. 2007). Comparative analysis for *C. reinhardtii* genome with then limited number of other algal genomes as well as non-algal genomes revealed 186 cilia-specific protein (CiliaCut) families, which are represented by 195 proteins in *C. reinhardtii.* A subset (125) of the CiliaCut were predicted to be flagellar structural components (16%), to mediate protein-protein interactions (26%), transduce signals (11%), and function in GTP-binding (6%) and trafficking (6%). Of the CiliaCut proteins, 133 are absent in the nonmotile *Caenorhabditis*, hence potentially specific to motility, whereas 62 shared with *Caenorhabditis* and hence are predicted to have structural, sensory, or assembly roles in the cilium.

In comparison, the moss *Physcomitrella* misses five of the outer dynein arm proteins, but contains inner dynein arm subunits IDS4 and DH2, central microtubule subunits, the radial spokes, and the dynein regulatory complex. In contrast, the motile flagella of centric diatoms lack the central pair of microtubules, radial spoke, and inner dynein arm proteins, although 69 of the 195 CiliaCut proteins were predicted to occur in *Thalassiosira*, including most of the outer dynein arm proteins. These comparative results suggest that the central pair and radial spokes function as a unit with the inner arms, but are not essential for generating motility by the outer arms (Merchant et al. 2007).

Outer and inner dynein arms generate force for the bending motion of the flagella. Electron cryotomographic analysis of *C. reinhardtii* flagella indicated that the ATPase domains of the dynein heavy chains, which are the ATP-driven motors, move toward the distal end of the chains. The N-terminal tails bend sharply during product release, thereby driving the adjacent microtubule to cause a sliding motion (Movassagh et al. 2010).

Green Autofluorescence

Various species of microalgae (e.g., dinoflagellates, diatoms) have been reported to emit green autofluorescence (GAF) (Tang and Dobbs 2007 and references therein). GAF can occur in the cell body of vegetative cells or cysts, spines, or flagella, depending on the species. In *Dunaliella tertiolecta*, both the flagella and the cell body emit GAF; however, while that in the flagella is inducible by Cd, low pH, and high light exposure, that in the cell body is not (Pavlinska et al. 2020). A light-sensitive flavin-like compound has been suspected to be the cause of GAF (Kawai 1988), and a fluorescent flavoprotein has been identified (Fujita et al. 2005). However, the exact identities of the fluorescent molecules in different species of algae and the mechanism of their biosynthesis have yet to be resolved. Additionally, the evolutionary mechanism and the fitness benefit of GAF remain to be deciphered.

Aquaporins (AQPs)

One of the most significant biological discoveries of the last century was about the existence of aquaporins (AQPs) as water channel (Agre et al. 1993), a breakthrough recognized with the Nobel Prize in chemistry in 2003. Although aquaporins were named in 1993, their existence was known earlier (e.g., Benga et al. 1986). AQPs are water channels across cell membranes that allow transport of small solutes as well as water in both plants and animals (Laloux et al. 2018). Initially identified as water channels, AQPs were later found to conduct small solutes such as urea, ammonium, and glycerol. Consequently, these transporters are collectively referred to as Membrane Intrinsic Proteins (MIPs), a term used interchangeably with AQPs. Both plants and animals have multiple types of MIPs, including Plasma membrane intrinsic Proteins (PIPs), Tonoplast Intrinsic Proteins (TIPs), Nod26-like MIPs (NIPs), small basic intrinsic proteins (SIPs), and unknown intrinsic proteins (XIPs) in plants (Maurel et al. 2008, Laloux et al. 2018).

In algae, AQPs have been reported in only few lineages (Table 2.2). In the chlorophyte *C. reinhardtii*, an AQP identified as a glycerol channel (aquaglyceroporin, abbreviated as GLP) was demonstrated to conduct glycerol (Anderca et al. 2004). AQP gene homologs have also been identified in the diatoms *P. tricornutum* (PtAQP) and *T. pseudonana* (TpAQP). in *P. tricornutum*, Green Fluorescence Protein (GFP)-AQP fusion proteins were localized to the plasma membrane, the Chloroplast Endoplasmic Reticulum (CER), and the tonoplast, while in *T. pseudonana*, they were localized to the vacuole membrane and CER (Matsui et al. 2018). Furthermore, PtAQP gene transcription was induced by NH_3 (PtAQP1 and PtAQP2), high CO_2 (PtAQP2). Constitutive overexpression of GFP-tagged PtAQPs significantly increased CO_2 and NH_3 permeability and exhibited elevated nonphotochemical quenching under high light, suggesting they are channels of CO_2/NH_3 and provide photoprotection.

An AQP was recently identified in the charophyte *Klebsormidium nitens* and was functionally characterized, using *Xenopus laevis* oocytes as the experimental system, as conduit of water and ions (McGaughey et al. 2021). A GLP was identified from the genome of the putative coral symbiont *Fugacium kawagutii* (formerly *Symbiodinium kawagutii*), presumably serving to translocating glycerol to coral hosts (Lin et al. 2015). However, using AQP sequence of *A. thaliana* as query, blast analysis yielded significant hits in a wide range of algae, including chlorophytes, rhodophytes, dinoflagellates, diatoms, haptophytes, pelagophytes, ochrophytes, and cryptophytes, suggesting that this transporter is widespread and conserved in algae (Table 2.2). Their functions, however, remain to be experimentally demonstrated.

Table 2.2. Aquaporin genes identified in algae.

Phylum	Species	Gene name	Function	Gene ID or NCBI accession no.	References
Charophyta	*Klebsormidium nitens*	PIP-like	water and ions transport*	GAQ80027.1	McGaughey et al. 2021
	Chara longifolia	PIP	water transport	CL_NODE_ 34384	Phipps et al. 2021
	Chara brauni	TIP	water and others transport	98238.t1	Phipps et al. 2021
Chlorophyta	*Chlamydomonas reinhardtii*	putative aquaporin	glycerol transport*	AAP33478.1	Anderca et al. 2004
	Monoraphidium neglectum	aquaporin	glycerol transport	XP_013897215.1	Bogen et al. 2013
	Coccomyxa subellipsoidea C-169	CcPIP4;1	water transport	XM_005647020.1	Anderberg et al. 2011
Rhodophyta	*Galdieria sulphuraria*	aquaglyceroporin related protein	glycerol transport	XP_005702633.1	Schönknecht et al. 2013
	Pyropia yezoensis	PyAQP1	glycerol transport	KT735045	Kong et al. 2017
	Gracilaria domingensis	aquaporin TIP4-1	Unknown	KAI0562366.1	Nakamura-Gouvea et al. 2022
Diatoms	*Fragilaria crotonensis*	glycerol channel	glycerol transport	KAI2513945.1	Zepernick et al. 2022
	Phaeodactylum tricornutum	PtAQP1	NH_3 transport	BBB86745.1	Matsui et al. 2018
	Phaeodactylum tricornutum	PtAQP2	CO_2 and NH_3 transport	BBB86746.1	Matsui et al. 2018
Dinoflagellates	*Fugacium kawagutii*	aquaglyceroporin	glycerol transport	Skav220488	Lin et al 2015
	Karlodinium veneficum	aquaglyceroporin-like	glycerol transport	KvenUN24968.p1	SAGER (http://sampgr.org.cn/)
	Symbiodinium sp.	Symb-GlpF	glycerol transport	UXG07099.1	Teng et al. 2022
	Cladocopium sp.	Clad-GlpF	glycerol transport	QTF34003.1	Teng et al. 2022
	Durusdinium sp.	Duru-GlpF	glycerol transport	QTF34004.1	Teng et al. 2022
	Polarella glacialis	MIP-like	water transport	evm.model.CCMP1383_ scf7180000595586.1	SAGER (http://sampgr.org.cn/)
Haptophyta	*Emiliania huxleyi*	AQP2	water and CO_2 transport	75635	Mackinder et al. 2012
	Chrysochromulina tobinii	mip family channel protein	glycerol transport	KOO26607.1	Hovde et al. 2014
Cryptophyta	*Guillardia theta*	MIP	Unknown	XP_005826812.1	Curtis et al. 2012
	Cryptophyta sp.	MIP	Unknown	KAJ1479813.1	Kuo et al. 2018

* Functions have been verified

3

Genome Evolution, Plasticity, and Adaptation

Variations and Evolutionary Trend of Genome Size

Genome size variation is ubiquitous. However, understanding the driving forces behind this variation has proven challenging. Multiple hypotheses have been proposed to explain the patterns of genome size, but a consensus explanation has yet to emerge. Any attempt to address this question should begin with an analysis of the trends and patterns of genome dynamics.

Macroevolutionary and Microevolutionary Trends of Genome Size

The range of genome size across domains of life is vast (Table 3.1). Viral genomes range from 2 to 400 Kbp. Bacterial genomes typically range from 0.5 to 9 Mbp, while mitochondrial and chloroplastic genomes range from 6-490 and 35-190 Kbp, respectively. Some exceptionally large plastid genomes exist; for instance, the chloroplast genome of the single-celled Ulvophyte *Acetabularia acetabulum* exceeds 1 Mbp. Archaeal genomes range from 490 Kbp to 5.75 Mbp. Free-living eukaryotes possess nuclear genomes that span from 10 to 250,000 Mbp. There is a clear trend of genome expansion from viruses to bacteria and Archaea (20-fold), and from these prokaryotic domains to eukaryotes (up to 20,000-fold).

The evolutionary trend of genome size expansion from simple to complex forms of life does not hold at microevolutionary scale (e.g., between lineages within a phylum). For instance, in dinoflagellates, after the group branched off from Perkinsids, currently recognized species range from the ancestral parasitic lineages such as *Amoebophrya*, then to the bioluminescent genus *Noctiluca*, and proceed toward the more derived lineages such as Suesiales and *Kryptoperidinium*. Along this evolutionary sequence, their genomes underwent repetitive cycles of expansion and contraction (Fig. 3.1). For example, the coral symbiotic lineage Symbiodiniaceae has the smallest genomes among documented dinoflagellates, despite its later-diverging position in the phylogenetic tree (Janouškovec et al. 2017). The generally large genome sizes of dinoflagellates (compared to related lineages such as apicomplexans and diatoms) and the wide range of genome sizes within dinoflagellates suggest multiple independent events of genome expansion in the evolution of Dinophyta. This lack of phylogenetic trend of genome size variation is widely found in algae as well as other groups of organisms.

Table 3.1. Genome size ranges in different groups of organisms (Kbp).

Free-living Eukaryotes	Symbiotic Eukaryotic	Parasitic Eukaryotes	Bacteria	Archaea	Mitochondria	Chloroplasts	Viruses
10689.2–250,000,000	13,300–48,800	551.3–87,700	580.1–9,949.9	490.9–5,751.5	6.0–491.5	27–1,350	2.4–407

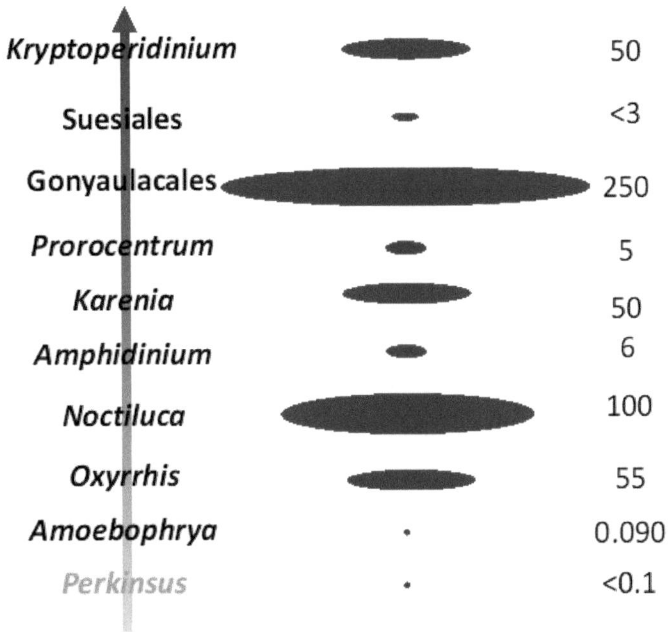

Figure 3.1. Evolutionary dynamics of dinoflagellate genome size (Gbp) showing the lack of a phylogenetic trend.

Genome Size Variation by Lifestyle

There is a clear difference in genome size between symbiotic organisms and their free-living relatives. Bacterial genomes range from 0.58 Mbp to 9.95 Mbp, larger than their symbiotic counterparts, the mitochondrial genomes (6-491 Kbp) and plastid genomes (27-1,350 Kbp). Genomes of eukaryotes also show clear dichotomic divergence with lifestyle: parasitic lineages have genomes of 0.55 to 9.71 Mbp, while free-living lineages have genomes of 10 Mbp to over 200 Gbp. Parasitic eukaryotes have undergone dramatic genome reduction (by 15 to 2,600 folds) compared to their free-living counterparts. In dinoflagellates, free-living species have genomes ranging from 6 Gbp to 250 Gbp, whereas the parasitic species *Amoebophrya ceratii* has a genome size of 87.7Mbp. Genome reduction is also common in symbiotic species. For example, the smallest genomes of dinoflagellates documented are from the family of Symbiodiniaceae, which are the source of symbiotic algae for corals and many other invertebrates. The genome size of the free-living symbiodiniacean *Effrenium voratum* is 2.67 Gbp, and that of *Polarella glacialis* is about 6 Gbp, while the genomes of symbiotic species in this family range from 1.0 to 2.2 Gbp.

Genomic Divergence Between Closely Related Species

Algal genomic studies have revealed unsuspected divergence in overall genome structure between closely related species. For instance, when genomes of symbiodiniacean species of dinoflagellates were compared, fewer than 2% of reads from one genome could be aligned with the assembly of the other, although their proteomes were much more similar (Lin et al. 2015, González-Pech et al. 2021). Remarkable genomic differences also occur between intraspecific populations. For example, only ~ 60% of a reference genome created from a Chinese strain (CNS00066) of *Phaeocystis globosa* could be mapped to the genome of a Thai strain (CNS00080) of the same species (Chen et al. 2024). Similar marked differences were found between *Emiliania huxleyi* strains (Read et al. 2013).

Molecular Mechanisms Underpinning Genome Evolution

The most efficient way of genome expansion is Whole Genome Duplication (WGD) or segmental duplication (Wolfe and Shields 1997, Crow and Wagner 2005). However, segmental duplication is also a frequent mechanism leading to genome expansion. Furthermore, transposon and retrotransposon can contribute significantly to genome expansion. On the other side of the coin, gene loss is an important mechanism of genome contraction. In addition, Horizontal Gene Transfer (HGT) serves as a crucial mechanism in genome evolution, driving both functional innovation and genome size dynamics through expansion and contraction.

Whole Genome Duplication (WGD)

WGD is widespread in eukaryotes facilitating gene family expansion, genetic innovation, and escape of extinction (Wendel 2000, Crow and Wagner 2005). WGD is particularly common in plants. In algae, report of WGD is sparse. In chlorophytes, relatively recent WGD was detected through genome analysis in the calcareous coral reef species *Halimeda opuntia* (Zhang et al. 2024). *H. opuntia* is one of the coenocytic algae, which is unicellular multinucleate and a macroscopic (with cactus-like morphology). K-mer analysis revealed four subgenomes in the cell. For dinoflagellates, their remarkable genome size increase than perkinsids (< 100 Mbp) and apicomplexans (< 130 Mbp) could have resulted from WGD. However, K-mer profile and scaffold synteny analyses for *F. kawagutii* did not show evidence of WGD (Lin et al. 2015). Nevertheless, the genome of *Polarella glacialis* (~ 3 Gbp) is about double the size of that in its phylogenetically close relative Symbiodiniaceae. Its k-mer profile shows double peaks, characteristic of diploid genomes (Stephens et al. 2020), suggesting a genome duplication. For another dinoflagellate, *Pyrocystis lunular*, "diploid" like chromosome pairs have been observed (Seo and Fritz 2006). There is a good chance that WGD is widespread in large dinoflagellate genomes.

Retroposition

Besides WGD, genome expansion can result from other processes such as proliferation of mobile DNA elements, including transposons and retrotransposons. The role of transposable elements (mobile DNA) in remodeling a genome has been demonstrated in plants (Chen et al. 2020). Among algae, there is evidence that transposons play a role in genome expansion, for instance, in dinoflagellates (John et al. 2019). Furthermore, retroposition is a process in which transcripts of a gene are reverse transcribed into cDNA, which is inserted into the genome. Retrotransposons are prevalent in eukaryotes, but direct evidence is generally challenging to obtain. In dinoflagellates, spliced leader (DinoSL) *trans*-splicing phenomenon provides a handy tool for detecting retroposed genes because the inserted cDNA would retain the entire or partial DinoSL (Fig. 3.2). By searching in the sequenced genomes for genes with DinoSL sequence at the upstream region, two massive

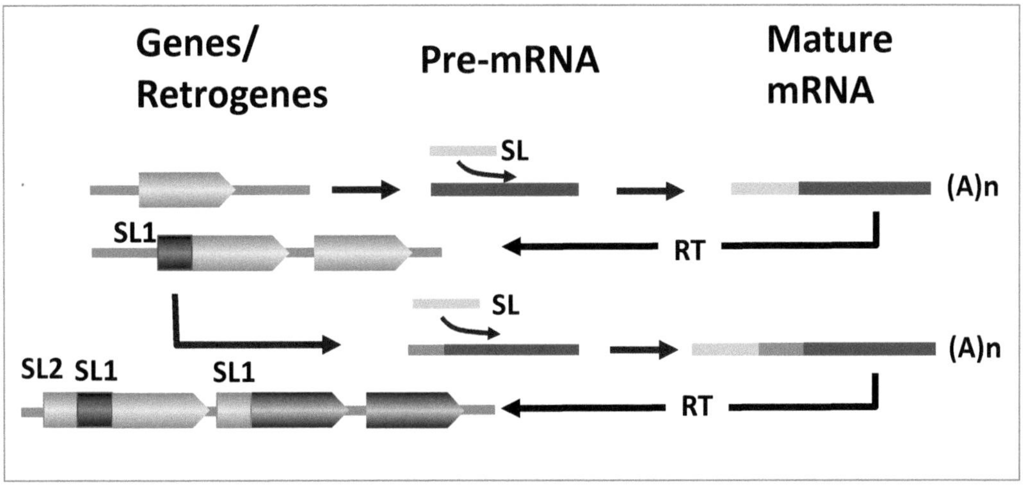

Figure 3.2. Schematic of retroposition.

episodes of retroposition were identified in two species in the symbiotic family of dinoflagellates, Symbiodiniaceae (Song et al. 2017).

Gene Family Expansion and Contraction

Gene family expansion can result from genome duplication or gene duplication. By the same token, genome reduction is often associated with gene loss. Gene family expansion and contraction can be detected by comparing contemporary species with ancestors for gene family numbers and sizes via phylogenetic analysis. The phylogenetic tree can be reconstructed using an alignment-based or alignment-free approach. The latter is not only promising for reconstructing the evolutionary trajectory, but it also enables one to probe differential evolutionary trends and paces among different components of the cell (Lo et al. 2022). For instance, an alignment-free phylogenomic analysis of the dinoflagellate family Symbiodiniaceae showed that the protein-based tree was the closest to the rRNA gene tree, followed by the whole genome tree, whereas trees inferred from repeat-masked whole genome, coding sequences, repeats, or introns showed different deviations relative to the rRNA reference tree. These variable deviations are indicative of differential evolutionary pressures acting on these distinct genomic regions (Lo et al. 2022).

Repetitive DNA

Repetitive DNA is also a significant contributor to genome size variations. As can be found in the tables in each chapter listing sequenced genomes, the proportion of the genome accounted for by repetitive DNA differs between phyla. In dinoflagellates, for instance, a study using FISH (Cuadrado et al. 2022) revealed clusters of dense trinucleotide repeats in the bean-shaped nuclei of *Karenia selliformis* and *K. mikimotoi*. Based on genome data, there appears to be a trend of larger dinoflagellate genomes containing more abundant repetitive elements (Stephens et al. 2020). For instance, repetitive elements account for 23.8% of the 0.1156-Gbp *Amoebophrya* A120 genome, 13.1% of the 0.1160-Gbp *Amoebophrya* A25 genome, 27.9% of the ~ 1.10-Gbp *Symbiodinium microadriaticum* genome, and 16% in the ~ 1.2-Gbp *Fugacium kawagutii* genome (Liu et al. 2018). For *P. glacialis*, with a genome size of ~ 3 Gbp, the level of repetitive elements increases to ~ 68% (Stephens et al. 2020). *Crypthecodinium cohnii* has a ~ 7 Gbp genome (Bhaud et al. 1991), and 55–60% of its genome is composed of repetitive DNA (Allen et al. 1975). A survey of the ~ 115 Gbp *Alexandrium ostenfeldii* genome indicated that repetitive DNA accounted for ~ 58%

(Jaeckisch et al. 2011). Overall, these results suggest that repetitive elements are significant contributors to genome size evolution in algae such as dinoflagellates.

Gene Transfer from Organelles/Symbionts, Bacteria, and Viruses

Endosymbiotic gene transfer (EGT)

Gene transfer from endosymbiont to the host genome (EGT) is widespread, resulting in nuclear genome expansion and organellar genome reduction. Numerous nuclear genes in eukaryotic algae originate in the chloroplasts and mitochondria. Free-living bacteria harbor 1500–7500 genes, and archaea have 1500–2700 genes, whereas parasitic bacteria contain 500–1200 genes in their genomes (Gregory 2005). In comparison, mitochondrial genomes only code for 3–89 genes (Hou and Lin 2009) and plastid genomes code for 13–200 genes (Table 3.2). Dinoflagellates represent one of the extreme EGT cases, resulting in only three protein-coding genes in the mitochondria and 16-17 genes in the chloroplast. To maintain the function of the organelles, the cytoplasmic proteins encoded by the EGT-genes need to be transported back to the organelles. In the case of plastids, protein import involves translocases of the outer (TOC) and inner (TIC) chloroplast envelope membranes.

Table 3.2. Algal plastid and cyanobacteria genome size ranges.

Phylum	Plastid genome size	Protein-coding genes	Conserved genes set	References
Cyanobacteria	1.44 Mbp–14.1 Mbp	1678–3291	559	(Larsson et al. 2011, Simm et al. 2015, Hirose et al. 2021)
Chlorophyta	37 kbp–1.35 Mbp	85–200	97	(Brouard et al. 2010, Fang et al. 2021, Ren et al. 2021)
Rhodophyta	150 kbp–200 kbp	> 149–197	135	(F Costa et al. 2016, Lee et al. 2016, Van Beveren et al. 2022)
Cryptophytes	77 kbp–107 kbp	> 82–150	143	(Kim et al. 2017)
Haptophyte	95 kbp–116 kbp	108-119	110	(Hovde et al. 2014, Ha et al. 2023)
Heterokonts	69 kbp–150 kbp	127–132	86	(Rogers et al. 2007, Brembu et al. 2014)
Dinophyceae	27 kbp–62 kbp	13	12	(Zauner et al. 2004, Barbrook et al. 2014)

Horizontal Gene Transfer (HGT) Between Eukaryotic and Prokaryotic Algae

Horizontal Gene Transfer (HGT) occurs rampantly in prokaryotes (up to 80% of genes in *Escherichia coli*). Although much more rarely, HGT is still remarkable in eukaryotes, accounting for 0.10 to 6.49% of total genes, depending on the species, with an average of ~1% (van Etten and Bhattacharya 2020). HGT facilitates the gain of adaptive functions and sometimes lifestyle transitions. As is true in most eukaryotes, many genes in algae have been acquired from bacteria. Diatoms are one of the algal lineages that have acquired a large number of genes from bacteria: 587 for *Phaeodactylum tricornutum* (5.6%; Bowler et al. 2008) and 252 for *Pseudo-nitzschia multistriata* (2.1%; Basu et al. 2017). One of the most notable impacts of HGT on diatom physiology is the evolution of the ornithine-urea cycle—a unique metabolic pathway derived from bacterial genes integrated into diatoms (Dorrell et al. 2021). Some diatoms can sequester iron in ferritin, a trait acquired due to HGT of its encoding gene from bacteria, which enables these diatoms to bloom in iron-limited waters (Marchetti et al. 2009).

Another notable example of HGT is from dinoflagellates. Sequencing the genomes of both the Antarctic (CCMP1383) and Arctic (CCMP2088) strains of *Polarella glacialis* and comparing them with other dinoflagellate genomes revealed potential bacteria-to-dinoflagellate horizontal gene transfer (HGT) for ice-binding domain-containing protein Lin et al. 2015).

Horizontal Gene Transfer (HGT) Mediated By Viruses

Viral invasion is active in both cyanobacteria and eukaryotic algae. Cyanophages (viruses infecting cyanobacteria) are highly abundant, affecting approximately 1 to 5% of picocyanobacterial cells in the ocean at any one time (Sullivan et al. 2006, Carlson et al. 2022). Cyanophages are typically tailed double-stranded DNA viruses that belong to the Myoviridae, Podoviridae, and Siphoviridae families. Their genomes are often integrated into host genomes, greatly influencing cyanobacterial ecology and evolution. For instance, some core photosystem II genes are present in the majority of marine cyanophages, with 88% of cyanophages containing psbA, and 50% containing both psbA and psbD; the psbA gene was found in all myoviruses and *Prochlorococcus* podoviruses, but not detected in *Prochlorococcus* siphoviruses or *Synechococcus* podoviruses, nearly all the phages that encode both psbA and psbD have broad host ranges (Sullivan et al. 2006). These photosystem genes may be employed to augment photosynthesis and maximize energy production during infection. Meanwhile, cyanophages carry Auxiliary Metabolic Genes (AMGs), and their expression modifies central carbon metabolism in the host, repressing carbon fixation during infection (Puxty et al. 2016).

Considered to be inconsequential for a long time, viral infection has also facilitated gene transfer and innovations in eukaryotic genomes, particularly impacting the algal cell wall (Irwin et al. 2022). The transferred genes can be co-opted, supplementing or supplanting host cellular components, or providing entirely novel functionality. For instance, histones in dinoflagellates have been replaced by viral proteins (Gornik et al. 2012). A complete pathway of sphingolipid biosynthesis has been transferred between the haptophyte *Emiliania huxleyi* and its DNA virus (Monier et al. 2009). In algae such as Phaeophyceae, double-stranded DNA viruses from Phycodnaviridae can persist as integrated proviruses in the host genome. Fragments of the viral genome are interspersed as repeats and pseudogenes throughout the host genome (Benites et al. 2022). The viral genome remains latent until it is activated in reproductive cells by light and temperature stimuli.

It is worth noting that viruses can play an important role in mediating cross-domain HGT, even from eukaryotic algae to blue-green algae (Godde et al. 2018). In this eukaryote-to-*Prochlorococcus* HGT event, an ATP sulfurylase (ATPS) gene and a class I Fructose Bisphosphate Aldolase (FBA I) gene appear to have been transferred from either a green or red algal origin, respectively.

Evolutionary Mechanisms of Genome Size Variation: Neutral or Selected For?

The evolutionary driving force of genome size variation is under debate. Two major classes of hypotheses have historically been set forth to account for the huge variation of genome size among organisms. One posits that genome size is a neutral trait, and the other asserts that genome size is subject to selection (Blommaert 2020). In essence, the neutral hypothesis regards genome size as a product of genetic drift, implying that selective pressures play no or minimal role in the gain/loss of DNA. There are two theories under the neutral hypothesis: the Mutational Hazard Hypothesis (MHH) and the Mutational Equilibrium Hypothesis (MEH). The nearly neutral MHH theory suggests that "extra" DNA is slightly deleterious and that larger genomes have higher mutation rates (Lynch and Conery 2003). It predicts that organisms with larger effective population sizes (Ne) will have smaller genomes, as selection against slightly deleterious DNA is more effective in larger populations. Conversely, the MEH proposes that genome size reflects a balance between insertion and deletion rates (Petrov 2002). It suggests that genome expansion occurs in bursts through duplications or transposon activity, while a constant rate of small deletions leads to genome shrinkage. Both theories have received some data support, but opposing data also exist, with the MEH being somewhat favored (Blommaert 2020). The MEH aligns with the common practice in genomics research of attributing genome dynamics to active transposition followed by gradual deletions (Canapa et al. 2016).

Consistent with the "selection" hypothesis, many argue that genome size has adaptive significance (Cavalier-Smith 2005). Major data support comes from correlations between genome size and parameters that represent fitness, such as cell size and gene content. The functional importance of cell size has been debated, particularly for multicellular organisms. However, the importance of cell size has been demonstrated in various cell types in vascular plants (e.g., stomatal guard cells) and some animal cell types (e.g., blood cells), as reviewed in Cavalier-Smith (2005). The adaptive implications of both increasing functional elements in expanded genomes and shedding "junk" DNA to streamline genomes are perhaps more obvious and hence easier to demonstrate. From the ecological perspective, the adaptive significance is particularly important, and thus will be elaborated with data from algae in the next section.

Adaptive Significance of Genome Streamlining and Expansion

On the basis that cell size is important, a positive correlation between cell size and genome size would provide support for the adaptive hypothesis of genome size. Such a correlation has been widely found, including in animal cells (Beaulieu et al. 2008), diatoms (Connolly et al. 2008), and dinoflagellates (Fig. 3.3A). Further supporting data come from a positive correlation between rRNA gene number and genome size, because rRNA is essential for protein synthesis and cellular functions (Fig. 3.3B). Additionally, a positive correlation between protein-coding gene number and genome size provides more direct support (Fig. 3.3; Hou and Lin 2009).

The genome size-gene number relationship is log-log hyperbolic instead of linear because in most eukaryotes, the larger genome size, the smaller the proportion (%) of protein-coding DNA (Fig. 3.4A, B, C). The trend holds true across domains of life, i.e., small eukaryotic genomes behave similarly to large prokaryotic genomes in terms of coding percentage (Fig. 3.4B, C). However, at the upper end of the gene size range, there is a break point of genome size, where the slope of the trend shifts. Below the break point, the genome is gene-dense, and the log genome size-log gene number linear regression has a markedly greater slope, whereas above the break point, the genome is expansive and the slope of the regression line decreases dramatically. The break point is at about 15 Mbp. Despite the quantitative variations, the increasing trend of functional potential (protein-coding gene number) with genome size is invariable.

A positive correlation also exists between genome size and the copy number of specific functional genes. For instance, there is a significant correlation between the copy number of the gene encoding Proliferating Cell Nuclear Antigen (PCNA) and genome size (Hou et al. 2019). Although the PCNA-encoding gene usually exists in one or two copies in eukaryotes when genome size is below about 4 Gbp, its copy number in dinoflagellates leaps to orders of magnitude higher as genome size increases beyond 4 Gbp (Fig. 3.5). The PCNA gene family expansion appears to have resulted from retroposition (Hou et al. 2019). This gene duplication is likely selected for because *pcna* paralogs within species differ considerably (up to 21%) and the divergence is dominated by nonsynonymous substitutions (Hou et al. 2019). As such, PCNA gene duplication is probably functionally driven. Interestingly, the enormous genome size in dinoflagellates does not significantly prolong the S phase of the cell cycle. These findings suggest that PCNA gene proliferation may potentially facilitate the operation of multiple replication origins and acceleration of DNA duplication in dinoflagellates.

The functional and adaptive significance of genome size evolution can be further demonstrated by comparing genomes and functional gene differences among different species of algae. This can be illustrated by the case of cyanobacteria *Prochlorococcus* and *Synechococcus*, as shown below.

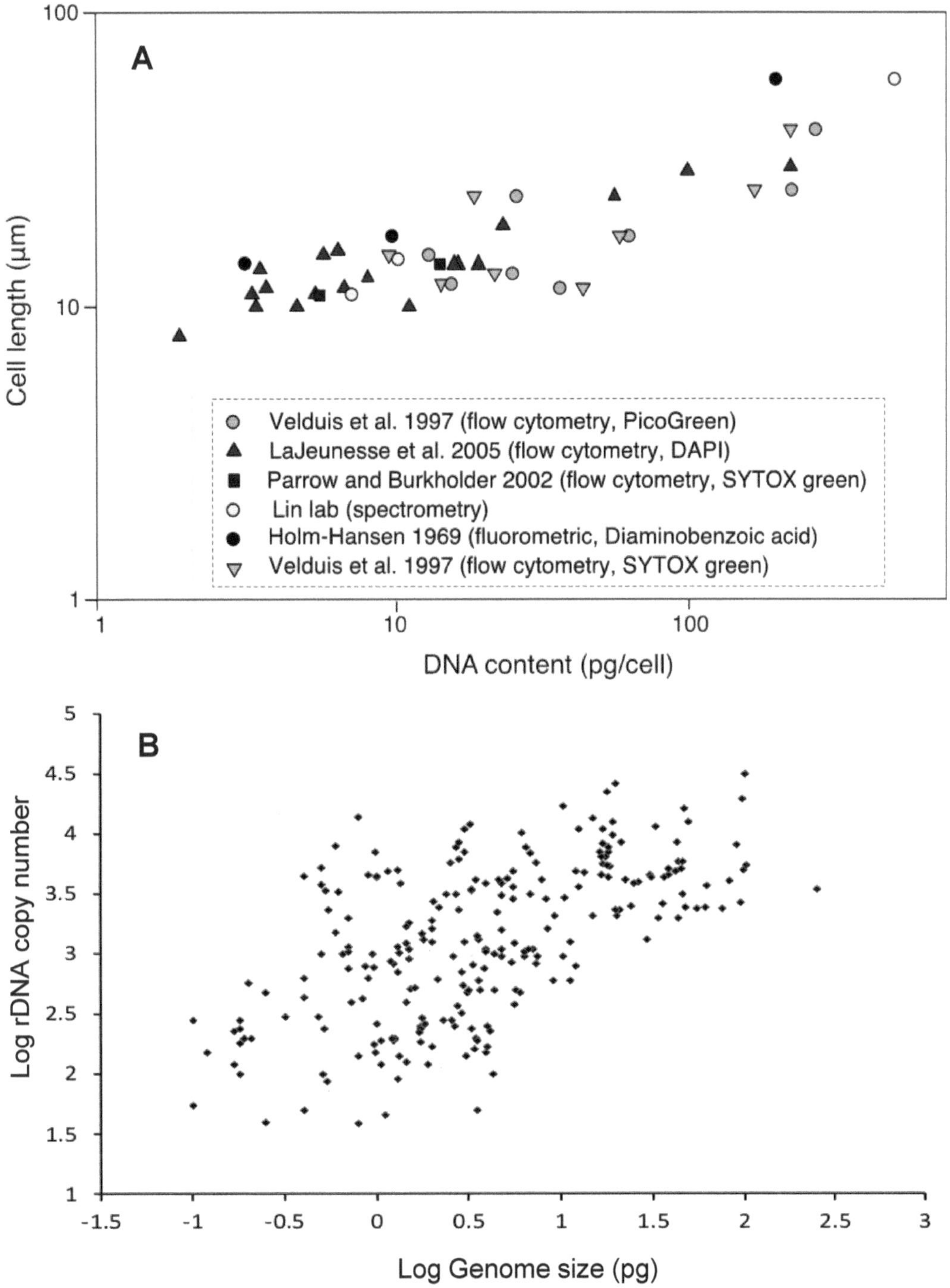

Figure 3.3. Correlations between cell size and genome size (A) and between genome size and rDNA copy number in eukaryotes (B).

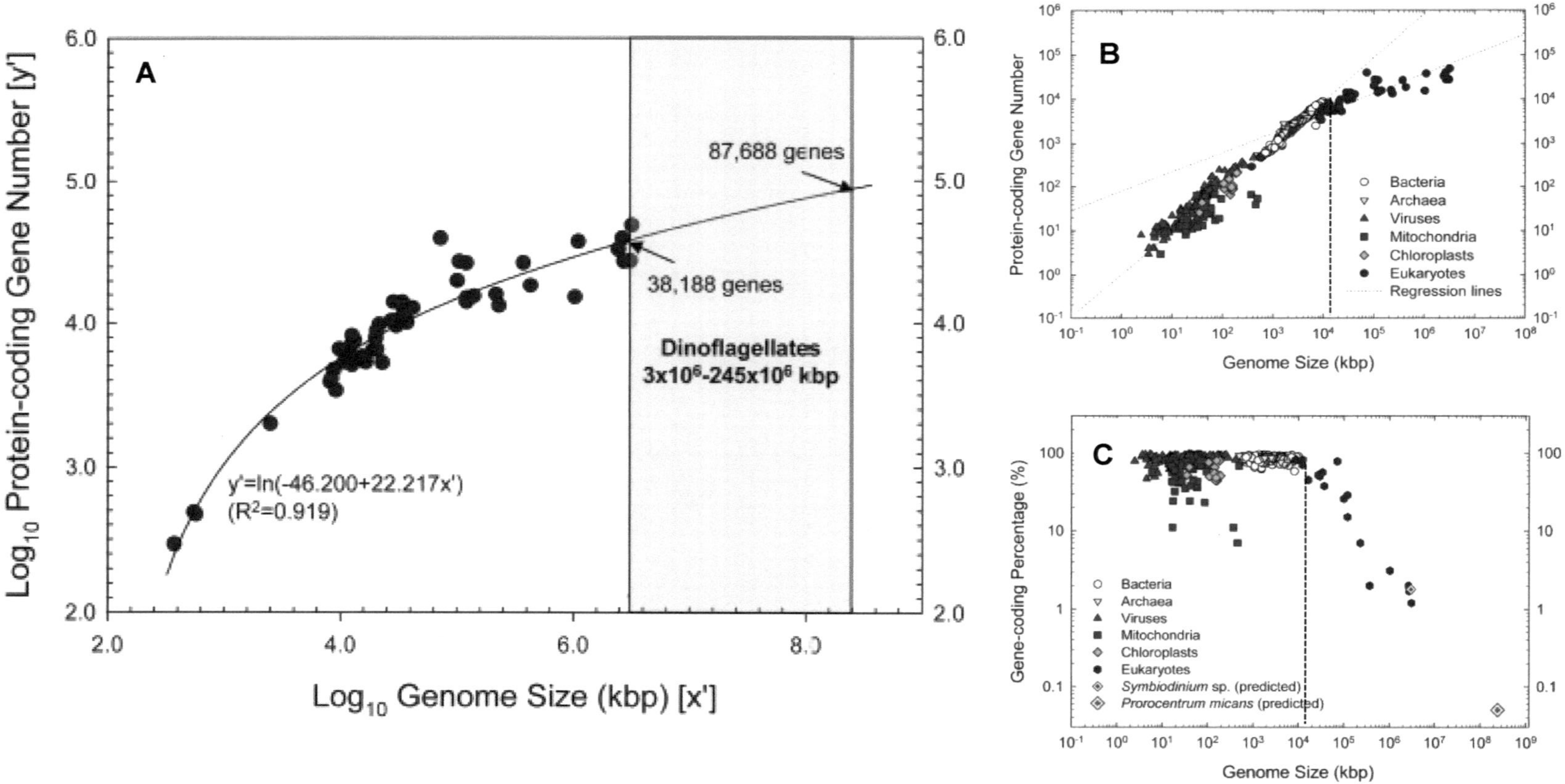

Figure 3.4. Genome number-genome size correlation and coding percentage of the genomic DNA. (A) Hyperbolic relationship between log protein-coding gene number and log genome size. Grey area predicts number of genes for dinoflagellate genomes. (B) Separate regressions with distinct slopes for small-sized genomes (including prokaryotes, viruses, organelles, and small eukaryotes) and large-sized genomes (larger eukaryotes). (C) Coding percentage. Dashed vertical lines in B and C indicate the inflection point (~ 15Mbp) between the two gene size groups. Reproduced from Hou and Lin 2009 under the terms of the Creative Commons Attribution License.

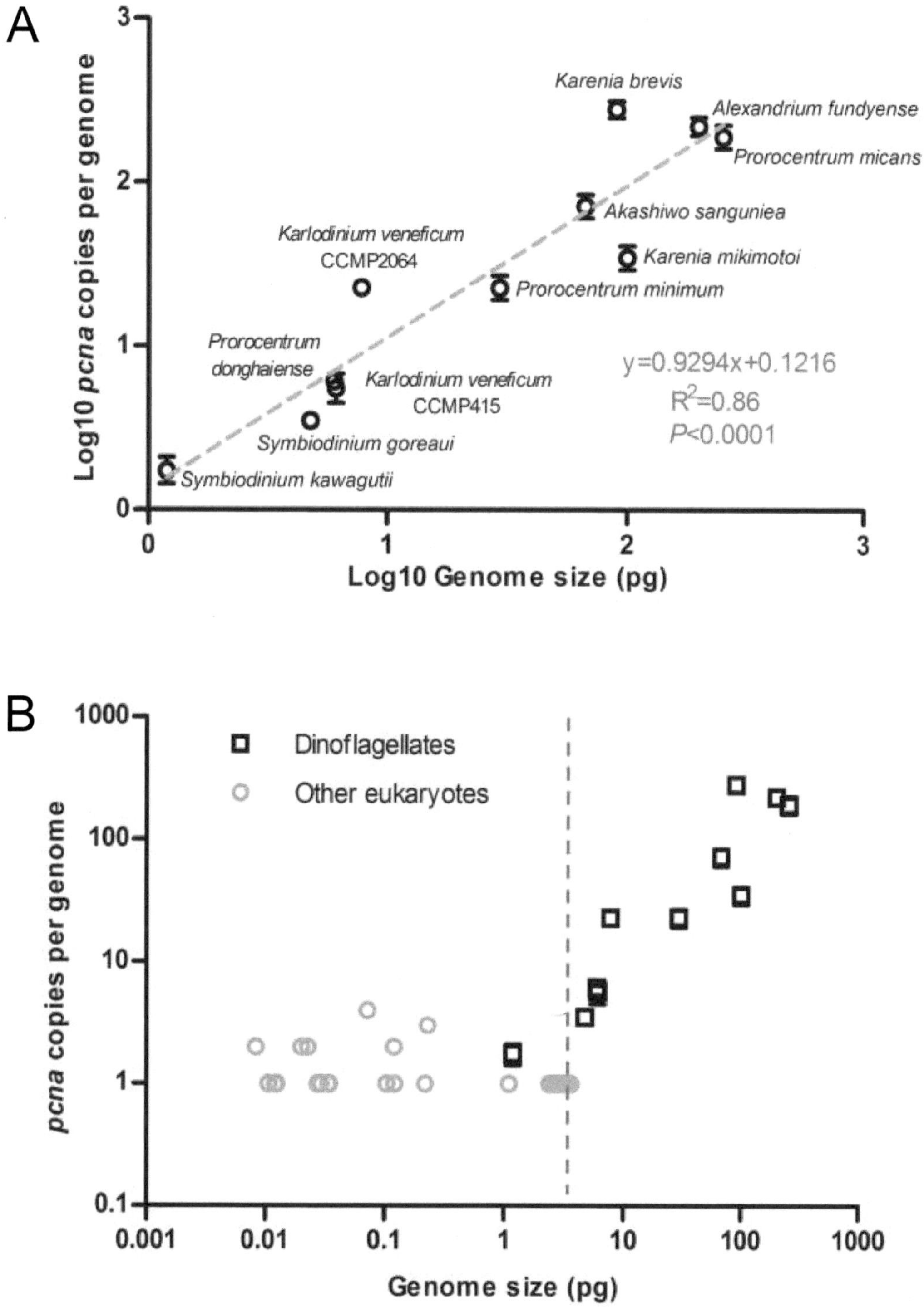

Figure 3.5. PCNA gene copy number and genome size relationship in eukaryotes. (A) Positive linear regression between log10 *pcna* copy number and log10 genome size in dinoflagellates. Dashed line is regression line (n = 38). (B) No relationship between *pcna* copy number and genome size in non-dinoflagellate eukaryotes (n = 29). The vertical dashed line indicates the hypothetical 4 pg genome size threshold for *pcna* multiplication. *Pcna* copy number shown are mean values. Reproduced with permission from Hou et al. 2019.

Adaptive Significance of Genome Size Evolution

There must be advantages and costs associated with having a reduced or enlarged genome. A large genome holds more functional genes, many of which may be duplicates, and thus renders the species more versatile, but it imposes higher energy costs for maintenance. For instance, dinoflagellates have markedly larger genomes than other algae, and they generally grow significantly more slowly (Tang 1996); however, they are extremely successful ecologically, indicative of versatility as demonstrated in their diverse habitats, trophic strategies, and lifestyle. In contrast, genome streamlining in a free-living organism conceivably saves energy for genome maintenance and duplication, but this advantage comes at the cost of physiological flexibility and versatility in resource utilization. Small genome size and fewer gene duplications may provide an adaptive advantage to life in the oligotrophic ocean (Swan et al. 2013), where poor nutrient and abundant light conditions prevail compared to the dynamic coastal areas.

Adaptive genome size evolution can also be demonstrated by the functional differentiation of cyanobacteria that have different genome sizes in adaptation to the environmental conditions of their habitats (Fig. 3.6). Comparison of *Prochlorococcus* and *Synechococcus* genomes (Rocap et al. 2003, Palenik et al. 2003, Dufresne et al. 2003, Fuhrman 2003) shows that *Prochlorococcus* MED4 and *Synechococcus* SS120 live in the stable top surface (high light, low nutrient) and dark lower euphotic zone (low light, rich nutrient), respectively, and they both have small genomes (1.7 Mbp). For inorganic nitrogen nutrient utilization, both only possess genes encoding ammonium transporters. In contrast, *Prochlorococcus* MIT9313 and *Synechococcus* WH8102 are both distributed in the subsurface where the nutrient conditions fluctuate due to mixing and upwelling. In adaptation to the variable environment, these two strains of cyanobacteria possess markedly larger genomes (2.4 Mbp). For N nutrition, they both possess genes encoding transporters of ammonium and nitrite, and *Synechococcus* WH8102 has a nitrate transporter in addition.

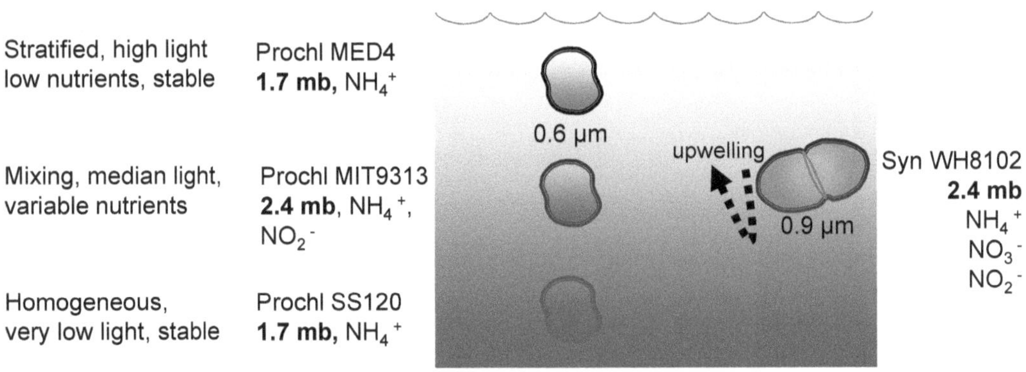

Figure 3.6. Illustration of genome size and gene repertoire evolution in cyanobacteria in adaptation to habitat environmental conditions. Strains with smaller genomes are distributed in more stable surface and deep areas with fewer genes for nitrogen nutrient uptake whereas species with larger genomes live in the more variable mid-depth, with more genes involved in taking up more forms of nitrogen nutrients available at that depth.

Additional evidence of adaptive genome size evolution comes from the massive retroposition found in the genomes of *F. kawagutii* and *Breviolum minutum* discussed earlier. Genome-wide analysis revealed two episodes of retroposition (Song et al. 2017). Genes duplicated by this mechanism were enriched in functions related to the evolution of symbiosis, with the first episode promoting sugar and fatty acid metabolism, oxidative stress response, and transport, whereas the second episode strengthened photosynthesis and adhesion (Fig. 3.7).

Swan et al. (2013) used Single Amplified Genome (SAG) sequencing technology and analyzed planktonic bacteria from the Gulf of Maine and the Mediterranean Sea, and the subtropical gyres

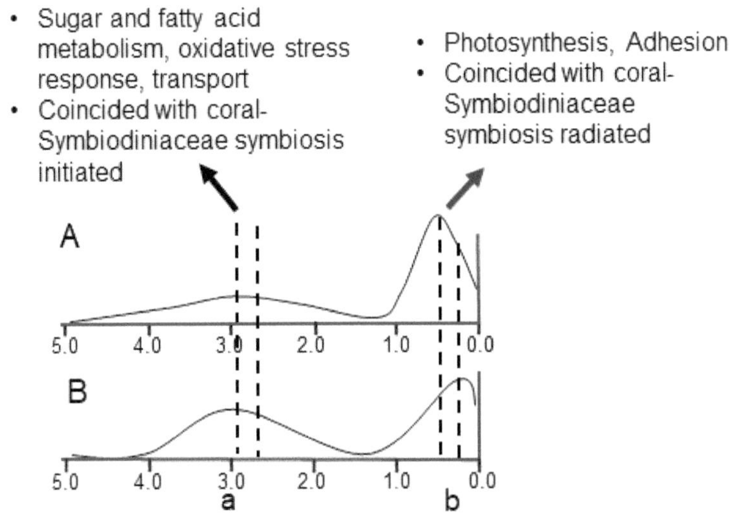

Figure 3.7. Two episodes of retroposition discovered in Symbiodiniacea genomes coincident with two major periods (a, b) of symbiosis evolution. (A) *Fugacium kawagutii*. (B) *Breviolum minutum*. Based on Song et al. (2017).

of the North Pacific and South Atlantic Oceans. They compared culture-independent genome datasets to previously reported genomes of cultured bacteria as well as environmental bacteria. The comparison indicates that natural bacterioplankton have smaller genomes, fewer gene duplications, reduced abundances of guanine and cytosine, noncoding nucleotides, and fewer genes related to transcription, signal transduction, and those coding for non-cytoplasmic proteins. These findings demonstrate that genome reduction and oligotrophy are prevalent features among diverse, free-living bacterioplankton, whereas existing laboratory cultures consist primarily of copiotrophs, suggesting that the apparently ubiquitous metabolic specialization and mixotrophy predicted from the data may explain why the majority of bacterioplankton are difficult to cultivate. In addition, the data show that the global distribution of surface ocean bacterioplankton correlates with temperature and latitude and is not limited by dispersal at the time scale of speciation determined by rRNA gene divergence.

The trade-off of benefits and costs in having large and small genomes is in line with the MEH theory discussed above. It might underlie the diversity of organisms living in a community. This may explain part of the "Paradox of the Plankton", which presents the irreconcilability between the apparent homogeneity of the marine environment and the observed high biodiversity in the plankton under the theory of competition (Hutchinson 1961).

Genome Size Dynamics Shaped by Physiological Conditions

There is evidence that genome size in some species is plastic in response to environmental and physiological changes. A cell's genome size typically changes by no more than two-folds throughout the cell division cycle, in which the genomic DNA is duplicated periodically. However, the genome size dynamics sometimes go beyond two-fold. In humans, DNA content can differ between cell types (Gillooly et al. 2015). In the zooplankton *Daphnia*, DNA content can change with nutrient or temperature conditions (Jalal et al. 2014). Besides, ploidy fluidity seems to be common in prokaryotes as well as eukaryotes. *Escherichia coli* genome copy number changes with growth rate and can reach 6–7 copies (Pecoraro et al. 2011). *Epulopiscium* and *Halobacterium* have been reported to show changes in ploidy. Among algae, the *Synechocystis* PCC 6803 wild-type strain was reported to have 218 genome copies in the exponential phase and 58 genome copies in the linear or stationary phase (Griese et al. 2011).

Genome size dynamics may reflect, or be contributed to by, the variations of gene copy number. This is evident from a close look at the correlation between rRNA gene (rDNA) copy number and genome size (Fig. 3.3B). A positive log-log correlation between DNA content (genome size) and rDNA copy number in a cell is commonly observed in eukaryotes (Prokopowich et al. 2003). However, the correlation is not very tight as data scatter substantially (Fig. 3.3B and Prokopowich et al. 2003). The data scatter suggests that rDNA copy number can change beyond the constraints of the genome size. Interestingly, non-inherited, cyclic, polyploidy and rDNA multiplication (e.g., in the form of extrachromosomal rDNA circles) occur in various eukaryotes (Parfrey and Katz 2010), in part explaining the data scatter in the genome size-rDNA copy number regression. In *Daphnia*, rDNA copy number varies with nutrient and temperature conditions (Jalal et al. 2014). In budding yeast, rDNA cluster length is a determinant of replicative lifespan (Hotz et al. 2022). Large changes in rDNA copy number have been reported in some dinoflagellates. For instance, in a strain of *Alexandrium pacificum,* rDNA copy number was reported to vary from 8.1×10^4 to 1.72×10^7 copies cell^{-1} (Ruvindy et al. 2023).

Functional genes are generally known to respond to growth conditions via changes in their expression levels (transcriptional, translational, or both) and post-translational modifications of the encoded proteins. However, changes in the copy number have also been reported, albeit rarely under normal conditions. In humans, gene Copy Number Variations (CNVs) are not uncommon between individuals and in association with diseases (Zhang et al. 2009). Consistent with the notion that gene copy number can be modulated by growth or physiological conditions in an organism, there have been reports about positive correlations between gene copy number and gene expression levels in human diseases (Shao et al. 2019). CNVs of functional genes have been reported for algae, albeit rarely. One conspicuous example concerns a gene involved in saxitoxin synthesis in dinoflagellates. SxtA is one of the key genes in the saxitoxin biosynthesis pathway. Among its four variants (SxtA1 to SxtA4), sxtA4 has been considered a marker of saxitoxin production. Differences in sxtA4 gene copy number have frequently been reported between *Alexandrium* species; for instance, ~ 180–325 copies cell^{-1} in *A. pacificum* and *A. catenella* (Murray et al. 2011, 2019, Geffroy et al. 2021), 1.5–11 copies cell^{-1} in *A. minutum* and *A. ostenfeldii* (Stuken et al. 2015, Savela et al. 2016), and none in many species which do not produce saxitoxins. Within-species variations can be high as well, e.g., ~ 8 to 609 copies cell^{-1} between strains of *A. pacificum* (Ruvindy et al. 2023). While the interspecific and inter-strain differences should not be surprising, it is rather striking that even within a population of *A. pacificium*, sxtA4 copy number could change from 43 to 609 copies cell^{-1} (Ruvindy et al. 2023). These variations were found to correlate with the level of saxitoxin production, thus exhibiting a gene dosage effect on toxin production (Stuken et al. 2015, Savela et al. 2016, Geffroy et al. 2021, Ruvindy et al. 2023). This is one of the most reported examples in which the copy number of a functional gene changes in proportion to a physiological activity.

Adaptive Significance of GC Content Variation and Codon Bias

GC Content

Genomic GC content, the percentage of (G+C) out of total nucleotides (%GC), varies greatly among taxa. The causes behind this diversity are a matter of debate, with various selective and neutral factors potentially at play. Nonetheless, researchers have suggested that the production of G and C is energetically more costly and hence these nucleotides are less available in the cell compared to A and T/U (Rocha and Danchin 2002). Furthermore, guanine and cytosine contain more nitrogen. Hence, lower genomic GC content may be advantageous in oligotrophic environments, while its impact on copiotroph ecology and evolution is diminished in resource-rich environments (Okie et al. 2020).

G/C pairing involves a triple bond, whereas A-T pairing involves a double bond. As such, G/C dissociation requires higher energy than A-T dissociation. Therefore, high GC content is often predicted to confer greater thermal tolerance. However, hot spring microbes do not seem to have remarkably higher GC content than mesothermal microbes. Therefore, the correlation between thermal tolerance and GC content has been questioned. A recent study analyzing a large set of genome data (681 bacteria and 155 archaea) revealed a positive correlation between optimal growth temperature and GC content of genomes, both in bacteria and archaea (Hu et al. 2022). This study also found that the conflicting results from past studies were due to small sample sizes that caused the insignificance of the correlation.

Codon Bias

Eighteen of the 20 proteinogenic amino acids can be encoded by more than one of the 61 proteinogenic codons, leading to genetic code redundancy. However, these synonymous codons exhibit distinct kinetic properties, such as varied translation rates and probabilities of mistranslation (Higgs and Ran 2008). In highly expressed genes that are crucial for rapid growth, like those encoding ribosomal proteins, there is likely heightened selection pressure favoring specific synonymous codons to enhance translation accuracy and speed, particularly in fast-growing organisms (Hershberg and Petrov 2008, Plotkin and Kudla 2011, Vieira-Silva and Rocha 2010). Therefore, copiotrophic environments may favor organisms that have higher codon usage bias in their ribosomal protein genes, whereas codon usage bias may play a smaller role in oligotrophic environments (Okie et al. 2020).

Investigations into the genomic GC content and codon bias outlined above have unveiled associations between these traits and growth patterns as well as trophic strategies across diverse eukaryotic and prokaryotic species (Vieira-Silva and Rocha 2010), These studies have begun to elucidate the mechanisms through which these traits impact fitness, suggesting that they may play an important role in ecology (Lauro et al. 2009, Weider et al. 2005, Okie et al. 2020).

Adaptive Significance of rRNA and tRNA Gene Copy Number

To grow quickly, organisms must synthesize proteins fast enough. As ribosomes facilitate protein synthesis, an enhanced ability to replenish depleted ribosomal resources or sustain substantial reserves promotes growth. As such, fast-growing organisms may benefit from higher rRNA gene copies. Indeed, in bacteria and eukaryotes, correlations between rRNA operon copy number and growth rate or habitat nutrient conditions have been reported (Condon et al. 1995, Klappenbach et al. 2000, Lauro et al. 2009). Thus, microbes that are competitive in growth-favoring (e.g., nutrient-rich) environments are likely to have a higher number of copies of the rRNA gene cluster.

Since faster-growing organisms require larger cellular quotas of rRNA, they are expected to have higher phosphorus contents, given that rRNA is rich in phosphorus. Consequently, they require more phosphorus to construct these rRNA pools (Okie et al. 2020). These observations, identified as the 'growth rate hypothesis' (Elser et al. 2000), have garnered substantial empirical backing from comparative studies. However, research into their applicability as a predictive trait for community ecology has been limited.

Likewise, as the concentrations of tRNAs influence the pace of protein translation, more copies of tRNA genes in the cell, along with a wider range of anticodons among these genes, may assist a cell in sustaining larger reserves of tRNAs (Kanaya et al. 1999, Higgs and Ran 2008). This enables the maintenance of accelerated translation rates essential for faster growth (Higgs and Ran 2008, Rocha 2004). Indeed, there is evidence indicating a negative correlation between the total number of tRNA genes and generation time (Higgs and Ran 2008, Rocha 2004).

4

Genomes of Cyanobacteria

Cyanobacteria belong to the domain of Bacteria, and their cell walls resemble those of gram-negative bacteria. They are also known as blue-green algae because they often appear bluish green due to chlorophyll (green) and phycocyanin (blue). They occur in marine, freshwater, and terrestrial environments, including hot springs and polar regions. Based on morphological characters and molecular analyses, 5,185 species have been identified, but the total species number may reach 8,000, which are affiliated with seven orders (Allaf and Peerhossaini 2022). Phylogenetic analyses have grouped cyanobacterial taxa into two clades: α-cyanobacteria and β-cyanobacteria. Alpha-cyanobacteria are a monophyletic group comprising the picoplanctonic genera *Prochlorococcus*, *Synechococcus*, and *Cyanobium*, featuring a proteobacterial form 1A RubisCO and α-carboxysomes obtained by Horizontal Gene Transfer (HGT). Beta-cyanobacteria, including all the remaining species, contain a form 1B RubisCO and β-carboxysomes (Badger et al. 2002, Marin et al. 2007). Cyanobacteria possess chlorophyll (Chl) *a*, as do all chloroplasts. Some cyanobacterial lineages (e.g., *Prochlorococcus*, *Prochloron*, *Prochlorothrix*) also have Chl *b*, but others (e.g., *Synechococcus, Trichodesmium*) lack this pigment but have phycobilipigments to absorb light and transmit energy to Chl *a*, covering the role of Chl *b*. Phycobilipigments are bound by proteins and hence are often called phycobiliproteins. Chlorophylls are bound by Chl-binding proteins, which are translated by membrane-bound ribosomes for proper and rapid integration into the membrane (Sobotka 2014).

Cyanobacteria are one of the most morphologically diverse groups of prokaryotes. Their growth forms range from unicellular to filamentous or colonial (Fig. 4.1). The unicellular species are tiny (0.5–1.5 μm) and belong to picoplankton (defined as < 2 μm), e.g., *Prochlorococcus*, *Synechococcus*, and *Crocosphaera*. In contrast, some species such as those from the genus *Microcystis* form colonies composed of numerous cells linked by mucilage known as extracellular polymeric substance. Additionally, many species occur as long filaments (100 μm or longer) of many cells, and the filaments may aggregate as a bundle or radial mass, as in *Trichodesmium* spp. The filaments and colonies are visible to the naked eye, e.g., *T. erythraeum* blooms appearing as sawdust.

The emergence of cyanobacteria ushered in the planet's oxygenic era, making it of high interest to understand when and how this occurred. As elucidated in Chapter 2, oxygen appeared in the atmosphere about halfway through the Earth's 4.543-BY history, which has been credited to the emergence of the crown group Cyanobacteria. However, it is worth noting that recent data from structural biology and molecular phylogeny strongly suggest that cyanobacteria arose after oxygen production (through water oxidation) had been ongoing for quite a long time (Sánchez-Baracaldo and Cardona 2019).

Research on trait evolution has indicated that ancestral Cyanobacteria were likely to have inhabited low salinity and terrestrial environments (Blank and Sánchez-Baracaldo 2010) and existed in unicellular as well as small-sized forms (Larsson et al. 2011, Sánchez-Baracaldo 2015).

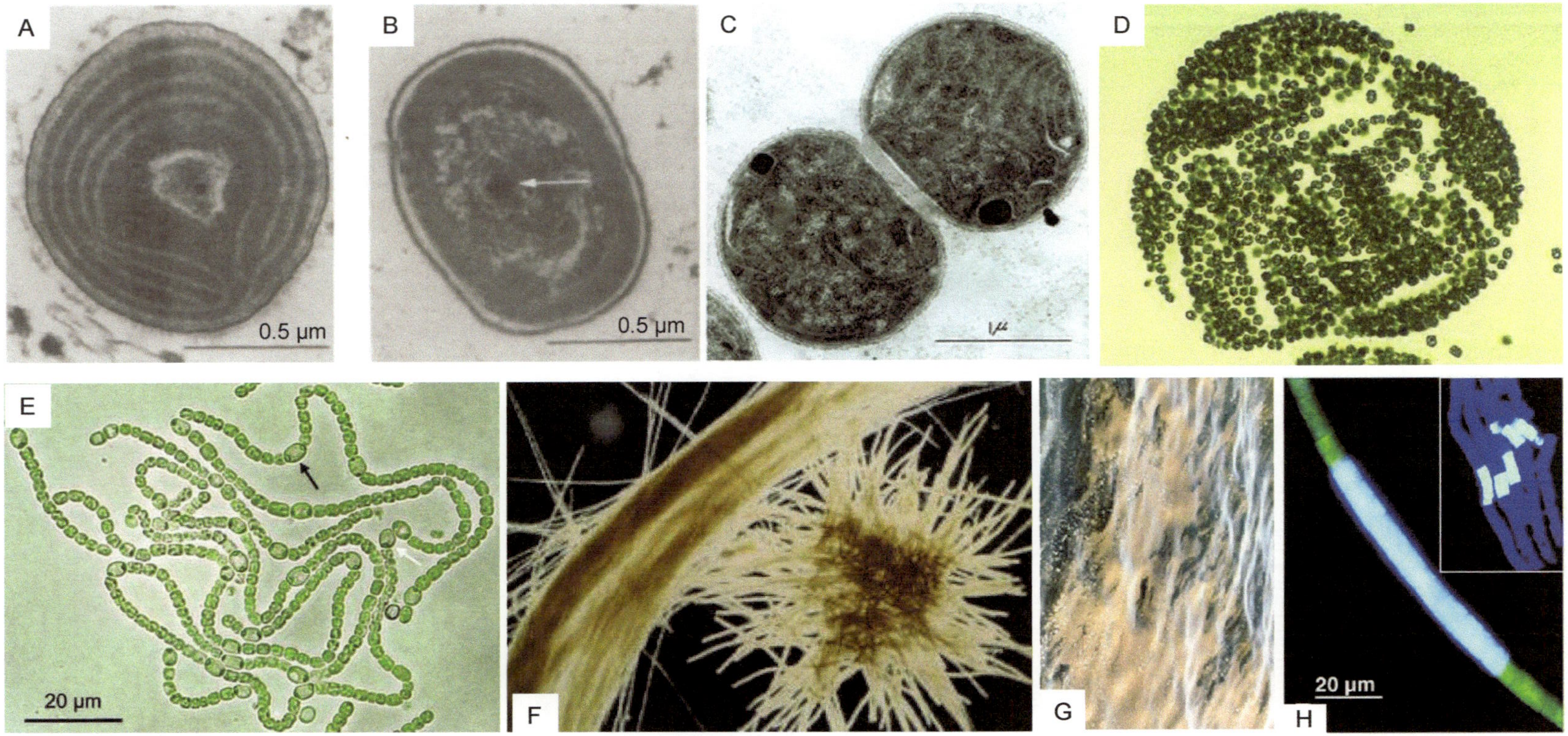

Figure 4.1. Morphotypes of cyanobacteria. A-C are single-celled species, *Synechococcus* sp. (A), *Prochlorococcus marinus* (B), and *Crocosphaera watsoni* (C). D, colony of *Microcystis* cells. E-I are filamentous species. (E), *Anabaena* cells in chains, in which some cells are specialized for N_2 fixation named heterocyst (arrow). F-H are micrographs of *Trichodesmium erythraeum*, in bundle and fusiform (F) colonies, that appear like sawdust on the surface of the ocean (G), and trichomes of *T. erythraeum* with central cluster of cells containing nitrogenase (diazocytes, the light blue fluorescence) (H) comes from immunostaining. Sources of the micrographs: Chisholm et al. (1988) for A and B, Zehr et al. (2019) for C, Backer and McGillicuddy (2006) for D, Herrero et al. (2016) for E, Bergman (2014) for F, Mohanty et al. (2010) for G, and Berman-Frank et al. (2001) for H, all with permissions from respective publishers.

Filamentous forms appeared shortly afterward and likely resembled extant *Pseudanabaena* lineages (Schirrmeister et al. 2011, Sánchez-Baracaldo 2015). Phylogenomic analyses have further shown that marine planktonic lineages are derived taxa and are polyphyletic, with some lineages being sister to unicellular freshwater taxa (e.g., *Synechococcus*, and *Cyanothece*), filamentous freshwater taxa (e.g., Nostocales) and benthic marine mat formers (e.g., *Hydrocoleum*) (Sánchez-Baracaldo et al. 2014, Sánchez-Baracaldo 2015). This phylogenetic pattern indicates independent incidents of colonization in open ocean habitats at different times during evolution (Sánchez-Baracaldo et al. 2014, Sánchez-Baracaldo 2015, Cornejo-Castillo et al. 2016, Sánchez-Baracaldo et al. 2019).

Sequenced Cyanobacterial Genomes

Over 650 genomes have been sequenced for cyanobacteria. The latest common ancestor of cyanobacteria had a genome size of ~ 4.5 Mbp and between 1678 and 3291 protein-coding genes, 4-6% of which are unique to cyanobacteria today. The evolution of cyanobacteria has led to reductions of genome size (Larsson et al. 2011). In contemporary cyanobacteria, most genomes range ~ 2 to 3 Mbp, but 4–8 Mbp genomes exist, and few species have exceptionally large genomes (Table 4.1). For instance, the chromosomal DNA of *Acaryochloris* spp. is > 6 Mbp, and in addition, they contain plasmids of > 7 Mbp (Yamamoto et al. 2022). Hundreds of genes appear to be correlated with the adaptation to various environments, including those with different light wavelengths, trophic states, and salinity. For terrestrial strains, gene correlation is also linked to the availability of water.

Larsson and colleagues analyzed the genomes of 58 contemporary cyanobacteria that were first sequenced by the early 2000's to identify shared and unique orthologs (Larsson et al. 2011). They found 404 protein families present in all cyanobacterial genomes, two of which were unique to the phylum: one corresponding to an AbrB family transcriptional regulator and the other to an unannotated gene with a conserved genomic neighborhood among all the species examined. Orthologs unique to cyanobacteria encode specific characteristics, including filament formation, heterocyst differentiation, diazotrophy, and symbiotic competence. The study revealed a mix of gene gains and losses in the evolution of the clade encompassing complex cyanobacteria but a single event of reduction in a clade dominated by unicellular cyanobacteria. Genome sizes and gene family copy numbers evolved at a higher rate in the complex cyanobacterial clade, and multi-copy genes were predominant in large genomes.

The different rates of genome-size evolution and multi-copy gene abundance led Larsson and colleagues to suggest two routes of genome development in the history of cyanobacteria (Larsson et al. 2011). The first is gene-family enlargement to generate a broad adaptive potential, and the second is genome streamlining that imposes adaptations to highly specific niches characterized by corresponding specific functional capacities. Several genomes are rich in non-coding nucleotides, which is thought to have resulted from initial expansion of genomes/gene copy number to gain adaptive potential, followed by loss-of-function drifts due to a lifestyle shift to a highly specific niche (e.g., symbiosis).

Genome Size Evolution and Adaptive Significance

Many epipelagic prokaryotic microbes have streamlined genomes (typically < 2 Mbp) with short intergenic regions, low %GC content, and few paralogous genes, as natural selection in nutrient-depleted environments favors cellular economization (Martinez-Gutierrez and Aylward 2022). This selection pressure causes loss of paralogous gene copies arising from gene duplication, particularly for marine lineages with streamlined genomes (Giovannoni et al. 2014). Despite the generally small size, variations occur in genome size among species. As demonstrated in Chapter 3, there is strong evidence that genomic size variations among cyanobacterial strains are associated with habitat adaptation. Terrestrial cyanobacteria, living on tree barks, rocks, soil, and any water-containing space, tend to have expanded genomes. This expansion is an evolutionary adaptation

Table 4.1. Genome features of cyanobacteria.

Species	Genome size (Mb)	Predicted genes	Coding gene number	Coding percentage (%)	%GC	References
Acaryochloris marina strain MBIC10699	7.6	6878	6800	82.3	47	Yamamoto et al. 2022
Acaryochloris marina strain MBIC11017	8.3			84.1	47	Swingley et al. 2008
Anabaena PCC7120	7.2118	6214	6132	82.41	41.2	Oliveira et al. 2015, Videau et al. 2014
Anabaena variabilis ATCC29413	7.1058	5805	5746	82.51	41.4	Thiel et al. 2014
Gloeobacter violaceus PCC 7421	4.659	4478	4430	89.31	62	Nakamura et al. 2003
Halomicronema hongdechloris	5.578		5273	83.5	54.6	Chen et al. 2019
Nostoc punctiforme PCC73102	9.0592	7465	7364	80.91	41.38	Moraes et al. 2017
Prochlorococcus marinus MED4	1.658	1756	1716	88.48	30.8	Tagwerker et al. 2012
Prochlorococcus marinus MIT 9313	2.4109	2324	2273	82.28	50.7	Zhaxybayeva et al. 2009
Prochlorococcus marinus strain MIT 9211	1.839	2168	2123	89	39.7	Rocap et al. 2003
Prochlorococcus marinus strain MIT 9301	1.6419	1963	1907	90	31.3	Gupta et al. 2010
Prochlorococcus marinus strain MIT 9303	2.6827	3136	2997	84	50	Gupta et al. 2010
Prochlorococcus marinus strain MIT 9312	1.709	1855	1809	89	31.2	Gupta et al. 2010
Prochlorococcus marinus strain NATL2A	1.843	1938	1890	85	35.1	Gupta et al. 2010
Prochlorococcus marinus subsp. marinus CCMP1375	1.7511	1926	1882	89.13	36.4	Dufresne et al. 2003
Prochlorococcus marinus subsp. marinus str. SS120 (CCMP1375)	1.751	1930	1882	88	36.4	Dufresne et al. 2003
Prochlorococcus marinus subsp. pastoris str. CCMP1986 (MED4)	1.668	1760	1712	87	30.8	Rocap et al. 2003
Prochlorococcus sp. CC9902 (coastal)	2.2348	2372	2321	90.31	54.2	Dufresne et al. 2008
Synechococcus elongatus BP-1	2.5939	2520	2475	89.8	53.9	Nakamura et al. 2002
Synechococcus elongatus PCC7942	2.695	2664	2611	88	55.47	Chen et al. 2012
Synechococcus sp. CC9605	2.511	2749	2638	86	59.2	Zhaxybayeva et al. 2007
Synechococcus sp. JA-2-3B'a(2-13)	3.047	2942	2862	85	58.5	Prabha et al. 2013

Table 4.1 contd. ...

...Table 4.1 contd.

Species	Genome size (Mb)	Predicted genes	Coding gene number	Coding percentage (%)	%GC	References
Synechococcus sp. JA-3-3Ab	2.933	2897	2760	84	60.2	Prabha et al. 2013
Synechococcus sp. PCC 6301	2.6963	2576	2525	87.98	55.5	Sugita et al. 2007
Synechococcus sp. RS9917	2.58	2820	2770	92	64	Fuller et al. 2003
Synechococcus sp. WH 5701	3.043	3401	3346	87	65	Dufresne et al. 2008
Synechococcus sp. WH 7805	2.62	2394	2883	90	60	Honda et al. 1999
Synechococcus sp. WH8102	2.4344	2576	2526	90.45	52.4	Zhaxybayeva et al. 2007
Trichodesmium erythraeum IMS101	8.1914	4801	4759	58.12	34.2	Dyhrman et al. 2006

to the fluctuating conditions of the terrestrial environment. In contrast, marine cyanobacteria have contracted genomes as an adaptation to the oligotrophic conditions in the open ocean where they dominate (Chen et al. 2021). Among marine lineages, genome size variation and functional gene distribution have also evolutionarily adapted to the light and nutrient conditions in their residing vertical zones (Chapter 3, see Fig. 3.4). Comparative genomic analyses by Chen et al. (2021) reveal that horizontal gene transfer is an important source of genes that provide cyanobacteria with selective advantages for living in their habitats.

When a gene is important for adaptation to an environment, it either occurs only in the strain living in that environment or at a significantly higher copy number in the strains living in that environment. Chen et al. (2021) reported three marine-enriched genes involved in osmolyte biosynthesis, which are involved in the biosynthesis of mannosylglycerate, glucosylglycerol, and mannosylglucosylglycerate. Present in most marine cyanobacterial genomes, these genes are missing in many non-marine cyanobacterial genomes. Also widely distributed in marine cyanobacteria is an ion transporter gene putatively involved in sodium effect for salt tolerance (yrbG). Interestingly, there is also a differentiation in the type of superoxide dismutase (SOD), with nickel-dependent SOD being prevalent in marine lineages but iron-dependent counterpart dominates the nonmarine lineages. Furthermore, for the acquisition of nutrients such as sulfate and nitrate/nitrite, marine cyanobacteria preferentially use monomer permease and transporter whereas non-marine strains use ABC transporters.

Horizontal Gene Transfer Diversifies Cyanobacterial Genomes

As is true in other branches of the tree of life, HGT is one of the mechanisms in cyanobacteria that cause genome innovation and metabolic diversification. HGT impacts photosystems and the nitrogen fixation pathway. A comparative genomics study on picocyanobacteria from metagenomes along a salinity gradient showed evidence that phycoerythrin (PE) genes were lost in the ancestor of phycocyanin-containing coastal picocyanobacterial and that multiple HGT events have re-introduced PE genes into brackish-adapted populations (Larsson et al. 2014). The cosmopolitan *Microcoleus chthonoplastes* is a filamentous, non-heterocystous cyanobacterium associated with microbial mats. Although typical cyanobacterial mats can fix N_2, the ability of *M. chthonoplastes* to fix N_2 was contested in part because no cyanobacterial *nif*H could be detected. A molecular study then revealed a complete *nif*-gene cluster in the genome of *M. chthonoplastes* PCC 7420 and *nif*HDK gene cluster in various axenic strains of *M. chthonoplastes* from distant locations. However, *nif*HDK phylogenetically clustered with the Deltaproteobacteria rather than cyanobacteria, even though the *nif* operon is flanked by typical cyanobacterial genes, suggesting that it was acquired from a deltaproteobacterial origin through horizontal gene transfer and has been integrated into the *M. chthonoplastes* genome (Bohuis et al. 2010).

Core Genome

Shared genomes across lineages are considered the core genome. Based on 14 cyanobacterial genomes, a comparative genomic analysis revealed 1,572 gene families as the core gene repertoire for *Synechococcus*, which accounts for as low as 52% of the total genome of strain WH5701 to as high as 67% in CC9902 (Dufresne et al. 2008). Most of the core gene families (93.4%) contain only one gene from each strain. When three *Prochlorococcus* strains were added to the comparative analysis, the core genome was reduced to 1,228 gene families. When both marine and freshwater cyanobacteria were included, the core genome was composed of 892 families of orthologs. Comparing the two major lineages of cyanobacteria *Synechococcus* and *Prochlorococcus*, only 70 gene families of the marine *Synechococcus* core genome were not present in any of the three *Prochlorococcus* genomes examined, including 23 related to photosynthesis (Dufresne et al. 2008). Notably, these include gene families encoding allophycocyanin and phycocyanin components indicating that *Prochlorococcus* has lost all phycobilisome genes except those encoding phycoerythrin. Also missing in *Prochlorococcus* are low-affinity carbon transporter genes ndhD4, ndhF4, and chpX.

Genomic Basis of Ecotypic Differentiation

The genera *Synechococcus* and *Prochlorococcus* are the two most dominant lineages of cyanobacteria in the ocean. Despite their close phylogenetic relationship, they have distinct niche specializations and genomic differentiations (Martinez-Gutierrez and Aylward 2022). *Synechococcus* prefers coastal and temperate waters, whereas *Prochlorococcus* is specialized in warm-oligotrophic waters. In adaptation to their distinct habitats, *Synechococcus* has greater phenotypic flexibility and regulatory capacity, whereas *Prochlorococcus* displays strong specialization to nutrient-depleted environments (Moore et al. 1995, Rocap et al. 2003, Partensky and Garczarek 2010). This ecotypic differentiation has genomic marks, including the loss of large extrinsic antenna complexes (phycobilisomes) in *Prochlorococcus* but their presence in *Synechococcus*, and the occurrence in *Prochlorococcus* of simpler compact antennae within thylakoid membranes in direct contact with the photosystems (Berube et al. 2018, Partensky and Garczarek 2010, Sánchez-Baracaldo et al. 2019). Furthermore, high-light adapted *Prochlorococcus* ecotypes have undergone a reduction in genome size and %GC content as part of streamlining and adaptation to the oligotrophic environment, while these changes do not seem to occur in *Synechococcus* (Rocap et al. 2003, Partensky and Garczarek 2010). The differences between the two lineages may be due to stronger purifying selection in *Prochlorococcus* than in *Synechococcus* (Wolf and Koonin 2013, Hu and Blanchard 2009). In addition, within *Prochlorococcus*, different ecotypes exist with respect to adaptation to light abundance; the High-Light adapted ecotype (HL) is more abundant in surface waters and tends to have small genomes and low %GC content whereas low-light adapted *Prochlorococcus* ecotypes are dominant in deeper waters and tend to have larger genomes with higher %GC content (Rocap et al. 2003, Kettler et al. 2007). Even within each ecotype, further differentiation is observed by growth temperature (Rocap et al. 2003).

Endosymbiotic Cyanobacteria

Various cyanobacteria form endo- and exo-symbiotic relationships with a wide of hosts in diverse environments (Mutalipassi et al. 2021). The hosts of cyanobacterial symbionts include diatoms, haptophytes, dinoflagellates, ciliates (Tintinnids), radiolarians, macroalgae, seagrasses, sponges, cnidarians, and ascidians (Mutalipassi et al. 2021). The best studied endosymbiotic cyanobacterium is arguably the heterocystous *Richelia intracellularis*, which resides inside diatoms such as *Hemiaulus* and *Rhizosolenia*. This was known as the only diazotrophic endosymbiotic cyanobacterium until the discovery of *Calothrix* in a symbiotic relationship with *Chaetoceros* sp. (Foster and Zehr 2006) and solitary-cell species such as UCYN-A living inside haptophytes (Thompson et al. 2012). Additionally, cyanobacteria in the dinoflagellate *Ornithocercus magnificus* were sequenced using single-cell sequencing technology (Nakayama et al. 2019). Cyanobacterial endosymbiosis will be discussed in more detail in Chapter 20.

Diversification of Pigments and Light Niche Space

Although the core photosystems (13 and 30 subunits in PSII and PSI, respectively) do not deviate much from their eukaryotic counterparts, cyanobacteria arguably have the most diverse and variable pigment complements among any photosynthetic group of organisms. Not only do cyanobacteria have the most diverse complements of chlorophylls, but their accessory pigments, phycobilipigments, are also most highly diverse in structure and variable in responding to light condition changes. These characteristics underlie the wide light niche space occupied by cyanobacteria collectively and the high-resolution differentiation between strains (Ting et al. 2002). Cyanobacteria also possess carotenoids as photoprotective pigments, including xanthophyll (zeaxanthin) and β-carotene.

Chlorophylls

Like all other oxygenic photosynthetic organisms, cyanobacteria use Chl *a* as the major photosynthetic pigment, which serves as the reaction center chlorophyll to ignite the photosynthetic electron transfer. While some cyanobacteria also use Chl *b*, like green chloroplasts, others use phycobilipigments instead to capture photoenergy and transmit it to the reaction center. In most cyanobacteria, Chl *a* and *b* (both being monovinyl) occur as divinyl Chl *a* and *b* (DVChl *a* and *b*) (Chisholm et al. 1992). Monovinyl Chl lineages possess the 3,8-Divinyl chlorophyllide 8-Vinyl Reductase (DVR), which reduces DVChl to monovinyl Chl, whereas the DVChl lineages do not (Nagata et al. 2005, Ito et al. 2008). DVChl harvests blue light more efficiently than Chl *a*, enabling DVChl lineages such as *Prochlorococcus* to grow in deep layers of the water column where light is dim and dominantly blue. The Chlorophyll-Binding Proteins (CBPs) have co-evolved as Chl *a* changed to DVChl *a*.

The change of Chl to DVChl was accompanied by the co-evolution of PS proteins. Comparing CBPs of the DVChl genus *Prorochlorococcus* with those of Chl lineages revealed two substitutions in the D1 protein (Ito and Tanaka 2011). V205 and G282 in D1 are conserved in Chl lineages but replaced with M205 and C282 in the DVChl group. When either or both of V205 and G282 of *Synechocystis* D1 protein were artificially replaced with M205 and C282, and the mutated gene construct was introduced into *Synechocystis* mutant with DVR disrupted (Δ*dvr*), the transgenic strain acquired light tolerance and grew at medium-light, and the tolerance increased when both residues were replaced, whereas Δ*dvr* could not survive (Ito and Tanaka 2011). However, when DVR genes in cyanobacteria and *Arabidopsis* were mutated, DVChl accumulated, but the mutants could not survive, both chlorophyll and their binding proteins need to co-evolve.

Some cyanobacteria use Chl *d* or Chl *f* to replace Chl *b* as the auxiliary chlorophyll to extend light absorption to red. The genus *Acaryochloris* possesses both Chl *a* and Chl *d*, but > 90% of its total chlorophyll is composed of Chl *d*, and the chlorophyll special pair in the photosynthetic reaction centers is also occupied by Chl *d* (Miyashita et al. 1996, Hu et al. 1998). The Chl *a*/*d* ratio is as low as 0.044 in strain MB10699 and 0.052 in strain MB11017 (Yamamoto et al. 2022). As such, *Acaryochloris* spp. are the only example of a Chl *d*-centered photosystem known to utilize far-red spectra for oxygenic photosynthesis (Hu et al. 1996, Tomo et al. 2007). Chl *d* is widely distributed in the low-light ecological niches enriched with far-red spectra, but it has only been documented in one cyanobacterial species, *A. marina*. Chl *f*, in contrast, occurs with Chl *a* and has been reported from several genera of cyanobacteria (Chen 2019). Additionally, Chl *f*-containing cyanobacteria were found to perform substantial near-infrared light-driven photosynthesis in the natural environment (Kühl et al. 2020). Interestingly, a recent study indicated that the cyanobacterial Chl *f* could substitute Chl *a* in soluble antenna of dinoflagellates, as reconstituted peridinin-Chl *f* protein complex (Chl *f*-PCP) showed a stoichiometry ratio of peridinin:Chl *f* at 4:1, consistent with the peridinin:Chl *a* ratio from native PCP complexes, and its excitation featured a broad peak at 453 nm and a shoulder at 511 nm while emission occurred at a wavelength of 712 nm, confirming energy transfer from peridinin to Chl *f* (Hernández-Prieto et al. 2022).

Phycobilisome

Possession of phycobilisome is one of cyanobacteria's signature characteristics. Other groups of algae that have phycobilisomes are glaucophytes, rhodophytes, and cryptophytes. Phycobilisomes are giant protein complexes (up to 600 polypeptides), bound by bilins, anchored to thylakoid membranes, and exposed to the stroma side. They are mainly composed of three types of phycobiliproteins: phycoerythrin (PE), phycocyanin (PC), and allophycocyanin (APC), together with their associated linker proteins. These proteins covalently bind to their corresponding chromophores, phycoerythrobilin, phycocanobilin, and phycourobilin, which are basically open-ring tetrapyrrole molecules (Fig. 4.2). These phycobiliproteins are organized like radiating antennas, with APC being situated next to chlorophylls in the reaction center and connected to PC on the other side. Next to PC is PE (PEII at the extremity followed by PEI), which captures light and passes it to chlorophylls through the PE-PC-APC cascade (Fig. 4.2). The energy transfer efficiency can reach 90%. Depending on the relative abundance of PE, PC, and APC and their organization,

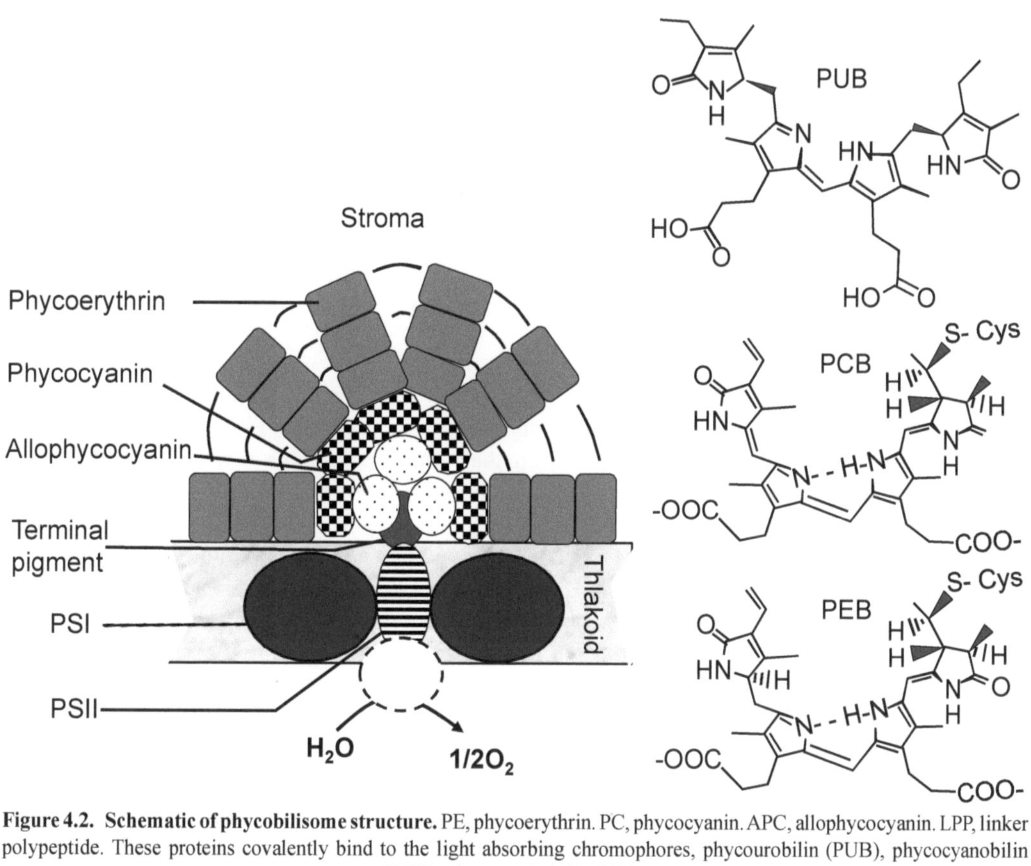

Figure 4.2. Schematic of phycobilisome structure. PE, phycoerythrin. PC, phycocyanin. APC, allophycocyanin. LPP, linker polypeptide. These proteins covalently bind to the light absorbing chromophores, phycourobilin (PUB), phycocyanobilin (PCB), phycoerythobilin (PEB).

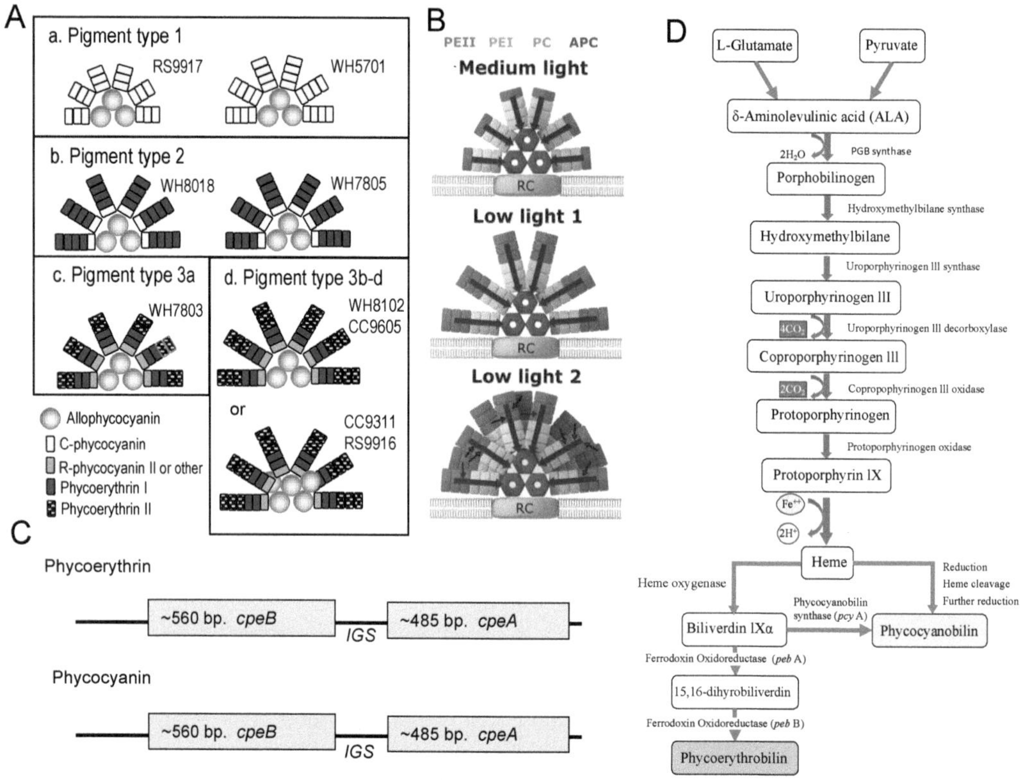

Figure 4.3. The diverse structures (A), light-induced changes (B), gene operon organizations (C) and biosynthetic pathway of phycobilins (D). (A) Schematic representation of the phycobilisomes in different strains of picocyanobacteria. Allophycocyanin (APC) core is attached to photosystem I or II present in the thylakoids membrane and it transfers the light energy absorbed by the rods to the photosynthetic reaction centers. Attached to the APC core are the rods that harvest the light energy. Close to the core, discs consisting of hexameric complexes of phycocyanin proteins with phycocyanobilin as the pigment (a). The next discs are composed of hexameric complexes with phycoerythrin (PE) I (b) and the most distal discs contain PE II proteins (c and d). In general, PE I binds only phycoerythrobilin (b), while PE II binds PUB (c and d). Based on Six et al. 2007. (B) Phycobilisome structure under different light intensities. Under low light the PBS is larger, with longer rods that have additional PEII hexamers (with higher PUB content). Energy absorbed at the extremities travels longer distance to the reaction centers (RCs). Reproduced with permission from Kolodny et al. 2021. (C) The operon structure encoding the alpha and beta subunit genes of the phycoerythrin (cpe) and phycocyanin (cpc) proteins. IGS, intergenic spacer. (D) Schematic of the biosynthesis pathway of phycobilins.

the architecture of phycobilins varies among cyanobacteria lineages (Fig. 4.3A; Six et al. 2007). The architecture in a species can change in response to light conditions (Fig. 4.3B; Kolodny et al. 2021). Salinity of the habitat can induce changes in the pigment phenotype as well as phycobilisome gene clusters, with unique gene organization found in the brackish-water Baltic Sea (Larsson et al. 2014). The biosynthesis pathway and the organization of major gene operon of phycobiliproteins are shown in Fig. 4.3C,D. Most genes involved in the biosynthesis and regulation of these proteins are located in a dedicated genomic region called the PBS rod region (Grébert et al. 2022). A recent study indicated that APC alpha-subunit in cyanobacterial phycobilisome is a carbon dioxide receptor; when binding to CO_2, it regulates Electronic Energy Transfer (EET) and its fluorescence quantum yield in the cyanobacterium *Synechocystis* sp. PCC6803, enhancing EET in the cell, potentially as a crucial mechanism to balance energy capture and carbon fixation (Guillén-García et al. 2022).

As alluded to earlier, while most cyanobacteria possess the signature pigment, phycobilipigment (PBP), some lineages have lost this trait. The PBP-less lineages such as *Prochlorococcus* use Chl *b* instead for capturing light energy. The variation in this trait can occur even within species. In the Chl *d*-containing species *Acaryochloris marina*, strain MBIC11017 has PBP whereas strain

MBIC10699 does not (Yamamoto et al. 2022). It is believed that strain MBIC11017 has reacquired PBP synthesis genes because they are located in the plasmids rather than chromosomes (Yamamoto et al. 2022). Additionally, although most of the protein-coding genes are identical between these two strains, genes in their plasmids are quite different. This evolutionary dynamics within species is striking and unprecedented.

Regulation of N Nutrition

Despite species- or strain-specific differentiation of nutrient niches, as discussed in Chapter 3, cyanobacteria collectively can take up nitrate, nitrite, ammonium, and urea as the source of N-nutrients. Cyanobacteria sense the deficiency of nitrogen nutrients through two central regulators of N assimilation, PII and NtcA, involving 2-oxoglutarate (2-OG). 2-OG is the substrate in ammonium assimilation into glutamate, which involves glutamine and glutamine-oxoglutarate aminotransferase (GOGAT). When N supply is insufficient, the assimilation slows down, resulting in more abundant free 2-OG. The binding of 2-OG stimulates the activity of NtcA, the main transcriptional regulator of N assimilation. When 2-OG is abundant (N is scarce), PipX switches from binding to PII to interacting with NtcA, enhancing the binding affinity of the PII-PipX-NtcA complex to the promoters of N assimilation genes. In the model cyanobacterium *Synechocystis* sp. PCC6803, NtcA directly activates 51 genes and represses 28 other genes after 4 h of N starvation. Among the NtcA-activated genes are the co-transcribed genes NblA1 and NblA2, which initiate degradation of phycobiliproteins and other physiological responses. The coordinated degradation of phycobilisomes involves a regulatory factor, NblD, a cysteine-rich small protein (Krauspe et al. 2020).

Cyanophycin as Storage of N

Cyanophycin, also known as Cyanophycin Granule Polypeptide (CGP) or multi-L-arginyl-poly (L-aspartic acid), is a non-protein, non-ribosomally produced amino acid polymer composed of an aspartic acid backbone and arginine side groups. It can be found in most cyanobacteria and a few heterotrophic bacteria such as *Acinetobacter* sp. In cyanobacteria, cyanophycin is used as a nitrogen- and possibly carbon-storage compound and also serves as a dynamic buffer for fixed nitrogen in cyanobacterial heterocysts. Cyanophycin granules can be stored in akinetes as well. It is largely insoluble under physiological conditions and is accumulated in the form of granules in the cytoplasm during phosphate or sulfur starvation, generally in the early and mid-stationary phase. Cyanophycin is synthesized from arginine and aspartate in an ATP-dependent reaction catalyzed by a single enzyme, cyanophycin synthetase (Ziegler et al. 1998). Nitrogen and carbon are mobilized from cyanophycin by intracellular cyanophycinase in the form of aspartate-arginine dipeptides. In *Raphidiopsis raciborskii*, the synthesis of cyanophycin appears to be regulated by the abundance of nitrogen nutrients in the environment (Lu et al. 2022), consistent with the notion that cyanophycin is synthesized to store N when environmental N nutrients are copious and is mobilized when environmental nitrogen is depleted. Hence, cyanophycin serves to ensure sufficient nutrient supply for bloom development (Lu et al. 2022).

Nitrogen Fixation

The most important characteristic of cyanobacteria is arguably the ability to fix N_2 to ammonium in many species. For a long time, this ability was ascribed exclusively to the filamentous species such as *Trichodesmium* (Fig. 4.1). Later, single-celled diazotrophic cyanobacteria were identified in the ocean (Zehr et al. 2001). Currently, N_2-fixing potential has been recognized in various microbial groups beyond cyanobacteria, including heterotrophic bacteria and archaea, in plankton and deep-sea cold seep sediments (Dong et al. 2022), but the magnitude of nitrogen fixation and contribution

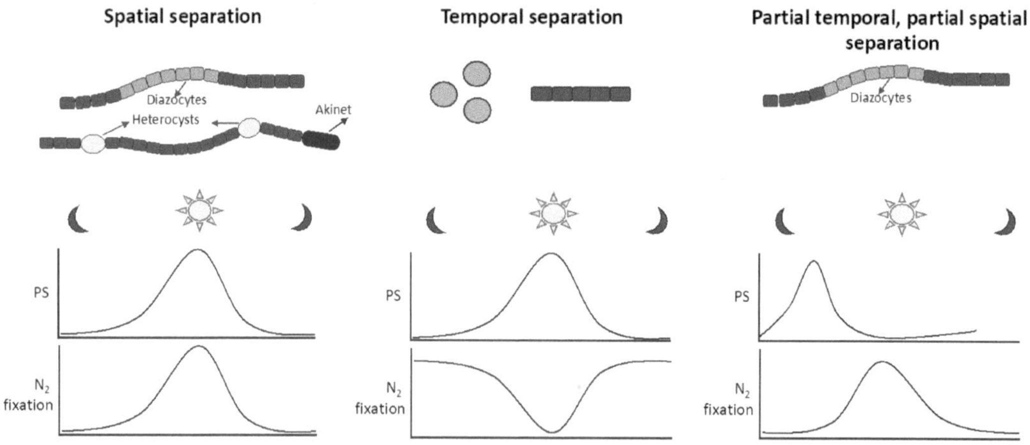

Figure 4.4. Three major strategies of diazotrophic cyanobacteria to separate photosynthesis (PS) and N_2 fixation to avoid O_2 inhibition of nitrogenase.

to the marine nitrogen budget by these bacteria and archaea remains unclear. As such, cyanobacteria are still regarded as the most important contributors of biogenic nitrogen nutrients in the ocean.

Nitrogen fixation is an energy-demanding biochemical reaction because of the triple bond between the two N atoms, which is tough to break. It consumes six NADPH and up to 12 ATP to reduce N_2 to NH_3. This biochemical reaction is catalyzed by an enzyme system name nitrogenase, which is a complex of two metalloproteins: dinitrogenase ($\alpha 2\beta 2$ heterotetramer) that contains the active site for N_2 reduction to NH_3 and dinitrogenase reductase ($\gamma 2$ homodimer). Dinitrogenase is encoded by *nif*D (α) and *nif*K (β) whereas dinitrogenase reductase is encoded by *nif*H. Nitrogenase is sensitive to O_2, making N_2 fixation and photosynthesis incompatible. Diazotrophic cyanobacteria have evolved at least three strategies to circumvent this problem (Fig. 4.4).

Heterocysts: cell differentiation to separate photosynthesis and N_2 fixation

In heterocystous species of diazotrophic cyanobacteria such as *Anabaena* spp., some cells are differentiated into heterocysts, which are distributed intermittently along the filament of cells, with about 10 vegetative cells in between heterocysts. Heterocystous cells have a thick cell wall to resist O_2 diffusion and a reduced photosystem to lower O_2 production. Therefore, heterocysts are specialized in N_2 fixation and share the fixed N with neighboring vegetative cells, while vegetative cells are specialized in photosynthesis and C fixation and share the fixed organic carbon with heterocysts.

Heterocyst development and pattern formation involve many processes, including environmental N-nutrient conditions and a host of genes (Herrero et al. 2004, Zhang et al. 2006, Flores and Herrero 2010, Herrero et al. 2013, 2019). Under N-rich conditions, pro-heterocystous cyanobacteria such as *Anabaena* develop no heterocysts, and their filaments consist only of vegetative cells performing photosynthesis. However, in response to various environmental stresses, some vegetative cells can differentiate into heterocysts capable of fixing N_2 (Casanova-Ferrer et al. 2022). The HetR gene is believed to be the central regulator of heterocyst differentiation (Khudyakov and Golden 2004, Casanova-Ferrer et al. 2022). A diffusible peptide 13–17 residues long, named PatS, is an inhibitory regulator of hetR (Yoon and Golden 1998, Casanova-Ferrer et al. 2022). Besides, PatX and HetN seem to influence the production of the diffusible peptide (Borthakur et al. 2009), whereas PatA interacts with hetF, influencing the activation of hetR (Corrales-Guerrero et al. 2014). The function of PatS and HetN creates a concentration gradient of hetR in the cell (Risser and Callahan 2009). The PatS/HetN::hetR regulatory core interacts with the other components in a sequential or complementary manner to furnish two major regulatory points in heterocyst differentiation: the early inhibitory mechanism and the late inhibitor mechanism (Muñoz-García and Ares 2016).

Much of the experimental evidence of the regulatory network of heterocyst differentiation came from *Anabaena* as the model, and gene mutation was used as the experimental method. For instance, in *Anabaena* sp. strain PCC 7120, researchers found that overexpressing patS inhibited heterocyst differentiation. Conversely, a patS null mutant displayed an increased frequency of heterocysts and an irregular pattern. Using a patS-gfp reporter, the expression of patS was localized to developing pro-heterocysts. Interestingly, the introduction of a synthetic peptide corresponding to the C-terminal five amino acids of PatS suppressed heterocyst development. These findings suggest that PatS plays a crucial role in regulating heterocyst pattern formation (Yoon and Golden 1998).

Anabaena sp. strain PCC7120 possesses other genes functionally similar to patS, asr1734 and patA (Yoon and Golden 2001). The overexpression of asr1734 inhibited heterocyst development in several strains, including the PatS null mutant (Wu et al. 2007). An Asr1734 knockout produced phenotypes of forming 5% heterocysts after a nitrogen shift from ammonium to nitrate, and forming 15% heterocysts and a weak multiple contiguous heterocyst (Mch) formation after step-down to medium lacking combined nitrogen. PatA was found to regulate terminal heterocyst differentiation in *Anabaena* (Casanova-Ferrer et al. 2022). The study showed that a patA knockout mutant of *Anabaena* successfully developed heterocysts at the end of the cyanobacteria termini.

Temporal and Spatial Segregation of Photosynthesis and N_2 Fixation

In non-heterocystous species, no cells are differentiated into O_2-proof N_2-fixing cells. Two strategies have evolved to separate photosynthesis and N_2 fixation. For species like *Synechoccocus* sp., photosynthesis and N_2 fixation are separated in time to avoid oxygen poisoning of nitrogenase. Photosynthesis is active during daytime whereas N_2 fixation is active at night. *Trichodesmium* spp. exhibit cryptic cellular differentiation for nitrogen fixation, as these specialized cells lack distinct morphological features that would allow their visual identification. In these species, a cluster of cells within the filament of cells (trichome) contain nitrogenase whereas the rest of the filament has little or no nitrogenase, as shown through immunofluorescence (Fig. 4.4; Lin et al. 1998). A ^{14}C-labeling experiment indicated that the nitrogenase-containing cells had depressed C-fixing activity. These functionally heterocyst-like cells are called diazocytes (Sandh et al. 2012). This spatial differentiation allowed simultaneous active operation of both C and N_2 fixation in the trichome (Berman-Frank et al. 2001).

More recent studies using nanometer-scale secondary ion mass spectrometry (Finzi-Hart et al. 2009) and modeling (Zhang et al. 2022) revealed a partial temporal and partial spatial segregation between C- and N_2-fixing activities. Photosynthesis is most active in the "morning" to fix C and store the resulting organic C. At "noon", respiration is elevated dramatically and photosynthesis declines, creating a low-O_2 environment that favors N_2 fixation. This slight mismatch between photosynthesis and N_2 fixation is coupled with the diazocyst specialization to facilitate both C and N_2 fixation without the conflict between O_2 and nitrogenase.

Transcriptional regulators ChlR and CnfR were found to be essential for coordinating photosynthesis and N_2 fixation in nonheterocystous filamentous cyanobacteria such as *Leptolyngbya boryana* (Tsujimoto et al. 2014). In this diazotroph, there is a 50-kb nitrogen fixation (nif) gene cluster that encodes 50 genes, including 14 *nif* genes, the regulators ChlR and CnFR, and 28 *nif*-related proteins. By analyzing growth and N_2-fixing activities of ChlR and cnFR mutants, it was found that these regulators activated the transcription of genes responsible for chlorophyll biosynthesis and those involved in N_2 fixation.

Regulation of P Nutrition

Cyanobacteria are versatile and efficient in utilizing different forms and low concentrations of P nutrients, enabling them to thrive in the oligotrophic open oceans. Besides phosphate uptake that involves low- and high-affinity transporters, cyanobacteria use Alkaline Phosphatase (AP) to scavenge diverse phosphoester types of Dissolved Organic Phosphorus (DOP). Mirroring heterotrophic bacteria, cyanobacteria possess different forms of AP, including PhoA, PhoD, and PhoA (Liu and Wu 2012). These APs contain characteristic domains. PhoA-709 has a conserved domain of alkaline phosphatase (alkPP), both PhoD1-709 and PhoD2-709 contain two conserved metallophosphatase domains (MPP), which are associated with YfcE-like phosphodiesterase activity, and PhoS-709 has a conserved PQQ-ABC-repeats (PQQ) domain. In addition, also like heterotrophic bacteria, cyanobacteria are able to utilize phosphonate compounds, which are characterized by C-P bonds (Dyhrman et al. 2006). Cyanobacterial genomes code for the gene operon of phosphonate utilization genes, including those encoding transporters and lyases of phosphonates.

As is true for N-nutrient utilization, cyanobacteria exhibit species- or strain-specific differentiation of P nutrient utilization. For example, *Synechococcus* species are ecologically diverse occurring in numerous environments. Thermophilic species are known to engage in photosynthesis and aerobic respiration during daytime but switch to fermentative metabolism and nitrogen fixation at nighttime. From Octopus Spring in Yellowstone National Park, several thermophilic strains were isolated (Gomez-Garcia et al. 2011). The genome of *Synechococcus* OS-B' contains a phn gene cluster encoding a phosphonate (Phn) transporter and a C-P lyase. In contrast, the closely related *Synechococcus* OS-A lacks this cluster but has genes encoding putative phosphonatases, which appear active only in the presence of Phn substrates. Both isolates can grow on various Phns as their sole phosphorus source. Notably, *Synechococcus* OS-B' can utilize the organic carbon backbones of Phns for heterotrophic growth in the dark, releasing organic carbon as ethane or methane in the light, depending on the specific Phn. *Synechococcus* OS-A lacks these capabilities. These differing metabolic strategies for phosphorus and carbon assimilation in the two *Synechococcus* strains suggest niche-specific adaptations and syntrophic relationships within the hot-spring microbial mats.

Life Cycle

Our knowledge of the life cycle of cyanobacteria is quite limited, especially for unicellular species (i.e., picocyanobacterial). For filamentous cyanobacteria, their life cycle includes vegetative cell stage (active) and akinete stage (dormant). These two stages are ecologically significant as they connect the water column and the sediment. Akinetes, formed from vegetative cells, stay dormant in the sediment under unfavorable conditions. When the environment becomes favorable again, they germinate and move to the water column, where they grow into vegetative cells.

Akinetes are essentially enlarged vegetative cells (Fig. 4.5) that develop a thickened wall in response to adverse environmental conditions. Akinetes are crucial for the survival and ecological resilience of cyanobacteria, allowing them to endure unfavorable conditions such as cold, desiccation, and nutrient scarcity. These specialized cells can remain dormant for extended periods (years in the sediment), resuming growth when environmental conditions improve. Akinetes also help cyanobacteria cope with extreme stressors such as freezing, oxidative stress, and lysozyme attack, through the protection conferred by their unique glycolipid-rich envelopes (Sukenik et al. 2012, Garg and Maldener 2021). This ability to withstand stress and facilitate long-term survival gives cyanobacteria a significant ecological advantage, ensuring their persistence and proliferation in various habitats (Hori et al. 2003, Kaplan-Levy et al. 2010).

During akinete differentiation, cyanobacterial cells increase in size and accumulate storage compounds like glycogen, cyanophycin globules, and nucleic acids (Garg and Maldener 2021). These cells develop thickened cell walls and multilayered extracellular envelopes composed of polysaccharides, mucilage, and glycolipids, which are absent in vegetative cells. Stored materials are consumed during maturation, facilitating the formation of the complex envelope, and sustaining the cell during dormancy (Perez et al. 2016). Several transcriptional regulators involved in heterocyst differentiation also play roles in akinete formation. For instance, the overexpression of devR enhances akinete differentiation in *Nostoc punctiforme*, while hetR and hepA are essential for

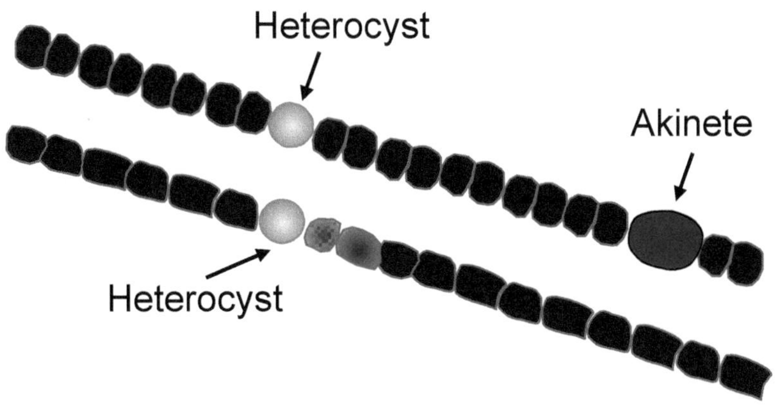

Figure 4.5. Schematic of akinete and heterocysts in filamentous cyanobacteria.

normal envelope formation and differentiation in various species (Leganés et al. 1994, Campbell et al. 1996, Wong and Meeks 2002). The regulation involves common genetic components, suggesting similar developmental processes for akinetes and heterocysts (Bell et al. 2017, Garg and Maldener 2021).

Akinete germination begins under favorable conditions such as adequate light, moderate temperatures, and nutrient availability. Initially, cell size increases, followed by the first cell division inside the akinete envelope within 18–24 hours. Subsequent cell divisions build pressure, eventually rupturing the envelope and releasing a short filament that resumes the vegetative cycle (Kaplan-Levy et al. 2010). In nitrogen-free medium, the terminal cell differentiates into a heterocyst early in cell division, indicating nitrogen sensing. This heterocyst likely fixes nitrogen to support the growing filament as nitrogen storage compounds are depleted during maturation (Perez et al. 2018). Germination involves both respiration and photosynthesis for energy, with glycogen respiration powering initial cell division and photosynthesis sustaining later stages. In species like *Anabaena cylindrica*, light, oxygen, and phosphate are required for efficient germination, and the rate is temperature-dependent, highlighting the intricate regulation of this process (Garg and Maldener 2021).

5

Genomes of Green Algae

Green algae (Fig. 5.1) are photosynthetic eukaryotes with plastids bound by double-layered membrane. Their plastids contain chlorophylls *a* and *b* as major pigments, while beta carotene and xanthophylls serve as accessory pigments. These algae store their photosynthetic products as starch within the plastids (Lewis and McCourt 2004). Green algae encompass the chlorophytes, prasinophytes, and part of the Streptophyta (charophytes), which is dominated by land plants (embryophytes). According to AlgaeBase, there are about 22,000 species of green algae (Guiry and Guiry 2017). Green algae are with or without flagella and cell walls and if present cell walls are composed of cellulose. Most green algae are unicellular and microscopic, typically several to ten microns in diameter, but there is a variety of macroscopic and multicellular species that can be over a meter long (Chapter 11). Most of the microscopic green algae are motile and planktonic but there are nonmotile species that are attached to substrates (benthic) or associated with lichen. Some unicellular species form colonies each containing as many as thousands of cells (e.g., *Volvox*). Cells in colonies may be joined by gelatinous strands or share a common parental wall.

Green algae are distributed in aquatic environments, soil, rock surface, tree trunks, deserts, and on surface of other organisms. Most green algae live in freshwater (lakes, ponds, rivers). Two major marine lineages include the macrophyte Ulvales and the prasinophytes, which are probably most ecologically conspicuous green algae in current ocean. The marine prasinophyte *Ostreococcus* (< 1 μm) and the more recently identified freshwater trebouxiophyte *Medakamo hakoo* (~ 1 μm) are the smallest eukaryotic cells known to science whereas *Micromonas* is one of the most common prasinophytes in the ocean, distributed from polar waters to the tropics. Although most green algae appear green in color due to the dominance of chlorophylls in pigment composition, there are species that appear red due to the high abundance of carotenoid pigments (e.g., snow algae *Chlamydomonas nivalis* and birdbath alga *Hematococcus pluvialis*; Fig. 5.1W-Z).

Interesting, some single-cell species of chlorophytes have multicellular species-appearance, being macroscopic (Fig. 5.2). This includes *Acetabularia*, *Halimeda*, and *Caulerpa*. Among these, some lineages contain numerous nuclei in a cell (e.g., *Halimeda*, *Caulerpa*). These multinucleated algae are also known to be siphonous or coenocytic. *Halimeda* has a cactus-like morphology, with segmented thalli, hence commonly known as cactus algae. Furthermore, *Halimeda* spp. occur in the coral reef ecosystem and their thalli are calcified.

Evolution of Green Algae

The successful colonization of the terrestrial landscape by green plants is one of the most significant biological events. The search for the algal ancestor of land plants has been an unceasing scientific inquiry. The reconstruction of phylogenetic relationships across green plants using transcriptomic or genomic data have provided evidence that unicellular, often scaly, flagellate organisms might be the common ancestor of all green plants. Identification of an extant relative of such a flagellate has proven to be challenging. *Mesostigma viride* is one of the candidate species that have been studied, which turns out to be an early-diverging member of the streptophytes. Another candidate is Prasinococcales, which could not be definitively positioned in phylogenetic trees, and the macroalgal genera *Verdigellas* and *Palmophyllum*, which belong to the class Palmophyllophyceae, the earliest divergence in Chlorophyta. Recently, a phylogenomic study established a new phylum, the Prasinodermophyta, representing the earliest divergence of the Viridiplantae. This phylum, comprising two classes and featuring the pico-eukaryotic species *Prasinoderma coloniale*, marks the dawn of green plant emergence (Li et al. 2020). As a result, the Viridiplantae is composed of three instead of two phyla.

The evolution within Chlorophyta remains a topic of debate, as multiple lineages do not form monophyletic clades in phylogenetic analyses. Based on chloroplast genome studies, a "Core Chlorophyta" has been proposed, encompassing the well-supported grouping of Chlorophyceae, taxa of Trebouxiophyceae and Ulvophyceae, as well as the classes Chlorodendrophyceae and Pedinophyceae (Fucikova et al. 2014).

Genomes of Chlorophytes

Thirteen genomes have been sequenced in the phylum of Chlorophyta (Table 5.1). Most chlorophyte algae have a diploid genome, but siphonous species (e.g., *Caulerpa lentillifera*) are multinucleate and appear to be polyploid (Zhang et al. 2024).

Chlamydomonas reinhardtii

Chlamydomonas reinhardtii is a soil-dwelling unicellular chlorophyte that can grow in the dark using an organic carbon source while maintaining a functional chloroplast. Its two anterior flagella make the species motile, a trait more commonly associated with animals. The genome of *C. reinhardtii* (121 Mbp) was sequenced to understand its evolutionary position relative to plants and animals (Merchant et al. 2007). This species has 17 chromosomes with a high GC content (64%). The genome encodes 15,143 protein-coding genes, including 1,226 gene families, 26 of which have 10 or more members. There are 798 two-gene families, with 317 arranged in tandem, indicating extensive gene duplications. On average, protein-coding genes have 8.3 exons each, with a longer average intron length than many eukaryotes, and only 8% of genes lack introns. This makes *C. reinhardtii* more intron-rich than many other unicellular eukaryotes and vascular plants. The intron number and size are more similar to multicellular lineages than protists.

C. reinhardtii has large families of kinases and P-450 proteins, with the largest being the 51-member class III guanylyl and adenylyl cyclase family, which is absent in land plants but present in animals for synthesizing the secondary messengers cGMP and cAMP. Cyclic nucleotides are crucial for mating processes, flagellar function, and acclimation to changing nutrient conditions. The genome encodes a P450 called CYP55B1, similar to fungal Nitric Oxide (NO) reductase CYP55A1, shown to encode NO reductase (Gong et al. 2018).

A
B
C
D
E
F
G
H
I
J
K
L
M
N
O
P
Q
R
S
T
U
V
W
X
Y
Z

Figure 5.1. Representative species showing morphological diversity of green algae. A: *Pterosperma* (Pyramimonadales), a marine prasinophyte characterized by quadriflagellate unicells (photo by Bob Andersen, reproduced under license from microscope.mbl.edu). B: *Nephroselmis* (Nephroselmidophyceae), a prasinophyte with bean-shaped cells and two unequal flagella occurring in marine or freshwater environments. C: *Palmophyllum* (Palmophyllales) forming lobed crusts composed of coccoid cells embedded in a gelatinous matrix, growing in deep-water marine habitats (photo by Jordi Regas). D: *Tetraselmis* (Chlorodendrophyceae), quadriflagellate unicells from marine or freshwater habitats. E: *Chlorella* (Trebouxiophyceae, Chlorellales), coccoid cells, endosymbiontic inside the single-celled protozoan Paramecium (Photo by Antonio Guillén). F: *Oocystis* (Trebouxiophyceae, Oocystaceae), small colonies of coccoid cells within a thin mucilaginous envelope from freshwater habitats (image copyright Microbial Culture Collection, NIES). G: *Haematococcus* (Chlorophyceae, Chlamydomonadales), a freshwater biflagellate unicell (photo by William Bourland, reproduced under license from microscope.mbl.edu). H: *Pediastrum* (Chlorophyceae, Sphaeropleales), a coenobial colony of non-motile cells arranged in a circular plate, occurring in freshwater habitats. I: *Bulbochaete* (Chlorophyceae, Oedogoniales), branched filaments with terminal hair-cells. J: *Chaetophora* (Chlorophyceae, Chaetophorales), highly branched filaments, occurring in freshwater habitats (H-J: photos by Jason Oyadomari). K: *Ulothrix* (Ulvophyceae, Ulotrichales), unbranched filaments from marine or brackish areas (photo by Giuseppe Vago). L: *Ulva* (Ulvophyceae, Ulvales), sheet-like plants, mainly from marine habitats (photo by Tom Schils). M: *Cladophora* (Ulvophyceae, Cladophorales), branched filament with cells containing numerous chloroplasts and nuclei (photo by Antonio Guillén). N: *Boergesenia* (Ulvophyceae, Cladophorales), plants composed of giant, multinucleate cells, from tropical marine habitats (photo by HV). O: *Acetabularia* (Ulvophyceae, Dasycladales), siphonous plants (i.e., single-celled) differentiated into a stalk and a flattened cap, with a single giant-nucleus situated at the base of the stalk; typically found in subtropical marine habitats (photo by Antoni López-Arenas). P: *Caulerpa* (Ulvophyceae, Bryopsidophyceae), siphonous plant differentiated into creeping stolons anchored by rhizoids and erect photosynthetic fronds, containing millions of nuclei; occurring in (sub)tropical marine waters (photo by FL). Q: *Klebsormidium* (Klebsormidiophyceae), unbranched filamentous charophyte, mainly from moist terrestrial habits (photo by Jason Oyadomari). R: *Spirotaenia*, a unicellular charophyte with typical spiral chloroplast; generally growing in acidic freshwater habitats (photo by Antonio Guillén). S: *Nitella* (Charophyceae), morphologically complex charophyte from freshwater habitats, consisting of a central stalk and whorls of branches radiating from nodes that bear oogonia and antheridia (photo by Nadia Abdelahad). T: *Micrasterias* (Zygnematophyceae, Desmidiales), characterized by non-motile unicells constricted in two parts with ornamented cell wall; generally found in acidic, oligotrophic freshwater habitats (photo by Antonio Guillén). U: *Coleochaete* (Coleochaetophyceae), branched filaments adherent to form a disc-like, parenchymatous thallus; found in freshwater habitats, often as epiphytes on submerged vascular plants (photo by CFD). A-U, reproduced from Leliaert et al. 2012 with permission from Taylor and Francis. V: *Volvox*; W: birdbath algae *Hematococcus pluvialis*, credit: Creative Commons license CC BY-NC-SA 2.0; X: birdbath; Y: snow algae *Chlamydomonas nivalis* (credit: Stefanie Lutz from EurekAlert!); Z: photo of snow with red spots containing snow algae, credit: Y. Zhao, https://news.cgtn.com/news/3d3d514e3341444d32457a6333566d54/index.html.

Figure 5.2. Single-cell green algae that grow to macroscopic bodies. (A) *Acetabularia crenulate*, a unicellular, uninucleated, alga, also known as "Mermaid's Wineglass". (B) *Halimeda discoida*, also known as "cactus algae". Image credit Francois Rebufat; source of the image: http://pictolife.net/pages/popup.php?espece=Algues&id=9&menu=12. (C) *Caulerpa lentillifera*, also known as "Sea grapes". Image credit: Sommail, through https://www.flickr.com. Both *H. discoida* and *C. lentillifera* are multi-nucleated unicellular algae.

The genome contains 486 membrane-localized transporters, including 61 ion channels, 124 primary ATP-dependent transporters, 293 secondary transporters, and eight unclassified ones. There are 12 phosphate transporters (PiT) and 11 ammonium transporters (AMT). Several transporters relocate intracellular Ca^{2+} in response to environmental cues like light. Changes in Ca^{2+} concentrations can affect flagella activity, a feature found in animals but absent in vascular plants.

C. reinhardtii shares 706 protein families with humans, none of which are shared with *Arabidopsis*. These genes either underwent loss or divergence beyond recognition in green plants and are notably enriched for sequences encoding cilia and centriole proteins. However, 2,489 *C. reinhardtii* proteins are homologous to those in both *Arabidopsis* and humans, suggesting a common ancestor. Most *Chlamydomonas* proteins are slightly more similar to *Arabidopsis* than to human proteins, but many, particularly in the flagellar and basal body proteomes, are more similar to animal homologs.

Merchant et al. (2007) used comparative genomics to define specific protein sets for various functional groups of phototrophs. They identified the GreenCut, PlantCut, and DiatomCut protein sets, which include proteins involved in plastid and thylakoid membrane biogenesis, photosynthetic electron transport, carbon fixation, antioxidant generation, and other primary metabolic processes. This approach also identified previously uncharacterized photosynthesis-associated proteins. With more algal genomes available, further comparative analysis can refine these protein sets for each functional group.

Chlorella variabilis

Chlorella variabilis NC64A, a unicellular green alga from the Trebouxiophyceae class, is an intracellular photobiont of the ciliate *Paramecium bursaria*. Its 46-Mbp nuclear genome was sequenced, revealing 9,791 protein-coding genes (Blanc et al. 2010). The genome contains meiosis-specific proteins and a subset of proteins found in flagella, despite *Chlorella* species being traditionally considered asexual and nonmotile. Additionally, a survey of phytohormone pathways in this and other chlorophyte algae identified algal orthologs of *Arabidopsis thaliana* genes involved in hormone biosynthesis and signaling. The detected pathways include those for abscisic acid, cytokinin, brassinosteroids, and polyamines, but lack gibberellin biosynthesis pathway genes,

Table 5.1. Published genomes of green algae.

Species	Genome size (Mbp)	Number of chromosomes	Predicted genes	% Annotated	Coding percentage	%GC	References
Chlorophytes							
Auxenochlorella protothecoides	22.9	6	7,039	82.9	96.6	63	Fan et al. 2017, Gao et al. 2014
Bryopsis sp.	90.7		14,034			45.9	Ochiai et al. 2024
Caulerpa lentillifera	29		9311	82.1	93.8	40.4	Arimoto et al. 2019
Chlamydomonas reinhardtii	121	17	17741			64	Merchant et al. 2007, Blanc et al. 2010, Arimoto et al. 2019
Chlorella variabilis NC64A	46.6	12	9,791			67.2	Fan et al. 2017
Chloroidium sp. UTEX 3007	52.5	16	8,153				Nelson et al. 2017
Coccomyxa subellipsoidea C-169	48.8	20	9,851			53	Fan et al. 2017, Nelson et al. 2017, Gao et al. 2014
Medakamo hakoo	15.8	16	7,857		97.1	73	Kato et al. 2023
Monoraphidium minutum	68.17		15,341	96.8		72.1	Calhoun et al. 2022
Picochlorum sp.	13.3		7,367			46.1	Fan et al. 2017, Foflonker et al. 2015
Prasinophytes							
Bathycoccus prasinos RCC1105	15	19	7847			48	Moreau et al. 2012
Micromonas pusilla CCMP1545	21.9	19	10,575			65	Worden et al. 2009
Micromonas pusilla RCC299	20.9	17	10,056			64	Worden et al. 2010
Ostreococcus lucimarinus	13.2	21	7,651			59	Palenik et al. 2007, Lanier et al. 2008
Ostreococcus tauri	12.6	20	8,166		94	59.2	Arimoto et al. 2019, Blanc-Mathieu et al. 2014
Pedinophyte YPF701	27.9		7,940			70	Repetti et al. 2021
Pycnococcus provasolii CCAP190/2	32.2	44	12,671			51.7	Green et al. 2023
Pycnococcus provasolii NIES2893	22.7	43	11297			52.3	Makita et al. 2021, Green et al. 2023
Streptophytes							
Penium margaritaceum	4,700		52,333				Jiao et al. 2020
Prasinoderma coloniale	25	22	7139			70	Li et al. 2020
Spirogloea musicola	163		52,529		51.7	57.7	Cheng et al. 2019

although putative orthologs to some *Arabidopsis* hormone receptors were found. *Chlorella* also exhibits the ability to produce chitinous cell walls, likely acquired through horizontal transfer from algal viruses, prokaryotes, or fungi.

Chloroidium sp.

To investigate the phenomic and genomic traits enabling green algae to survive in deserts, researchers characterized *Chloroidium* sp. UTEX 3007, a ubiquitous species isolated from various locations in the United Arab Emirates (UAE) (Nelson et al. 2017). Metabolomic analyses revealed that *Chloroidium* sp. UTEX 3007 accumulates a broad range of carbon sources, including several desiccation tolerance-promoting sugars and unusually large stores of palmitate. Growth assays showed its ability to grow in salinities from zero to 60 g/L and heterotrophically on more than 40 distinct carbon sources. Genomic assembly and annotation yielded a 52.5 Mbp genome with 8,153 functionally annotated genes. Comparative analysis with other sequenced green algae identified unique protein families involved in osmotic stress tolerance and saccharide metabolism, supporting phenomic findings. These results highlight the robust and flexible biology that allows this green alga to thrive in a desert coastline environment.

Chromochloris zofingiensis

Chromochloris zofingiensis belongs to the chlorophyte order Sphaeropleales in the class Chlorophyceae. It is a small-sized (~ 4 μm) unicellular coccoid green alga containing multiple mitochondria occurring in tubular network and a single interconnected chloroplast that occupies ~ 40% of the cell volume. Unlike typical chlorophytes, no flagella or pyrenoids have been observed. Similar to *C. reinhardtii*, this species undergoes multiple fission with temporal separation between cell growth and cell division. It primarily divides into two or four daughter cells, but can also divide into 16, 32, or 64 cells, in uniform size. Also like *C. reinhardtii*, nucleus division precedes chloroplast in *C. zofingiensis*. Strikingly, *C. zofingiensis* has an extremely high photoprotective capacity and can dramatically increase production of lipids and secondary carotenoids such as astaxanthin under certain conditions. As such, *C. zofingiensis* is considered one of the most promising biofuel feedstocks and source of astaxanthin for commercial production.

Roth et al. (2017) sequenced the genome of *C. zofingiensis* (strain SAG211-14) and achieved chromosome-level assembly using the combination of Illumina short reads, PacBio long reads, and whole-genome optical mapping (OpGen). The ~ 58 Mbp genome was assembled into 19 chromosomes, with canonical telomeric sequence of (CCCTAAA)n and noncanonical telomeric sequence of (CCTAAA)n and (CCCTGAA)n. Through a transcriptomic and functional genetic approach (gene disruption), the biosynthesis pathway of astaxanthin was characterized. Two genes encoding β-ketolase (BKT), the key enzyme synthesizing astaxanthin, were identified and found to be upregulated under high light growth condition. Gene mutation analysis indicated that one of these BKT1 genes is required for the production of astaxanthin.

Microglena sp.

Microglena sp. YARC is a psychrophilic unicellular green alga isolated from sea ice in the Antarctic. Psychrophilic microalgae thrive in the extreme and highly variable polar ecosystems, which are crucial for food webs and nutrient cycling. Their ecological success is due to adaptive evolution. *Microglena* sp. YARC has a large genome, probably as a result of gene duplication for polar adaptation. Genome sequencing for this strain indeed shows a significant increase in protein-coding genes, with duplications estimated to occur when the circumpolar Southern Ocean was formed (Zhang et al. 2023). Duplicated genes are enriched in pathways related to photosynthesis, DNA repair, and fatty acid metabolism. Additionally, 657 gene families, including collagen-like proteins, are unique to *Microglena*. Divergence in expression patterns of duplicated and species-specific genes indicates sub- and neo-functionalization during stress acclimation. The observed gene duplication and functional novelty may contribute to polar algae adaptation.

Medakamo hakoo

Medakamo hakoo, first identified in 2015, is a freshwater unicellular alga and one of the smallest algae known to science (~ 1 μm). It contains one mitochondrion and one chloroplast in a cell. During cell division, one-cell, two-cells, and four-cells can be found (Kato et al. 2023). Using PacBio RS II system, aided by RNA-seq, Kato et al. (2023) analyzed the genome. Sequencing reads were assembled into a 15.8 Mbp genome, which is one of the smallest known genomes in the Viridiplantae. Evolutionary analyses suggest that this species belongs to a new genus within the class Trebouxiophyceae. The genome encodes 7,629 genes, with relatively few genes associated with genetic information processing, basal transcription factors, and RNA transport. The *M. hakoo* genome has a GC content of 73%, markedly higher than the genome of other ultrasmall alga (notably, the thermoacidophilic hot-spring red alga *C. merolae* has a 55% GC content).

Monoraphidium minutum

Monoraphidium minutum 26B-AM, a cold-tolerant freshwater green microalga, is a top performer in biomass production during winter season screens. Its genome was sequenced, generating a 68.17 Mbp assembly, and a multi-omics approach was applied to profile this high-potential strain under high salt and cold temperature conditions (Calhoun et al. 2022). Annotation of the genome assembly predicted 15,341 genes. Comparative analysis with other green algae in the class Chlorophyceae revealed gene families unique to the genus *Monoraphidium*, including a desaturase gene linked to cold tolerance in plants. This desaturase contains a sphingolipid delta-4 desaturase (DES4) domain, unique to *Monoraphidium* within Chlorophyceae. Under salt stress, osmolytes such as trehalose, proline, and betaine accumulated, corresponding to the upregulation of genes involved in their biosynthesis. Genome annotation enabled the reconstruction of a metabolic model, providing a detailed map of metabolic pathways and allowing for simulations of growth and reaction fluxes.

Caulerpa lentillifera

Caulerpa lentillifera is a siphonous chlorophyte, known as sea grape. Its nuclear genome has been sequenced and assembled to ~ 26 Mbp in length, revealing 9,311 protein-coding genes (Arimoto et al. 2019). Comparative genomic analysis revealed that the TALE superclass of homeobox genes, which has expanded in land plants, also underwent lineage-specific duplications in *C. lentillifera*. Additionally, plant hormone signaling components exhibited similar lineage-specific expansions. The diversification of nuclear pore proteins, indicated by the spatially varied expression of expanded transport regulators (fronds versus stolons), suggests that the structural patterning strategy of a multinucleate cell relies on this diversification.*C. lentillifera* has the largest mitochondrial genome (209,034 bp) among green algae. The genome encodes 17 proteins, 20 tRNA, three rRNA, 42 putative ORFs, and 29 introns, which has homologs in green algal mtDNAs, with a gene density of 65.2% (Zheng et al. 2018).

Bryopsis sp.

Bryopsidales are commonly known as green feather algae. Species in this order exhibit a unique life cycle characterized by repeated nuclear division without cytokinesis, resulting in a coenocytic thallus (> 100 mm) with distinctive feather-like morphology. *Bryopsis*, a representative coenocytic alga, demonstrates exceptional regeneration ability, with extruded cytoplasm rapidly aggregating in seawater to form protoplasts. A sequencing effort (Ochiai et al. 2024) generated a high-quality assembly and annotation of the *Bryopsis* sp. nuclear genome (90.7 Mbp, 27 contigs, N50 = 6.7 Mbp, 14,034 protein-coding genes). Comparative genomic analyses revealed heavy duplication of genes encoding BPL-1/Bryohealin, an aggregation-promoting lectin, in *Bryopsis*, while homologous genes are absent in other ulvophyceans. This suggests a genetic basis for *Bryopsis*' regeneration capability (Ochiai et al. 2024).

Bryopsis sp. possesses over 30 kinesins but only a single myosin, contrasting with other green algae that have multiple myosin gene types. Consistent with this biased motor toolkit, bidirectional chloroplast motility in the cytoplasm was observed to be dependent on microtubules but not actin in *Bryopsis* sp. Most genes required for cytokinesis in plants are present in *Bryopsis*, including those in the SNARE or kinesin superfamily. However, a kinesin crucial for cytokinesis initiation in plants (NACK/Kinesin-7II) is hardly expressed in the coenocytic part of the thallus, potentially explaining the lack of cytokinesis in this portion. Among the three major myosins, Myosins-XI, -XIII, and -VIII, existing in land plants, Myosin-VIII, regulator of microtubule-actin crosslinking required for cell tip growth, branching and cytokinesis in moss, was found missing in *Bryopsis* sp. (Ochiai et al. 2024).

Halimeda opuntia

Halimeda opuntia is, like Bryopsis, a giant multinucleate unicellular chlorophyte, which is characterized by interutricular (IUS) calcification. It is commonly found in tropical and subtropical oceans, where it plays crucial roles in coral reef ecosystems. Its genome was recently sequenced, achieving a 169.53-Mbp assembly (Zhang et al. 2024). The genome contains 47,597 protein-coding genes, 94.9% functionally annotated. Transposable elements accounts 44.4% of the genome, indicative of high genome fluidity. The genome assembly reveals four distinct subgenomes, reflecting polyploidization events that coincide with historical climate and sea level fluctuations. Like other siphonous chlorophytes, *H. opuntia* lacks myosin VIII genes, suggesting this absence may facilitate multinucleation. The genome exhibits expansion of gene families of actin, alpha tubulin, dynein light chain, as well as kinesin-like proteins KIN-7D and KIN-13A, likely genetic basis to sustain the species' large thallus size. Genomic analysis revealed mechanisms underlying the alga's remarkable regenerative capacity following fragmentation, including genes encoding wound-inducible factors auxin-responsive GH3 and respiratory burst oxidase, and genes involved in tissue healing and regeneration such as Act and bryohealin, as well as expansion of concanavalin A-like lectin/glucanase and glycine-rich protein. In addition, the genome displays remarkable expansion of Ca^{2+}-ATPase genes, likely responsible for transporting intracellular Ca^{2+} into the IUS, providing insights into its unique extracellular calcification process and its integration. Furthermore, the organism's ability to modify its proteomic alkalinity profile in response to pH changes may explain its resilience to ocean acidification.

Genomes of Prasinophytes

Prasinophytes are morphologically diverse, ranging from non-motile coccoid form to flagellates with one to eight flagella, and while some species carry organic body scales others are naked. Their high genetic diversity (18S rRNA gene or rDNA) divides species in this lineage into nine clades. Prasinophytes represent the smallest free-living eukaryote cells and genomes and mixotrophic picoeukaryotic species (e.g., the Antarctic species *Pyramimonas gelidicola* capable of phagocytosis). Eight nuclear genomes are available in prasinophytes (Table 5.1), some of which will be briefly discussed below. Among the prasinophyte species, the number of genes increases linearly with genome size, but the number of chromosomes varies significantly, from fewer than 20 to over 40, suggesting that chromosome length varies with chromosome number (Fig. 5.3).

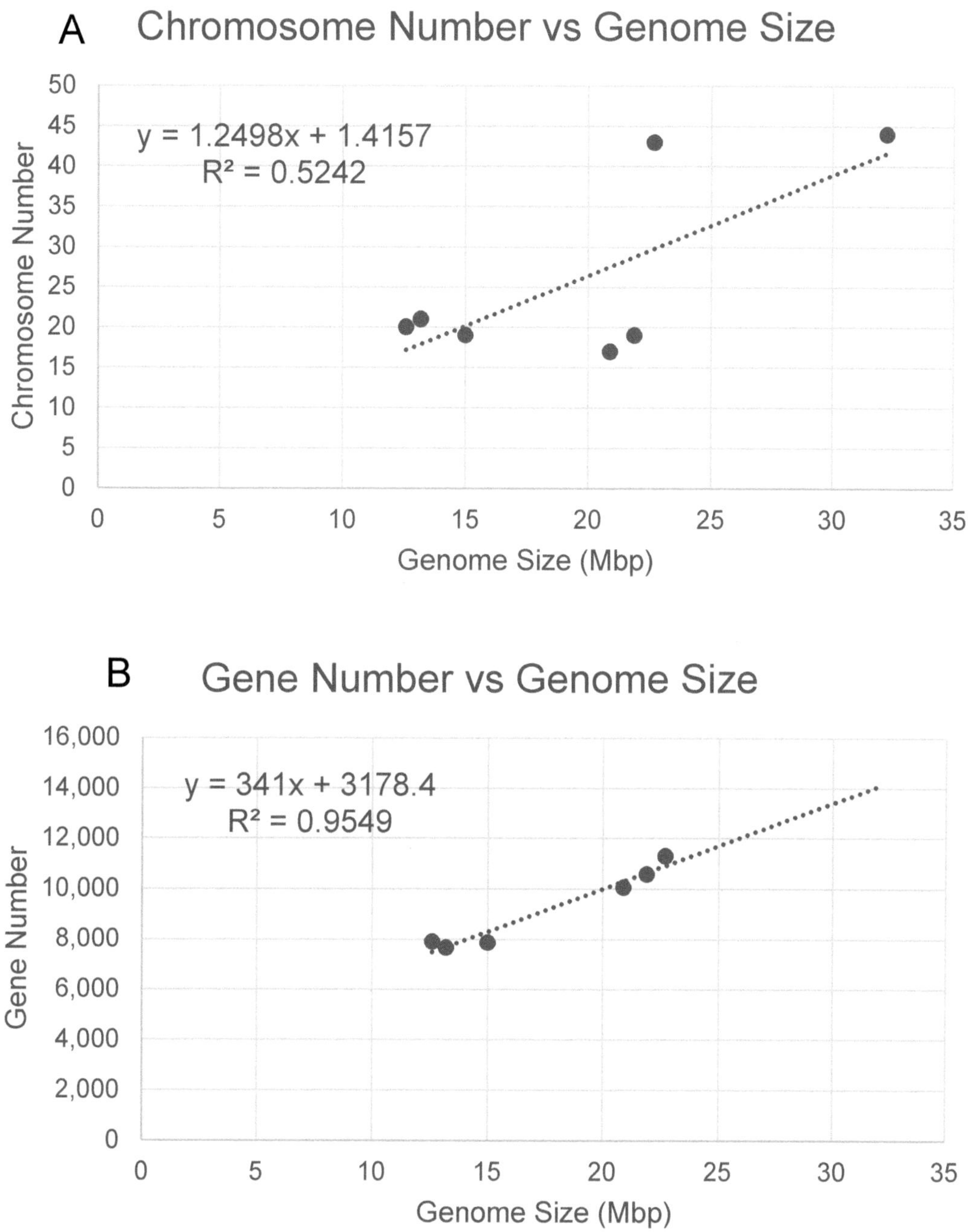

Figure 5.3. Genome features of prasinophytes. (A) Chromosome number varies more than gnome size accounts for. (B) Genome number varies linearly with genome size.

Ostreococcus spp.

Ostreococcus spp. represent the smallest eukaryotic cells (< 1 μm) with small genomes (~ 12 Mbp). The 12.6-Mbp nuclear genome of cosmopolitan *O. tauri* was sequenced to address the genomic basis of how a green alga adapts to endosymbiosis (Derelle et al. 2006). Sequencing output was assembled into 20 superscaffolds corresponding to 20 chromosomes. Among these 20 chromosomes, two chromosomes (#2 and #19) differ structurally from the other eighteen and contain most (77%) of the Transposable Elements (TEs) present in the genome. The G+C content of this genome is more similar to that of the rhodophyte *Cyanidioschyzon merolae* than to that of plants, diatoms, and fungi. From the genome 8,166 protein genes were predicted, making *O. tauri* potentially the most gene rich free-living eukaryote. The high gene density is due in part to extensive reduction of intergenic regions as well as gene fusion and other forms of compaction. The genome contains few repeated sequences, but has an internal duplication of 146Kbp on chromosome 9 (Chr 9), apparently a recent occurrence due to > 99% identity. It also exhibits unprecedented levels of heterogeneity for a eukaryote. Chr 2 genes display unique codon usage and splicing patterns, but analysis suggests no alien gene origin; these and the abundant TEs suggest that this might be a sexual chromosome. In contrast, many Chr 19 genes are distinct from those in the green lineage, with a focus on cell surface processes. Besides evidence of downsizing of many gene families, the genome harbors remarkable features such as diversified resource acquisition machinery (complete set of genes for uptake and assimilation of ammonium, nitrate, and urea), an unusual photosynthesis apparatus ('hybrid' of bacterial and eukaryotic pairs of photosynthetic genes), and C4 photosynthesis pathway. Notably, the compact genome contains four ammonium transporter genes, two being of chlorophyte origin and the other two being prokaryote like.

Palenik and colleagues (2007) sequenced the genome of *O. lucimarinus* and compared it with that of *O. tauri* (Palenik et al. 2007). The *O. lucimarinus* genome is 13.2 Mbp long, distributed in 21 chromosomes, encoding 7,651 genes, highly similar to the genome of *O. tauri*. However, the comparation of the two genome datasets revealed striking dissimilarities among orthologous chromosomes in the two species, ranging from highly syntenic chromosomes in most instances to chromosomes showing minimal similarity in other cases. Species divergence in these phytoplankton seems to occur through various mechanisms, acting disparately on different chromosomes, and likely involves the acquisition of new genes via horizontal gene transfer. About 20% of all genes in both genomes exhibit a multiexon structure, with the majority situated on Chr 2, characterized by introns of unusual size and structure. The 6,753 ortholog pairs identified between the two *Ostreococcus* species show 93% coverage and 70% amino acid identity, indicating greater divergence than similar synteny levels in *Saccharomyces* species.

Both *Ostreococcus* genomes show unique nutrient uptake mechanisms (Palenik et al. 2007). They lack genes common in unicellular free-living eukaryotes functioning in iron acquisition such as the coupled activity of a ferric reductase, multicopper oxidase, and a ferric permease. Instead, both genomes contain genes with significant sequence similarity to prokaryotic siderophore-iron uptake system, suggesting these algae either use siderophores secreted by bacteria or they synthesize their own siderophores. Both *Ostreococcus* genomes also encode for a gene of flavodoxin, a Fe-free protein known to replace the Fe-containing ferredoxin in diatoms when iron is limited. Both species rely on bacteria for vitamin B_{12} because their genomes lack B_{12} biosynthesis pathway, but their methionine synthases are exclusively B_{12}-dependent. Also, *O. lucimarinus* genome harbors a large quantity of selenocysteine-containing proteins, which are more catalytic active than similar selenium-lacking proteins, thus requiring less of that protein. These, along with novel fusion proteins and the absence of certain major protein families, particularly those linked to chromatin, appear to be significant adaptations that contribute to achieving a smaller cell size.

Intriguingly, chromosome 21 of *O. lucimarinus* contains segments from chromosome 9 and Chr 13. The high identity between these duplicated regions suggests this chromosome arose through a recent fusion event, likely via recombination between Chr 9 and Chr 13. Furthermore, intrachromosomal duplication has occurred in multiple chromosomes. Chr 2 is so dissimilar to its counterpart in *O. tauri* that they are unlikely to pair in meiosis, potentially constituting a sex barrier between the two species. The analysis also revealed the first example of heme-handling component genes such as CcsA and Ccs1 and thiol-metabolizing components like CcdA transferred from organelle genomes (the usual location in all other plant and algae) to the nucleus. Additionally, an EST study on this species revealed a high number of introns in highly expressed genes (Lanier et al. 2008), a striking characteristic for a compact genome in a tiny eukaryote.

Micromonas pusilla

M. pusilla is widely distributed in the ocean spanning from tropical to polar regions and may serve as sentinel organisms, indicating biogeochemical fluxes in contemporary oceans amid climate change. Analyses of the genome for *M. pusilla* were conducted to illuminate the divergence in evolution from land plants and to understand the "sentinel" behavior of this algae (Worden et al. 2009). While *Micromonas* isolates exhibit significant identity in their 18S ribosomal RNA genes, genomes from two isolates shared only 90% of their predicted genes. This indicates independent evolutionary trajectories, underscored by unique riboswitch arrangements. Additionally, one isolate exhibited intronic repeat elements, observed in metagenomic data, which are absent in other isolates, further highlighting divergence. This divergence seems to have been driven by selective pressures and acquisition processes, shaping a set of genes that are mutually exclusive between the two isolates, distinct from the core genes.

Pycnococcus provasolii

P. provasolii is a marine green alga first described in the early 1990's (Guillard et al. 1991) and was among the first planktonic eukaryotes found below the pycnocline. It is widely distributed covering different marine regions and from surface to 100 m deep. The high flexibility of *P. provasolii* to light conditions is consistent with its unique pigment composition. Besides Chl *a* and Chl *b* shared with other green algae, *P. provasolii* contains Mg 2,4-divinylphaeoporphyrin a_5 monomethyl ester, and prasinoxanthin as a major xanthophyll. Furthermore, the species possesses a dual-waveband photoreceptor (DUC1) that includes a phytochrome (PHY) and a cryptochrome (CRY) (Makita et al. 2021).

Makita et al. (2021) sequenced the ~ 22.7 Mbp genome of *P. provasolii* strain NIES-2893 using paired-end and mate-pair short reads and Nanopore long reads. The reads were assembled into 43 scaffolds. From the genome 11,297 genes were predicted, with 89.1% (270/303) BUSCO match. Of these genes, 47.8% were conserved across seven prasinophytes compared, five CRYs and one DUC1 were identified but no PHY was detected. Based on PacBio single-molecule HiFi long reads coupled with high-throughput chromosome conformation capture (Hi-C), Green et al. (2023) sequenced the genome of strain CCAP190/2 and assembled it to 32.2 Mbp in 44 scaffolds with N50 of 0.8 Mbp and 86.9% BUSCO completeness. The study also assembled the mitochondrial genome (24.3 Kbp) and plastid genome (80.2 Kbp). Telomeres were identified on both ends of 36 of the 44 chromosomes, indicative of chromosome-scale assembly. The 44 chromosomes range from 0.26 to 1.49 Mbp in length. The larger genome size and one more chromosome detected in this strain suggests remarkable genomic differences between intraspecific strains.

P. provasolii also possesses homologs of vascular plant photoreceptors and light signal transduction genes, including PHOT, COP1, CK2alpha, signalosome subunits, and HY5 homologs, while others like PIF, SPA1, SPA3, FHY1, FHY3, CIB1, and BIC are absent. The LAF1 amino acid sequence shared 44% identity with *Arabidopsis* LAF1, challenging the suggestion that this gene is angiosperm-specific. The light signal transduction gene set was found to be conserved across 13 chlorophyte genomes, including the absence of PIF, CIB1, and BIC in all these genomes. Regarding photosynthesis, the genome of *P. provasolii* harbors a large gene family coding for prasinophyte-specific light-harvesting complex (Lhcp). This includes 16 Lhcp genes, with 10 clustered on chromosome 29, showing almost identical sequences and forming a monophyletic group. Although similar *lhcb* gene duplications are seen in Mamiellophyceae, those in *P. provasolii* seem to have occurred independently.

Bathycoccus prasinos

B. prasinos is cosmopolitan coccoid picoeukaryotes, like the other prasinophytes *Ostrecoccus* and *Micromonas*. Morphologically it has intricate spider's web patterned scales whereas *Ostreococcus* and *Micromonas* are well-less. *Micromonas* possesses a long flagellum whereas the other two are immotile. *B. prasinos* has two identical rRNA genes, and both 18S rRNA genes contain an identical 433-bp long group I intron. Phylogenomic analysis has indicated that the genus *Bathycoccus* is closer to *Ostreococcus* than to *Micromonas*. *B. prasinos* strain RCC1105 has a genome of 15 Mbp distributed in 19 chromosomes. Genome sequencing revealed 7,847 genes (Moreau et al. 2012).

Genomes of Streptophyte Algae

Spirogloea muscicola and Mesotaenium endlicherianum

To understand how the transition from algae to land plants occurred, research effort on streptophyta algae has increased rapidly in recent years. Cheng et al. (2019) sequenced *Spirogloea muscicola* and *Mesotaenium endlicherianum*, which belong to the class Zygnematophyceae, and compared them with land plants. Analysis identified 902 genes in 22 gene families that are shared by the two semiterrestrial algae with land plants but are absent in other algae, suggesting their role in adaptation to the terrestrial environment. Of these, GRAS (GIBBERELLIN-INSENSITIVE, REPRESSOR of GAI, and SCARECROW) and PYR/PYL/RCAR (PYRABACTIN RESISTANCE, PYRABACTIN RESISTANCE-LIKE, and REGULATORY COMPONENTS of ABA RECEPTORS) are known to enhance resistance to biotic and abiotic stresses in land plants. These genes represent those that just evolved before the divergence between the algae and land plants (~ 580 Mya). Two of the shared gene families are functionally related to desiccation resistance and stress responses. These genes are also present in soil bacteria, but not other organisms, suggesting that they were acquired by the streptophytes through horizontal gene transfer from bacteria.

Penium margaritaceum

Sequencing of the streptophyte *Penium margaritaceum* revealed molecular mechanisms for terrestrial adaptation, such as a large number of genes for cell wall synthesis and photoprotective proteins (Jiao et al. 2020). Specifically, the species has a huge genome (4.7 Gbp) relative to most green algae, with 52,333 protein-coding genes predicted. The genome contains a high proportion of repeat sequences linked to extensive segmental gene duplications, which likely facilitate neofunctionalization. A large set of transcription factors (935) and transcriptional regulators (454) were identified from the genome. A notable feature is the exceptionally large size of the GRAS gene family, consisting of 291 genes. As indicated earlier, expansion of the GRAS gene family is also a feature of *Spirogloea muscicola* and *Mesotaenium endlicherianum*, indicating its importance to land plants and some zygnematophycean algae. In land plants, GRAS genes have diversified to regulate crucial processes for multicellular structures, such as meristem development, cell division and expansion, vascular development, seed maturation, and stress responses.

Compared to earlier diverging algal lineages, *P. margaritaceum* has expanded gene families, signaling networks, and adaptive responses, underscoring its evolutionary trajectory toward terrestrialization. These include various physiological processes and protective cellular features, such as flavonoid compounds and large families of modifying enzymes involved in cell wall biosynthesis, assembly, and remodeling. Transcriptome profiling has further clarified the adaptations, responses, and selective pressures associated with *P. margaritaceum*'s semi-terrestrial habitat, where a simple body plan is advantageous. These features in the charophyte genome suggest that *P. margaritaceum* is one of the closest green algal relatives to land plants.

Prasinoderma coloniale

Recently, a novel phylum within green plants, Prasinodermophyta, was identified, which diverged before the split between Chlorophyta and Streptophyta (Li et al. 2020). The genome of *Prasinoderma coloniale* (25.3 Mb) is intermediate in size between the small genomes of picoplanktonic Mamiellophyceae and early-diverging streptophyte algae. Analysis of the minimal core genome revealed an ancestral roadmap highlighting transcription factors, flagellar proteins, an expansion of light-harvesting proteins, and a reduction of early light-induced proteins. Additionally, the genome showed the evolution of a distinct type of C4 photosynthesis and carbon-concentrating mechanism, the synthesis of the metal-complexing metabolite picolinic acid, and auxotrophy for vitamins B1, B7, and B12.

Organellar Genomes

At least 12 plastid genomes have been sequenced in the prasinophyte lineage (Lemieux et al. 2014). Their sizes ranged from 64,335 bp (*Prasinophyceae* sp. CCMP1205) to 200,801 bp (*Nephroselmis olivacea* NIES 484), coding for 94 (*Monomastix* sp. OKE-1) to 128 genes (*Nephorselmis olivaceae* NIES 484). There was a major reduction in gene content and restructuring of the chloroplast genome in conjunction with cell organization in at least two prasinophyte lineages and uniquely shared derived gene clusters within the cpDNA. Despite both *Nephroselmis* and *Picocystis* chloroplast DNAs featuring a large inverted repeat encoding the rRNA operon, the distribution of genes among the single copy regions differs significantly. Surprisingly, all three species from clade VI (Prasinococcales) were found to harbor chloroplast genes not previously documented for chlorophytes (ndhJ, rbcR, rpl21, rps15, rps16, and ycf66), and *Picocystis* contains a *trans*-spliced group II intron. Phylogenetic analyses based on chloroplast DNA-encoded proteins closely mirror those based on 18S rDNA, robustly resolving all six examined prasinophyte lineages, except for the Pycnococcaceae, with strong support. Overall, the plastid genome data highlight the significant variability in plastid genome structure among prasinophyte lineages, indicating a strong selection pressure to retain a small and condensed chloroplast genome in picoplanktonic species. The presence of a distinct set of six chloroplast genes in the Prasinococcales supports their ancestral position within the prasinophytes. The distinct characteristics observed in members of clade VII (*Picocystis* and *Prasinophyceae* sp. CCMP 1205) align with their classification as separate lineages in the chloroplast phylogeny.

Chloroplast and mitochondrial genomes of the symbiotic green algae *Chlorella heliozoae, Chlorella variabilis*, and *Micractinium conductrix* were sequenced (Fan et al. 2017). The results showed no evidence of functional genomic reduction. Instead, these genomes were larger and contained more introns compared to their free-living trebouxiophyte counterparts.

Shared Minimal Proteome among Microalgae

Due to the small cell size and genome size, prasinophytes have been used in comparative genomic analyses to identify the minimal proteome shared by microalgae. One of the analyses revealed 1,263 common orthogroups (gene families) shared among 15 ultrasmall algae from distinct phylogenetic groups (Kato et al. 2023). The small genome sizes of prasinophytes and species from other marine or extremophilic microalgae such as *Cyanidioschyzon merolae* suggest that maintaining a small genome and small cell is advantageous for survival not only in seawater and extreme environments, but also in common freshwater environments (Kato et al. 2023).

Repetti et al. (2021) generated a high-quality nuclear genome for the pedinophyte YPF701 and performed a comparative analysis with 25 other genomes from Viridiplantae. Their findings suggest positive selection for genome streamlining in Pedinophyceae, distinct from the genome minimization observed in prasinophyte lineages. The study emphasizes multiple independent signals of genome minimization within Chlorophyta and highlights genomic innovations predating the diversification of the UTC clade (classes Ulvophyceae, Trebouxiophyceae and Chlorophyceae) of Chlorophyta, which may contribute to the success of this species-rich clade in diverse habitats.

Multinucleation

Multinucleation occurs in coenocytic or siphonous algae. When these algae grow, their nuclei divide without inner cell wall and cytoplasm separation. Coenocytic species are present in diverse and unrelated groups of algae, including Xanthophyceae (e.g., *Vaucheria*), rhodophytes (e.g., *Griffithsia*), and green algae. Coenocytic green algae include some species in the streptophyte (internodal cells of *Chara*) and chlorophyte orders Bryopsidales and some Dasycladales. Extant species in Bryopsidales are morphologically diverse ranging from simple branched siphons (*Boodleopsis*, *Bryopsis*, *Caulerpa*, *Chlorodesmis*, *Derbesia*, *Pseudochlorodesmis*, *Ostreobium*) to more complex, multiaxial thalli (*Codium*, *Halimeda*, *Rhipilia*, *Rhipiliopsis*, *Udotea*). The Bryopsidales comprise about 564 species (Guiry and Guiry 2017) distributed from tropical to Arctic marine waters (Kerswell 2006), with a single genus (Dichotomosiphon) found in freshwater habitats. They are among the major primary producers on both coral reefs and rocky shores, as well as in embayments, lagoons, and seagrass beds. As a result of cell growth without division, siphonous algae are usually macroscopic.

These unicellular, multinucleate, macroscopic algae typically have a large central vacuole surrounded by a layer of protoplasm, containing nuclei and chloroplasts, which line the cell wall. They have as many as millions of nuclei in one cell. This incredible cytological characteristics places these in an evolutionary stage transitioning from unicellularity to multicellularity. Recent genomic studies revealed the loss of myosin VIII (Myo8) in the siphonous species *Halimeda opuntia* and *Bryopsis* sp., a protein important in cytokinesis (Ochiai et al. 2024, Zhang et al. 2024). Additionally, gene families related to mitosis, spindle organization and assembly, as well as microtubule structures are expanded in the *H. opuntia* genome (Zhang et al. 2024). The expansion of mitosis-related gene families and the absence of Myo8 likely facilitate the development of multinucleate cells by preventing cytokinesis completion. This modified cellular division mechanism appears to be an adaptive trait that enables multinucleation.

Polycistronic Gene Expression

Chlorophytes are now known to be one of the few eukaryotic lineages that express polycistronic genes. This was first found in *C. reinhardtii* and *C. zofingiensis*, in which hundreds of gene loci were shown to be polycistronic, and several other chlorophyte species (Gallaher et al 2021). In this mode of expression, multiple proteins are translated from a single gene transcript expressed from a single gene. Polycistronic genes are a hallmark of prokaryotic genomes and are rare in eukaryotes. Trypanosomes and nematodes (*Caenorhabditis elegans*) are the main exception in which polycistronic gene occur (Blumenthal 2004, Michaeli 2011). The discovery of numerous polycistronic genes in chlorophytes is potentially useful for using the green algae as experimental systems for research and industrial applications.

6

Genomes of Rhodophytes

Rhodophyta is one of the three founding groups of eukaryotic photosynthetic organisms (Archaeplastida), which emerged from primary endosymbiosis. Hence, the phylum shares the cyanobacterial progenitor with the other two groups, Chlorophyta and Glaucophyta. The bangiophyte taxon *Bangiomorpha* (resembling the extant marine species *Bangia*) represents the oldest taxonomically resolved multicellular eukaryote in the fossil record (1.2 BYA). The genus *Gloeomargarita* is the closest extant cyanobacterial relative documented to date of the common ancestor (Sanchez-Baracaldo et al. 2017).

Rhodophytes (red algae) are globally distributed and genetically diverse. There are estimated 6,100–14,000 species in this phylum (Bhattacharya et al. 2013). Phylogenetically, Rhodophyta is divided into seven classes, including two classes of red seaweeds, Bangiophyceae and Florideophyceae, and five classes of red microalgae. The red seaweed group includes the edible cultured seaweeds *Porphyra yezoensis*, *Chondrus crispus*, and *Kappaphycus alvarezii*. The unicellular rhodophytes include classes Cyanidiophyceae, Compsopogonophyceae, Porphyridiophyceae, Rhodellophyceae, and Stylonematophyceae. As seaweeds will be treated separately in Chapter 11, emphasis will be placed on unicellular rhodophytes here, although some common characteristics will be presented herein.

Most rhodophyte species are marine, with only ~ 5% living in freshwater, and these are mostly multicellular species. Few rhodophytes live in soils, probably because an evolutionary bottleneck has led to the loss of plasticity. Some rhodophytes tend to live in warm waters. For instance, the freshwater taxon *Compsopogon* is more abundant in warm waters with maximum photosynthetic rate at 30–35°C. The unicellular species *Cyanidioschyzon merolae* and *Galdieria sulphuraria* thrive in hot and acidic springs. Some rhodophytes are calcifying species that deposit calcium carbonate in the cell wall. These include all members of the Corallinales and some members of the Nemaliales (*Liagora*, *Galaxaura*). The Corallinales deposit $CaCO_3$ as calcite whereas the calcified members of the Nemaliales produce aragonite. The red algae that have the highest rates of calcification also have the highest rates of photosynthesis and are usually found in waters less than 20 m deep (Goreau 1963).

Rhodophytes exhibit a unique hybrid light-harvesting system: their Photosystem II utilizes phycobiliproteins as antenna proteins, while their Photosystem I employs Light-Harvesting Complex proteins (Bhattacharya et al. 2013). This arrangement contrasts with cyanobacteria, which use phycobiliproteins for both photosystems, and green algae and land plants, which exclusively use Light-Harvesting Complex proteins. *In situ* structures of PBS-PSII-PSI-LHC megacomplexes from *Porphyridium purpureum* was resolved at near-atomic level using cryogenic electron tomography (You et al. 2023). The structure reveals several unidentified and incomplete proteins and their role in the assembly of the megacomplex, including a gigantic and sophisticated pigment network. Red algal plastid genomes are considered ancestral and evolutionary stable for their largest repertoire of genes (166–235) and compact organization among algae. Some of the rhodophyte species have exceptionally large plastid genomes. For instance, the plastid genome size of the sea sand-dwelling unicellular Rhodellophycean *Corynoplatis japonica* is 1.13 Mbp, the largest plastid genome ever sequenced (Munoz-Gomez et al. 2019). The remarkable plastid genome expansion in *C. japonica* seems to have been driven by the massive proliferation of introns, which account for > 63.9% of total ptDNA with at least 311 introns. Besides, plastid genomes as well as mitochondrial or in some cases even nuclear genomes of rhodophytes contain plasmid DNA and mobile DNA (Lee et al. 2016). However, no evidence of plasmid DNA was found in the plastid genome in *C. japonica*.

Rhodophytes have a complex life cycle, with vast variations in alternations of sexual and asexual reproduction. It is interesting to note that crustaceans such as the isopod *Idotea balthica* can play a role in the "pollination" of the female gametes of red seaweed like *Gracilaria gracilis* (Lavaut et al. 2022). Another distinct trait of rhodophytes is the ability of cells in red seaweed to fuse as part of a wound-repair process or even under normal vegetative morphogenesis (review by Borg et al. 2023).

Many rhodophyte species are cultivated through aquaculture for human consumption as or fodder in mariculture. In China, *Gracilariopsis lameneiformis* is farmed as a food source. Another widely utilized genus, *Pyropia* (=Neopyropia), has tremendous nutritional values, with *P. yezoensis* and *P. haitanensis* being the most popular species. Like *Porphyra*, *Pyropia* spp. thrive in cold shallow coastal waters, typically between the upper intertidal zone and splash zones. The edible *P. yezoensis* is particularly renowned for its economic and medicinal importance. Research has shown that ethanol extract of *P. yezoensis* exhibit neurotrophic and neuroprotective effects in primary hippocampal neuron cultures. These extracts particularly their active component taurine, promote neuronal development and maturation, reduce apoptosis, and enhance reactive oxygen species (ROS) scavenging (Mohibbullah et al. 2016).

Sequenced Genomes

Among single-cell species, *Cyanidioschyzone merolae*, *Galdieria sulphuraria*, and *Porphyridium purpureum* have been sequenced. Sequenced macroscopic rhodophyte species include *Chondrus crispus*, *Porphyra umbilicalis*, *Pyropia yezoensis* (Table 6.1).

Cyanidioschyzon merolae

The 16.52 Mbp genome of the unicellular rhodophyte *C. merolae* that inhabits sulfate-rich hot springs (pH 1.5, 45°C) was sequenced in 2004 (Matsuzaki et al. 2004). *C. merolae* is a tiny (2 μm) eukaryote with a minimal set of small membrane-bounded compartments, including a microbody (peroxisome), a single Golgi apparatus with two cisternae, coated vesicles, a single endoplasmic reticulum, and several lysosome-like structures, as well as a small volume of cytosol. The highly reduced genome represents the smallest of all photosynthetic eukaryotes. Its G+C content is 55%, not particularly high given that it lives in hot spring, but comparatively the dinucleotide CpG (1.151) is exceptionally high. The genome is composed of 20 chromosomes and was found to contain 5,331 genes, only 26 of which contain introns (0.5%). The streamlined genome only harbors three copies of ribosomal DNA units and 30 tRNAs. While a large gene family consisting of more than 10 members encoding functionally diverse dynamins with diverse membrane pinching roles exist in other organisms, the *C. merolae* genome has only two dynamin genes, which are involved only in the division of mitochondria and plastids. The genome contains only one set of tubulin, no myosin gene or dynein motor domain-containing gene, and its actin gene is not expressed, suggesting a simpler system of cytokinesis.

Standard sets of photosystem genes, including phycobilisome encoding genes, were found in the *C. merolae* genome. Most photosystem genes (11 PSI and 17 PSII) are plastid-encoded while

Table 6.1. Sequenced genomes of rhdophytes.

Species	Cellularity	Sequenced ploidy	Genome size (Mbp)	# of chromosomes	GC%	# of coding genes	% Repetitive DNA	References
Chondrus crispus	Multicellular	Both haploid and diploid stages	105	19	52	9,606	73	Collén J et al. 2013
Cyanidioschyzon merolae	Unicellular	Haploid	16.52	20	55	5,331	28	Matsuzaki et al. 2004
Cyanidium caldarium	Unicellular	Haploid	14.67–22.005	5(7)	54	5,331	1	NCBI Bioproject SAMN29217976
Galdieria sulphuraria	Unicellular	Haploid	13.71	2	55	7,174	1.5	Muravenko et al. 2001, Kanika et al. 2015
Gracilariopsis chorda	Multicellular		92.1		49.26	10,806		Lee et al. 2018
Gracilaria changii	Multicellular		35.8		50.6	10,912		Ho et al. 2018
Porphyra umbilicalis	Multicellular	Haploid	87.7	3–5	65.8	13,125	43.91	Brawley et al. 2017
Porphyridium purpureum	Unicellular	Diploid	19.70	8.00–10.00	55.50	8355	9.50	Bhattacharya et al. 2013, Hammel et al. 2024
Pyropia yezoensis	Multicellular	Haploid	43.48		63.6	10,329	1.4	Nakamura et al. 2013
Pyropia yezoensis	Multicellular	Haploid	108	3	65.9	12,855	48	Wang et al. 2020

PsbO, P, U, Z and the distant PsbQ homolog are encoded in the nuclear genome. PsaH, N, X, and S as well as NDH genes were not found, essentially lacking various mechanisms for dissipating excess light energy. Apart from Rubisco, known to have been acquired through horizontal gene transfer in rhodophytes, the origin of other Calvin-Benson-Bassham (CBB) cycle enzymes is essentially identical as in the land plant *Arabidopsis thaliana*. The conserved mosaic origin of CBB cycle enzymes in this red alga and in green plants supports the hypothesis of the existence of single primary plastid endosymbiosis and indicates that no crucial changes have occurred since the separation of the two lineages.

The *C. merolae* genome encodes several blue light receptor (cryptochrome) genes, but no phytochromes or phototropins. Because in cyanobacteria, phytochrome was found only in large genomes, the lack of this photoreceptor in *C. merolae* suggests that the ancestor of plastids might be an ancestral cyanobacterium without phytochrome, and this receptor in land plants might not be cyanobacteria-originated. In addition, *C. merolae* does not code for complete two-component signaling systems (having only histidine kinase genes but no response regulator genes) and its MAPK pathway is limited to a dozen MAP kinase-related molecules.

Porphyridium purpureum

The genome of *P. purpureum* was reported in 2013 (Bhattacharya et al. 2013). The 19.7-Mbp genome contains 8,355 gene models with only 235 splicesomal introns (2.8% genes with introns, 0.03 intron per gene on average). Analysis of the sequence reads revealed 26,383 single-nucleotide polymorphisms in contigs with > 10x average depth coverage, mostly (94.6%) represented by two variants, suggesting that the strain is a diploid. Further analysis showed that at least 453 genes were impacted by horizontal gene transfer, implicating that ancient red algae might act as mediators of horizontal gene transfer between prokaryotes and photosynthetic eukaryotes (Bhattacharya et al. 2013). The genome possesses gene sets encoding phycobiliproteins for PSII and LHC genes (Lhcr1 ad Lhcr2), a complex set of carbohydrate-active enzymes (CAZymes), methylerythritol phosphate pathway of isoprenoid biosynthesis and proteins regulating sexual reproduction. About 38% of the 107 CAZymes show prokaryotic provenance. Of note is the sharing of reproductive and cell development functions between this species and the Plantae supergroup, evidence of vertical inheritance from the common ancestor.

Common Genome Characteristics and Evolutionary Trends

Small Genome Size, High GC Content, and RNA-Dependent Methylation

Red algae have relatively small genome sizes and gene repertoires among eukaryotic algae (Table 6.1). The haploid genomes range from about 10 to 108 Mbp (Table 6.1), and protein-coding genes spans 4,803–13,215 across species (Lee et al. 2018). Among rhodophytes, red seaweeds have the largest genomes, but they are still no larger than about 100 Mbp. Comparative genomics of rhodophytes indicates that there is no evidence of polyploidization, and genome size variability among rhodophyte lineages is primarily due to the variability of Transposable Elements (TEs), which are dominated by long terminal repeat elements and DNA repeats. As can be seen from Table 6.1, rhodophytes generally have high GC contents, which can be as high as nearly 66%, with few exceptions.

In *Gracilariopsis chorda*, about 8.3% of the genome undergoes cytosine methylation, and 71.5% of TEs contain methylated DNA, with 57% of these regions being associated with small RNA (sRNA). These attributes suggest that genome size evolution in red algae might have resulted from the combined action of TE spread and epigenetic suppression, the latter of which is facilitated by RNA-dependent DNA methylation aided by TE-associated sRNA (Lee et al. 2018). A recent genome and transcriptome reanalysis revealed that gene duplication is the highest in the seaweed lineages, Florideophytes and Bangiophytes, and lowest in the microalgal lineages *Porphyridium* and cyanidiophytes (McKinnie et al. 2023). A similar trend to that of TE expansion and gene duplication might have contributed to the trend of genome expansion.

Genome Expansion and Contraction

Rhodophytes have undergone two phases of large-scale genome reduction early in their evolution. The first occurred in the stem lineage of Rhodophyta and the second in the ancestor of the extremophilic *Cyanidiophytina* (Lee et al. 2018). These genome reductions were accompanied by the loss of various enzymes and proteins in red algae. For instance, among the lost genes were those encoding enzymes of the glycosyl-phosphatidylinositol (GPI) anchor biosynthesis pathway, autophagy proteins, and most flagellar proteins (Qiu et al. 2015, Brawley et al. 2017). Red algae also have reduced cytoskeletons, with complete absence of dynactin complex Arp 1, dynein motor, and in most red algae, myosin (Brawley et al. 2017). These genome reduction events and gene losses explain why rhodophyte genomes are generally smaller than other eukaryotic algae. However, within rhodophytes, some lineages have larger genomes than others (e.g., Florideophyceae vs. Bangaiophyceae), which is due to the expansion of TEs and duplication of genes. Rhodophyte species share many duplicated genes that encode key enzymes and functions, including core metabolic processes. Notably, genes associated with ATP bindg and protein binding, are highly duplicated (McKinnie et al. 2023).

Furthermore, the comparative genome/transcriptome analysis also indicated that specific functions and enzymes have undergone more extensive gene duplication in specific lineages of rhodophytes. Cyanidiophyceae seems to have a higher rate of duplication for transmembrane transporters, particularly the Major Facilitator Superfamily (MFS) transporters, and significantly higher in the Galdieriales and *C. caldarium* than in the Cyanidioschyzonales. In the Florideophytes, haem peroxidases are highly duplicated, most prominently in the Gracilariales and *Asparagopsis*. The Gracilariales and Asparagopsis also underwent a notable duplication of the von Willebrand factor domain (IPR002035), while *Porphyridium* exhibits a rampant duplication of the integrase catalytic core domain (IPR001584).

Innovation of Biosynthesis Pathways

The phytosterol biosynthesis module M00917 represents the biosynthesis pathways for the phytosterols campesterol and sitosterol from squalene 2,3-epoxide. Both pathways are largely identical from a gene ortholog context, only differing by one additional ortholog in the sitosterol pathway, although their sequences of reactions do differ. A study reanalyzing genomes and transcriptomes of 72 species from all seven classes of rhodophytes, encompassing 410 pathway modules, revealed that the phytosterol pathway module was incomplete in all assemblies. However, each assembly contained at least four KEGG orthologues required for the pathway (McKinnie et al. 2023). Similarly, the cholesterol and ergosterol biosynthesis modules were largely incomplete across the red algal assemblies. This contradicts *in vitro* evidence of sterol production in red algae, indicating that red algae may have novel genes involved in the biosynthesis of sterols that have not been identified in the KEGG database.

Sulfated galactans are common to the cell wall of rhodophytes. These polysaccharides comprise a linear backbone of galactose residues linked by alternating β-1,4 and α-1,3 glycosidic bonds. The rhodophyte seaweeds either produce agars or carrageenans, which differ in the stereochemistry of the bicyclic α-3,6-anhydrogalactose moiety in their galactan chain. It is in the L-configuration in agars and in the D-configuration in carrageenans (Rees 1969). Based on the existing experimental and genomic data, Ficko-Blean et al. (2015) proposed two alternative carrageenan biosynthesis pathways. It is strikingly, however, that no sulfatase was found, suggesting that rhodophytes use alternative steps to bypass the reaction catalyzed by sulfatase.

Adaptive Differential Evolution Between Rhodophyte Lineages

Cyanidiophytes exhibit extensive duplication of transmembrane transporters and possess unique orthologs across various pathways that are absent in other rhodophyte algae (McKinnie et al. 2023). Furthermore, evidence from genomes indicates that Cyanidiophyceae has lost genes involved in urea utilization. This is consistent with the previous observation that Cyanidiophyceae diverged early from other rhodophyte classes (Yoon et al. 2004). However, divergence also occurs within Cyanidiophyceae, for instance among the three orders Galdieriales (*Galdieria* spp.), Cyanidiales (*Cyanidium caldarium*), and Cyanidioschyzonales (*Cyanidioschyzon merolae* and *C. yangmingshanensis*). Galdieriales and *C. caldarium* differ consistently from Cyanidioschyonales in several ways: *C. merolae* has a larger proportion of DNA transposons, Galdieria and *Cyanidium* have higher transmembrane transporter duplications, and their ortholog presence patterns vary. Despite all three orders being polyextremophiles in hot and acidic environments, Galdieria and *Cyanidium* spp. have greater salt tolerance and possess a rigid cell wall, unlike *C. merolae*. This correlates with the increased duplication of transmembrane transporters in *Galdieria* and *Cyanidium*.

The Florideophytes also show systemic differences to other classes due to gene duplications, with heme peroxidases being highly duplicated across the class. The expansion of this gene family seems to be an adaptation to life in marine environments where halogenated compounds are common and utilized in cellular defense and secondary metabolism (Collén et al. 2013). Heme peroxidases are diverse enzymes involved in oxidation reactions using hydrogen peroxide. They play roles in metabolite biosynthesis, cell signaling, and defense in plants and algae. Gene duplication in red algae, particularly Florideophytes, has led to the subfunctionalization of these enzymes. Many duplicated genes were annotated as animal heme peroxidases or peroxinectin, both belonging to the peroxidase-cyclooxygenase superfamily. Some species also show duplication of eosinophil peroxidase genes (Fig. 6.1), involved in bromide ion oxidation. This gene duplication, especially of halide-associated heme peroxidases, suggests an evolutionary trend towards the production of halogenated secondary metabolites in red seaweeds, particularly brominated compounds in Florideophytes (McKinnie et al. 2023).

Response to Nutrient Variations

Rhodophytes requires a high amount of nitrogen from the environment and contains abundant proteins in the cell because of the abundant phycobiliproteins. As such, rhodophytes is sensitive to Nitrogen (N)-nutrient deficiency. A study using *Gracilariopsis lameneiformis* showed that the availability of N and Phosphorus (P) nutrients in the environment triggers intracellular resource reallocation through metabolic reprogramming (Li et al. 2022). As shown in Fig. 6.2, under high N and P nutrient (H-NP) conditions, ammonium transporter genes (AMT1-1 and AMT1-2) were downregulated whereas nitrate transporter gene (NRT2) and phosphorus transporter were upregulated. There was a differential expression pattern among the N assimilation genes. Nitrite reductase (NiR), Glutamine Synthetase (GS) and ferredoxin-dependent glutamine oxoglutarate amine transferase (fd-GOGAT, plastidic) were all regulated by N and P nutrient conditions. There are two isoforms of GS: GS1 localized in the cytosol and GS2 in the plastid, which catalyz the ATP-dependent condensation of ammonium and glutamate to form glutamine. GS2 and fd-GOGAT were upregulated in high-NP cultures, but GS1 and Nir were not. Under this H-NP condition, the synthesis of N-rich amino acids such as glutamate, glutamine, aspartic acid, asparagine, and arginine increased. On the contrary, gene coding for Alkaline Phosphatase (AP) and 5'-nucleotidase (5'-NT), the essential enzymes for utilizing dissolved organic phosphorus, were expressed at high levels under low-NP conditions. In particular, *G. lameneiformis* stores N in the form of arginine under the nutrient rich condition.

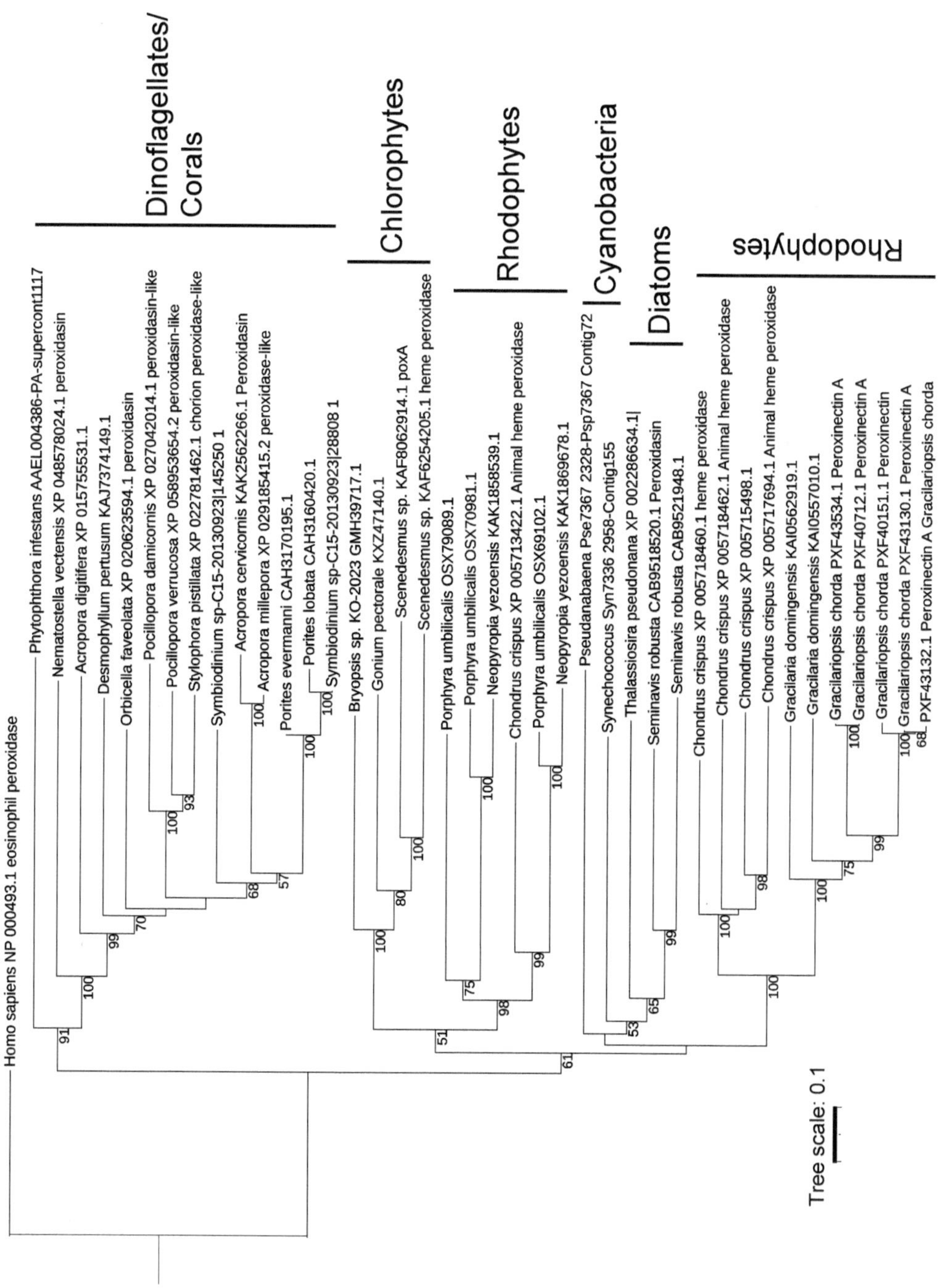

Figure 6.1. Nitrogen nutrient utilization genes in representative rhodophyte species compared to those in other algae.

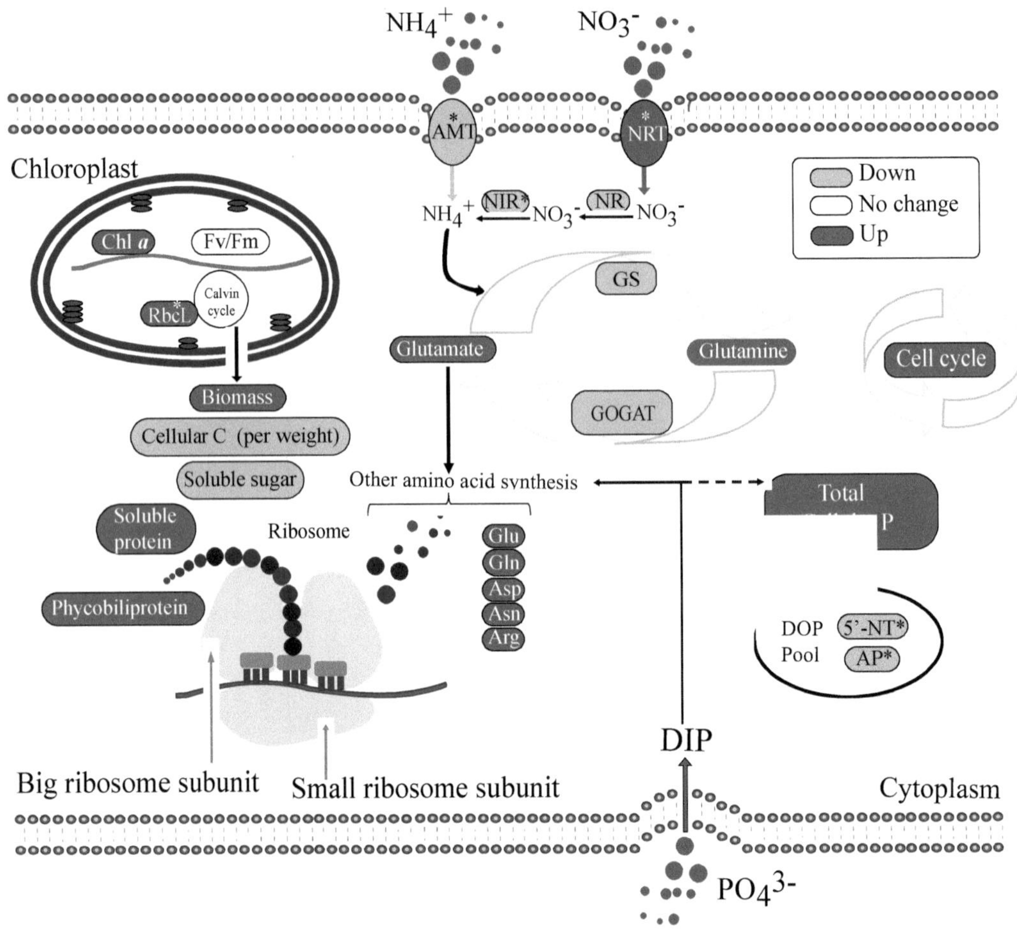

Figure 6.2. Schematic representation of biological pathways in *G. lemaneiformis* under high the N- and P-enriched environments. Schematic representation of biological pathways in *G. lemaneiformis* under high the N- and P-enriched environments. AMT: ammonium transporter, NRT: nitrate transporter, NIR: nitrite reductase, NR: nitrate reductase, GS: glutamine synthetase, GOGAT: glutamate synthase, AP: alkaline phosphatase, 5′-NT: 5′-nucleotidase, Glu: glutamic acid, Gln: glutamine, Asp: aspartic acid, Asn: asparaginate, Arg: arginine. Represents the data from RT–qPCR. Based on Li et al. 2022.

Under variable N and P nutrient conditions, *G. lameneiformis* exhibits a negative correlation between the soluble sugar and soluble protein contents. Phycoerythrin seems to be the main biologically active N-compound used by *G. lameneiformis* for growth. The findings of the study indicate that N and P nutrient abundances shape the metabolic landscape and in turn influence C, N, and P stoichiometry in *G. lameneiformis*.

It is intuitive to expect that rhodophytes may have expanded the gene families related to N-nutrient uptake and assimilation. When *Gracilariopsis* genes for AMT, NRT, GS, and fd-GOGAT were used as queries in BLAST analyses, the results were surprising in that these genes in rhodophytes were not duplicated more than in other lineages of algae (Table 6.2). This result indicates that a physiological capacity is not necessarily determined by the family size of the underlying genes.

Table 6.2. Nitrogen nutrient utilization genes in representative rhodophyte species compared to those in other algae.

Species	AMT	NRT	GS1	GS2	GOGAT2
Rhodophyta					
Chondrus crispus	1	1	1	3	1
Cyanidinium caldarium	2	1	0	0	5
Cyanidioschyzon merolae	4	1	0	5	1
Galdieria sulphuraria	2	1	0	1	1
Gracilariopsis chorda	3	1	1	1	2
Gracilariopsis lemaneiformis	5	1	3	5	3
Porphyra umbilicalis	7	1	0	2	1
Porphyridium purpureum	2	4	1	2	2
Other phyla					
Chlamydomonas reinhardtii	10	3	1	5	2
Phaeodactylum tricornutum	8	6	0	1	2
Fugacium kawagutii	30	14	0	5	9
Prorocentrum shikokuense	23	3	0	1	3
Polarella glacialis CCMP1838	29	31	0	4	5

7

Genomes of Diatoms

Diatoms (Bacillariophyta) are one of the major lineages of photosynthetic eukaryotes on Earth, ubiquitous in the plankton and benthos of marine and freshwater ecosystems and are also abundant in temporarily wet terrestrial environments. Diatoms contribute around 20% of global primary productivity, or 40% of global ocean productivity. Diatoms are also important contributors in the biological carbon pump as their silicate frustules ballast the cells for sinking and carbon export and as such facilitate paleo-oceanographic research as they remain in the sediment over an extended period of time. Therefore, diatoms are often considered the most important players in marine ecological and biogeochemical processes.

Diatoms are diploid and their cells commonly have a frustule as part of the cell protective surface, which are composed of two silicate valves, an epitheca or epivalve (larger) and a hypotheca or hypovalve (smaller). The two valves fit together like a petri dish, with the side band named girdle band or cingulum. Despite this basic common feature, diatoms have numerous morphologies. They are divided into two major classes based on their modes of symmetry, the centric (radially or bipolarly symmetric) and the pennate lineages (laterally symmetric). Phylogenetic analyses indicate that the centric lineage is ancestral to the pennate lineage (Medlin 2016), indicating a paraphyletic relationship (Fig. 7.1).

Emerging about 240 million years ago (Mya) based on molecular clock (Kooistra and Medlin 1996, Medlin et al. 1997) or about 190 Mya based on fossil records, diatoms have radiated into one of the most diverse groups of eukaryotes. Earlier species number estimates for diatoms were wide ranged, from 1,800 to 200,000 earlier (Sournia et al. 1991, Mann and Droop 1996), and 12,000 to 30,000 more recently (Mann and Vanormelingen 2013, Guiry and Guiry 2017). Based on the V9 region of the 18S rRNA gene, the *TARA Oceans* expedition retrieved ~ 12 million diatom ribotypes from 293 size-fractionated plankton communities collected at 46 sampling sites across the global ocean euphotic zone (Malviya et al. 2016). These ribotypes were binned into 4,748 Operational Taxonomic Units (OTUs), representing the oceanic planktonic diatom flora dominated by the genera *Chaetoceros*, *Fragilariopsis*, *Thalassiosira*, *Corethron*, *Leptocyclindrus*, *Actinocyclus*, *Psedo-nitzschia*, and *Proboscia*. However, the latest estimate of diatom species number is slightly over 10,000 species, of which 3,221 (~ 30%) are marine species and 7,349 (~ 70%) are freshwater species (Nakov et al. 2019). Overall, 1,800 species (17%) are planktonic and 8,770 (83%) are benthic species, and the planktonic-benthic species split is the same for freshwater and marine groups.

Although many species occur in chains or colonies of many cells connected by silicate spines, organic fibers, or mucilage, diatoms are exclusively single-celled eukaryotes. Their cell sizes range from < 1 μm to several mm. *Ethmodiscuss rex* is one of the largest diatoms, up to 2 mm in size, which is typically found in the open ocean. The large cell size makes the species useful for cytological and autecological studies to understand its diel vertical migration for nutrient scavenging (Villareal et al. 1999), cell cycle progression with the nucleus migrating between the end and the middle of the

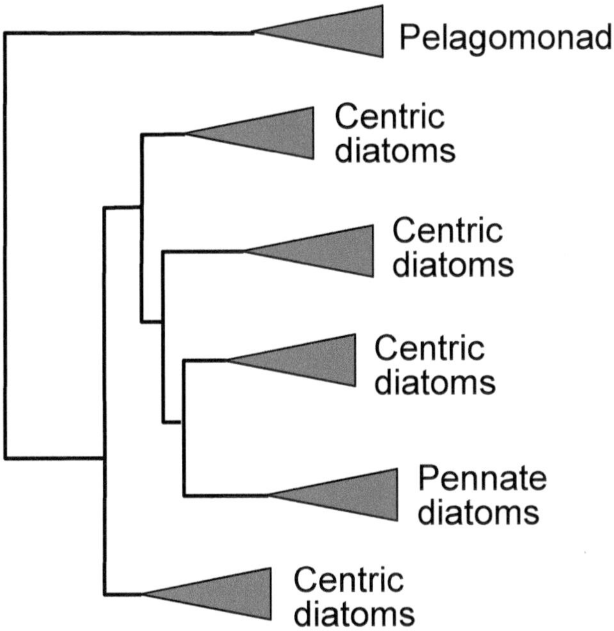

Figure 7.1. Schematic diatom phylogeny indicating that centric diatoms are ancestral to pennate taxa. Note the paraphyletic relationship between centric and pennate lineages.

cell (Lin and Carpenter 1995), among other attributes. Chain formation can enhance nutrient uptake under turbulent condition by counteracting the diffusion boundary limitation and promoting cell movement across turbulent current (Dell'Aquila et al. 2017).

One distinct feature of diatoms is their siliceous frustules. With few exceptions (e.g., *Phaedactylum tricornutum*), diatoms require silicate in their growth medium in the stoichiometric ratio with N of 1:1. The porous frustule gives diatoms morphological support and allows for nutrient uptake. Diatoms possess most, if not all, of the photoreceptors known to operate in plants (aureochrome, phytochrome, cryptochrome) and some diatoms also have proton-pump rhodopsins (Andrew et al. 2023). Diatoms are relentless competitors for iron and have long been a model for phytoplankton iron nutrition research (Gao et al. 2021).

In addition, diatoms are the algae in which the Ornithine-Urea Cycle (OUC) was first found outside animals (Armbrust et al. 2004, Allen et al. 2011). Mechanisms underlying these and other important capabilities of diatoms have increasingly come to light thanks to the accelerating pace of advances in algal genomics.

Sequenced Diatom Genomes

Over a dozen diatom species have been sequenced to date (Table 7.1), including *Thalassiosira pseudonana*, *Phaeodactylym tricornutum*, *Fragilariopsis cylindrus*, *Thalassiosira weissflogii*, *Thalassiosira oceanica*, *Fistulifera solaris*, *Mayamaeaa pseudoterrestria*, *Minidiscus variabilis*, *Nitzschia putrida*, *Plagiostriata* sp., *Pseudo-nitzschia multiseries*, *Pseudo-nitzschia multistriat*a, *Skeletonema costatum*, and *Synedra acus*. The sizes of these genomes range from 27 to 231 Mbp. The large differences between species within the same genus, e.g., between *Pseudo-nitzschia multiseries* (~ 219 Mbp) and *P. multistriata* (59 Mbp), are notable. Based on the comparison of *C. weissflogii*, *Thalassiosira rotula*, and three species of *Pseudo-nitzschia*, similar GC contents were found, suggesting that diatoms have high GC content homogeneity (Li et al. 2021). It is rather intriguing that although diatoms are highly plastic in response to light, temperature, and nutrient fluctuations, most of their genomes are small, contrary to the general notion that larger genomes

Table 7.1. Sequenced genomes of diatoms.

Species	Sequenced ploidy	Genome size (Mbp)	# of chromosomes	GC%	# of coding genes	# of exons per gene (% gene with intron)	# of rRNA gene	% Repetitive DNA	References
Conticribra weissflogii CCMP1336	Diploid	231.5	30	47.5	19,531	2.13	195		Li et al. 2021
Fistulifera solaris JPCC DA0580	Allodiploid	24.9(49.7)	42(84)	46%	11,448 (20,455)	1.76		16	Tanaka et al. 2015
Mayamaea pseudoterrestris		30.6			11,019			19	Suzuki et al. 2022
Fragilariopsis cylindrus	Diploid	61.1	11	39.80%	21,066	2.08	163	35	Kamikawa et al. 2022; Mock et al. 2017
Nitzschia putrida	Diploid	35(47)		47.60%	15,003(20,461)	1.69(1.68)			Kamikawa et al. 2022
Phaeodactylum tricornutum	Diploid	27.4	33	47	10,402	0.79	3		Bowler et al. 2008
Plagiostriata sp. CCMP470		23		51%	8,970				Sato et al. 2020
Pseudo-nitzschia multiseries	Diploid	218.7		41%	19,703	2.38		16	JGI*; Sato et al. 2020
Pseudo-nitzschia multistriata	Diploid	59	45	46%	12,008	1.87		3	Basu et al. 2017; Sato et al. 2020
Seminavis robusta	Diploid	125.77	27	48.32%	36,254	2.02		22.64	Osuna-Cruz et al. 2020
Skeletonema costatum	Diploid	46.9		45.10%	16,449	6.3	2	2	Ogura et al. 2018
Synedra acus		98.4			11,184				Galachyants et al. 2015
Thalassiosira oceanica	Haploid	81.6		53%	34,642	2.29		3	Lommer et al. 2010
Thalassiosira pseudonana	Diploid	32.4	24	48	11,776	1.52	43	3	Armbrust et al. 2004

* https://mycocosm.jgi.doe.gov/Psemu1/Psemu1.info.html

confer plasticity (Chapter 3). Comparison of the multiple diatom genomes has revealed common attributes of diatoms, mainly oxidation-reduction processes, responses to cadmium, and nucleus/cell division regulation.

Thalassiosira pseudonana

The first sequenced diatom genome was that of the centric diatom *T. pseudonana* (Armbrust et al. 2004). With 24 chromosomes, its nuclear genome is approximately 34 Mbp in size. The species has a plastid genome of 129 Kbp (31% GC, 144 protein genes) and a mitochondrial genome of 440 Kbp (30.5% GC, 40 protein genes). Despite its small size, the nuclear genome is gene-rich: with 11,242 protein-coding genes, the genome shows an average of one gene per 2.9 kilobases of DNA. The genome has a relatively low repeat content, with only about 3% of the genome composed of transposable elements.

T. pseudonana brought to light a number of gene families that were not found in other diatoms or in other eukaryotic organisms at the time. For example, it encodes a novel family of genes involved in silicic acid transport and biosynthesis of silica-based cell walls (silaffins, frustulins, prolyl 1-4 hydroxylases, silica deposition vesicle). The genome revealed biosynthetic enzymes for several types of polyunsaturated fatty acids, transporters, and enzymes for utilizing inorganic and organic nitrogenous compounds as N nutrients, including transporters of NO_3^- and NH_4^+, transaminases, glutamate dehydrogenase, uricase, and allantoicase. Among these, the OUC discovered in this species changed the long-standing view that OUC only occurs in metazoans. The genomic data also illuminate the role of arginine not only in OUC but also in the synthesis of Nitric Oxide (NO) via NO synthase for signaling (see Chapter 13 for more information on NO biosynthesis pathways) and generation of the energy storing molecule creatine phosphate through an OUC branch. The genome also sheds light on biosynthesis and degradation of chitin fibers, potentially as a mechanism to regulate flotation and nutrient delivery. Also unveiled by the genomic data were mechanisms of photoreception and photoprotection, two pathways for β oxidation of fatty acids (one mitochondrial and the other peroxisomal), and sophisticated iron uptake (including the bacteria-like siderophore) and homeostasis regulation machinery (metallothioneins and phytochelatin synthases). Overall, the genome of *T. pseudonana* has provided important insights into the biology of diatoms and the evolution of eukaryotic algal genomes.

Phaeodactylum tricornutum

The pennate diatom *P. tricornutum* was the second diatom species that was sequenced at full scale (Bowler et al. 2008). The size of this genome is approximately 27 Mbp, even smaller than that of *T. pseudonana*. Similar to *T. pseudonana*, the genome of *P. tricornutum* has a high density of genes, with an average of one gene per 3.5 kilobases of DNA. *P. tricornutum* has a higher repeat content than *T. pseudonana*, with about 12% of the genome consisting of transposable elements. While the *T. pseudonana* genome seems to be influenced more by intron gain, the *P. tricornutum* genome is contributed more by diatom-specific copia-retrotransposable elements. *P. tricornutum* also has many genes involved in photosynthesis, reflecting its importance as a primary producer in marine ecosystems. *P. tricornutum* has a unique ability to synthesize large amounts of lipids, and its genome contains a great number of genes involved in lipid metabolism and storage. *P. tricornutum* is adept at perceiving the environment and responding to environmental stresses such as nutrient limitation, oxidative stress, and fluctuating light levels. Its genome contains genes encoding a high variety of receptors and stress response pathways, such as those involved in the synthesis of antioxidants and the repair of DNA damage. Both *P. tricornutum* and *T. pseudonana* genomes encode abundant heat shock transcription factors: 69 in *P. tricornutum* and 89 in *T. pseudonana*, suggesting a high capacity to respond to thermal and other stresses. Notably, this diatom possesses the bacteria-like histidine kinase-based phosphorelay two-component signaling system.

Although the pennate and centric lineages have only been diverging for 90 million years, their genome structures are dramatically different and a substantial fraction of genes (~ 40%) are not shared by the two best studied representatives of the two lineages, *T. pseudonana* and *P. tricornutum*. Comparative analyses including yeasts and metazoans revealed rapid rates of gene diversification in diatoms, spanning from selective gene family expansions, differential losses and gains of genes and introns to differential mobilization of transposable elements. *P. tricornutum* stands out among sequenced algal genomes in its large number of bacterium-derived genes apparently acquired through Horizontal Gene Transfer (HGT). More than 300 of these bacteria-originated genes were also found in *T. pseudonana*, indicative of their ancient origins, and many of these are likely to enable functional innovation for metabolite management and for perception of environmental signals. Among the bacteria-originated HGT genes, many code for organic carbon and nitrogen utilization, OUC, polyamine metabolism related to diatom cell wall silicification, and unorthodox mechanisms of DNA replication, repair, and recombination. The genome also encodes a distinct cell cycle regulation machinery. Overall, findings in the diatom genomes go a long way towards explaining the incredible diversity and success of the diatoms in contemporary oceans and provide important insights into the potential applications of diatoms in biotechnology.

Fragilariopsis cylindrus

The pennate diatom *Fragilariopsis* from the Southern Ocean is a cold-adapted ecologically successful taxon in the polar region. The species *F. cylindrus* is frequently found to form large populations in both the bottom layer of sea ice and the wider sea-ice zone, including open waters. The genome of *F. cylindrus* (61.1 Mbp) has many imprints of adaptations to living in polar environments, including genes involved in cold acclimation, light sensing, and nutrient uptake (Mock et al. 2017). The study showed that divergent alleles were differentially expressed across environmental stressors of low iron, darkness, freezing temperatures, elevated temperatures, and increased CO_2. This was associated with 24.7% of the diploid genome. Alleles with higher ratios of non-synonymous to synonymous nucleotide substitutions indicated selection imposed on the allele differentiation and more pronounced condition-dependent expression, suggesting a correlation between diversifying selection and allelic differentiation. *F. cylindrus* produces various bioactive compounds, including polyunsaturated fatty acids, and the study of its genome has provided insights into the biosynthesis of these compounds, which could have biotechnological applications. Overall, the study of the *F. cylindrus* genome has provided a better understanding of the biology and ecology of this important polar diatom species and revealed many potential applications for its unique genetic features.

Conticribra weissflogii

Sequencing of the *C. weissflogii* (strain CCMP1336; formerly *Thalassiosira weissflogii*) genome yielded a 231 Mbp draft assembly, predicting 22,823 gene models, with 66.01% BUSCOs coverage (Li et al. 2021). Comparative analysis of the species with other diatoms revealed an enrichment of the gene ontology category of photosynthesis-light harvesting and the existence of the C4 pathway as part of the carbon concentrating mechanism, suggesting that this is probably a shared characteristic of diatoms. Furthermore, the genomic data augment the transcriptome data (Cheng et al. 2021), revealing the biosynthesis pathway of chitin in *C. weissflogii*.

Thalassiosira oceanica

The 92.9-Mbp genome of *T. oceanica* (strain CCMP1005) was sequenced to understand the adaptive mechanisms in an oceanic diatom to cope with chronic iron limitation (Lommer et al. 2012). The genome encodes a 34,642-protein proteome. The combination of the genome with transcriptomes and proteomes reveals a high metabolic flexibility in response to iron availability, involving cellular retrenchment and remodeling of bioenergetic pathways to reduce iron-rich photosynthetic proteins (i.e., photosynthetic electron transfer complexes) while preserving iron-rich mitochondrial proteins. The iron economy is enhanced by substituting three metal-containing fructose-bisphosphate aldolases involved in metabolic conversion of carbohydrates for enzymes that contain no metals. Besides, this oceanic species is equipped with a robust iron uptake system, featuring duplicated genes (including *ISIP1A*, *ISIP1B* and the flavodoxin genes *FLDA1* and *FLDA2*) and domains (such as *CREGx2*). Its adaptation to low-iron environments is further evidenced by high genomic plasticity, with many iron-related genes acquired through horizontal gene transfer. The horizontally transferred genes originated from diverse prokaryotic and eukaryotic taxa, most abundantly from the prasinophycean genus *Micromonas*.

Fistulifera solaris

F. solaris is an oil-rich marine diatom, which has been studied for biofuel exploitation. Its genome was sequenced using the Oxford Nanopore platform MinION, yielding a chromosome-scale assembly of the ~ 50-Mbp (~ 25 haploid) genome, with 41 out of the 44 chromosomes reaching telomere-to-telomere completeness (Maeda et al. 2022). Putative centromere regions were predicted in 16 chromosomes, revealing consensus motif sequences. This discovery marks the first reported characterization of consensus centromere motifs in diatom. The genome also shows the existence of multi-copy mini-chromosomes and evidence of segmental genome duplication. Furthermore, the genome also displays gene duplication and tandem repeat arrangement of lipogenesis genes related to the oleaginous phenotype. Although major metabolic pathways are shared with nonoleaginous diatoms, transcriptomic analysis showed unique expression patterns, particularly with concomitant upregulation of fatty acid/triacylglycerol biosynthesis and fatty acid degradation (β-oxidation) in concert with ATP production (Tanaka et al. 2015).

Heterotrophic Diatoms *Nitschia* spp.

Secondary loss of photosynthesis widely occurs in nearly all major groups of eukaryotic algae, although its occurrence in diatoms remain relatively unexplored. So far, *Nitzschia* represents the only diatom genus with heterotrophic species characterized at the genomic level (Onyshchenko et al. 2021, Kamikawa et al. 2022). Studies of *Nitzschia* sp. NIES-3581 reveal a complex pattern of metabolic retention and loss in its non-photosynthetic plastid. Transcriptomic, subcellular localization, and phylogenetic analyses show the loss of isopentenyl pyrophosphate biosynthesis and Rubisco-based carbon fixation. However, the plastid has retained various proteins for metabolic pathways such as amino acid biosynthesis and a portion of the Calvin-Benson-Bassham (CBB) cycle including glycolysis/gluconeogenesis and the reductive Pentose Phosphate Pathway (rPPP). Plastid glycolysis, gluconeogenesis, and rPPP might serve to feed indispensable substrates to plastid amino acid biosynthesis pathways, explaining the retention of plastid glycolysis/gluconeogenesis and rPPP proteins in multiple independent lineages of nonphotosynthetic algae/plants (Kamikawa et al. 2017).

Nitzschia Nitz4

This heterotrophic strain isolated from a shallow, warm, mangrove-lined coastal waterway is another non-parasitic non-photosynthetic diatom whose genome has been studied. Genome sequencing resulted in a 27.6-Mbp assembly, from which 9,478 protein coding genes were predicted (Onyshchenko et al. 2021). Most of these genes are shared with other diatoms, with 8,285 (87.5%) assigned to orthogroups that include other diatoms, suggesting that diatoms share a conserved 'core' metabolic space. The small number of unique genes (374 or fewer) suggests massive loss of nuclear genes, as supported by the Nitz4-specific loss of 10 gene families and reduction of several others.

The lost genes are largely enriched in functions directly or indirectly related to photosynthesis. Among these lost genes is the calmodulin-lysine N-methyltransferase family of proteins, with a role in photosynthesis suggested by the previous observation of upregulation of plastid-localized calmodulin-dependent protein kinases at low CO_2 in *Thalassiosira pseudonana* (Clement et al. 2017) and the previous conclusion that calcium is an important regulator of photosynthesis in plants and algae (Hochmal et al. 2015). More importantly, the genome study (Onyshchenko et al. 2021) revealed a large-scale loss of plastid genome, with complete loss of intergenic DNA and retention of only 90 of the 122 core diatom photosynthetic genes, accompanied by a decrease in GC content (from > 30% in photosynthetic diatoms to 22.4% in Nitz4). The gene losses eliminate isoprenoid biosynthesis in the plastid and lead to the remodeling of mitochondrial glycolysis to maximize ATP yield. However, the genome contains a beta-ketoadipate pathway that may allow Nitz4 to metabolize lignin-derived compounds. In addition, because diatom plastids lack an oxidative Pentose Phosphate Pathway (oPPP), leaving photosynthesis as the primary source of NADPH to support biosynthetic pathways in the plastid, the plastid loss in Zitz4 limits available sources of NADPH.

Nitschia putrida

This is another non-photosynthetic diatom, with its diploid genome sequenced (Kamikawa et al. 2022). *Nitschia putrida* is an osmotrophic species. About 15,003 protein-coding genes were predicted from the 35-Mbp genome assembly, and many diverged alleles were identified from the genome. This gene repertoire misses over 500 orthogroups of nucleus-encoded plastid proteins that are usually shared between photosynthetic diatoms, including proteins for pigment synthesis, light harvesting antenna, photosystems, and the CBB cycle. The species retains the OUC gene set, suggesting that N cycling is as important in the heterotrophic diatom as in photosynthetic species. The species does not have the gene set required by photorespiration (including Rubisco), but appears to operate peroxisomes, probably for the production of malate or glyoxylate to feed mitochondrial respiration. Photoreceptors such as aureochrome and cryptochrome-DASH/CPF2 are present in the nuclear genome, probably playing a role in light-dependent gene expression regulation. Further comparative analyses revealed multiple potential evolutionary signatures marking the transition from phototrophy to free-living secondary heterotrophy. These signatures encompass gene loss, accumulation of genes involved in organic carbon degradation, a unique secretome, and rapid divergence of conserved gene families associated with cell wall and extracellular metabolism. Together, these changes provide insight into the molecular adaptations underlying this significant ecological shift.

Skeletonema costatum

The genus *Skeletonema* has been thought to contain only one species, *S. costatum*, and the clear morphological differences visible under the light microscope have been attributed to population-specific variations. This nominal species is distributed in global coastal waters where it typically forms blooms in spring, causing harmful effects as its high biomass clogs fish gills and consumes oxygen when dead cells are decomposed. Not until the 21st century, when molecular techniques were easily accessible, was the "single species" notion changed. Eleven species have been recognized so far (Kooistra et al. 2008), and multiple species often coinhabit an ecosystem. A strain of *S. costatum* isolated from Japanese coastal waters, where this species is a dominant diatom that causes harmful algal blooms, was sequenced (Ogura et al. 2018). The genome was estimated to be ~ 51.365 Mbp in size and contain 14,669 protein coding genes, and 65.4% of its orthologous genes are shared with *T. pseudonana*. Comparative genomic and transcriptomic analyses revealed mechanisms of oxidative stress responses and evidence of cytokinin promoting the proliferation of diatoms. Specifically, diatoms causing harmful algal blooms have gained multi-copy genes governing responses to oxidative stress and cytokinin presence, suggesting the importance of oxidative stress response and cell cycle promotion in bloom development.

Ornithine-Urea Cycle and its Diversified Function in Diatoms

In the discovery of OUC in *T. pseudonana* changed the dogma that OUC only occurs in metazoans (Armbrust et al. 2004). OUC was then also found in the pennate diatom *P. tricornutum* (Bowler et al. 2008). All six genes in the OUC known in animals were detected, including ornithine carbamoyltransferase (OCT), Carbamoyl Phosphate Synthetase (CPS), argininosuccinate synthase (ASuS), and argininosuccinate lyase (ASL), arginase (ARG), and N-acetylglutamate synthase (NAGS). As shown in Fig. 7.2, OCT catalyzes the conversion of ornithine and carbamoyl phosphate into citrulline, a reaction that initiates the cycle. CPS then synthesizes carbamoyl phosphate from ammonia and bicarbonate, providing the substrate for the subsequent reactions. ASuS and ASL further transform citrulline into arginine, and ARG subsequently converts arginine to ornithine and urea. NAGS catalyzes the formation of N-acetylglutamate, an essential allosteric activator of CPS. These enzymatic conversions enable diatoms to recycle nitrogenous compounds for essential cellular processes.

OUC is a crucial metabolic pathway that enables diatoms to regulate intracellular nitrogen flux (Fig. 7.2). Diatoms are known for their ability to thrive in diverse aquatic environments, often characterized by fluctuating nutrient concentrations. In adaptation to variations in nitrogen availability, diatoms utilize OUC to distribute nitrogen to different metabolic pathways. Experimental studies indicated that OUC in diatoms was coupled with the TCA cycle and the glutamine synthetase/glutamate synthase cycles, and various parts of the interactive network were evolutionarily derived from bacterial or secondary endosymbiotic hosts through horizontal gene transfer (Allen et al. 2001). Furthermore, the role of OUC in diatoms as a central hub for inorganic carbon and nitrogen distribution and repackaging sets it apart from OUC in metazoans, and enables diatoms to recover from N-nutrient stress quickly.

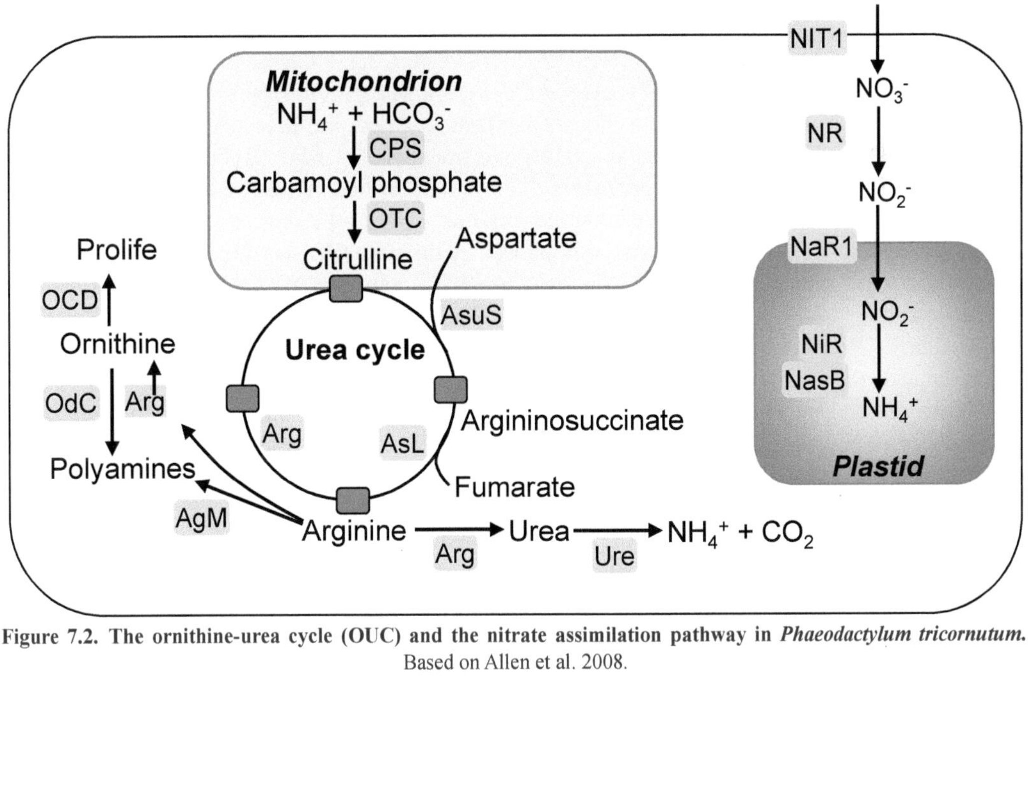

Figure 7.2. **The ornithine-urea cycle (OUC) and the nitrate assimilation pathway in *Phaeodactylum tricornutum*.**
Based on Allen et al. 2008.

Carbon Concentrating Mechanism (CCM)

In the genomes of *P. tricornutum* and *T. pseudonana*, researchers identified genes that encode aquaporins, the membrane channels for water and small-molecule solutes and gases, some of which have been shown to regulate entry of CO_2 and NH_3 into cells (Matsui et al. 2018). Due to the low CO_2 concentration in the modern-day ocean, diatoms, like other algae, have evolved a carbon (CO_2) concentrating mechanism (CCM) to elevate the concentration of CO_2 around the CO_2-fixing enzyme Rubisco (see Chapter 2). There are two main types of CCMs, one biophysical and the other biochemical, which have been best studied in diatoms among eukaryotic algae (Reinfelder 2011). The biophysical CCM, identified and characterized in many diatoms (Hopkinson et al. 2011, Matsuda et al. 2017), involves active transport of CO_2 and/or HCO_3^- across biological membranes, accompanied by inter-conversion between CO_2 and HCO_3^- catalyzed by carbonic anhydrases (CAs) (Nakajima et al. 2013, Kikutani et al. 2016). Cellular modeling has illustrated how CCM facilitates the uptake of HCO_3^- from the environment, the storage of absorbed CO_2 and HCO_3^-, and the enrichment of CO_2 in the chloroplast (Hopkinson et al. 2011). HCO_3^- transporter (SLC4, where SLC stands for solute carrier) genes have been reported in *P. tricornutum* (Nakajima et al. 2013) and *T. pseudonana* (Armbrust et al. 2004). Diatoms are equipped with numerous common (α, β, γ) and unusual (δ, ζ) families of CAs, and some of them are extracellular and serve to convert HCO_3^- to CO_2 on the cell surface for CO_2 uptake, as demonstrated in *T. pseudonana* (Samukawa et al. 2014). In *P. tricornutum*, plastidial CAs, especially those located in the pyrenoid (the β-CA PtCA1 and PtCA2, and the thylakoid luminal θ-CA Pt43233) (Tachibana et al. 2011, Kikutani et al. 2016), are critical for CCM function to enhance photosynthetic efficiency.

Biochemical CCMs are known as C4-type CCM or C4 pathway. As demonstrated in Chapter 2, some plants in tropical and arid environments use the C4 pathway to prevent water loss and enhance the efficiency of the Rubisco-mediated carbon fixation. While the prevalence and importance of the pathway in algae largely remain elusive these have been more extensively explored in diatoms. There is evidence that the C4 pathway (Reinfelder et al. 2000, 2004, Clement et al. 2017) or C3-C4 intermediate type (Roberts et al. 2007) occur in *Conticriba weissflogii*. However, in *T. pseudonana*, transcriptomic and physiological data did not confirm or reject the presence of this mechanism (McGinn and Morel 2008, Kustka et al. 2014), and tracer pulse-chase (Roberts et al. 2007) and localization (Tanaka et al. 2014) experiments did not support occurrence of a functional C4-type biochemical CCM. Recent data showed that the activities of key C4 enzymes phosphoenolpyruvate (PEP) carboxylase (PEPC) and pyruvate orthophosphate dikinase (PPDK) were lower in cells under low CO_2 than in those under high CO_2, opposite to what is expected for C4 metabolism (Clement et al. 2016).

Similarly, *P. tricornutum* possesses all the genes required for C4-type CCM, but the intracellular distribution of C4-related enzymes does not support the presence of a C4 pathway in the diatom (Ewe et al. 2018). Furthermore, knockdown of PPDK (Haimovich-Dayan et al. 2013) and PEP carboxykinase (PEPCK) caused no significant effect on the inorganic carbon acquisition and photosynthetic activity under ambient CO_2 conditions (Yang et al. 2016), and the transcript abundance of PEPCK and PEPC did not respond significantly to CO_2 concentration (McGinn and Morel 2008), thus raising questions about whether a C4 metabolism really operates in *P. tricornutum*. The absence of a C4 mechanism was also suggested in *Navicula pelliculosa* and the freshwater species *Asterionella formosa* (Clement et al. 2017). Interestingly, a recent study on *P. tricornutum* suggests that mitochondrial PEPC might catalyze biochemical carbon fixation when ambient inorganic carbon concentrations are low (Yu et al. 2022). Whether a modified C4 pathway involving mitochondria operates in diatoms warrants further investigation.

Similar to what was observed in *T. pseudonana*, the transcript and protein abundances of PPDK (NCBI Accession PRJEB34512) were significantly elevated (Jin et al. 2022a, b) in long-term

(8–24 months) high CO_2-selected populations of *P. tricornutum*. Its transcript abundance was also raised in short-term high $NaHCO_3$ treatment (Huang et al. 2023), indicating that PPDK might be involved in the long-term adaptation and short-term acclimation to higher inorganic carbon. Knockdown or knockout of PPDK in *P. tricornutum* inhibited the growth and the maximal photosystem II electron transport rate, and the growth inhibition became stronger when $NaHCO_3$ concentration increased from 1 to 8 mM. PPDK was found to localize in the plastid stroma at lower $NaHCO_3$ but both in stroma and periplasmic space (PPS) at higher $NaHCO_3$ (Huang et al. 2023). Under the high inorganic carbon condition, a large amount of HCO_3 transported by SLC4 can be converted into CO_2 by CA around the pyrenoid for photosynthesis. In addition, PPS-localized PPDK produces PEP, the substrate of PEPC. Oxaloacetate produced from the C4 pathway can be transported into the mitochondria and then enter the tricarboxylic acid cycle or be converted into α-ketoglutarate by aspartate aminotransferase to provide carbon skeletons for amino acid synthesis. In conclusion, due to the lack of the chloroplastic decarboxylases (Ewe et al. 2018), the decarboxylation of fixed carbon by PEPC will not take place in the plastid, and the C4 pathway might not function as a true CCM. Rather, the C4 cycle may consume excess inorganic carbon both from the plastid and extracellular environment to maintain pH homeostasis or dissipate excessive ATP in *P. tricornutum* (Haimovich-Dayan et al. 2013).

Genomic Basis of Cold Adaptation

In the genome of *Fragilariopsis cylindrus*, a cold-adapted diatom originally isolated from near Antarctica (the Southern Ocean), nearly 25% of the 61-Mbp diploid genome is composed of genetic loci with highly divergent alleles. These divergent alleles show differential expression across environmental conditions such as darkness, low iron, freezing, elevated temperature, and increased CO_2. Approximately 66% of the diverged alleles show greater than fourfold significant differential expression relative to optimal nutrient-replete growth, and about 45% of divergent alleles display over fourfold unequal bi-allelic expression between allele 1 and allele 2 in at least one RNA-seq experiment. Furthermore, the alleles with the highest ratios of nonsynonymous to synonymous nucleotide substitutions also exhibit the most pronounced condition-dependent expression pattern, suggesting that the allelic differentiation is a result of selection (Mock et al. 2017).

When grown in prolonged darkness, mimicking the dark Antarctic winter season, the diatom expressed a higher percentage of genes that do not map to predicted genes, potentially regulatory RNA molecules or unknown genes. Genes involved in starch, sucrose, and lipid metabolism were also strongly upregulated. In addition, this genome contains disproportionately higher gene dosage for Zn-binding domains, copper-binding domains, proton pump rhodopsin, and chlorophyll a/c Light-Harvesting Complex (LHC) proteins, including 11 members of the Lhcx clade, which is involved in stress response.

Membrane Energization

Instead of using proton motive to energize membranes as in vascular plants, bryophytes, charophytes, and fungi, marine diatoms seem to utilize Na^+ (Brownlee et al. 2022). Evidence includes the following. First, measured resting membrane potentials of approximately –60 mV in diatoms studied to date give a calculated inwardly directed Na^+ electrochemical gradient of 12 kJ mol^{-1}, and membrane energization similar to animal cells. Second, diatoms can generate fast action potentials involving Na^+/Ca^{2+} influx through voltage-gated channels. In addition, *in silico* and physiological studies suggest the presence of Na^+-dependent co-transporters of Si $(OH)_4^-$, NO_3^-, NH_4^+ and inorganic phosphate at the diatom plasma membrane. It is less clear for freshwater diatoms.

Morphotype Regulation

The regulatory mechanism of cell morphology in algae is poorly understood. Some breakthrough has been made on morphotype alteration in *P. tricornutum*. This species exists in three morphotypes: fusiform, triradiate, and oval, all of which often coexist in a strain although one or two of them tend to dominate. Using CRISPR/Cas9 gene editing and overexpression methodologies, a study found that one of the uncharacterized genes (Pt722015) in *P. tricornutum* genome (PHATRDRAFT_47103) was involved in morphotype regulation (Gao et al. 2022). Pt722015 knockout did not change the dominance of the fusiform morphotype, but the Pt722015-overexpression strain caused the predominance of the triradiate form (Fig. 7.3) and appeared to confer resistance to grazing by micrograzers. Besides, this strain grew at the same rate as the wild type, but accumulated more

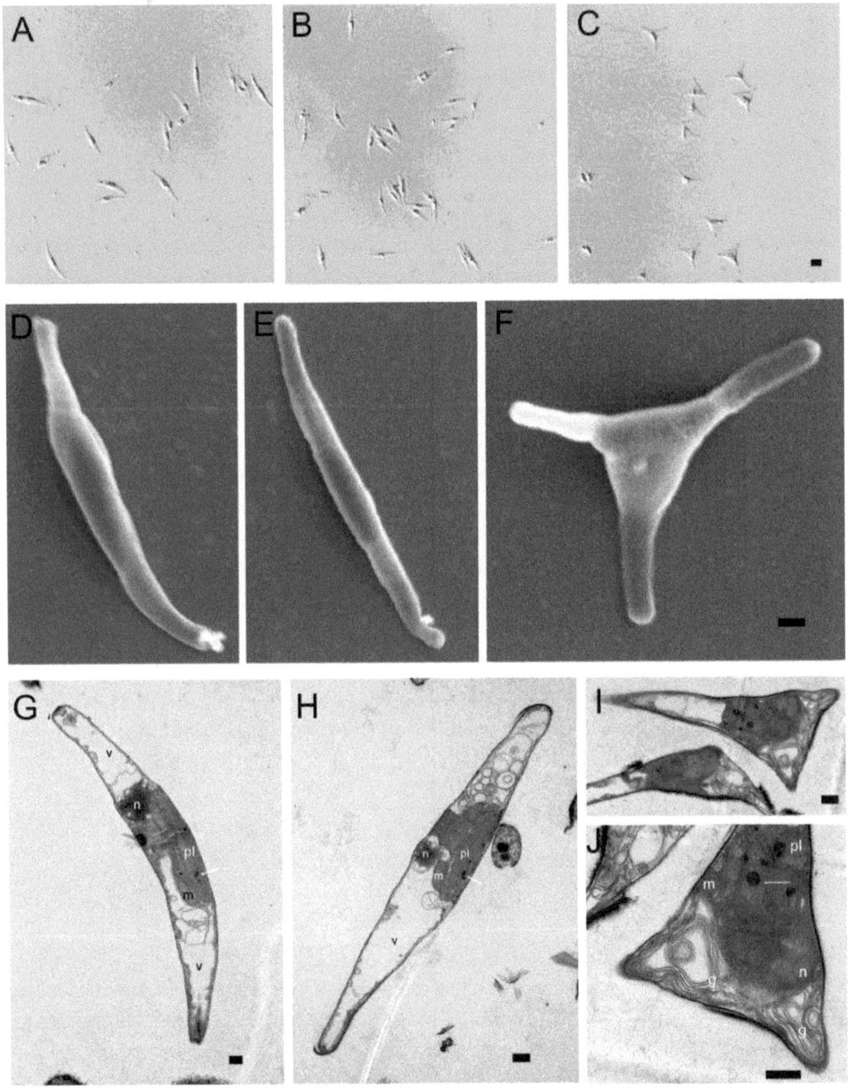

Figure 7.3. Micrographs of the wild type (WT), Pt2015 knockout strain (2015KO), and the Pt2015 overexpression (oeT) strains. Differential interference contrast micrographs of the WT (A), 2015KO (B), and oeT (C) strains (scale bar: 5 μm in C applies to A and B as well). Scanning electron microscopy (SEM) micrographs of the WT (D), 2015KO (E), and oeT (F) strains. The scale bar is 0.5 μm. Transmission electron microscopy (TEM) micrographs of the WT (G), 2015KO (H), and oeT (I and J) strains. White arrows indicate lipid droplets. The scale bar is 0.5 μm. Reproduced with permission from Gao et al. 2022 under the Creative Commons CC BY 4.0 license.

lipids than the wild type, making this gene a promising candidate biomarker for strain screening or creating a strain for biofuel production.

Chloroplast-Photoprotection

Having arisen from a secondary endosymbiosis event involving a eukaryotic cell and a red algal ancestor (Bhattacharya et al. 2007), diatom chloroplasts have a four-membrane envelope and thylakoids organized in a three-band structure, lacking the grana-stroma organization (Keeling 2010). Photosystems have evolved in photosynthetic organisms with pigments, reaction centers, and antenna complexes adapted to the different environmental conditions of their habitats (Blankenship 2010). Except for anoxygenic species, photosystem (PS) cores are highly similar between prokaryotes and eukaryotes, whereas the peripheral light-harvesting complexes are more variable in adaptation to their respective ecological niches (Calvaruso et al. 2020).

Diatoms possess 12 subunits in PSI and 28 subunits in PSII (Fig. 2.2 in Chapter 2). Photosynthetic pigments in diatoms include Chl *a* and *c*, fucoxanthin as the major carotenoid, and diatoxanthin/diadinoxanthin serving in the xanthophyll cycle. Even though diatoxanthin/diadinoxanthin are the main functional xanthophylls for photoprotection, the green plant-type xanthophylls, violaxanthin/zeaxanthin, also exist and function in diatoms (Lohr and Wilhelm 1999).

A Cryptochrome Functioning in High Temperature Acclimation

Members of the Cryptochrome/Photolyase Family (CPF) are widely distributed across all kingdoms of life and, as blue/UV light sensitive proteins, they typically serve as either photoreceptor or DNA repair enzyme. While photolyase was thought to be ancestral and the highly similar cryptochrome was derived, animal and plant cryptochromes seem to have lost DNA repair activity (Coesel et al. 2009). Cryptochromes in plants are involved in the regulation of growth and circadian rhythms, and those in mammals and insects function as transcriptional repressors that control the circadian clock. However, in *A. thaliana*, CPFs were reported to be involved in regulating responses to high temperature, with Cry1 repressing high temperature-induced hypocotyl elongation and Cry2 regulating blue light-dependent thermosensory flowering (Li et al. 2021, Zhao et al. 2022). This may not be as surprising as it sounds because temperature is influenced by light intensity.

In *P. tricornutum*, one of the CPFs (PtCPF1, Phatr2_27429) was found to exhibit DNA repair activity and serve as a transcriptional repressor of the circadian clock (Coesel et al. 2009). A follow-up study showed that PtCPF1 in *P. tricornutum* was expressed at a much higher transcriptional level at 26°C (thermal stress) than at 20°C (optimal temperature), suggesting a role for this protein in high temperature acclimation (Gao et al. 2024). Consistent with this postulation, Gao et al. (2024) also reported that homologs of PtCPF1 were widely distributed in the global ocean covered by the *Tara Oceans* expedition, where their transcript abundances were positively correlated with ambient seawater temperature. Furthermore, knockout of the PtCPF1 gene made the species more susceptible to high temperature with greater decreases in growth rate at 26°C than at 20°C. Additionally, PtCPF1 gene knockout caused the disruption of the expression of genes encoding two phytotransferrins (ISIP2A and ISIP1) and two Na+/P co-transporters at 26°C, suggesting the high temperature acclimation regulating role of PtCDF1 might be associated with iron and phosphorus nutrient uptake (Gao et al. 2024).

Rampant Horizontal Gene Transfer and Remarkable Genomic Innovation

Endosymbiotically (EGT) and horizontally transferred (HGT) genes are common in algae. It has been estimated that about 1% of genes in any given eukaryote have been recruited by HGT (van Etten and Bhattacharya 2020). Diatoms seem to champion the HGT activity among algae. Based on the genomes of the centric diatom *T. pseudonana* and the pennate diatom *P. tricornutum*, > 400 genes were acquired from a red algal origin, and between 100 and > 1,500 genes from a green algal origin. While the genes of red algal origin can be interpreted as having resulted from EGT from the diatom's red algal ancestor, the imprint of green algae in diatom nuclear genomes hints at a more complex evolutionary scenario. This could signal the possibility that diatoms once harbored a green plastid and recruited its genes into the nuclear genome before replacing it with the red alga-derived secondary plastid we know today (Moustafa et al. 2009, Deschamps and Moreira 2012).

Initial phylogenomic analysis identified nearly 5000 such green-originated proteins, but a more stringent phylogenetic examination reduced the number to 16. Even with this small number, the finding casts a significant question on the 'universal' red algal origin of secondary plastids that group ochrophytes, dinoflagellates, haptophytes, and cryptophytes together and calls for further research to examine this question.

Diatoms and other ochrophytes seem to have acquired large numbers of bacterial genes through HGT. Using ochrophyte genome and transcriptome data, employing automated and manual phylogenetic annotation, *in silico* modeling, and experimental techniques, Dorrell et al. (2021) found that continuous bacterial HGT was the predominant source of recently arrived genes in *P. tricornutum*. They observed a probable elevation of bacterial HGTs at foundational points in diatom evolution, following their divergence from other ochrophytes. Notably, they found that throughout ochrophyte evolutionary history, bacterial HGTs have been enriched in genes encoding secreted proteins.

Silicon Bioprocesses

Diatom cell walls contain amorphous hydrated silicon dioxide (silica), which is embedded in small amounts of organic matter. Therefore, silicate is essential to most diatoms. Its absence can cause cell cycle arrest in the G1 phase alone (e.g. *T. pseudonana* [Hildebrand et al. 2007]) or in both the G1 and G2/M phases (e.g., *T. weissflogii* [Vaulot et al. 1987]), depending on the species and the severity of silicate deficiency (Brzezinski et al. 1990 and references therein). Few exceptions to this include *P. tricornutum*, some strains of which have no silica cell wall. While silicate is clearly required for the formation of the new valve during cytokinesis, it is also required for DNA synthesis, although a mechanistic understanding is still lacking. It was hypothesized early on that DNA polymerase in diatoms might need silicate to be active; however, experiments to verify the hypothesis did not provide support.

A strong silicon uptake mechanism has evolved in diatoms. The first silicon transporter protein (SIT) gene was identified in the marine pennate diatom *Cylindrotheca fusiformis* (Hildebrand et al. 1997). SIT genes have since been identified in many species, thanks to the growing diatom genome data. Based on the primary sequence, diatom SIT genes are diverse (Fig. 7.4), currently classified into five clades (A to E). SIT genes in diatoms are also highly duplicated (see multiple SIT sequences from the same species in Fig. 7.4 and the phylogenetic tree in Alverson 2007). Despite the overall sequence diversity, the SIT protein consistently contains 10 transmembrane domains, and, as shown in *Synedra*, contains two highly conserved motifs: CMLD and DXDID (Marchenkov et al. 2018). In species such as *Synedra acus* subsp. radians, multiple SIT-Coding Units (CUs) are tandemly organized in an open reading frame (three CUs in one array) and expressed as a polyprotein. The polyprotein is then proteolytically cleaved, at the CMLD motif, into individual mature SITs. The polyprotein mode of gene expression is rarely documented in diatoms, but occurs in viruses and some eukaryotic genes (see Chapter 1). This mode of gene organization and expression can probably enhance expression efficiency of the genes to produce a large quantity of proteins in a short time. Furthermore, the sequences of the silicon transporters (SITs) show strong purifying selection, in both marine and freshwater diatoms (Alverson 2007). As such, the previously observed lower silicon uptake efficiency in freshwater diatoms is not due to transporter sequence divergence. All these observations in concert indicate the importance of SIT genes to diatoms.

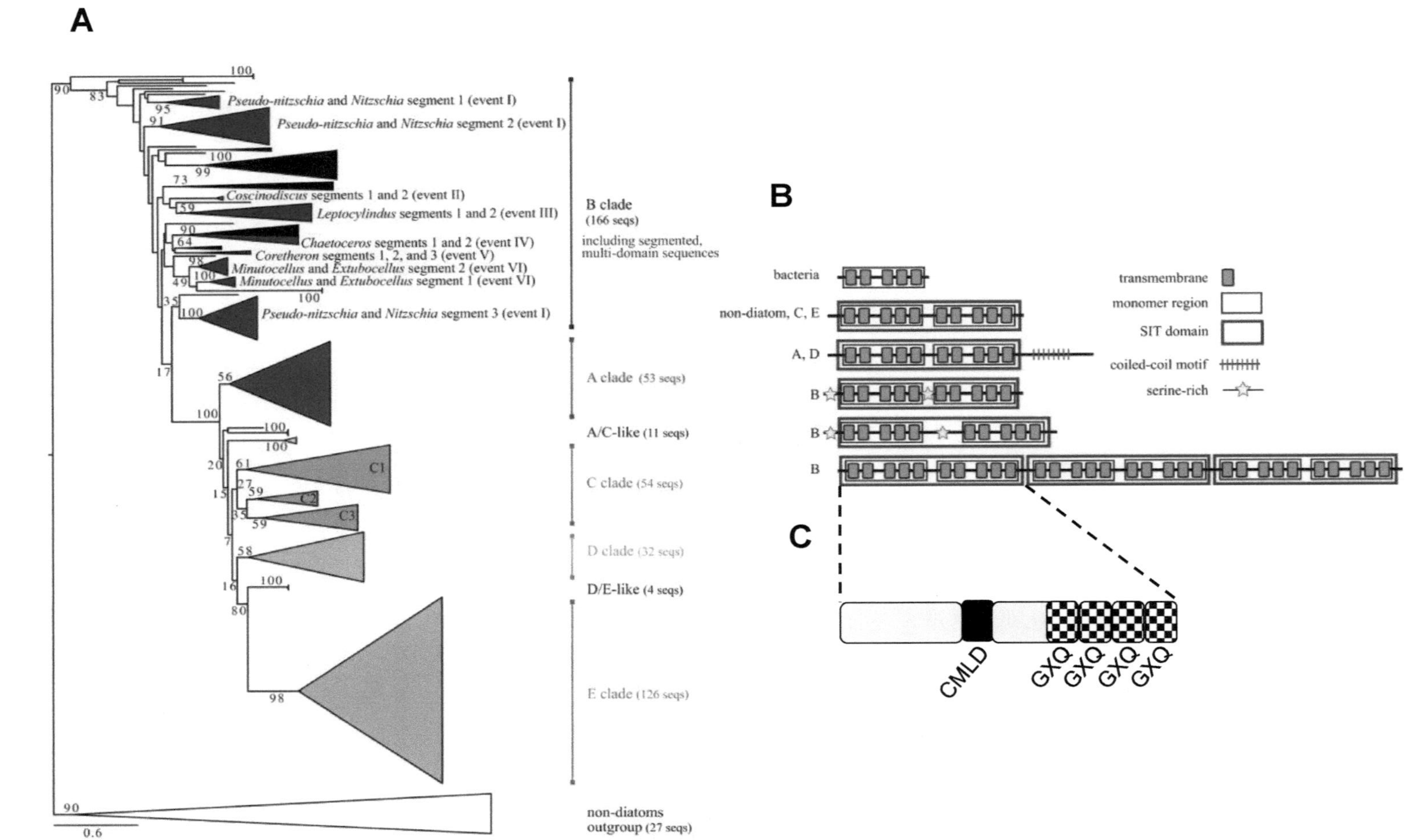

Figure 7.4. Diversity and extensive duplication of diatom silicon transporter genes (SITs) with conserved motifs. (A) Phylogenetic tree showing diversity and duplication of SITs. Each SIT is identified by species name and the last three digits of its GenBank accession ID. (B) Schematic of SIT gene organization. (C) Schematic of a mature SIT coding region, which contains a CMLD motif and four GXQ motifs. (A and B) is reproduced with permission from Durkin et al. (2016) under the Creative Commons CC BY-NC-ND 4.0 license and C is based on Marchenkov et al. 2018.

There is evidence that proteins other than SITs may be involved in silicon processing in diatoms. Using whole-genome oligonucleotide tiling arrays and tandem pass spectrometry analysis of transcriptomes, Mock et al. (2008) identified about 3000 new gene transcripts that had not been described, some of which were noncoding and antisense RNA molecules. From these, 75 genes were induced only under low concentrations of silicon but not under low concentrations of nitrogen or iron, alkaline pH, or low temperatures. Most of these induced gene products were predicted to contain secretory signals and/or transmembrane domains but matched no known proteins. In addition, a set of 84 genes were induced by both silicon and iron limitations, suggesting shared pathways in silicon and iron processing or iron-dependency of silicon processing.

Iron Nutrition

Iron is one of the essential trace elements for algae and, like many other trace metals, is often limited, especially in oceanic waters. Algae have evolved a diverse range of mechanisms to take up the limited amount and even usually unusable forms of iron. In addition, algae have developed strategies to cope with iron deficiency. These mechanisms and strategies will be examined in greater depth in Chapter 17. Discussion of this topic here will be limited to what has been uniquely found or characterized in diatoms. Diatoms are one of the best investigated algal models for iron metabolism (Gao et al. 2021). *P. tricornutum* is particularly well-suited to low-iron environments and can thrive at iron concentrations 50 times lower than the minimum requirement of *T. pseudonana*, resembling species found in the High Nutrient, Low Chlorophyll (HNLC) regions of the oceans (Sutak et al. 2020).

Proteins Inducible By Iron Deficiency

Iron Uptake Proteins

Ferric Reductases (FREs) are enzymes that reduce Fe^{3+} complexes such as siderophore-bound Fe^{3+} to Fe^{2+} complexes. The reduction either facilitates transport of iron via ferrous iron transporters or release of iron from siderophore after import into cells by TonB-dependent transporters (Cain and Smith 2021). In general, there are two types: soluble and membrane bound. Analysis of *P. tricornutum*'s transcriptome unveiled that under iron-deprived conditions, the expression of FRE genes (FRE1-FRE4) is induced. Interestingly, FRE2 in *P. tricornutum* is positioned nearby a highly iron-responsive gene encoding a putative Ferrichrome-Binding Protein (FBP1), divergently oriented with approximately 1.5 kb separating them (Allen et al. 2008). This is reminiscent of the bacterial FhuD protein found in the ferrichrome transport system, which is also known as the Fe^{3+} hydroxamate uptake system.

Although FRE2 and FBP1 transcript levels are consistently lower than those of canonical diatom iron stress biomarkers (such as flavodoxin and ISIP genes), their expression is correlated and significantly responsive to iron concentration (Allen et al. 2008). Furthermore, the role of FBP1 in this siderophore-iron uptake machinery was experimentally verified by the loss-of-function knockout and functional complementation by reintroduction of the native FBP1 as well as a homolog from the marine actinomycete *Salinispora tropica* (Coale et al. 2019).

Another iron deficiency-inducible protein is the IRT-like protein of the ZIP (Zrt/Irt-like-protein) family. The ZIP family is a group of membrane proteins that transport divalent metals, such as zinc, manganese, and iron, across cell membranes. These proteins are involved in maintaining homeostasis of these metals at the cellular and systemic levels. ZIPs occur in both prokaryotes and eukaryotes. *P. tricornutum* possesses an IRT-like protein of the ZIP family, which exhibits induction under iron limitation, potentially involved in ferrous iron uptake (Allen et al. 2008).

Iron Starvation-Induced Proteins (ISIPs)

ISIPs are a group of iron-regulated proteins, which occur not only in *P. tricornutum*, but also in other low-iron-quota diatoms (Allen et al. 2008). The expression of ISIP genes is inversely correlated with iron availability, as shown in *TARA Oceans* data (Caputi et al. 2019), making them useful markers of iron limitation (Sutak et al. 2020). ISIP2A has been shown to concentrate ferric iron at the cell surface (Morrissey et al. 2015), and a knockdown cell line displayed lower Fe (III) uptake than the wild type, but showed no effect on Fe(II) uptake. All research results suggest that ISIP2A contributes to a thermodynamically regulated iron uptake process, alongside the reductive, kinetically controlled mechanism of iron uptake (Morrissey et al. 2015). Furthermore, ISIP2A shows distant relation to the FEA1 protein of *C. reinhardtii*, which has been hypothesized (although not conclusively demonstrated) to aid iron uptake by concentrating iron near the plasma membrane (Narayanan et al. 2011).

Flavodoxin and Other Proteins

Diatoms are probably the earliest and best studied algal model for understanding how flavodoxin (FLD) is induced under iron deficiency to substitute its iron-containing functional equivalent, ferredoxin (FED). It has been reported for diatoms and dinoflagellates that under iron deficiency, the ratio of FLD abundance to FED abundance increased (Erdner et al. 1999).

Besides ISIPs and FLD, other common core low iron-induced genes in diatoms include class I fructose-bisphosphate aldolases, copper-dependent oxidases, ferric reductases, and permeases (Lommer et al. 2012, Mock et al. 2008, Cohen et al. 2018).

Iron Storage: Ferritin and Vacuole

Depending on the species, diatoms generally operate two iron storage mechanisms: the protein ferritin and vacuolar storage. Ferritin (FTN) is an iron-binding protein. Although present across all four diatom lineages, FTN appears to be largely absent from centric diatoms, including many species of *Thalassiosira* and *Chaetoceros*, two prominent genera in the ocean. Furthermore, diatom ferritins do not seem to all serve the same functional role. In a study involving iron addition and removal incubations for natural phytoplankton communities from varying iron environments, it was found that among the predominant diatoms, *Pseudo-nitzschia* were favored by iron removal and displayed unique ferritin expression consistent with a long-term storage function (Lampe et al. 2018). The iron storage was exceptionally high both under steady-state high and low iron conditions, as well as following iron resupply to iron-limited cells. In contrast, *Chaetoceros* and *Thalassiosira* gene expression patterns were as expected for vacuolar storage mechanisms. These results suggest that different diatom taxa use different iron storage mechanisms, and that ferritin utilization may provide an advantage in areas of prolonged iron limitation with pulsed iron inputs. As such, changes in iron distributions and availability may cause shifts in diatom community composition.

Endocytosis-Mediated Iron Uptake and Cell Surface-To-Chloroplast Trafficking Pathway

Iron uptake by endocytosis is believed to be common in mixotrophic protists when iron is limited in the environment (e.g., Liu et al. 2015). Although diatoms with frustules were thought to be incapable of endocytosis (phagotrophy), recent studies have indicated that diatoms can take in iron and silicon in seawater through pinocytosis to support their growth (Thamatrakoln and Kustka 2009, Kazamia et al. 2018, Annenkov et al. 2020). An environmental metatranscriptomic study revealed expression of endocytosis-related genes in diatoms in a subtropical marginal sea, in positive correlation with the expression of the Calvin-Benson-Bassham cycle genes, suggesting that these diatoms deploy pinocytosis to take up iron to support photosynthetic carbon fixation (Li et al. 2024).

Iron Starvation Induced Protein 1 (ISIP1) is responsible for endocytosis of siderophore-bound iron. Diatom ISIP1 proteins have a short C-terminal intracellular domain with a putative endocytosis motif. Additionally, diatoms have convergently evolved a high-affinity ferric (Fe^{3+})-binding protein, phytotransferrin (pTF), the functional analog of human iron delivery proteins transferrins (Cheng et al. 2004, McQuid et al. 2018). pTF is localized to the cell surface and presumably endosomal vesicles (Morrissey et al. 2015, McQuid et al. 2018). In *P. tricornutum*, pTF was initially identified as iron starvation induced protein 2A (ISIP2A), which was shown to concentrate iron on cell surface (Morrissey et al. 2015), consistent with pTF being a cell surface protein. The iron-bound phytotransferrin is then transported into the diatom cells via endocytosis (McQuid et al. 2018).

The downstream proteins involved in the intracellular trafficking of iron were identified using engineered ascorbate peroxidase APEX2-based subcellular proteomics to catalog proximal proteins of phytotransferrin in the model diatom *P. tricornutum* (Turnšek et al. 2021). This study identified three previously uncharacterized proteins, two of which were localized near the chloroplast.

Production of Domoic Acid (DA) and B-Methylamino-L-Alanine (BMAA)

Both DA and BMAA are metabolites of diatoms that are toxic to humans. DA is produced by *Pseudo-nitzschia* spp. and *Nitzschia* sp. When the diatoms are transferred in the food chain, the toxin is accumulated, and when it reaches humans via contaminated shellfish seafood, it causes short-term memory loss, a syndrome named Amnesic Shellfish Poisoning (ASP). BMAA is a non-proteinogenic amino acid known to be produced by cyanobacteria and diatoms. BMAA is a neurotoxin and has been linked to various neurodegenerative disorders including Alzheimer's disease (Koksharova and Safronova 2022). Interestingly, the biosynthesis of both toxins in diatoms is promoted by iron stress. It has thus been proposed that diatoms produce DA to serve as an iron chelator and enhance iron uptake.

DA synthesis starts with acetyl CoA and has two potential pathways, one involving the TCA cycle followed by the metabolism of glutamate and the other involves oxidation of 3-hydroxyl-methylglutaryl CoA to 3(R)-mevalonic acid followed by production of geranyl pyrophosphate (Fig. 7.5). In both proposed pathways, additional steps are required, which are currently unclear. In *P. multiseries*, DA synthesis seems to involve a four-gene cluster encoding a terpene cyclase (*dabA*), a hypothetical protein (*dabB*), a-ketoglutarate-dependent dixoygenase (*dabC*), and CYP450 (*dabD*) (Brunson et al. 2018). For BMAA biosynthesis, a transcriptomic investigation indicated that cellular processes such as the inhibition of ubiquitination and autophagy function, as well as reduced accuracy and efficiency of COP vesicle transport, could potentially contribute to the biosynthesis of BMAA-containing proteins in diatom cells (Li et al. 2023).

Chitin and Floating

Chitin, a polymer of 1,4-linked ß-D-N-acetylglucosaminyl residues, is the most abundant polymer in the ocean and the second most abundant polymer on Earth, surpassed only by cellulose (Durkin et al. 2009). Diatoms such as *Conticribra/Thalassiosira* and *Cyclotella* have been shown to contain chitin (McLachian and Craigie 1966). Planktonic diatoms, like other phytoplankton, require some mechanisms to keep the cell mass floating to receive light and sometimes sinking to explore nutrients. This seems to be mediated in part by extruded chitin fibers.

T. pseudonana genome contains at least 22 putative genes of Chitin Synthase (CS), the enzyme for chitin biosynthesis, suggesting that CS gene family expansion might be a genomic innovation in this taxon in adaptation to the planktonic lifestyle (Armbrust et al. 2004). The deduced amino acid sequences contain CS domain IPR004835 (also annotated as fungal chitin synthase domain IPR004834). According to Pfam domain prediction, the CS protein is localized in the cell periphery or cell septum, involved in the cell wall biosynthetic process.

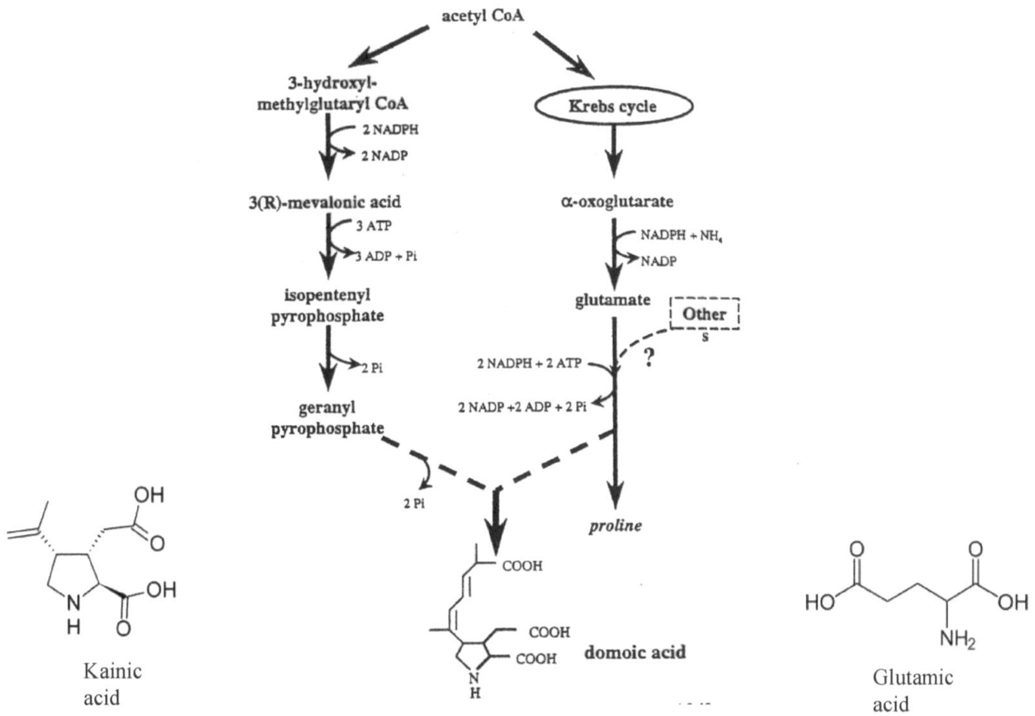

Figure 7.5. Hypothetical pathway of DA biosynthesis in diatoms.

This CS-coding gene has since been identified in more centric as well as pennate diatoms, with divergent sequences forming three major clades (Durkin et al. 2009), indicating a widespread and potentially differential ability to synthesize chitin in diatoms. A species can have genes affiliated with different clades, indicating early divergence of this gene family during diatom evolution. Interestingly, while *T. pseudonana* CS genes are similar to Division II fungal CS genes, homologs in *P. tricornutum* are similar to Division I fungal CS genes (Li et al. 2021). Among the three clades of CS genes (A, B, and C), clade B in *T. pseudonana* exhibited marked upregulation, whereas the other two clades showed no change, in cells experiencing depletion of silicic acid or iron in the medium.

Genomic data of *C. weissflogii* (Li et al. 2021), along with transcriptome data (Cheng et al. 2021), reveal the chitin biosynthesis pathway in the species, which has the highest content of chitin (up to 34% of the total cell mass) among diatoms examined so far. Comparative analysis for diatom genomes also indicated a divergence in the chitin metabolic pathway between the centric and pennate diatoms.

8

Genomes of Dinoflagellates

Dinoflagellates are one of the most important and unique groups of protists, renowned for their ecological success and many intriguing characteristics (Hackett et al. 2004, Lin 2011). Believed to have emerged in the Silurian period (~ 400 million years ago), the Dinoflagellata phylum comprises over 2,300 extant species and roughly 2,000 fossil species (Fensome et al. 1993, Gomez 2012). Morphologically diverse (Fig. 8.1), dinoflagellates are generally classified into two major morphotypes: thecate, with cellulose-filled vesicles (amphiesma) beneath the cell membrane, and athecate (naked), lacking or with minimal amphiesma.

Ecologically, dinoflagellates adapt to a wide range of habitats, from small ponds and rivers, lakes, and estuaries, to the ocean, and from the Antarctic to the Arctic, with lineages evolutionarily migrating between freshwater and marine environments (Logares et al. 2007, Annenkova et al. 2020). Numerous genetically diverse species and strains in the Symbiodiniaceae family are crucial endosymbionts of reef-building corals. Many species from various lineages cause harmful algal blooms, devastating marine ecosystems and coastal economies as well as posing public health risks. Additionally, dinoflagellates live in diverse trophic modes (from autotrophic, heterotrophic, to mixotrophic) and lifestyles (from parasitic, symbiotic, to free-living), with photoautotrophic lineages (~ 50% of species) contributing substantially to photosynthetic carbon fixation and oxygen production and heterotrophic/mixotrophic species being important trophic links in the food chain. Dinoflagellates are the only group of protists that produce bioluminescence, which occurs in both photosynthetic (e.g., *Alexandrium* spp., *Lingulodinium polyedrum*) and heterotrophic (e.g., *Noctiluca scintillans*) species.

From an evolutionary perspective, dinoflagellates encompass parasitic (Syndiniales) lineages at the base followed by the typical dinoflagellate lineages composed of athecate and thecate lineages (Fig. 8.1). The parasitic lineage is phylogenetically basal, leaving the photosynthetic ancestor that is believed to exist according to the current evolutionary theory unidentified. Most studied species retain the ancestral plastid type, which contains the light-harvesting pigment peridinin as the major accessory pigment, and these species are often referred to as typical dinoflagellates. However, a substantial number of dinoflagellate lineages have replaced plastids through serial secondary or tertiary endosymbiosis events (see Chapter 2). The archetypical plastid genome exists in more than ten plasmid-like circles (minicircles), each carrying one to two genes except for some being geneless, and only coding for 16–17 proteins in total (Zhang et al. 1999, Dorrel and Howe 2015; Fig. 8.2). In *Amphidinium carterae*, psbA and petB/atpA minicircles can be transcribed into antisense RNA (Dorrel and Howe 2019).

At the cellular and genomic levels, dinoflagellate possess many peculiar characteristics (Lin 2011, Wisecaver and Hackett 2011). Chromosomes are permanently (or nearly so) condensed, with cholesteric liquid crystalline state of DNA and lack of nucleosomes. Many species contain numerous chromosomes, which appear to wrap quite closely into a "spaghetti ball" (Fig. 8.3).

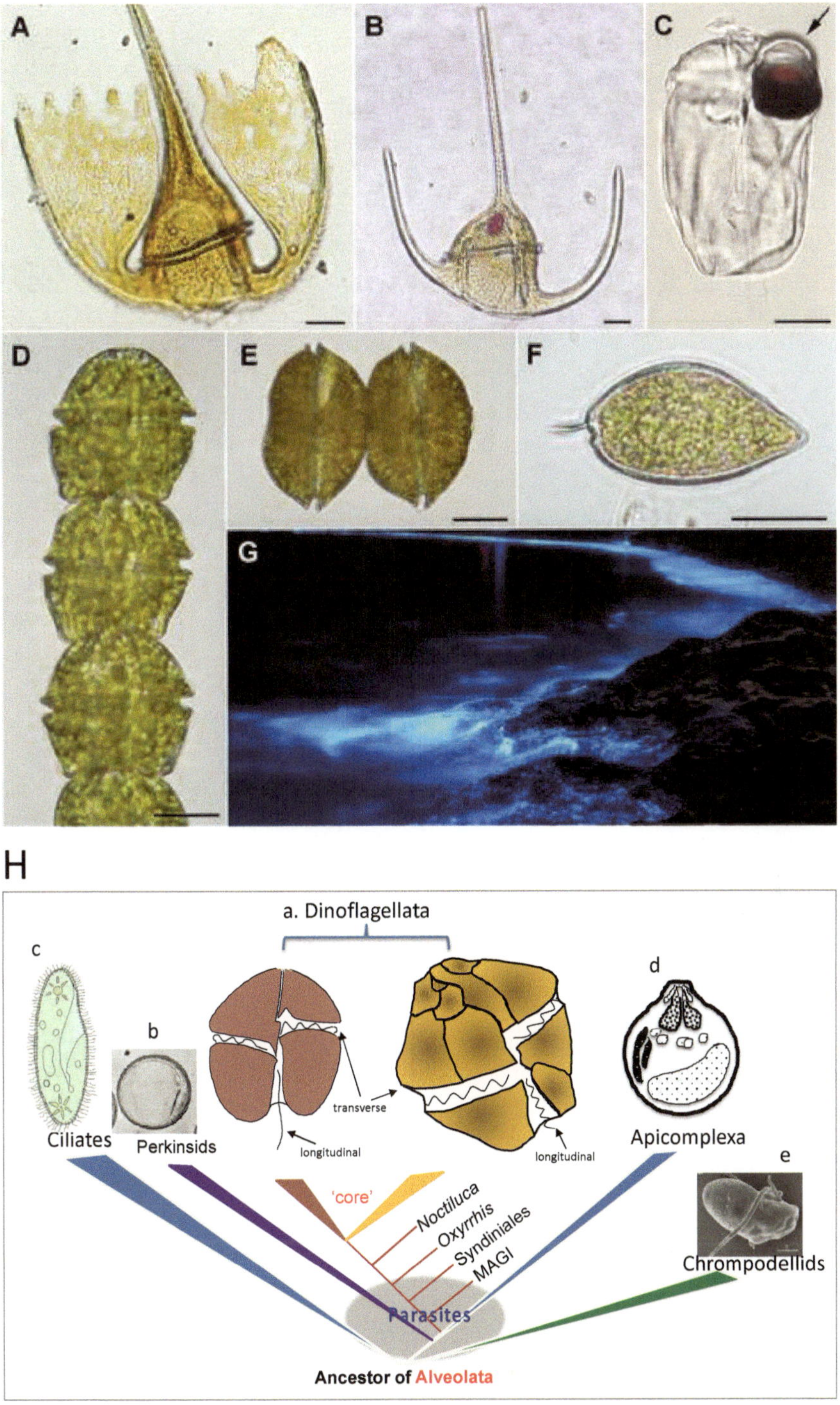

Figure 8.1. Light micrographs of dinoflagellates and a bioluminescence scene (A-G) and phylogenetic relationship of dinoflagellates, apicomplexa, and other alveolates (H). A. *Tripos platycorne*. B. *Tripos muelleri*. C. *Erythropsidinium agile*. The arrow points the ocelloid. D. *Gymnodinium catenatum*. E. *Gessnerium* (=*Alexandrium*) *monilatum*. F. *Prorocentrum micans*. Scale bars = 20 µm. G. Bioluminescence by *Noctiluca scintillans* at Zhejiang, China. Images A-F courtesy of Dr. Fernando Gómez. Image G courtesy of Dr. Hao Luo. In H), core dinoflagellates (a) include naked dinoflagellate with two flagella shown (longitudinal and transverse) and thecate dinoflagellate with two flagella (longitudinal and transverse) and the conspicuous thecal plates. Non-core dinoflagellates include Marine Alveolate Group I (MAGI, euduboscquellids), Syndiniales (also known as MAGII), and the basal lineages *Oxyrrhis* and *Noctiluca*. Reproduced from Lin (2024) under the Creative Commons CC BY 4.0 license.

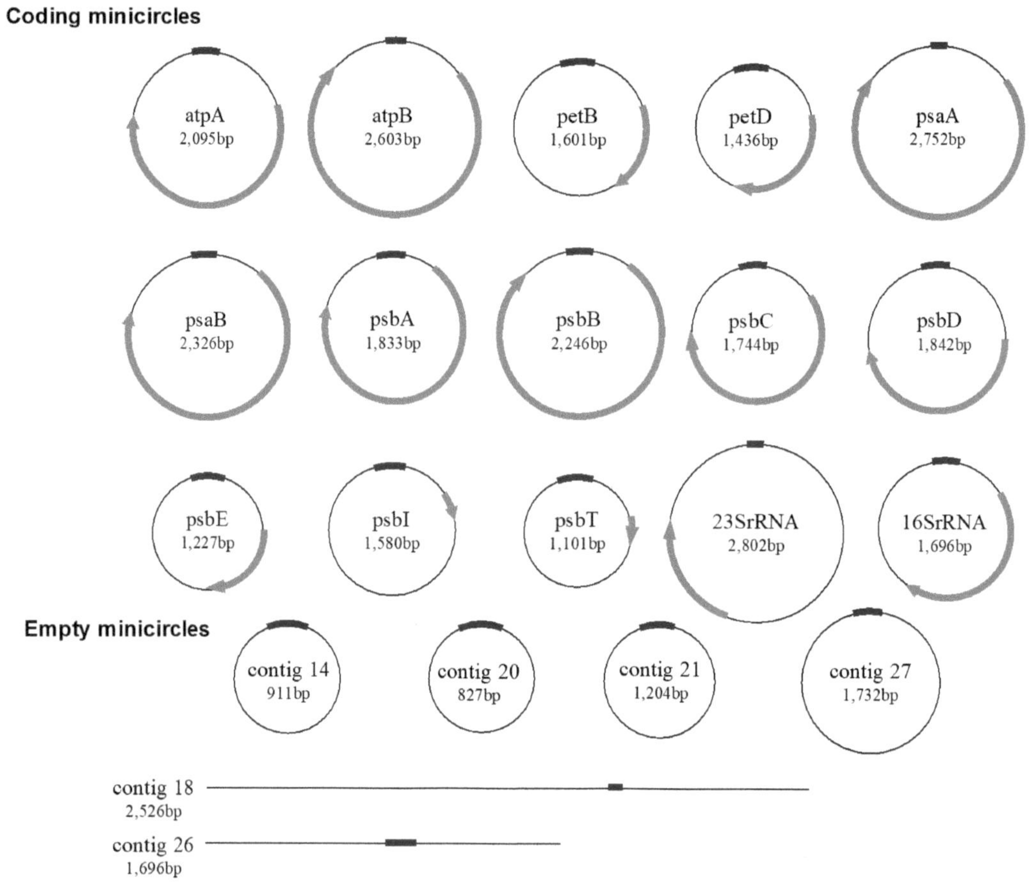

Figure 8.2. Minicircles of chloroplast genome of *Fugacium kawagutii*. In each circle, there are core region (thick black line on the top), non-coding region (thin line), and, in most circles, a coding region (grey thick line with arrow). In this species, contigs 18 and 26 do not form circles. Reproduced with slight modification from He et al. 2024 under the Creative Commons CC BY 4.0 license.

Cells perform closed mitosis, with extranuclear mitotic spindle. DNA in dinoflagellates is extensively methylated, containing hydroxymethyluracil (5-hmU) in place of some thymine bases besides the more common cytosine methylation. How dinoflagellates regulate DNA duplication in the cell division cycle and regulate gene expression, given the permanently condensed DNA state, remains perplexing and poorly understood.

Sequenced Nuclear Genomes

Due to their large genome sizes (up to 250 Gbp), dinoflagellate genome sequencing has lagged behind that of other algal phyla. Most of molecular and genomic insights about dinoflagellates have come from EST (Expressed Sequence Tag) or similar analyses of transcriptomes and cloning-based sequencing of genes (Lin 2011, Wisecaver and Hackett 2011). However, recent advances in sequencing technologies and bioinformatics, coupled with a precipitous drop in sequencing costs, ushered in an era of thriving dinoflagellate genomics, yielding significant milestones over the past decade. With the exponential growth of transcriptome and metatranscriptome data, the dinoflagellate omic (Dinomic) data has expanded into a substantial database (Table 8.1). Although biased towards smaller genomes, this database provides rich insights into dinoflagellate biology. Recent findings have challenged long-standing notions about dinoflagellate nuclear biology and molecular genetics, while revealing unexpected physiologies.

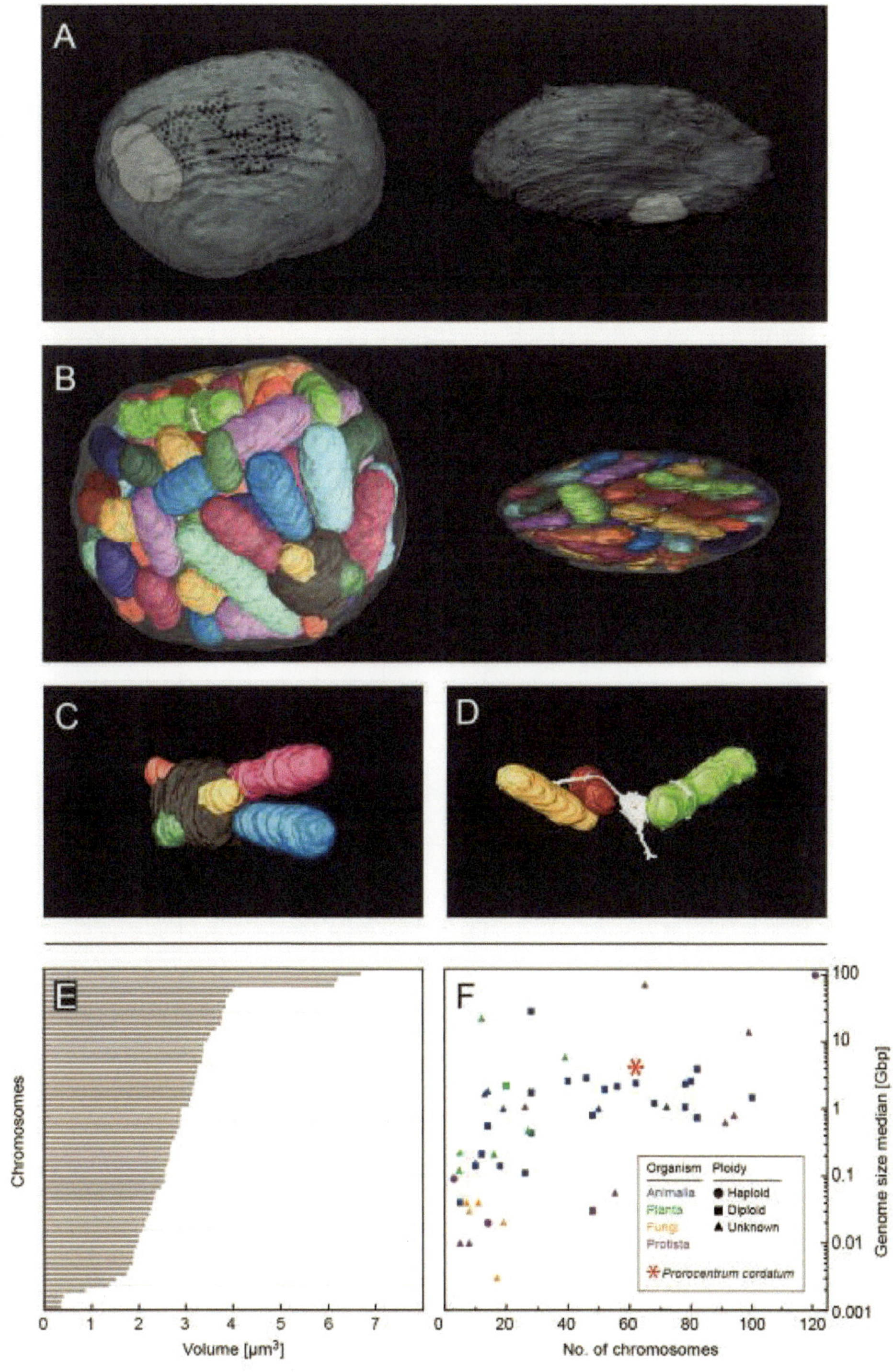

Figure 8.3. Three-dimensional reconstruction of the nucleus of *Prorocentrum cordatum* based on FIB/SEM images. (A) Distribution of nuclear pores across the nuclear envelope. Left panel, patch with high number of pores proximal to the nucleolus; right panel, pore-poor region. (B) Tight packing of chromosomes in the nucleus. Left panel, top view; right panel, side view. Chromosomes are arbitrarily colored, the nucleolus is marked dark grey and the nuclear membrane displayed transparently. (C) Focus on nucleolus with interacting chromosomes. (D) Conspicuous structure (white, probably extension of endoplasmic reticulum) interacting with several chromosomes. (E) Volume distribution of detected chromosomes. (F) Comparison of P. cordatum genome size and chromosome numbers with reported literature data. After Kalvelage et al. (2023) with the permission of mSphere under the Creative Commons CC BY 4.0 license.

Table 8.1. Published genomes of dinoflagellates.

Species	Assembled genome size (Mbp)	Assembled rate (%)	Gene No.	Scaffolds N50 (kbp)	GC (%)	Repetitive DNA (%)	References
Amoebophrya sp. ex *Karlodinium veneficum*	~ 130	94.62	41,936	21	>58		Bachvaroff 2019
Amoebophrya ceratii AT5.2	87.7	75.86	19,925	83.9	55.9	23.8	John et al. 2019
Amoebophrya ceratii A120	115.5	92.21	26,441	9,243	51.2	13.1	Farhat et al. 2021
Amoebophrya ceratii A25	116	97.8	28,091	1,082	47.8		Farhat et al. 2021
Amphidinium gibbosum	6300	98.44	85,139	166.4	47.1	29.9	Beedessee et al. 2020
Breviolum minutum	616	41.07	41,925	126.2	43.6		Shoguchi et al. 2013
Cladocopium goreaui (original)	1,030	85.55	35,913	98	44.8		Liu et al. 2018
Cladocopium goreaui (improved)	1,200		45,322	354	44.4	36.5	Chen et al. 2022
Cladocopium spp.	704.8		65,832		43		Shoguchi et al. 2021
Durusdinium trenchii	670		30,054	97.5	47.4		Shoguchi et al. 2021
Effrenium voratum RCC1521	1,200	85.71	32,108	720	50.8		Shah et al. 2024
Effrenium voratum CCMP3420	1,300		39,878	252	50.6		Shah et al. 2024
Effrenium voratum CCMP421	1,100	57.89	32,615	304	50.9		Shah et al. 2024
Fugacium kawagutii CCMP2468	935	79.24	36,850	381			Lin et al., 2015
Fugacium kawagutii CCMP2468	937	79.41	45,192	13,533.5			Li et al. 2020
Fugacium kawagutii CCMP2468	1,050	88.98	26,609	268.8		16	Liu et al., 2018
Hematodinium sp.*	4,769		29,510	17,235			Gornik et al. 2015
Polarella glacialis CCMP1383	2,980		58,232	170.3	45.91	68	Stephens et al. 2020
Polarella glacialis CCMP2088	2,760		51,713	129.2	46.15	68	Stephens et al. 2020
Prorocentrum cordatum	4,750	87.37	85,849	349.2	59.7		Dougan et al. 2023
Symbiodinium linucheae CCMP2456	695	75.96		58.1	50.36		González-Pech et al. 2021
Symbiodinium microadriacticum CCMP2467	808	73.45	49,109	573.5	50.5	34.8	Aranda et al., 2016
Symbiodinium microadriacticum CassKB8	813	72.65		43	51.91		González-Pech et al. 2021

Symbiodinium microadriaticum 04-503SCI.03	775	73.62*		50	50.46	27.9	González-Pech et al. 2021
Symbiodinium natans CCMP2548	762	100		610.5	51.79		González-Pech et al. 2021
Symbiodinium necroappetens CCMP2469	768	76.26		11.4	50.85		González-Pech et al. 2021
Symbiodinium pilosum CCMP2461	1,089	54.64		62.4	48.21		González-Pech et al. 2021
Symbiodinium spp. clade C	767		69,018		50		Shoguchi et al., 2018
Symbiodinium spp. clade A	705		65,832		43		Shoguchi et al., 2018
Symbiodinium tridacnidorum CCMP2592	1,103	85.71		651.3	51.01		González-Pech et al. 2021

* Information came from transcriptome and incomplete draft genome.

Dinoflagellate Genomics Landscape and Incredible Inter-Species Dissimilarities

To date, 28 dinoflagellate genomes have been sequenced (with two sequenced more than once), mostly from the family Symbiodiniaceae (Table 8.1). Sequenced taxa other than Symbiodiniaceae include *Amoebophrya* sp. ex *Karlodinium veneficum* (Bachvaroff 2019), *Amoebophrya ceratii* (John et al. 2019, Farhat et al. 2021), *A. gibbosum* (Beedessee et al. 2020), *Hematodinium* (Gornik et al. 2015), *Polarella glacialis* (Stephens et al. 2020), and *Prorocentrum cordatum* (formely *P. minimum*) (Dougan et al. 2023). These taxa cover a wide range of environmental conditions and different lifestyles. Species from the family Symbiodiniaceae are endosymbionts of tropical corals. *Amoebophrya* species are parasitic infecting other dinoflagellates. *A. gibbosum* and *P. cordatum* form harmful algal blooms and *A. gibbosum* is toxigenic. *P. glacialis* thrives in both the Antarctic and Arctic, can withstand temperature shocks up to 15°C (Zheng et al. 2012), and DNA barcoding suggests its presence, along with closely related species, in the temperate estuary Long Island Sound (Lin et al. 2009).

Most of these genome assemblies were based on Illumina Hi-Seq technology and hence are largely fragmentary. However, using PacBio to generate long sequence reads and Hi-C to establish linkages between scaffolds, chromosome level assemblies have been achieved for part of the *F. kawagutii* genome (> 120 Mbp; Li et al. 2020) and the genomes of *Symbiodinium microadriaticum* (Nand et al. 2021) and *Breviolum minutum* (Marinov et al. 2021).

Chromosome numbers in dinoflagellates vary widely, ranging from several to over 200 (Sigee 1986). *F. kawagutii* has 7-11 chromosomes (Lin et al. 2015, Zaheri et al. 2019), while *S. microadriaticum* and *B. minutum* have 94 and 91 chromosomes, respectively (Nand et al. 2021, Marinov et al. 2021). This inter-species disparity is remarkable given the similar genome sizes among these symbiodiniacean taxa.

Chromosome lengths also differ significantly. *S. microadriaticum* (16 Kb to 16.6 Mb, median 6.6 Mb; Nand et al. 2021) and *B. minutum* (median 6.7 Mb, longest ~ 11 Mb; Marinov et al. 2021) have shorter chromosomes than *F. kawagutii* (maximum > 120 Mb; Li et al. 2020). This suggests an evolutionary trend of decreasing chromosome numbers and increasing lengths from the basal *S. microadriaticum* to the later-diverging *F. kawagutii* (Lin et al. 2021).

Based on the existing data, a pattern of genome size and chromosome number can be inferred. Small non-parasitic dinoflagellate genomes (e.g., Symbiodiniaceae) contain approximately 1-3 Gbp DNA in 10-100 chromosomes of 100-10 Mbp each. The largest dinoflagellate genomes contain about 250 Gbp DNA, likely distributed in 200-1000 chromosomes of 1.25 Gbp-250 Mbp each.

Genome sequencing combined with fluorescent *in situ* hybridization (FISH; Fig. 8.4A,B) revealed the telomere sequence in *F. kawagutii* as repeats of TTTAGGG, depicted as $(TTTAGGG)_n$, which is identical to typical plant telomere (Richards and Ausubel 1988). The same telomeric repeat has been documented for other dinoflagellates (Alverca et al. 2007, Fojtová et al. 2010). The length of the telomere tract was estimated at 25-80 Kbp in *Karenia papilionacea* and *Crypthecodinium cohnii* (Fojtova et al. 2010) but is unclear in other species. Using non-denaturing fluorescent *in situ* hybridization, Cuadrado et al. (2022) revealed bipolar distribution of clustered multiple foci of telomeric DNA: one in the nucleolus organization area and the other peripherally along the convex side of the nucleus, suggesting that chromosome configuration changes during cell division.

Dinoflagellate genomes are highly divergent. This is reflected in the genomes of Symbiodiniaceae, among which a > 98% dissimilarity is consistently found by mapping reads from one genome to the assembly of the other genome (Lin et al. 2015, González-Pech et al. 2021). Whether this extreme divergence reflect rapid genome evolution in dinoflagellates in general or severe isolation between symbiodiniacean species by their facultative symbiotic lifestyle remains to be unveiled.

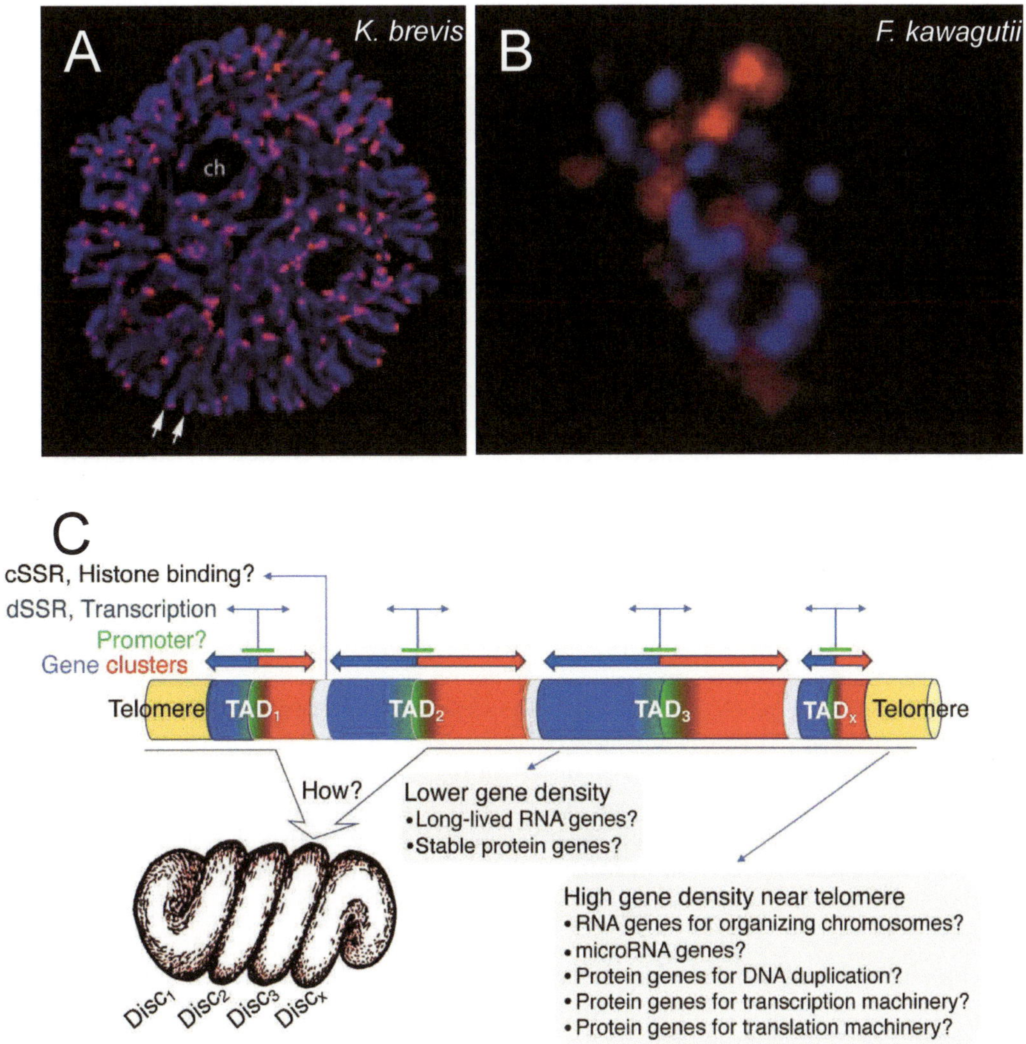

Figure 8.4. Distribution pattern of telomeres in dinoflagellates (A,B) and schematic of TADs in dinoflagellate chromosomes and associated gene distribution and expression machinery (C). Reproduced with permission from Cuadrado et al. (2019) under the Creative Commons CC BY 4.0 license for (A), Lin et al. 2015 for (B), and Lin et al. 2022 for (C).

Topologically Associated Domains (TAD)

Dinoflagellate genomes form topologically associated domains (TADs; Fig. 8.4C), a feature widely found in eukaryotic genomes (Nand et al. 2021, Marinov et al. 2021). Between the two species that have been examined, *S. microadriaticum* have about 450 TADs with a median size of 1.8 Mb whereas *B. minutum* contains 580 TADs with a median size of 5.7 Mb). These are markedly larger than mammalian TADs (~ 0.18 Mb) and bacterial CIDs (~ 0.17 Mb), and do not show interactions with more distant regions of the chromosome in the Hi-C maps, suggesting the absence of long-range looping interactions (Naumova et al. 2013). Furthermore, in these two genome assemblies, genes are distributed along the chromosomes as an alternating series of unidirectional clusters (Nand et al. 2021, Marinov et al. 2021). As shown in Fig. 8.4C, the TADs show two types of Strand Switch Regions (SSR). One, where values of the DNA contact probability plot are closest to the diagonal axis in Hi-C interaction maps, corresponds to the end of transcription (a convergent strand switch region, cSSR). The other, where regions of the TADs show DNA interactions furthest from the diagonal corresponds to the initiation of transcription (a divergent SSR, dSSR). Roughly half the genes in the genome are found in groups of at least nine all sharing a common direction. The TAD boundaries are also associated with decreased GC contents.

The newly documented dinoTAD structure raises the intriguing possibility that transcriptional promoters might be found uniquely in the central region of the TADs where genes are oriented outward in both directions (Fig. 8.4C). The limited number of these sites in the genome could represent sites for histone attachment, which could thus reconcile the conservation of histone sequences with the paucity of the proteins (Roy and Morse 2012). It would be of interest to test for the presence of DNA sequences in these potential promoter regions with DNA immunoprecipitated by anti-histone antibodies (ChIP). A closer examination on sequences in these regions might be fruitful in detecting common promoter motifs.

The average interaction frequency between two regions of the chromosome decreases when the distance between them increases. The trend in *S. microadriaticum* (Nand et al. 2021) suggests that DNA in this species is organized as linear rods. Comparing the trend between different organisms indicates that although Dinoflagellate nuclei presumably use Viral Nuclear Protein (DVNP) and bacterium-originated histone-like proteins more than histones, the chromosome organization appears more like eukaryotic chromosomes, with the highest similarity to mitotic chromosomes in metaphase (Lin et al. 2021) when the chromosomes are highly condensed (Nand et al. 2021).

The observation by Marinov et al. (2021) and Nand et al. (2021) that when the transcription was blocked, the TADs disappeared reveals a critical role for transcription in maintaining the TADs in dinoflagellate chromosomes. This is reminiscent of an old experiment in which treatment of dinoflagellate chromosomes by ribonuclease, but not pronase, led to the relaxation of condensed structure (Soyer and Haapala 1974a). It is also consistent with experiments in bacteria, where treatment with rifampicin largely eliminated the CID (Le et al. 2013). It has been suggested that RNA or RNA-protein complexes play an important role in stabilizing the super helical loops of dinoflagellate chromosomal DNA (Soyer and Haapala 1974b). However, the data from Marinov et al. (2021) and Nand et al. (2021) suggest it might not be RNA as much as the transcription activity (i.e., the transcription bubble or the machinery) that is involved in stabilizing the domain structure observed in the Hi-C experiments.

Mechanisms of Genome Expansion: Duplication, (Retro)Transposition, and Repetitive DNA

Dinoflagellate genome size varies vastly among species, from < 0.1 in the parasitic taxon *Amoebophrya*, < 3 Gbp for most coral symbionts from the family Symbiodiniaceae, to ~ 250 Gb for free-living taxa (Veldhuis et al. 1997, LaJeunesse et al. 2005, John et al. 2019). There is no phylogenetic trend, however, in the genome size variation (Fig. 8.5A). For instance, the coral symbiotic lineage Symbiodiniaceae has the smallest genomes among documented dinoflagellates (LaJeunesse et al. 2005) despite its later-diverging position in the phylogenetic tree (Janouškovec et al. 2017). The generally large sizes (compared to phylogenetically related lineages such as *Plasmodium falciparum* [23.3 Mbp] and *Perkinsus marinus* [86.8 Mbp]) and the wide range of dinoflagellate genomes suggest multiple independent events of genome expansion in the evolution of Dinophyta. Alternatively, genome contraction might have occurred occasionally, which is not necessarily due to symbiosis lifestyle (Shah et al. 2024).

Whole Genome Duplication (WGD) is the most efficient way to chromosome multiplication and genome expansion (Wolfe and Shields 1997, Crow and Wagner 2005). The immense dinoflagellate genomes might have resulted from WGD. However, K-mer profile and scaffold synteny (proxy of chromosome synteny) analyses for *F. kawagutii* did not show any sign of WGD (Lin et al. 2015). Nevertheless, the genome of *Polarella glacialis* (~ 3 Gbp) may have been duplicated, as it is about double the size of that in Symbiodiniaceae, and its k-mer profile shows double peaks characteristic of diploid genomes (Stephens et al. 2020) (Fig. 8.5B). "Diploid" like chromosome pairs have been

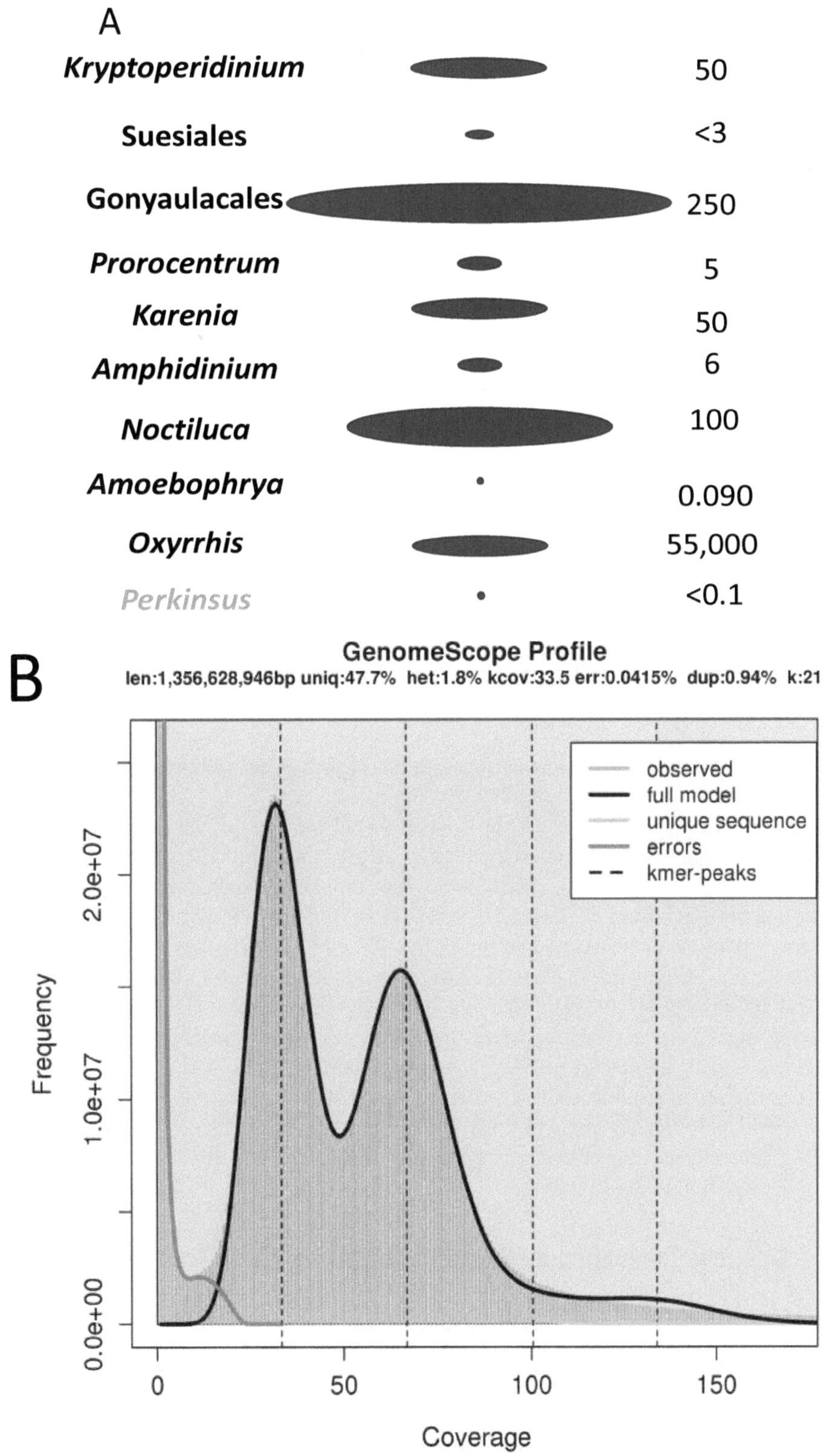

Figure 8.5. Evolutionary dynamics of dinoflagellate genome size showing no phylogenetic trend (A) and double peak in 21-mer profile of *Polarella glacialis* showing evidence of genome duplication (B). Part B is reproduced from Stephens et al. 2020 under the Creative Commons CC BY 4.0.

reported for *Pyrocystis lunular* as well (Seo and Fritz 2006). There is a good chance that WGD is widespread in the large dinoflagellate genomes.

Besides WGD, genome expansion can result from other processes such as proliferation of mobile DNA elements, including transposons and retrotransposons. The role of transposable elements in remodeling a genome has been demonstrated in plants (Chen et al. 2020). Indeed, there is evidence that transposons (John et al. 2019) and retrotransposons (Slamovits and Keeling 2008, Hou et al. 2019) play a role in dinoflagellate genome expansion. The role of retrotransposition is corroborated by the prevalent gene paralogs and DinoSL (complete or truncated) at the upstream region of genes which account for > 20% of total genes (Song et al. 2017), as well as the large families of reverse transcriptase and integrase genes in the genomes (Lin et al. 2015, Nand et al. 2021). In contrast, transposition is rare in the small genome (< 0.1 Gbp) of the parasitic species *Amoebophrya ceratii* (John et al. 2019). González-Pech et al. (2021) compared genome features of *S. tridacnidorum* and *S. natans*, and the high abundance of retrotransposition related genes led them to postulate that retrotransposition might be a major driving force for the two genomes to diverge.

Besides, Stephens et al. (2020) detected a substantial proportion of Long Terminal Repeat (LTR) elements (~ 12%) in *P. glacialis* genomes. Transposable elements (such as LTRs) commonly comprise up to 80% of the genomes in plants and are induced by genome shock and polyploidization, resulting in genome restructuring (Galindo-González et al. 2017). The abundance of LTRs in *P. glacialis* and the role of LTRs in genome restructuring may provide partial explanation for the difference in genome size between the two populations. In *S. microadriaticum*, heat stress-activated Ty1-copia-type LTR retrotransposons were detected, and evidence of recent expansion was documented (Chen et al. 2018), suggesting proliferation of retrotransposable elements for adaptation to hot habitats. The survival of the retroposed genes may in part be promoted by their methylation (de Mendoza et al. 2018), as DNA methylation is thought to silence and promote the transposition (Hu et al. 2020). Similar TE domestication by epigenetic regulation seems to be quite common in eukaryotes (Collens and Katz 2021), and methylation of transposal elements is presumably a major driving force of genome expansion in all eukaryotes (Zhou et al. 2020).

The abundance of repetitive DNA can contribute to genome size variations too. A work using FISH (Cuadrado et al. 2022) revealed clusters of dense trinucleotide repeats in the bean-shaped nuclei of *Karenia selliformis* and *K. mikimotoi*. Based on genome data, there seems to be a trend of larger dinoflagellate genomes containing more abundant repetitive elements (Stephens et al. 2020). For instance, repetitive elements account for 23.8% of the 0.1156-Gbp *Amoebophrya* A120 genome, 13.1% of the 0.1160-Gbp *Amoebophrya* A25 genome, 27.9% of the ~ 1.10-Gbp *S. microadriaticum* genome and 16% in the ~ 1.2-Gbp *Fugacium kawagutii* genome (Liu et al. 2018). For *P. glacialis* with genome sizes of ~ 3 Gbp, the level of repetitive elements increases to ~ 68% (Stephens et al. 2020). *Crypthecodinium cohnii* has a ~ 7 Gbp genome (Bhaud et al. 1991), 55–60% of its genome is composed of repetitive DNA (Allen et al. 1975). A survey of the ~ 115 Gbp *Alexandrium ostenfeldii* genome indicated that repetitive DNA accounted for ~ 58% of the genome (Jaeckisch et al. 2011). Overall, these results suggest that repetitive elements are significant contributors of genome size evolution in dinoflagellates.

Genomic Innovation by Rampant Horizontal Gene Transfer (HGT)

HGT might be an important driver of gene innovation in dinoflagellates (Wisecaver et al. 2013). Horizontal Gene Transfer (HGT) occurs widely in eukaryotes, accounting for 0.1–6.4% of total genes in the genomes, depending on species (van Etten and Bhattacharya 2020). Although the complete range of HGT in dinoflagellates awaits the sequencing of large genomes, there is ample evidence of HGT in the currently available data. Using EST data, Norsenko and Bhattacharya (2007) identified 16 proteins that were acquired through ancient HGTs in the common ancestor of the genera *Karenia* and *Karlodinium* and one protein derived through a more recent HGT. Eight of these appear to have resulted from independent HGTs in several eukaryotic lineages. These HGT-gained genes are highly concentrated in energy metabolism. Most of these were transferred directly from bacteria but some were originated from another chromalveolate that had acquired it from a bacterium. Some of the HGT-derived genes serve essential functions in dinoflagellates and deserve some in-depth discussion.

Genome sequencing for both the Antarctic (CCMP1383) and the Arctic (CCMP2088) strains *of Polarella glacialis* and comparison with other dinoflagellate genomes unveiled putative bacteria-to-dinoflagellate HGT for ice-binding domain-containing proteins (Stephens et al. 2020). Several bacterium-originated genes were identified in the plastid genomes of *Pyrocystis lunula* and *Ceratium horridum*, including FtsY, Rpl28, Rpl33, Yef16, and Ycf24 (Mackiewicz et al. 2013). *F. kawagutii* genome analysis also revealed 56 putative bacterium-originated HGT genes (0.12% of total gene models), of which 22 are flanked by eukaryotic genes at least on one side (Table S17, Lin et al. 2015). These included erythromycin esterase superfamily gene, radical SAM protein, and Phosphoadenosine phosphosulphate (PAPS) reductase. Erythromycin esterases (IPR007815) disrupt erythromycin by hydrolyzing the macrolactone ring, thereby confers antibiotic resistance (InterPro). Radical SAM proteins (IPR007197) catalyze diverse reactions, such as unusual methylations, isomerization, sulfur insertion, ring formation, anaerobic oxidation, and protein radical formation in the biosynthesis and in biodegradation pathways of DNA precursor, vitamin, cofactor, antibiotic and herbicide (InterPro). PAPS reductase (IPR002500) is part of the adenine nucleotide alpha hydrolases superfamily that also includes N type ATP PPases and ATP sulfurylases; this enzyme uses thioredoxin as an electron donor for the reduction of PAPS to phosphor-adenosine-phosphate (PAP) (InterPro).

Form II Rubisco

One of the most notable bacterium-originated HGT genes in dinoflagellates is Form II Rubisco (Rubisco II). Rubisco II only exists in peridinin-containing dinoflagellates besides anaerobic bacteria (Morse et al. 1995). Perhaps because of the anaerobic bacterial origin, Rubisco II is more sensitive to O_2 and is less efficient for CO_2 fixation than Form I Rubisco that is prevalent in oxygenic photosynthetic organisms. Rubisco II in dinoflagellates is encoded in the nuclear genome by arrays of tandem repeated genes, which are expressed as polyprotein, as demonstrated in *P. cordatum* and *P. shikokuense* (Zhang and Lin 2003, Shi et al. 2013).

Proton Pump Rhodopsin (PPR). PPR (also known as proteorhodopsin) is another remarkable bacterium-originated HGT genes in dinoflagellates. Initially found in prokaryotes from a metagenomic library (Béjà et al. 2000), PPR was later found to be widespread in the domain of bacteria throughout the global ocean (Béjà et al. 2001, Rozenberg et al. 2021), and now known to exist in various lineages of protists, including dinoflagellates (Lin et al. 2010), diatoms (Marchetti et al. 2012), and many other lineages. As a photoreceptor harvesting light energy, this is a transmembrane protein carrying an all-*trans* retinal as the chromophore. Dinoflagellate PPR has been shown to increase in response to dim light and starvation or phosphorus (P)-nutrient deficiency (Guo et al. 2014, Shi et al. 2015, 2017b). In *P. shikokuense*, PPR expression was upregulated by 40-fold in a bloom outbreak, when P nutrient was depleted (Zhang et al. 2019). Results from different studies consistently point to the role of PPR in providing energy for the dinoflagellates to cope with nutrient stress.

PPRs in dinoflagellates have been shown to be localized either on plasma membrane or endosomal membrane. PPR genes from *Prorocentrum shikokuense* (formerly *P. donghaiense*) and *A. catenella* fused with Green Fluorescence Protein (GFP) were expressed in HEK293 cell line, and the fusion protein showed cell membrane localization (Ma et al. 2020). *Oxyrrhis marina* possesses both proton pump and sensory rhodopsin (Slamovits et al. 2011, Guo et al. 2014). The PPR in this heterotrophic dinoflagellate has unique distribution; antibody staining showed its localization on endosomal membranes (Slamovits et al. 2011) as well as plasma membrane and membranes of vacuoles and other intracellular compartments (Rhiel et al. 2022, Westermann et al. 2023).

The multifocal distribution suggests that PPR has multiple roles, including ATP production and proton or pH modulation. In addition, some dinoflagellate genomes encode a variety of microbial rhodopsin genes that belong to different functional types. One of these is a giant ion channel encoded by a rhodopsin gene fused with a bestrophin gene encoding an ion channel, forming gigantic pentameric ion channels (Rozenberg et al. 2022).

Dinoflagellate Viral Nuclear Protein (DVNP)

Frequent viral infection (Nagasaki 2008, Correa et al. 2013, Veglia et al. 2023) and colonization in a non-lytic fashion (Wang et al. 2023) provides copious opportunities for invasion into dinoflagellate genomes. The recent detection of a Clustered Regularly Interspaced Short Palindromic Repeat (CRISPR)-associated Protein (Cas) in *Symbiodinium pilosom* (Li et al. 2024) suggests that dinoflagellates might have evolved an anti-viral defense mechanism similar to the CRISPR/Cas system in bacteria. Virus-to-dinoflagellate HGT can occur. The acquisition of dinoflagellate/viral nuclear protein or DVNP (Gornik et al. 2012) is hitherto the most prominent case of virus-to-dinoflagellate HGTs.

DVNP acquisition coincided with the disappearance of nucleosome, which occurred early in dinoflagellate evolution (Gornik et al. 2012). Heterologous expression of a DVNP in *Saccharomyces cerevisiae* showed that DVNP expression inhibited growth, diminished histones, disrupted nucleosomes, and impaired transcription (Irwin et al. 2018). The detriment was relieved through histone depletion. These results suggest that histone abandonment might have been an adaptive response to the ‘colonization’ of the ‘invading’ viral protein. The probable functional replacement of DVNP for histones in structuring chromatin (Gornik et al. 2019) testifies for the remarkable impact of HGT on the biology of the host.

Loss of Nucleosome and Gains of Nuclear Proteins

Chromosome structure is influenced by a suite of factors, most importantly DNA binding proteins, but also cations and RNA. Divalent cations, especially Mg^{2+} and Ca^{2+}, have been shown to be important in dinoflagellate DNA configuration by neutralizing the strong negative charges of the “naked DNA” (with little proteins) and achieving maximal chromosome compaction and structural configuration (Kearns and Sigee 1980, Herzog and Soyer 1983, Levi-Setti et al. 2008). Besides, RNA may be involved in DNA packaging (Levi-Setti et al. 2008, Soyer and Haapala 1974). Abundant RNA binding proteins were detected in the extracts of purified chromatin by high throughput proteomic analysis in *Lingulodinium polyedra* (Beauchemin and Morse 2017).

Dinoflagellate nuclei are characterized by 10-fold lower nuclear protein to DNA ratio (1:10) than typical eukaryotes (1:1), lack of nucleosome (Rizzo 1991) despite presence of core histones (Lin et al. 2010, Roy and Morse 2012). In typical eukaryotes, DNA wraps arounds a Histone (H) octomer (two each of H2A, H2B, H3, and H4), with the linker histone H1 serving as a “stake” fixing the DNA to the histone octomer, and with a 60-nucleotide interval (“sting”) between neighboring octomer units (“beads”). This “beads on a sting” architecture also occurs in most archaea but is absent in most bacteria. In few bacteria, histone proteins do occur, but they bind to DNA “head on” instead of being wrapped around by DNA (Hocher et al. 2023). Whether dinoflagellates do away with nucleosome completely, albeit never held to question previously, might need to be reexamined more closely. Some studies have suggested the occurrence of low amount of nucleosome (Marinov and Lynch 2016, Marinov et al. 2024).

Dinoflagellates appear to have also lost histone H1 (Wang et al. 2024). HCc is about 10% similar to H1, implying that HCc might be a functional replacement of H1 (Wong et al. 2003). The similarity could cause misidentification of HCs as H1, potentially explaining why H1 presence has been reported (Marinov and Lynch 2016). Stringent *in silico* studies suggest that H1 is absent in dinoflagellates (Wang et al. 2024). Experimental evidence is required to nail the issue.

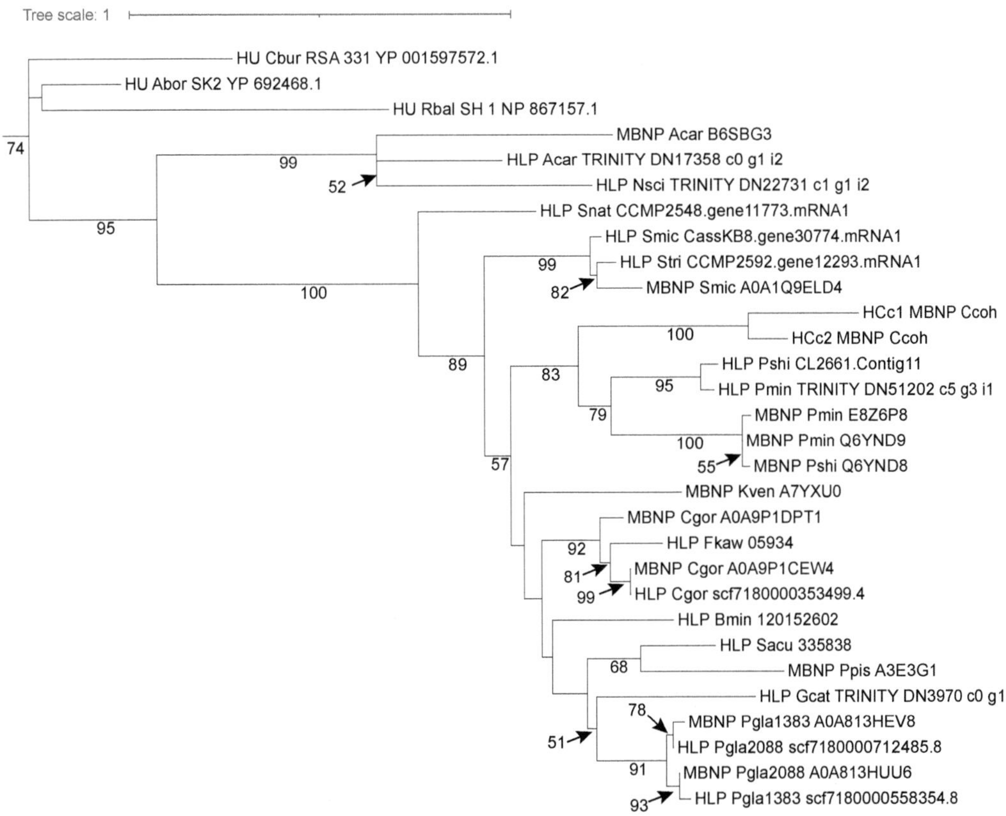

Figure 8.6. Phylogenetic tree of major basic nuclear protein (MBNP), histone-like protein (HLP), and bacterial DNA binding protein (HU)-like protein (Hcc).

While abandoning (entirely or not) the canonical role of histones in chromosome organization, dinoflagellates have acquired alternative proteins through Horizontal Gene Transfer (HGT). One of these proteins is the Major Basic Nuclear Protein (MBNP), the coding gene of which is one of the most highly expressed genes in dinoflagellates (Lin 2011). Two related proteins have been abundantly reported: Histone-Like Proteins (HLP) and the bacterial DNA binding protein (HU)-like protein originally identified in the *Crypthecodinim cohnii*, hence named Hcc, which was credited for a role in supporting liquid crystalline chromosome structure (Wong et al. 2003, Chan and Wong 2007, Wong 2019). Whether these are synonyms of the same gene family was recently examined (Lin 2024). A phylogenetic analysis including all three proteins (Fig. 8.6) shows that, they do not form monophyletic groups and hence cannot be separated. This suggests that MBNP, HLP, and HCc belong to the same protein family, and their divergence have stemmed from duplications followed by mutations. Thus, unifying the names is warranted. The use of MBNP to represent this gene family has been proposed (Lin 2024).

Gene Structure: Unidirectionality and Intron Dynamics

In bacteria, genes are largely organized in operons and thus tend to be unidirectionally oriented. Eukaryotic genes, in contrast, are organized bidirectionally due to recombination (López et al. 2010). Unlike typical eukaryotes, dinoflagellate genes are dominantly unidirectionally arranged in the genome, forming co-oriented clusters (Shoguchi et al. 2013, Lin et al. 2015, Dougan et al. 2023). This characteristic is widespread in dinoflagellates, including the basal taxa *Amoebophrya* spp. (Farhat et al. 2021). Such an organization might facilitate transcription in unison and in foci on certain areas of the TADs (Nand et al. 2021, Marinov et al. 2021, Lin et al. 2021).

Contrary to previous belief, dinoflagellate genes are not intron-depleted. Even in streamlined parasitic *Amoebophrya* spp. genomes, 67–71% of genes contain introns, with 64 to 99% in Symbiodiniaceae species (Table 8.1). The ~ 6.4 Gbp *A. gibbosum* genome has fewer introns per gene but longer total intronic regions compared to smaller dinoflagellate genomes. Average intron lengths vary: 335–345 bp in *Amoebophrya* spp., 505, 517, and 893 bp for *S. microadriaticum*, *B. minutum*, and *F. kawagutii*, respectively (Farhat et al. 2021). Notably, a 26,372 bp intron was detected in *C. cohnii* (Mendez et al. 2015).

Intron-exon boundaries in dinoflagellates also differ from typical eukaryotes (Farhat et al. 2021). Besides the canonical GT-AG boundary, dinoflagellates use GCIGA-AG (25 to 74% in Symbiodiniaceae; < 3% in *Amoebophrya*), and other boundaries (up to 67% in *Amoebophrya*). Interestingly, 99.98% of introns in *Amoebophrya* strain AT5 have canonical GT-AG splice sites, while over 60% in strains A25 and A120 have non-canonical splice sites.

Intron creation may be driven by introners (specialized transposable elements). Analysis of five dinoflagellate genomes revealed: (1) evidence of historical intron creation, (2) recently active introners in four of the five species, (3) dynamic intron gain and loss, with higher gain rates than most eukaryotes, and (4) In *P. glacialis*, 12,253 introns from recent introner insertion, with 15 distinct families accounting for at least 100 introns each (Roy et al. 2023).

High Gene Content, Complex Organization, and Abundant "Dark" Genes

High Gene Content

Using genome information available in 2008, which was limited to genomes smaller than ~ 2 Gbp, a predictive correlation between the number of protein-coding genes and genome size was developed: log (gene number)=LN (–46.200+22.217*log[genome size]), where genome size is in Kbp (Hou and Lin 2009). This predictive equation provides a tool for predicting gene numbers and a baseline with which observed gene numbers can be compared to assess trends of gene density in genomes. This equation predicts 38,188 to 87,688 protein-coding genes for dinoflagellate with genomes of ~ 2 to 250 Gbp. However, the number of identified genes already reached 85,139 even though the largest sequenced dinoflagellate genome was only 6.3 Gbp (Table 8.2). Even for the parasitic species *Amoebophrya ceratii* AT5.2, the equation predicts only 14,180 genes for its 87.7 Mbp genome, yet genome annotation identified 19,925 genes, 26% above the predicted number. Most of the genomes examined encode more genes than predicted, except for *Hematodinium*, which encodes 30% less than predicted (Table 8.2). These discrepancies suggest that dinoflagellate genome expansion disproportionately favored gene increase compared to other eukaryotes.

Shared and Unique Genes and Gene Families

Current genome sequencing efforts for dinoflagellates have predominantly focused on species with smaller genomes, which hinders the comprehensive identification of genes that are ubiquitous across the entire phylum. Nevertheless, comparative genomic analyses still provide interesting insights into the differential dynamics and diversification of dinoflagellate genomes. González-Pech et al. (2021) analyzed 15 Suessiales genomes (Symbiodiniaceae and *Polarella glacialis*) and found 555,682 predicted protein sequences in total, 42,539 of which formed gene families. Out of these families, 2,500 (5.88%) have homologs in all 15 Suessiales isolates. This result indicates that most gene families are specific to certain families or genera. Interestingly, more gene families (8,828) were found to be specific to sequenced isolates of *Symbiodinium* than to sequenced isolates of all the other Symbiodiniaceae combined. This may be because substantially more gene families have been gained (or preserved) in *Symbiodinium* than in the other three genera, but it may also be due to the undersampling of genera outside *Symbiodinium*. When the comparison was narrowed down to within Symbiodiniaceae, they found 3,445 unigenes (non-redundant) genes shared by the eight

Table 8.2. Comparison of identified gene number with predicted gene number based on Hou and Lin (2009).

Species	Genome size (Mbp)	Detected gene no.	Model-predicted gene no.	% deviation*
Parasitic				
Amoebophyra ceratii AT5.2	120	19,925	15834	26%
Amoebophyra ceratii A25	116	28,091	15656	79%
Amoebophyra ceratii A120	115.5	26,441	15633	69%
Hematodinium sp.*	4800	29,510	42412	-30%
Symbiotic				
Breviolum minutum	616	41,925	25830	62%
Cladocopium goreaui	1,200	45,322	30710	48%
Durusdinium trenchii	670	30,000	26419	14%
Fugacium kawagutii	935	45,192	28828	57%
Symbiodinium microadriacticum CCMP2467	808	49,109	27758	77%
Free-living				
Effrenium voratum RCC1521	1,200	32,108	30710	5%
Effrenium voratum CCMP3420	1,300	39878	31328	27%
Effrenium voratum CCMP421	1,100	32,615	30046	9%
Polarella glacialis CCMP1383	2,980	58232	38147	53%
Prorocentrum cordatum	4,750	85,849	42316	103%

* (detected-modeled)/modeled x 100

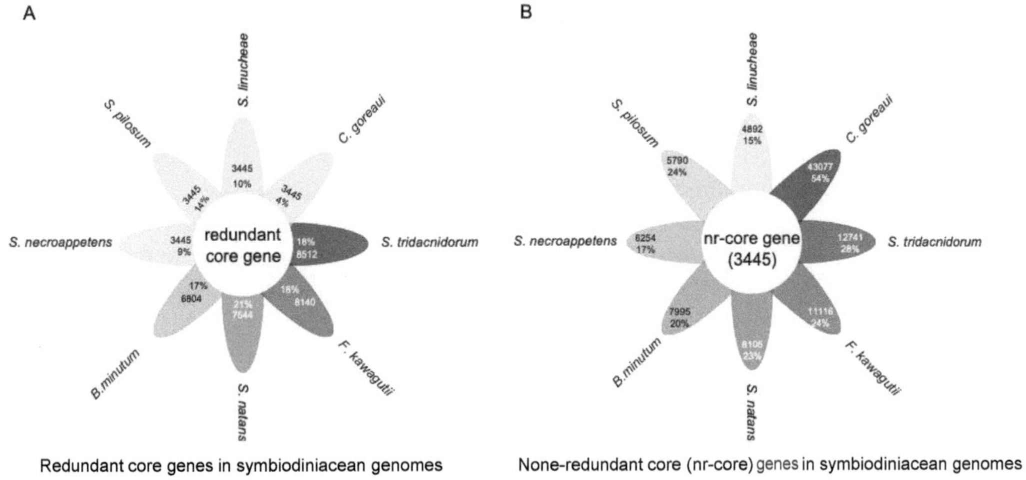

Figure 8.7. Shared and unique genes in Symbiodiniaceae genomes. (A) Core genes (3,445 after de-redundancy) in proportion of total genes in each species. (B) Unique genes in each species. Darker colors indicate a larger number of genes.

symbiodiniacean species covered in the analysis (Fig. 8.7). These genes may represent the core genes of Symbiodiniaceae. These common genes are enriched in 219 KEGG pathways, including various basic metabolism such as ribosome, signal transduction, photosynthesis, and DNA repair.

Highly Expressed Genes

Previous transcriptomic data from several dinoflagellate species showed that the most highly expressed genes included MBNP, glyceraldehyde-3-phosphate dehydrogenase (GAPDH), Peridinin-Chl *a*-binding Protein (PCP), heat shock protein (HSP) 90, HSP70, HSP40, elongation factor EF-1α,

S-adenosyl-L-methionine (SAM) synthetase, A-adenosyl-homocysteine hydrolase (SAHase), and calmodulin (Lin 2011). With the accumulation of more data, it has become evident that the highly expressed genes exhibit more inter-specific differences (Table 8.3). Now that MBNP is unified with Hcc and HLP, its high expression will likely be found more widespread.

Pfam analysis for *A. gibbosum* genome revealed Leucine-Rich Repeat (LRR), Ankyrin, Tetratricopeptide (TPR), and Pentatricopeptide repeat (PPR) domains as the most abundant domains in this species (Beedessee et al. 2020).

Tandem Gene Arrays and Polyprotein Expression

Numerous genes in dinoflagellates occur in multiple copies (paralogs) arranged in tandem arrays (Lin 2011). These arrays can form through unequal crossing over during meiosis, replication slippage, retroposition, DNA transposons, and gene conversion (Zhang 2003). Some tandem-arrayed genes are transcribed into multi-gene transcripts that are then *trans*-spliced with a 22-nt highly conserved spliced leader (DinoSL; Zhang et al. 2007), producing individual mature mRNA with a DinoSL at the 5' end and a poly A tail at the 3' end. This type of genes is polycistronic. Other arrays are expressed as polyproteins: each array is transcribed as a poly-gene transcript, which is translated into a polyprotein peptide, and the polyprotein is then cleaved into individual proteins. Genes expressed as polyproteins in dinoflagellates documented to date include the chlorophyll *a-c* binding protein (Hiller et al. 1995), luciferase (Li et al. 1997), as well as Rubisco II (Zhang and Lin 2003, Shi et al. 2013). In *P. cordatum*, each Rubisco polyprotein contains four Rubisco proteins, which is guided to the chloroplast by a signal peptide at the upstream N-terminus. At the chloroplast, the polyprotein is cleaved into individual Rubisco proteins, each with a transit peptide to facilitate plastid entry (Zhang and Lin 2003, Lin 2011). A similar structure is found in *P. shikokuense* Rubisco II (Shi et al. 2013). Recent genome sequencing has also identified polyprotein gene arrays (Dougan et al. 2023). Despite the prevalence of tandem repeats, polycistronic expression is rare, as shown in a transcriptomic analysis of *L. polyedrum* (Beauchemin et al. 2012).

Polyprotein genes such as Rubisco II likely resulted from retroposition, because the gene cluster contains only one regulatory system and lack introns, characteristic of retroposed genes (Zhang 2003). Polyprotein expression, though less common in eukaryotes than viruses, has notable examples like the TY1 retrotransposon in yeast, certain plant storage proteins, and mammalian prohormones. Viruses can express both polycistronic and polyprotein genes; some viruses use polycistronic mRNAs, while others use polyproteins. Further research is needed to understand the evolution and functional distinction of tandem repeats and polyprotein genes in dinoflagellates.

"Dark" Genes

In sequenced dinoflagellate genomes, about 50% or fewer of the predicted genes match functionally annotated genes in genomic databases. The remaining genes fall into the "unknown function" category. This is consistent for both the smallest dinoflagellate genomes, like those in *Amoebophrya* strains (over 50% dark genes; Farhat et al. 2021), and the largest ones sequenced so far, such as *P. cordatum* (52.2% dark genes; Dougan et al. 2023) and *A. gibbosum* (51.7% by TrEMBLE annotation; Beedessee et al. 2020). While some of these "dark" genes are common across dinoflagellate taxa, most are specific to individual genera or species (Stephens et al. 2018, González-Pech et al. 2021).

Some dark genes may have crucial functions related to bloom formation or other ecological and biological phenotypes. For example, a bloom metatranscriptomic study identified six unknown genes that were differentially expressed between bloom and non-bloom stages (Zhang et al. 2019). Four of these genes exhibited a 10- to over 100-fold upregulation during the bloom stage, suggesting they may encode key enzymes promoting cell proliferation. Conversely, two dark genes were significantly downregulated during the bloom stage, indicating they might act as negative regulators of growth.

Table 8.3. Ten most highly expressed genes in dinofalgellates.

Species	Ranking											References
	1	2	3	4	5	6	7	8	9	10	11	12	
Non-Symbiodiniaceae	MBNP	GAPDH	PCP	HSP90	HSP70	HSP40	EF-1α	SAMS	SAHase	CaM			Lin et al. 2011
Cladocopium goreaui	PCP	atpE	unk1,2	CaM	MBNP	FAM135A	unk3	PSMG1	psbD	unk4	unk5	RPL28e	Huang and Lin unpubl
Durusdinium trenchii	PEX16	EFG	GAPDH	UBE	USP7	PCP	unk8	psbJ	EF-1α	petJ	unk9	TUBB	Huang and Lin unpubl
Effrenium voratum	PCP	psbA	petJ	unk10	petF	unk11-12	LHCP	unk13	RPL	unk14-16	HSP40	Ub	Li T et al. 2021
Fugacium kawagutti	psbA	petF	NRT2.5	PCP	unk18-19	psbV	rbcL	PGK	psbC	psaA	LHCP		Li et al. 2020
Karenia mikimotoi	HLP	MBNP	LHCP9	LHCP12	Ub	unk20	LHCP4	LHCP	LHCP8	unk21	actin	unk22	Lin et al. 2022
Karenia mikimotoi	unk23	psbA	unk24	SAG-L	MBNP	HLP	psbD	psaB	atpH	petB	rbcL	psbV	Lin et al. 2022
Prorocentrum shikokuense	SAG-L	unk25-26	LSU-rRNA	unk27-28	SKP1	MBNP2	psbA	psbE	psbD	PCP	unk29-30	petB	Zhang et al. 2019
Prorocentrum shikokuense	unk25	SRm300	psbA	unk31-32	psbD	unk33-34	MBNP2	RPS15A	RPS30	unk35-37	RPL36	RHO	Yu et al. 2020
Prorocentrum shikokuense	MBNP2	SRm300	PCP	unk41	FCP	unk42	atpE	unk34	RHO	MBNP	HSP40	petB	Li H et al. 2021

Gene Expression and Regulation

Dinoflagellates are generally known to have limited transcriptional regulation (5–27% of genes; for review see Lin 2011). However, significant transcriptional changes occur in response to important environmental and physiological fluctuations such as nutrient changes and diel cycles. For example, alkaline phosphatase gene expression and enzyme activity increase under phosphate limitation (Lin et al. 2012, Morey et al. 2011, Shi et al. 2017b, Luo et al. 2017). Nitrogen deficiency elicits significant upregulation of acquisition genes and downregulation of nitrogen associated protein gene (NAP50) (Morey et al. 2011, Zhuang et al. 2015). In addition, Rubisco, antenna protein, and some cell cycle protein genes show strong diel dynamics of transcript abundance (Shi et al. 2013, 2017a). Earlier studies indicated that several percentages of genes exhibited circadian rhythms of transcript abundance at 2- or more fold variations (Okamoto and Hastings 2003). These findings suggest that dinoflagellates may have more dynamic transcriptional regulation than previously thought, particularly in response to environmental changes and daily cycles.

General Gene Expression

General transcriptional machinery in eukaryotes includes RNA polymerase II transcription initiation factors (TFII), TFIIA, TFIIB, TFIID, TFIIE, and TFIIH. These factors assemble on promoter DNA alongside polymerase II to form a multiprotein-DNA complex crucial for precise initiation of basal-level transcription (Zaheri and Morse 2022). Additionally, transcriptional activators may bind to the enhancer region and coactivators may mediate interactions between enhancer-bound activators and promoter-bound gene-specific Transcription Factors (TFs), regulating transcription responses to environmental cues (Nikolov and Burley 1997).

Of the five TFII genes, only one TFIID and three TFIIH genes were identified in the *F. kawagutii* genome (Lin et al. 2015, Zaheri and Morse 2022). Dinoflagellates lack the typical eukaryotic TATA box (consensus sequence TATAAA) that interacts with TATA-Binding Protein (TBP), but instead use TTTT interacting with a TBP-like protein (TBL), as previously proposed and confirmed by genome data (Guillebault et al. 2002, Lin et al. 2015).

Cold Shock Domain (CSD)

CSD-containing proteins (CSPs) were initially identified in *Escherichia coli* during cold shock stress and are now known to occur in all major groups of organisms. In prokaryotes, CSPs act mainly as RNA chaperones and can regulate transcription by binding to the gyrA promoter. In eukaryotes, CSPs usually contain additional domains and play roles beyond cold stress response, such as coping with nutrient limitation and regulating growth. In plants, CSPs are essential for freezing tolerance and translation regulation during cold stress, seed germination, and flowering (Sasaki and Imai 2012).

CSPs are ubiquitous in dinoflagellates. In *L. polyedrum*, CSP genes were the only gene family with higher transcriptional levels than in other protists (Beauchemin et al. 2012), suggesting their significant role. Some LpCSPs contain only the CSD domain while others have additional domains, but neither showed cold induction (Beauchemin et al. 2016). In *F. kawagutii*, at least eight of 121 identified transcription factors (TFs) contain CSD (Skav204535, Skav204536, Skav217190, Skav218283, Skv218284, Skv223430, Fkav224338, Skv231215), suggesting that CSPs may be the functional domains of TFs in dinoflagellates.

CSPs were identified in *Scrippsiella acuminata*, showing increased expression in resting cysts (Deng et al. 2018). *P. cordatum* has two CSD genes that respond significantly to low temperatures (Wang et al. 2021). Dinoflagellate CSPs contain a ZF domain and a glycine-rich motif besides the conserved CSD, with varying ZF and GR order between species. This organization resembles bacterial and plant CSPs, suggesting possible bacterial origin through HGT.

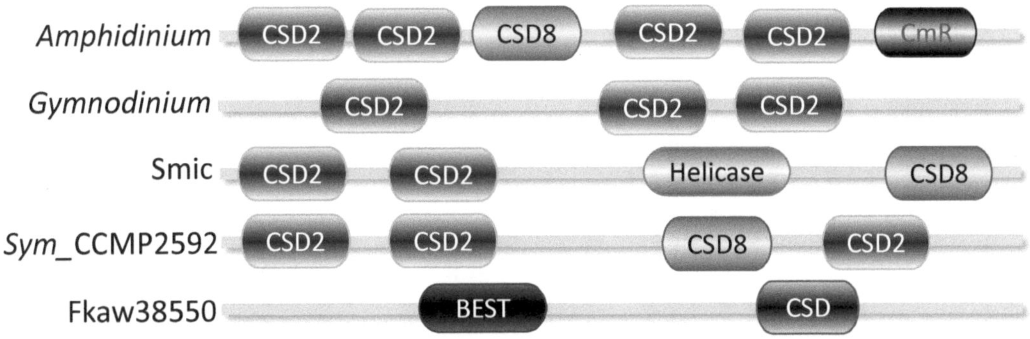

Figure 8.8. Organization of cold shock domain (CSD) in various proteins in dinoflagellates. *Amphidinium, A. massarti*, data source MMETSP0689; CSD2, domain PS51857; CSD8, domain SM00357; CmR, Calcium regulated mRNA-binding domain (IPR052069); *Gymnodinium, G. catenatum*, data source MMETSP0784-20121206; Smic, *Symbiodinium microadriaticum*, data source CAE7767176.1 dbp2, Helicase = helicase superfamily ATP-binding domain; Sym_CCMP2592, *Symbiodinium* strain CCMP2592; Fkaw38550, *Fugacium kawagutii*, BEST = bestrophin.

DNA and RNA binding assays showed dinoflagellate CSD proteins prefer binding RNA over DNA, and single-stranded over double-stranded DNA, without sequence specificity (Zaheri and Morse 2022). This suggests they are unlikely to act as transcription factors but may be involved in DNA unwinding. Some CSPs contain calcium-regulated mRNA-binding or helicase ATP-binding domains in certain species (Fig. 8.8).

MicroRNA (miRNA)

The scarce and weak transcriptional regulation in dinoflagellates suggests other mechanisms must exist. MicroRNA (miRNA) is one of the widely existing post-transcriptional epigenetic mechanisms in eukaryotes. MiRNA repertoires have been documented in the transcriptomes of *S. microadriaticum* (Baumgarten et al. 2013) and *P. shikokuence* (Shi et al. 2017b, Yu et al. 2020) and the genome of *F. kawagutii* (Lin et al. 2015). Functional miRNA machinery in dinoflagellates is evidenced by effective RNA silencing (Zhang and Lin 2019). *A. gibbosum* genome analysis suggests miRNA involvement in regulating responses to nitrogen and phosphate starvation (Beedessee et al. 2020).

In silico analyses of *F. kawagutii* miRNAs have indicated potential influence on coral host gene expression (Lin et al. 2015). The genome contains a putative RNA transporter homologous to Systemic RNA Interference Deficiency–1 (SID-1) in nematodes and mammals. The similarity between Symbiodiniaceae and cnidarian homologs suggests possible HGT (Fig. 8.9A). Potential target genes of symbiodiniacean miRNA were identified in coral hosts, showing broad interactive protein network (Fig. 8.9B) and GO terms (Fig. 8.9C).

These findings suggest Symbiodiniaceae microRNA may cross over into the invertebrate host and modulate gene expression, potentially impacting symbiosis mechanisms like nutrient uptake and photosynthate translocation. While remaining to be experimentally demonstrated, this putative host-symbiont cross-regulation mechanism is consistent with findings in other symbiotic and pathogenic systems. *Vibrio fischeri* produces noncoding small RNA that controls host gene expression in squid light organs (Moriano-Gutierrez et al. 2019, 2020). Pathogen-to-host miRNA transfer silences host immunity genes in plants (Weiberg et al. 2013).

Alternative Splicing

Genome analyses have detected alternative splicing in dinoflagellates. In *A. gibbosum*, a high number of alternative splicing events were identified, potentially responsible for gene diversification, particularly in secondary metabolite biosynthesis pathways (Beedessee et al. 2020). The majority (77%) of the alternative splicing events occur by skipping exons, and the rest uses Alternative

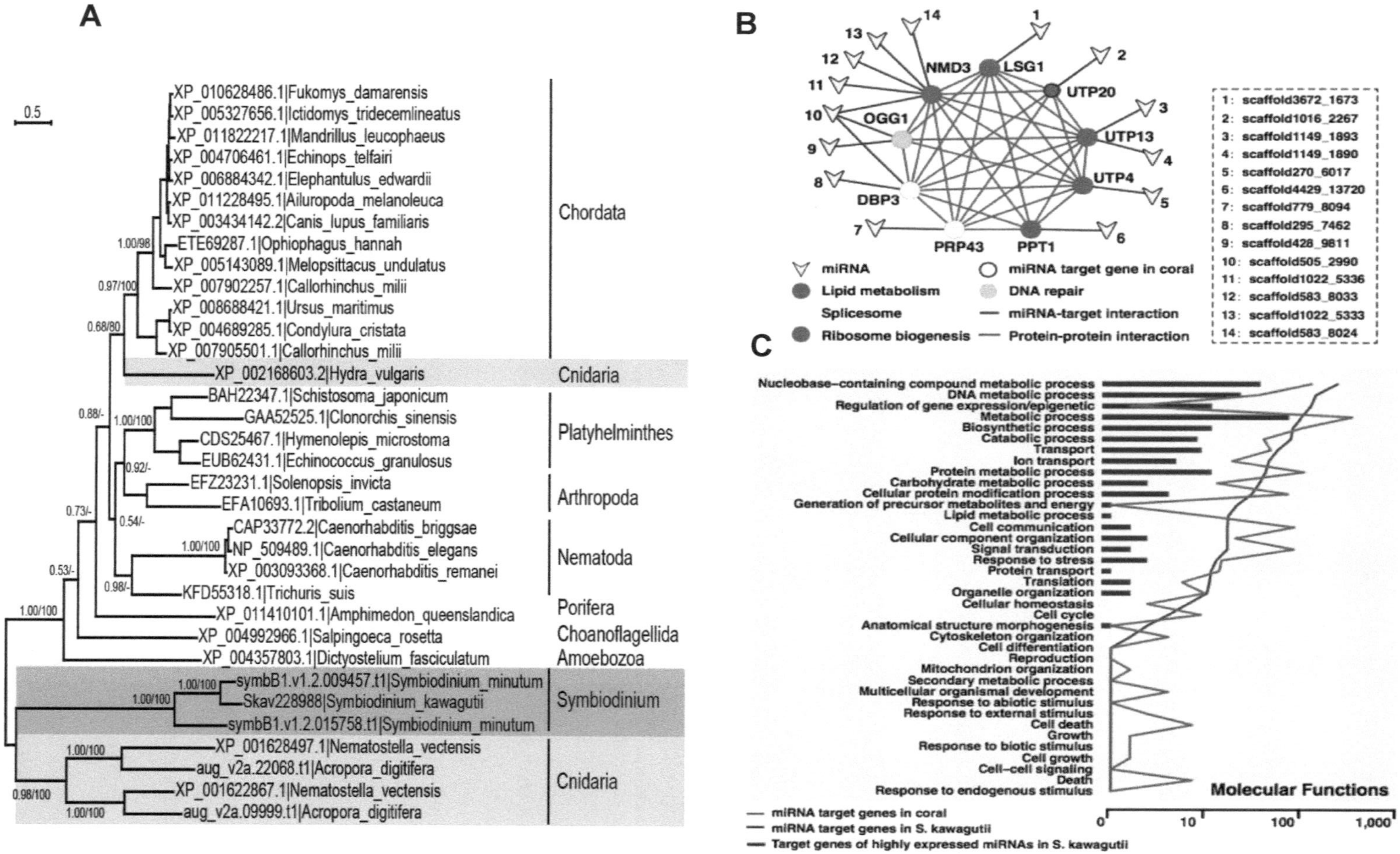

Figure 8.9. Evidence that Symbiodiniaceae miRNA may interact with coral host to modulate host gene expression. (A) Phylogenetic tree of double-stranded RNA channel (SID-1). The close relationship between the homologs in *Symbiodinium* and the cnidarians suggests horizontal gene transfer between the two lineages. (B) Protein interaction network and metabolic pathways predicted to be impacted by miRNA. C) GO categorization of predicted miRNA target genes in both *F. kawagutii* and coral. Reproduced with permission from Lin et al. (2015).

3′Splice Sites (A3SS) or Alternative 5′Splice Sites (A5SS) (6.8 and 11.3%, respectively). These events mainly impact ion transport, nucleic acid metabolism, and RNA metabolic process. Notably, polyketide synthase (PKS) genes have alternatively spliced isoforms and the genes are expressed as polycistronic. Alternative splicing is also suspected in *F. kawagutii* and other dinoflagellates (Stephens et al. 2018, Li et al. 2020). However, the extent and function of alternative splicing in dinoflagellates remain largely unknown.

Translational Regulation

Post-transcriptional gene regulation is widely suspected for dinoflagellates but experimental data are sparse. A recent study using ribosome-protected mRNA profiling revealed pulses of synchronous translation of gene clusters throughout the diel cycle (Bowazolo et al. 2022), indicating circadian or light/dark modulation of gene expression. Pearson correlation analysis of rRNA count versus mRNA count using data reported in that study shows that on average, translation of an mRNA requires 3.6 ribosomes (Fig. 8.10). Deviations from this may indicate above or below average translational efficiency, with exceptionally high translation activity occurring at LD20 (four hours before light-on).

A study of 11 core dinoflagellate transcriptomes revealed diversity in the eIF4E family (Jones et al. 2015). Each species contained 8–15 (average 11) different eIF4E family members, phylogenetically segregated into three clades. Dinoflagellate eIF4E-1 likely contains at least one translational initiation factor. eIF4E-2 lacks mRNA cap-binding residues and may have a non-essential regulatory role. eIF4E-3 genes share attributes with metazoan Class I eIF4E. The three clades were further divided into nine subclades, six containing members from all eleven species, suggesting ancestral duplication.

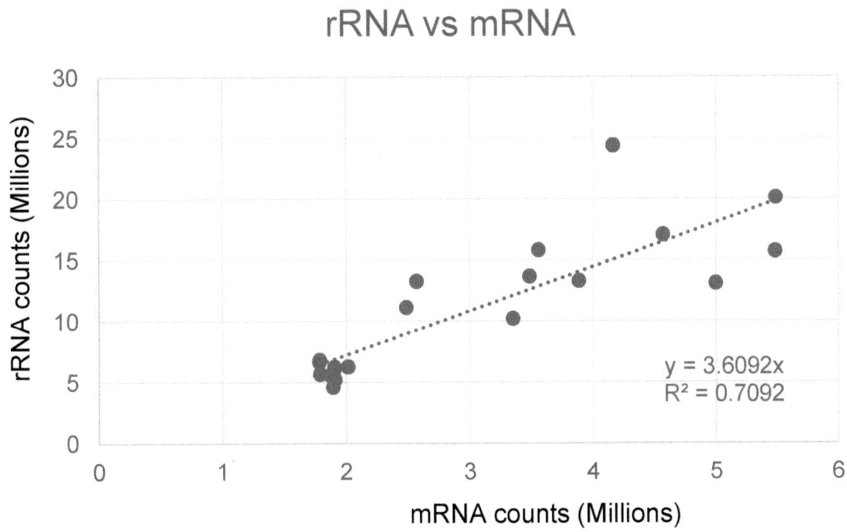

Figure 8.10. Correlation between ribosomal rRNA and ribosome-bound mRNA abundances in *Lingulodinium polyedra*. Data from Bowazolo et al. (2022).

Powerhouse of RNA Editing

RNA editing is a gene modification process where single or multiple nucleotides in a gene are altered during or after transcription. This can be identified by comparing the nucleotide sequence of a gene with its corresponding mRNA. RNA editing occurs in mRNA, tRNA, and rRNA, involving mitochondrial, plastid, and nuclear genes (Gray 2012).

In dinoflagellates, RNA editing was first documented in mitochondrial protein-coding gene transcripts, showing extensive and mechanistic diversity (Lin et al. 2002). Mitochondrial mRNA editing exhibits an increasing trend in terms of the percentage of impacted nucleotides throughout dinoflagellate evolution (Lin et al. 2008). Most nucleotide changes are synonymous and do not affect the encoded proteins but tend to increase the GC content of the impacted genes. This may enhance translation efficiency since dinoflagellate mitochondria import cytoplasmic tRNA for translation, which may not function well with the AT-rich mitochondrial genes (Lin et al. 2008). This explanation, however, would not apply to rRNA editing.

The field of RNA editing in dinoflagellates has advanced significantly in the last decade. Following the discovery in mitochondrial mRNA, editing was also found in chloroplast-encoded transcripts, including mRNA and rRNA (Dang and Green 2009, Mungpakdee et al. 2014). For coral symbiotic Symbiodiniaceae, RNA editing was even detected in *in-hospite* populations (Shoguchi et al. 2020). Interestingly, mitochondrial gene transcripts appear to be more frequently edited than plastid genes (Shoguchi et al. 2020). RNA editing occurs not only in typical peridinin-containing dinoflagellate plastids but also in Chl *b*-containing plastids such as that in *Lepidodinium* spp. and fucoxanthin-containing tertiary plastids of haptophyte origin such as that in *Karlodinium veneficum* (Matsuo et al. 2022, Dorrell and Howe 2012, Jackson et al. 2013). No RNA editing has been reported for plastid genes in closely related Chl *b*-containing algae or haptophytes. This, along with the accelerated evolution of plastid protein-encoding genes in peridinin- and fucoxanthin-containing species, indicates a significant host effect on plastid proteome (He et al. 2024).

Additionally, using high-throughput sequencing, mRNA editing has been detected in nuclear genes. Up to 10,486 and 69,953 putative RNA editing sites were identified in the nuclear genomes of the coral symbiont *D. trenchii* CCMP2556 and the free-living bloom-forming species *P. cordatum* CCMP1329, respectively (Dougan et al. 2023). These events included all 12 possible types of RNA edits, with more transitions than transversions and a dominance of A-to-T transversion in noncoding regions, many of which were condition-specific. Furthermore, A-to-T editing within untranslated regions appears to be associated with the upregulation of edited genes under heat stress (Chen et al. 2024). If experimentally verified, the high number of RNA editing events suggests that RNA editing may be a more influential molecular mechanism regulating gene expression in dinoflagellates than previously thought.

RNA Methylation

RNA methylation is one of the > 170 post-transcriptional RNA modifications (Boccaletto et al. 2021, Wiener and Schwartz 2021), including methylation of adenosine, cytosine, and guanosine (Liu and Pan 2015). Methylation protects the 3'-UTR of mRNA from hydrolysis, enhancing stability and lifespan. It also stabilizes the 3'-endo conformation, improving interactions with ribosomal and RNA-binding proteins, thus influencing mRNA translation (Wang et al. 2014).

While methylation can occur at different positions on cytosine (m^5C) and guanosine (m^7G), adenosine methylation at N^1, N^2, N^6, or the 2' oxygen of ribose is most common (Zaccara et al. 2019, Boulias and Greer 2023). These forms of RNA methylations are regulated by proteins called writers, erasers, and readers that recognize modification sites, affecting RNA fate (Zhu et al. 2020). Proteins that bind modified bases are readers. YTH-domain proteins function as readers in animals and yeast by selectively removing meiosis-specific transcripts in mitotic cells (Harigaya et al. 2006). In plants, YTH-domain proteins like ECT2, 3, 4, and 8 serve as m^6A readers (Arribas-Hernández et al. 2018). YTHDF1 binding to elF3 or direct interactions with METTL3, a methyltransferase-like protein. METTL3 was found to be responsible for m^6A methylation and plays a role in developmental processes in *Arabidopsis*.

Both m6A and m1A methylations were recently detected in dinoflagellates and found to probably affect translation efficiency under nutrient stress (Li et al. 2024). BLAST searches against the existing dinoflagellate genomes revealed widespread presence of ECT8 like protein genes, encompassing species from parasitic ancestral taxon *Amoebophrya* (Gene ID AD7963638), the Gymnodiniales species *Karenia brevis* (Karenia-brevis-CCMP2229-20130916|33997_1), the Prorocentrales *Prorocentrum shikokuense* (Pdh.Unigene68085_All), to the symbiodiniacean taxa *Cladocopium* (CAL1140513) and *Durusdinium* (CAK9055763). The roles of their encoded enzymes and the presence of other genes responsible for RNA methylation remain to be investigated.

Chloroplast Genome

Nucleomorphs

Nucleomorphs, relic nuclei of eukaryotic endosymbionts, seem to occur in some dinoflagellate chloroplasts. Previously known only in the phylum Cryptophyta (Curtis et al. 2012), a recent study discovered nucleomorphs in two newly described dinoflagellate species with green chloroplasts (Sarai et al. 2020). These chloroplasts, like those in *Lepidodinium* spp., originate from prasinophytes, although *Lepidodinium* spp. do not contain nucleomorphs.

Minicircles and Macrocircles

Typical dinoflagellate chloroplasts contain a genome composed of minicircles (Zhang et al. 1999). Each of these circles mostly contains one or two genes but some are empty. This type of dinoflagellate chloroplast genomes encodes a total of 16 or 17 genes (Fig. 8.2). However, the operation of the chloroplast requires between 100 and 200 proteins, indicating that the majority of the plastid genes have been lost to the nuclear genome. How many proteins are involved in the function of the typical dinoflagellate chloroplasts is still unclear.

The plastid genomes of two dinotoms, *Durinskia baltica* and *Kryptoperidinium fociaceum*, have been sequenced and found to be circular genomes of 116.470 kbp and 140.426 kbp, respectively (Imanian et al. 2010). The organization and gene content of the *D. baltica* plastid genome are similar to those of the pennate diatom *Phaeodactylum tricornutum*, whereas the *K. foliaceum* plastid genome has undergone more extensive reorganization and is more similar to the plastid of centric diatoms.

Photosystem Structure and Novel Proteins

Dinoflagellates have conserved photosystem cores (PSI and PSII) but varied light-harvesting complexes (LHCI and LHCII). Dinoflagellates have 12 subunits in PSI, ten of which are shared by all photosynthetic organisms, and 24 in PSII, 20 of which are shared by all photosynthetic organisms (Fig. 2.2 in Chapter 2). Dinoflagellate LHCs consist of a thylakoid intrinsic Chl *a*-Chl *c2*-peridinin-protein complex (acpPC) and a water-soluble peridinin-Chl *a*-protein (PCP) complex (Hofmann et al. 1996, Jiang et al. 2014). PCP exists in two forms, differing in absorption and fluorescence properties (Sharples et al. 1996, Ilagan et al. 2006).

A study using single-particle electron microscopy on PSI-LHCI supercomplexes in *B. minutum* revealed 25 LHCs, including LHCF and LHCR families, and indicated a role in photoprotection via nonphotochemical quenching (Kato et al. 2020). Recent cryo-EM studies identified two novel PSI subunits (PsaT and PsaU) and suggested enhanced electron transport and energy quenching efficiency (Li et al. 2024, Zhao et al. 2024).

Photochemical experiments have shown high quenching efficiency in dinoflagellates such as symbiodiniacean species. Under stress conditions, *Cladopodium* cells activate a "super-quenching" mechanism, transferring excess excitation energy to PSI and converting it into heat, potentially triggering symbiont expulsion or bleaching (Slavov et al. 2016).

Mitochondrial Genomes in Dinoflagellates

Mitochondrial (Mt) genomes in dinoflagellates remain poorly understood and underexplored. Pulse field gel electrophoresis experiments have suggested dinoflagellate Mt genomes consist of multiple linear chromosomes up to ~ 30 Kbp in size (Nash et al. 2008). Limited sequencing efforts have revealed that these genomes are highly derived, showing extensive amplification and recombination compared to their close relatives in Apicomplexa (Jackson et al. 2007, 2012). This evolution has resulted in multiple copies of genes and gene fragments linked in various combinations (Jackson et al. 2012). Typically, dinoflagellate Mt genomes encode three proteins: cytochrome B (cob) and cytochrome C oxidase subunits I (cox1) and III (cox3), along with two rRNA genes. Notably, a conventional stop codon occurs in cox3 but not in cox1 and cob, where the stop signal is often absent and recoded to TGG for tryptophan through mRNA editing (Lin et al. 2002).

Comparing parasitic species with later-branching taxa can offer insights into dinoflagellate mitochondrial evolution. Initially, no mitochondrial genome was found in *Amoebophrya ceratii* (John et al. 2019), but a highly reduced, cryptic mtDNA was later identified in conspecific strains (Kayal and Smith 2021). In *Hematodinium* sp., Mt genome sequencing revealed an inflated, heavily recombined genome with multicopy genes and gene fragments, similar to later-diverging dinoflagellates (Jackson et al. 2012). Noncoding sequences are also duplicated and rearranged, often containing inverted repeats capable of forming secondary structures (Norman and Gray 2001, Jackson et al. 2007, Nash et al. 2007). The extensive recombination and duplication make assembling genomic fragments challenging, leaving the ultimate size and organization of dinoflagellate mtDNA uncharacterized (Waller and Jackson 2009).

Cell Wall

The protective bound of dinoflagellates is composed of cytoplasmic membrane and the underlying amphiesmal vesicles that constitute the cell wall. Whether the amphiesmal vesicles are filled with cellulose dinstinguishes thecate and athecate (naked) taxa. A cell wall-bound cellulase is required for and coupled to cell cycle progress as observed in the heterotrophic dinoflagellate *Crypthecodinium cohnii* (Kwok and Wong 2010). Using $Na^{125}I$ labeling of *L. polyedrum* cell surface proteins exposed to the growth medium, Bertomeu et al. (2003) identified a ~ 43kDa protein (P43). The protein, deduced from cDNA, contains a signal peptide at the N-terminus. Search into dinoflagellate genomes and the transcriptomes of microeukaryotes reveals presence of this protein in various species of dinoflagellates but also in some haptophytes and other lineages (Fig. 8.11). This protein contains a conserve motif, GKYGPMGEDLFAQ. Dinoflagellates also share with other alveolates a cell surface protein named alveolin (Gould et al. 2008). Cell surface proteins are potentially useful for developing probes to detect species and their physiological conditions.

Flagella

Most dinoflagellates have two flagella: one wraps around the cell body, propelling it forward in a spiral motion, while the other extends backward from the sulcus (a vertical slit on the ventral side), steering the cell's movement. Besides enabling cell motility through dynein-dependent beating, flagella in eukaryotes like *Chlamydomonas reinhardtii* also perform four beating-independent motility functions (Kozminski et al. 1993). Three of these involve glycoprotein movements in the flagellar membrane plane, while the fourth, intraflagellar transport (IFT), involves bidirectional movement of multi subunit protein particles within the flagella. IFT is crucial for flagellum assembly, transporting building blocks (e.g., tubulin, dynein arms) from the base to the growing tip and back (Taschner and Lorentzen 2016). IFT also supports cell stimulus perception and sensory functions; defects in IFT can cause developmental disorders in mammals.

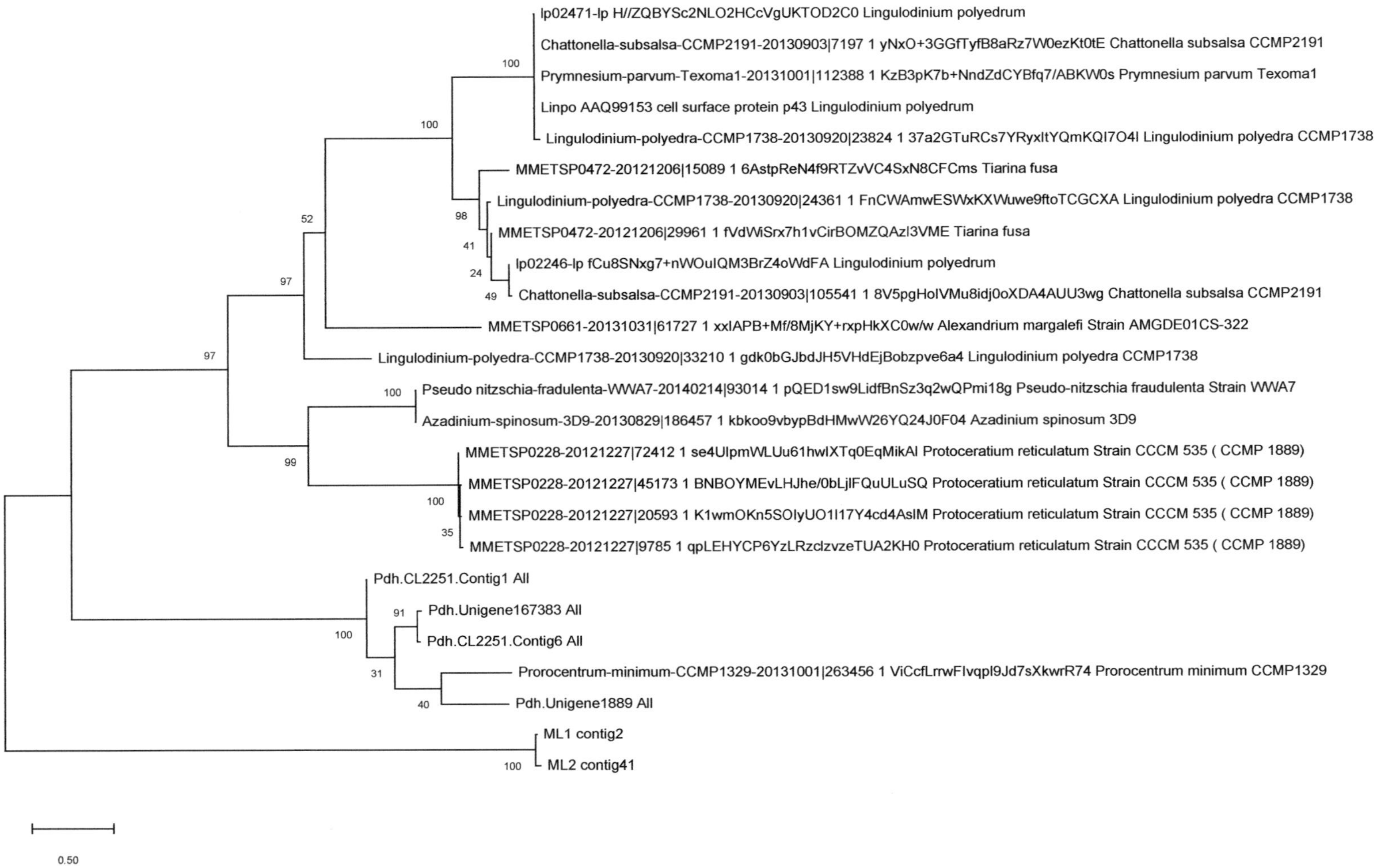

Figure 8.11. Maximum Likelihood tree of P43 of dinoflagellates and homologs identified in other algal lineages.

The IFT machinery comprises two complexes (IFT-A and IFT-B) in contact with microtubule doublets and the flagellar membrane. IFT-A includes six proteins (IFT144, 140, 139, 122, 121, 43), and IFT-B includes 16 subunits (IFT172, 88, 81, 80, 74, 70, 57, 56, 54, 52, 46, 38, 27, 25, 22, 20), featuring interaction domains like tetratricopeptide repeats (TPRs), WD40, and coiled coils. IFT-A handles retrograde (backward) transport, and IFT-B manages anterograde (forward) transport. Khalili and Usup (2015) used transcriptomic analysis to identify genes involved in flagellar synthesis, assembly, and functioning in *Alexandrium minutum*, finding two complex A components (IFT122, IFT140) and ten complex B components (IFT20, 27, 46, 52, 57, 74, 80, 81, 88, 172).

Adaptation to Symbiotic Lifestyle

Comparative genomic analyses reveal that genes involved in nutrient uptake, photosynthate transport, stress response, and infection are present and expanded in symbiodiniacean dinoflagellates (Lin et al. 2015, Aranda et al. 2016). In the genomes of *B. minutum* and *F. kawagutii*, genes important for mutualism, such as those related to sugar and fatty acid metabolism, oxidative stress, transport, photosynthesis, and adhesion, have undergone duplication during two major retroposition events (Fig. 8.12; Song et al. 2017). Comparison of symbiodiniacean genomes with those of other algae, the land plant *Arabidopsis thaliana*, and the parasite malaria *Plasmodium falciparum* identified genes unique to *Plasmodium* and Symbiodiniaceae, such as those involved in cell recognition (e.g., merozoite surface protein 1) and genes with unknown functions (e.g., uncharacterized protein PF11_0213, protein dpy-19 homolog 1) (Lin et al. 2015). In *Plasmodium*, merozoite surface proteins are crucial for red blood cell invasion (Beeson et al. 2016).

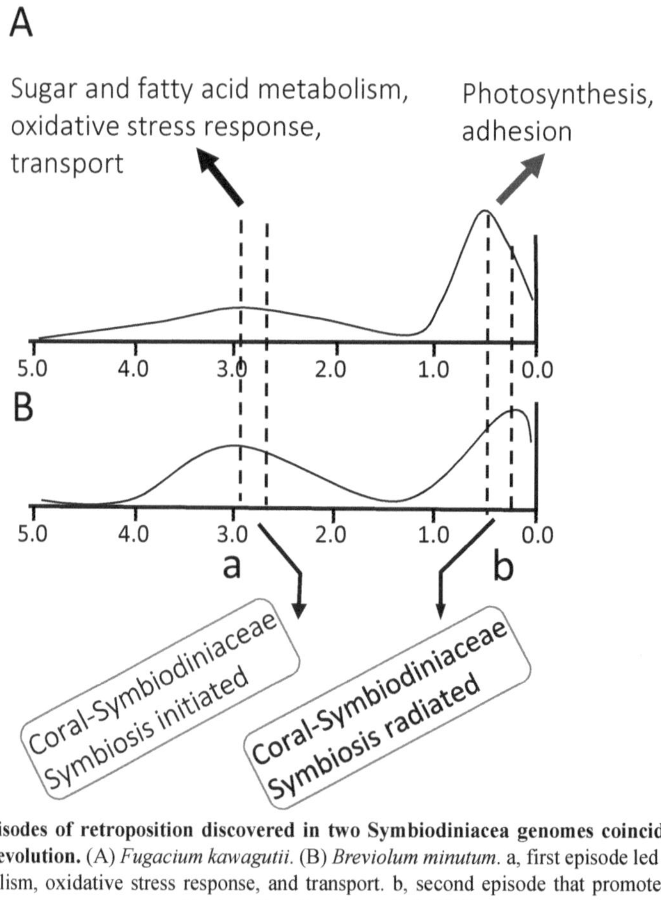

Figure 8.12. Two episodes of retroposition discovered in two Symbiodiniacea genomes coincident with two major periods of symbiosis evolution. (A) *Fugacium kawagutii*. (B) *Breviolum minutum*. a, first episode led to expansion of sugar and fatty acid metabolism, oxidative stress response, and transport. b, second episode that promoted photosynthesis and adhesion. Based on Song et al. 2017.

The *F. kawagutii* genome encodes 16 putative circumsporozoite protein (CiSP)-coding genes and a CiSP-like gene with a forkhead domain, potentially regulating vesicle-mediated transport and CiSP secretion. In the chlorophyte symbiont *Symbiochlorun hainanensis*, a CiSP-like protein is upregulated by acidification and heat stress (Gong et al. 2020). Protein secretion via the T2S system, encoded by at least 12 genes, is a major virulence mechanism in bacterial infections and may participate in bacteria-eukaryote mutualism (Sandkvist 2001, Cianciotto and White 2017, Epstein and Tiffin 2021).

UV protection is crucial in coral-Symbiodiniaceae mutualism. Mycosporine-like amino acids (MAAs) prevent UV damage, initially believed to be synthesized only by the basal lineage (clade A) of Symbiodiniaceae (Banaszak et al. 2000). Genome analyses confirmed the presence of MAA biosynthesis genes in *S. tridacnidorum* and absence in later-diverging lineages like *F. kawagutii* and *Cladocopium* (Lin et al. 2015, Shoguchi et al. 2018), although an improvement of the genomes later revealed partial MAA biosynthesis genes in *F. kawagutii* (Li et al. 2020). Furthermore, the MAA gene cluster was found to be present in *D. trenchii*, in addition to *Symbiodinium*, but absent in *B. minutum*, consistent with the original notion of ancestral synthesis ability with secondary loss (Shoguchi et al. 2021). In this context, *F. kawagutii* may represent an intermediate state of MAA biosynthesis pathway.

Insights into mutualism were also gained from green *Noctiluca scintillans*, which harbors prasinophyte *Protoeuglena noctilucae* as endosymbionts (Wang et al. 2016). This mixotrophic dinoflagellate forms massive blooms in the Arabian Sea, feeding on phytoplankton and producing ammonium for symbiont photosynthesis (Luo et al. 2022). During the bloom investigated in this study, symbiont photosynthesis was active, but cell division was suppressed, suggesting that the host promotes photosynthesis while inhibiting symbiont proliferation to maximize photosynthate translocation to the host.

Adaptation to Parasitic Lifestyle

Amoebophrya species are intracellular parasites of marine dinoflagellates, radiolarians, ciliates, and other *Amoebophrya* strains. The genomes of *Amoebophrya ceratii* strains AT5.2, A25, and A120 are significantly streamlined, with sizes (~ 120 Mbp) much smaller than other dinoflagellates (~1-250 Gbp). They have few transposable elements, short introns and intergenic regions, and a limited number of gene families, encoding 19,925, 28,091, and 26,441 genes, respectively (John et al. 2019, Farhat et al. 2021). These genomes are gene-rich, with 26, 27, and 80% more genes than predicted for AT5.2, A25, and A120, respectively, comparable to other dinoflagellates (7–89% above model prediction; Table 8.2).

Most genes in all three strains (56, 63.7, and 59%) lack functional annotation in public databases, suggesting some may be specific to parasitism. The intron density (about 1.3, 1.47, and 1.42 introns per kb of coding sequence in AT5.2, A25, and A120, respectively) is typical of dinoflagellates, but the introns are mostly non-canonical, including repeated introns with highly variable splicing motifs. The canonical GT-AG splice sites account for only 34.02% and 30.41 in strains A25 and A120, respectively, with the majority (>65%) being various other boundaries. In these strains, 11-30% of the non-canonical introns contain 8-20 nt inverted repeat motifs and 3–5 nt direct repeat motifs, spreading similarly to Transposable Elements (TEs).

High gene block synteny (> 49%) but low gene sequence similarity (< 50%) was observed among the *Amoebophrya* strains, suggesting strong selection pressure for protein changes due to parasitic interaction with different hosts. All three strains exhibit organelle reduction, including loss of the plastid and potential loss of mitochondrial genome and functions, indicating adaptive evolution. This is consistent with findings in *Hematodinium*, another parasitic dinoflagellate, which infects shellfish and other organisms (Lin et al. 2015).

Transcriptome sequencing of *Hematodinium* revealed 29,510 unique ORFs with no evidence of nucleus-encoded plastid-targeted protein genes or plastid gene homologs. The same BLAST workflow revealed 94 putative mitochondrial proteins. Genes involved in fatty acid, lysine, and tetrapyrrole synthesis lack organelle-targeting signal peptides, suggesting these pathways occur in the cytosol. Interestingly, the type I Fatty Acid Synthase (FAS) responsible for cytosolic FA synthesis appears to have taken the form of plastid type II FAS by fusing multiple protein genes (Gornik et al. 2015).

Highlight of Newly Found Genes and Physiologies

The rapidly growing dinoflagellate genomic data constitute a rich resource for exploring unique physiologies or metabolic features of dinoflagellates. The many functionally unknown genes predicted from their genomes will also serve as a target of future studies. A few of these genes, which are highly interesting but have not received adequate attention, deserve some highlight here. They are involved in acquisition and storage of nutrients, defense (toxin production), response to stress, and life cycle.

Use of Cyanate as a Nitrogen Nutrient

Alexandrium catenella (=*A. fundyense* or "Atama" complex Group I) actively expresses cyanate lyase (cyanase) during toxic bloom outbreaks. Zhuang et al. (2015) observed cyanase expression in a bloom outbreak in Northport Harbor, Long Island Sound in 2008. Metatranscriptomic data derived from samples collected throughout a 24-hour cycle at two-hour intervals showed downregulation of Nitrogen-Associated Proteins (NAPs) and active expression of urea and nickel transporters (nickel is a cofactor of urease) and urease as well as ammonium and nitrate transporters. These findings suggest that the bloom population was experiencing nitrogen stress due to rapid proliferation and that *A. catenella* utilizes all nutrient acquisition mechanisms, including cyanate metabolism. In a follow-up study mining metagenomics data, cyanase was found to be globally prevalent in marine plankton, underscoring the role of cyanate metabolism in the nitrogen biogeochemical cycle (Mao et al. 2022).

Storage of N: Uric Acid or Polyguanine?

Dinoflagellates have long been thought to store nitrogen in uric acid crystals (Clode et al. 2009). NanoSIMS and transmission electron microscopy studies confirmed that Symbiodiniaceae species store nitrogen as uric acid crystals following sudden environmental nitrogen increases (Kopp et al. 2013). Pulses of ammonium, nitrate, or aspartic acid promoted uric acid crystal accumulation, which formed quickly and were remobilized within 24 hours. In accordance, genome analysis of *F. kawagutii* revealed genes promoting uric acid biosynthesis, such as xanthin dehydrogenase/oxidase (Lin et al. 2015).

Recent studies, however, indicated that dinoflagellates also synthesize guanine polymers. Using cryo-SEM, cryo- FIB SEM and Raman spectroscopic imaging, Jantschke et al. (2019) found crystals of polymer of guanine close to chloroplasts in *Calciodinellum operosum*, and postulated that these polyguanine crystals might function to scatter light into chloroplast. Using similar methods, Mojzes et al. (2020) detected guanine crystals in symbiotic as well as free-living dinoflagellates and other algae. This study showed that in *A. carterae*, the stored guanine supported growth as the sole source of N-nutrient for several new generations. The authors suggested that the function of guanine ($C_5H_5N_5O$) crystals as N storage in these eukaryotic algae was similar to that of cyanophycin ($C_{10}H_{19}N_5O_5$) in cyanobacteria.

Phagotrophy Gene Repertoire

Phagotrophy, a mode of nutrition involving particle ingestion through endocytosis, is common among dinoflagellates. This process includes four major steps: endocytosis, phagosome formation, fusion of phagosomes with lysosomes, and autophagy (McKie-Krisberg et al. 2018, Uribe-Querol and Rosales 2020). Phagotrophy is an ancient feature of protists, including photosynthetic lineages (Mitra et al. 2016, Villanova and Spetea 2021, Li et al. 2023), because the evolution of algae involved a heterotrophic eukaryote (Raven et al. 2009) according to the prevailing endosymbiosis hypothesis of chloroplast evolution initially postulated by Lynn Margulis (Sagan 1967).

Dinoflagellates, comprising 50% heterotrophic species, exhibit mixotrophy with phagotrophic capabilities (Stoecker 1999, Jeong et al. 2010). The molecular mechanisms of phagotrophy in dinoflagellates are still being uncovered. Genes involved in autophagy and endocytosis, such as ATG3, ATG8, ATG12, PIP5K, PLD, AP2 adaptor complex, dynamin, epsin, Eps15, EH domain-containing protein 1, VPS4, VPS35, VPS45, Rab8, and Rab11, have been documented in laboratory cultures and natural assemblages (Li et al. 2020, Zhang et al. 2019, Yu et al. 2020, 2023). In a natural bloom of *Prorocentrum shikokuense*, genes involved in endocytosis, phagosome, peroxisome, and lysosome were actively expressed, particularly at night, suggesting increased feeding activity during nighttime (Yu et al. 2020).

These findings highlight the complexity and diversity of nutrient acquisition and storage mechanisms in dinoflagellates, emphasizing their adaptability and ecological significance.

Polyketide Synthase (PKS)/Non-ribosomal Peptide Synthetase (NRPS) Enzyme System and Secondary Metabolites

PKS/NRPS enzymes are crucial for producing specialized metabolites, including toxins, in dinoflagellates (Kellmann et al. 2010). PKSs consist of an Acyl-Transferase (AT) domain, an Acyl-Carrier Protein (ACP), and a ketosynthase (KS) domain and are classified into Type I, II, and III based on their organization (Hertweck 2009). Type I PKS has all catalytic domains on a single polypeptide, which is used in a processive fashion for chain elongation, as in animals and fungi. Type II PKS is a multiprotein complexe where each catalytic domain is found on a separate polypeptide. Type III PKSs, functioning iteratively similarly to Type II, are known from the plant chalcone/stilbene synthases (CHS/STS) producing compounds such as naringenin chalcone. NRPSs synthesize non-ribosomal peptides and include adenylation (A-domain), thiolation (T-domain), and condensation (C-domain) domains (Marahiel et al. 1997). These enzymes can work together to produce hybrid natural products (Kellmann et al. 2010). PKS and NRPS enzymes are found in toxin-producing dinoflagellates like *Karenia brevis, Gambierdiscus* spp., and *Amphidinium gibbosum* (Lopez-Legentil et al. 2010, Bachvaroff et al. 2015, Verma et al. 2019, van Dolah et al. 2020, Beedessee et al. 2020) and in Symbiodiniaceae species (Aranda et al. 2016, Beedessee et al. 2019).

Dinoflagellates are unique in possessing both canonical Type I PKS and an unusual Type I sequence single-domain PKS (e.g., KS) (Monroe and van Dolah 2008, van Dolah et al. 2017) while significantly expanding the PKS family relative to other protists (Koli et al. 2016). A study scanning 47 dinoflagellate transcriptomes for modular synthase domains and their co-occurrence with thiolation domains revealed widespread presence, but in particularly high abundance in Gymnodiniales, of the thiolation domains (Williams et al. 2021). Some of these occur alongside tetratricopeptide repeats (especially hexa- and hepta-repeats), which are unique to dinoflagellates.

Cyclophilins

Cyclophilins are a family of proteins known for binding cyclosporin A (CsA), an immunosuppressant drug in humans (Singh and Argáez 2018). These proteins possess peptidyl-prolyl cis-trans isomerase activity, essential for correct protein folding. Cyclophilins have diverse functions, including protein

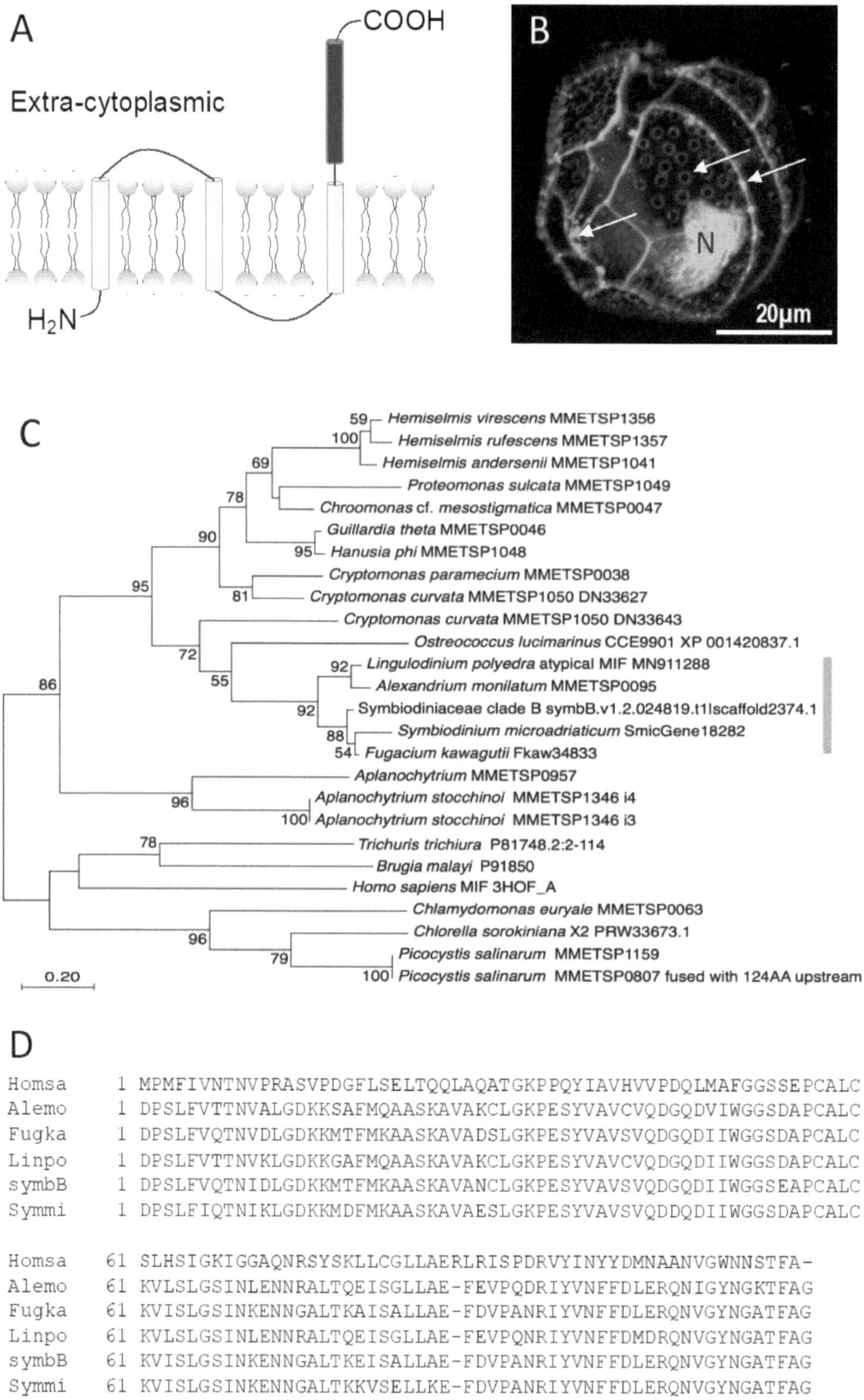

Figure 8.13. Macrophage migration inhibitory factors (MIF) detected in dinoflagellates. (A) Schematic of MIF structure. (B) Localization of MIF at the apical plates and the thecal pores shown by immunofluorecence (arrows). Reproduced from Jaouannet et al. (2020) under the Creative Commons CC BY 4.0. license. (C) Alignment of amino acid sequences from dinoflagellates and humans.

folding, signaling, transcriptional regulation, cell cycle control, and stress response (Singh et al. 2020). In algae, cyclophilins have been reported in *Chlamydomonas* (Vallon 2005) and *Ulva* (Wu et al. 2009), where they are implicated in immunity and stress response. For dinoflagellates, cyclophilin genes were detected over two decades ago (Lin and Zhang 2010), but their functions remain unclear. Recent studies indicate cyclophilins in *P. cordatum* are induced by copper and polychlorinated biphenyl exposure, suggesting a role in stress response (Ponmani et al. 2015). Cyclophilin B in *Margalefidinium polykrikoides* is implicated in environmental stress responses (Abassi et al. 2017).

Macrophage Migration Inhibitory Factors (MIF)

MIFs are cytokines and chemokines essential for the vertebrate immune system, mediating innate immune responses (Calandra and Roger 2003). These proteins are conserved across the tree of life, serving various cellular functions (Michelet et al. 2019). In free-living organisms, MIFs activate immune responses, promote immune cell proliferation, and inhibit p53-mediated apoptosis (Jung et al. 2008). Parasitic species produce multiple MIF proteins to modulate host immune responses, facilitating infection (Augustijn et al. 2007, Miller et al. 2012, Ajonina-Ekoti et al. 2013, Naessens et al. 2015). A homolog of MIF (PmMIF) was identified in the cyanobacterium *Prochlorococcus marinus*, showing high structural homology with mammalian MIFs but lacking the Cys-X-X-Cys motif necessary for oxidoreductase activity (Wasiel et al. 2010).

A MIF-like protein was detected in *L. polyedrum* (Jaouannet et al. 2020). However, while typical MIF is a soluble cytoplasmic protein, the protein in *L. polyedrum* (LpMIF) is a transmembrane protein localized in cell wall-associated membranes, vesicular bodies, and membranes of extracellular vesicles accumulating at the secretory pores of the cells (Fig. 8.13A,B). Tblastn analysis using human MIF as query against the MMETSP database (E-value e-20 cutoff) yielded 92 hits, 28 from dinoflagellates including *P. shikokuense*, *Alexandrium* spp., Symbiodiniaceae species, *K. brevis*, *K. veneficum*, and *Heterocapsa triquetra*. Phylogenetic analysis shows it is closely related to a prasinophyte homolog within a clade containing cryptophytes (Fig. 8.13C). Most dinoflagellate hits align with human MIF (hsMIF) from position 2 to 115, with some shorter due to incompleteness (Fig. 8.13D).

Pfam search confirms these dinoflagellate MIF sequences contain the MIF functional domain. KEGG analysis indicates they are phenylpyruvate tautomerase (EC 5.3.2.1), the second member of the MIF superfamily, involved in Tyrosine and Phenylalanine metabolism. Under biotic stress, *L. polyedra* shows decreased MIF gene expression and reduced protein levels on the surface, suggesting this MIF may be involved in intercellular communication (Jaouannet et al. 2020) or the sensing of environmental stimuli.

RACK1 and Hemerythrin-Like Protein

Islas-Flores et al. (2021) constructed yeast two-hybrid system cDNA libraries for *S. microadriaticum* using DinoSL and SMART technology. The DinoSL-based libraries yielded significantly more cDNAs than the SMART-based libraries, highlighting the advantage of DinoSL for creating full-length cDNA libraries in dinoflagellates. Using a yeast strain harboring *Symbiodinium* RACK1 (SmicRACK1) as bait and the DinoSL-based library as prey, they demonstrated an interaction between SmicRACK1 and a hemerythrin-like protein, suggesting the latter as the ligand of the former.

RACK1, a WD-repeat protein similar to the G-protein β subunit (Gβ), features a seven-bladed β-propeller structure that facilitates protein binding (Adams et al. 2011). These proteins typically serve as scaffolds for protein complexes, interacting with multiple partners to coordinate signaling events and contribute to cellular organization. RACK1 is involved in shuttling, anchoring, and stabilizing proteins, interacting with ribosomal machinery, cell surface receptors, and nuclear proteins, influencing various cellular pathways (Adams et al. 2011).

Hemerythrin is an oxygen-carrier non-heme protein responsible for oxygen transport in marine invertebrates and worms (Li et al. 2015, Alvarez-Carreño et al. 2016), and also functions in innate immunity and tissue regeneration in certain worms (Coates and Decker 2017). The hemerythrin gene is found in numerous bacterial, archaeal, and eukaryotic genomes (Alvarez-Carreño et al. 2016). Besides oxygen transport, hemerythrin proteins can be involved in signal transduction, phosphorelay regulation, abiotic resistance, and protein binding (Li et al. 2015). The specific roles of SmicRACK1 and the hemerythrin-like protein in *S. microadriaticum* remain to be determined.

Abscisic Acid 8'-Hydroxylase

Abscisic acid 8'-hydroxylase (EC 1.14.13.93) belongs to the family of oxidoreductase, and it catalyzes the oxidative degradation of abscisic acid (ABA). As a member of the cytochrome P450 monooxygenase family (Krochko et al. 1998), this enzyme is involved in the regulation of germination and dormancy of plant seeds (Footitt et al. 2011) and participates in carotenoid biosynthesis. In dinoflagellates, the encoding gene of this enzyme was shown to be upregulated in cysts, suggesting its role in encystment and maintenance of dormancy (Deng et al. 2017). Besides, upregulated expression of this gene was found during the initial response to heat stress in a *Cladocopium* species (Rosic et al. 2010).

Small Heat Shock Protein

sHSPs are ancient, universal molecular chaperones involved in cellular stress response (Haslbeck et al. 2015, 2019). They have a molecular weight of 12-43 kDa and a tripartite domain architecture, consisting of a conserved Alpha-Crystallin Domain (ACD) and variable C and N terminal extensions (Waters and Vierling 2020).

Recent research has revealed new information on sHSP gene sequences and stress-dependent expression profiles in dinoflagellates (Deng et al. 2021). Using DinoSL as a selective tool, Deng et al. (2021) constructed dinoflagellate-specific full-length cDNA libraries from marine sediment-derived RNA samples. They found sHSPs with A-X-X-X-D/N/S/H/R-G-V-L in the typical conserved A-X-X-X-N-G-V-L motif. Active expression of these sHSPs in sediment suggests their potential role in maintaining dinoflagellate cyst dormancy.

Meiosis Genes

Sexual reproduction is widespread among eukaryotes and serves as a mechanism for chromosome recombination, thereby generating genetic diversity (Vesteg and Krajcovic 2011, Goodenough and Heitman 2014, Speijer et al. 2015). Dinoflagellates typically have a haplontic life cycle, consisting of haploid asexual and diploid sexual stages. Meiosis is one of the hallmarks of sexuality, and its regulating genes only began to be identified in dinoflagellates. From genomes and transcriptomes, a range of genes (31-51) with homology to meiosis genes in model organisms have been identified (Chi et al. 2014, Lin et al. 2015, Morse 2019). These include most of the 12 core meiosis genes previously developed for eukaryotes (Schurko and Logsdon 2008) and homologs of yeast meiosis regulating genes such as MEI2 and several meiotically upregulated genes (Table S8 in Lin et al. 2015). However, securing, which normally serve in preventing the transition from metaphase to anaphase by inhibiting the protease separin, and some components of the cohesion complex with the function to hold homologous chromosomes together are absent (Morse 2019). The absence of cohesion complex elements is consistent with more recent findings suggesting that dinoflagellates may have lost the canonical synaptonemal complex (Shah et al. 2020).

Syngamy is another important hallmark of sexuality. In a transcriptomic study of *P. cordatum*, HAP2 was identified, along with MEI2, MutS2, and several other meiosis-associated genes

(Berdieva et al. 2021). HAP2 (hapless 2) participates in the fusion of cell membranes (Wong and Johnson 2010).

In dinoflagellates, sexual reproduction is known to occur under adverse environmental conditions and lead to encystment for survival ('sex for encystment'). However, increasing laboratory observations indicate that sexual reproduction can also produce vegetative cells, bypassing encystment. Recent natural bloom transcriptomic studies suggest that sexual reproduction in dinoflagellates may aid in bloom proliferation or extension. Genes specific to or associated with meiosis were significantly upregulated during blooms, unexpected for the 'sex for encystment' scenario (Lin et al. 2022, Luo et al. 2022). Integrating transcriptomic data from natural blooms and laboratory cultures of *P. shikokuense*, *Scrippsiella acuminata*, and *K. mikimotoi*, 28 meiosis genes were identified and found to be highly expressed during blooms or cyst germination. Key genes include MEI2, initially identified in yeast for initiating meiosis. RAD21, SMC3, and MEI2 are linked to bloom-promoting sex, HOP2 and MSH4 to cyst germination, while SPO11, MND1, and DMC1 are common to both cyst-forming and non-encysting sex (Lin et al. 2022). Verification of these findings could provide markers for distinguishing different sexual reproduction ecotypes in dinoflagellates.

Response to Temperature Variation

Temperature responses have not received as much research attention as nutrient responses for molecular understanding. Recently, Zhang et al. (2022) investigated proteomic and metabolomic changes in the HAB species *P. shikokuense* across four temperatures: optimal (25°C), supra-optimal (28°C), and sub-optimal (19°C and 22°C). They observed decreases in Particulate Organic Carbon and Nitrogen (POC, PON) as temperature increased, with an increased POC/PON ratio at higher temperatures, except at 28°C. These results suggest a higher cell division rate and photosynthetic carbon fixation at higher temperatures. However, at the supra-optimal temperature, cells increased production of light harvesting, photoreaction, and protein homeostasis proteins. Sub-optimal temperatures upregulated glutathione (an antioxidant), transcription, and lipid biosynthesis, compensating for decreased translation efficiency and cell membrane fluidity. At 19°C, there was an increase in nitrate reduction and nitrogen flux towards asparagine, aspartic acid, and glutamine, with the accumulation of glutathione, glutarate semialdehyde, and 5-KETE.

Interestingly, their metabolomic data showed a significant increase in a sulfinpyrazone-like metabolite at 19°C compared to 25°C (13-fold) and 28°C (26-fold). Sulfinpyrazone (C23H20N2O3S) is a medicine to treat gout patients as it inhibits reabsorption (transport) of uric acid and enhances its urine excretion (Margulies et al. 1980). Whether this compound blocks intracellular cross-membrane trafficking of organic nitrogen molecules or is an accumulated intermediate at lower temperatures remains to be investigated.

Another study on *P. cordatum* revealed differential gene cluster responses between mRNA editing and alternative splicing (exon usage) to temperature changes between 20°C and 26°C (Dougan et al. 2023). Further verification could indicate a novel mechanism of gene regulation. Additionally, *P. glacialis* is an ideal model for understanding cold and darkness adaptation, having acquired genes from bacteria for ice-binding proteins. This species shows enhanced transcriptional responses through unidirectional, tandem duplication of single-exon genes involved in cold survival and low-light adaptation (Stephens et al. 2020).

9

Genomes of Haptophytes

Haptophytes (also known as prymnesiophytes) are characterized by the presence of a unique flagellum-like structure called the haptonema, which is located between the two flagella (Fig. 9.1) and is known to be involved in prey capture among other functions (Kawachi et al. 1991). With the haptonema, photosynthetic haptophytes are naturally mixotrophic. Most haptophytes are single-celled flagellates or planktonic nonflagellate protists, but some species form colonies (e.g., *Phaeocystis globosa*). Based on molecular clock estimates, the Haptophyta phylum likely emerged between 800–1000 million years ago (Liu et al. 2010). Currently, Haptophyta comprises 312 morphologically described species (Edvardsen et al. 2016).

Haptophyte taxa are generally divided into two functional groups, the calcifying group (coccolithophorids) and the non-calcifying group. Coccoliths, composed of calcium carbonate plates (calcite) fit together as an intricate housing of the cell, are believed to protect cells from grazing pressure, viral infection, and photodamage, but calcification is energetically costly (Monteiro et al. 2016). Typically, non-calcifying haptophytes bear organic scales. The organic scales of Prymnesiophyceae are elaborate and ornamented, resembling coccoliths (de Vargas et al. 2007). The earliest geological records indicate that the order Coccolithales emerged at least approximately 220 million years ago. They are most abundant and diverse in marine ecosystems, but also occur in freshwaters; they are distributed from the equator to polar regions, but are more abundant in tropical and subtropical regions, and in open oceans than coastal waters. Molecular analyses have suggested that tropical waters were the original center of haptophyte diversification, which subsequently spread into higher latitudes (Liu et al. 2009). Liu et al. (2009) conducted DNA barcoding based on the D1-D2 region of rRNA the large subunit (LSU), using the forward primer Hapto_4 (5′-atggcgaatgaagcgggc-3′) and the reverse general eukaryote primer Euk_34r (5′-gcatcgccagttctgcttacc-3′) for subpolar and subtropical oceanic samples. Their results revealed very high diversity in noncalcifying lineages, suggesting that many new haptophyte species that had been missed by classical PCR using eukaryote-universal primers remain to be described.

Another signature characteristic of haptophytes is their haplo-diploid life cycle. Both the haploid and diploid stages are capable of independent asexual reproduction. In the prymnesiophytes, the morphology of the cell covering differs between the two phases. In the coccolithopores, diploid phase is non-motile and typically produces heterococcoliths whereas the motile haploid phase produces holococcoliths or nannoliths or does not calcify at all; whether the absence of calcification or the production of nannoliths also occurs cryptically in diploid stages remains to be unraveled (de Vargas et al. 2007).

Haptophytes are a globally distributed biogeochemically important group of phytoplankton in the ocean. Their contribution to the marine primary production can be as high as 20% in some systems, and their biomass can account 30–50% of total chlorophyll *a* in the oceans

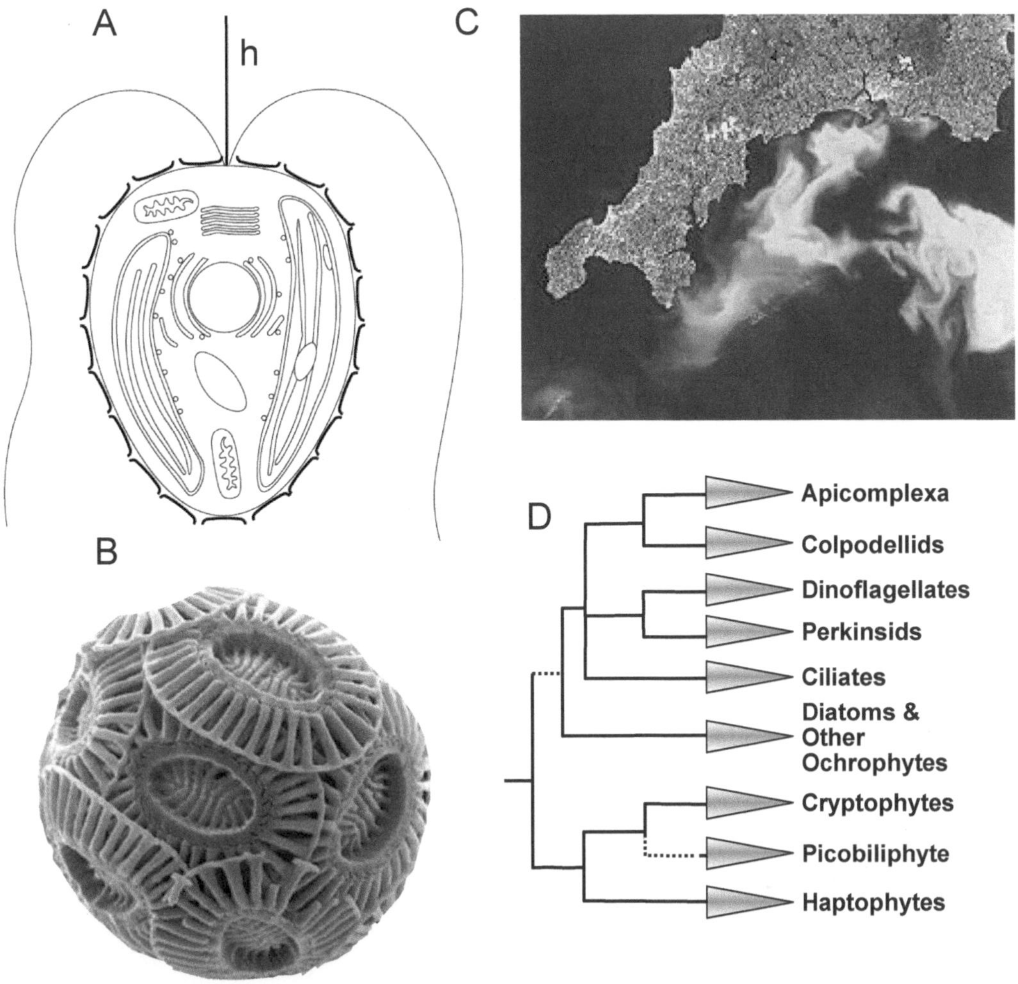

Figure 9.1. Haptophyte cell morphology, bloom, and phylogenetic placement. (A) drawing of a typical haptophyte cell with a haptonema (h). (B) an electron microscopic image of *E. huxleyi*. (C) Satellite image of *E. huxleyi* bloom; credit: Wikimedia Commons/Steve Groom, Landsat. (D) Schematic phylogenetic tree of algae highlighting the position of haptophytes; based on Keeling 2009; dashed line indicates uncertain branching point.

(Liu et al. 2009). Blooms of coccolithophorid species such as *Emiliania huxleyi* often cover > 100,000 km^2 and are conspicuous on satellite images (Fig. 9.1). The bloom appears as milky patches in the surface often, due to the high light reflectance (albedo) of the coccoliths. Many species of haptophytes also form harmful algal blooms, some of which produce lipolytic toxins that disrupt cell membranes (e.g., on fish gills).

Two other signature features make haptophytes biogeochemically significant. First, the coccolithophores are important carbon exporters, because their coccoliths ballast the cells to sink quickly, thus conveying carbon from sea surface to the interior for sequestration for thousands of years. Despite the long-term benefit in terms of carbon sequestration, the formation of coccoliths ($CaCO_3$) releases CO_2 increasing partial pressure of CO_2 on ocean surface and weakening or reversing the role of the ocean as carbon sink. Second, haptophytes play a significant role in marine sulfur cycle. They produce dimethylsulfoniopropionate (DMSP), which is readily converted to dimethyl sulfide (DMS) by bacteria and some algae via DMS lyases.

Third, related to the last point, DMSP-producing haptophytes are climate feedback relevant. DMS gas forms aerosols, which serve as nuclei of cloud formation, thus potentially reducing radiation heat on earth surface (Charlson et al. 1987, Curson et al. 2018). As such, DMSP production in haptophytes (as well as other groups of algae) may act as a climate negative feedback system.

Sequenced Haptophyte Genomes

Eight species in the Haptophyta phylum have been sequenced (Table 9.1). These represent different morphotypes and ecotypes, including the coccolithophorid species *Emiliania huxleyi* (recently renamed as *Gephyrocapsa huxleyi*) and *Chrysotila roscoffensis*, colonial species *Phaeocystis globose,* toxigenic species *Prymnesium parvum,* the aquaculture prey species *Tisochrysis lutea* and *Diacronema lutheri,* and species from the coastal genera *Chrysochromulina.* Their genomes range in size from 43.5 to 674 Mbp with nearly 12,000 to over 38,000 encoded genes.

Emiliania huxleyi

Emiliania huxleyi is a cosmopolitan and highly abundant coccolithophore in modern oceans, making it a model organism for studying coccolithophores and the entire haptophyte lineage (Wheeler et al. 2023). The genomes of *E. huxleyi* strain CCMP1516 and an additional 13 strains were sequenced in 2013 as the first haptophyte reference (Read et al. 2013). The haploid genome of strain CCMP1516 is estimated to contain 141.7 Mbp DNA, while genomes of the other 13 strains range from 99 Mbp to 133 Mbp. The assembly was 97% complete based on conserved eukaryotic single-copy genes. The genome codes for 30,569 genes, corresponding to a 21.9% coding percentage. Over 64% of the genome sequence is dominated by repetitive elements, much higher than seen in sequenced diatoms.

Table 9.1. Published genomes of Haptophyta.

Species	Genome size (Mb)	Assembled genome size	Number of chromosomes	Predicted genes	Protein genes	% Annotated	% GC	References
Chrysochromulina parva	65.77	65.77			28,138		63.58	Hovde et al. 2019
Chrysochromulina tobin	59.07	59.07		16,777	6,710	61.4	63.37	Hovde et al. 2015
Chrysotila roscoffensis	674	636		23,341				Meng et al. 2021
Diacronema lutheri	43.50	43.50			14,446		73.25	Hulatt et al. 2021
Emiliania huxleyi	141.7	167.7		38,549	30,569		64.5	Read et al. 2013
Phaeocystis globosa	144	129.7	23	32,618			64.6	Chen et al. 2024
Prymnesium parvum (CCMP 3037)	98	98			19,426		57.4	Jian et al. 2024
Prymnesium parvum (UTEX2797)	107.32	107.33			20,578		57.6	Jian et al. 2024
Tisochrysis lutea	82	82		20,582	11,763	47		Berthelier et al. 2018, Carrier et al. 2018

A high level of diversity was found, suggesting that no single strain can represent all levels of genetic diversity. The genome reveals diverse iron uptake mechanisms, with many iron-containing proteins present in both the core and variable parts of the genome.

Chrysotila roscoffensis

In 2022, the genome of another coccolithophorid species, *Chrysotila roscoffensis*, was reported (Meng et al. 2022). This species was originally isolated in 1934 from Roscoff, France, and was formally classified as *S. roscoffensis* in 2015 (Andersen et al. 2015). The sequenced strain (NMBjih026-8) was isolated from coastal water in Xiangshan Bay (N 29°32'48.44", E 121°48'34.62") near Ningbo, Zhejiang, China, in January 2009. Its presence in the North Sea and the East China Sea suggests wide distribution of this species. The genome was sequenced combining Illumina, SMRT (PacBio Sequel), and 10X Genomics GemCode. K-mer analysis (k=17) of the sequencing data resulted in an estimate of the genome size as 674.07 Mbp and its heterozygosity at 0.64%, markedly lower than that of the *E. huxleyi* genome.

An ~ 636 Mbp assembly was achieved, covering 94.35% of the genome. The genome encodes 23,341 genes, while repetitive elements account for ~ 59% of the genome. Core Eukaryotic Genes Mapping Approach (CEGMA) analysis (Parra et al. 2007) indicated that the assembly captured 81.05% of the conserved core eukaryotic genes, of which 75% were complete. Compared with *E. huxleyi*, 22 gene families were significantly expanded, and 39 families contracted. The expanded gene families are enriched in 60 GO terms, including those associated with the cytoskeleton, cellular motility and morphology, and ion transport. The expansion of the cytoskeleton gene family may boost the calcification capacity of the species.

Phaeocystis globosa

Phaeocystis globosa occur in unicells (3–10 μm in size) as well large spherical colonies. Colonies range from millimeters to centimeters in diameter, in which cells are embedded in gelatinous materials. The species is responsible for increasing frequent harmful algal blooms that, among other devastating environmental and ecological impacts, threaten the operation of nuclear power plant cooling water system, as has happened in China. *P. globosa* genome was sequenced recently using PacBio HiFi and Hi-C technologies (Chen et al. 2024). The sequencing output was assembled into a ~ 144Mb genome composed of 23 scaffolds that are of chromosome scale, some of which contain terminal telomeric repeats (CCCTAA). The genome has a GC content of 64.6%. The analyses of 4dTv plot, gene repeats, and reasonable level of synteny between assembled blocks provided evidence that *P. globosa* has likely undergone two rounds of segmental genome duplication. Nine copies of the ribosomal RNA genes were found, eight being localized on chromosome 23 interspersed by DNA of 14.4-290 Kbp and the other being localized on chrome 16.

The results shed light on the potential molecular basis of colony formation of *P. globosa*, highlighting the expansion of gene families related to the biosynthesis of glycan and polysaccharide chains, and the metabolism of amino sugar and nucleotide sugar. This finding is consistent with that of an earlier comparative transcriptomic study indicating the potential role of glycosaminoglycan in colony formation of the species (Zhang et al. 2020). The genome data also revealed dual pathways of dimethylsulfoniopropionate (DMSP) biosynthesis, as opposed to other DMSP-producing organisms that usually have only one of the pathways. Besides, enzymes responsible for the degradation of DMSP to produce dimethyl sulfide (DMS) were identified in the genome.

The genome also reveals 19 rhodopsin genes of proton pump type and anion channel type. One of them is a proton pump rhodopsin similar to that of diatoms and dinoflagellates. Several of them resemble proton/chloride pump rhodopsin of *Halorubrum* and *Halobacterium*. The rest are unclassified types (Chen et al. 2024). The diverse phylotypes suggest that some of these might have been acquired through horizontal gene transfer.

Prymnesium parvum

Prymnesium parvum is cosmopolitan and ecologically successful in part because it is euryhaline and eurythermal, tolerating salinities from 3 to 30 PSU and temperatures from 2 to 30°C (Wagstaff et al. 2018). Genome sequencing of several strains of *P. parvum* revealed extreme genome diversity and existence of cryptic species (Wisecaver et al. 2023). The three cryptic species identified (A-, B-, and C-type clades) are reproductively isolated with markedly varying genome sizes (115-845 Mbp). These strains include haploid, diploid, and polyploid cells, and show differences not attributable to genome copy number variations. At the same time, there are variations of up to 243 Mbp in haploid genome size among strains of different chemotypes. Syntenic and phylogenetic analyses suggest that UTEX2797, a widely used laboratory strain isolated from Texas, is in fact a hybrid containing two genetically distinct haplotypes. Total repetitive genes ranged in values from 22.9%–64% reported in other haptophytes. The number of protein-coding genes ranged from 26,458 (CCMP3037) to 32,339 (UTEX 995). The examination of gene families across different strains revealed various functional categories linked to metabolic and genome size differences in *P. parvum*. These include genes associated with the biosynthesis of toxic metabolites and the proliferation of transposable elements.

Diacronema lutheri

D. lutheri is frequently used in aquaculture as prey because it has high lipid contents. Hulatt et al. (2021) sequenced the genome of strain NIVA-4/92, generating a 43.50-Mbp assembly. Its GC content (73.25%) is very high. Compared to other haptophyte species, *D. lutheri* has more intron-free (i.e., single exon) genes, consistent with its smallest genome size among haptophyte genomes sequenced so far. The genome has relatively low level (22.9%) of repetitive DNA, encoding for 14,446 protein genes. The genome provides the needed foundation for further investigating the nuclear genome architecture of the class Pavlovales.

High Intraspecific Genomic Divergence

Extraordinary genetic divergence was observed based on the sequenced haptophyte genomes. Not only the genomic differences are large between different lineages within the phylum Haptophyta, but the differences are also substantial even between strains within species. For instance, in *P. globosa*, only about 60% of the genome from a strain isolated from Beibu Bay in China (CNS00066) can be aligned with reads of the genome (CNS00080) in a strain isolated from Thailand (Chen et al. 2024). Similar levels of genomic divergence have also been reported for intraspecific strains of *E. huxleyi* (Read et al. 2013) and *P. antarctica* (Luxem et al. 2017).

Haptonema

A haptonema is a unique microtubule-based motile organelle in haptophytes, extending from the base between two flagella and reaching over 100 μm in some species (Gregson et al. 1993). It serves various functions, including attachment, gliding, food aggregation and transport, and mechanical stimulus reception (Nomura et al. 2019).

As observed in *Chrysochromulina hirta*, food capture and transport occur during forward swimming with extended haptonema and backward-beating flagella. Particles adhere to the haptonema, accumulate at a specific point, and move to its tip. The cell stops, and the haptonema bends to deliver food to the cell's lateral posterior surface for phagocytosis (Kawachi et al. 1991).

The haptonema shaft typically contains up to eight singlet microtubules arranged in a ring or crescent, surrounded by endoplasmic reticulum (Graham et al. 2020). Upon mechanical stimulation, it coils rapidly within milliseconds, changing the circular microtubule arrangement to a crescent shape. An electron-dense structure is found in the microtubule ring's center, with structures linking adjacent microtubules (Gregson et al. 1993). No motor protein-like structures have been identified, suggesting that haptonema coiling is not motor-dependent.

Nomura et al. (2019) demonstrated that coiling begins with left-handed twisting, followed by helical writhing from the distal tip. Uncoiling occurs slowly from the proximal region. Electron microscopy showed seven mostly parallel microtubules, with one often winding around others in the extended state. Paclitaxel inhibited coiling and induced right-handed twisting without Ca^{2+}, while adding Ca^{2+} caused planar bends near the proximal region. This suggests that microtubule conformation changes, possibly aided by Ca^{2+}-binding proteins, drive rapid haptonema coiling.

Coccolith Formation

Coccolithophores produce two types of coccoliths, the heterococcoliths and holococcoliths. Non-coccolithophorid taxa, e.g., *Prymnesium* spp., often have small non-mineralized organic scales (Graham et al. 2022). While holococcoliths are made of identical plates and formed internally by the Golgi apparatus, heterococcoliths are composed of plates of different sizes and shapes and are formed externally. The coccolith vesicle, where calcification occurs, has been found to be associated with calcium organelles, polysaccharides linked to some novel low-complexity proteins, and some 68 coccolith-associated proteins (Wheeler et al. 2023). While the roles of these elements in the formation of coccoliths are still unclear, the voltage-gated H^+ channel has been shown to be important in maintaining H^+ homeostasis required for calcification, which is disrupted by environmental acidification (Kottmeier et al. 2022).

Studies have shown that cytoskeleton proteins are important for the correct formation of coccoliths. In several coccolithophorid species and the silicifying species *Prymnesium neolepis*, the disruption of cytoskeleton was shown to prevent the secretion of coccoliths, resulting in the production of malformed coccoliths (Langer et al. 2010, Durak et al. 2017).

Besides, the calcification process requires silicate, with its transporter existing in both silicifying and calcifying haptophytes (but absent in *E. huxleyi*) and being upregulated during calcification (Durak et al. 2016, Ratcliffe et al. 2023). Although cellular quota of Si in haptophytes is low, often below the detection limit, Si incorporation in the coccolith or organic scale has been demonstrated (Fig. 9.2). The differential Si requirement among haptophyte species has profound ecological implications in the ocean where Si bioavailability is highly variable.

Calcification is a dynamic process in which *E. huxleyi* cells continually produce and shed coccoliths (Paasche 2001). The shed coccoliths have high absorption efficiency. Remarkably, *E. huxleyi* can "recruit" free-floating coccoliths and incorporate them to their own coccoliths, causing changes in the morphology of the living coccoliths (Johns et al. 2023). In addition, the free coccoliths can also absorb viruses (creating a living entity named "virolith"), which promote viral infection of the living *E. huxleyi* cells (Johns et al. 2023).

Regulation of Colony Formation

Species like *P. globosa*, *P. antarctica*, and *P. pouchetii* alternate between their two morphotypes: solitary free-living cells and colonial cell groups. During their bloom outbreaks, the colonial form often dominates. In the past decade, there has been an increasing trend of *P. globosa* blooms in subtropical and tropical seas, causing devastating impacts on the ecosystem as well as fisheries and aquaculture industries, because the colonies deter grazing due to their large sizes and disrupt the food chain. In addition, the colonies can clog cooling water inlets of nuclear power plants, posing nuclear security threats. The colonies are formed by non-flagellate cells embedded in mucilaginous masses. The proportion of colonial cells in a *Phaeocystis* assemblage is variable. Potential triggers of the transition from solitary cells to colonies include, according to various studies, nutrient availability, light intensity, temperature, mechanical stress, grazing cues, and viral infection.

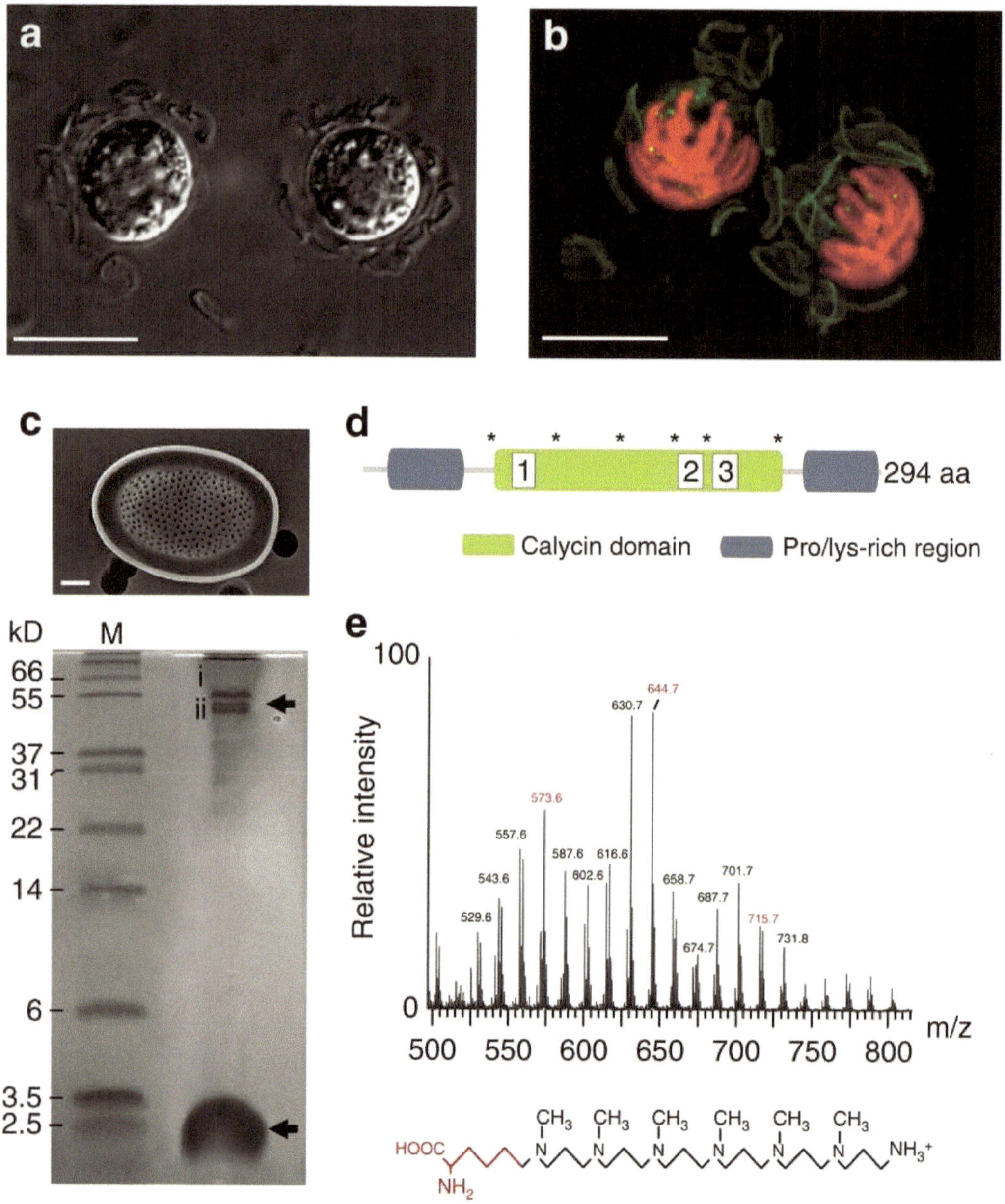

Figure 9.2. Molecular mechanisms of silica scale production in *Prymnesium neolepis*. (a) Differential interference contrast (DIC) micrograph of *P. neolepis* cells displaying the loose covering of silica scales. Scale bar, 10 µm. (b) Confocal microscopy of a *P. neolepis* cell showing incorporation of the fluorescent dye HCK-123 into newly formed silica scales (green). Chlorophyll autofluorescence is shown in red. Scale bar, 10 µm. (c) Tricine/SDS–PAGE of organic components released after dissolution of silica scales with NH_4F. A SEM image of an isolated silica scale is also shown (Scale bar, 1 µm). The higher molecular weight component around 50 kDa is a single protein that runs as two bands (i, ii), whereas the low-molecular-weight components around 2.5 kDa are long-chain polyamines (LCPA). (d) Domain organization of the lipocalin-like protein (LPCL1) identified from both protein bands in NH_4F extracted silica scales. The approximate positions of the proline/lysine-rich regions and the calycin domain (IPR012674) are shown. Also shown are the positions of six highly conserved cysteines (asterisk) that may be involved in the formation of disulphide bridges. (e) Long-chain polyamines (LCPAs) from *P. neolepis* silica scales. The proposed structure of the LCPAs in *P. neolepis* is shown with the putative lysine residue highlighted in red. Reproduced from Durak et al. (2016) under the Creative Commons CC BY license (https://creativecommons.org/licenses/).

Transcriptomic data generated for *P. globosa* strain CCMP1528 revealed marked differences in gene expression profiles between colonial cells and solitary cells (Brisbin and Mitarai 2019). The results indicated that most of the differentially expressed genes were downregulated in the colonial form, providing evidence of a metabolic shift to reallocate resources towards colony production. Chen et al. (2024) conducted a comparative analysis of genomes from strains of the colony-forming species *P. globosa* and other algae, and found the expansion of 926 gene families, impacting 184 KEGG pathways, many of which are associated with colonial formation during *P. globosa* blooms. These include the biosynthesis of various types of N-glycans, O-glycans, polysaccharide chains, amino sugar and nucleotide sugar metabolism, glycolysis/gluconeogenesis, galactose metabolism, and the amino acid metabolism-related pathways encompassing cysteine and methionine metabolism and glycine, serine, and threonine metabolism. In accordance, the expanded genes showed elevated expression during bloom development when colonial densities were high, suggesting that the expansion of gene families facilitates gene expression, which in turn promotes colony formation, favoring bloom development.

Haploid-Diploid Life Cycle of Coccolithophorids

Most haptophytes possess a haplodiplontic life cycle, with distinct heteromorphic differentiation between two ploidy levels, as demonstrated for coccolithophores (Fig. 9.3). Both life cycle stages are capable of independent asexual growth. The diploid (2N) phase is calcified non-motile, whereas the haploid (1N) phase is composed of non-calcified biflagellated cells bearing organic scales. How the complex life cycle is regulated or what factors trigger the transition from one phase to the next is not entirely understood. Interestingly, a study showed that viral infection triggered the shift in *E. huxleyi* from diploid cell type to the more virus-resistant haploid type (Frada et al. 2008), suggesting that viral infection or other similar biotic interactions might be an important influential factor.

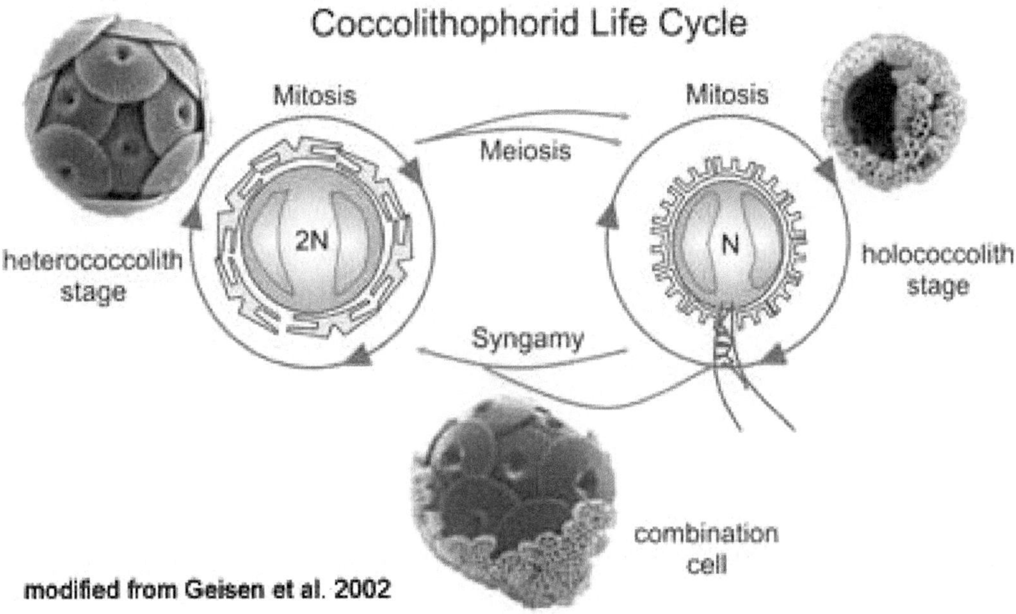

Figure 9.3. Haploid-diploid life cycle of coccolithophores. http://www.geo.uni-bremen.de/cocco/images/stories/life_cycle.jpg.

Viral Infection of Haptophytes

Haptophyte-virus interactions, particularly involving *E. huxleyi*-virus, are among the best studied alga-virus models, in part because the infection cycle can be induced in culture. The collapse of *E. huxleyi* blooms is attributable to the infection by a specific large double-stranded DNA (dsDNA) virus (EhV). Given the complex life cycle characterized by variations in coccoliths, flagellate, and ploidy, it has been proposed that some of these forms, such as the haploid cells of *E. huxleyi*, non-calcified biflagellated cells bearing organic scales, are resistant to viral infection (Frada et al. 2008). However, subsequent studies revealed that the haploid morphotype could be infected by EhV without causing host cell lysis (Mordecai et al. 2017). Viral infection not only influences the dynamics of *E. huxleyi* blooms, but also affects microbial interactions and biogeochemical processes, potentially strengthening or weakening carbon export, playing roles known as "viral shuttle" and "viral shunt", respectively (Vincent et al. 2023).

EhV has developed intricate interactions with its host cells, allowing it to reprogram the host's metabolic network and create the virocell metabolic condition that meets its specific requirement (Rosenwasser et al. 2016). The term "virocell", coined by Patrick Forterre, refers to the 'living form' of the virus within infected host cells, which exhibits active metabolic activities, contrasting with the passive 'free-living' virions existing in the environment (Forterre 2013). EhV belongs to the Phycodnaviridae family, a group of common algae-infecting dsDNA viruses with relatively large genomes (0.3-2.5 Mb), often referred to as giant viruses or Nucleo-Cytoplasmic large DNA Viruses (NCLDV). Sequencing of the EhV86 viral genome revealed 472 genes, including a complete pathway for sphingolipid biosynthesis (Wilson et al. 2005), which plays a direct role in the viral infection cycle (Vardi et al. 2009).

Using high-throughput single-molecule messenger RNA *in situ* hybridization (smFISH) of both virus and host mRNAs, Vincent et al. (2021) quantified active viral infection in the bloom forming coccolithophorid species *E. huxleyi*. In this study, *psbA* which encodes D1 protein, a key component of the photosystem II complex involved in the first step of the light reaction in photosynthesis, was used to monitor photosynthetic activity of *E. huxleyi*, while the major capsid protein MCP-encoding gene (*mcp*) was used to monitor viral replication activity as it is expressed at the late stage of viral infection. The smFISH technique allowed the authors to track the viral infection and host photosynthesis dynamics in the course of infection.

In the early stage, most of the cells were $mcp^-/psbA^+$, indicating active photosynthesis and low viral activity. At four hours postinfection (hpi), 24% of cells were $mcp^+/psbA^+$, indicating simultaneous viral infection and photosynthesis. At 24 hpi, 60% of cells were $mcp^+/psbA^-$, signaling *psbA* transcription shutdown or rapid *psbA* mRNA degradation, while 20% were $mcp^-/psbA^-$. Intriguingly, viral release increased in the $mcp^+/psbA^+$ stage rather than later. The results indicated that in natural assemblages, no more than 25% of the cell population was infected, despite synchronous bloom demise. Furthermore, the work revealed that photosynthetically active cells can chronically release viral particles through nonlytic infection whereas viral-induced cell lysis can occur without viral release, challenging the classical paradigm of viral infection cycle.

In a study of a *Prymnesium parvum* bloom, Wagstaff and colleagues discovered a new virus infecting this algal species (Wagstaff et al. 2018a). This bloom occurred in Spring 2015 on the Norfolk Broads, eastern England, and caused massive fish kills. The identified virus, belonging to the Megaviridae family, had an average capsid diameter of 221 nm, and a genome of approximately 500 Kbp.

Finally, sialic acids (Fig. 9.4) have been shown to mediate viral infection in haptophytes such as *E. huxleyi* and *P. parvum* (Rose et al. 2014, Fulton et al. 2014, Wagstaff et al. 2018). These acidic sugars, comprising a family of over 50 structurally distinct molecules, play crucial roles in cell-cell and host-pathogen interactions on vertebrate cell surfaces. While absent in plants, a structurally related C8 acidic sugar, 3-deoxy-d-manno-2-octulosonic acid (Kdo), is found in plant cell walls.

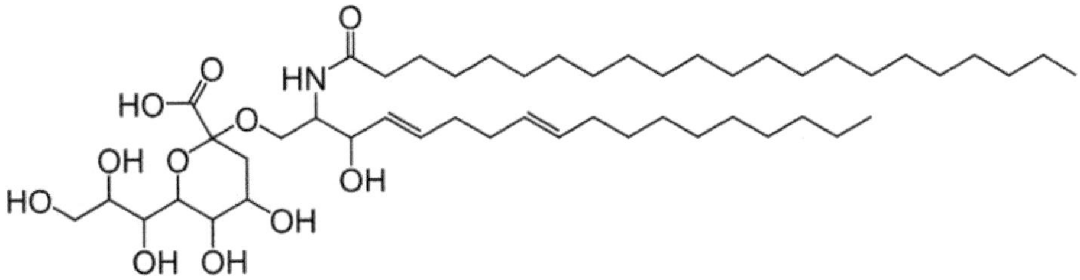

Figure 9.4. Tentative structure of a sialic acid-containing sphingolipid from the haptophyte *E. huxleyi*.

In microalgae, sialic acids were first discovered as Kdn-linked sphingolipid in the lipid rafts of *E. huxleyi*, which are membrane microdomains important for host-virus interactions (Rose et al. 2014). The biosynthesis of the deaminated sialic acids, specially 3-deoxy-d-glycero-d-galacto-2-nonulosonic acid (Kdn), has been biochemically characterized in *P. parvum* (Wagstaff et al. 2018). Transcriptomic analysis of this alga identified two genes in this pathway, Kdn-9-P synthetase (CAMPEP_0191217894) and CMP–Kdn synthetase (CAMPEP_0191219004). The gene for Kdn-9-P phosphatase, involved in the second step of the pathway, was not found in the transcriptome, although the corresponding biochemical reaction was detected.

Production of Dimethylsulfoniopropionate (DMSP) and Dimethylsulfide (DMS)

Haptophytes are one of the most important algal contributors of dimethylsulfoniopropionate (DMSP) in the ocean, which is readily transformed into DMS, a volatile sulfur-containing molecule that enhances cloud formation, thus having a cooling effect on the climate (Graham and Wilcox 2000). Many lineages of algae (including dinoflagellates and diatoms) can produce DMSP. Some algae can also degrade DMSP to DMS, as they possess DMSP lyase-encoding genes (e.g., *E. huxleyi* [Alcolombri et al. 2015]), but others release DMSP to the environment, where bacteria or other organisms (e.g., zooplankton) are responsible for DMSP degradation and DMS production. Some haptophyte species produce high amounts of DMSP and can convert it to DMS, and are thus the group of algae that has been best studied regarding DMSP/DMS production.

Based on differences in one of the biosynthesis steps, two pathways are proposed (Fig. 9.5). Both pathways involve methionine transamination and 4-methylthio-2-oxobutyrate (MTOB) reduction, but then diverge in two ways for converting 4-methylthio-2-hydroxybutyrate (MTHB) to dimethylsulphonio-2-hydroxybutyrate (DMSHB). In this methylation step, one pathway involves the enzyme DSYB and operates in high DMSP producers (HiDP) whereas the other involves the enzyme MT2 and operates in low DMSP producers (LoDP). *P. globosa* not only possesses both the HiDP and LoDP pathways, but also has duplicated DYSB genes, suggesting the evolutionary reinforcement of the ability to produce DMSP (Chen et al. 2024). The conversion of DMSP to DMS involves Alma. The *P. globosa* genome harbors two to three Alma genes (depending on the strain), again indicating the strengthening of DMSP degradation, similar to that of DMSP production, in this colonial haptophyte (Chen et al. 2024).

Synthesis of Unique Lipids

Fatty acids, sterols, alkenones, alkenoates, alkenes are rich in haptophytes (Volkman et al. 1980). As discussed earlier, some of the haptophyte species are highly rich in fatty acids, making them ideal food for animals. Furthermore, alkenones and alkenoates have been used as biomarkers to track plankton populations over time in paleo-oceanographic and paleoclimate studies (Volkman et al. 1995). The utility of alkenones and alkenoates stems from the correlation between their degree of unsaturation and temperature at which the algae grow (Prahl and Wakeham 1987, Volkman et al. 1995).

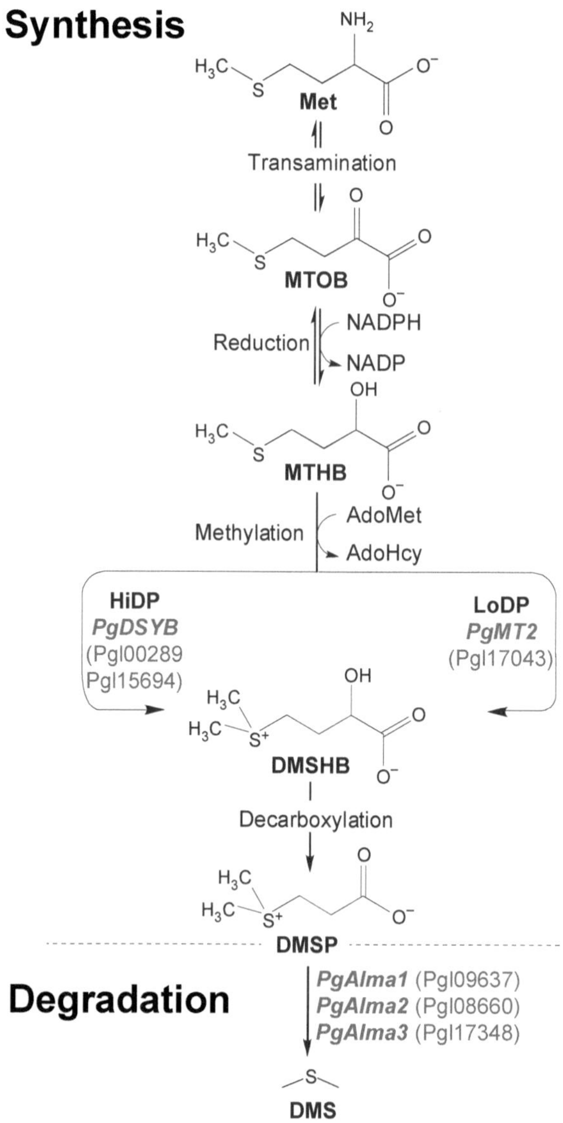

Figure 9.5. Two pathways for DMSP biosynthesis and degradation to DMS in algae based on information on the haptophyte *Phaeocystis globosa*. Reproduced with permission from Chen et al. (2024) under the Creative Commons CC BY 4.0 license.

Some aspects of the alkenone and alkenoate biosynthesis machinery have been elucidated. A proteomic investigation revealed several abundant proteins co-occurring with the alkenone bodies (AB) in *Tisochrysis lutea* (Shi et al. 2015). These AB-associated proteins include V-ATPase subunit V_A, V-ATPase subunit V_d, hypothetical protein EMIHUDRAFT_465,517, coccolith scale associated protein-1, cycloartenol-c-24-methyltransferase 1-like and SPFH domain-containing protein. Except SPFH domain-containing protein, which remained unchanged, the genes encoding these proteins were upregulated when AB size increased in *T. lutea* grown under N-nutrient deficient conditions (Shi 2019). Among these genes, the expression of the coccolith scale associated protein-1 gene was up-regulated by 50-650 fold. These results suggest that these vacuole ATPases may be involved in the biosynthesis of alkenones in haptophytes.

Because alkenones are long-chain neutral lipids, their biosynthesis must involve desaturation reactions. A gene, named Akd1, was identified from *T. lutea* that appears to encode for one of these desaturases (Endo et al. 2018). The artificial overexpression of this gene led to increases in the conversion of di-unsaturated C_{37}-alkenone to tri-unsaturated molecule by alkenone desaturation. In the wild type cells exposed to cold stress, which induced elevation of desaturation, Akd1 exhibited drastic upregulation.

Sialic acid in *E. huxleyi* is part of the lipid raft that mediates viral infection, as indicated earlier. As demonstrated in *P. parvum*, the biosynthesis of deaminated sialic acid, 3-deoxy-d-glycero-d-galacto-2-nonulosonic acid (Kdn) involves Kdn-9-P synthetase and CMP-Kdn synthetase that convert mannose-6-P to CMP-Kdn. Bioinformatic analysis revealed the occurrence of the genes coding for these enzymes in *P. parvum* (Wagstaff et al. 2018). However, whether this biosynthesis pathway occurs in other haptophyte species remains to be investigated.

10

Genomes of Cryptophytes and Other Algal Phyla

For the many other algal phyla, genomic research has not advanced as far as the major phyla treated in previous chapters. Due to the small amount of genomic information currently available and the scope of the book, they are combined in one chapter. These include the "red-line" lineages Cryptophyta, Pelagophyta, and Raphidophyta, the "green-line" lineage Euglenophyta, and the lineage that represents the latest primary endosymbiosis known to date, *Paulinella* from the phylum of Cercozoa (Fig. 10.1). The genomes from these phyla that have been sequenced are listed in Table 10.1.

Cryptophytes

Cryptophytes are a small group of unicellular algae, with estimates of 12–23 genera and approximately 200 species equally split between marine and freshwater taxa (Graham and Wilcox 2000). They are estimated to have emerged anywhere from 500 to 200 Ma (Zauner et al. 2000). The asymmetric cell body with a furrow-gullet system at near the apical point makes cryptophyte morphologically unique (Fig. 10.1). They seem to favor oligotrophic, temperate, and high latitude waters, typically becoming abundant in winter and early spring. Cryptomonads are a favorite food source of heterotrophic or mixotrophic protists in the estuaries such as the dinoflagellates *Pfiesteria* spp. probably due to their high protein (phycobiliproteins) content. Some of these protistan grazers can retain the ingested cryptomonad cells (symbionts) or their chloroplasts alone (kleptoplastidic, 'plastid-stealing') for temporary photosynthesis. For instance, *Pfiesteria* spp. and *Dinophysis* spp. feed on cryptophytes and retain their chloroplasts for temporary photosynthesis (Feinstein et al. 2002, Qiu et al. 2011). However, kleptoplastids quickly lose their functions and need to be replenished continually, apparently because the hosts do not have all the genes needed to maintain and regulate the chloroplast (Wisecaver and Hackett 2010). As elaborated in Chapter 20, ciliates represent another notable example of cryptophyte-dependent protists. *Mesodinium rubrum* feeds on cryptomonads such as *Teleaulax*, *Geminigera*, or *Plagioselmis*, and retains the cells or all organelles of the cryptomonads such as *Teleaulax*, *Geminigera*, or *Plagioselmis* (Johnson et al. 2007, Qiu et al. 2016). Another ciliate, *Strombidinium*, also steals plastids from cryptophytes (McManus et al. 2018). A single-cell transcriptomic analysis on *Strombidium* cf. *basimorphum* showed that this species was also able to retain the chloroplast and transcriptionally active genetic materials from the cryptophyte prey (Maselli et al. 2021).

One unique feature of cryptophytes is the presence of nucleomorph, the remnant of the nucleus of the red algal endosymbiont that gave rise to the plastid (Gillott and Gibbs 1980). Cryptophyte

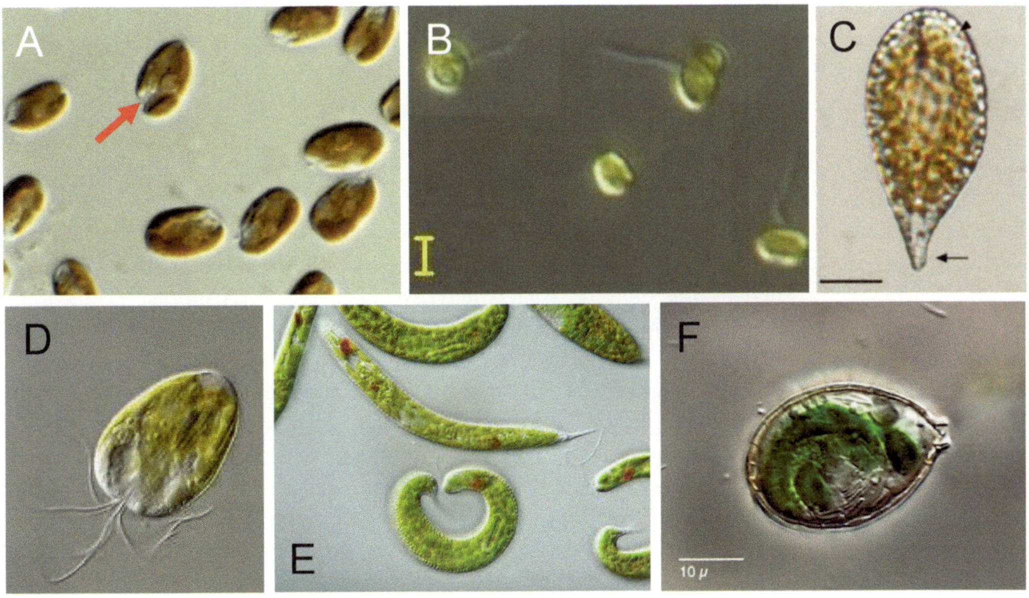

Figure 10.1. Morphologies of cryptophytes, prasinophytes, euglenophytes, and *Paulinella*. (A) Cryptophyte cells showing the asymmetric cell body and the furrow-gullet system unique for cryptophyte (arrow). Courtesy of By CSIRO, CC BY 3.0, https://commons.wikimedia.org/w/index.php?curid=35486525. (B) *Pelagomonas calceolate* (strain RCC0978), scale bar = 2 μm. Image is a courtesy from the Roscoff Culture Collection). (C) Raphidophyte cells. After Branco et al. 2019 with permission from Phycol Res. https://doi.org/10.1111/pre.12377. (D) Prasinophyte *Pyramimonas amylifera*. Resproduced with permission of SciencePhotoLibrary (https://www.sciencephoto.com/keyword/prasinophytes). (E) *Euglena* sp. reproduced with permission from https://www.britannica.com/science/Euglena. (F) *Paulinella chromatophora*. Reproduced with permission from https://www.plingfactory.de/Science/Atlas/KennkartenProtista/source/Paulinella_chromatophora.html.

Table 10.1. Sequenced genomes of Cryptophyta, Pelagophyte, Raphidophyta, Euglenophyta, and Paulinella.

Species	Genome size (Mb)	N0. of chromosomes	No. of coding genes	Coding percentage (%)	%GC	References
Cryptophytes						
Guillardia theta	87.2	3	24,840	42.73	53	Curtis et al. 2012
Pelagophytes						
Pelagomonas calceolata	32.4	6	16,667	77.16	63.6	Guérin et al. 2021, Sibbald et al. 2021
Aureococcus anophagefferens	56	22	11,570	30.99	67.4	Gobler et al. 2011, Moniruzzaman et al. 2018
Chrysophaeum taylorii	71		10,810	22.84		Davison et al. 2023
Raphidophytes						
Chatonella antiqua		29	67,823		42	Shikata et al. 2019
Chatonella subsala			41,237		42	Shikata et al. 2019
Heterosigma akashiwo			45,293		52	Shikata et al. 2019
Euglenaophytes						
Euglena gracilis	500		36,000	< 1	50	Ebenezer et al. 2019
Paulinella						
Paulinella chromataphora	707		32,361	6.87	44	Lhee et al. 2021

plastids are bound by four layers of envelope membranes, two inner and two outer. The outermost membrane is continuous with the endoplasmic reticulum, which is connected to the outer membrane of the nuclear envelope. The nucleomorph is located between the two pairs of inner and outer plastid membranes.

Guillardia theta

Curtis et al. (2012) sequenced the genome of *Guillardia theta* and revealed 24,840 protein genes in this 87-Mbp nuclear genome. *G. theta* nuclear genome is gene rich relative to other secondary plastid-bearing algae such as the diatoms *Phaeodactylum tricornutum* (Armbrust et al. 2004) and *Thalassiosira pseudonana* (Bowler et al. 2008) and the filamentous brown alga *Ectocarpus siliculosus* (Cock et al. 2010). Besides, *G. theta* has 145 genes in the plastid, 36 in the mitochondria, and, remarkably, 487 in the nucleomorph. Importantly, *G. theta* is genetically and biochemically highly mosaic, with both host- and endosymbiont-derived genes servicing the mitochondrion, the host cell cytosol, the plastid and the remnant endosymbiont cytosol (Curtis et al. 2012). There is evidence that mitochondrion-to-nucleus gene transfer is still occurring in *G. theta*, but plastid-to-nucleus and nucleomorph-to-nucleus gene transfers is not, indicating that the plastid evolution is complete in this lineage.

Goniomonas avonlea

Sequencing the genome of the plastidless cryptophyte *Goniomonas avonlea* resulted in a ~ 92 Mbp genome assembly and predicted 33,470 protein genes (Cenci et al. 2018). Comparative genomics analyses have indicated that some metabolic pathways, such as fatty acid biosynthesis, which are predicted to occur in the plastid and periplastidal compartment of *Gu. theta*, operate in the cytoplasm of *Go. avonlea*. This suggests that metabolic redundancies were generated during secondary plastid evolution. Conversely, certain cytosolic pathways present in *Go. avonlea* are not found in *Gu. theta*, indicating secondary loss in *Gu. theta* and other plastid-bearing cryptomonads. Interestingly, no algal endosymbiont (plastid)-derived genes were identified in the *Go. avonlea* genome, suggesting that the absence of a plastid is not due to secondary loss but indicates that goniomonads are ancestrally non-photosynthetic (Cenci et al. 2018).

Pelagophytes

Pelagophytes (Heterokontophyta) are a morphologically diverse class of marine algae united by DNA sequences (Wetherbee et al. 2021). Initially found in marine plankton, recent studies revealed occurrence of sand-dwelling taxa. The number of species has grown rapidly from the initial recognition of *Aureococcus anophagefferens* and *Aureoumbra lagunensis*, known for their 'brown tide' blooms, to the inclusion of multiple species in the new genera *Gazia* and *Glomerochrysis*. High-pressure freezing of samples revealed a distinctive perforated theca in all pelagophyte genera, suggesting this as a unique morphological feature of this lineage (Wetherbee et al. 2021). Some of these species (e.g., *Ga. saundersii*) form colonies of thousands of cells.

Aureococcus anophagefferns

A. anophagefferens is a small nanoplanktonic unicellular alga (~ 5 μm in diameter) that can outcompete co-occurring phytoplankton in estuaries with elevated levels of DOM and turbidity and low levels of DIN. Its genome, at 56 Mbp, is larger with markedly more genes involved in light harvesting and the utilization of organic carbon and nitrogen nutrients than the other co-occurring phytoplankton species (Gobler et al. 2011). Additionally, the genome encodes genes for the synthesis of microbial deterrents, likely reducing mortality during blooms and facilitating species population proliferation. Furthermore, it contains abundant genes for metal-dependent proteins and has the highest number of selenoprotein genes reported so far among phytoplankton (Gobler et al. 2013), indicating a high capacity to cope with stress. The abundance of selenoprotein genes in this small alga mirrors that found in the prasinophyte *Ostreococcus* (Palenik et al. 2007), which will be discussed later in this chapter.

To assess responses to stressors for dense phytoplankton blooms for *A. anophagefferens*, a transcriptome profiling was performed for strain CCMP 1850 (Frischkorn et al. 2014). *De novo* assembly using Trinity yielded approximately 54,000 transcripts, while assembly based on ABySS resulted in around 31,000 transcripts. On average, 93% of the significantly upregulated transcripts identified through genome mapping were also found in the significantly upregulated pool detected by both *de novo* assembly methods. Transcripts associated with nitrogen transport and metabolism exhibited elevation of expression in the low nitrogen treatment, while those encoding enzymes involved in DOP hydrolysis or arsenic detoxification showed upregulation in the low P treatment. Additionally, transcripts related to organic compound catabolism, lipid membrane restructuring, and sulfolipid biosynthesis were upregulated in the low light treatment. Comparing this transcriptome to the nutrient-regulated transcriptional response of strain CCMP 1984 revealed shared responses between the two strains. These analyses unveil the transcriptional mechanisms underlying physiological changes that may enhance the ecological adaptation of this species in its natural environment, including organic matter processing, metal detoxification, lipid restructuring, and turnover of the photosynthetic apparatus.

Pelagomonas calceolata

P. calceolate was the first described species of the stramenopile class of Pelagophyceae, which has been shown to be widely distributed in the global ocean (Worden et al. 2012). The cell has a single flagellum that is anteriorly directed and has two rows of tripartite hair. Studies have shown that it can adapt to different environmental conditions, exhibiting high plasticity to light fluctuations with rapid activation of the xanthophyll cycle and non-photochemical quenching observed in laboratory experiments (Dimier et al. 2009). It also shows strong responsiveness to iron fertilization documented in the natural sea (Caputi et al. 2019). The iron-responsive trait is consistent with the findings of genes associated with low-iron adaptation in pelagophytes throughout the global ocean (Carradec et al. 2018). Besides, when grown under low N-nutrient conditions in the laboratory, *P. calceolate* upregulated the expression of genes coding for cathepsin, urease, arginase, which function in utilizing dissolved organic nitrogen (Kang et al. 2021).

Guérin et al. (2022) sequenced the 31 Mbp genome of *P. calceolate* (strain RCC100) using the Oxford Nanopore Technologies (ONT). The ONT long reads were assembled into six nuclear contigs (32.4 Mbp), one circular plastid contig (90 kbp), and one circular mitochondrial contig (39 kbp). The authors detected two large and highly similar duplicated regions (> 99% identity) at the extremity of contigs 1 and 5 (393 kbp) and the extremities of contigs 3 and 6 (192 kbp). Contigs 2, 3, 4, and 6 end at both extremities with the telomeric repeats (TTAGGG)n, indicating that these four contigs represent complete chromosomes. A total of 16,667 genes was predicted on the genome, which shows 94% completeness with BUSCO. Fifty-eight of the predicted proteins (9,812) are homologous with at least one gene in a stramenopile genome, but surprisingly, only 16% of these annotated genes (2,631) are shared with the pelagophyte *A. anophagefferens*. Further analysis showed that the average GC content was 63%, and regions encompassing centromeres had a GC of 52%, a low GC content attributable to the suppression of meiosis in centromeres (Nambiar and Smith 2016).

The *P. calceolata* genome encodes a repertoire of 18 meiosis genes, including the double-strand DNA break (DSB) initiator SPO11; the DSB-binding proteins RAD50, RAD52, and MRE11; proteins involved in pairing and invading the homologous strand HOP2, MND1, DMC1, and

RAD51; and proteins involved in the synthesis-dependent strand annealing pathway MSH2, MSH3, PMS1, and MSH6; and the protein essential to non-interfering (class II) cross-over MUS81.

Using metagenomic data (instead of rRNA data, which could cause biases due to copy number variations between species) from the *Tara Oceans* expedition, Guérin and colleagues (2022) found that *P. calceolate* accounted for 1.39% in surface samples and 2.67% in DCM samples for the 0.8–5 μm size fraction. The numbers decreased to 1.01% for surface and 1.56% for DCM samples in the 0.8-2000 μm size fraction. Comparing with environmental conditions under which the samples originated, the authors discovered that *P. calceolate* was most abundant in temperate, low-light and iron-poor environments. *P. calceolata* has five genes coding for the phytotransferrin ISIP2A, which is involved in Fe^{3+} uptake via endosomal vesicles and two putative iron-storage protein ISIP3. Mining the *Tara Oceans* metatranscriptomic data reveals that three ISIP2A and one ISIP3 were overexpressed in iron-poor environments. This result indicates that, similar to diatoms, *P. calceolata* upregulated ISIP2 following iron starvation, potentially to improve cell growth in low-iron environments. Furthermore, *P. calceolata,* like diatoms, possesses both flavodoxin (11 genes) and ferredoxin (17 genes), an iron-free and iron-containing protein, respectively, and replaces ferredoxin with flavodoxin when iron is depleted in the environment. Additionally, eight zinc/iron permeases potentially involved in iron uptake form the environment were identified from the genome. Two of these were upregulated in high-iron conditions and one in low-iron conditions. Interestingly, the iron permease FTR1, the iron storage ferritin, and the iron-starvation induced protein ISOP1 (involved in endosytosis of siderophores in diatoms) are missing in the *P. calceolate* genome. In comparison, the coastal pelagophyte *A. anophagefferens* lacks ISIP3 genes, has fewer zinc-iron permeases, and possesses five flavodoxin genes and nine ferredoxin genes.

The presence of three ferroportin genes in *P. calceolata* is interesting, as these transmembrane iron export proteins play a significant role in iron homeostasis in multicellular organisms. While the function of ferroportin in microalgae is unknown, it could potentially act to export iron from endosomes to the cytoplasm (Turnšek et al. 2021). In the green alga *Chlamydomonas reinhardtii,* a ferroportin gene is overexpressed under low Fe conditions (Urzica et al. 2012). However, no significant changes were found in the expression of the three *P. calceolata* ferroportin genes according to iron concentration. The function of these genes in iron nutrition remains to be studied.

The *P. calceolaae* genome also contains genes necessary for utilizing nitrate and nitrite (nitrate/nitrite transporter, formate/nitrite transporter, nitrate reductase, nitrite reductase), ammonium (ammonium transporter), urea (urea transporter, urease), and cyanate (cyanate lyase). Similar to diatoms, this pelagophyte has a complete urea cycle. Additionally, it possesses a higher number of genes involved in the assimilation of ammonium than other species: five Glutamine Synthetase (GS) and four glutamine oxoglutarate amine transferase (GOGAT) genes. As observed in other algae, the nitrogen transporter genes are upregulated in low-N nutrient environments, while the reductase and GS/GOGAT genes exhibit different response patterns to nitrogen availability. Notably, cyanate lyase is upregulated in low-nitrate environments, suggesting that cyanate is an important alternative source of nitrogen for *P. calceolata.*

The *P. calceolata* genome harbors three genes that possess the nitrate and nitrite sensing (NIT) domain (IPR013587). One of these genes has the NIT-sensing domain flanked by two transmembrane domains, suggesting its potential for external nitrate/nitrite sensing. The other two genes contain a protein kinase domain (IPR000719), suggesting a phosphorylation-based signal transduction mechanism that is dependent on intracellular nitrate or nitrite concentration. Interestingly, the gene associated with potential external NIT sensing is significantly overexpressed in low-nitrate environments.

Chrysophaeum taylorii

Chrysophaeum taylorii is a colonial pelagophyte species known to accumulate bioactive natural products, including hormothamnione and the chrysophaentin family of antibiotics. To gain insights into the molecular mechanisms underpinning this remarkable capability and expand genomic studies of pelagophytes to a colonial species, researchers sequenced the ~ 71 Mbp genome of *C. taylorii* using a combination of Illumina HiSeq, Oxford Nanopore, and Pacific Biosciences (Davison et al. 2023). The genome data revealed a wide range of secondary metabolite biosynthetic genes. Interestingly, the biosynthesis of chrysophaentins appears to involve non-canonical polyketide synthase (PKS) genes. Thirteen unique PKS genes were identified from the genome, showing significant divergence from known fungal and bacterial counterparts.

The study by Davison et al. (2023) provides valuable insights into the genomic characteristics of *C. taylorii* and its capacity for secondary metabolism. The research highlights the extensive complement of secondary metabolite biosynthetic genes in the genome and suggests the involvement of non-canonical PKS genes in chrysophaentin biosynthesis. These findings contribute to our understanding of the molecular mechanisms underlying the bioactive natural product accumulation in *C. taylorii*.

Raphidophytes

Raphidophytes are a small group of eukaryotic microalgae that belong to the supergroup heterokonts (Stramenopiles). The were previously considered part of the division Xanthophyta (yellow-green algae). However, they are currently classified as a class of the larger division Ochrophyte, which also includes diatoms and Phaeophyta (Graham and Wilcox 2000). It is important to note that the classification may vary depending on the systematics scheme. Raphidophytes consist of approximately five genera and 10 species. They are widely distributed in both marine and freshwater environments. In marine environments, raphidophytes often form Harmful Algal Blooms (HABs). Two common genera associated with HABs in the ocean are *Chattonella* and *Heterosigma*. Raphidophyte cells have no cell wall, with two unequal flagella. Primary pigments include Chl a, c_1, and c_2, and primary accessory pigments include β-carotene, diatoxanthin, diadinoxanthin, and vaucheriaxanthin.

Chattonella marina

Chattonella marina has been reported to possess the ability to undergo Extracellular Electron Transfer (EET) when exposed to light (Li et al. 2015). Initially, EET was primarily associated with anaerobic biofilms at mineral surfaces, but it was later discovered in photosynthetic microbes, such as cyanobacteria (Pisciotta et al. 2011). Among these microbes, cable bacteria are notable champions of EET, as they can conduct electron transfer over centimeter distances. While only a few species of eukaryotic algae have been shown to generate current in a light-dependent manner (McCormick et al. 2011), bacteria, in which EET is well-characterized, are believed to mediate the reaction through a chain of proteins. These proteins first produce electrons in the inner layer of the cell membrane and then relay them to the outer surface of the cell membrane, where electron receptors like Fe (III) and DMSP exist in the hydrosphere (Jiang et al. 2019). For oxygenic phototrophs like *C. marina*, the electrons are likely produced in PSII. However, the specific mechanisms by which these electrons are transferred to the extracellular area remain to be unveiled through genomics and functional genetic studies.

C. marina has not had its genome sequenced. However, researchers conducted RNA-seq for another species, *C. antiqua*, to identify genes related to the dynamics of the noxious red tide caused by this lineage (Shikata et al. 2019). RNA sequencing of daytime and nighttime *C. antiqua* cDNA libraries produced 186 million paired-end reads (2 × 101 bp). Assembly of these reads yielded 67,823 contigs, from which 39,031 coding sequences were discovered. The transcriptomic analysis of *C. antiqua* revealed the presence of orthologous genes involved in nutrient uptake, photosynthesis, and ichthyotoxicity. Notably, genes of photoreceptors such as aureochrome, cryptochrome/photolyase (CRY/PHR), and phytochrome were retrieved. These genes play a crucial role in regulating photoresponses and various other physiological processes, including flagellar motion and the cell cycle, as observed in model organisms. Regarding ichthyotoxicity, the study also unveiled the existence of two types of NADPH oxidase. These oxidases produce superoxide on the cell membrane, leading to respiratory burst. One type has 5–6 transmembrane domains, while the other has 11 membrane domains.

Chattonella species have a need to protect their own cells from the oxidative stress they impose to grazers and coexisting phytoplankton. Studies have found that *C. marina* exhibits a high level of peroxiredoxin (Prx) production during the exponential growth stage (Shimasaki et al. 2021). This has implications in its antioxidative role, promoting its bloom while suppressing competing algal species. The development of blooms in *Chattonella* species is often triggered by nutrient input and promoted by diurnal vertical migration to exploit nutrients in deep water at night and photosynthesize in surface on daytime. For these species, nutrient uptake, photoreception, and the ability to rapidly respond to changes in light are crucial. Additionally, they possess a strong capability to produce photoprotective and oxidation-protecting elements. The transcriptomic data obtained from *C. antiqua* (Shikata et al. 2019) serve as a foundation for further exploration into the specific mechanisms underlying *Chattonella* blooms.

Heterosigma akashiwo

Heterosigma akashiwo is distributed globally and able to form harmful algal blooms and produce ichthyotoxin that kills wild and farmed fish (Honjo 1992). The growth and HAB formation of *H. akashiwo* has been ascribed to light, temperature, nutrient, pCO_2, salinity, and interacting organisms. It has been found the rising CO_2 during the ongoing climate change has strong effects on the growth of *H. akashiwo* (Hennon et al. 2019).

The nuclear genome of *H. akashiwo* was initially estimated to be about 3 Gbp (Cattolico et al. 1976). The large size of the genome has prevented its sequencing so far, but there have been transcriptomic studies. A set of transcriptomes was generated for *H. akashiwo* cultures grown under elevated temperature and CO_2 concentration (Thangaraj and Sun 2023). The data revealed, among other gene expression characteristics, the presence and expression of cyclophilins G and D. Between the two types, cyclophilin G was dominant. Another transcriptome sequencing was conducted for *H. akashiwo* cultures with or without viral infection (Sato et al. 2023). It generated 84,693,530 high-quality deduplicated short-read sequences, which was assembled into 144,777 contigs with N_{50} being 1,085 bp. From these contigs, 82,716 gene models were predicted, 60,877 of which were coding genes. The most abundant types of pfam domains in the transcriptome were ankyrin repeats, tetratricopeptide repeat, AAA ATPase domain, AAA domain, ABC transporter, EF hand, Zinc finger, and various kinases.

To create a reference transcriptome to facilitate ecological genomic studies on *H. akashiwo*, Ji et al. (2023) conducted a full-length transcriptome sequencing using the PacBio Sequel sequencing platform. To maximize the coverage of the genome, diel samples were collected, and the RNA extracts were pooled for the generation of the Single-Molecule Real-Time (SMRT) cDNA libraries for sequencing. Sequencing runs produced 48.66 Gbp, from which 16,668 unigenes were obtained, with an N_{50} of 2007 bp. A total of 8,666 (52.00%) unigenes were successfully annotated, of which 6478 (38.86%) and 6551 (39.30%) were enriched in the GO terms and the KEGG database, respectively. From the data, major uptake and assimilation pathways of nitrogen and phosphorus nutrient were assembled for the species. Besides, 4,392 long non-coding RNA and 300 putative transcription factors were predicted.

A metatranscriptomic study of a *H. akashiwo* bloom event revealed significant metabolic shifts to promote the growth of the bloom species (Ji et al. 2018). RNA-seq analysis identified 3,378 differentially expressed transcripts, with 2,826 differing between the pre-bloom and bloom samples. Genes involved in nutrient uptake, such as ATP-binding cassette (ABC) transporters and phosphatases, were upregulated in the bloom. Additionally, (ppGpp)ase, an enzyme involved in the degradation/synthesis of pp(G)pp—a stress response nucleotide—was also upregulated. Genes related to phagocytosis were upregulated in the morning of the bloom, while photosynthesis and cell division-related genes were upregulated in the afternoon. These findings suggest that *H. akashiwo* possesses a metabolic programming system for nutrient utilization, photosynthesis, and cell proliferation.

Euglenophytes

The phylum Euglenozoa comprises four main groups: euglenids (including the phototrophic Euglenophyceae), kinetoplastea (parasitic flagellates), diplonemids (mostly found in marine environments), and Symbiontida (inhabiting hypoxic marine environments and hosting epibiotic bacteria). The phylum exhibits distinct molecular genetic features, including polycistronic transcription, *trans*-splicing, mRNA editing of mitochondrion-encoded genes, and, in some cases, the absence of introns. These characteristics resemble those of dinoflagellates, suggesting convergent evolution between the two protist lineages (Lukeš et al. 2009). Euglenophyte genomes are also known for possessing the noncanonical base "J" (β-D-glucopyranosyloxymethyluracil), a hypermodified nucleobase replacing approximately 1% of all thymidines (Dooijes et al. 2000).

Within the class Euglenophyceae, approximately 3,000 species have been described (Bicudo and Menezes 2016). Most of these species are free-living, unicellular flagellates with one to several plastids, engaging in phototrophic activities. Euglenozoa is one of the two groups that possess secondary green plastids, the other being chlorarachniophytes. Euglenoid plastids originated from a close relative of *Pyramimonas*, which acted as an endosymbiont within phagotrophic euglenids (Zakryś et al. 2017). These plastids have three bounding membranes and contain chlorophyll *a* and chlorophyll *b*. However, this phylum also includes colorless species that have lost their ability to photosynthesize and lack the photosensory apparatus associated with cilia (known as secondary osmotrophic species). One unique member of the phylum is *Rapaza viridis*, which possesses a functional chloroplast but cannot survive without the presence of a prasinophyte alga (Yamaguchi et al. 2012).

Euglena gracilis

To gain an understanding of the complex biology and evolutionary processes of Euglenophytes, a draft genome was created for *Euglena gracilis* (Ebenezer et al. 2019), followed by a chromosome-level assembly (Chen et al. 2024). The assembly, approximately 2.37 Gbp in size, was organized into 42-46 linear chromosomes and has a GC content of around 50%. The genome encodes 39,362 protein genes, with 80.39% BUSCO completeness, accounting for less than 1% of the genome. Findings suggest similarities among gene families involved in nuclear architecture and protein trafficking. There are also indications of large lineage-specific paralogs, suggesting significant flexibility in environmental monitoring and adaptation. Notably, most protein expression is post-transcriptionally regulated, with little transcriptional regulation despite the presence of conventional introns. The genome data reveal a highly sophisticated endomembrane system, RNA processing mechanisms, and a nuclear genome contributed by several photosynthetic lineages. This finding is consistent with earlier studies using Expressed Sequence Tags (EST), which found genes acquired via eukaryote-to-eukaryote transfer from algal sources other than green algae, the progenitor algae of euglenids (Maruyama et al. 2011). The eukaryote-to-eukaryote gene transfer might be an imprint of Endosymbiotic Gene Transfer (EGT) from a secondary or tertiary endosymbiosis or gene acquisition from phagocytosed algae.

Ebenezer et al. (2019) found that several gene families, potentially comprising signal transduction elements, have undergone significant expansion. Post-transcriptional regulation, rather than transcriptional regulation despite the presence of conventional introns, governs changes in protein abundance between light and dark conditions, exhibiting striking similarities to trypanosomatids. The extensive gene family expansion and gene acquisition, along with unique regulatory mechanisms, have important implications for the organism's extreme adaptability.

Supporting the evidence of secondary loss of photosynthesis among euglenid species, a transcriptomic study of the colorless species *Euglena longa* confirmed the absence of genes for photosynthetic mechanisms but showed plastid-localized proteins bearing N-terminal bipartite topogenic signals (BTSs) (Zahonova et al. 2018). Furthermore, the study unveiled salient aspects of plastid infrastructure, such as the plastidial targeting of several C-terminal translational fusions with BTS-bearing proteins. Investigating whether such unique gene structures exist in photosynthetic euglenid species, such as *E. gracilis*, will be insightful.

Paulinella

The genus *Paulinella*, belonging to the phylum Cercozoa, consists of both freshwater and marine species. These organisms have cells covered by rows of siliceous scales and utilize filose pseudopods to move across the substrate in the benthic zone. The best-known members of this genus are three photosynthetic species: *P. chromatophora*, *P. micropora*, and *P. longichromatophora*. The first two species are found in freshwater environments, while the third species is a marine form.

Belonging to the phylum of cercozoan amoeboids, these phototrophic species are believed to have recently (~ 124 million years ago) acquired a cyanobacterium as an endosymbiont. This endosymbiont serves as a novel photosynthetic organelle called a chromatophore or cyanelle. The chromatophore derived from an α-cyanobacterium, which is distinct from the β-cyanobacterium that gave rise to common primary plastids approximately 1.6–2.1 billion years ago (Marin et al. 2005, Yoon et al. 2006, Gabr et al. 2022). The size of the chromatophore genome is about one-third the size of closely related cyanobacteria genomes but ten times larger than most plastid genomes (Zhang et al. 2017). Thus, *Paulinella* provides living evidence of a second event of primary endosymbiosis (Marin et al. 2005, Gabr et al. 2022).

Furthermore, the chromatophore is still in the early stage of integrating into a fully functional plastid. The first line of evidence is the limited Endosymbiont-to-host Gene Transfer (EGT) and the presence of genes encoding major subunits of PSI and PSII, ATP synthase, phycobilisome, and the cytochrome b6f complex in the chromatophore genome. Additionally, certain photosynthetic genes, such as the PSI protein PsaI and the carboxysome shell protein Csos4A, are encoded in both the chromatophore and nuclear genomes (Zhang et al. 2017).

Chromatophore Genome

The chromatophore genome of *Paulinella* was sequenced to elucidate its metabolic capabilities and integration with its host (Nowack et al. 2008). The data revealed a significant reduction in the genome compared to cyanobacteria genomes, with a single circular chromosome of 1.02 Mb encoding 867 protein-coding genes, more than many chloroplasts. The chromatophore lacks genes not essential for intracellular lifestyle and crucial pathways involved in amino acid and cofactor synthesis. A study comparing the chromatophore genomes of *P. longichromatophora* and *P. micropora* to that of *P. chromatophora* showed that only 35% of the ancestral orthologous gene families from the cyanobacterial endosymbiont remained in chromatophore DNA after the basal split among photosynthetic *Paulinella* species approximately 60 million years ago. The examination of nucleotide substitution and insertion/deletion occurrences across various *P. micropora* strains revealed ongoing and dynamic processes of chromatophore evolution, characterized by increasing AT-content and genome reduction (Lhee et al. 2019). The chromatophore, as a photosynthetic entity, is dependent on its host for growth and survival.

Nuclear Genome

Prior to any whole-genome sequencing, a reference transcriptome dataset was generated for *P. chromatophora* (Nowack et al. 2011). The data revealed several key points: (1) an estimated 0.3–0.8% of the nuclear genes might have been obtained through endosymbiotic gene transfer (EGT); (2) the transferred genes tend to encode small proteins involved in photosynthesis and photoacclimation; (3) the host cells have established control over the expression of transferred genes; and (4) Horizontal Gene Transfer (HGT) from food organisms, though to a lesser extent than EGT, might have also contributed to shaping the nuclear genome.

To understand the rules that govern organelle integration into cellular metabolism, a draft genome assembly was achieved from the first bacteria-free *P. chromatophora* culture (Nowack et al. 2016). Among 229 nuclear genes acquired via HGT from various bacteria, 25% likely arose through EGT from the chromatophore genome. Many of the HGT-derived bacterial genes encode proteins that fill gaps in critical chromatophore pathways and processes, highlighting HGT's role in compensating for organelle genome reduction.

The nucleus encodes not only proteins with chromatophore-targeting signals but also a substantial number of short Chromatophore-Targeted Proteins (sCTPs; fewer than 90 amino acids) lacking identifiable targeting signals. Three putative DNA-binding (DB)-sCTPs were isolated as recombinant DB-sCTPs for functional exploration. All three proteins possess antimicrobial activity by permeabilizing bacterial membranes and binding to bacterial lipids *in vitro*. The membrane binding of DB-sCTPs may facilitate their import into chromatophores, where they could potentially influence genetic information processing under nuclear control (Macorano et al. 2023).

Numerous genes derived from EGT encode proteins involved in photosynthesis and photoprotection, including an expanded family of high-light-inducible (ncHLI) proteins. Homologs of ncHLI proteins in cyanobacteria aid in cell viability under excess light. Zhang et al. (2017) examined the impact of light on *P. chromatophora* and found it lacked light-induced transcriptional regulation and was thus light-sensitive. However, several ncHLI genes have reestablished light-dependent regulation, and the gene family has expanded. This regulation may reflect the light and oxidative stress experienced by *P. chromatophora* due to the incomplete integration of host and chromatophore metabolisms.

There is evidence that some of the nucleus-encoded photosynthesis-related genes in *Paulinella* have undergone remarkable duplication and pseudonization (Calatrava et al. 2022). PSI system protein PsaE gene, for instance, exists in three copies in *P. micropora* (PsaE1, PsaE2, PsaE3), which are 53, 46, and 72% identical, respectively, to PsaE of *Synechococcus* sp. WH5701, the closest cyanobacterial relative of the chromatophore. The more divergent PsaE1 and PsaE2 are incomplete, missing either 3' or 5' terminus, each with a seven-residue deletion compared to PsaE of *Synechococcus* sp., indicating that they are pseudogenes. In contrast, PsaE3 is full length and identical to the *Synechococcus* sp. WH5701 homolog. It has a single exon, and is associated with a downstream poly(A) tract and retrotransposon domains (LTR reverse transcriptase, ribonuclease H, and integrase), features characteristic of gene retrocopies. Furthermore, similar retrocopy features were found for PsaI and the highlight inducible protein (HLIP) in *P. micropora*. These results suggest that the domestication of chromatophore into a plastid involves extensive retrotransposition, and that this widely occurring innovation mechanism may be a major facilitator for host-endosymbiont integration and evolution of organelles (Calatrava et al. 2022).

11

Genomes of Seaweeds and Multicellularity

Seaweeds are treated here in a separate chapter for the convenience of discussion about their common ecological (living in shallow water) and morphological (multicellular) features. They are genetically diverse, phylogenetically belonging to Rhodophyta, Chlorophyta, and Phaeophyta (Fig. 11.1). Chlorophyta and Rhodophyta contain many unicellular microscopic species. Phaeophyta does not have a unicellular representatives, but it belongs to the supergroup of Ochrophyta (also known as Stramenopiles or Heterokontophyta), which includes diatoms and many other unicellular algae. Seaweeds are benthic algae *per se*, attached to rocks or sediment by holdfasts, although some species can float on the surface (e.g., *Ulva* spp.). Due to their need for light, their habitats are restricted to shallow coastal areas, from the upper intertidal zone to the subtidal zone, an area subject to strong atmospheric, terrestrial, and tidal impacts and highly fluctuating and often stressful conditions. Some species of phaeophytes (e.g., kelp) can grow to tens of meters long and form large populations, creating "forests."

The genomes of microalgae from Chlorophyta and Rhodophyta have been examined in Chapters 5 and 6, respectively. This chapter will focus on their macroscopic counterparts as well as phaeophytes (Table 11.1).

Chlorophytes

Ulva

Ulva is a widely distributed macroscopic chlorophyte, found in oceans ranging from temperate to tropical regions. The genus is an iconic representative of the green seaweeds. *Ulva's* rapid and abundant growth makes it a key contributor to coastal biogeochemical cycles (de Clerck et al. 2018). Its role in marine sulfur cycles is particularly important due to its high production of dimethylsulfoniopropionate (DMSP), the main precursor of volatile dimethyl sulfide (DMS). The rapid growth of *Ulva* also makes it increasingly a driver of nuisance blooms known as "green tides". For instance, *U. prolifera* often forms massive blooms covering aerial areas exceeding 30,000 km^2 in regions such as Qingdao, China, where its proliferation nearly disrupted the Olympic sailing events in 2008. These blooms typically emerge from macrophyte communities initially consisting of several *Ulva* species alongside other seaweed groups. On the other hand, *Ulva* spp. is edible (called "Sea Lettuce"), and their rapid growth makes them attractive biomass feedstock. *U. lactuca* is among the most widely consumed species, with the Latin word *lactuca* meaning "lettuce". Species with hollow, one-layered thalli were formerly included in the genus *Enteromorpha* to separate them

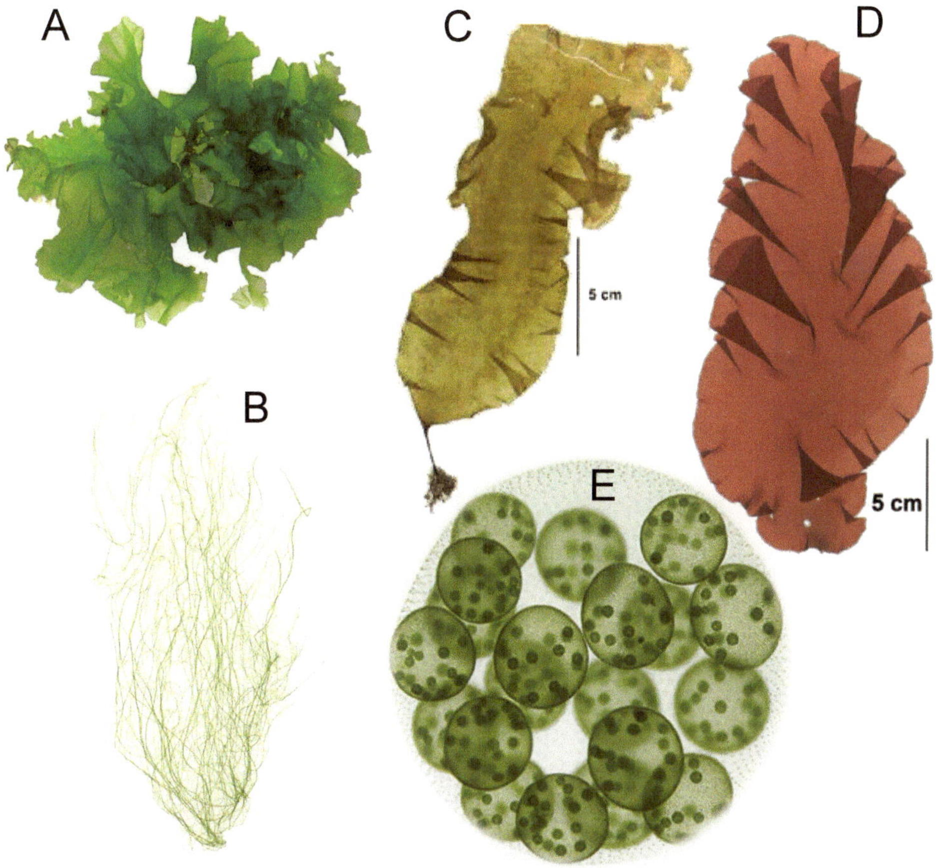

Figure 11.1. Morphotypes of seaweed and colonial unicellular algae.

from *Ulva* species, which are flat, lettucelike blades two cell layers thick. However, it is now widely accepted that such species should be included in *Ulva* because the two morphotypes cannot be separated in molecular phylogenetic trees and both morphotypes can occur within species (Tan et al. 1999, Hayden et al. 2003, Spoerner et al. 2012). Interestingly, their morphogenesis is critically influenced by an assortment of bacterial species (Spoerner et al. 2012).

U. mutabilis genome was the first among *Ulva* spp. that was sequenced (de Clerck et al. 2018). The genome is 98.5 Mbp in size, from which 12,924 genes were predicted, covering 92% of eukaryotic BUSCO. Comparison with other chlorophyte genomes indicated that *U. mutabilis* lacks ten families of Transcription Factors (TFs) and two families of Transcriptional Rregulators (TRs) that are present in other green algae (De Clerck et al. 2018). This apparent "reduction in complexity" trend in *U. mutabilis* also includes the absence of the retinoblastoma (RB)/E2F pathway and associated D-type cyclins (CYCDs), the co-option of which is a recognized hallmark of multicellularity in volvocine algae (Prochnik et al. 2010, Hanschen et al. 2016). Candidate genes, including some that arose from horizontal gene transfer from chromalveolates, are present for the transport and metabolism of DMSP.

Few transcription factors (TFs) were expanded in *U. mutabilis*, contrary to the expectation for a multicellular organism, but a notable exception is CONSTANS-like (CO-like) TFs. The CO-like transcription factors are among the families potentially involved in the development of complex multicellularity in green algae and land plants (Lang and Rensing 2015). *U. mutabilis* has five CO-like TF genes, whereas other sequenced algae have 0–2. This species exhibits a significantly higher diversity of protein domains associated with the extracellular matrix and cell surface

Table 11.1. Sequenced genomes of macroalgae.

Phylum	Species	Genome size	Number of chromosomes	Predicted genes	Coding%	GC%	References
Chlorophyta	*Caulerpa lentillifera**	28.7		9,311	48.66	40.4	Arimoto et al. 2019
	*Klebsormidium nitens**	117.1		16,063	20.58	52.4	Hori et al. 2014
	Ostreobium quekettii	151.9		10,663	10.53	52.4	Iha et al. 2021
	Ulva compressa	80.8		19,207	35.66	57.3	Osorio et al. 2022
	Ulva mutabilis	98.5		12,924	19.68	57.2	Osorio et al. 2022
Rhodophyta	*Chondrus crispus*	105	14	9,606	13.72	8	Collen et al. 2013
	Gracilariopsis chorda	92.1		10,806	17.60	49.26	Lee et al. 2018
	Gracilaria changii	35.8		10,912	45.72	50.6	Ho et al. 2018
	Pyropia yezoensis	43	6	10,327	36.02	63.6	Nakamura et al. 2013
	Pyropia yezoensis	108	3	12,855		65.9	Wang et al. 2020
	Porphyra umbilicalis	87.7	3–5	13,125	22.45	65.8	Brawley et al. 2017
Phaeophyta	*Cladosiphon okamuranus*	130		13,000	15.00	54	Wang et al. 2020
	Ectocarpus siliculosus	214	25	16,256	11.39	56	Cock et al. 2010
	Macrocysitis pyrifera	537	30	25,919	7.24	50.37	Diesel et al. 2023
	Nemacystus decipiens	154		15,156	14.76	54	Wang et al. 2020
	Saccharing japonica	537	62	18,733	5.23	49.6	Wang et al. 2020, Ye et al. 2015
	Sargassum fusiforme	394.4		20,222	7.69	48.4	Wang et al. 2020
	Sargassum horneri	385		58,211	22.68	30.6	Wang and Wu 2023
	Undaria pinnatifida	539	30	14,178	3.95	50.14	Wang et al. 2020, Osorio et al. 2022

* coenocytic (a.k.a. siphonous) and macroscopic species, which are unicellular with multi-nuclei.

compared to its sequenced sister taxa. *U. mutabilis* has 36 genes coding for heme peroxidases, 21 of which are predicted to be extracellular, while peroxidases may have important functions in cell wall modification and in plants may anchor to pectins and cross-link extensins to stiffen the primary cell wall. Coincidentally, the *Ulva* genome harbors abundant genes encoding extensins and pectin-like polysaccharide. The genome also contains the complete dimethylsulfoniopropionate (DMSP) biosynthesis pathway, specifically the MAT pathway, one of the three major pathways of DMSP biosynthesis and DMSP lyase (see Chapter 14 for more information), indicating that this *Ulva* has the capability to produce and regulate DMSP.

The genome of *U. compressa* has also been analyzed to identify protein coding genes (Osorio et al. 2022). The analysis revealed genes encoding enzymes involved in the synthesis of antioxidant enzymes as well as ascorbate and glutathione, showing similarities to plant and bacterial counterparts. Additionally, signal transduction protein kinases such as MAPKs, CDPKs, CBLPKs, and CaMKs detected showed significant similarities to homologs in plants, green microalgae, and bacteria. Regulatory transcription factors such as ethylene- and ABA-responsive factors (MYB, WRKY, HSTF) were also found, and they showed similarity to plant and green microalgae transcription factors.

Gonium pectoral

Gonium pectoral is an undifferentiated colonial alga, and as such can serve as an effective system for studying the genetic basis of multicellularity, including the initial formation of cooperative cell groups. The genome of *G. pectoral* was sequenced (Hanschen et al. 2016). The analysis of the genome suggested that group formation of this species evolved through the co-option of the retinoblastoma cell cycle regulatory pathway. Notably, expressing the *Gonium* retinoblastoma cell cycle regulator in unicellular *Chlamydomonas* induces colonial behavior. The presence of these changes in undifferentiated *Gonium* indicates extensive group-level adaptation during the initial stage of multicellularity evolution. These findings highlight the evolution of cell cycle regulation as a crucial step in the transition to multicellularity, offering insights into other multicellular innovations and evolutionary transitions.

Rhodophytes

Chondrus crispus

Metabolic features of *Chondrus crispus* were studied by analyzing its 105-Mbp genome (Collen et al. 2012). The genome is characterized by gene-dense regions surrounded by repeat-rich regions. From the genome, 9,606 gene models were identified, 36% lower than the genome size-based estimate (15,138) by the existing predicting model (Hou and Lin 2009). This gene sparsity coincides with gene compactness, with only 1.32 exons on average per gene and 88% of genes being intron free. This genome has an extraordinary level of repetitive DNA (73%), which is dominated by transposable elements (class I LTR accounting for 58 Mbp). The small gene repertoire has resulted from the loss of numerous genes, such as those for selenoproteins, the machinery for DNA methylation, sulfatases, core components of the endocytic machinery, heterotrimeric G proteins, and flagella-specific genes (red algae have no flagella in any stage of the life cycle), and the genome has only one photoceptor (a cryptochrome).

Despite the small gene repertoire, the *C. crispus* genome harbors many unknown genes, potentially unique to this species. Genes related to halogen metabolism, oxylipins, and multicellularity (microRNA and transcription factors) were identified. Of particular interest are features related to carbohydrate metabolic processes, including a minimalistic gene set for starch biosynthesis, which was found to have existed prior to primary endosymbiosis. An occurrence of mannosylglycerate synthase was also noted and suggested to have come from a marine bacterium by Horizontal GeneTtransfer (HGT).

Pyropia yezoensis

The genome of *Pyropia yezoensis* was sequenced to better understand its biological properties (Nakamura et al. 2013). A 43-Mbp assembly was achieved, and 10,329 genes were predicted. Similar to that of *C. crispus*, the genome of *P. yezoensis* has low intron content, with 60% of genes free of introns. From the gene repertoire, the authors identified B12-dependent and independent methionine synthase. Light-harvesting genes involved in the phycobilisomes were predicted. In addition, the nuclear genome also harbors a homolog of the phycobilisome-degradation gene (Ycf18), which is usually encoded in the chloroplast. The homolog of this gene in cyanobacteria is called non-bleaching A (NblA), which is upregulated under nutrient starvation and involved in phycobilisome degradation (Collier and Grossman 1994). The nuclear Ycf18 was upregulated in ammonium medium but not in nitrate medium, and its function remains to be investigated.

Another strain was sequenced. Using homozygote cells, the sequencing output was assembled into three large scaffolds, totaling 108 Mbp, which is more than double the above-mentioned size (Wang et al. 2020). The huge difference in genome size might be due to ploidy and bacterial contribution. Nearly half of the genome (48%) was contributed by repetitive elements, predominated by LTR, similar to *C. crispus* (Collen et al. 2012). Horizontal gene transfers from bacteria and the expansion of gene families (e.g., Carbonic Anhydrase [CA], anti-oxidative related genes) were observed, with many showing gametophyte-specific expression or significant up-regulation in gametophytes in response to dehydration. A total of 24 CA genes were identified, dominated by αCA (15), but βCA (6) and γCA (3) were also found. In conchocelis, carbonic anhydrase facilitates the release of HCO_3^- from shells, providing a source of inorganic carbon for the algae that may experience CO_2 inadequacy while residing inside the shell of bivalves. An experiment on conchocelis co-cultured with ^{13}C-labeled $CaCO_3$ produced results in support of that hypothesis. The genome also revealed expansion of gene families involved in antioxidative stress and detoxification, adaptive features for inhabitants of highly dynamic and often stressful tidal environments.

Porphyra umbilicalis

An international research team led by Susan Brawley sequenced the 87.7-Mbp haploid genome of *P. umbilicalis* to illuminate adaptive evolution in intertidal rhodophytes (Brawley et al. 2017). The genome has 65.8% GC content and 13,125 gene loci. Comparative analyses revealed innovative features in the genome shared by other red algae related to cytoskeleton, calcium signaling, the cell cycle, and stress-tolerance mechanisms including photoprotection. Of cytoskeletal motor proteins, *Porphyra* only possesses a small group of kinesins, which seem to be the only cytoskeletal motors commonly present in red algae. *Porphyra* possesses representatives of several kinesin motor subfamilies, including kinesins 5, 7, and 14. These kinesins are anticipated to play roles in spindle assembly, kinetochore function, and regulation of microtubule dynamics, respectively. However, dynein motors are absent, and most rhodophytes, including *Porphyra*, lack myosin. This sequenced genome encodes four closely related actin genes and actin-related protein Arp4, but no other Arp proteins. In eukaryotes, Arp1 is part of the dynactin complex, and Arp2/3 nucleates the formation of branched microfilaments that mediate amoeboid motion. This notably minimal cytoskeleton can explain why cells and multicellular structures in rhodophytes are generally smaller than in most other multicellular lineages. The lack of dynein motor suggests that *Porphyra* and other red algae have other motor devices to fulfill its functions such as intracellular transport of membranes (Brawley et al. 2017).

The genome analysis indicated strong stress tolerance in intertidal red algae. Whereas *P. umbilicalis* and other red algae lost flagellar and cytoplasmic dynein motors, intermediate chains, and most light chains, they retain a particular light chain that is also conserved in flowering plants, which independently lost flagellar motility. Remarkably, this dynein light chain is expressed under abiotic stress and phytohormone treatments in plants, suggesting that it may have a role in the stress tolerance of *Porphyra*. As an intertidal organism, *Porphyra* is subjected to a diverse array of stresses, including periodic desiccation, high light and UV light exposure, osmotic stress, and nutrient depletion. Molecular mechanisms found in the genome to cope with these stressors will be elaborated in a subsequent section.

Gracilariopsis chorda

The genome of *G. chorda* was sequenced (Lee et al. 2018) to address three questions: (1) What is the role of repeats and TEs in explaining Rhodophyta genome size variation? (2) What is the history of genome duplication and gene family expansion/reduction in these taxa? and (3) Is there evidence for TE suppression in red algae? Red algae genome sizes are found to be typically small compared to land plants, with no signs of polyploidization. Size variation was attributed to TE expansion, with long terminal repeat elements and DNA being the major source. Around 8.3% of the *G. chorda* genome is methylated among the gene bodies, promoters, and TEs, and 71.5% of TEs contain methylated DNA, 57% of which is associated with small RNAs (sRNAs). The latter results suggest the involvement of TE-associated sRNAs in RNA-dependent DNA methylation to facilitate gene silencing. The results of the genomic work suggest that the evolution of genome size in red algae results from the combined TE spread and the concomitant emergence of epigenetic suppression.

Gracilaria changii

Poor genomic understanding of agar-producing red seaweeds motivated the genome study of *G. changii* (Ho et al. 2018). From the 35.8 Mbp genome, 10,912 genes were predicted, 60.5% being intron-free. The gene repertoire contains genes for amylopectin/glycogen branching enzyme (EC 2.4.1.18) and amylopectin/glycogen phosphorylase (EC 2.4.1.1). Furthermore, genes encoding glycogenin-1 (UDP-α-d-Glc: glycogen α-d-glucosyltransferase, EC 2.4.1.186) and glycogen synthase (UDP-Glc:glycogen 4-α-d-glucosyltransferase, EC 2.4.1.11) were identified. These enzymes are likely specific to UDP-Glc and play roles in initiating and elongating floridean starch, an α-linked glucose polymer that serves as a storage glucan found in cytoplasmic grains or granules, and is typically the main repository for fixed carbon from photosynthesis in rhodophytes. These enzymes might have replaced the canonical floridean starch biosynthesis genes such as UDP-Glc: 1,4- α-D-glucan-4- α-D-glucosyltransferase, which was not found in the genome. As the genome is not complete, this gene might have been missed.

Phaeophytes

Only a few eukaryotic lineages have independently evolved complex multicellularity. Brown algae are arguably the most developmentally complex lineage after animals and land plants. This lineage of algae demonstrates remarkable size and complexity, as exemplified by the giant kelp *Macrocystis*, which can grow up to 50 meters in length (Bourdareau et al. 2021).

Ectocarpus siliculosus

This filamentous seaweed is a model organism for phaeophytes. Its haploid male genome is 214 Mbp in size and contains 16,256 protein genes as well as 22.7% repeated sequences including DNA transposons, retrotransposons and helitrons (Cock et al. 2010). Small RNAs are preferentially mapped to transposons, suggesting their involvement in silencing these mobile elements, even in the absence of detectable levels of cytosine methylation in the genome. Notably, a copy of a large DNA virus, closely related to the *Ectocarpus phaeovirus* EsV-1, was found integrated in the sequenced genome, consistent with the prevalence of viral infection in natural populations of this brown alga. The viral DNA contains many genes known to be important for the viral life cycle. The genome and its comparison with genomes of other multicellular organisms reveal gene families related to signal transduction and receptor kinases, which are crucial to multicellular life forms.

Bourdareau et al. (2021) investigated histone modifications during the life cycle of *Ectocarpus*. The genome of *E. siliculosus* encodes 34 core histone and nine H1 histone genes. Four of the core histone genes code for isoforms, including probable CenH3, H2A.X and H2A.Z proteins. A set of histone-modifying enzymes (47 in total) was identified from the genome, including acetyltransferases, methyltransferases, deacetylase and demethylase. This set of post-translational modifications includes the novel mark H2AZR38me1, yet *Ectocarpus* lacks both H3K27me3 and the major polycomb complexes that are widespread in eukaryotes. Polycomb repressive complexes play crucial roles in regulating diverse developmental processes in animals and land plants. These results suggest that epigenetic regulation plays a very important role in *Ectocarpus*. Overall, stable patterns of histone modifications occur throughout the life cycle (Bourdareau et al. 2021).

Saccharina japonica

The genome of *Saccharina japonica* was sequenced and 98.5% of the genome (537-Mbp) was assembled (Ye et al. 2015). From the assembly, 18,733 genes were predicted. The analysis results showed that gene families related to cell wall synthesis, halogen concentration, development, and defense systems were expanded. Furthermore, mannuronan C-5-epimerase and haloperoxidase gene families have been diversified in this genome, indicating adaptive evolution of polysaccharide biosynthesis and iodine antioxidation. Additional sequencing of seven cultivars and nine wild individuals showed a higher genetic diversity within the wild populations compared to the cultivars.

Brown algae use mannitol and laminarian for carbon storage, and alginates and sulfated fucans as cell wall polysaccharides, instead of cellulose and trehalose found in land plants and other algae. Carbohydrate metabolism pathway reconstruction in *S. japonica* and 14 other algal genomes revealed similarities with *E. siliculosus* but differences from *Aureococcus*, *Nannochloropsis*, and diatoms in various pathways. Notably, starch and sucrose pathways are absent in all four stramenopile species examined. In *S. japonica*, gene expansion and duplication events were identified in key nodes of the alginate and sulfated fucan pathways. Alginate, the primary matrix component of brown algal cell walls, enhances rigidity in the stipe and holdfasts while providing flexibility to the blades. The matrix and physicochemical properties of alginate depend on the final epimerization step, catalyzed by MC5Es27. *S. japonica* has 105 MC5E genes, in contrast to only 28 in *E. siliculosus*. The expansion of MC5E genes in *S. japonica* likely contributes to its exceptionally high alginate content (up to 45% of dry weight) and enhanced bending flexibility.

Macrocysitis pyrifera

The genome of *Macrocysitis pyrifera* was sequenced to better understand this specific giant kelp species and inform utilization for commercial, pharmaceutical, and food stock usage (Diesel et al. 2023). The assembled genome is 537 Mb with a GC content of 50.37%. From the genome assembly, 25,919 genes were predicted, a coding level almost exactly as predicted by the formular of Hou and Lin (2009), suggesting that the gene density is typical of eukaryotes. Genetic variation was identified by analyzing 48 diploid giant kelp sporophytes from three separate southern California populations.

Sargassum spp.

The genome of *Sargassum fusiforme* was analyzed to further the understanding of the order Fucales and provide insight into evolutionary trajectories of the Chromista kingdom (Wang et al. 2020). The genome assembly was 394.4 Mbp, from which 20,222 genes were predicted, which covered about 90% of the BUSCO dataset.

Another sequenced species in this genus is *Sargassum horneri*, one of the species that have caused "Golden Tide", a destructive bloom formed due to rapid growth of the species. Its genome was sequenced to identify the molecular basis of bloom formation (Wang and Wu 2023). The effort led to an assembly of a 385-Mbp draft genome, composed of 30.6% repetitive sequences and 58,211 predicted gene models. A gene encoding an alkaline phosphatase family protein was found to exhibit positive selection in *S. horneri*. This evolution of alkaline phosphatase may enhance *S. horneri*'s ability to utilize organic phosphorus, aiding its population dominance and contributing valuable molecular insights for studying the *S. horneri* blooms.

Additionally, the authors also characterized putative genes involved in fucoidan biosynthesis, which is a significant pathway in *S. horneri*. Specifically, these included 3 genes homologous to GDP-mannose 4,6-dehydrogenase (GM46D) and GDP-fucose pyrophosphorylase (GFPP) from *Saccharina japonica*; 2 genes corresponding to GDP-L-fucoidase synthase (GFS) and Fucose kinase (FK) from *E. siliculosus*; 7 genes matching FT from *E. siliculosus* (4 from Glycosyl Transferase family 23 [GT23] and 3 from GT10); and 12 genes matching Sulfotransferase (9 from *E. siliculosus* and 3 from *S. japonica*). In contrast, *S. fusiforme* displayed 3 genes homologous to GM46D, GFS, and GFPP from *S. japonica*; 1 gene with FK from *E. siliculosus*; 9 genes matching FT from *E. siliculosus* (8 from GT23 and 1 from GT10); and 8 genes matching ST (7 from *E. siliculosus* and 1 from *S. japonica*).

Evolution of Multicellularity

The transition from unicellular to multicellular organization is one of the major innovations in the evolution of eukaryotes. This multicellularity has been foreshadowed by cyanobacteria, where cell differentiation occurs in filamentous taxa in the order Nostocales, including *Anabaena* and *Nostoc* (Garg and Maldener 2021). Under favorable conditions, these aquatic forms grow in filaments of hundreds of cells connected by cell-cell junctions, facilitating communication and rapid responses to stress signals with a division of labor among different cell types. For example, about 10% of cells differentiate into nitrogen-fixing heterocysts during nitrogen starvation, supporting the filament with nitrogen through septal junctions. Additionally, many species form spore-like resting cells, called akinetes, in response to unfavorable conditions like cold winters. Akinetes differentiate from vegetative cells to withstand harsh environments. Their differentiation is triggered by environmental factors such as light intensity, light quality, temperature, and nutrient deficiency, varying in a species-specific manner.

Chlorophyte and streptophyte algae form a large and diverse clade with multiple independent transitions from unicellularity to multicellularity. Colonial forms such as that in *Volvox* and multinucleate single-celled (coenocytic or siphonous) algae such as green algae from Bryopsidales and Dasycladales are considered intermediate stages of evolution from unicellular to multicellular lineages. Furthermore, algae from the charophytes within the phylum Streptophyta are the closest relatives of land plants known to date and encompass the transition from unicellularity to simple multicellularity (Umen 2014). These green algae encompass unicellular, simple colonial, and obligate multicellular taxa with germ-soma differentiation (Fig. 11.2), and hence represent an excellent model for studying the evolution of multicellular macroscopic organisms.

Intuitively, multicellular organisms should have more complex cellular regulatory machinery and a larger gene repertoire than unicellular relatives. However, "complexity" does not depend on more genes or larger genome sizes (Hahn and Wray 2002). Growing evidence suggests that the expansion of *cis*-acting regulatory elements such as the number of transcription factors is a more important predictor of organismal complexity than the number of protein-coding genes (Taft et al. 2007). Prochnik et al. (2010) sequenced the 138-Mbp genome of the colony-forming chlorophyte *Volvox carteri* and compared its ~ 14,500 predicted proteins to those of its unicellular relative *Chlamydomonas reinhardtii*. Surprisingly, the two species have highly similar proteomes,

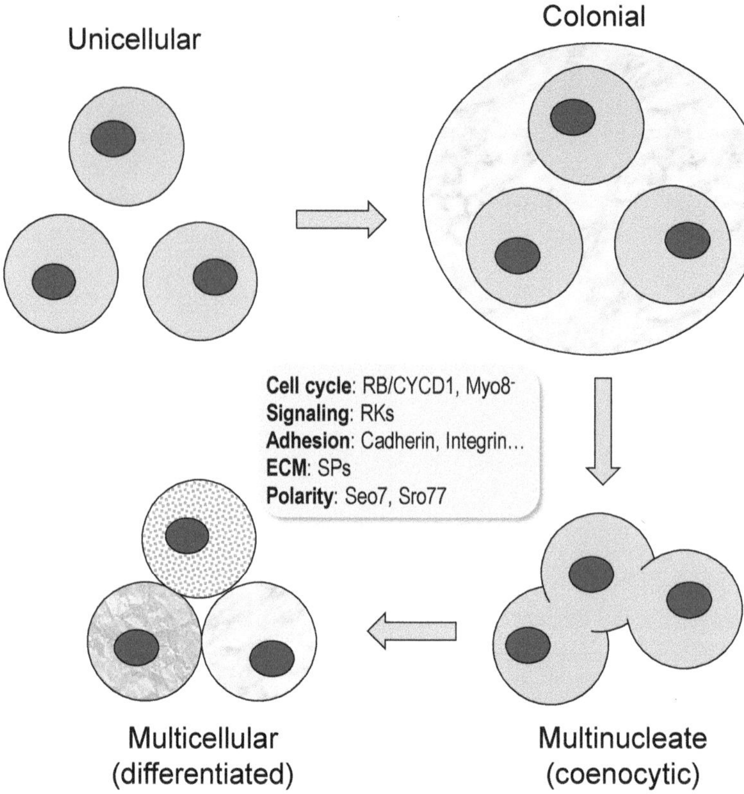

Figure 11.2. Hypothetical evolutionary trajectory of algal cell organization from unicellular to multicellular. Pathways and example genes involved are shown in the center. Cell cycle, signaling, extracellular matrix (ECM), and cell polarity have been shown to be the key processes involved in evolution to multicellularity. RB, retinoblastoma; CYCD1, D-type cyclin; Myo8-, loss of myosin VIII; RKs, receptor kinases such as serine/threonine receptor kinases; SPs, sulfated polysaccharides.

with few lineage-specific genes. Even gene families involved in key developmental processes showed essentially no difference between the two taxa, including protein secretion and membrane trafficking (potentially involved in cytoplasmic bridge formation via incomplete cytokinesis), cytoskeleton, as well as ECM (extracellular matrix) and cell-wall proteins that are involved in ECM expansion, sexual differentiation, and morphogenesis. The defining differences between unicellular and multicellular algae likely lie in subtle genetic differentiation in combination with divergence in regulatory mechanisms.

A study sequenced the genomes of 110 macroalgae from diverse lineages and climate zones to identify key genomic features distinguishing macroalgae from their microalgal relatives (Nelson et al. 2024). This effort covered 65 strains from the phylum Rhodophyta, 11 from Chlorophyta, and 36 from Phaeophyta. Comparative genomic analyses across this extensive range of taxa revealed genes common to all three major macroalgal phyla, defining the genetic foundation for multicellularity. These included genes involved in cell adhesion, ECM formation, cell polarity, transport, and cell differentiation. There are 222 macroalgae-specific PFAM domains conserved in all three seaweed phyla, which are significantly enriched in 59 GO terms, including terms related to morphology, ECM formation, cell-cell adhesion, the establishment of cell polarity, and signaling. These features, along with cell cycle regulation (Hanschen et al. 2016), are recognized as hallmarks of multicellularity.

Adhesome

Macroalgae exhibit complex cell-to-cell adhesion, a hallmark of multicellular development, which is mediated by proteins and polysaccharides that are collectively named the adhesome (Nelson et al. 2024). The macroalgal adhesome is composed of cadherin, integrin, armadillo (ARM), lectin, F actin-binding calponin homology, Kringle, Pleckstrin Homology (PH), Src Homology 2 (SH2), and LIM PFAMs. Besides the adhesome, the function of multicellularity also involves other enzymes such as guanine nucleotide exchange factors and GTPase activating proteins, protein tyrosine, serine/threonine kinases and phosphatases, and E3 ligases and proteases, which regulate adhesion by post-translational modification and do not directly adhere to other proteins. About 1,154 adhesome PFAMs/genome were identified in the macroalgal genomes, 294 of which differ between phyla, some (129) even showing differences between phyla within their respective habitats. The data in Nelson et al. (2024) also indicate geographic (climate zone) differences, with a minimal adhesome gene set occurring in subpolar algae. The geographic variations suggest that environmental cues may shape adhesome development in a convergent fashion, and fewer adhesome proteins are needed in subpolar regions.

Extracellular Matrix (ECM)

In the comparative genomics study for *V. carteri* and *C. reinhardtii,* Prochnik et al. (2010) identified *Volvox*-specific genes including the expanded and highly compartmentalized extracellular matrix. One of these is the gametolysin/VMP (*Volvox* matrix metalloprotease) family, whose substrates are cell-wall/ECM proteins. *Volvox* has 49 pherophorin genes and 42 VMP genes compared to 27 and 8, respectively, in *Chlamydomonas*. Pherophorins have dual roles, one in ECM structure and the other as a sex-inducer glycoprotein, indicating co-option of an ECM protein for sexual signaling, mirroring sexual agglutinins of *Chlamydomonas* related to ECM proteins.

A notable feature of macroalgal-specific gene metabolic maps is the abundance of polysaccharide metabolism pathways. These pathways contribute to complex, tissue-specific glycoprotein patterns that shape organismal morphology and cell development. Macroalgae are known to utilize diverse sugars and sugar derivatives for specific ECM components (Domozych and Domozych 2014, Kloareg et al. 2021, Mazéas et al. 2022). ECM polysaccharides play crucial roles in cell adhesion mechanics and provide protection for macroalgae from their surrounding environment (Nelson et al. 2024). Components of seaweed ECMs, such as mannose derivatives, sulfated polysaccharides (SPs), and acidic-residue-rich linking proteins like integrins, are indispensable for macroalgae and set them apart from microalgae and land plants. SPs, which are particularly structurally diverse and unique to some marine plants and macroalgae, comprise numerous building blocks for tissue-specific ECMs. The work by Nelson and colleagues found that genes involved in mannose were conserved across all three macroalgal phyla, suggesting their significance for multicellular algae. In contrast, macroalgae lack glycosyltransferase activity required in complex-type N-glycan biosynthesis (Yoshiie et al. 2012). Complex N-linked glycans differ from high-mannose and hybrid glycans by having additional GlcNAc residues at both the α-3 and α-6 mannose sites. Unlike high-mannose glycans, complex glycans do not contain mannose residues outside of the core structure.

Cell Polarity

Cell polarity is a fundamental aspect of multicellular organisms, and is essential for the formation of tissues and organs. It is defined as an asymmetry in the molecular composition or structure of a cell's two sides, which creates a polarity axis that regulates cellular processes. This asymmetry allows cells to perform specialized functions, such as migration and asymmetric cell division. Spatio-temporal orientation of cells is important for coordinating cell movement, migration, and adhesion (Niklas et al. 2019). Cell polarity evolved before the emergence of multicellular organisms. For instance, in the yeast *Saccharomyces cerevisiae*, cell polarity proteins Seo7 and Sro77 (homologous to *Drosophila meganogaster* Lgl) are involved in regulation of polarization of the actin cytoskeleton and vesicle exocytosis (Hattendorf et al. 2007). In the cross-phyla comparative genomics analysis in Nelson et al. (2024), out of the macroalgae-specific GO terms, terms related to the establishment of cell polarity were recovered, along with others critical for multicellularity.

In certain macroalgal taxa, sexual reproduction polarity is established during the zygotic phase (Kropf 1997), leading to the development of polarized embryos after zygote fertilization (Charrier et al. 2017). One of the key features setting macroalgal cells apart from microalgae is the extensive cell-cell communication networks in macroalgae. Genomic data suggest that these networks partially arose due to gene gains compared to their unicellular ancestors (Nelson et al. 2024).

Receptor Kinases

Comparative genomics analysis for *Ectocarpus siliculosus* and other multicellular as well as unicellular algae revealed the gain of a family of membrane-spanning receptor kinases in the multicellular algae. Receptor kinases have been shown to play crucial roles in developmental processes such as differentiation and cellular patterning in both animals and green plants. These include Receptor Tyrosine Kinases (RTKs) in animals and serine/threonine receptor kinases in green plants. The emergence of receptor kinases is regarded as a key event in the evolution of multicellularity (Cock et al. 2002).

Developmental and Cell Cycle Reprogramming

It is expected that the transition to multicellular life forms involves construction of a life cycle at the level of the cell group (Maliet et al. 2015). Studies including genomic analyses in the volvocine algae have shown that cell cycle genes are among the earliest diversifying gene families (Prochnik et al. 2010, Hanschen et al. 2016). Among others, these studies revealed the expansion of cyclin D1 (CYCD1) genes in the colonial volvocine *Gonium pectoral* and *Volvox carteri* relative to the unicellular *Chlamydomonas reinhardtii*. In most eukaryotes including the volvocines, D-type cyclins interact with retinoplastoma (RB) gene (Müller et al. 1994). The Volvocine RB homolog (a.k.a. MAT3) plays a crucial role in the formation of colonial multicellularity, as evidenced by the phenotypic outcome of introducing the *G. pectorale* RB gene into the MAT3/RB-negative *C. reinhardtii*, which was to form colonies (Hanschen et al. 2016). However, as indicated earlier, *U. mutabilis* genome shows exceptions in lacking the RB/CYCD genes (De Clerck et al. 2018), suggesting the presence of multiple evolutionary trajectories to multicellularity. It is of interest to note that CYCD1 has two forms in humans, the nuclear form that regulates G1 to S cell cycle transition and the cytoplasmic membrane-associated form that induces cell proliferation and colony formation (Chen et al. 2020) as well as cell invasion (Fusté et al. 2016). Another significant distinction between the multicellular and unicellular division programs lies in the fact that the maximum number of divisions (n) is genetically predetermined in multicellular taxa but not in *C. reinhardtii* (Kirk 2005).

Featherston et al. (2017) sequenced one of the most morphologically simple species of this lineage, the four-celled colonial *Tetrabaena solialis,* whose genome size is 135.7 Mbp. Using conservative estimates of gene family expansions, the authors identified a minimal set of expanded gene families associated with the origin of multicellularity. These gene families are rich in genes involved in developmental processes, including DNA repair (Rad51-type), DNA damage surveillance and cancer genes, structural maintenance of chromosome 3 proteins, MutL C-terminal dimerization domain containing proteins and a family of proliferating cell nuclear antigen domain containing proteins. Comparison with genomes of three other complete volvocine nuclear genomes revealed lineage-specificity of a subset of these gene families. One of these was CYCD1, as found previously from comparing *Volvox* and *Chlamydomonas*.

Alteration of Cytokinesis

Research on multinucleate single-celled macroscopic algae, known as coenocytic or siphonous algae, provides insights into the evolution of multicellularity. Species from the chlorophyte orders Bryopsidales and Dasycladales and several species from Rhodophyta and Phaeophyta are notable examples of this type of algae. Among them genera *Halimeda* and *Caulerpa* are commonly found in tropical and subtropical oceans, where they play vital roles in biogenic reef ecosystems. Coenocytic algae undergo multiple rounds of nuclear division as the cell enlarges but without undergoing division (cytokinesis). This process results in the formation of large cells with numerous nuclei.

Cytokinesis, the last step of cell division after nuclear division, involves the formation of the septum (bacteria) or phragmoplast (plants), a cytoplasmic structure made of microtubules and actin filaments (Smertenko et al. 2017), or a functional equivalent in non-model organisms like algae. In eukaryotes, cytokinesis entails myosins serving as molecular motors for organelle movement or tension facilitator in the formation of new cell wall plate. Specifically, in plants, myosins of class VIII (Myo8) and class XI (Myo11) are involved in this process (ShuZon and Bezanilla 2014, Haraguchi et al. 2018). Myo11 proteins in *Arabidopsis thaliana* have tissue-specific roles, such as facilitating cytoplasmic streaming and promoting pollen tube growth (Haraguchi et al. 2018). Myo8 proteins are linked to microtubule ends and actin, contributing to guiding phragmoplast expansion in moss (ShuZon and Bezanilla 2014).

Comparative analyses of genomes of siphonous algae provide clues about how multinucleation is regulated. From the 90.7-Mbp genome of *Bryopsis* sp., 14,034 protein genes were predicted, with a chlorophyte BUSCO completeness of 95.5% (Ochiai et al. 2024). Only a single myosin (myo11) was found, in contrast to non-siphonous green algae that have multiple types of myosin genes. In addition, even though *Bryopsis* sp. possesses over 30 kinesin genes, NACK/Kinesin-7II, which is crucial for cytokinesis initiation in plants, is hardly expressed in the coenocytic part of the thallus. Similarly, from the recently sequenced genome of *Halimeda opuntia*, myosin XI (Myo11) was found, but despite an extensive search in both genome and transcriptome datasets, no myosin VIII (Myo8) was detected (Zhang et al. 2024). Importantly, previous studies have demonstrated that Myo8 depletion in moss leads to cytokinesis abnormalities, resulting in a multinucleate phenotype (ShuZon and Bezanilla 2014). Besides, cytokinesis-defective plant mutants exhibit a multinucleate phenotype, suggesting that nuclear division can occur despite incomplete cytokinesis (Söllner et al. 2002). Thus, the loss of myo8 and downregulated expression of kinesin-7II may represent an evolutionary innovation that contributed to multinucleation in these species.

Ubiquitin Proteasomal Pathway (UPP)

There are multiple lines of evidence that modifications to the Ubiquitin Proteasomal Pathway (UPP) are associated with the beginning of coloniality (Featherston et al. 2017). In multicellular volvocines, genes experiencing positive or accelerating selection were discovered to be enriched in components of the UPP (ubiquitin-proteasome pathway), and gene families acquired during the emergence of multicellularity include components of the UPP.

Roles of Bacteria

The *P. umbilicalis* genome contains three GH16 enzymes, which cluster together with those from other red algae as well as with agarases and porphyranases from marine bacteria (Brawley et al. 2017). These enzymes are likely involved in cell wall remodeling and are not found in unicellular red algae, suggesting a potential role of bacteria in the evolution of multicellular algae (Miranda et al. 2013).

Roles of Viruses

It has been hypothesized that the interaction of early eukaryotes with viruses favored the evolution of multicellularity as a defense mechanism against infection (Iranzo et al. 2014). Based on the extensive comparative genomics analysis, Nelson et al. (2024) found remarkable viral imprints distinct in the macroalgal genomes, particularly in phaeophytes. *Ectocarpus siliculosus* Virus-1 (EsV-1) is a lysogenic dsDNA virus affiliated with the superfamily of nucleocytoplasmic large DNA viruses (NCLDV). EsV-1-7 domain-containing proteins are most abundant in phaeophytes among the three macroalgal phyla. The EsV-1-7 domains in macroalgae are connected to various proteins necessary for a multicellular phenotype, suggesting that the acquisition of lysogenic virus-sourced genes may be essential to genomic architecture in phaeophytes. Intriguingly, some of these EsV-1-7 domain-containing proteins also contain elements originated from other viruses. One example is the EsV-1-7 domain-FNIP motif protein. The FNIP motif is not encoded by EsV-1 virus but found in viruses from the Megaviridae. In *Megavirus baoshan,* there are many FNIP repeat-containing protein genes, most of which have terminal F-box domains, potentially involved in exploiting the ubiquitin-proteasome pathways of its host during infection. In macroalgae, EsV-1-7 domain-FNIP motif proteins may play a role in cell type differentiation.

Grazer Pressure

Grazing might have played some roles in the evolution of multicellularity. For instance, in the presence of the heterotrophic protist *Ochromonas vallescia*, *Chlorella vulgaris* cultures form inheritable algal colonies (Boraas et al. 1998). When *C. reinhardtii* were cultivated with the presence of a single-celled filter-feeding predator, the ciliate *Paramecium tetraurelia*, after 50 weeks of weekly transfers (~ 750 asexual generations), the algal cells divided and stayed in groups (Herron et al. 2019). The newly evolved multicellular traits seem to curb predation by the ciliate, suggesting the role of grazing in the latter part of multicellularity evolution.

Adaptation to Intertidal Habitats

Intertidal macroalgae are subject to highly variable and harsh environmental conditions due to tidal changes. The wide range of stressors includes desiccation, high light and UV exposure, osmotic shock, heat stress, and nutrient deprivation. Conceivably, the tolerance and adaptive mechanisms may include common strategies as well as lineage- or habitat-specific ones. For instance, oxidative stress is prevalent and stems from a wide range of stressors; accordingly, increases of antioxidants are largely common mechanisms. In this context, it is no surprise that *Porphyra* and *Pyropia* genomes exhibit expansion of the catalase gene family (Brawley et al. 2017). Another common stress is heat stress, which induces expression of heat shock proteins (HSPs). Using microarray technology, Collén et al. (2007) investigated transcriptomic responses in *Chondrus crispus* to exposure to high light, high temperature, hypo- and hyper-osmotic conditions in the laboratory and to natural habitat stress of different loads. The study identified a host of key stress genes that showed responses to multiple stressors, including Early Light-Inducible Proteins (ELIPs), HSP70, glutamate-cystine ligase, zinc finger protein, and a protein kinase. The microarray results also revealed potential marker genes for specific stress conditions, e.g., antioxidative proteins that were upregulated in high-light stress and high natural stress; HSPs that were increasingly expressed under high temperature and high light stresses. Not all lineages have been studied to the same degree, however; therefore, much of the discussion will be biased toward individual lineages, although the hope is to provide insights for all groups of seaweed. Besides, the discussion here will be focused on stresses unique to seaweeds that inhabit the intertidal zones but will inevitably involve some common stresses that are elaborated in Chapter 13.

Desiccation Tolerance

Intertidal seaweeds are subjected to desiccation during low tide and must withstand desiccation and be able to recover upon rehydration during rising tide. Depending on the habitat location (high, middle, and low intertidal zone), algal taxa have different levels of tolerance of desiccation. A study comparing species of different habitat types found high activities of antioxidant enzymes and low ROS production in higher intertidal species, recovery of photosynthesis capacity during rehydration only in higher intertidal species, and absence or low level of stress tolerance in lower intertidal species (Flores-Molina et al. 2014). Different taxa may use different mechanisms too for desiccation tolerance. The chlorophyte *Ulva prolifera* was shown to use PSI cyclic electron flow to survive and quickly recover from dehydration (Gao et al. 2011). In this species, PSII activity declined progressively in the course of desiccation, with decreasing maximum quantum yield and effective quantum yield. Meanwhile, the electron transport rate of PSII approached zero at severe levels of desiccation. In contrast, PSI activity fluctuated markedly in the course of desiccation and remained active even at severe desiccation. During rehydration, the electron transport of PSI recovered more rapidly than that of PSII. These observations suggest that the linear electron flow was impaired in desiccated *Ulva*, whereas cyclic PSI activity was elevated and could recover rapidly upon rehydration, as a mechanism to withstand periodic desiccation in the intertidal zone.

In *Porphyra*, the extracellular matrix (ECM) component of the cell wall is primarily composed of a highly hydrated agar known as porphyran. Porphyran, with porphyranobiose as the primary constituent with varying percentages of agarobiose, serves to restrict desiccation and enhances the flexibility of the cell wall under osmotic stress conditions. The *P. umbilicalis* genome contains a family 2 carbohydrate sulfotransferase (ST2), which likely functions as a L-galactose-6-O-sulfotransferase in the biosynthesis of porphyran. This red algal sulfotransferase are found in metazoans (*Nematostella vectensis*, sea anemones), with the closest characterized enzymes being dermatan and chondroitin 4-O-sulfotransferases. Similarly, the *Porphyra* GT7 is highly similar to metazoan GT7s (*Acropora digitifera*, staghorn corals), with chondroitin sulfate N-acetylgalactosaminyltransferase 2 (*Homo sapiens*) being the closest characterized enzyme. These findings provide support for the ancestral nature of sulfated polysaccharide biosynthesis in eukaryotes. This sulfated polysaccharide-containing ECM present in marine organisms is absent in terrestrial plants and freshwater organisms, suggesting its crucial roles in high-salinity environments.

Osmotic Regulation and Ca^{2+} Signaling

The *P. umbilicalis* genome encodes enzymes that support floridoside and isofloridoside synthesis, low molecular weight carbohydrates that act as compatible solutes. Taurine is a likely osmolyte, and enzymes such as cysteine dioxygenase, implicated in metazoan taurine biosynthesis, were found in the genome. Also found in the genome are a range of Ca^{2+}-permeable membrane channels that could play a role in osmotic stress signaling, including two OSCA, which were identified in land plants as Ca^{2+} channels. The Ca^{2+} sensor kinases through which *Porphyra* senses and responds to cytosolic Ca^{2+} rises seem to be distinct from land plants and other eukaryotes. The kinases in *Porphyra* contain two to three Ca^{2+}-binding EF-hand domains at the N terminus and a kinase domain at the C terminus, and as such, they belong to the Tyrosine Kinase-Like (TKL) class of kinases compared to the kinases in most eukaryotes that belong to the CAMK group. The Ca^{2+}-dependent TKLs represent a novel class of Ca^{2+}-regulated kinases apparently unique to red algae.

The Sucrose nonfermenting-1–Related Kinase (SnRK) family of serine/threonine kinases plays crucial roles in mediating metabolic and stress signaling across various organisms, from fungi (yeast) to land plants. The *P. umbilicalis* genome encodes 31 proteins similar to *Arabidopsis* SnRK proteins. Phylogenetic analysis places the *Porphyra* SnRK group in two clades that include *Arabidopsis* SnRK1 and SnRK3/CIPK, and a third expanded clade containing 19 unique members (Brawley et al. 2017). Several members of the SnRK3/CIPK clade contain motifs homologous to the C-terminal domain II of SnRK2, which mediates interaction with PP2C in the abscisic acid (ABA) signaling pathway, which appears to occur in *Porphyra*. These findings raise the possibility that the SnRK family is important in stress responses including ABA-mediated responses in *Porphyra*.

Photoprotection

Porphyra also harbors 11 genes encoding "high light-induced" or "One-Helix" Proteins (OHPs), which have essential roles for photoacclimation and survival under stressful environmental conditions (Sturm et al. 2013). The *Ectocarpus* genome harbors an extended set of Light-Harvesting Complex (LHC) and pigment biosynthesis genes, including a cluster of 11 genes highly similar to the LI818 family of light-stress related LHCs. The genome also encodes a light-independent protochlorophyllide reductase (DPOR), allowing efficient synthesis of chlorophyll under dim light. These features reflect the species' ability to cope with highly variable light conditions in its habitat, and new metabolic processes such as halide metabolism, potentially underpinning this tidal environment.

Porphyra is well known for its high productivity of Mycosporine-like Amino Acids (MAAs) that act as "sunscreen", with porphyra-334 being the major compound. Four enzymes are involved in the biosynthesis of MAA, MysA, MysB, MysC, and MysD, although alternative enzymes are used in some lineages of algae. The *Porphyra* genome contains a gene encoding a MysA and MysB fusion protein that also occurs in several other MAAs-producing red algae. Interestingly, while *Porphyra* and related species contain a MysC-MysD fusion protein and MysA-MysB and MysC-MysD fusion genes are next to each other but transcribed on opposite DNA strands with adjacent 5'-ends, the counterparts in *Chondrus* are organized with 3'-ends being adjacent (Fig. 11.3). The preservation of the MAA gene cluster in *Porphyra* and *Chondrus*, along with two gene fusion events, indicates that this configuration offers a selective advantage and efficient biosynthesis of MAAs for red algae, which are often exposed to intense UV radiation. In addition, the *P. umbilicalis* genome harbors ancestral enzymes for sulfation of the hydrophilic galactan-rich cell wall, evidence for mannan synthesis that originated before the divergence of green and red algae, and a high capacity for nutrient uptake, which confer bangiophytes stress tolerance while living in the intertidal zone (Brawley et al. 2017).

Homologues of flavonoid pathway genes were found in the *Ectocarpus* genome although they are completely absent from diatom or green algal genomes. This is consistent with the high abundance of phenolic compounds in brown algae thought to protect against ultraviolet radiation in a manner analogous to flavonoids in terrestrial plants. In addition, a diverse complement of enzymes involved in the metabolism of reactive oxygen species are encoded in the genome.

Efficient Nutrient Uptake

Porphyra has genes for Fe uptake mechanisms that likely enable high-affinity assimilation during the narrow tidal window when blades are underwater (Brawley et al. 2017). The *Porphyra* genome encodes various putative Fe^{2+} transporters from the NRAMP and ZIP family proteins, which

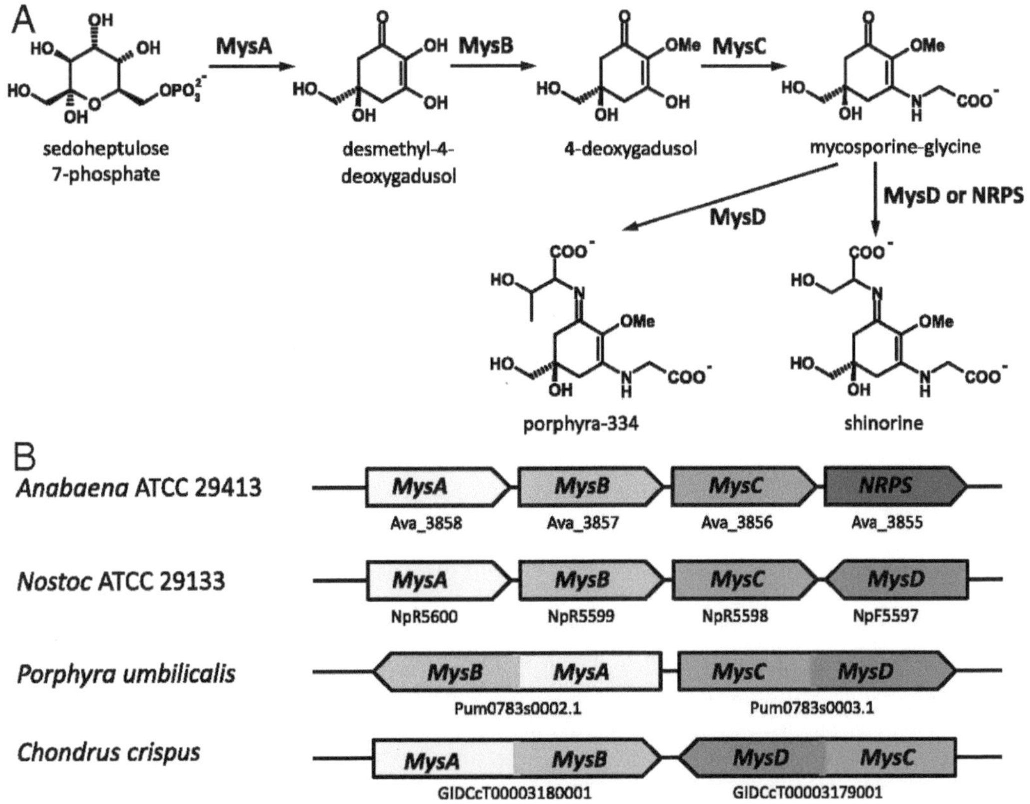

Figure 11.3. Conserved gene clusters involved in MAA biosynthesis in *P. umbilicalis* and related red algae. (A) Biosynthetic pathway from sedoheptulose 7-phosphate to shinorine in cyanobacteria and proposed pathway to porphyra-334 in red algae. (B) Comparison of gene clusters and gene fusions in the cyanobacteria *Anabaena* and *Nostoc* and the red algae *P. umbilicalis* and *C. crispus*. Reproduced with permission from Brawley et al. (2017) under the Creative Commons CC BY 4.0 license.

typically function as broad-specificity divalent metal transporters, and a high-affinity iron transport complex. This complex includes a permease (FTR1), a multicopper oxidase (FET3 in yeast/FOX1 in green algae), and members of the FEA and ISIP2A families. The latter, as discussed in Chapter 7 on diatoms, are algal-specific protein families comprising secreted soluble proteins involved in iron assimilation and membrane-bound iron-uptake facilitators. While both iron-uptake strategies are observed in microalgae, it was unexpected to find FTR1 and FOX1 homologs in a macroalga, given the absence of this complex in higher plants and animals. Phylogenetic distribution analysis revealed that FTR1 homologs were present in all three main Archaeplastida lineages, but they were lost after the transition to land, when vascular plants evolved true roots.

12

Sensing of Environmental Stimuli and Signal Transduction

Like other organisms, algae must develop mechanisms to sense environmental conditions and make appropriate responses. This starts with sensing molecules that detect extracellular stimuli, as in the case of photoreceptors intercepting light signals. These interactions result in changes in the sensing (or ligand-binding) molecules, which in turn trigger a biochemical cascade, a series of biochemical events known as a signaling pathway. The series of molecular events mostly involve protein phosphorylation catalyzed by protein kinases, ultimately leading to a cellular response. The "osmotic stress-MAPKKK-MEK-ERK-TF-Lipid production" signaling cascade (Fig. 12.1) is an example of such a pathway.

When multiple signaling pathways interact, they form networks that coordinate cellular responses through combinatorial signaling events (van Straalen and Roelofs 2012). At the molecular level, these responses involve changes in gene transcription or translation, post-translational modifications, and protein conformational and locational shifts. These molecular events are fundamental mechanisms for controlling cell growth, proliferation, metabolism, and other processes. In multicellular organisms, signal transduction pathways regulate cell communication in various ways.

Each component of a signaling pathway is classified by its role in the initial stimulus response. Ligands, which are mostly soluble molecules from the extracellular medium binding to cell surface receptors, are termed first messengers. Receptors act as signal transducers, which then activate primary effectors, typically proteins that link to secondary messengers. These secondary messengers can then activate further effectors, creating a cascade of signaling events.

The structure and operation of signal transduction pathways in algae are not well understood, but research is advancing rapidly due to the increasing availability of genome data and functional genetic tools. Models like diatoms and the chlorophyte *Chlamydomonas reinhardtii* have been extensively studied for environmental sensing and signal transduction. Diatoms use Ca^{2+} signaling to perceive abiotic and biotic stimuli such as water motion, osmotic pressure, and nutrients (Falciatore et al. 2000, Vardi et al. 2006, Helliwell et al. 2019).

While most eukaryotes use four-domain voltage-gated Na^+/Ca^{2+} channels (4D-Cav/Navs) to regulate Na^+/Ca^{2+}-based action potentials, eukaryotic phytoplankton (including diatoms, dinoflagellates, cryptophytes, haptophytes, and pelagophytes) seem to use single-domain voltage-gated Na^+ channels, termed EukCats. These channels are grouped into EukCatA, EukCatB, and EukCatC based on their taxonomic divergence. Diatoms possess the 'A' group, which comprises fast voltage-gated Na^+- and Ca^{2+}-permeable channels essential for voltage-activated Ca^{2+} signaling and membrane excitability (Helliwell et al. 2019).

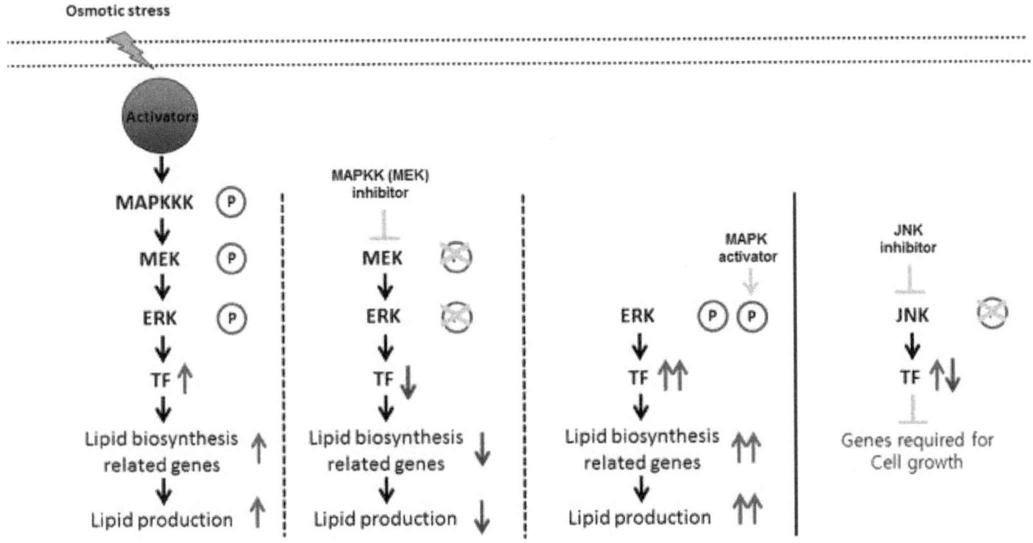

Figure 12.1. Proposed of MAPK signaling pathways in *Chlamydomonas* regulating lipid biosynthesis and cell growth. Reproduced from Yang et al. (2018) under the Creative Commons CC BY 4.0 License.

By transforming *P. tricornutum* cells with jellyfish-derived cDNA encoding the calcium-sensitive photoprotein aequorin (NCBI accession number AAA27720), researchers identified systems for detecting and responding to fluid motion (shear stress), osmotic stress, and iron supply (Falciatore et al. 2000). The cytosolic Ca^{2+} concentration transiently increases when transformed diatom cells face hypoosmotic shock and mechanical stimulus, with a recovery response after desensitization. Interestingly, this sensing system responds to iron but not to nitrogen, silicate, or phosphate nutrients.

Two-Component Response Regulator

A two-component system (TCS) typically comprises a membrane-bound Histidine Kinase (HK) that detects specific environmental stimuli and a corresponding Response Regulator (RR) that governs the cellular response, primarily by altering the expression of target genes (West and Stock 2001). The histidine kinase serves as the sensor, detecting environmental stimuli through its input domain and autophosphorylating on a conserved His residue within the kinase domain. This phosphoryl group is then transferred to a conserved Asp residue on the receiver domain of the response regulator, triggering downstream responses via the output domain. In some cases, the sensor kinase and the response regulator are carried by a single polypeptide, categorizing the TCS as a hybrid system (HY).

Two-Component Systems (TCS) have long been recognized as important in bacteria for sensing and responding to environmental changes, yet they are much less understood in eukaryotes. Extensive research on TCS in cyanobacteria has revealed a higher abundance and diversity of TCS proteins compared to other bacteria, indicative of a complex evolutionary history (Ashby and Houmard 2006, Tang et al. 2023). Interestingly, the abundance of TCS genes is positively correlated with genome size (Tang et al. 2023), suggesting that the need to respond to environmental conditions may exert selection pressure on genome size. While the separate histidine kinase/response regulator (HK/RR) system is prevalent in cyanobacteria (Table 12.1), hybrid types have also been reported in thermophilic lineages (Tang et al. 2023).

Research on algal TCS is just catching up as genomic data has grown explosively. In most eukaryotes (and some bacteria), TCS uses a multistep His-Asp-His-Asp phosphotransfer process for signal transduction. TCS genes can be found in various eukaryotic algae (Table 12.1). Both HK

Table 12.1. Two-component system (TCS) genes in algae.

Group	Species	Gene name (domain code)	Gene ID*
Cyanobacteria	*Prochlorococcus marinus*	His kinase	CAE19800.1
	Prochlorococcus marinus	Response regulator	CAE19572.1
	Synechococcus elongatus	His kinase	BAD80692.1
	Synechococcus elongatus	Response regulator	BAD80587.1
	Prochlorococcus marinus	Phosphate sensor kinase	KZR77567.1
	Prochlorococcus marinus	Phosphate response regulator	KGG08159.1
	Synechococcus elongatus	Phosphate sensor kinase	BAA02454.1
	Synechococcus elongatus	Phosphate response regulator	BAD78723.1
Diatoms	*Phaeodactylum tricornutum*	His kinase	Phatr3_EG02387.t1
	Phaeodactylum tricornutum	Response regulator	EEC48840
	Phaeodactylum tricornutum	HK-RR hybrid	XP_002185350
	Skeletonema marinoi	His kinase (GO: 0000155)	KAK1735225
	Fragilariopsis cylindrus	His kinase kinase (GO: 0000155)	OEU23738
	Pseudo-nitzschia multisetriata	Response regulator (IPR001789)	VEU39019.1
Dinoflagellates	*Prorocentrum shikokuense*	HK-RR hybrid (IPR003594, 001789)	Pshi_cluster_contig8886.p1
	Durusdinium trenchii	HK-RR bydrid (IPR003661,001789)	CAK9098629.1
	Polarella glacialis	Sensor kinase/phytochrome (IPR052162)	CAK0900951.1
	Prorocentrum shikokuense	Response regulator (IPR001789)	comp98388_c9, contig12287.p1
	Prorocentrum cordatum	His kinase (IPR003594, 003661)	evm.Pminimum_s226_g6658
	Prorocentrum cordatum	Response regulator (IPR001789)	evm.Pminimum_s329_g8825
	Effrenium voratum	His kinase (IPR003594, 003661)	EvCCMP421_s9628_g29036
	Effrenium voratum	Response regulator (IPR001789)	CAJ1373479.1
	Fugacium kawagutii	His kinase (IPR003594, 003661)	Fkaw39661
	Fugacium kawagutii	Response regulator (IPR001789)	Skv223432
Cryptophytes	*Rhodomonas salina*	His kinase (IPR003661, 003594)	YP_001293515
	Guillardia theta	His kinase (GO: 0000155)	XP_005821110
Chlorophytes	*Chlamymonas eustigma*	B-type AAR	GAX77779.1
	Cocccomyxa sp.	B-type AAR	BBK07877
Prasinophytes	*Ostreococcus lucimarinus*	B-type AAR	XP_001415508.1

* GenBank accession number or gene ID in genomes

and RR are present in various dinoflagellate species. Interestingly, in *Prorocentrum shikokuense* and symbiodiniacean species such as *Durusdinium trenchii*, protein-coding genes with both a HK domain and a RR domain were identified, indicating that these species employ a hybrid form of TCS. Furthermore, in the polar dinoflagellate *Polarella glacialis*, one gene (CAK0900951.1) contains a sensor kinase domain that appears to be located in the same region as a phytochrome-like domain, whereas another has a histidine kinase-response regulator hybrid system plus a phytochrome-like domain. The multi-domain proteins may function to sense and relay light signals. This is striking given that phytochrome has not been reported for dinoflagellates. Response regulators in dinoflagellates belong to the CHEY superfamily and are characterized by the presence of an OmpR/PhoB-type DNA binding domain. In the genome of *Symbiodinium* sp., a larger protein (GenBank CAE6969219.1) is predicted. This protein not only contains the OmpR/PhoB-type DNA binding domain, but also features a peptidase (trypsin-like serine protease) domain in the N terminal region and a PAS domain in the C terminal region.

In diatoms, the presence of both Histidine Kinase (HK) and Response Regulator (RR) components has been observed in the *P. tricornutum* genome. However, in species such as *Skeletonema marinoi*, *Fragilariopsis cylindrus*, and *Pseudo-nitzschia multisetriata*, only one of these components (WalR-like) has been identified. This discrepancy is likely due to the incomplete nature of the database, which may not encompass the full range of genetic information for these species. In the genome of *P. tricornutum*, not only are separate Histidine Kinase (HK) and Response Regulator (RR) protein genes present, but there is also a gene encoding a hybrid HK-RR polypeptide. This hybrid polypeptide contains both the HK and RR domains within a single protein (Table 12.1). It is interesting to note that many algal sensor kinases are characterized by the presence of multiple repeats of PAS domains. Furthermore, the type-A *Arabidopsis* Response Regulator (ARR) proteins are a family of 10 genes that exhibit rapid induction in response to cytokinin. These proteins bear a striking resemblance to bacterial two-component response regulators. Notably, putative homologs of ARR have been identified in chlorophytes and prasinophytes, as indicated in Table 12.1.

G Protein-Coupled Receptor (GPCR) Signaling Pathway

The G-Protein Coupled Receptor (GPCR) signaling pathway in known to play an essential role in responding to external stimuli in mammalian cells. This pathway is widely present in eukaryotes, including animals, plants, and some simpler eukaryotes like yeast. However, research on the GPCR pathway in algae has only recently begun, and the available information is still limited. Despite the limited information, GPCRs have been identified in some algae based on sequence similarity and the presence of the domain IPR002455 (Table 12.2). For example, the cyanobacterium *Gloeocapsa* and the prasinophyte *Ostrecoccus* are equipped with a rhodopsin-like GPCR. In the dinoflagellate *Alexandrium monilatum* strain CCMP3105, a protein with a type IV pili methyl-accepting chemotaxis transducer N-terminus has been found. In the phaeophyte *Ectocarpus*, an abscisic acid GPCR has been identified. Additionally, a GABA_B receptor containing a 5'-nucleotidase C in the N-terminus has been identified in the diatom *Skeletonema costatum*. A comparative genomic investigation on *Pseudo-nitzschia multiseries*, *Thalassiosira pseudonana*, *Phaeodactylum tricornutum*, and *Fragilariopsis cylindrus* has revealed the presence of core components of GPCR in these species, including proteins with similarity to GPCRs, human G protein α- and β-subunits, and downstream effectors (Port et al. 2013). In *T. pseudonana*, a G protein γ-subunit also has been identified, suggesting the presence of a full heterotrimeric G protein in this species. Phylogenetic analysis indicates that the GPCR primary sequences are relatively conserved and cluster along taxonomic boundaries (Fig. 12.2), suggesting that the GPCR system is ancient and has diverged with speciation.

MAPK Pathway: One of the Most Common Signaling Pathways

The Mitogen Activated Protein Kinase (MAPK) pathway is a highly conserved signaling cascade present across eukaryotic organisms. It orchestrates cellular responses to diverse external stimuli, including biotic and abiotic stresses (Huang et al. 2012). The pathway consists of a sequential three-tiered kinase cascade invoving MAP kinase kinase kinase (MAP3K/MAPKKK), MAP kinase kinase (MAP2K/MAPKK), and MAP kinase (MAPK). Each kinase in the cascade phosphorylates specific serine and threonine residues on its downstream target. In plants, these phosphorylation sites are typically separated by five amino acid residues, following a Ser/Thr-X-X-X-X-X-Ser/Thr motif (Neupane et al. 2019).

Information about the MAPK pathway in algae is limited, although many sequenced algal genomes reveal its existence. MAPK in plants and algae have evolved differently from animals, developing an expanded network of Extracellular signal-Regulated Kinase (ERK)-like MAPKs

Table 12.2. G-protein coupled receptor (GPCR) in algae.

Phylum	Species	Name of receptor	GenBank accession number/ Genome gene ID
Cyanobacteria	*Gloeocapsa* sp. PCC7428	Rhodopsin-like orphan GPR	AFZ31342.1
Phaeophyta	*Ectocarpus siliculosus*	Abscisic acid GPCR	CBN79133.1
Prasinophyta	*Ostreococcus tauri*	Rhodopsin-like GPR	OUS45439.1
Diatoms	*Phaeodactylum tricornutum*	GPCR_3	Phatr44133-3Phatr2
	Thalassiosira pseudona	GPCR	XP_002290529.1
	Ditylum brightwellii	GABA-B Receptor	MMETSP1063-20121207|2424_1
	Nitzschia inconspicus	GPCR 157-related, Serpentine type 7 TM GPCR chemoreceptor	KAG7362405.1
	Nitzschia inconspicus	rhodopsin-like GPR	KAG7358806.1
	Skeletonema costatum	GABA-B receptor	MMETSP0013_2-20120614|4301_1
	Skeletonema marinoi	GABA-B receptor	MMETSP1040-20121108|4099_1
Raphidophyta	*Heterosigma akashiwo*	GABA-B Receptor	Heterosigma akashiwo-CCMP2393-20130911|13962_1
Dinoflagellates	*Alexandrium monilatum*	Type IV pili methyl-accepting chemotaxis transducer N-term	PhyloDB: Alexandrium-monilatum-CCMP3105-20140214|407280_1
	Amphidinium gibbosum	GPCR	g46052.t1
	Gonyaulax spinifera	GPCR	MMETSP1439-20131203|127569_1
	Karenia brevis	GABA-B Receptor	PhyloDB: Karenia brevis-SP3-20130916|256282_1
	Oxyrrhis marina	GABA-B Receptor	PhyloDB: Oxyrrhis marina-20131105|4905_1
	Prorocentrum shikokuense	GABA-B Receptor	Pshi_cluster_contig24273.p1
	Symbiodinium microadriaticum	GABA-B Receptor	CAE7431219.1

(Kalapos et al. 2019). This ERK-like MAPK pathway (also known as Ras-Raf-MEK-ERK) transmits signals from cell surface receptors to the nucleus and regulates various cellular processes through crosstalk with other signaling pathways, such as hormone signaling (Jagodzik et al. 2018). These processes include cell division cycle (Kalapos et al. 2019), lipid metabolism (Yang et al. 2018), chemotaxis (Wang et al. 1998), stress response (Koga et al. 2000), cell survival (Bonni et al. 1999), and osmoadaptation (Hohmann 2002).

Documentation of the MAPK pathway in algae is most extensive for chlorophytes and prasinophytes (Kalapos et al. 2019). In *Chlamydomonas reinhardtii* (Fig. 12.1), 17 MAPKs, 2 MAP2Ks and 108 MAP3Ks (11 MEKK-, 94 RAF- and 3 ZIK-type) in the MAPK signaling cascade have been identified (Gomez-Osuna et al. 2020). The high sequence similarities among chlorophyte MAPKs indicate conservation and purifying selection. The selection pressure appears to be higher for more distant taxa, as reflected by higher Ka/Ks ratios. Algal MAPKs can contain TTY, TSY and TPY motifs, diverging from the classical TEY or TDY signatures (Gomez-Osuna et al. 2020).

In *C. reinhardtii*, *Chlorella vulgaris*, and *Hematococcus pluvialis*, MAPK and cyclic adenosine monophosphate (cAMP) play an inhibitory role in regulating neutral lipid accumulation (Choi et al. 2015). *C. reinhardtii* responds to osmotic stress by upregulating ERK and c-jun N-terminal kinase (JNK) activities, leading to increased cellular lipid content (Yang et al. 2018). Additionally, the MAPK pathway phosphorylates Nitrate Reductase (NR) and regulates nitrate assimilation. There is evidence suggesting a link between the MAPK pathway, NO production, and the action of NR

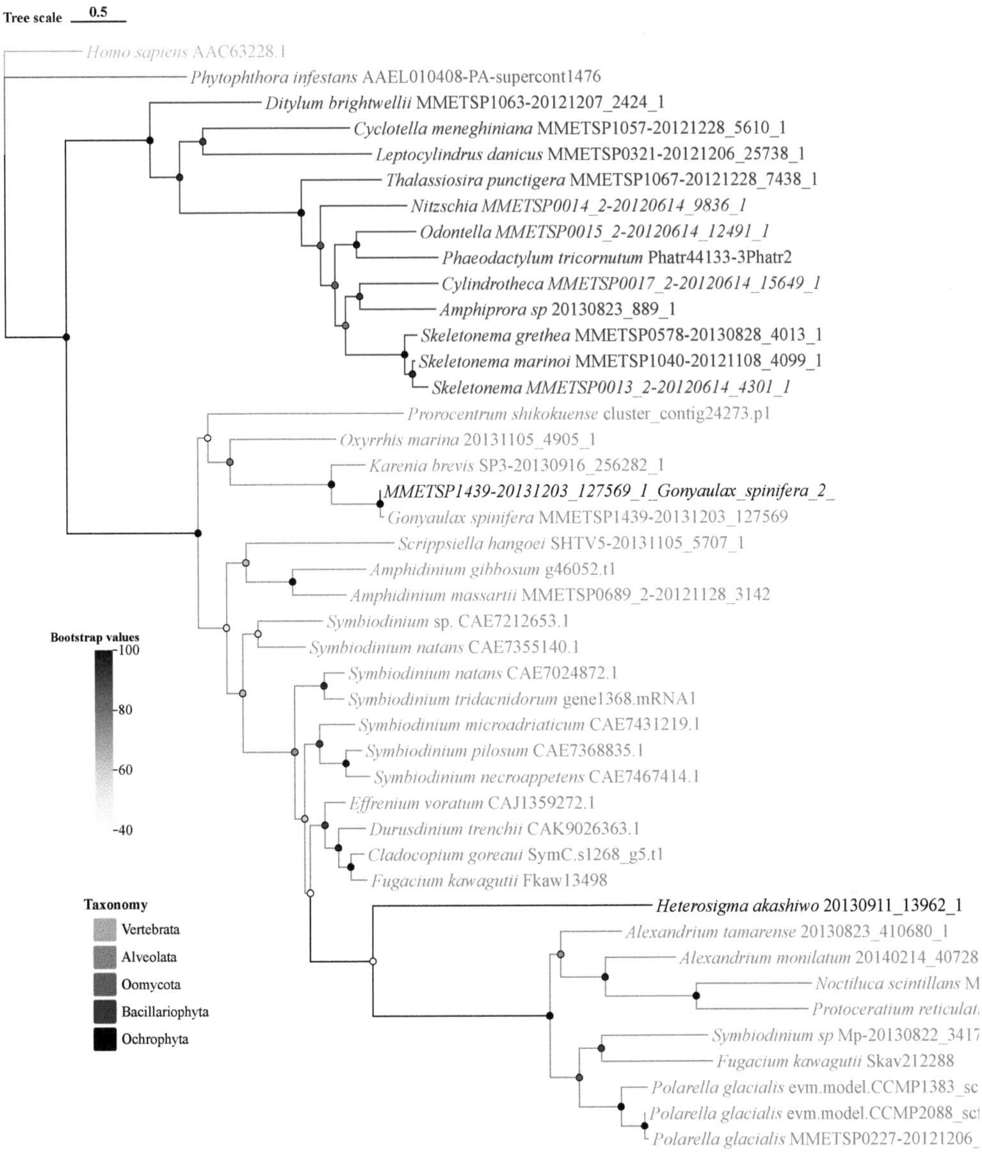

Figure 12.2. Phylogenetic tree of G protein coupled receptors in algae indicating prevalence and conserved nature. Human and Phytophthora GPCR are used as outgroup.

with its accessory proteins, NOFNiR (a molybdoenzyme containing a Moco domain) and THB1 (a truncated hemoglobin) (Gomez-Osuna et al. 2020).

In *Skeletonema marinoi*, the perception of external cues such as the presence of copepod grazers involves GPCR and its downstream MAPK signaling pathway (Amato et al. 2018). A study on the green seaweed *Ulva compressa* showed that the MAPK pathway plays a role in copper detoxification (Celis-Plá et al. 2019). In dinoflagellates, a MAPK homolog was cloned and sequenced for the heterotrophic dinoflagellate *Pfiesteria piscicida* (Zhang and Lin 2003). This MAPK is characterized by the signature motif LTDYVVTRWYRAPELM, which resembles that in land plants such as MAPK11 of *Tanacetum cinerariifolium* (GFD04998.1). The MAPK in *P. piscicida* shows growth-related expression dynamics.

Cell-to-Cell Communications

Bacteria use quorum sensing to measure and respond to population density changes through the production, release, and detection of autoinducers, such as acylated homoserine lactones (AHLs) (Bassler and Losick 2006). Detected autoinducers can initiate the activation of intracellular second messenger signaling systems, such as Ca^{2+} and cAMP, thereby regulating gene expression (Werthén and Lundgren 2001). This allows bacteria to function collectively, similar to multicellular organisms, and gain benefits unattainable individually (Bassler and Losick 2006).

In algae, density-dependent physiological responses are well-documented in diatoms, dinoflagellates, cyanobacteria, and other lineages, regulating processes like resting spore formation, sexual reproduction, and nutrient competition (Vardi 2008, Oliver et al. 2010, Pereira and Giani 2014, Russo et al. 2020, Pelusi et al. 2020). However, the chemical cues and molecular signaling mechanisms remain poorly understood.

Recently, genome analysis identified a calcium exchanger, PtSLC24A (XP_002176308.1 or Gene ID_7205024 in NCBI), in the diatom *P. tricornutum* as crucial for cell density-driven signaling (Liu et al. 2024). The SLC24 gene family, known in animals for Na^+/Ca^{2+}-K^+ exchangers, serves to extrude Ca^{2+} from the cytosol (Dipolo and Beauge 2006). The PtSLC24A sequence resembles other eukaryotic and prokaryotic Na^+/Ca^{2+} exchangers, with a specific Ca^{2+} binding site predicted within the transmembrane pore-forming segment (Liu et al. 2024). Molecular and physiological analyses suggest a potential Ca^{2+}-dependent intracellular signaling mechanism related to cell density. PtSLC24A facilitates Ca^{2+} efflux upon sensing population density signals, maintaining specific intracellular calcium levels. This intracellular transduction of density signals regulates physiological responses such as cell apoptosis, influencing the fate of algal blooms (Liu et al. 2024).

PAS Domain: A Common Sensing Mechanism

PAS domains (Fig. 12.3) are structural modules found in proteins across all life forms (Vreede et al. 2003). They act as universal signal sensors and interaction hubs in signal transduction proteins (Möglich et al. 2009). Initially identified in the *Drosophila* clock protein Per (period circadian protein), the mammalian transcription factor Arnt (aryl hydrocarbon receptor nuclear translocator protein), and the insect protein Sim (single-minded protein), PAS domains typically function as sensory modules. They detect a variety of stimuli, such as oxygen tension, small molecules, ions, gases, redox potential, or light intensity. Alternatively, they facilitate protein-protein interactions or bind small ligands (Vreede et al. 2003). PAS domains regulate processes like nitrogen fixation in rhizobia, phototropism in plants, circadian behavior in insects, and ion channel gating in vertebrates (Möglich et al. 2009). Thus, PAS domains play a critical role in monitoring changes in the environment, enabling organisms to adapt accordingly.

PAS domain-containing proteins function in a modular fashion. PAS sensor (input) domains detect a wide range of physical and chemical stimuli and regulate the response activity of effector (output) domains, such as catalysis or DNA binding. Despite the diversity of sensed stimuli and functional outputs, different PAS domains share low sequence similarity. However, their three-dimensional structure is generally conserved, consisting of a five-stranded antiparallel β-sheet and several α-helices. Typically, PAS domains encompass 250–300 amino acids, with two highly degenerate 50 amino acid subdomains termed the A and B repeats (Gu et al. 2000).

In algae, PAS domain research is limited. However, studies have identified PAS in proteins such as phytochrome, bZIP transcription factors, bHLH transcription factors, and aureochrome, which contains the LOV (light, oxygen, voltage) domain (Annunziata et al. 2019, Coesel et al. 2021).

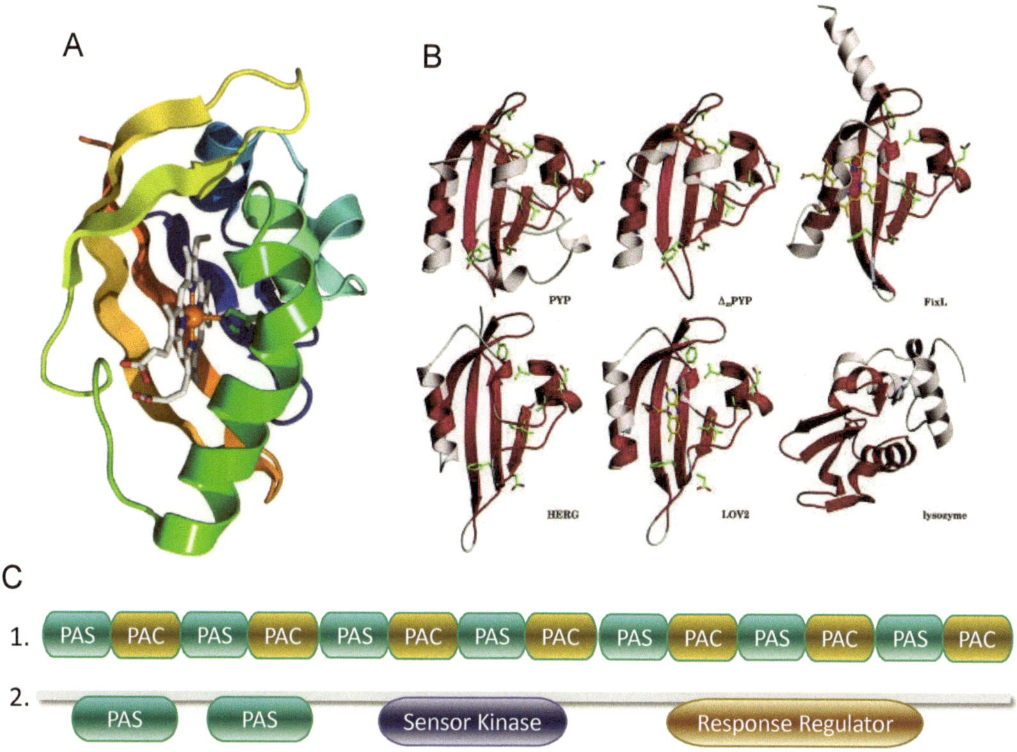

Figure 12.3. PAS domain structure and presence in different proteins. (A) Rendering by Boghog2 (https://en.wikipedia.org/wiki/PAS_domain#/media/File:FixL_1y28.png). (B) Ribbon representations of the structures of wtPYP, Δ25PYP mutant, FixL, HERG, LOV2, and turkey lysozyme (Protein Data Bank code 135L). Residues that align structurally are highlighted in red to facilitate comparison. The cofactors, if present, are drawn in a yellow-colored stick model. Homologous residues at similar structural positions are depicted in green. Reproduced with permission from Vreede et al. (2003) under the Creative Commons CC BY 4.0 license. (C) Example PAS domain organization in dinoflagellates; (1) PAS/phytochrome in *Polarella glacialis* (CAK0900951.1); (2) Two-component system in *Durusdinium trenchii* (CAK9003307.1).

Photoreceptors and Light-Dependent Pathways

Photosynthetic organisms contain multiple families of photoreceptors, with at least five classes identified in the plant model *Arabidopsis* (Bian et al. 2022). Phytochromes (phyA-phyE) are bilin-binding photoreceptors that sense red and far-red light spectra. Cryptochromes (CRY1 and CRY2), phototropins (PHOT1 and PHOT2), and ZEITLUPE family members (ZTL, FKF1, and LKP2) perceive ultraviolet (UV)-A and blue light, while UV-B RESISTANCE LOCUS 8 (UVR8) absorbs UV-B. These photoreceptors detect light signals and transmit this information to downstream signaling networks, precisely controlling plant growth and development.

Interestingly, the prasinophyte *Pycnococcus provasolii* has a phytochrome-cryptochrome chimeric gene (dualchrome1, DUC1) that detects light in the orange/far-red and blue wavebands, an adaptive innovation for expanding the absorption spectrum (Makita et al. 2021). Algae from various lineages also contain aureochromes, which have flavin mononucleotide (FMN) as the chromophore. An aureochrome contains one basic region/leucine zipper (bZIP) domain and one Light–Oxygen–Voltage (LOV)-sensing domain. Additionally, various algae lineages, such as dinoflagellates and diatoms, possess rhodopsin.

Phytochrome

Phytochromes (Fig. 12.4) are a class of photoreceptors found in plants, bacteria, fungi, and algae (Legris et al. 2019). In algae, phytochrome-encoding genes are present in streptophytes (*Mougeotia*), chlorophytes (*Tetraselmis*), cryptophytes (*Cryptomonas*), haptophytes (*Pavlova*), and cyanobacteria (*Nodularia*). No phytochrome genes have been identified in diatoms, dinoflagellates, and lineages in the supergroup Ochrophytes (Fig. 12.5). Phytochromes are sensitive to either the far-red (type I) or red spectrum (type II), depending on the type.

In plants, phytochromes control development under a wide range of environmental conditions and throughout the entire life cycle. Light-induced conformational changes enable phytochromes to interact with signaling partners, particularly transcription factors or proteins that regulate them, leading to extensive transcriptional reprogramming. Phytochromes also regulate promoter usage, mRNA splicing, and translation through less defined routes.

In land plants like *Arabidopsis*, phytochromes sense red and far-red light and exist in two forms. They are synthesized in the inactive Pr state, which is converted by light absorption to the active Pfr conformation. Pfr is inactivated upon Far-Red (FR) light absorption or through thermal relaxation, a temperature-dependent process known as dark or thermal reversion. Phytochromes function as dimers, in three possible combinations: Pr–Pr, Pfr–Pr, and Pfr–Pfr. Pr and Pfr have

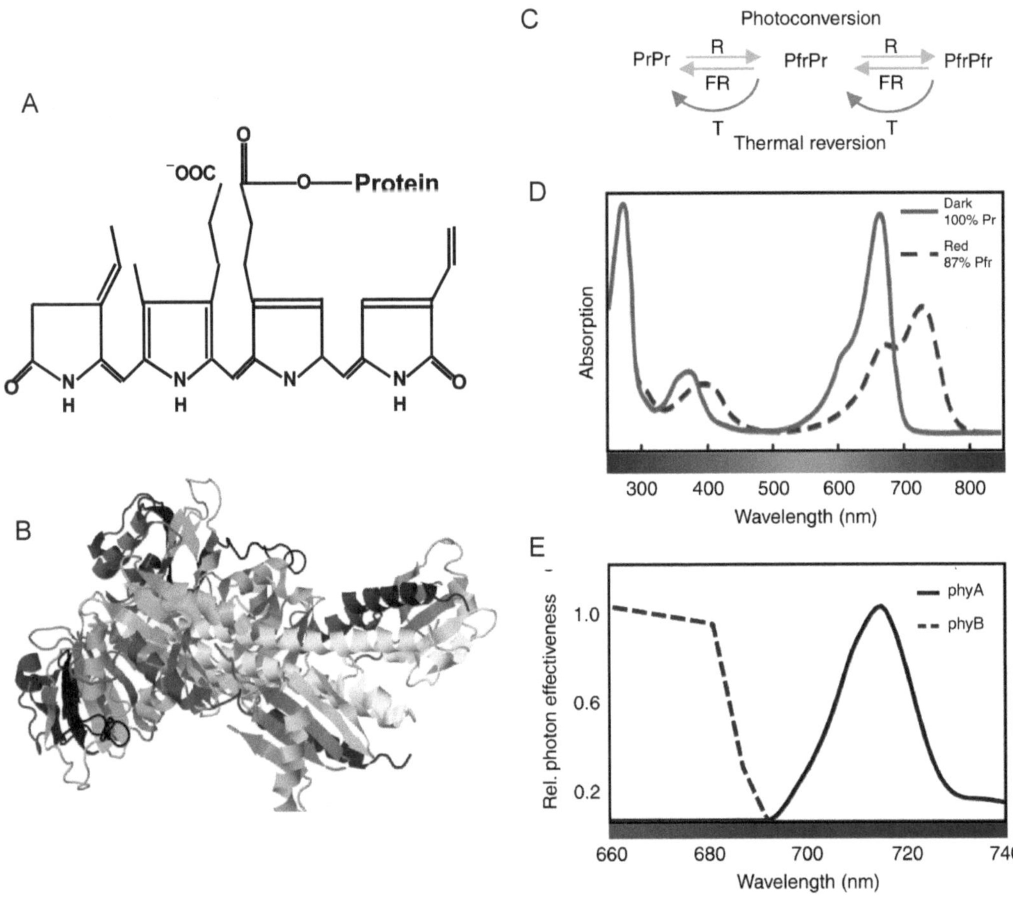

Figure 12.4. Phytochromes in plants. (A) bilin. (B) bilin-containing phytochrome. (C) Conversion between different dimer forms (PrPr, PfrPr, and PfrPfr induced by light and temperature. (D) Absorption spectra of Pf and Pfr-dominant forms. (E) Absorption of the two types of phytochromes (phyA and phyB). (B) is a courtesy of Jmol Development Team under GNU General Public License. (C-E) are reproduced from Legris et al. 2019 under the Creative Commons CC BY license.

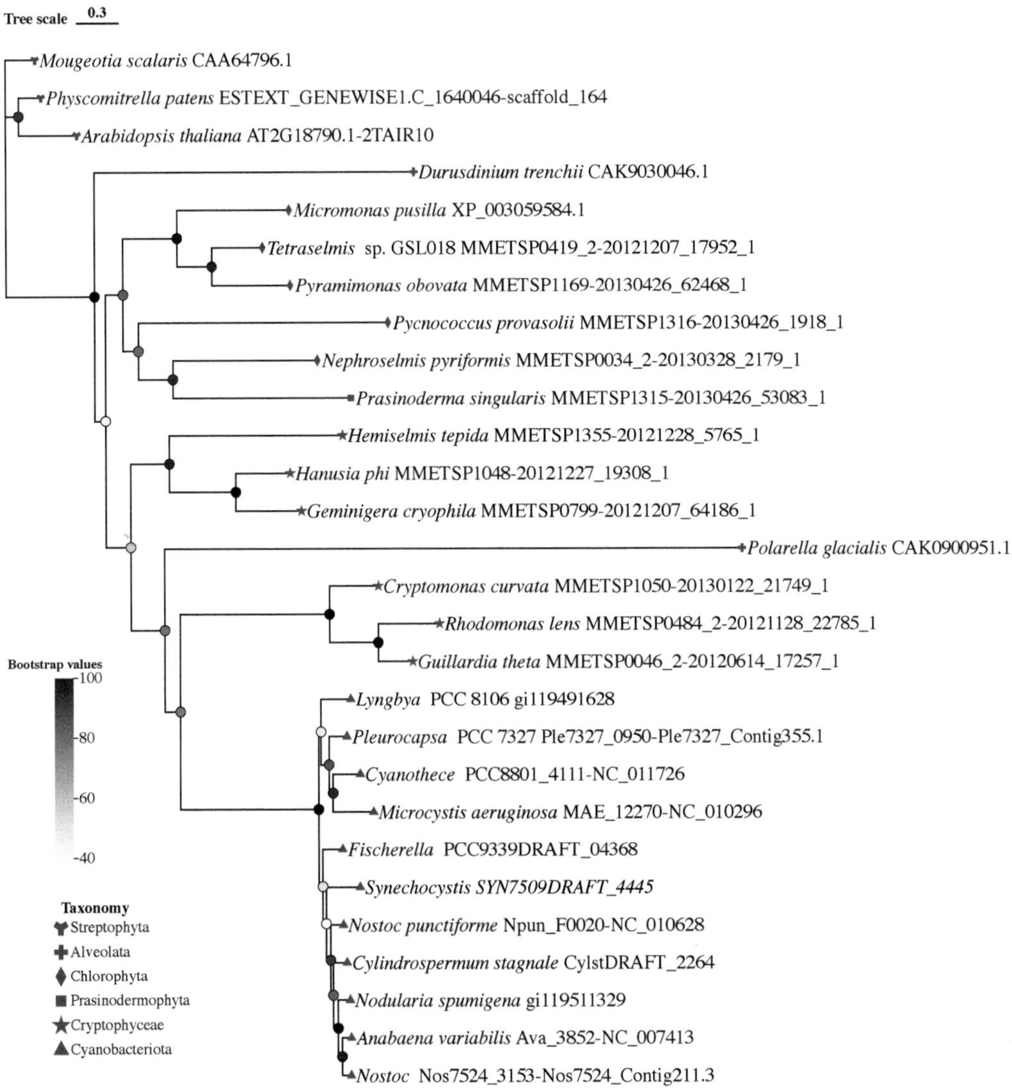

Figure 12.5. Phytochrome in algae. Note that *Durusdinium* and *Polarella* with long branches are sequences that contain two-component system besides the phytochrome domain.

different absorption maxima, but due to overlapping spectra, both conformers are constantly present in the light, except when prolonged darkness converts all phytochrome to Pr (Fig. 12.4). Since phytochrome responses depend on the proportion of Pfr conformers, signaling is influenced by a combination of light quantity, color, and temperature. Whether algal phytochromes function similarly to their Viridiplantae counterparts remains to be investigated.

Cryptochrome

Cryptochromes are a class of flavoproteins sensitive to blue light, found in a wide range of animals, plants, and algae (Fig. 12.6). Although they are highly similar in sequence to DNA photolyases, cryptochromes lack DNA repair activity. In animals, cryptochrome (CRY) is a key element of the circadian system. In plants, CRY acts as a blue or ultraviolet (UV-A) light receptor, entraining circadian rhythms and mediating various light responses such as the regulation of flowering and

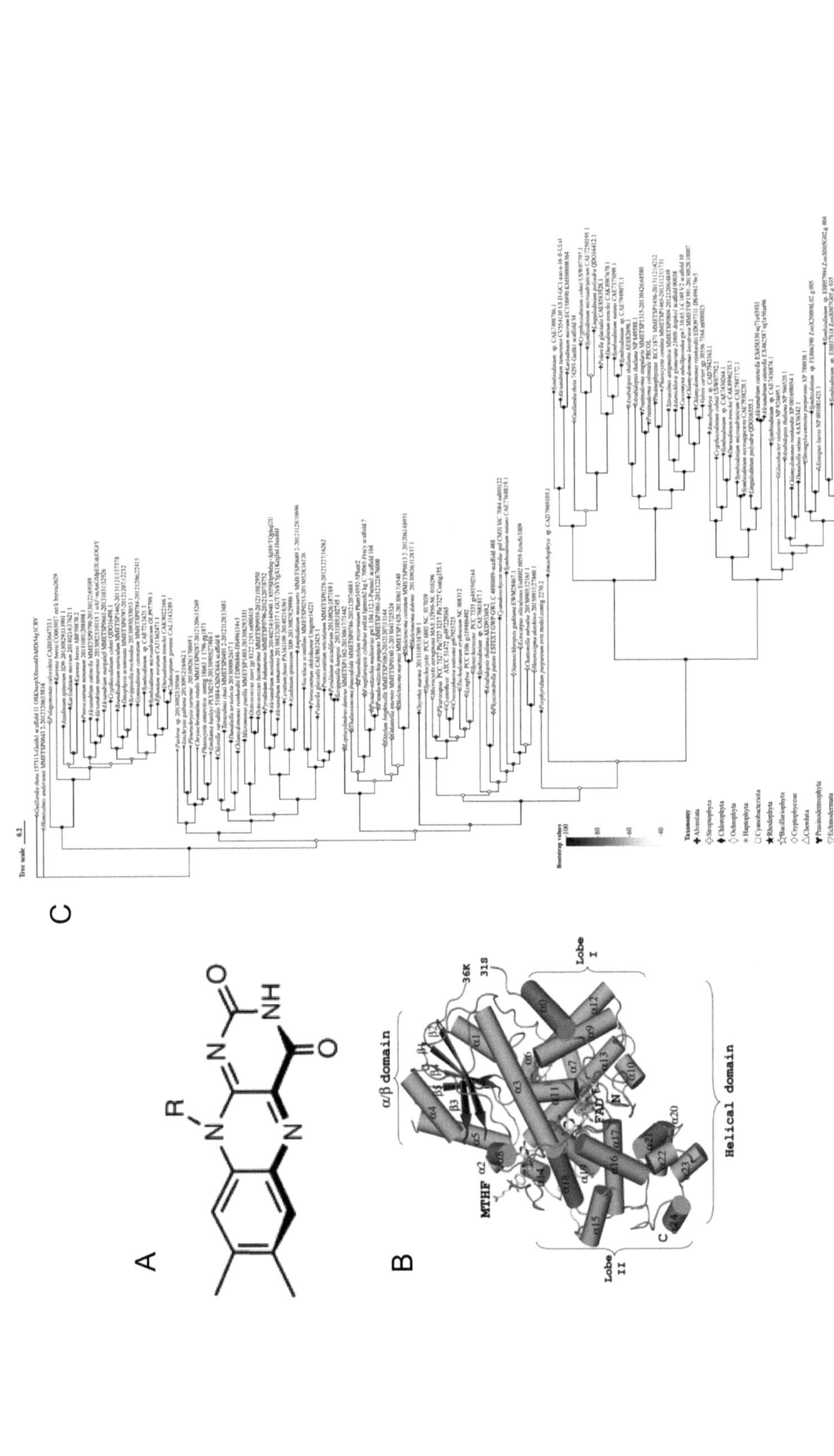

Figure 12.6. Cryptochromes in plants and algae. (A) Flavin. (B) Flavin-containing cryptochrome. (C) Phylogenetic tree showing the wide distribution and distinct groups of cryptochrome in algae as well as model organisms.

seedling growth (Li and Todo 2005). Additionally, cryptochrome functions as a magnetoreceptor underlying the light-dependent magnetic compass in migratory birds (Liedvogel and Mouritsen 2010). Interestingly, in the plant model *Arabidopsis thaliana*, cryptochrome also exhibits linkage to magnetic field effects. Processes involving cryptochrome signaling, such as hypocotyl growth inhibition, are enhanced under a magnetic field (Ahmad et al. 2007).

The cryptochrome 1 protein (CRY1) in *Arabidopsis* is found in both the nucleus and cytoplasm of cells, with each pool having different functions (Wu and Spalding 2007). Nuclear CRY1 is responsible for blue light-induced hypocotyl growth suppression and petiole elongation, while cytoplasmic CRY1 promotes primary root growth and cotyledon expansion in blue light, which are inhibited by nuclear CRY1. Additionally, anthocyanin production in response to blue light is strongly stimulated by nuclear CRY1 and, to a lesser extent, by cytoplasmic CRY1.

Cryptochromes are present in all major lineages of algae, including dinoflagellates and diatoms, based on current genomic and transcriptomic data (Fig. 12.6), showing high diversity (multiple clades in the phylogenetic tree). In the neurotoxin-producing dinoflagellate *Karenia brevis* has a cryptochrome homolog with a molecular mass of approximately 55 kDa. Comparative sequence analysis indicates it is a blue-light receptor, resembling the DASH type of cryptochrome, and is localized in the chloroplast, similar to CRY DASH in *Arabidopsis thaliana* (Brunelle et al. 2007). DASH (*Drosophila*, *Arabidopsis*, *Synechocystis*, human)-type cryptochromes (cry-DASHs) form one subclade of the Cryptochrome/Photolyase Family (CPF). Among the algal cyptochromes, some cluster with animals and land plants, likely the DASH group, but others cannot be classified with current data (Fig. 12.6).

In diatoms, two types of cryptochromes have been identified and characterized: CPF1 and CryP (König et al. 2017). The *Phaeodactylum tricornutum* genome contains seven cryptochrome genes, including PtCPF1 and CryP. CPF1 is related to animal-type cryptochromes, functioning as both photolyase and photoreceptor, and is involved in clock-related processes and gene transcription. PtCPF1 (gene ID: 27429) in *P. tricornutum* also plays a role in high-temperature acclimation (Gao et al. 2023).

CryP, on the other hand, is distantly related to plant-type cryptochromes and influences light-harvesting protein accumulation. It also impacts gene transcription, as revealed by comparative analyses of CryP knock-down mutants and wild-type subjected to prolonged darkness and blue light exposure, which showed massive transcriptomic alterations (affecting 70% of expressed genes). These genes were influenced in different manners: some only under blue light, others independently of light conditions, and some only in darkness (König et al. 2017). More recently, a study indicated that the knockout of CryP in *P. tricornutum* led to increases in fucoxanthin content and fucoxanthin chlorophyll a/c-binding protein levels (Yang et al. 2022).

Aureochrome

Aureochromes are photoceptors initially found in the stramenopiles *Vaucheria frigida* (Xanthophyte) and *Fucus distichus* (Phaeophyte) and have been identified in eight algal species (Table 12.3). Each aureochrome contains a basic region/leucine zipper (bZIP) domain for DNA binding and one light-oxygen-voltage (LOV)-sensing domain and a flavin mononucleotide as a chromophore (Takahashi et al. 2007; Fig. 12.7). The DNA binding domain indicates that aureochromes are transcription factors. Studies have shown the role of aureochrome in regulating high light acclimation and the cell cycle. They are essential for light-dependent morphogenesis in these algae.

The *P. tricornutum* genome hosts four aureochrome genes (PtAUREO1a, 1b, 1c, and 2). TALEN-mediated knockout of PtAUREO1a and PtAUREO1b resulted in loss of photoacclimation ability, highlighting their crucial role in this process (Mann et al. 2017). PtAUREO1a and PtAUREO1b are not fully functionally redundant, as their respective knockouts exhibit distinct phenotypes. However, both are crucial for photoacclimation under blue light conditions. In wild-type *P. tricornutum*, a shift from red to blue light triggers strong (affecting about 75% of genes) and rapid

Table 12.3. Aureochrome identified in algae.

Group	Species	AUREO gene #	References
Diatoms	*Phaeodactylum tricornutum*	5	Takahashi et al. 2007
	Thalassiosira pseudonana	+	Takahashi et al. 2007
Phaeophytes	*Cladosiphon okamuranus*	7	Wu et al. 2022
	Ectocarpus siliculosus	5	Wu et al. 2022
	Fucus distichus		Takahashi et al. 2007
	Nemacystus decipiens	9	Wu et al. 2022
	Saccharina japonica	9	Deng et al. 2012
	Undaria pinnatifida	8	Wu et al. 2022
Pelagophytes	*Aureococcus anophagefferens*	+	Ishikawa et al. 2009
Raphidophytes	*Chattonella antiqua*	+	Ishikawa et al. 2009
	Heterosigma akashiwo	4	Ji et al. 2017
Chrysophytes	*Ochromonas danica*	+	Ishikawa et al. 2009
Xanthophytes	*Vaucheria frigida*	2	Takahashi et al. 2007

+ presence reported without copy number

(within 10 to 60 minutes) transcriptional changes, whereas PtAUREO1a knockout lines showed a dramatically reduced response to this light shift, indicating that PtAUREO1a plays a key role in blue light-induced gene regulation (Mann et al. 2020).

Further functional genetic experiments (Im et al. 2024) demonstrated that PtAUREO1a and 1b proteins can activate the genes *diatom-specific cyclin 2*, *PtAureo1a*, and *PtAureo1c* under both light and dark conditions, and that PtAUREO1a regulates the expression of PtLHCF15, which is essential for red light acclimation. The promoters of these genes targeted by PtAUREO were identified.

Rhodopsin

Rhodopsin is a transmembrane protein (opsin) that contains retinal as its chromophore and features seven transmembrane domains. It is found in all three domains of life albeit in different forms, serving various functions. In animals with eyes, rhodopsin is the primary photoreceptor molecule for vision. Upon absorbing light, the chromophore 11-*cis*-retinal is photoisomerized to its all-*trans* form, activating the receptor and initiating a highly amplified signal transduction cascade.

In prokaryotic (bacterial and archaeal) and eukaryotic microbes, rhodopsin also acts as a light receptor with different downstream functions (Figs. 12.8 and 12.9). They create motive for flagellar mobility, serve as pumps or ion channels, and regulate phototaxis (both positive and negative). A notable subgroup is proton pump rhodopsins (PPRs), which provide a photosystem-independent mechanism for converting photoenergy to ATP, as detailed further in Chapter 16.

Microbial rhodopsins, such as bacteriorhodopsin in *Halobacterium halobium*, harvest light energy to drive chloride pumps (Oesterhelt and Stoeckenius 1971). In other microbes, rhodopsins capture photoenergy to drive flagellar movement. In *Chlamydomonas reinhardtii*, there are two rhodopsins: Channelrhodopsins 1 and 2 (CSRA and CSRB), which mediate phototaxis and photophobic swimming responses. CSRA primarily contributes to the photophobic response at high light intensities, while CSRB mediates phototaxis at low light intensities through an efficient amplification cascade (Govorunova et al. 2004).

In some species, PPRs create an influx of protons. For instance, the deep-ocean marine bacterium *Parvularcula oceani* has three rhodopsins, one functioning as a light-driven inward H+ pump (Inoue et al. 2016). Mechanistic analyses have revealed that small differences in interactions at the active center determine the direction of H^+ transfer.

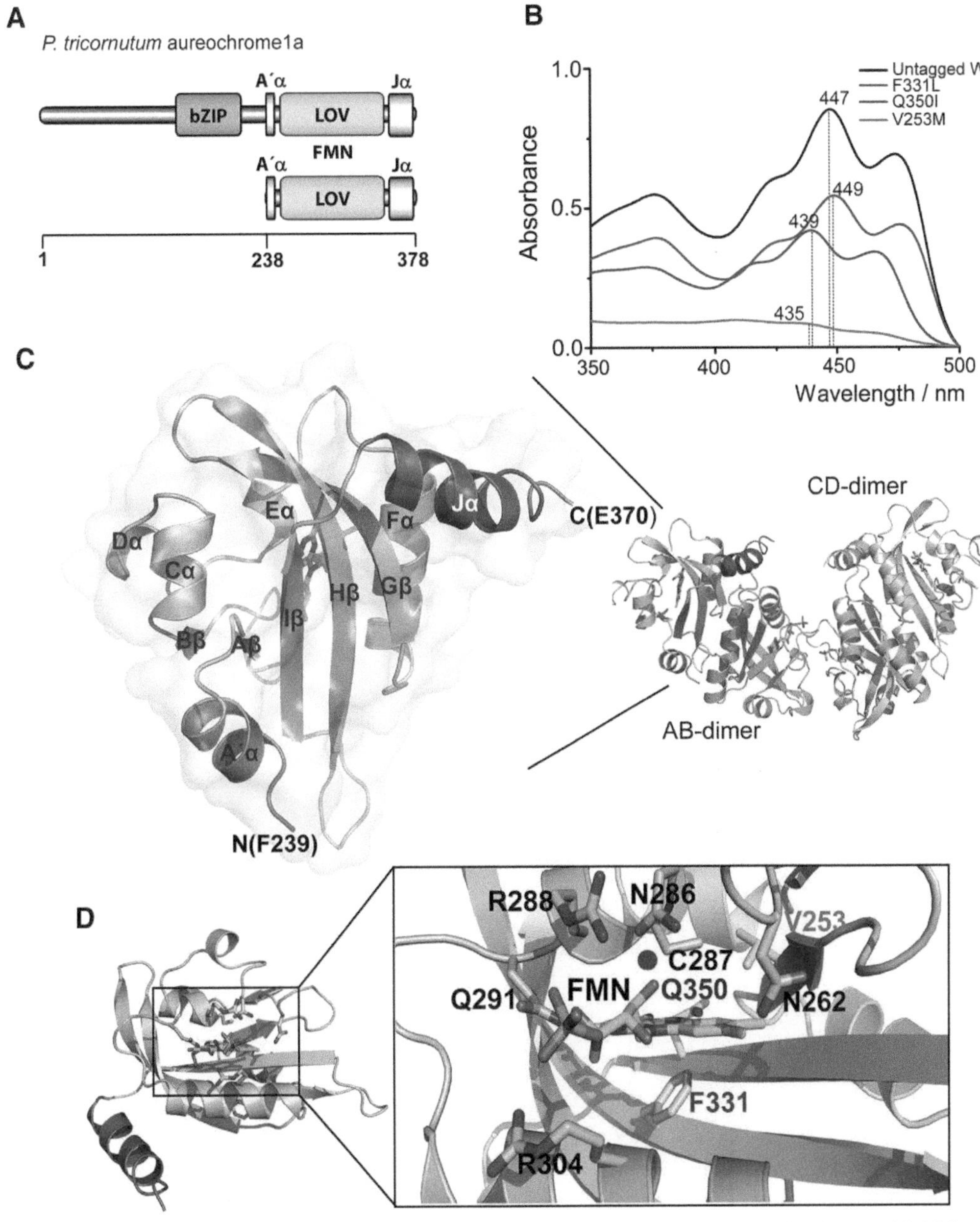

Figure 12.7. Crystal structure of A′α-LOV-Jα in the Dark-Adapted State of aureochrome in diatoms. (A) Molecular architecture of full-length *PtA*UREO1a and its A′α-LOV-Jα module. (B) UV/Vis spectra of recombinant A′α-LOV-Jα and *PtA*UREO2 LOV domain-based mutants with their bound flavin cofactors. WT, wild-type. (C) The overall fold of the A′α-LOV-Jα module (molecule A) is shown on the left; the four molecules in the asymmetric unit as found as AB and CD dimers are depicted on the right. (D) The FMN-binding site of PtAUREO1a A′α-LOV-Jα. Residues crucial for binding of the FMN chromophore are highlighted as stick models. N286 interacts with FMN via a bridging water molecule (sphere). Reproduced with permission from Banerjee et al. 2016.

Some dinoflagellate species possess both sensory and proton pump rhodopsins. In the heterotrophic species *Oxyrrhis marina*, sensory rhodopsin is involved in phototaxis, while the other rhodopsin functions in photoenergy harnessing (Hartz et al. 2011, Guo et al. 2014).

Retinal, the chromophore of rhodopsin, occurs as 11-*cis* retinal in vision rhodopsin and all-trans retinal in proton pump rhodopsin (Pushkarev et al. 2018). The precursors of both forms are

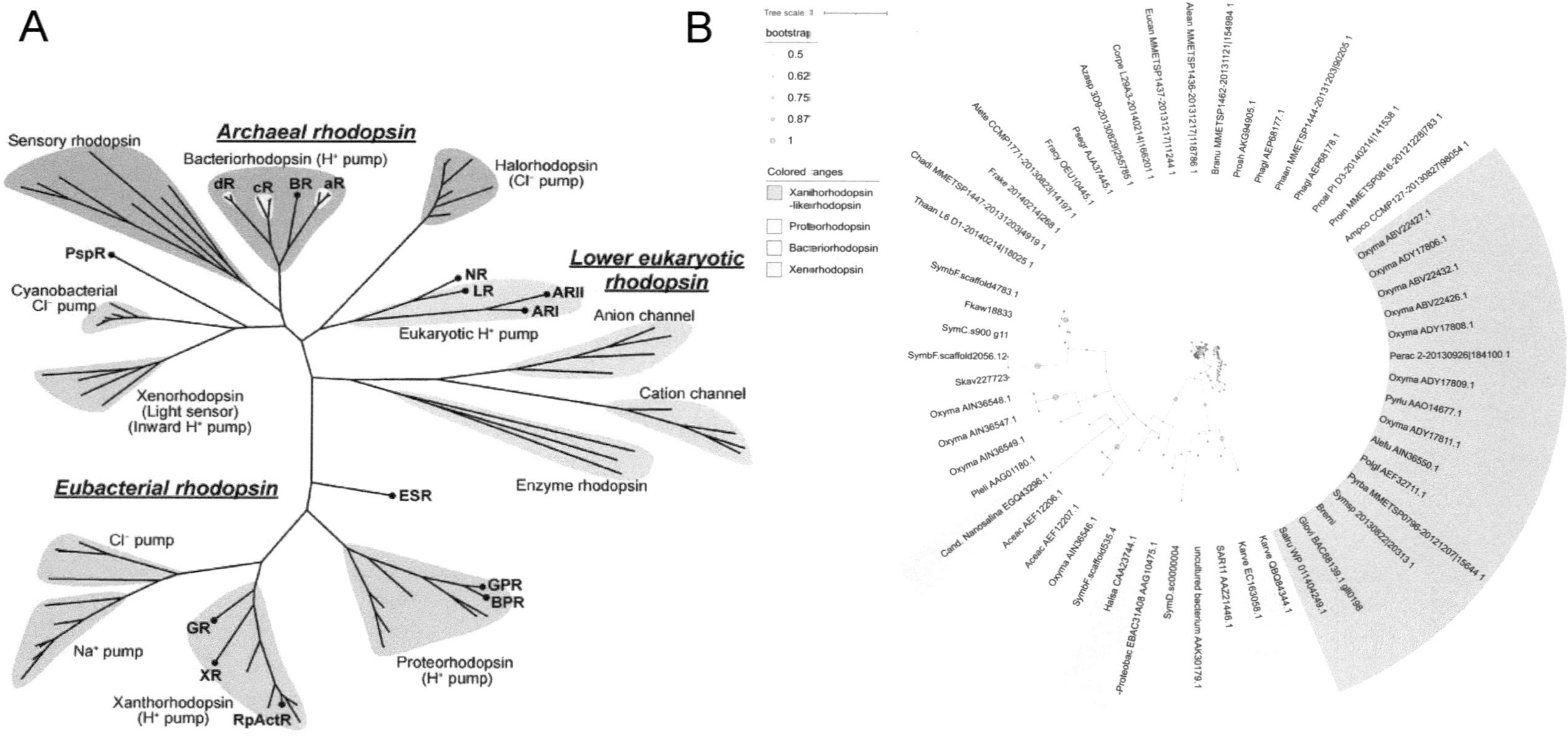

Figure 12.8. Microbial rhodopsins in all three domains of life. (A) Microbial rhodopsins in archaea, eubacteria, and lower eukaryotes. After Tamogami and Kikukawa 2021 under the Creative Commons CC BY 4.0 license. (B) proton pump rhodopsins in dinoflagellates that are clustered into three major groups, including xanthorodopsin, bacterorhodopsin, and proteorhodopsin.

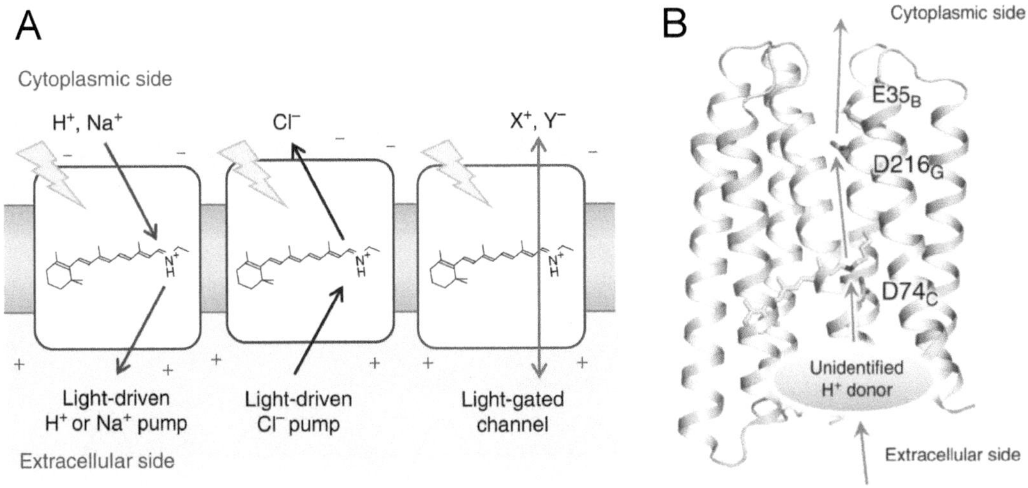

Figure 12.9. Schematic showing the mechanistic about the functions of different types of light-driven rhodopsin-based ion pumps. (A) Light-driven outward cation pumps (left) and inward anion pumps (middle) function as active transporters, while light-gated channels conduct cations or anions in a passive manner (right). (B) The H+ transfer pathway in *Po*XeR suggested by the results of the present study. The primary H+ transfer occurs from the Schiff base to D216, and E35 affects this reaction. The Schiff base is not directly reprotonated from the aqueous phase, but there is an internal H+ donor, yet to be identified. Reproduced from Inoue et al. (2016) under the Creative Commons CC BY 4.0 license.

carotenes or xanthophyll. Animals must incorporate these from their diet using β-carotene 9’,10’ oxygenase 2 (BCO2) (Vila et al. 2019), whereas rhodopsin-possessing microbes use the blh-encoded β-carotene 15,15′ dioxygenase to convert β-carotene to retinal (Sabehi et al. 2005). In cyanobacteria, retinal is produced by the diox1 gene (Ruch et al. 2005). Some bacteria lack these genes but still have functional rhodopsins, suggesting alternative pathways for synthesizing or acquiring retinal (Nakajima et al. 2020, Chazan et al. 2022).

Whether rhodopsin-carrying eukaryotic algae synthesize retinal or obtain it externally remains unclear. Recent reports indicate that heterotrophic protists, including *Thecamonas trahens*, *Oxyrrhis marina*, and *Noctiluca scintillans*, contain carotenoid biosynthesis pathways and genes related to retinal generation (Rius et al. 2023). However, no homologs of these genes have been found in photoautotrophic protists, warranting further investigation.

Circadian Rhythm

Circadian rhythmicity is a universal feature in living organisms, including prokaryotic algae (cyanobacteria) and eukaryotic algae, regulated by an endogenous clock. Daily rhythms in RNA abundance, for instance, are common across algal lineages, except for dinoflagellates (Jadhav et al. 2022). Circadian rhythms and the underlying clock are aligned when both are entrained by the light-dark cycle. However, a cell population can lose the circadian rhythm while the clock in individual cells function as normal. This can happen under conditions such as constant light or darkness. This loss of detectable rhythm is due to desynchronization among individual cells, not a loss of the clock's function or individual cells' rhythmicity.

Synechococcus elongatus PCC 7942 serves as a model for studying circadian rhythms (Fig. 12.10). In this cyanobacterium, circadian regulation affects 65% of total genes via a post-translational clock mechanism (Markson et al. 2013). The core oscillator is encoded by the genes kaiA, kaiB, and kaiC, controlling gene expression patterns, chromosome compaction, cell division timing, and metabolic switches (Schmelling et al. 2021). Environmental stimuli integration and synchronization involve interactions between the core clock and redox-sensing proteins CikA and LdpA, which relay cellular redox conditions to the core clock.

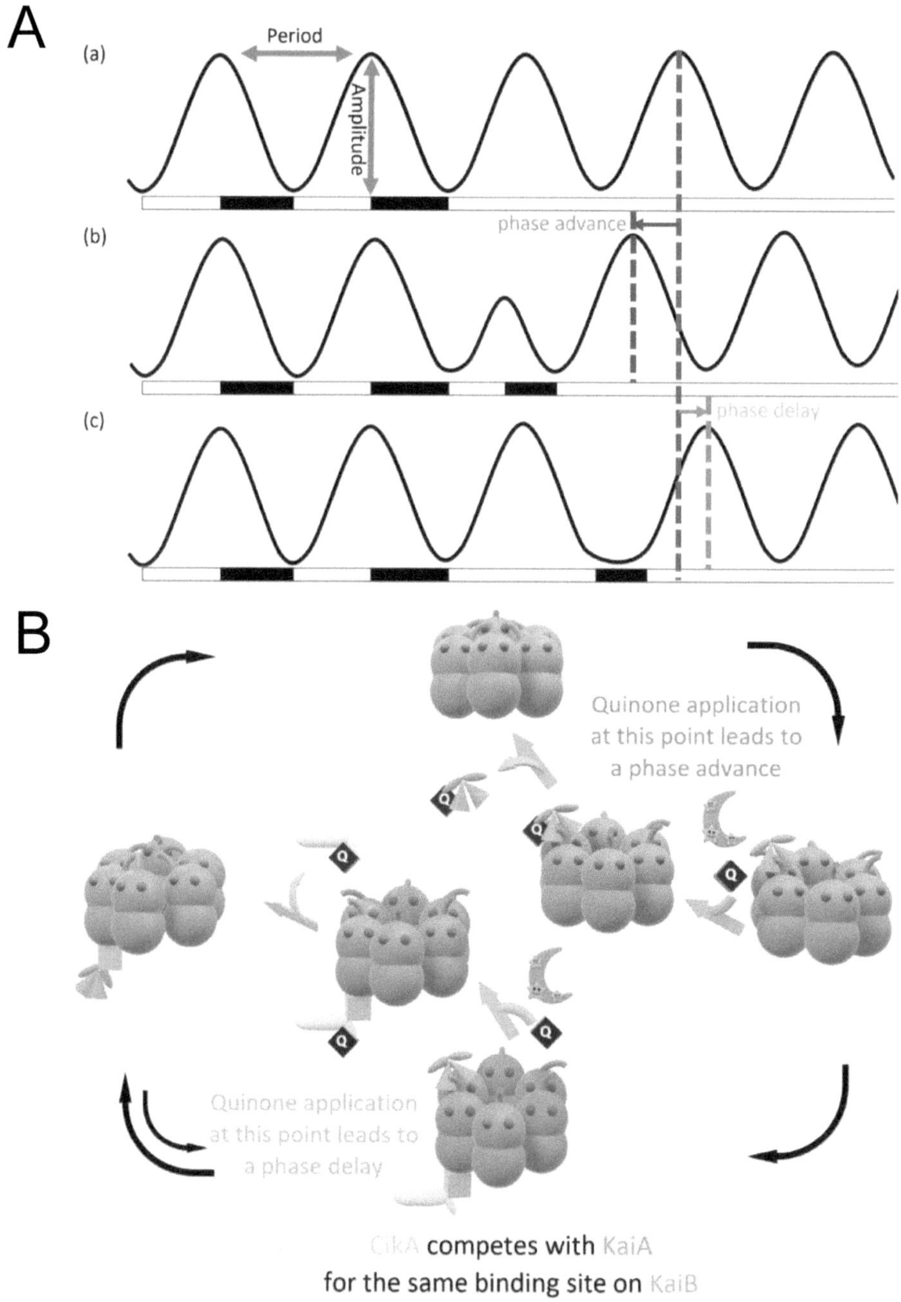

Figure 12.10. The entrainment of cyanobacterial circadian rhythm with a dark pulse (A) and molecular mechanism (B). (A) The white part on the x-axis represents light or daytime; the dark area represents either the application of a dark pulse *in vivo* or the application of ADP or oxidized quinone in vitro. The y-axis represents the circadian gene expression pattern or KaiC phosphorylation rhythm in cyanobacteria. A free-running rhythm that still runs even under a continuous light condition (a), a phase advance caused by a dark-related cue during the rising phase of the clock (b), and a phase delay that is caused by a dark-related cue during the falling phase (c). (B) Oxidized quinone ("Q" in a black rhombus) becomes acutely abundant at night as a proxy for darkness. As KaiC autophosphorylation takes place, quinone application at this point inactivates KaiA, leading to premature dephosphorylation and a phase advance (upper right arrows). On the other hand, in the in vitro mixture containing CikA and KaiABC, as fully phosphorylated KaiC enters the autodephosphorylating phase, quinone application at this point inactivates CikA, adding an extra time frame for the KaiA sequestration to take place, delaying the phase (lower left arrows). Reproduced from Kim et al. 2020 under the Creative Commons CC BY 4.0 license.

Another pathway involves the transcription factor RpaB, which interacts with the core oscillator's output components, histidine kinase SasA, and its response regulator RpaA. RpaA is a major transcription factor essential for genome-wide rhythmic gene expression (Iwasaki et al. 2000, Takai et al. 2006). Interestingly, *Prochlorococcus* has a simplified timing system lacking KaiA and several central input and output factors, compared to the more complex setup in *Synechocystis* (Schmelling et al. 2021).

Circadian clocks in eukaryotic algae have been studied in model organisms such as *Chlamydomonas reinhardtii*, *Ostreococcus tauri*, and dinoflagellates. In chlorophytes, some clock genes are homologous to those in vascular plants. For instance, *C. reinhardtii* has genes similar to *Arabidopsis* CCA1 and LHY, which help maintain the circadian clock's rhythmicity. A proteomic study on *O. tauri* revealed distinct diurnal and circadian protein expression rhythms, showing high coordination between plastid-encoded and nucleus-encoded plastid-localized proteins (Kay et al. 2021). However, the genetic network architectures of algal clocks remain largely unknown.

Cryptochrome (CRY) proteins play a role in regulating the circadian clock in algae and other organisms. Algae may have additional clock-associated proteins, including transcription factors, kinases, and phosphatases, which participate in feedback loops to maintain circadian oscillations. In dinoflagellates, circadian rhythms govern processes like bioluminescence, vertical migration, aggregation, cell division, photosynthesis, nitrogen assimilation, and expression of numerous proteins and enzymes (Jadhav et al., 2022). Interestingly, bioluminescence rhythms in dinoflagellates do not involve transcriptional regulation (Morse et al. 1989, Roy et al. 2014, 2018).

To identify clock components in dinoflagellates, Jadhav et al. (2022) analyzed existing omic data and found cryptochrome (CRY), chalcone synthase (CHS), and casein kinase 2 (CK2), with significant similarities to plant clock counterparts. The CK family is the only clock gene shared with other eukaryotes and appears to be a universal clock gene.

Pseudo-Response Regulator proteins (PRRs) are central to the plant circadian oscillator. PRRs bind intranuclearly to Phytochrome Interacting Factor (PIF) family members, repressing their capacity to activate Direct Target Genes (DTGs) (Zhang et al. 2020). Signal transduction from the photoactivated receptor phytochrome (Phy) to the cellular response network involves rapid translocation of the far-red absorbing PHY form (Pfr) into the nucleus, where it binds to PIF proteins, leading to their phosphorylation, ubiquitination, and degradation. BLASTP analysis with plant PRR as a query yielded no homologs in algae, suggesting that the PRR clock component likely does not exist outside vascular plants. Instead, top hits were two-component response regulator (ARR) proteins from chlorophytes and prasinophytes, with no matches in non-green algae.

Adaptation to Light Fluctuations

RpaB is an OmpR-type Transcription Factor (TF) that controls various photosynthesis-associated genes under different light conditions. This TF binds to a DNA motif with imperfect 8-nt direct repeats (G/T)TTACA(T/A)(T/A) separated by two random nucleotides. Binding sites close to the Transcriptional Start Site (TSS) mediate repression, while sites about 50 nt upstream mediate activation (Riediger et al. 2019). RpaB was first discovered in the cyanobacterium *Synechocystis* strain PCC6803, where it regulates the distribution of energy from phycobilisomes between PSI and PSII (Ashby and Mullineaux 1999). This energy redistribution, known as "state transition", occurs under fluctuating light conditions. In *Synechococcus elongatus* PCC7942, RpaB is involved in transcriptional regulation during dark-to-light transitions and light intensity increases, highlighting its broader regulatory role (Yasuda et al. 2020). RpaB is essential and as such cannot be deleted by conventional methods (Riediger et al. 2019).

RpaB is found not only in the cyanobacteria core genome but also in the chloroplast genomes of the charophyte alga *Chlorokybus atmophyticus* and all non-green algae, including rhodophytes, glaucophytes, cryptophytes, haptophytes, and raphidophytes, indicating its significant and conserved functions (Riediger et al. 2019).

Diatoms, known for their adaptation to fluctuating light environments, show significant transcriptional responses under different light regimes. In *Phaeodactylum tricornutum*, 2,673 genes (25% of total expressed genes) showed transcriptional changes, including 497 transcription factors with higher expression under light fluctuations (Zhou et al. 2022). Seven of these TFs contain light-sensitive PAS domains, and ten are predicted to regulate genes related to light-harvesting complex proteins. Whether these are part of the RpaB regulon remains to be examined, but so far, no RpaB has been reported in diatoms. Severe light fluctuation was shown to divert the metabolic flux of assimilated carbon preferentially towards fatty acid storage over sugar and protein.

Temperature

Response and adaptation to temperature elevation and global warming are detailed in Chapter 14. This section focuses on temperature sensing, both cooling and warming. In plants, cellular membranes, calcium (Ca^{2+}) channels, and the rice CHILLING TOLERANCE DIVERGENCE 1 (COLD1) gene function as low-temperature sensors. Conversely, warm temperature sensing involves the red-light photoreceptor PhyB, its interacting partner PIF7, and EARLY FLOWERING 3 (ELF3), a core component of the plant circadian clock (Bian et al. 2022). Notably, several light signaling components are involved in temperature-mediated development, indicating crosstalk between light and temperature signaling to regulate plant growth and development. This integration is mediated by the liquid-liquid phase separation of PhyB. PhyB forms liquid-like droplets driven by its C terminus self-association and modulation by the intrinsically disordered N-Terminal Extension (NTE), coordinating light and temperature signaling networks to optimize plant growth during daily and seasonal changes (Chen et al. 2022).

In green algae, cold response mechanisms resemble those in plants, involving calcium-based signaling and ubiquitin-mediated degradation (Fig. 12.11). Small heat shock proteins (sHSPs) were identified in the basal red alga *Cyanidioschyzon merolae* and the green alga *Chlamydomonas reinhardtii* (Kohayashi et al. 2014). Two sHSP genes in *C. merolae* were drastically and rapidly upregulated (> 1,000 fold) in response to heat shock treatment (~ 50°C). Their upregulation appears to be triggered by absolute temperature rather than temperature changes. This suggests that sHSPs may serve as precise molecular thermometers in algae, with the threshold temperature to trigger gene expression varying among algal species (e.g., ~ 36°C for *C. reinhardtii*).

Temperature changes can significantly affect cell surface properties, including hydrophobicity and cell membrane stiffness or fluidity. Decreases in temperature cause membrane rigidification in algae like *Chlamydomonas*, which may serve as an initial sensor for cold stress (Ermilova 2020). To restore membrane fluidity, algae increase the amount of polyunsaturated fatty acids in their membranes upon cold exposure. This change in membrane properties can activate downstream cold signaling pathways, similar to mechanisms proposed in plants like *Arabidopsis*.

Some cyanobacteria and heterotrophic bacteria possess specialized cold sensor proteins called histidine kinases that enable them to detect and respond to low temperature conditions. These cold sensors, such as DesK of *Bacillus* (Saita and de Mendoza 2015) and Hik33 of *Synechocystis* (Murata and Los 2006), function by perceiving alterations in the physical properties of the cell membrane that occur upon exposure to cold temperatures. When the membrane properties change due to cold, these sensor kinases undergo conformational shifts that initiate downstream signaling cascades, allowing the species to adapt their cellular processes and physiology to cope with the low temperature stress.

Cold stress initiates signal transduction through secondary messengers like Ca^{2+} (Fig. 12.11), which in land plants can activate mechano-sensitive or ligand-activated Ca^{2+} channels, triggering an increase in cytosolic Ca^{2+} and stimulating the Mitogen-Activated Protein Kinase (MAPK) signaling pathway (Knight et al. 1996, Conde et al. 2011). While cytosolic Ca^{2+} increases in response to cold in *Chlamydomonas* remain unconfirmed, transcripts encoding Cyclic Nucleotide-Gated Channel (CNGC)-like proteins, Calcium-Dependent Protein Kinases (CDPKs), and cyclic nucleotides accumulate after cold treatment, suggesting their involvement (Li et al. 2020). Cold also induces

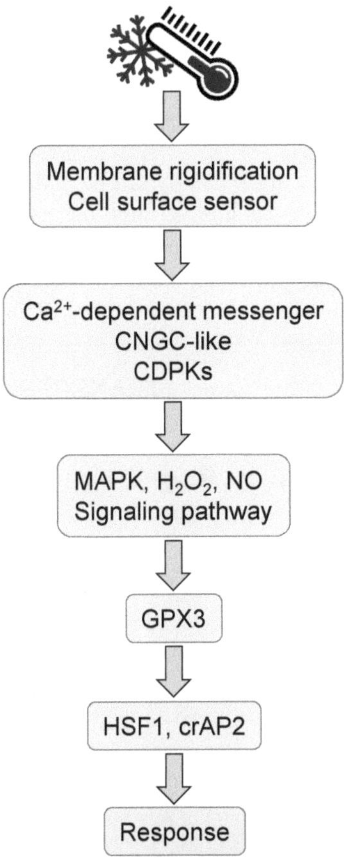

Figure 12.11. Summary of fundamental responses of *Chlamydomonas* during low-temperature exposure. Left and right circles represent key cold-induced components of *Chlamydomonas* and land plants, respectively.

H_2O_2 accumulation, a key signaling molecule mediating responses. In *Chlamydomonas*, the cold-responsive gene GPX3 appears to be regulated by H_2O_2 (Zalutskaya et al., 2019). H_2O_2 may coordinate with NO, which regulates cold-responsive genes and protein modifications in plants, though its role in algae is unexplored.

Major cold signaling regulators include dephosphorylating proteins and transcription factors (TFs). Potential TF candidates in *Chlamydomonas* include HSF1 (heat shock factor 1), crAP2 (APETALA2 [AP2]/ethylene-responsible element binding proteins [EREBPs] family of TF), and others from various families. Besides CNGCs and CDPKs (Li et al. 2020), kinases like SnRK2.2 (plant-specific serine/threonine kinase), which has a key role in sulfur deprivation response (Gonzalez-Ballester et al. 2008), may also participate (Unal and Cekic 2019). However, further studies are needed to fully elucidate the regulatory networks underlying the cold stress response in algae.

Another group of thermosensitive receptors is the transient receptor potentials (TRPs), transmembrane proteins that serve as transducers by altering membrane potential in response to changes in intracellular calcium concentrations (Samanta et al. 2018). In model organisms, TRPs function as cellular sensors that detect environmental stimuli such as temperature, pH, osmolarity, and mechanosensation (Clapham et al. 2003). Some of these TRPs are known as thermo-TRPs and act like microscopic thermometers in animals, sensing hot or cold temperatures (Himmel and Cox 2020). The TRPM (melastatin-related) family members TRPM2, TRPM3, TRPM4, and TRPM5 are activated by heat, TRPM3 responds to harmful heat, while TRPM8 responds to cold (Wang and Siemens 2015). TRP thermoreceptors might exist in algae, as my BLAST and domain analyses revealed multiple putative TRP sequences in dinoflagellates (e.g., CAE6972685.1 in *Symbiodinium*; CAK9104464.1 in *Durusdinium*). This is an open field for exploration.

Nutrients

Nutrients in the marine environment are typically scarce and variable. To maximize nutrient use efficiency, algae and plants have evolved sophisticated mechanisms to monitor both external and internal nutrient levels. These mechanisms allow them to dynamically adjust nutrient uptake, allocation and cellular compartmentation in response to changing nutrient abundances (Fig. 12.12).

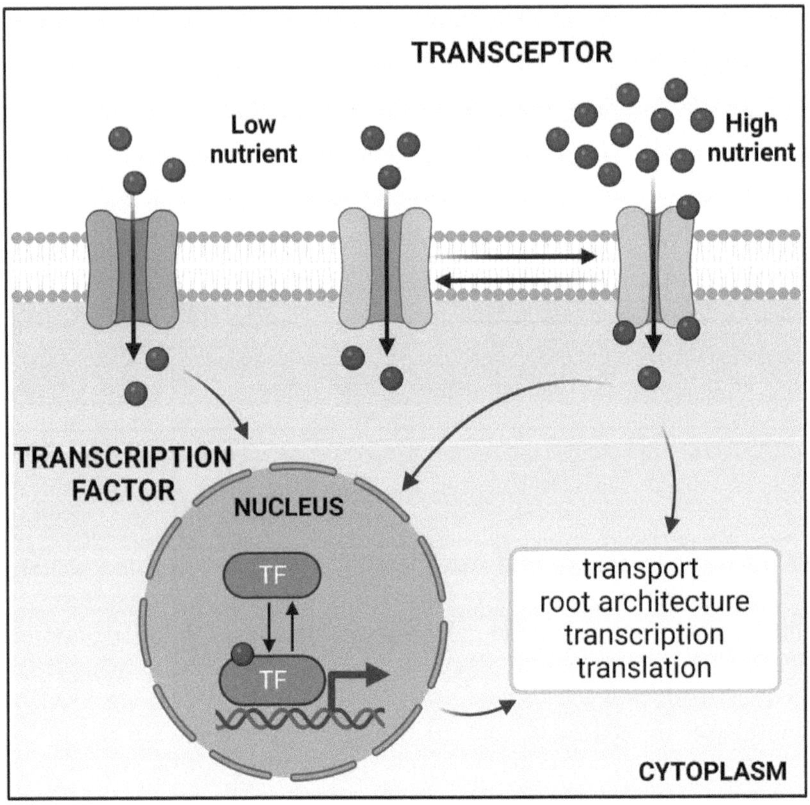

Figure 12.12. Schematic of nutrient sensing mechanism in plants. Reproduced from Podar and Maathuis (2022) under the Creative Commons CC BY 4.0 license.

Primary Sensors and Secondary Signaling

Sensing nutrient levels in cells involves primary sensors typically comprising either transceptors or transcription factors (Podar and Maathuis 2022). While the identification of primary sensors in plants for specific nutrients is just beginning, even less is known about nutrient sensors in algae. In plants, transporter networks can also function as nutrient sensors (Dreyer et al. 2022). Using computational cell biology simulations and laboratory experiments, Dreyer and colleagues demonstrated that these transporter networks translate changes in external nutrient concentrations into cytosolic H^+ and Ca^{2+} signals. This suggests that transporter networks not only fulfill their primary role of nutrient transport but also act as receptors without needing conformational changes in transporter proteins, exhibiting transceptor-like characteristics.

A dual function of transporters has also been proposed for the nitrate transporter NRT1 (Liu and Tsay 2003). NRT1 operates as both a high-affinity and low-affinity transporter, and its role in nutrient monitoring appears to involve cytosolic Ca^{2+} as a messenger to transduce nutrient signals (Krouk 2017). Whether algae use similar transceptors and transcription factors to monitor nutrient status and whether their nutrient transporters also serve dual functions remain to be investigated.

Target of Rapamycin Signaling Pathway

The Target Of Rapamycin (TOR) pathway integrates signals from growth factors, nutrients, stress, and cellular energy levels to control cell growth in animals and various other organisms. The Tor pathway mediates cell growth in response to nutrient availability, in part by inducing ribosomal protein (RP) gene expression (Rohde and Cardenas 2003). Rapamycin, an antifungal agent originally isolated from *Streptomyces hygroscopicus*, inhibits cell growth in many types of organisms (Inoki et al. 2005). TOR proteins are part of the phosphatidylinositol kinase-related kinase (PIKK) family. There are two TOR proteins, TOR1 and TOR2, each approximately 280 kDa. Mutation in one of these genes makes the yeast mutant resistant to rapamycin. Mechanistically, rapamycin forms a complex with FKBP12 (binding protein of the macrolide antibiotic FK506), and this complex binds to TOR, inhibiting its activity. The mammalian counterpart of yeast TOR is called mTOR.

TOR is an evolutionarily conserved protein, found in a wide range of organisms from yeast to mammals, with homologs in invertebrates, fungi, plants, and microalgae, including diatoms, prasinophytes, and chlorophytes (Mallén-Ponce et al. 2022). It also occurs in dinoflagellates such as *Prorocentrum shikokuense* and *Polarella glacialis* (Fig. 12.13). A study revealed that the kinase in the prasinophyte *Ostreococcus tauri* responded to TOR inhibitors (e.g., rapamycin and PP242) in a manner more similar to vascular plants than to the chlorophyte *Chlamydomonas*. The inhibitors induced cell growth arrest but not cell death and altered carbon reserve content. Additionally, the *OtTOR* gene was dynamically regulated by nutritional stress (Caló et al. 2022).

TOR associates with other proteins to form two distinct complexes: TOR complex 1 (TORC1) and TOR complex 2 (TORC2), defining two independent signaling branches. The core components of TORC1 include TOR kinase, LST8, and Raptor/Kog1, whereas TORC2 contains TOR, LST8, Sin1/Avo1l, and Rictor/Avo3, with TOR and LST8 shared by both. LST8 is essential for TOR activity, interacting with the kinase domain of TOR and required for full catalytic activity of both complexes (Aylett et al. 2016). TOR inhibitors used in experiments compete with the kinase domain for ATP.

The core proteins of TORC1 are found in nearly all eukaryotes, including microalgae and land plants (Fig. 12.13). Their presence in rhodophytes and chlorophytes indicates an early evolutionary origin in photosynthetic eukaryotes. However, LST appears to be absent in some microalgae including *Cyanophora paradoxa* and some rhodophytes (Fig. 12.13A). *Cryptophyceae* sp. CCMP2293 seems to lack both LST8 and Raptor but retains TOR. These absences might be due incomplete genome assemblies.

TORC2 is not conserved in algae (Mallén-Ponce et al. 2022). It likely arose in the Last Eukaryotic Common Ancestor (LECA) but was lost early in algal and plant evolution. In support of this, a Rictor homolog is present in *Bigelowiella natans* and *Ectocarpus siliculosus*.

TORC1 regulators Rheb and RAGs also are not conserved in green algae and plants but are found in other algal groups (Mallén-Ponce et al. 2022). This suggests they arose with TORC1 in the LECA but were lost in green algae. Other mammalian and yeast TORC1 regulators are absent in green algae and plants but present sporadically in some microalgae, suggesting unique TORC1 pathways in different algal lineages.

How the TOR pathway functions in algae remain poorly understood. TORC1 typically controls cell growth by promoting anabolic processes (protein synthesis and ribosome biogenesis) and inhibiting catabolic processes (e.g., autophagy). As such, nutrient deficiency is expected to

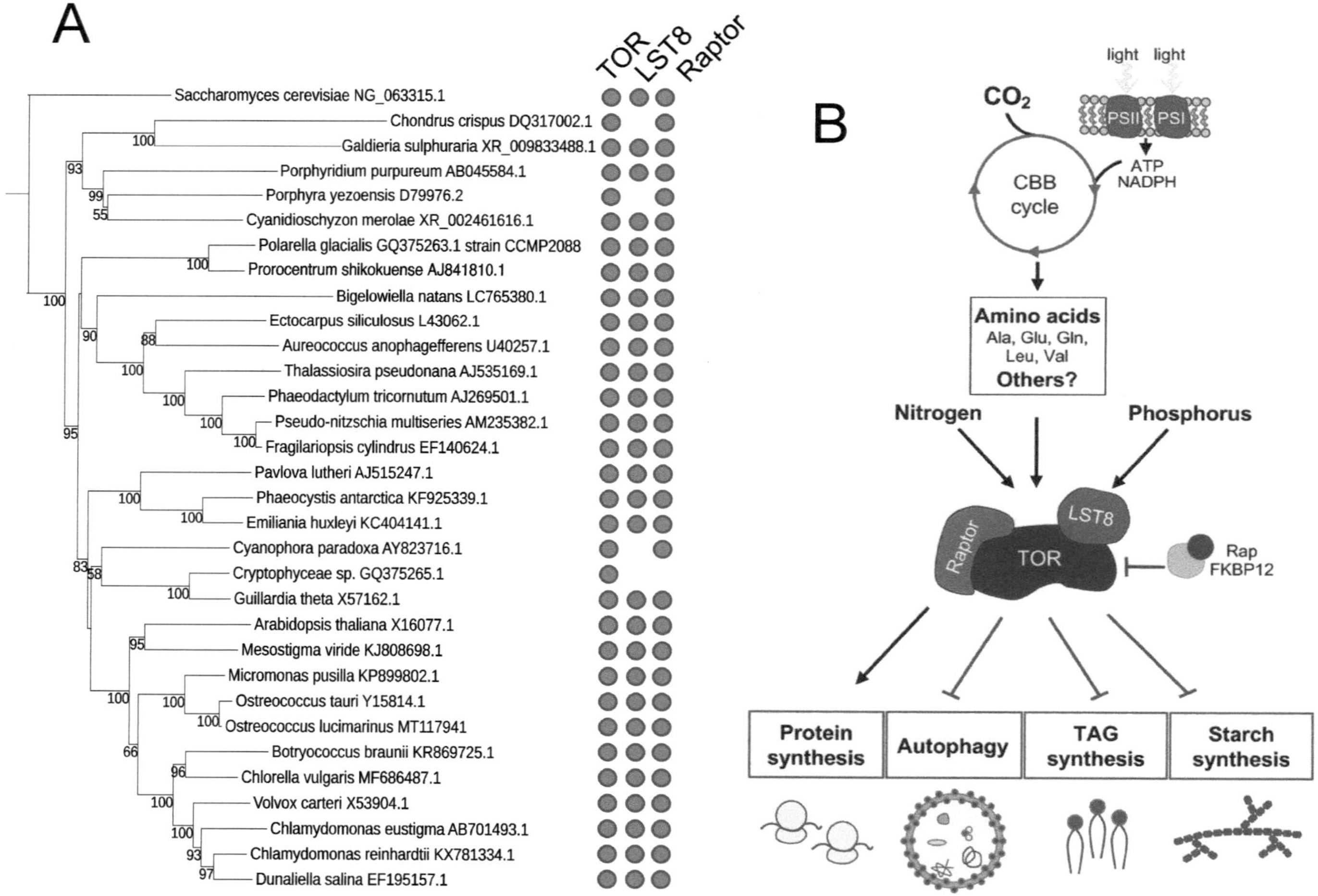

Figure 12.13. TOR pathway in algae. (A) Evolutionary conservation of core components TOR signaling pathway in microalgae (with *Saccharomyces cerevisiae*, *Arabidopsis thaliana*, and *Mesostigma viride* included as reference). The cladogram was constructed based on 18S rRNA sequences of the indicated species. (B) Proposed model for the TOR signaling pathway in the green model microalga *Chlamydomonas reinhardtii*. Cited from Mallén-Ponce et al. (2022) under the Creative Commons CC BY 4.0 license.

decrease the expression of TOR. The best studied algal model is *Chlamydomonas reinhardtii*, where rapamycin-sensitive TORC1 signaling regulates cell growth, autophagy, lipid metabolism, and central metabolic pathways. *C. reinhardtii*'s FKBP12 has a conserved rapamycin-binding pocket but lacks the glutamine in position 53 (Gln53), explaining its reduced rapamycin sensitivity.

Rapamycin sensitivity varies among algae. Some, like *C. reinhardtii*, are sensitive. When Gln53 was artificially generated to mimic the rapamycin-binding pocket of yeast and human FKBP12s, the expression of the mutants in *C. reinhardtii* led to increased sensitivity of the cells to rapamycin; on the contrary, FKBP12-deficient mutants of *C. reinhardtii* exhibited complete resistance to rapamycin (Crespo et al. 2005). Other sensitive algae include *Euglena gracilis* and the coccolithophorid alga *Emiliania huxleyi*, although their cell growth is not inhibited by this drug (reviewed by Pérez-Pérez et al. 2017).

In contrast, some other algae (as well as plants) are resistant due to the inability of FKBP12 to form a stable complex with rapamycin (reviewed by Pérez-Pérez et al. 2017). *Cyanidioschyzon merolae* is resistant, but when it was transformed with yeast FKBP12, the transgenetic strain exhibited sensitivity (Imamura et al. 2015). In *Ostreococcus tauri*, TOR transcript abundance increases first (two hours after nitrogen nutrient deprivation) and then decreases (Caló et al. 2022). This complicates the general notion that TOR gene expression is downregulated by nutrient deprivation. Therefore, both the molecular mechanisms and physiological roles of TOR in algae remain to be fully elucidated.

In the hypothetical model for *C. reinhardtii* (Fig. 12.13B), TOR forms complex with LST8 and Raptor, regulated by nutrients, including phosphorus (Couso et al. 2020) and nitrogen (Upadhyaya et al. 2020). Photosynthetic CO_2 assimilation can also activate TOR through the CBB cycle and the synthesis of key amino acids such as Ala, Leu, Val, Glu, and Gln (Mallén-Ponce et al. 2022). In nutrient-rich conditions, TOR promotes cell growth by activating translation and repressing nutrient stress-induced processes like autophagy and the synthesis of energy storage molecules (TAG and starch).

P-Ca^{2+} Signaling Pathway

Calcium ions (Ca^{2+}) function as universal second messengers in cellular signaling, regulating diverse processes including cell metabolism, growth, gene expression, and protein function through allosteric effects. There are diverse calcium sensors including calmodulin (CaM), calcium-dependent protein kinases (CDPKs), and calcineurin B-like proteins (CBLs), among which CaM is predominant (Kim et al. 2009). The regulatory role of calcium signaling occurs through direct modulation of protein activity, control of transcriptional processes, post-transcriptional modifications, and translational regulation (Aldon et al. 2018). For algae, two sensors involving calcium are particularly interesting.

In *P. tricornutum* transformed with jellyfish-derived aequorin cDNA, cytosolic Ca^{2+} responds differently to iron supply compared to osmotic and mechanical stimuli. Cytosolic Ca^{2+} increases slowly (detectable 5 min after iron addition), and the increase continues for an extended period (Falciatore et al. 2000). When *P. tricornutum* was transformed with a fluorescent Ca^{2+} biosensor (NCBI accession number AEO16866.1, R-GECO1), a P-Ca^{2+} signaling mechanism was discovered (Helliwell et al. 2021). This mechanism is sensitive to sub-micromolar concentrations of inorganic phosphate and various environmental P forms.

Recovery of the diatom from P limitation involves rapid and substantial increases in N assimilation and this process is dependent on the P-Ca^{2+} signaling. This suggests a role of this P-Ca^{2+} signaling in N/P stoichiometric homeostasis. The P-Ca^{2+} signaling protein is highly conserved across the phylogenetic spectrum of algae (Fig. 12.14), suggesting that this Ca^{2+}-based P signaling mechanism may operates in all algae. The relationship between the P-Ca^{2+} signaling pathway and a trypsin gene known to regulate N/P stoichiometric homeostasis in *P. tricornutum* (You et al. 2022) remains to be studied.

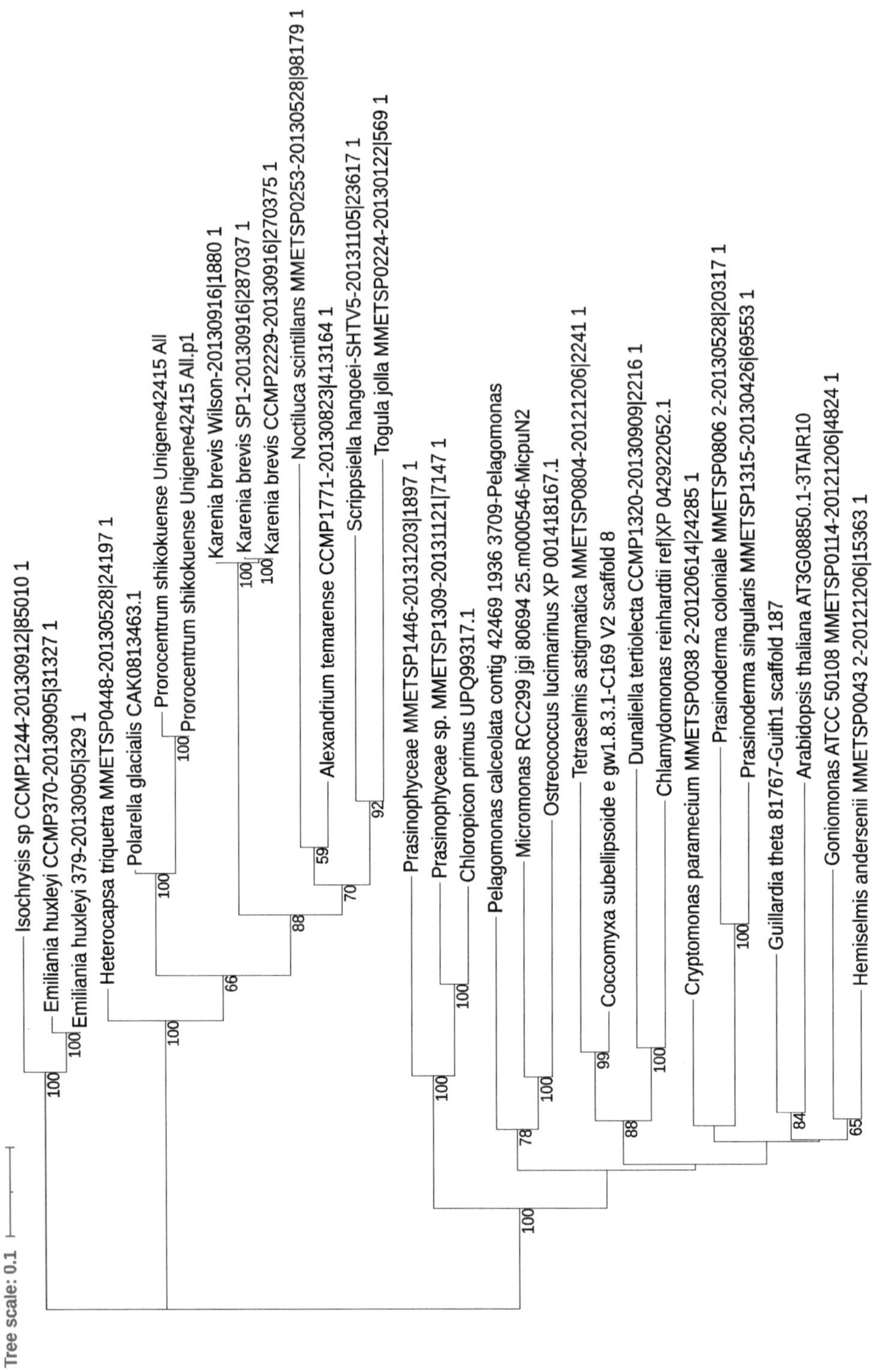

Figure 12.14. Phylogenetic tree of P-Ca2+ signaling protein wide occurrence of the protein in algae.

Transcription Factor ROC40

In the chlorophyte *Chlorella*, a quantitative proteomics study shows that under Nitrogen (N) starvation, the accumulation of triacylglycerol (TAG) is regulated (promoted) by the MYB-related transcription factor ROC40 (Goncalves et al. 2016). Consistent with this finding is earlier observation of a *Chlamydomonas* mutant in which ROC40 was impaired and TAG accumulation under N starvation was abolished. This impairment could be rescued when wild-type ROC40 is reintroduced and expressed. These in concert indicate that the transcription factor ROC40 plays a role in N starvation-induced lipid accumulation, at least in the green lineage of algae.

Sensing of Turbulence (Shear Stress)

The ocean is often turbulent, and algae constantly receive mechanical stimuli. These stimuli act on the cell surface in two forms: pressure (acting normal to the plane of cell membrane) and tension (acting on the plane of the cell membrane), with hypo-osmotic shock as an example of tension-imposing stimuli (Cox et al. 2019). Algae such as diatoms are tolerant of physical perturbance, whereas dinoflagellates are well known to be susceptible to turbulence, though sensitivity varies with species (García-Portela et al. 2019). In *Protoeratium reticulatum*, hydrodynamic shear forces inhibit growth, apparently due to the production of reactive oxygen species that damage lipid-dependent membranes (Rodríguez et al. 2009). Cultures of *Ceratocorys horrida* under constant stirring were found to grow to smaller cell size with shortened spines (/Zirbel et al. 2000). Small-scale turbulence can cause disruption of the cell cycle and cell death in *Alexandrium minutum* (Llaveria et al. 2009).

The sensitivity of dinoflagellates to mechanical stimuli is best demonstrated by their bioluminescence (Maldonado and Latz 2007). During blooms of the bioluminescent dinoflagellates, shear stress created by boating, swimming, or simply scooping of water with hands or a container could generate glowing halo or patch (Fig. 12.15). The mechanotransduction pathway activating bioluminescence in dinoflagellates involves a G-protein system and includes a calcium flux in the cytoplasm, a self-propagating action potential across the vacuole membrane, and a proton flux into the scintillon that activates the luminescent chemistry (Chen et al. 2007, Valiadi and Iglesias-Rodriguez 2013).

More generally, there are three major types of mechanosensitive sensors on the cell membrane: mechanosensitive (MS) ion channels, G Protein-Coupled Receptors (GPCRs), and integrins. These receptors directly or indirectly interact with cytosolic force-sensing elements such as microtubules and actin filaments. This interaction allows cells to progressively remodel their cortical cytoskeleton and cell membrane, enabling them to adjust to their mechanical niche until they are mechanically protected within the new stress limits (Luo et al. 2013). Consequently, these receptors can sense and transmit biochemical signaling from the cell surface to the nucleus to regulate gene transcription (Cox et al. 2019). Among the primary mechanosensory membrane proteins at the origin of cellular signaling cascades, mechanosensitive ion channels are quite unique, as they are the most rapid in converting mechanical stimuli into electrochemical intracellular signals (e.g., on the millisecond timescale) (Martinac and Cox 2017). MS ion channels consist of MscL and MscS subfamilies. While members of the MscL subfamily are largely confined to bacteria and archaea, members of the MscS subfamily are very diverse in terms of structure and function, as well as their presence throughout organisms of different evolutionary origins, including algae (Cox et al. 2019). Mechanosensitive ion channels in algae have been studied as early as in the 1990's (Yoshimura 1998). Falciatore and colleagues (2000) transformed *P. tricornutum* with the aequorin cDNA derived from the jellyfish *Aequorea victoria*, which had long been used as an indicator of Ca^{2+} because its binding to the cation triggers irreversible reaction with the production of light in the visible range. Experiments with different stimuli showed that this aequorin-transformed diatom exhibited the strongest response to shear stress compared to other stimuli in terms of intracellular Ca^{2+} concentration changes, suggesting that this mechanosensory transduction pathway is Ca^{2+}

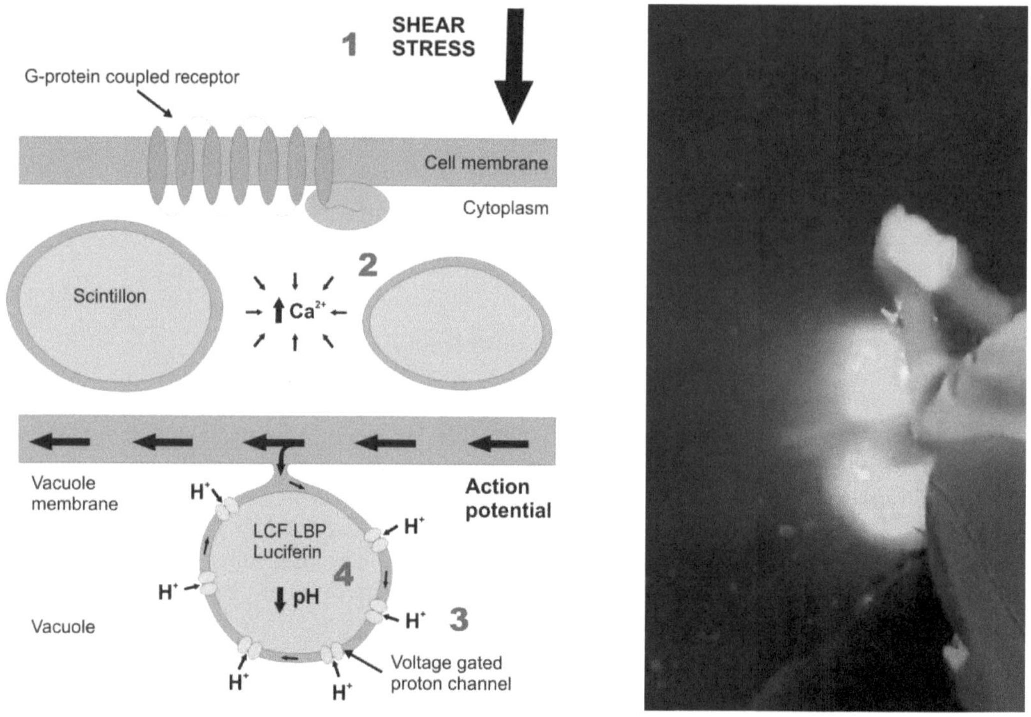

Figure 12.15. Mechanoreception pathway (A) and bioluminescence from a *Noctiluca scintillans* bloom in Xiamen, China (2021) induced by hand scooping of the bloom water (B). The diagram in (A) is reproduced from Valiadi and Iglesias-Rodriguez (2013) under the terms and conditions of the Creative Commons Attribution license (http://creativecommons.org/licenses/by/3.0/).

dependent. From the genomic data, dinoflagellates such as *Polarella glacialis* contain numerous mechanosensitive ion channel protein genes, suggesting that they likely have the MC ion channel-based mechanotransduction pathway.

Additionally, eukaryotes share the highly conserved Notch signaling pathway. This pathway is important in deciding cell fate in higher eukaryotes, but its roles include regulating cell polarity and mechanotransduction in the endocytosis process. It is possible that this pathway is involved in the mechanosensory function of algae.

Oxygen

Like other aerobic organisms, algae require O_2 for survival. When O_2 is limited (hypoxia), a response is required to reduce demand and/or improve supply. In plants, a hypoxic response mechanism is based on the stability of certain proteins with N-terminal cysteine residues regulated in an O_2-dependent manner by the Cys/Arg branch of the N-degron pathway (Weits et al. 2014). Among these are the Group VII ethylene response factors (ERF-VIIs), which can instigate adaptive responses to hypoxia. Their N-terminal cysteine residues undergo oxidation catalyzed by plant cysteine oxidases (PCOs), producing Cys-sulfinic acid at the N-termini of target proteins and leading to their destabilization under normoxia conditions; however, PCO inactivity in hypoxia leads to their stabilization. PCOs are sensitive to oxygen availability and function as plant oxygen sensors (White et al. 2018).

Among algae, PCO-Like sequences (PCOL) have been identified in the freshwater alga *Klebsormidium nitens* (NCBI accession number GAQ79467), exhibiting a similar function as counterparts in *A. thaliana* (Taylor-Kearney et al. 2022). Using a PCO sequence of *A. thaliana* as a

Table 12.4. Identification of plant cysteine oxygenase (PCO)-like genes in algae.

Phylum	Species	NCBI accession number
Charophyta	*Klebsormidium nitens*	GAQ79467.1
	Closterium sp.	GJP37486.1
	Chara braunii	GBG67584.1
Chlorophyta	*Chlorella desiccata*	KAG7669610.1
	Chloropicon primus	UPQ97168.1
	Chlamydomonas reinhardtii	XP_042927086.1
	Chlamydomonas reinhardtii	XP_042927085.1
	Coccomyxa subellipsoidea	XP_005646266.1
	Micromonas pusilla	XP_003057792.1
	Micromonas commoda	XP_002500146.1
	Ostreococcus lucimarinus	XP_001418893.1
	Pleodorina starrii	GLC63285.1
	Pycnococcus provasolii	GHP03147.1
	Volvox carteri	XP_002946534.1
Rhodophyte	Galdieria sulphuraria	XP_005707183.1
Ochrophyta	*Tribonema minus*	KAG5184539.1
Cryptophyta	*Guillardia theta*	XP_005839351.1

query, searches of algal genomes yielded significant hits in charophytes, chlorophytes, rhodophytes, ochrophytes, and a cryptophyte with highly conserved regions (Table 12.4). These algal PCOL all contain the functional domain of PCO (cupin_ADO-cd20289). Further research is needed to determine whether the PCOL in these groups of algae has the same function as in *K. nitens* and *A. thaliana*, and if so, how this oxygen sensor evolved in algae.

13

Stress Response and Adaptation

Environmental Changes and Stress

Environmental changes, whether natural or anthropogenic, can impose stress on organisms. The definition and classification of stress is thoroughly examined in the title edited by Rai, L. C. and Gaur, J. P. (2001), *Algal adaptation to environmental stresses: Physiological, biochemical* and *molecular mechanisms*. Those most relevant to the genomic theme are summarized below.

According to the Shorter Oxford English Dictionary (1933), stress is "the overpowering pressure of some adverse force or influence". Among researchers, stress has been defined as an external constraint "limiting the rates of resource acquisition, growth or reproduction of organisms" (Grime 1989), or "as an environmental condition that reduces Darwinian fitness when first applied" (Sibly and Calow 1989), or "any change in the environmental conditions is experienced as a stress that threatens the normal metabolic balance and produces a response designed to counter the effects of the disturbing influence" (Franks et al. 1990). Intuitively, the conditions of intense cold, desiccation, minimal or excessive light and mineral deficiency under which the green alga *Hemichloria antarctica* lives, endemic in the dry valleys of Antarctica, or the hot and acidic environment of the host spring that the rhodophyte *Cyanidioschyzon merolae* inhabit, are stressful. However, these species have survived and thrived there for thousands or millions of years. In this context, research on stress aims to understand how organisms cope with or adapt to their apparently 'abnormal' habitats in comparison to organisms that adapt to their own, 'normal', habitats.

Stressful conditions can be grouped into mechanical (e.g., waves, currents), physical (radiation, desiccation, microplastics), chemical (heavy metal pollution, organic pollution), and nutritional (e.g., deficiency of phosphate, nitrogen, or trace metals). Although every stress condition elicits a unique suite of metabolic and molecular responses, all eukaryotes are believed to share a set of conserved stress-related transcription factors (Table 13.1).

This chapter will briefly discuss common response pathways and specific stressor responses, excluding nutritional stress, which is covered in Chapter 17.

The response may involve different levels of biological organization, from the cell or individual, the population or species, to the community level (Fig. 13.1). The response occurs via two mechanisms: (1) the expression of phenotypic plasticity (the ability of one genotype to express varying phenotypes when exposed to different environmental conditions), and (2) evolution via selection for specific genotypes (which either exists in a pool of genotypes or arises from mutation induced by the environmental stress).

Plasticity at the cell/individual level reflects a wide niche breadth. This may involve epigenetic regulation of gene expression, which translates into biochemical changes and metabolic

Table 13.1. Overview of stress-related transcription factors and consensus sequences of their DNA binding sites (http://www.gene-regulation.com/pub/databases.html).

Transcription factor	DNA-binding site	Consensus sequence (5'→3')	Context
Activator protein 1(AP-1)	AP-1-binding element	TGA(C/G)T(A/C)A	General stress response
Heat-shock factor (HSF)	Heat-shock-responsive element (HSE)	NGAANNGAANNG AAN	Heat shock, general stress response
Nuclear factor erythroid 2-related factor 2 (Nrf2)	Antioxidant-responsive element (ARE), electrophile-responsive element (EpRE)	TGACNNNGC	Induction of antioxidant enzymes and ROS-scavenging systems
Metal-responsive-element-binding transcription factor (MTF-1)	Metal-responsive element (MRE)	TGC(A/G)CNC	Induction of metallothionein
Aryl hydrocarbon receptor (AhR)	Xenobiotic-responsive element (XRE), dioxin-responsive element (DRE), aryl hydrocarbon-responsive element (AhRE)	TGCGTGAGAAGA (human, mouse, rat, guinea pig)	Induction phase I and II biotransformation enzymes
Centromere-binding factor 1 (CBF1)	Centromere DNA element 1 (CDE1)	(A/G)TCAC(A/G)T G(yeast)	Regulation of sulphur amino acid biosynthesis pathway, induced by exposure to heavy metals
Msn2p, Msn4p	Stress-response element (STRE)	GATGACGTGT(Msn4) (yeast)	General stress defence and carbohydrate homeostasis in yeast
DREB family transcription factors	Dehydration-responsive element (DRE)	CCGAC	Induction of defence against cold and drought in plants
Abscisic acid-responsive element binding factor (ABF)	Abscisic acid-responsive element (ABRE)	(C/T)ACGTGGC	Water stress signalling in plants
Hypoxia-inducible factor (HIF-1)	Hypoxia responsive element (HRE)	TACGTGCT (human)	Metabolic switches related to hypoxia
Oestrogen receptor (ER-α)	Oestrogen-responsive element (ERE)	AGGTCANNNTGA CCT (human, mouse, cattle, clawed frog, chicken)	Activation of processes related to oogenesis and other female functions

reprograming. Plasticity can also manifest at the population/species level, which can be achieved through two mechanisms: the plasticity of the constituent cells/individuals or the "sorting" of different genotypes across the gradient of the environmental condition. Similarly, plasticity in the community level can stem from either the plasticity of all the constituent species or the reshuffling of species composition. In the latter case, one group of dominant species are succeeded by another that performs the same functions, demonstrating functionally redundant (Fig. 13.2). Adaptation, on the other hand, manifests as changes in allele frequencies and trait means in response to selection over multiple generations.

From a molecular perspective, the different levels of responses can be due to the involvement of different molecular strategies or reallocation of resources in the individuals or cells. Using phytoplankton as the example, there are bloomer and survivalist strategies (Arrigo 2005). Bloomers invest cellular resources primarily on cellular components controlling cell proliferation (e.g., ribosomes), whereas survivalists allocate resources heavily towards what is critical for survival, such as transporters for nutrient uptake or toxins for defense.

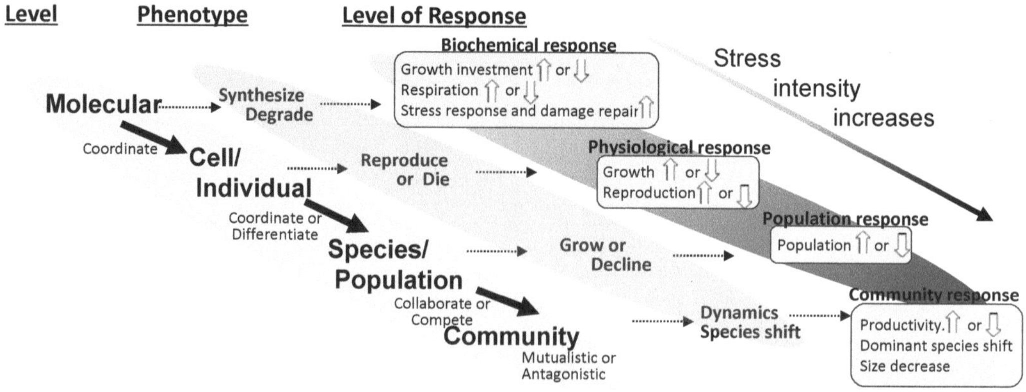

Figure 13.1. Schematic illustrating different levels of biological organization of responses to environmental changes (e.g., stress).

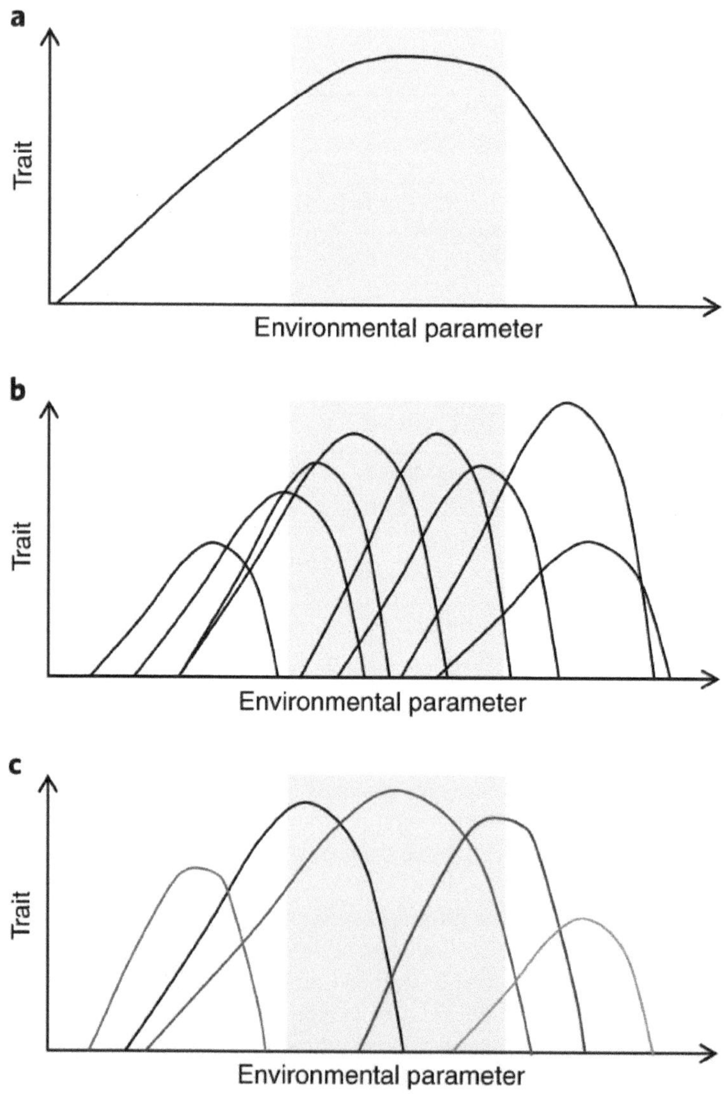

Figure 13.2. Schematic of potential mechanisms underlying resilience of an algal community and single species' differential responses to stressors.

Common Stress Response Pathways

Environmental stress signals are first sensed via the action of receptors on the cell surface. The signal is transduced, mediated by a signaling cascade, to the inside of the cell, ultimately reaching the nucleus, where the binding of a transcription factor to the promoter region of genes activates their expression. The final step can also involve a gene repressor, which inhibits the expression of genes through binding to gene promoter regions or interacting with activators to block transcription. Thus, transcription factors are critical elements of the stress response pathways. Many of the pathways are conserved among eukaryotes (Table 13.1).

The Integrated Stress Response

In response to various stress stimuli, eukaryotic cells activate the Integrated Stress Response (ISR) to restore cellular homeostasis. This pathway's key event is the phosphorylation of eukaryotic translation initiation factor 2 alpha (eIF2α) by one of four eIF2α kinases, reducing global protein synthesis and inducing the expression of specific genes, such as the transcription factor ATF4, to promote recovery (Pakos-Zebrucka et al. 2016). The ISR optimizes the stress response based on cellular context and the nature and intensity of the stress. While primarily a pro-survival mechanism, severe stress can shift ISR signaling toward cell death.

Heat Shock Protein Pathway

Elevated expression of Heat Shock Proteins (HSPs) is one of general stress response mechanisms. HSPs are molecular chaperones expressed during normal cell growth and induced under various stress conditions, including environmental changes, temperature fluctuations, and microbial attacks. Originally discovered in *Drosophila buskii* in 1962, the HSP family is now known as a universal stress-responsive and cell protecting mechanism. As ubiquitous and conserved protein families, HSPs occur in both prokaryotic and eukaryotic organisms, serving to maintain proteostasis and buffer against stresses. The function of HSPs is regulated by a complex signaling network and in an intricate negative feedback mode, primarily involving a heat shock factor (HSF, a transcription factor). Under normal conditions, HSFs are inactive in complexes with HSPs. Upon stress, they are released from the HSP-HSF complexes, allowing HSF to oligomerize, translocate to the nucleus, and bind to the highly conserved Heat Shock Elements (HSEs) in the promoter regions of the HSP genes, triggering their rapid transcription (Van Straalen and Roelofs 2012, Hu et al. 2020).

HSP families are classified based on their molecular masses, mainly including HSP100, HSP90, HSP70, and HSP60, which are designated as large HSPs, and small HSPs (sHSPs, 12-43 kDa). sHSPs feature a tripartite domain architecture with a conserved Alpha-Crystallin Domain (ACD), which is the functional core, and variable C and N terminal extensions, which may determine target specificity (Waters and Vierling 2020). sHSPs are known to suppress protein aggregation in an ATP-independent manner. Under normal conditions, HSPs may account for 5–10% of total cellular proteins and function in various cellular housekeeping processes such as folding newly synthesized polypeptides, refolding denatured proteins, assembling protein complexes, degrading misfolded proteins, and dissociating protein aggregates (Hu et al. 2022). In response to stresses, the heat shock response factors (HSFs) are activated, likely through a Ca^{2+}-mediated signaling pathway (see Chapter 14), leading to elevated expression of HSPs to mitigate the stress.

HSPs have been extensively documented in algal 'omic datasets. HSP90 and HSP70 are among the most highly transcribed genes in dinoflagellates (Lin 2011). The general lack of specificity of HSP response to types of stress renders HSPs less useful for ecological studies. However, it is noteworthy that functional differentiation may occur among the different HSPs. For instance, HSP70 (DnaK in prokaryotes) occurs in 17 copies in humans; while some of them are stress induced

(HSP70), others are housekeeping HSC70 (heat shock cognate) (Brocchieri et al. 2008). Similar functional differentiation among multiple HSPs in algae remains to be investigated.

Research on algal sHSPs is relatively limited, but recent studies have uncovered new sHSP gene sequences and stress-dependent expression profiles in various algae. Waters and Rioflorido (2007) surveyed the genomes of five phylogenetically distinct algae, the chlorophyte *Chlamydomonas reinhardtii*, the rhodophyte *Cyanidioschyzon merolae*, the prasinophytes *Ostreococcus lucimarinus* and *O. tauri*, and the diatom *Thalassiosira pseudonana* and identified diverse sHSP genes. They found that algal sHSP genes were distinct from plant counterparts and had a different evolutionary history from that of HSP70 genes. In the unicellular rhodophyte *Cyanidioschyzon merolae* genome there are two sHSP genes (*CmsHSP1*, *CmsHSP2*), located side by side in the chromosome and sharing a promoter region. These two encoded proteins likely target different cellular compartments: CmsHSP1 to the nucleus or cytoplasm and CmsHSP2 to chloroplasts and/or mitochondria. The differential localization may confer advantages for achieving rapid activation/delivery of sHSPs to multiple cellular compartments with minimum time and cost (Kobayashi et al. 2014). In rhodophyte seaweed genomes, 97 HSP20 genes were identified and classified into 13 subfamilies phylogenetically (Gao et al. 2022). These genes were shown to be polyphyletic, suggesting a complex evolutionary origin and trajectory, including duplication and purifying selection. The study also showed that HSP20 in *Pyropia yezoensis*, *P. haitanensis*, and *Porphyra umbilicalis* were highly conserved. Phytohormone-, light-, and stress-responsive as well as development-related *cis*-elements were identified in the rhodophyte HSP20 gene promoter sequences. For dinoflagellates, Deng et al. (2021) constructed dinoflagellate-specific full-length cDNA libraries from marine sediment RNA samples using DinoSL as the 5' primers and identified sHSPs. The work revealed a conserved A-X-X-X-N-G-V-L motif in these dinoflagellate sHSP genes. The active expression of these sHSPs in sediment suggests they may play a crucial role in the dormancy maintenance of dinoflagellate cysts.

Phenylpropanoid Pathway

The phenylpropanoid pathway used to be known only in terrestrial plants until recently when the pathway or core phenypropanoid building blocks as well as flavonoids were documented in streptophyte algae (Goiris et al. 2014, Maeda and Fernie 2021). The phenylpropanoid pathway produces precursors for thousands of metabolites, contributing around 40% of organic carbon on Earth (Vogt 2010). These metabolites include lignins, which serve as structural polymers, and various UV-protective compounds, including flavonoids and others (de Vries et al. 2021 and references therein). Phenylpropanoid derivatives also play crucial roles in plant responses to abiotic stress and defense against pathogens. Genomic analyses have revealed wide existence and expansion of the pathway in streptophyte and chlorophyte algae as well as land plants (de Vries et al. 2021).

Nitric Oxide-Mediated Stress Surveillance System

A stress surveillance system involving nitric oxide (NO) has been proposed for diatoms (Vardi 2008), which may exist and operate more broadly in algae (Fig. 13.3), as NO production or its genetic basis has been found widely in algae. NO is an essential signaling molecule involved in several physiological and biochemical processes. It was recognized by Science as the molecule of the year in 1992, and its American researchers Furchgott, Murad, and Ignarr were awarded Nobel Prize in 1998 for their discovery of NO as a signaling molecule in the cardiovascular system. This small gaseous molecule diffuses rapidly and acts as a messenger in diverse functions.

In mammals, sufficient NO levels are necessary for protecting internal organs from ischemic injury but sustained levels of NO production result in direct tissue toxicity and vascular collapse. The prevalence and function of NO production and NO-dependent signaling pathway is less understood in plants and algae, but research in this area has advanced quickly in recent years. In plants, NO

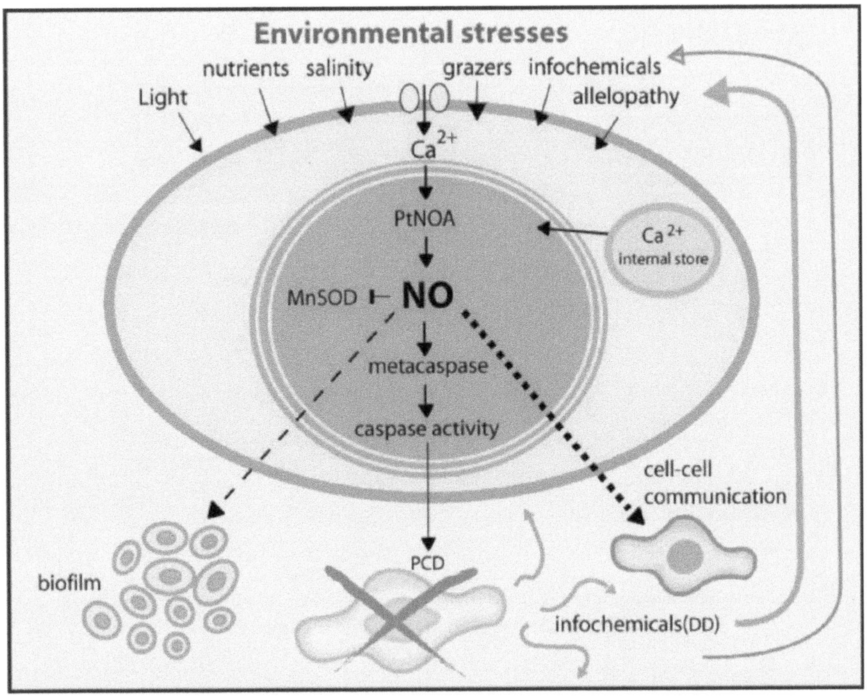

Figure 13.3. Schematic of stress surveillance pathway involving nitric oxide (NO) proposed for diatoms and potentially operating in all algae. Reproduced with permission from Vardi (2008).

is recognized as a potent signaling molecule (Jeandroz et al. 2016), with diverse physiological roles, including stress management and regulating seed germination in plants (Šírová et al. 2011, Hancock 2020, Allagulova et al. 2023). Furthermore, there is a crosstalk between NO and ROS (Reactive Oxygen Species), which is involved in the light- and hormone-specific regulation of seed development during the initiation of plant ontogenesis. The proper adjustment of NO and ROS levels is tightly linked to the mechanisms of plant pollen hydration, germination, tube growth, as well as pollen-stigma recognition. Besides, the interaction of NO with ROS and secondary messengers such as Ca^{2+}, cAMP and cGMP in pollen represent a common mechanism of NO signaling (Šírová et al. 2011).

Two main pathways are known to be responsible for NO production (Fig. 13.4). One is the reductive pathway, which is based on the reduction of nitrites to NO. In this pathway, nitrate is reduced in the cytosol by the assimilatory Nitrate Reductase (NR) to nitrite (with two-electron transfer), which is further reduced by the cytosolic NR (with one-electron transfer) or a mitochondrial electron transport chain to produce NO (Allagulova et al. 2023). The other is the oxidative pathway, which is dominantly based on the oxidation of L-arginine (Arg) by Arg-dependent NO synthase (NOS), a pathway well established in animals (Lapina et al. 2022). Besides, the oxidative formation of NO can involve polyamines with enzymes such as copper amine oxidases (Tun et al. 2006) or hydroxylamine with hydroxylamine reductase (Allagulova et al. 2023). For plants, the reductive pathway used to be considered the most relevant sources of NO synthesis, but increasing evidence suggests the existence of NOS-like activity (Hancock 2020, Lapina et al. 2022).

In algae, NO has been reported to regulate (promote at low and inhibit at high concentration) growth (Lehner et al. 2009, Tang et al. 2011), facilitate stress or toxicity tolerance (Qian et al. 2009, Li et al. 2013), induce programmed cell death (Bouchard and Yamasaki 2009, Yordanova et al. 2010) and serve in signaling (Chung et al. 2008). A study comparing HAB species (*Chattonella marina*, *C. ovata*, and *Heterosigma akashiwo*) and non-HAB species found NO production in the HAB species only, potentially linking NO production to growth promotion (Kim et al. 2008).

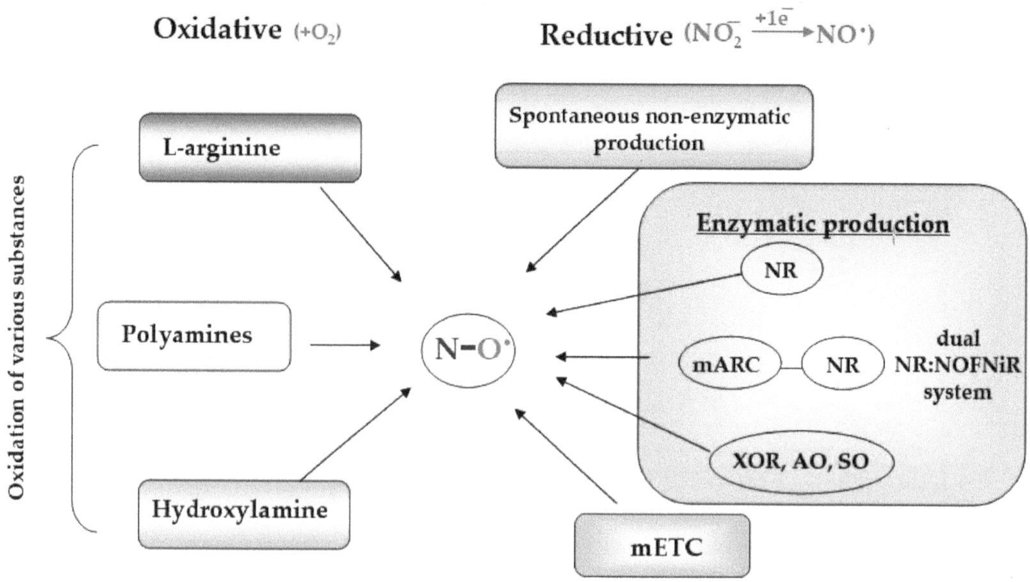

Figure 13.4. Schematic of the main pathways of NO formation in plants. Reproduced from Allagulova et al. (2023) under the Creative Commons CC BY 4.0 license.

In dinoflagellates, NO has been shown to trigger redox-related cyst formation (Ganini et al. 2013). The signaling role of NO in algae has been investigated in diatoms in which NO seems to act with and upstream of Ca^{2+} (Vardi et al. 2006), as in land plants. Both the reductive and oxidative pathways of NO production have been reported in algae (Table 13.2). A number of algae has been shown to possess homologs of animal NOS (Jeandroz et al. 2016), and the number is growing, including more species of diatoms and dinoflagellates (Table 13.2). A comparative genomic study led to the identification of NOS in *Pseudo-nitzschia multistriata* and some other diatoms (Di Dato et al. 2015). In the chlorophyte *Chlamydomonas reinhardtii,* NO is produced by the complex between NR and nitric oxide-forming nitrite reductase (NOFNiR) (Chamizo-Ampudia et al. 2016).

Oxidative Stress

Oxidative stress often emerges in hostile environments, which induce excess production of reactive oxygen species (ROS). For algae and plants, most oxidative stress results from ROS production in the location of metabolisms characterized as high redox potentials such as electron transport chains of chloroplasts and mitochondria (Wang et al. 2023). Under normal conditions, cells produce ROS that functions as signaling molecules, contributing to growth, development, defense mechanisms against pathogens such as hypersensitive reactions and systemic acquired resistance, the synthesis of stress hormones, acclimation processes, and programmed cell death (Tripathy and Oelmüller 2012). However, heightened levels of oxidizing agents within cells disturb cellular function by extracting electrons from vital organic molecules, resulting in excessive ROS production and an imbalance known as oxidative stress. This results in the perturbation of redox homeostasis and damages to the organism. The oxidative stress in photosynthetic organisms due to light-dependent generation of ROS is named photooxidative stress.

Reactive Oxygen Species (ROS)

ROS includes superoxide radical ion (O_2^{*-}), hydrogen peroxide (H_2O_2), the hydroxyl radical (OH^{*-}), singlet molecular oxygen ($^1O_2$), and organic hydroperoxides. Not all ROS are equally reactive: the

Table 13.2. Reductive and oxidative pathways of NO generation in alge (with a plant reference) inferred from existence of the responsible enzyme coding genes (+) or response to subdtrate or inhibitors of the responsible enzyme (*).

Phylum	Species	Reductive		Oxidative		References
		NR/NiR	NOS	Polyamine	Hydroxylamine	
Land plants	*Arabidopsis thaliana*	+	+	*	*	Tun et al. 2006, Allagulova et al. 2023
Cyanobacteria	*Anabaena doliolum*	*				Malllick et al. 1999
	Microcystis aeruginosa	*				Tang et al. 2011
	Synechoccocus leopoliensis		+			Picciano and Crane 2019
Chlorophyta	*Bolbocoleon piliferum*		+			Jeandroz et al. 2016
	Carteria crucifera		+			Jeandroz et al. 2016
	Carteria crucifera		+			Jeandroz et al. 2016
	Chaetosphaeridium globosum		+			Jeandroz et al. 2016
	Chlamydomonas reinhardtii	*				Sakihama et al. 2002
	Chlamydomonas cribrum		+			Jeandroz et al. 2016
	Chlamydomonas sp.		+			Jeandroz et al. 2016
	Chlorella sorokiniana	*				Tischner et al. 2004
	Cosmarium subtumidum		+			Jeandroz et al. 2016
	Golenkinia longispicula		+			Jeandroz et al. 2016
	Helicodictyon planctonicum		+			Jeandroz et al. 2016
	Leptosira obovata		+			Jeandroz et al. 2016
	Lobochlamys segnis		+			Jeandroz et al. 2016
	Lobochlamys segnis		+			Jeandroz et al. 2016
	Nephroselmis pyriformis		+			Jeandroz et al. 2016
	Ostreococcus tauri		+			Foresi et al. 2015
	Pandorina morum		+			Jeandroz et al. 2016
	Phacotus lenticularis		+			Jeandroz et al. 2016
	Planophila terrestris		+			Jeandroz et al. 2016
	Pleurastrum insigne		+			Jeandroz et al. 2016
	Polytomella parva		*			Lapina et al. 2022
	Pteromonas angulosa		+			Jeandroz et al. 2016
	Pteromonas sp.		+			Jeandroz et al. 2016
	Scenedesmus obliquus	*	+			Mallkick et al. 1999
	Scherffelia dubia		+			Jeandroz et al. 2016
	Volvox aureus M1028		+			Jeandroz et al. 2016
Rhodophyta	*Rhodochaete parvula*		+			Jeandroz et al. 2016
Diatoms	*Pseudo-nitzschia multistriata*		+			Di Dato et al. 2015
	Skeletonema marinoi strain FE7					Jeandroz et al. 2016
	Skeletonema marinoi strain FE60					Jeandroz et al. 2016
	Thalassiosira minuscula CCMP1093					Jeandroz et al. 2016
	Thalassiosira oceanica					Lommer et al. 2012
	Thalassiosira rotula strain GSO102					Jeandroz et al. 2016
Dinoflagellates	*Gambierdiscus caribaeus*					Price et al. 2016
	Prorocentrum micans		+			Jeandroz et al. 2016
Phaeophyta	*Sargassum muticum*		+			Jeandroz et al. 2016
Raphidophyta	*Chattonella marina*		+			Kim et al. 2006

most reactive species are the hydroxyl radical and singlet oxygen, which react immediately with a suitable molecule and so inflict injury mainly on local cellular structures; H_2O_2 and superoxide anion are more stable and can move through the cell by diffusion; H_2O_2 can even pass through cell membranes (van Straalen and Roelofs 2012). The damaging effect of ROS largely stems from their ability to generate hydroxyl radicals.

Many algae are known to produce detectable levels of ROS, including those produced extracellularly (Diaz and Plummer 2018, Cho et al. 2022). These ROS producing algae include raphidophyte species from the genera *Chattonella*, *Heterosigma*, and *Fibrocapsa*, dinoflagellates such as *Akashio sanguineum*, *Heterocapsa circularisquama*, *Karenia mikimotoi*, *Margelefidinium polykrikoides* (formerly *Cochlodinium polykrikoides*), *Prorocentrum cordatum* (formerly *P. minimum*), *P. micans*, and Symbiodiniaceae, diatoms such as *Thalassiosira pseudonana* and *T. weissflogii*, chlorophytes such as *Ulva* spp., rhodophytes such as *Chondrus crispus*, and cyanobacteria such as *Trichodesmium*. Remarkably, some toxin-producing HAB species also produce excess ROS. Examples include *Karlodinium veneficum* and *Alexandrium* spp., which are best known to produce karlotoxins and saxitoxins, causing Neurotoxic Shellfish Poisoning (NSP) and Paralytic Shellfish Poisoning (PSP), respectively (Mardones et al. 2015). Some cyanobacteria, such as *Trichodesmium*, are also known to produce ROS (Cho et al. 2022).

ROS can be produced in the photosynthetic apparatus, respiratory system, peroxisomes, endoplasmic reticulum, and cell membrane (Fig. 13.5). ROS production is widely accepted as

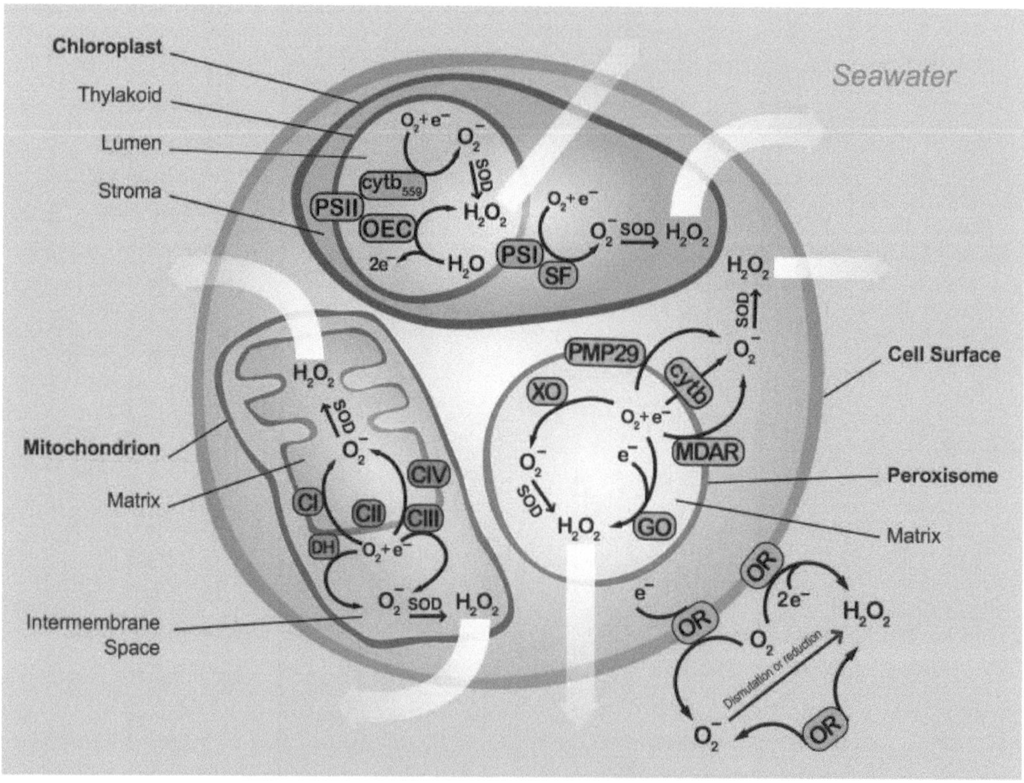

Figure 13.5. Mechanisms of superoxide (O_2^-) and hydrogen peroxide (H_2O_2) production in phytoplankton. Chloroplast: PSI—photosystem I, PSII—photosystem II, SF—stromal factor, OEC—oxygen-evolving complex, cytb559—cytochrome b559; Mitochondria: CI—complex I, CIII—complex III, DH—NAD(P)H dehydrogenase; Peroxisomes: GO—glycolate oxidase, XO—xanthine oxidase, MDAR—monodehydroascorbate reductase, cytb—cytochrome b, PMP29—peroxisome membrane polypeptide 29; Cell surface and cell-free environment: OR—oxidoreductase. Intracellular hydrogen peroxide can diffuse within and outside of cells (white arrows), but intracellular superoxide is unlikely to escape the cell. Reproduced with permission from Diaz and Plummer (2018) under the Creative Commons CC BY 4.0 license.

an inevitable outcome of the photosynthetic electron transport chain functioning in an oxygen atmosphere (Foyer et al. 1994). Light-dependent ROS generation is due to the oxygen-consuming processes associated with photosynthesis, including (a) the oxygenase reaction of the ribulose-1,5-bisphosphate carboxylase/oxygenase (Rubisco), which is the initiating reaction of photorespiration and (b) direct reduction of molecular oxygen by the photosystem I (PSI) electron transport chain. Besides, certain photosystem II (PSII) components can reduce molecular O_2 to a radical. Oxygen may also be reduced by a cyanide-insensitive respiratory pathway in chloroplasts that competes for electrons with photosynthetic electron transport.

In the respiratory system, during cellular respiration, oxygen (O_2) interacts with electrons from the Electron Transport Chain (ETC), generating superoxide ions (O_2^-). The primary sites where electrons are transferred to O_2 are the NADH-quinone oxidoreductase and coenzyme Q: cytochrome C oxidoreductase. O_2^- is created by the combination of electrons and O_2 in the intermembrane space or matrix. Being a charged species, O_2^- cannot freely cross the mitochondrial membrane. Hence, it appears that voltage-dependent mitochondrial anion channels aid in releasing intermembrane mitochondrial O_2^- into the cytosol, where O_2^- is rapidly converted to hydrogen peroxide (H_2O_2).

Besides, ROS can also be produced from the reduction-oxidation reactions catalyzed by cytochromes of the smooth endoplasmic reticulum, and the xanthine dehydrogenase pathway, which is involved with the degradation of purines to urate. Singlet oxygen is produced by so-called photo-sensitization reactions, in which light energy is absorbed by molecules such as riboflavin, chlorophyll, and retinol, and transferred to molecular oxygen. Other metabolic activities catalyzed by the cytochrome P450 system, xanthine oxidase, NAD(P)H oxidoreductases, and other enzymes may be involved as well.

NADPH oxidases (NOX) play a significant role in extracellular ROS production. NOX occur in algae as well as plants and animals. In plants they are encoded by Respiratory burst oxidase homolog (Rboh) genes and catalyze the transfer of electrons from cytosolic NADPH to extracellular O_2, leading to the production of extracellular superoxide anion (O_2^-) (Sagi et al. 2004). Extracellular superoxide anion production is widespread in algae, having been reported in green algae such as *Ulva pertusa* and *U. compressa*, the brown alga *Laminaria digitata*, the red alga *Chondrus crispus*, and red intertidal algae and the various microalgae indicated earlier (Anderson et al. 2016). NOX activity has been shown to be a significant component of light-dependent generation of electron flow in *Chlamydomonas reinhardtii,* exporting electrons across the plasma membrane to form superoxide anion from oxygen (Anderson et al. 2016). NOX enzymes and genes have also been reported in many species of algae (Table 13.3).

Excess amounts of ROS have detrimental effects on the cell membrane (via lipid damage), DNA, proteins, and enzymes. Lipid damage is due to a process known as lipid peroxidation, which is a chain reaction where an oxygen radical abstracts a hydrogen atom from an unsaturated bond in a fatty acid chain, followed by molecular rearrangement and uptake of oxygen, to form a lipid peroxyradical (LOO*), which can abstract H* from another lipid. Polyunsaturated fatty acid chains of membrane lipids are particularly sensitive to lipid peroxidation. Peroxidation causes loss of membrane flexibility, loss of activity of membrane-bound enzymatic processes, and, in its most severe form, lipid destruction fueled by the appearance of volatile alkanes, alkenes, and aldehydes (van Straalen and Roelofs 2012). Protein damage may be caused by oxidation of free thiol (SH) groups, leading to loss of function. DNA damage may be due to thymidine dimers and strand breaks.

Diverse Oxidative Stress Protection Mechanisms

Organisms have evolved a diversity of protective systems against ROS, which can be categorized into two groups. One is by scavenging ROS (by reaction with a reductant) and the other is by enzymatic transformation (dismutation or reduction) of ROS to a non-reactive form (Table 13.4). Superoxide dismutases (SODs) represent the first line of antioxidant defense (Alscher et al. 2002).

Table 13.3. NADPH oxidase enzymes (NOX) and genes reported in algae.

Algal Phylum	Algal species	NOX enzyme	NOX gene	Induction condition	References
Charophyta	*Chara vulgaris*		X		Fichman et al. (2023)
Chlorophyta	*Chlamydomonas reinhardtii*		X		Anderson et al. (2016)
Chlorophyta	*Chlamydomonas reinhardtii*		X		Fichman et al. (2023)
Chlorophyta	*ParaChlorella kessleri*		X		Petrova et al. (2024)
Diatoms	*Phaeodactylum tricornutum*		X	LL/LD	Laohavisit et al. (2015)
Rhodophyta	*Chondrus crispus*		X	The gene was over-transcribed in some situation.	Hervé et al. (2006)
		X	X		Anderson et al. (2011 Nov)
NOX family in the Eukarya domain		X			Massari, Marta et al. (2022)

Table 13.4. Major antioxidant systems protecting the cell against injury from radical oxygen species.

Antioxidant system	Primary localization	Actions
Copper- and zinc-containing superoxide dismutase (Cu/Zn SOD)	Cytosol, nucleus	Catalyses dismutation of O^-_2 to H_2O_2
Manganese-containing superoxide dismutase (Mn SOD)	Mitochondrion	Catalyses dismutation of O^-_2 to H_2O_2
Rubredoxin oxidoreductase	Only found in anaerobic bacteria and Archaea	Catalyses dismutation of O^-_2 to H_2O_2
Catalase (CAT)	Peroxisomes	Catalyses reduction of H_2O_2 to H_2O
Glutathione peroxidase (GSH-Px)	Cytosol, mitochondrion	Catalyses reduction of H_2O_2 to H_2O
Glutathione peroxidase (GSH-Px,m)	Lipid membranes	Catalyses reduction of liquid hydroperoxides
Glutathione reductase	Cytosol, mitochondrion	Catalyses reduction of low-molecular-mass disulphides
Ascorbate peroxidase	Chloroplast (only in plants)	Catalyses reduction of H_2O_2 to H_2O
Thioredoxin (TRX)	Cytosol, mitochondrion, nucleus	Restores redox state by reducing oxidized thiols, scavenges H_2O_2 and OH·
α-Tocopherol (vitamin E)	Lipid membranes	Scavenges H_2O_2, OH·, and LOO·, interrupts lipid peroxidation
β-Carotenoid (vitamin A)	Lipid membranes	Scavenges O^-_2 and peroxyl radicals
Ascorbate (vitamin C)	Throughout the cell	Scavenges O^-_2 and OH·, contributes to regeneration of vitamin E
Glutathione	Throughout the cell	Scavenges O^-_2 and organic free radicals; substrate in enzymatic reduction reactions

These are metalloenzymes (with metal ions at the active site) that catalyze the conversion of the highly reactive and potentially damaging oxygen radicals (O_2^-, OH^-) to the less toxic moieties such as hydrogen peroxide (H_2O_2). There are four classes of SODs, named for the metal in their reaction centers: iron SOD (FeSOD), manganese SOD (MnSOD), copper-zinc SOD (Cu/Zn SOD), and nickel SOD (NiSOD). The amino acid sequences and three-dimensional protein structure of FeSOD and MnSOD are very similar, suggesting a common ancestry, whereas those of Cu/Zn SOD and NiSOD are quite different from the other two and from each other, indicating remote evolutionary relationships.

Superoxide Dismutases (SODS) and Other Antioxidative Enzymes

Among the different classes of SODs, Cu/Zn SOD is generally distributed in the cytosol, MnSOD occurs in the mitochondria, and FeSOD occurs in the chloroplast (specifically, in PSI). It has been suggested that SOD in the cytosol is necessary for land plants because oxygen can diffuse into the cytosol from the atmosphere and can produce damaging radicals there (Lee 1999). Among green algae, only charophytes have the cytosol-localized Cu/Zn SOD. The exclusive presence of Cu/Zn SOD in Charophyceae and land plants within Viridiplantae supports their evolutionary relationship and explains the successful terrestrial colonization by charophycean descendants (Lee 1999).

Cu/Zn SOD and MnSOD are most common in prokaryotes and eukaryotes. FeSOD occurs in prokaryotes and eukaryotic photosynthetic organisms. NiSOD was better known in prokaryotes including cyanobacteria (Palenik et al. 2003), but is now known to exist in eukaryotic algae such as dinoflagellates, diatoms, cryptophytes, and prasinophytes (Zhuang et al. 2015, Sibbald et al. 2019). Interestingly, NiSOD sequences are quite distinct between the eukaryotic and prokaryotic (including cyanobacteria) lineages, and the eukaryotic algal NiSOD sequences are also divergent, although all contain the conserved nickel binding site (Ni Hook) with the sequence HCxxPCG(I/V)(F/Y) D(D/P). The latter are clustered into two clades, clade Euk I represented by different eukaryotic algal lineages and clade Euk II composed exclusively of dinoflagellates (Fig. 13.6). Furthermore, Euk I is characterized by fusion with ubiquitin in the N-terminus of the protein, as also found to be the case in the cryptophyte *Guillardia theta* and the chlorarachniophyte *Bigeloweilla natans* (Sibbald et al. 2019). In contrast, Euk II sequences instead contain a mitochondrion-targeting peptide, suggesting they are nucleus-encoded but mitochondria-localized. NiSOD in the paralytic shellfish poisoning (PSP) dinoflagellate *Alexandrium catenella* (formerly *A. fundyense*) contains Euk II NiSOD genes, which were actively expressed during a PSP outbreak (Zhuang et al. 2015). Strikingly, coral endosymbiotic dinoflagellates Symbiodiniaceae contain either Euk I or Euk II NiSOD, depending on strains (Fig. 13.6). Based on data currently available, chlorophytes that contain NiSOD seem to be restricted to the prasinophyte genera *Ostrecoccus* and *Micromonas*. The role of the Ubiq-NiSOD protein as antioxidant remains to be established.

Besides SOD, catalase, glutathione peroxidase, glutathione reductase, and ascorbate peroxidase (Table 13.4) also play important roles in transforming ROS to non-damaging or less damaging molecules.

Non-Enzyme Antioxidants

The non-enzymatic antioxidants include thioredoxin (TRX), α-tacopherol (vitamin E), β-carotenoid (vitamin A), ascorbate (vitamin C), thiamine (vitamin B1) and other members of the vitamin B family, and glutathione. Among these, glutathione is worth highlighting as it is both a ROS scavenger and a substrate of enzymatic reactions. As a γ-glutamyl-cysteinyl-glycine tripeptide, glutathione is a potent antioxidant with many crucial functions such as detoxification by conjugating lipophilic substances in xenobiotic biotransformation reactions. Besides its role in the control of ROS, glutathione is also involved in the detoxification of methylglyoxal (MG) in plants. Much like ROS, MG is produced at low levels by aerobic metabolism under normal conditions and is overproduced in response to stress (Dorion et al. 2021). In the reduced form (GSH), it has a free thiol group on the cysteine residue. Its conversion to the oxidized form (GSSG) involves two molecules of GSH, the thiol groups of which react with each other, losing two electrons, as in the reaction catalyzed by glutathione peroxidase. GSSG is then reduced back to GSH by involving glutathione reductase and NADPH.

Glutathione is synthesized in two ATP-dependent steps catalyzed by γ-glutamylcysteine synthetase (GSH1) and glutathione synthetase (GSH2) (Fig. 13.7). In the first step, Cys is

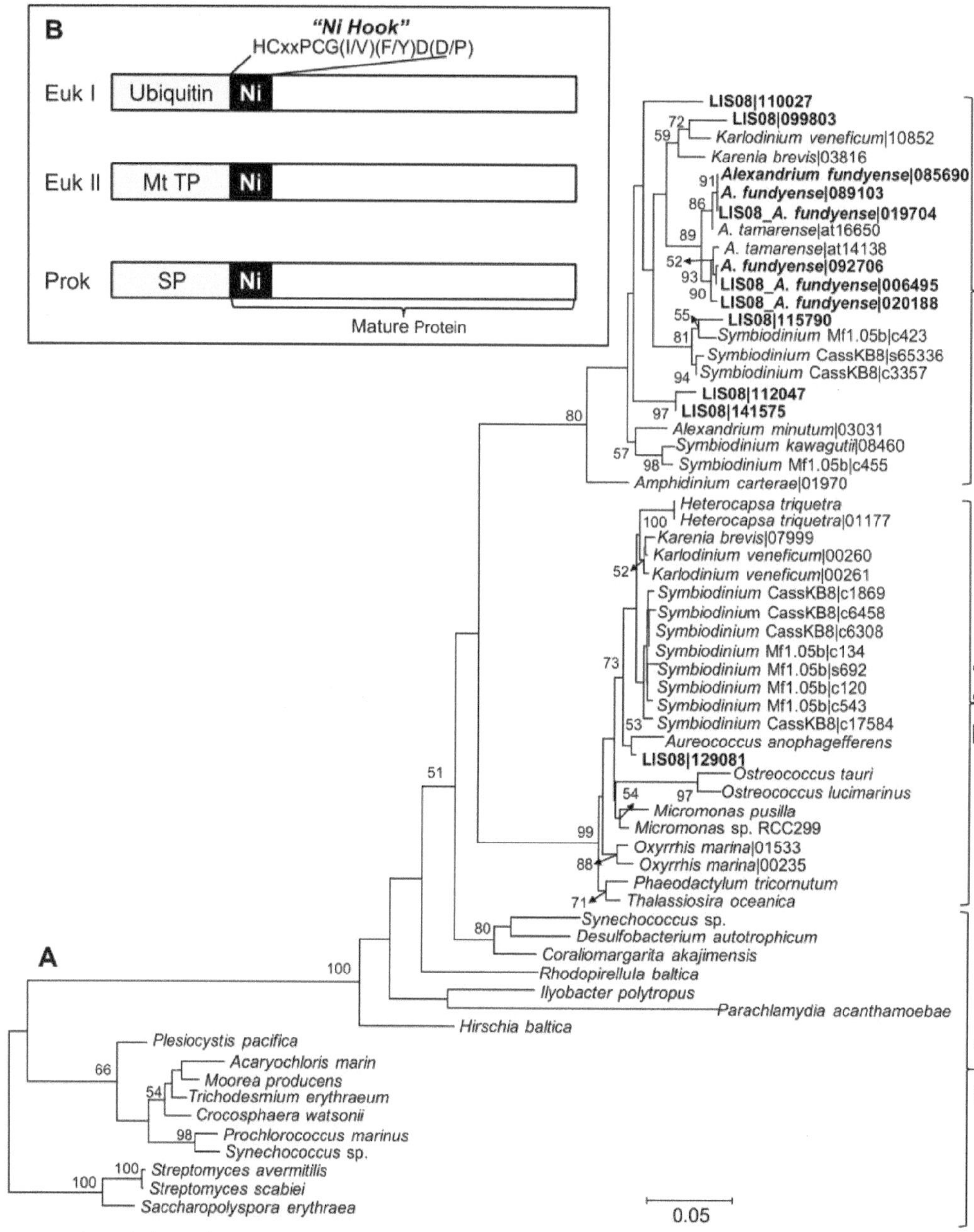

Figure 13.6. Phylogenetic trees of algal NiSOD protein sequences (A) and organization of the proteins (B). NiSOD from bacteria were used to root the tree. Note that eukaryotic algal NiSOD (Euk) are distinct from cyanobacterial and heterotrophic bacterial NiSOD (Prok) and classified into Euk I and Euk II, which contain a ubiquitin domain and mitochondrial targeted peptide (Mt TP), respectively. Reproduced with permission from Zhuang et al. (2015).

ligated with Glu to produce γEC by the action of GSH1. In the second step, GSH2 ligates Gly to γEC to yield GSH. The two GSH biosynthesis genes are conserved in all biological kingdoms (Tamaki et al. 2021). GSH1 exists in all phyla of algae examined except for the 'green' lineages: Chlorophyta, Prasinophyta, Euglenophyta, and Cyanobacteria (Fig. 13.8A). The absence of GSH1 in these phyla of green algae is surprising, particularly because it exists in land plants (but again surprisingly, not in *Arabidopsis* spp.). GSH1 gene structure seems to differ between dinoflagellates and diatoms, with the former containing a membrane-embedded region and a cytoplasmic domain,

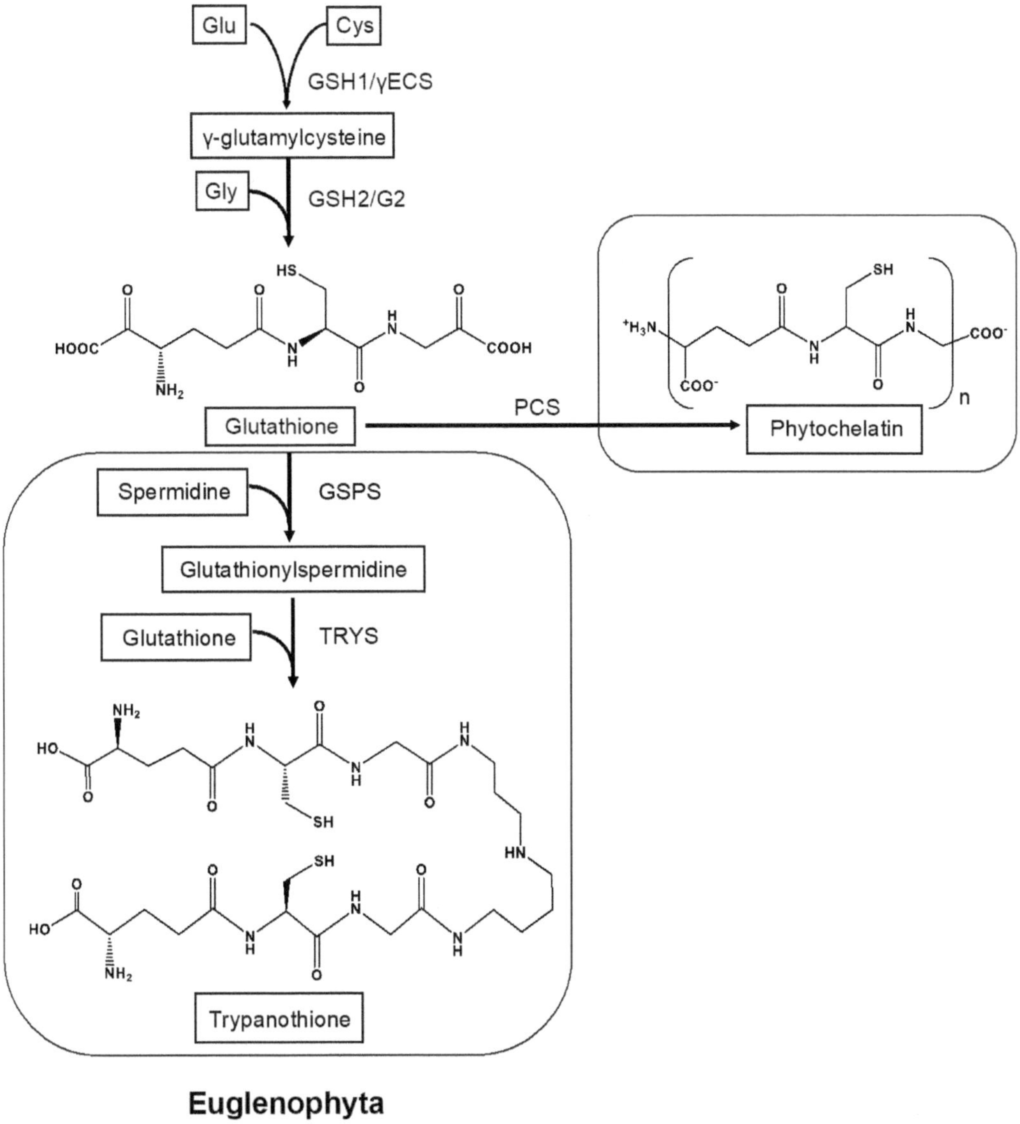

Figure 13.7. Structure and biosynthesis of glutathione, phytochelatins, and trypanothione in microalgae. Glutathione biosynthesis is conserved in all biological kingdoms. Phytochelatin biosynthesis occurs in most microalgae, whereas among photosynthetic organisms, trypanothione is detected only in Euglenophyta. GSH1/γECS, γ-glutamylcysteine synthetase; GSH2/GS, glutathione synthetase; PCS, phytochelatin synthase; GSPS, glutathionylspermidine synthetase; TRYS, trypanothione synthetase. Redrawn from Tamaki et al. 2021.

whereas the latter contains only one domain with unclear subcellular localization (Fig. 13.9). In contrast, GSH2 occurs in green algae and plants as well as other groups of algae (Fig. 13.8B). The polyphyletic topology of the GSH2 tree suggests a complex evolutionary trajectory of this gene in algae, including multiple independent acquisition and horizontal gene transfer.

Photoprotective Mechanisms

The fact that excess PAR can induce the formation of ROS poses a strong evolutionary pressure on photosynthetic organisms to develop potent photoprotective mechanisms. Like plants, algae have evolved a wide variety of photoprotective strategies that alleviate photoinhibition and are associated

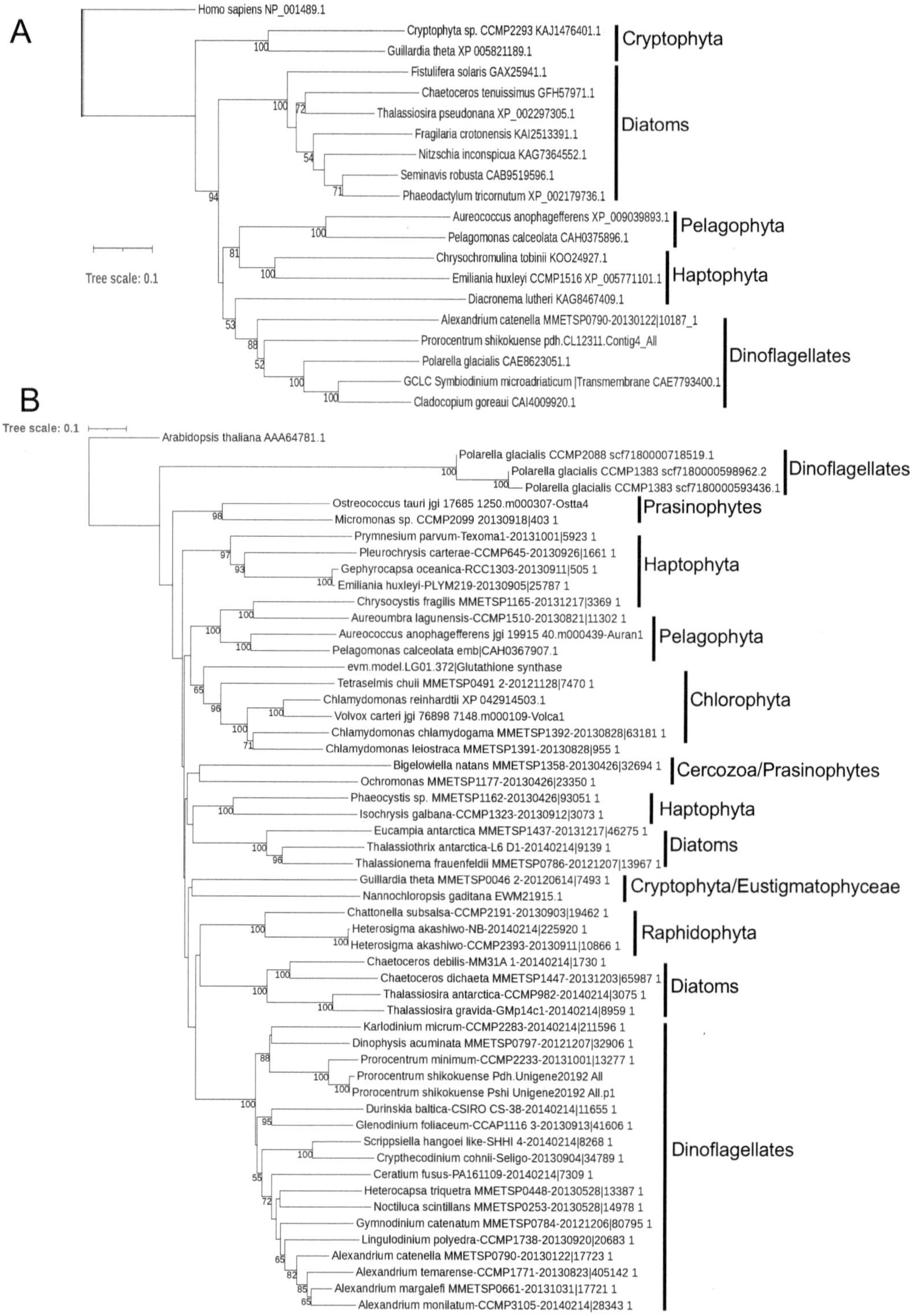

Figure 13.8. Phylogenetic distribution of the two key genes involved in glutathione biosynthesis. (A) γ-glutamylcysteine synthetase (GSH1). (B) Glutathione synthetase (GSH2).

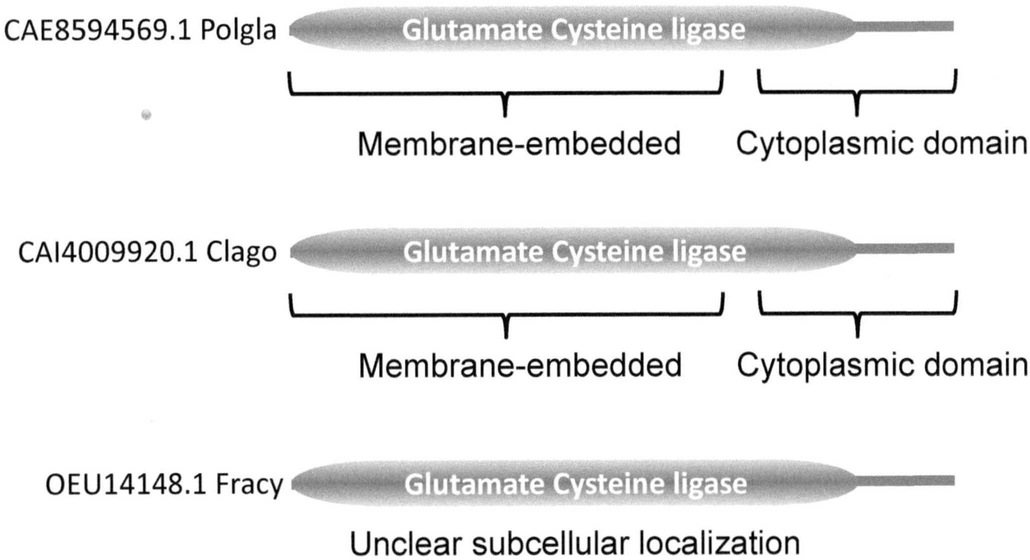

Figure 13.9. Different subcellular localization of glutamate-cysteine ligase (GSH1) in dinoflagellates compared to that of a diatom. Polgla and Clago depict the dinoflagellates *Polarella glacialis* and *Cladopodium goreaui*; Fracy depicts the diatom *Fragilariopsis cylindrus.*

with avoiding light absorption and successfully consuming or dissipating the light energy absorbed by light-harvesting pigments (Derks et al. 2015). Eukaryotic algae commonly possess the extended Light-Harvesting Complex (LHC) protein superfamily, which evolved from the cyanobacterial one-helix protein, with the chlorophyll-binding motif being strongly conserved across the entire extended LHC protein superfamily. The extended LHC protein superfamily comprises LHC, LHC-like, and PSBS protein families (Sturm et al. 2013). Besides, Non-Photochemical Quenching (NPQ) is a fast-induced photoprotection mechanism (time-scale of seconds to a few minutes) that facilitates the dissipation of excessive excitation energy as heat (Ruban 2016). In addition, all algae use the xanthophyll cycle or its equivalent (for cyanobacteria) to dissipate the excess energy harvested by the light-harvesting complex (Goss and Jakob 2010). The high-energy-state quenching (qE), known as the principal component of NPQ in plants, is related to the activity of a xanthophyll cycle.

Light-Harvesting Complex

The LHC protein family includes LHCR in rhodophytes ('R' for Rhodophyta), chl *a*/*c*-binding (CAC) proteins in algal groups with secondary plastids of red algal origin, which are called fucoxanthin CAC proteins (FCPs) or LHCF ('F' for fucoxanthin) in diatoms and brown algae, LI818 (also called LHCX) in diatoms, and a less known clade, LHCZ, in some algae with complex plastids. In chlorophytes, the LHC protein family comprises chl *a*/*b*-binding (CAB) proteins and LI818 (also called LHCSR). Besides LHC superfamily proteins, algae also use Chl *a*/*b* binding proteins or the IsiA chlorophyll-binding protein (in cyanobacteria) to handle excess PAR.

The LHC-like protein family consists of early light-induced proteins (ELIPs), Stress-Enhanced Proteins (SEPs, also named Light-Harvesting-Like [LIL] proteins), and One-Helix Proteins (OHPs, also called High Light-Induced Proteins HLIPs), and High Light (HL) intensity-inducible LHC-like 4 (LHL4) proteins. ELIPs and LHL4 are found only in the green lineage, whereas SEPs and OHPs are shared between red and green algae. OHPs occur in two types: the OHP1/HLIP-type in cyanophages, cyanobacteria, and eukaryotic algae and the OHP2-type exclusively in eukaryotic algae. The PsbS protein family only occurs in the green lineage. Besides, the red lineage possesses an additional, unique, member of the extended LHC protein superfamily, Chl *a*/*b*-binding-like proteins (RedCAPs) (Engelken et al. 2010).

Accumulation of ascorbic acid has been found to be an important photoprotective mechanism in the acidothermophilic red alga *Galdieria partita* against high light stress under mixotrophic culture with glucose supplement (Fu et al. 2020). Furthermore, ascorbate peroxidase was recently shown to play an important role in photoacclimation in the extremophilic obligate autotrophic red alga *Cyanidiococcus yangmingshanensis* (Fu and Wang 2023). Scavenging ROS is one important function of ascorbic acid in the photoprotective process. Ascorbic acid can directly quench singlet oxygen, superoxide, and hydroxyl radical, and participate in enzymatic ROS scavenging, in which superoxide dismutase (SOD) catalyzes the conversion of superoxide to hydrogen peroxide (H_2O_2). H_2O_2 then reacts with ascorbic acid through ascorbate peroxidase (APX) to produce monodehydroascorbate (MDHA) radical and water. Ascorbic acid can be regenerated from reduction of malondialdehyde (MDA) through MDA reductase (MDHAR) or from spontaneous disproportionation of MDA into dehydroascorbate (DHA) and ascorbic acid.

Xanthophyll Cycles

Xanthophyll cycle represents one of the important photoprotective mechanisms in photosynthetic organisms (Goss and Jakob 2010). There are two major types of xanthophyll cycles (Fig. 13.10). One is the violaxanthin (Vx) cycle (Fig. 13.10A), which occurs in chlorophytes, prasinophytes, and

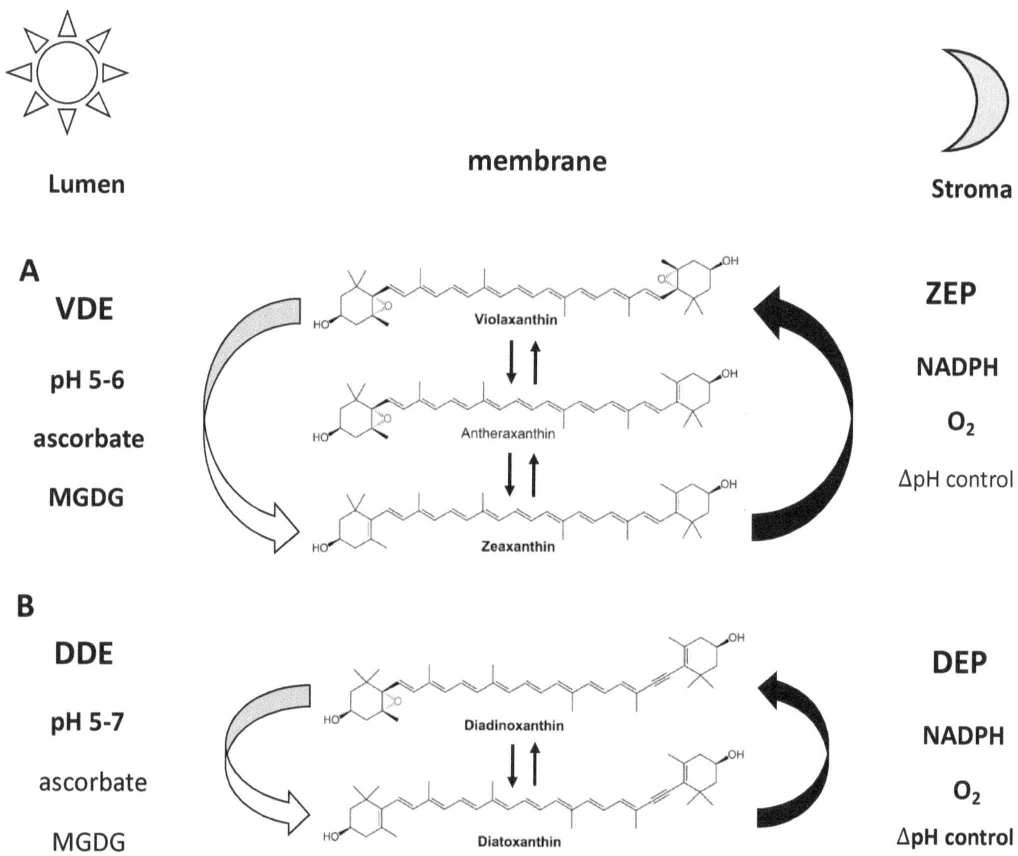

Figure 13.10. Xanthophyll cycles in algae. (A) Violaxanthin cycle in chlorophytes, prasinophytes, and phaeophytes, in which violaxanthin is converted to zeaxanthin through the intermediate antheraxanthin in the light, which is catalyzed by violaxanthin de-epoxidase (VDE). In the dark, zeaxanthin is oxidized back to violaxanthin with the aid of the enzyme zeaxanthin epoxidase (ZEP). (B) Diadinoxanthin cycle in diatoms, xanthophytes, haptophytes, and dinoflagellates in which diadinoxanthin is converted in the light to diatoxanthin by the enzyme diadinoxanthin de-epoxidase (DDE). In the dark, diatoxanthin is converted back to diadinoxanthin by the diatoxanthin epoxidase (DEP).

phaeophytes. Vx is converted to zeaxanthin (Zx) through the intermediate antheraxanthin in the light, which is catalyzed by Vx De-Epoxidase (VDE). In the dark, Zx is oxidized back to Vx with the aid of the enzyme Zx epoxidase (ZEP). The other is the diadinoxanthin (Ddx) cycle operating in diatoms, xanthophytes, haptophytes, and dinoflagellates (Fig. 13.10B). Ddx is converted in the light to diatoxanthin (Dtx) by the enzyme Ddx de-epoxidase (DDE). In the dark, Dtx is converted back to Ddx by the Dtx epoxidase (DEP). Interestingly, the Zx cycle also exists in the Ddx cycle-possessing lineages such as diatoms (Lohr and Wilhelm 1999), and seems to complement the function of the Ddx cycle in photoprotection (Kuczynska et al. 2020). Recent studies also reveal that the de-epoxidation of Ddx occurs in the dark.

The functional equivalent of the xanthophyll cycle in cyanobacteria includes a water-soluble carotenoid-binding protein, which specifically interacts with the phycobilisomes and large aggregates of the IsiA protein, which are formed under stress conditions and are characterized by a strong fluorescence quenching (reviewed by Goss and Jakob 2010). In the IsiA aggregates, a singlet–singlet excitation energy transfer between a Chl *a* and a xanthophyll molecule has been proposed as the structural basis for quenching in *Synechocystis* PCC 6803 (Berera et al. 2009). This resembles the mechanism proposed for zeaxanthin- and diatoxanthin-dependent quenching. In contrast to violaxanthin and diadinoxanthin cycle-containing algae, the quenching in *Synechocystis* PCC 6803 does not depend on active xanthophyll cycling, but is likely induced by the carotenoid echinenon, which interacts with Chl *a*.

Ultraviolet Radiation (UV)

Based on the damaging power, UV is divided into three spectra: UV-A (315–400 nm) and UV-B (280–315 nm), and UV-C (100–280 nm). The shorter the wavelength, the more damaging it is; however, UVC is completely filtered by the atmosphere and does not reach the earth's surface. UV-B (280–315 nm) is particularly destructive to DNA, causing mutations (with formation of pyrimidine dimers) and is also damaging to proteins and photosynthetic pigments. Due to ozone depletion resulting from anthropogenic use and release of chlorofluorocarbons, increasing UV radiation reaches the earth's surface.

Signaling of UV Stress

Terrestrial algae and land plants do not have the benefit of water efficiently absorbing UV and hence receive more extreme UV exposure than aquatic algae. Thus, more UV protection mechanisms are expected in terrestrial organisms. Plants use the UVR8 photoreceptor system to perceive UVB, and a similar system has been identified in the zygnematophyte alga *Serritaenia* (Busch et al. 2024). The UVR8 system appears to serve as a central component of its cellular reaction to harmful radiation in *Serritaenia.* It was found that *Serritaenia* upregulated two components of the UVR8 signaling network, Constitutive Photomorphogenic 1 (COP1) and Repressors of UVB photomorphogenesis (RUP) upon exposure to UV. In plants, UVR8 dimers split into monomers upon UVB exposure, which then interact with COP1. This reduces the ubiquitin-mediated breakdown of the transcription factor ELONGATED HYPOCOTYL 5 (HY5) by COP1 and induces the transcription of UVB-responsive genes responsible for UVB acclimation (Oravecz et al. 2006, Tilbrook et al. 2013, Liang et al. 2019). RUP, in contrast, maintains the photoequilibrium when the UVR8 dimer/monomer cycling rate increases by facilitating the re-dimerization of UVR8, and thereby reduces again the transcription of UVB-responsive genes (Liao et al. 2020). This appears to be a negative feedback loop that counterbalances UVB-induced signaling.

The *Serritaenia* study also showed that UV exposure led to the upregulation of components of the Jasmonic Acid (JA) mediated signaling pathway, including the digalactosyldiacylglycerol synthase 1 (DGD1; chloroplast), OPDA (12-oxo-phytodienoic acid) reductases (OPR2 and OPR3, chloroplast), "Novel Interactor of 334 JAZ" (NINJA; nuclear) and NAC transcription factors

(NAC019, NAC055, NAC072; 335 nuclear). This finding is consistent with the previous report that JA levels in *A. thaliana* increased upon UVB exposure, although algae have been thought to lack classical JA signaling systems (Rieseberg et al. 2023) and some important JA-related factors were not detected in *Serritaenia*.

In *A. thaliana*, transcriptional responses to UV stress involves promoter elements A and B, both of which are required to enable induction of tolerance to high light and UV light (Hayami et al. 2015). The signature sequences of these elements are TACACACC (element A) and GGCCACGCCA (element B). Element B occurs in the Early Light Induced Protein 2 (ELIP2) promoter, which is a target of the light signaling transcription factor Elongated Hypocotyl5 (HY5). It is unclear if similar elements occur in algae.

Stress Response to UV Exposure

Algae respond to UVB stress in different ways and by different mechanisms, depending on the species. Besides DNA damage, UVB exposure can induce lipid peroxidation, production of ROS, and photosynthesis inhibition. Accordingly, responses may involve DNA repair, upregulated production of antioxidants, and photosystem repair, and, as a common response mechanism, HSP increases.

Serritaenia transcriptomic analysis revealed upregulation of 11 HSP genes in response to UV B exposure, all predicted to act as molecular chaperones assisting in a wide range of folding processes of proteins (Busch et al. 2024). A rapid upregulation of HSPs under UVB stress has also been reported for *A. thaliana* (Swindell et al. 2007).

UV exposure also caused remarkable photosynthetic responses in *Serritaenia*, including the downregulation of almost all components of PSI and PSII, light harvesting complexes, cytochrome *b6f* complex, and photosynthetic electron transport chain except for the upregulation of genes encoding PsbA (D1 protein), PsbD (D2), PsbC (CP43) and PsbB (CP47) in the PSII reaction center. The differential gene expression pattern suggests that the UV exposure inflicted damages to PSII, which was undergoing repair by *de novo* synthesis of these proteins, and that the UV exposure did not significantly impact PSI. Besides, UV exposure induced in *Serritaenia* the upregulation of PSI cyclic electron transport, which is a mechanism to eliminate excess electrons resulting from the damage on PSII. *Serritaenia* expressed several ELIP genes, although most of them were not upregulated under UV exposure (Busch et al. 2024). Chlorophyll *a*/*b* binding proteins (ELIP: Early Light Induced Protein, SEP2: stress enhanced protein 2) have been found in chlorophyte algae such as *V. carteri* (Razeghi and Kianianmomeni 2019) and *C. reinhardtii* (Allorent and Petroutsos 2017) and might represent a universal mechanism in chlorophytes.

Protective Mechanisms

Antioxidative mechanism. While UV exposure could induce ROS production, enzymatic and non-enzymatic antioxidants increase in many algae. For instance, cytoplasmic catalase and an extracellular class III peroxidase were upregulated in the UV-exposed *Serritaenia* (Busch et al. 2024). Similarly, upregulation of enzymatic and non-enzymatic antioxidants occurs in the UVB-exposed rhodophyte *Neoporphyra haitanensis* (Xue et al. 2022) and chlorophyte *Ulva fasciata* (Shiu and Lee 2005).

Extracellular sunscreen. In plants, the oxidative enzymes act on extracellular phenolics to form polymeric lignin in the apoplast, which is mainly associated with the mechanic properties of plant tissues (Busch et al. 2024). The transcriptome of the UV-exposed *Serritaenia* surprisingly did not show a role of plant flavonoid biosynthesis pathways, but instead exhibited remarkable regulation of enzymes mapped on the shikimate and phenylpropanoid pathway, potential cross-membrane transporters of phenolics and oxidative enzymes targeted to the extracellular space. In embryophytes, the secreted class III peroxidase is known to function in apoplastic lignin formation. These findings suggest divergence between plants and close algal relatives in the biosynthesis of extracellular sunscreen. However, *Serritaenia* produces extracellular pigmented mucilage, which is inducible by UV, suggesting its role in protection against UV damage (Busch and Hess 2021). Interestingly, extracellular sunscreen was previously known only in cyanobacteria until its production by *Serritaenia* was recognized. Importantly, carotenoids in algae commonly play dual roles, one in harvesting light energy and the other serving as sunscreen.

Mycosporine-Like Amino Acids (MAAs). Algae have evolved the ability to produce MAAs to combat UV damaging effects. MAAs are a family of intracellular compounds with the basic structure of a chromophore ring with amino alcohol at the C1 position and a glycine moiety at the C3 position (Fig. 13.11A). They can absorb UV radiation and take up ROS to reduce cell damage. The ability to produce MAAs has been reported widely among algae, including cyanobacteria, chlorophytes, streptophytes, rhodophytes, diatoms, raphidophytes, and dinoflagellates (Karsten et al. 1998, Jeffrey et al. 1999, Jain et al. 2017, Sun et al. 2020, Raj et al. 2021). The biosynthesis of MAAs involves two pathways: the shikimate pathway and the pentose phosphate pathway (Fig. 13.11B), with the shikimate pathway being the predominant source of MAAs production. The shikimate pathway and pentose phosphate pathway are linked to each other (Fig. 13.11). The intermediates sedoheptulose 7-phosphate (SH-7P) from the pentose phosphate pathway can readily be transformed into erythrose 4-phosphate, an intermediate in the shikimate pathway, through the action of the trans-aldolase enzyme.

Cyanobacteria seem to use the pentose phosphate pathway for the biosynthesis of MAAs. It involves mys gene cluster (Balskus and Walsh 2010), starting with SH 7-P cyclase (2-epi-5-Epi-Valiolone Synthase (EVS, named mysA), together with an O-methyltransferase (OMT, named mysB), leads to the production of 4-deoxygadusol, a MAA precursor. These two genes seem to be absent in non-producers of MAA, suggesting that they are specific indicators of MAA production. mysA does not seem to be essential for MAA synthesis, however, at least in some cyanobacterial species. The other genes in the mys gene cluster include ATP-grasp ligase (mysC), d-Ala-d-Ala ligase (mysD) or a non-ribosomal peptide synthetase (NRPS-like) gene (mysE). Expressing cyanobacterial mysABCE and dehydrogenase genes in *Escherichia coli* resulted in exclusive production of shinorine, confirming that the cyanobacterial type mys cluster is accountable for shinorine production in these species, as corroborated by bioinformatic analysis (Dextro et al. 2023).

The shikimate pathway is responsible for the synthesis of aromatic amino acids, and the synthesis of MAAs starts from an intermediate point of the pathway. It involves four key enzymes, dehydroquinate synthase (DHQS), O-methyltransferase (O-MT), ATP-grasp, and non-ribosomal peptide synthetase (NRPS). This pathway converts the precursor 3-dehyroquinate into 4-deoxygadusol via gadusol (Fig. 13.11). Conceivably, UV exposure elicits the regulation of the shared pathways from 'normal' functions in aromatic amino acid synthesis or energy metabolism towards production of MAAs. However, the signaling pathway remains to be determined.

Consistent with their high UV exposure during low tides, red macroalgae inhabiting intertidal zones are the most MAA-rich lineage. Depending on the species, rhodophytes contain 1.5–12 mg MAAs g^{-1} Dry Weight (DW), mostly between 5–6 mg MAAs g^{-1} DW (Vega et al. 2021). Porphyra-334 is one of the major MAA compounds produced by rhodophytes. Interestingly, in *Porphyra umbilicalis*, mysA is fused with mysB and mysC with mysD, and the mysA-mysB and mysC-mysD loci are arranged next to each other in the genome but transcribed on opposite DNA strands with adjacent 5'-ends. The same fusion genes occur in the *Chondrus* genome but the two loci are on opposite strands with the 3'-ends of the genes adjacent. The conservation of the MAA gene cluster in *Porphyra* and *Chondrus* and the occurrence of two gene-fusion events indicate that this genetic arrangement offers a selective advantage and promotes efficient MAA biosynthesis for red algae exposed to high levels of UV irradiance (Brawley et al. 2017).

Among dinoflagellates, various species of *Alexandrium* have been reported to produce MAA when exposed to high light conditions (Carreto 2001). Some dinoflagellate species contain the same mysA-mysB fusion as found in rhodophytes, suggesting its origin in red algae. Due to the

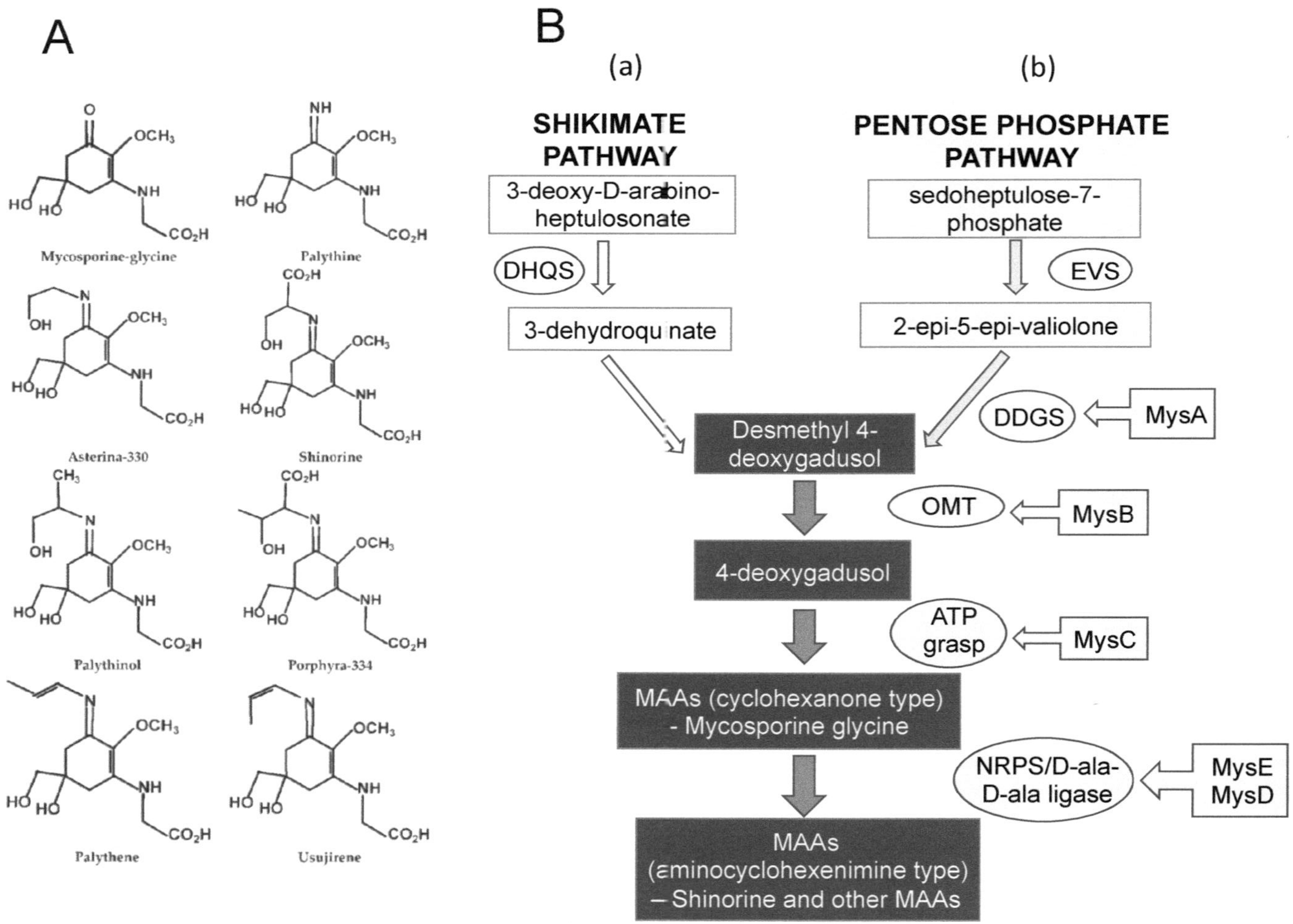

Figure 13.11. Chemical structure (A) and biosynthesis pathways (B) of mycosporine-like amino acids (MAAs) produced by algae. The Shikimate pathway and the Pentose phosphate pathway for MAA synthesis start with different substrates but merge at Desmethyl 4-deoxygadusol. Part B is redrawn from Raj et al. 2021.

importance of photoprotection in corals, there has been increasing research attention to MAA in the coral endosymbiotic lineage Symbiodiniaceae. Initially detected only in the ancestral genus *Symbiodinium* spp. (clade A), genome studies now have revealed the presence of the MAA biosynthesis gene set in some later diverging taxa (Shoguchi 2022). The adaptive significance of MAA evolution in the coral holobiont will be discussed in greater detail in Chapter 20.

Repair Mechanisms

UV exposure causes DNA damage, mainly the formation of Cyclobutene Pyrimidine Dimers (CPDs), thymine dimers (TTs), and pyrimidine (6-4) pyrimidone photoproducts (6-4PPs). One repair mechanism is based on DNA photolyases (PLs), enzymes activated by blue light to repair DNA photodamage (Sancar 2003). The photolyase family is part of the cryptochrome (CRY)/PL (a.k.a CRY/PHR) superfamily, which comprises 10 major subfamilies: CPD class I (CPD I), CPD II, CPD III, (6-4) photolyases, plant PHR2, CRY-DASH (*Drosophila*, *Arabidopsis*, *Synechocystis*, human), plant CRY (PCRY), mammalian-type CRY (MCRY), cnidarian-specific CRY (ACRY), and the putative magnetoreceptor CRY4 (Mei and Dvornyk 2015, Deppisch et al. 2022).

Cryptochromes are believed to have evolved from photolyases, and hence share structural similarity, but have gained new functions while losing their DNA repair capability. Photolyases contain conserved regions that bind to damaged DNA and chromophores that absorb light to drive the repair process, whereas cryptochromes have evolved distinct features such as the tryptophan triad and other tryptophan residues that are important for circadian regulation and light sensing (Mei and Dvornyk 2015). Photolyases were lost in higher animals, including most mammals and humans, which can only repair these DNA lesions through a process called nucleotide excision repair (Ramírez-Gamboa et al. 2022).

Among algae, photolyase occurrence has been reported in cyanobacteria (Aubert et al. 1999, Haney et al. 2022) and diatoms (Karentz et al. 1991, Coesel et al. 2009). In *Synechococcus*, the DNA photorepair capacity is attributed to multiple photolyase genes (Hanely et al. 2022). The photolyase of the diatom *Phaeodactylum tricornutum* (Phatr3_J34592.p1) demonstrates (6-4) photoproduct repair activity (Coesel et al. 2009). Furthermore, BLAST analysis into NCBI, SAGER, and PhyloDB databases reveals the prevalence of the gene, with high similarity (e value <–110) across a wide phylogenetic spectrum of algae, ranging from cyanobacteria to phaeophytes and dinoflagellates (Fig. 13.12A). Their functions, however, remain to be experimentally demonstrated.

Based on model organisms, the repair process (Fig. 13.12B) of Cyclobutane Primidine Dimers (CPDs) begins when the reduced cofactor $FADH^-$ absorbs light directly or receives energy transferred from an excited antenna chromophore. This excitation causes $FADH^{-*}$ to donate an electron to the CPD substrate, leading to the immediate splitting of the CPD into two thymine bases. Subsequently, the electron returns to the flavin cofactor, restoring the enzyme's activity (Zhang et al. 2016). Whether the same procedure operates in algae remains to be investigated.

Osmotic Stress

An algal cell in an aqueous medium must maintain a balance between intake and loss of water to avoid disruptive change in volume. An abrupt change in the osmotic potential of the medium will impose stress by plasmolysis, leading to inhibition of photosynthesis and other functions, on the one hand, or distention ultimately leading to bursting on the other. Global climate change alters precipitation and ice melt patterns, causing freshening of some marine areas and salinization of others. For instance, a study of ocean salinity since 1960 from the surface to 2000 m showed, based on the 0–2000 m averaged salinity trend, freshening in almost the entire Pacific Ocean, broad salinification in the low- and mid-latitude Atlantic, pronounced freshening in the North Atlantic, and contrasts between the north and south Indian Ocean (Cheng et al. 2020). In coastal oceans, the Baltic Sea, for instance, is projected to experience salinity declines of about 0.1–0.2 $g\,kg^{-1}$ per decade at

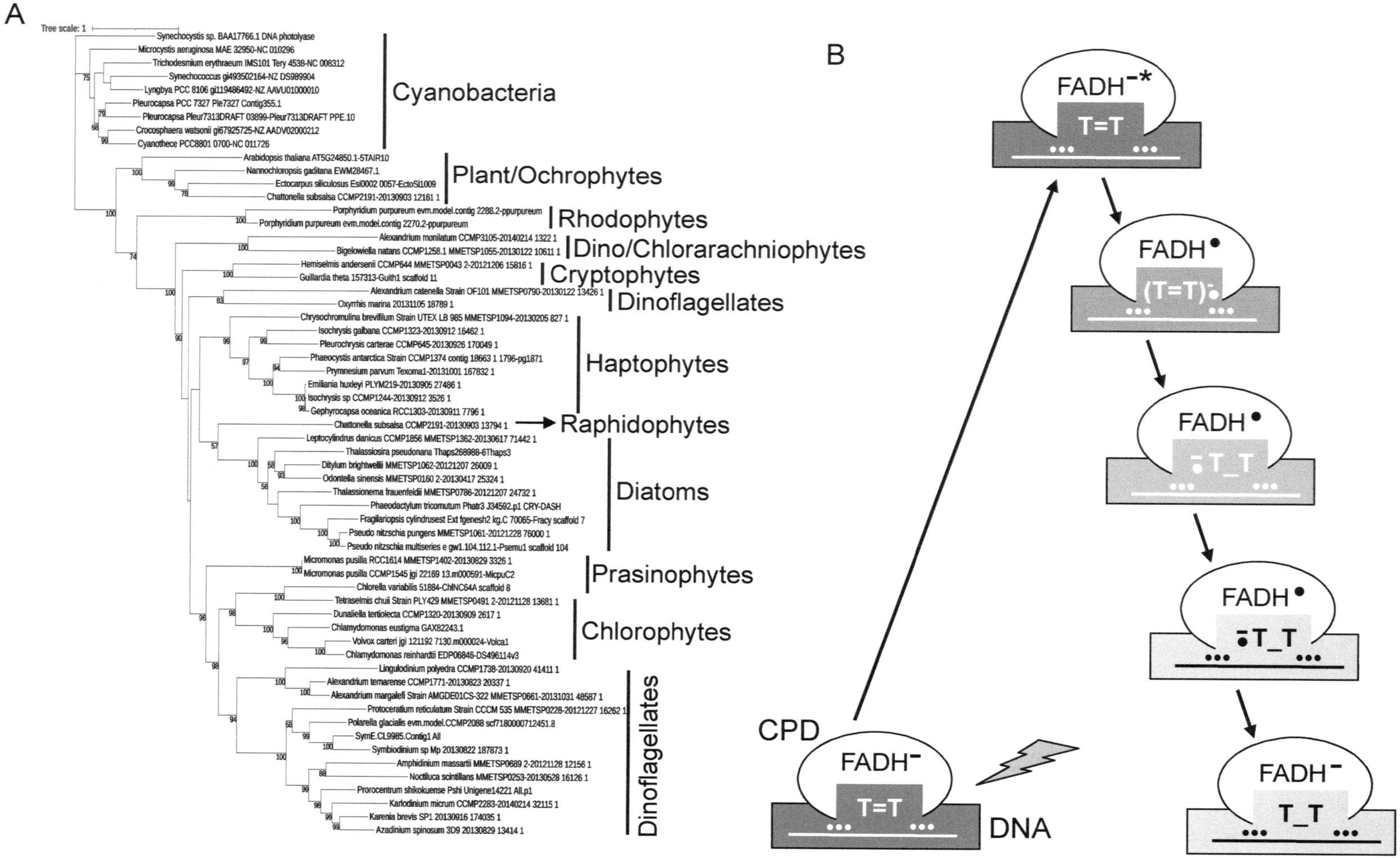

Figure 13.12. DNA Photolyase (PL) distribution in algae (A) and hypothetical procedure of photorepair of cyclobutene pyrimidine dimers (CPDs) by PL based on model organisms (B).

the surface (Lehmann et al. 2022). Meanwhile, freshwater salinization occurs globally too (Kaushal et al. 2018). These changes are likely to exert impacts on the growth, toxin production, and osmotic homeostasis of algae (Brown et al. 2006, Errera et al. 2010, Ochsenkühn et al. 2017).

Early forms of cyanobacteria (*Gloeobacter*) and Archaeplastida lineages (*Cyanidioschyzon*) appear to inhabit low-salinity (freshwater) environments (Sánchez-Baracaldo et al. 2017). The subsequent radiation of these lineages into the ocean represents a radical habitat shift. That evolutionary history might have afforded contemporary algae the plasticity to adapt to salinity variability, which is commonplace in the coastal ocean. Some algae or their heterotrophic relatives are euryhaline while others are stenohaline. For instance, the dinoflagellate *Pfiesteria piscicida* tolerates salinities from nearly 0 to ~ 30 (Lin and Zhang 2003). The marine chlorophyte *Chlamydomonas pulsatilla* was shown to grow best at 10% artificial seawater and its growth rate declined to 69% in 200% salinity (Shetty et al. 2019).

Roles of Aquaporins

In humans, AQP1 and AQP5 are involved in osmotic regulation, both being upregulated in intervertebral disk under hyperosmotic condition (Snuggs et al. 2021). In mouse oocytes, hyperosmosis generated by high concentrations of cryoprotectants such as DMSO, ethylene glycol and sucrose induced significant upregulation of AQP7, a response blocked by inhibition of the PI3K and PKC pathways (Tan et al. 2015). This suggests that the AQP7-based hyperosmotic response is mediated by PI3K and PKC pathways. AQPs in plants are classified into five groups, one of which is named Tonoplast Intrinsic Proteins (TIPs). TIPs are the most abundant in vacuolar membranes but also have important roles in cellular membranes (Maurel et al. 2008). BLAST analysis using plant TIP sequence as query yielded several significant hits in symbiodiniacean dinoflagellates (e.g., Fkaw21179 from *Fugacium kawagutii*). Their functions remain unclear.

Metabolic Reprograming in Response to Salinity Changes

A recent study based on a decade-long monitoring data indicated the increase of harmful algal blooms by the dinoflagellate *Karenia mikimotoi* in a Chinese coastal area. Physiological and omic investigation of an isolated strain showed a significant increase in photosynthetic efficiency and capacity, growth rate, and cellular protein content under lower salinity (25 [LoS] versus 33), along with a remarkable shift of metabolic pathway (Shi et al. 2024). Specifically, the transcriptomics data showed that LoS activated the glyoxylate shunt to bypass the decarboxylation reaction in the tricarboxylic acid cycle, hence diverting carbon flow from CO_2 generation to biosynthesis. This metabolic shift manifested in the enrichment of the Cysteine and Methionine Metabolism pathway by upregulated genes under LoS such as cysteine synthase, aspartate aminotransferase, and thiosulfate/3-mercaptophyruvate sulfurtransferase. This finding was corroborated by metabolomic data indicating a significant increase in S-adenosyl-L-methionine and 19 other metabolites. The omics data in concert showed that SAM and N-acetylputrescine were dominant metabolites in *K. mikimotoi* under LoS. Furthermore, the enhanced glyoxylate cycle could promote hydrogen peroxide metabolism, reducing reactive oxygen species (ROS). Consistent with this potential, experimental data showed a remarkable decrease under LoS in H_2O_2 and malondialdehyde (MDA), a lipid peroxidation marker produced by ROS attacks on polyunsaturated fatty acids. The findings in this study suggest that salinity declines can reprogram metabolism to enhance cell proliferation, potentially promoting bloom formation in HAB species like *K. mikimotoi*. Notably, *K. mikimotoi*'s global habitats range in salinity from 20.7 to 35.7 and the most common salinity for the strain used in this study is 33, suggesting that the species in the natural habitat frequently experience high-salinity stress and growth inhibition.

Desiccation Stress

Desert and intertidal algal lineages arguably represent the most desiccation-resistant algae. A transcriptomic study compared two desert (desiccation-tolerant) species and an aquatic (non-desiccation-tolerant) species from the family Senecesmaceae for their molecular responses to desiccation (Peredo and Cardon 2020). Results showed that both groups of the green algae upregulated desiccation protective genes such as those involved in photosystem repair, high light stress response, production of protective pigments, and metabolism of neutral lipids and triglyceride metabolism. Surprisingly, what defined the difference between the desiccation-tolerant species and the intolerant groups lie in the downregulation of a set of genes in the tolerant species. These included genes involved in 32 processes, 50% of which were upregulated in the aquatic species exposed to desiccation, and genes involved in photosynthesis, starch metabolism, and generation of precursor metabolites and energy. This finding suggests that the desiccation tolerance in the two desert species stems from their genetic capacity to reprogram their metabolic circuitry, thereby tuning down multiple metabolic processes in response to desiccation to save energy for coping with the desiccation stress.

A genomic study of Zygnematophyceae species (*Spirogloea muscicola* gen. nov. and *Mesotaenium endlicherianum*) revealed evidence that genes that promote resistance to desiccation and other stresses in land plants, such as GRAS andPYR/PYL/RCAR, are originated or expanded in the common ancestor of Zygnematophyceae and embryophytes, and they might have been acquired from soil bacteria through horizontal gene transfer (Cheng et al. 2019).

Heavy Metal Pollution

Heavy metals, inorganic substances, hydrocarbons, alcohols, and sterols are some of the common pollutants in the environment. Some of these are waste products, others used as fuels, pesticides, industrial reagents for drugs. Of these pollutant, heavy metals are the most widespread and diverse, including Cu, Zn, Pb, Cd, Cr, Hg, As, Ni, and Co (Xiao et al. 2023). The detrimental effects of heavy metals on algae involve physiological and biochemical processes, including inhibiting growth, limiting photosynthesis, altering cell ultrastructure, reducing protein content, impacting fatty acid composition, and inducing the production of ROS.

Copper

Plastocyanin and cytochrome c6, abundant proteins in photosynthesis, are indicators of cellular copper status in various algae (Merchant et al. 2020). Their accumulation is regulated by a transcription factor known as the copper response regulator (CRR1). Substituting the copper-containing plastocyanin with heme-containing cytochrome c6 conserves copper, allowing for the prioritized allocation of copper to cytochrome oxidase. In copper-rich conditions, the quota is determined by the abundance of different cuproproteins and is rigorously controlled, except in cases of zinc deficiency, where acidocalcisomes excessively accumulate Cu(I). CRR1 possesses a transcriptional activation domain, a zinc-independent DNA binding SBP-domain with a nuclear localization signal, and a C-terminal Cys-rich region that suppresses the zinc regulon. In *C. reinhardtii*, CRR1 activates over 60 genes through GTAC-containing CuREs, including genes that encode assimilatory copper transporters of the CTR/SLC31 family (Merchant et al. 2020).

Methylmercury

Mercury is toxic to humans exposed to it, primarily through consuming marine and estuarine fish. Mercury occurs in various chemical forms, but methylmercury is particularly concerning because it is the dominant form found in fish and is most efficiently taken up by organisms and bioaccumulated in the food chain. Algae are potentially the primary point of entry of MeHg into the food chain. Mercury in the environment is methylated by microbes, and the resulting methylmercury (MeHg) is transferred in the food chain via the "microbial loop" and is accumulated in various organisms. Some of the MeHg is released into the water, which can be taken up by algae leading to food chain transfer and accumulation. In the environment, MeHg forms complexes with organic ligands. The molecular mechanisms facilitating the uptake of MeHg by algae are not well understood. It is thought that small complexes of MeHg might enter the cells by diffusion. For large MeHg complexes, uptake may be mediated by the exchange of the complexing ligand. More specifically, this involves the replacement of the ligand bound to MeHg in the water by a ligand on the surface of the algal cell (Skrobonja et al. 2019).

Interactive Effects of Multiple Stressors

In their natural environments, algae are often subjected simultaneously or within a short space of time to more than one kind of stress. For instance, in the intertidal zone, seaweeds may undergo changes in mechanical stresses, aqueous to aerial environment, desiccation, intensity of both visible and UV radiation, temperature, and salinity, within one tidal cycle. Another example is in polar seas where algae incorporated in the ice are subjected to mechanical stress, decrease in temperature, and increase in salinity as brine is extruded during freezing.

The action of multiple stressors may have synergistic effects, i.e., the effect on the alga is greater than the sum of the effects produced by the two stresses applied separately. For instance, in *Scenedesmus acutus*, phosphorus deficiency sufficient to cause partial inhibition of photosynthesis renders this alga more sensitive to the toxic effect of copper. The brown kelp, *Laminaria saccharina*, subjected to a combination of heat and nitrogen deficiency, suffers more devastating effects, through impacts on Rubisco, RCII density, Fv/Fm, PSU size, and G3PDH), than when either stress is applied alone (Gerard 1997). On the contrary, nitrogen sufficiency enables algae to tolerate heat stress better (Gerard 1997, Fernández et al. 2020).

Apoptosis and Other Stress-Induced Death

In phytoplankton, numerous studies have highlighted that abiotic (e.g., osmotic shock, UV exposure, salt, heat, and nutrient deprivation) or biotic stressors (e.g., senescence, buildup of ROS, viral infection, and allelopathy) can induce apoptosis (a.k.a Programmed Cell Death or PCD)-like syndrome (Biddle 2016). The syndrome manifests as cell shrinkage, blebbing, chromatin condensation, and nuclear DNA fragmentation, culminating in cell autolysis. Besides PCD, several other types of cell death occur. One of this is necrosis caused by acute and irreversible cellular injury, which leads to general swelling of the plasma membrane, chromatin clustering and DNA strand breaks. Another is ferroptosis, an iron-dependent cell death. Excitotoxicity is glutamate-induced death, similar to but distinct from apoptosis. While PCD usually involves enzymes called caspase-like proteases as well as Ca^{2+} (Berges and Falkowski 1998), the signaling pathway that transmits the external stress into the cell and transduces it all the way to the execution of PCD is still not well understood.

In the coccolithophore *Emiliania huxleyi*, infection by EhVs induces production of bioactive viral-glycosphingolipids, which trigger algal PCD. In contrast, infection by the roseobacter *Phaeobacter inhibens* induces a highly specific type of PCD called apoptosis-like-PCD (AL-PCD). In this case, the presence of a pan-caspase inhibitor, which prevents the activation of caspase-like molecules, can abolish *E. huxleyi* cell death (Bramucci and Case 2019). This demonstrates that different death mechanisms may operate depending on the type of stressor.

Ferroptosis-inducing factors can directly or indirectly affect glutathione peroxidase, resulting in a decrease in antioxidant capacity and accumulation of lipid reactive oxygen species (ROS) in cells, ultimately leading to oxidative cell death (Li et al. 2020).

14

Temperature and Global Warming

Global Warming and General Effects on Algae

Mounting evidence shows that human activities, primarily fossil fuel combustion, are driving rapid changes in global climate (Fig. 14.1) and ocean temperature (Hoegh-Guldberg and Bruno 2010). Contemporary human CO_2 emissions to the atmosphere of approximately 10 billion metric tons of carbon per year result in an increase in atmospheric CO_2 of approximately 2 ppm/year or 0.5% per year. This rate of increase is nearly two orders of magnitude faster than what occurred during the large glacial-interglacial transitions and the present-day CO_2 level (~ 410 ppm, compared to the ~ 280 ppm preindustrial level) has not been experienced by life on Earth for several million years (Doney et al. 2020). Based on the current rate of increase, CO_2 concentration is projected to reach 750-1000 ppm at the end of the century.

As with any organisms, temperature changes affect the growth of algae. Different species of algae may have distinct reaction norms with temperature (also known as thermal performance curve), with different optimal temperature for growth and different temperature niche breadth (Fig. 14.2). Temperature changes may also cause cell death. For instance, the diatom *Phaeodactylum tricornutum*, was found to show temperature-independent mortality within its niche breadth, but mortality increased significantly at temperatures that approach or exceed its upper thermal tolerance (Baker and Geider 2021).

Temperature effects can occur at different scales of organization (from molecular to community), as elaborated in Chapter 13, and can be direct or indirect. The direct effect comes from temperature's influence on molecular kinetic energy (i.e., Maxwell-Boltzman energy distribution), which dictates the rate of fundamental processes such as enzyme reactions, diffusion, and membrane transport. Indirect effects of temperature on algae, particularly phytoplankton, stem from the intensification of water stratification that depresses the upwelling of nutrient-rich deep waters. As a result, a phytoplankton community faces the combined effects of heat and nutrient deficiency.

Warming may cause shifts in species composition in two major ways. First, small-sized phytoplankters are advantaged in warmer waters. This is not because size directly influences the capacity of the cells to tolerate or adapt to temperature rise (Huertas et al. 2011), but due to a higher cell surface area to volume ratio promoting nutrient uptake at low concentrations. Second, genetically more thermotolerant species may increase their dominance in the community due to greater thermal plasticity or genetic adaptation to higher temperatures (Daufresne et al. 2009, Moran et al. 2010, Huertas et al. 2011, Qu et al. 2022). At the community level, warming intensifies predation, promoting blooms of carnivoreous zooplankton such as jellyfish (Oviatt 2004), which emerged widely and rapidly in the global coastal oceans. The consequent decline in herbivorous zooplankton may lead to increased algal blooms.

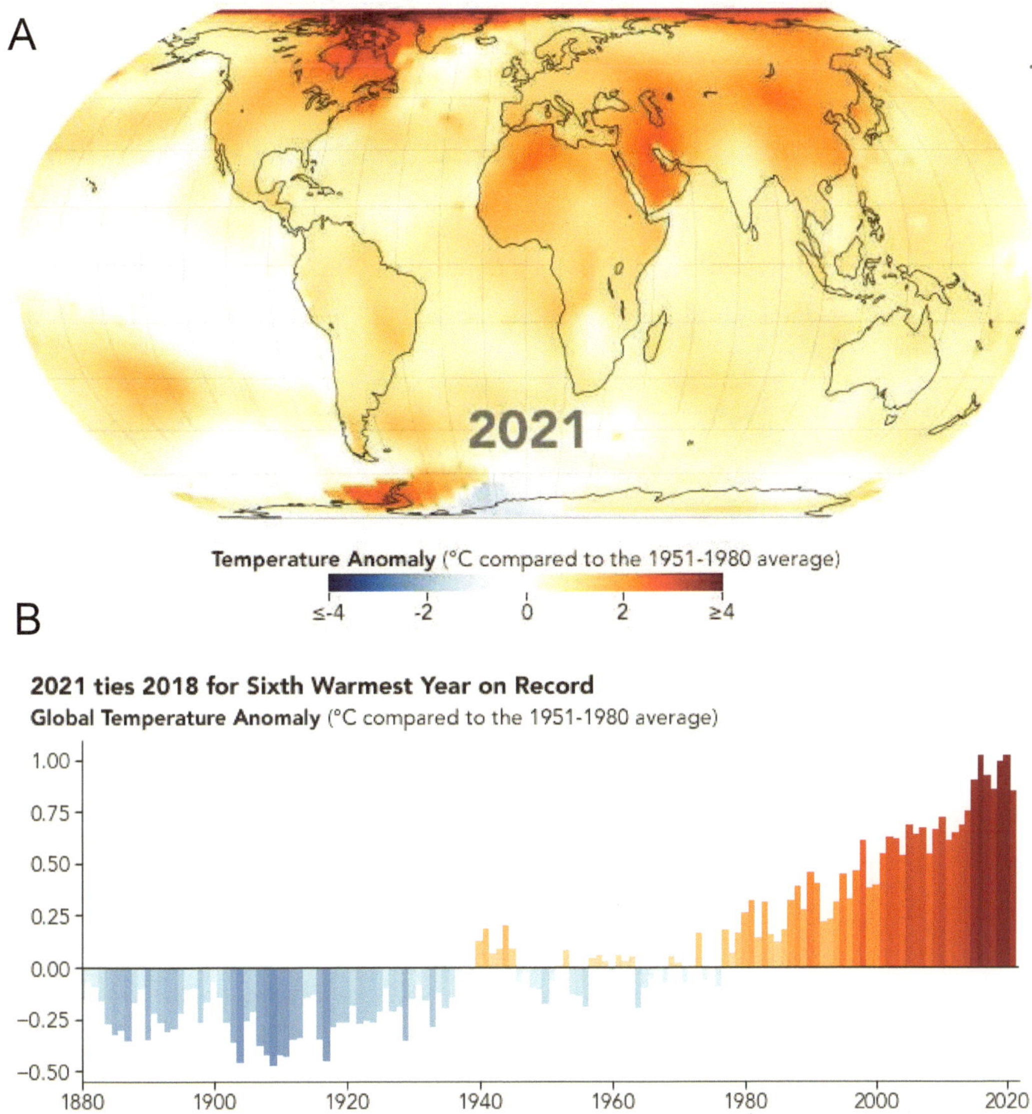

Figure 14.1. Warming trend of global climate. (A) Global map of temperature anomaly in 2021 relative to the average of 1951–1980. Note the very high increase in the Arctic and some areas of the Antarctic. (B) Temperature anomaly since 1880 until 2020 relative to the average of 1951–1980. From NASA Earth Observatory.

Impacts of Temperature Changes on Phytoplankton Population Dynamics

The physical condition in the ocean varies most in temperate regions, followed by polar regions, and least in tropical and subtropical regions. In temperate regions, phytoplankton abundance and chlorophyll concentration peak in spring/winter and autumn, with low biomass periods in between. This seasonality results from variations in nutrients and light. In summer, high sea surface temperatures cause stratification, reducing nutrient upwelling and phytoplankton growth. In winter, turbulent mixing brings nutrients up, but low light limits growth. Spring features calm seas and increasing light, while autumn sees nutrient delivery from depths and adequate light. In polar regions, phytoplankton growth is mainly limited by light, with a single growing season in summer. In tropical regions, stable light, temperature, and nutrient conditions keep phytoplankton levels stable but low.

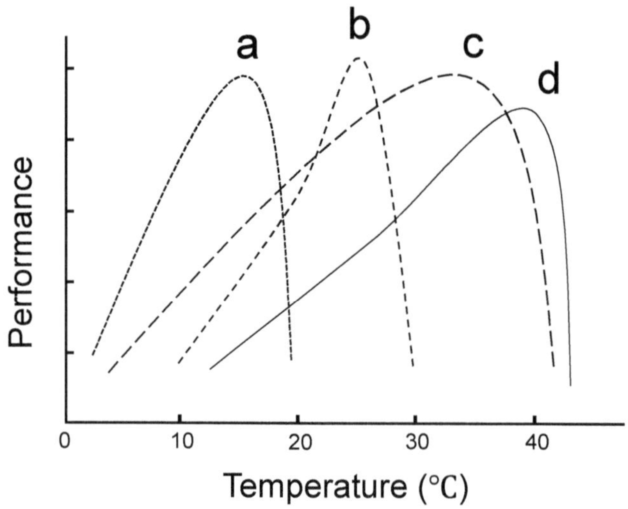

Figure 14.2. Differential temperature reaction norm in different species of algae. Species a has the lowest optimal temperature (T_{opt}), whereas species d has the highest T_{opt}, and species c has the widest thermal niche breadth.

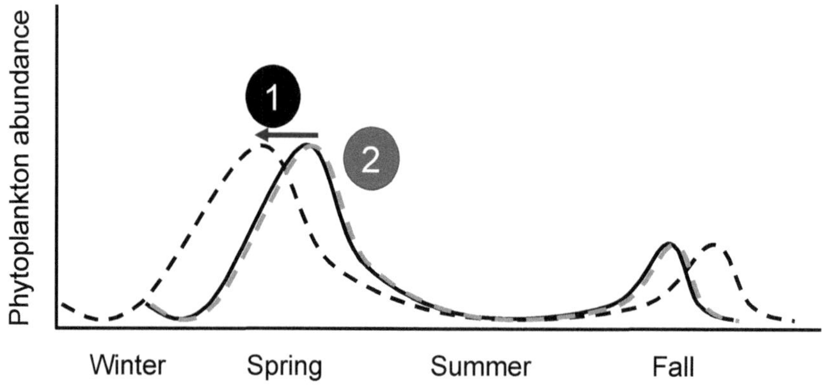

Figure 14.3. Hypothetical scenarios of phytoplankton population responses to temperature rises. Scenario 1, warmer temperature causes earlier spring bloom and later fall bloom without shift dominant species. Scenario 2, warmer temperature causes shift of dominant species to a high temperature-adapted species without causing phenological shift.

Global warming affects phytoplankton differently across regions (Behrenfeld et al 2006). In temperate regions, warming is expected to shift spring blooms earlier (Fig. 14.3). Global warming-induced temperature elevation may cause shifts in dominant species without changing overall phytoplankton abundance (Fig. 14.3). In polar regions, warming may boost primary production. However, a modeling study using methanesulfonic acid from GrIS ice cores showed a decline in subarctic Atlantic productivity over two centuries (Osman et al 2019). In tropical regions, rising temperatures will likely decrease primary productivity and abundance but may also change the timing, strength, and duration of phytoplankton peaks. For instance, in the northern Red Sea, temperature anomalies delay and weaken phytoplankton peaks, shortening their duration (Gittings et al 2018). Exceptions to this trend can occur, such as in parts of the Red Sea, where the interaction of the Arabian monsoon with prolonged positive phases of the multivariate El Niño-Southern Oscillation (ENSO) Index can promote horizontal nutrient transport from the Indian Ocean, elevating phytoplankton biomass (Raitsos et al 2015).

Impacts of Warming on Phytoplankton Species Diversity

Biodiversity results from the relative growth and competition of coexisting species. Ocean warming's effects on the seasonality and composition of algal communities have been widely reported. Using several models, Benedetti and colleagues (2021) indicated that global warming will significantly restructure marine plankton assemblages, increasing species richness by over 16% in most regions except the Arctic Ocean. The mechanism underlying this overall biodiversity increase was not explained.

Thomas et al. (2012) investigated temperature's direct effect on phytoplankton by fitting published data to temperature reaction norms for 194 phytoplankton strains from a wide latitudinal range. The resulting temperature-related traits (maximum growth rate, temperature optimum, and thermal niche width) revealed latitudinal trends in temperature optima and diversity. Biogeographical differences indicate increased susceptibility of tropical strains to further warming, with modeling predicting poleward shifts of tropical strains and a loss of phytoplankton diversity in the tropics within the next hundred years.

Experiments with ambient and elevated temperatures have documented species composition changes caused by warming. In a study using mesocosm set up in artificial ponds, warming's effect on phytoplankton species sorting and local adaptation within species (increasing C:P and N:P ratios while keeping the C:N ratio unchanged) led to seston stoichiometry changes, indicating warming's ripple effect on biogeochemistry through phytoplankton (Yvon-Durocher et al. 2017).

Temperature Effects on Metabolism and Growth

At the metabolic or cellular level, given adequate nutrients, higher temperatures within a species' thermal tolerance limit cause higher growth rates. Here, the carbon resource (photosynthate) available for growth is the difference between production by photosynthesis and consumption in various metabolic processes. Resource allocation to growth varies with temperature because enzymatic reactions and maintenance expenses for membrane structure are temperature dependent. For instance, at low temperatures, protein translation efficiency at the ribosomes is low, forcing algal cells to produce more ribosomes to compensate. This temperature effect may vary between genes, causing differential responses among different genes to temperature changes. For example, in the fission yeast *Schizosachromyces cerevisiae*, translation of mRNAs containing an AUG start codon upstream of the main ORF shows marked increases in efficiency at elevated temperatures compared to when UUG, ACG, or AUU is used as the start codon in the upstream ORF (Kulkarni et al. 2019).

Elevated expression of ribosomal RNA and a higher RNA/protein ratio at low temperatures have been demonstrated in phytoplankton in the polar ocean (Toseland et al. 2013, Pearson et al. 2015). Besides, at cold temperatures, cells shift lipid composition to increase nonsaturated fatty acids. In two strains of the diatom *Chaetoceros* sp. grown at different temperatures (6–30°C), transcriptomic data show that genes in the acetyl-CoA associated pathways, including lipid and branched-chain amino acid metabolism, balance photosynthetic light capture and metabolism with changes in growth temperature (Liang et al. 2019).

As respiration increases faster than photosynthesis with temperature, the amount of carbon available for growth (or carbon use efficiency) decreases as temperature increases (Fig. 14.4). When the photosynthetic rate is offset by the respiration rate, no growth is expected, but the population can potentially remain stable for an extended period, as observed for the polar dinoflagellate *Polarella glacialis*. *P. glacialis* shows noticeable growth at 4°C but a population decline at 10°C or 15°C. After about two weeks of acclimation, the cultures at both elevated temperatures remained stable (Zheng et al. 2012). During this population maintenance, Rubisco (Ribulose 1,5-bisphosphate carboxylase/oxygenase, a key enzyme in the CBB cycle) was degraded more quickly at 15°C than at

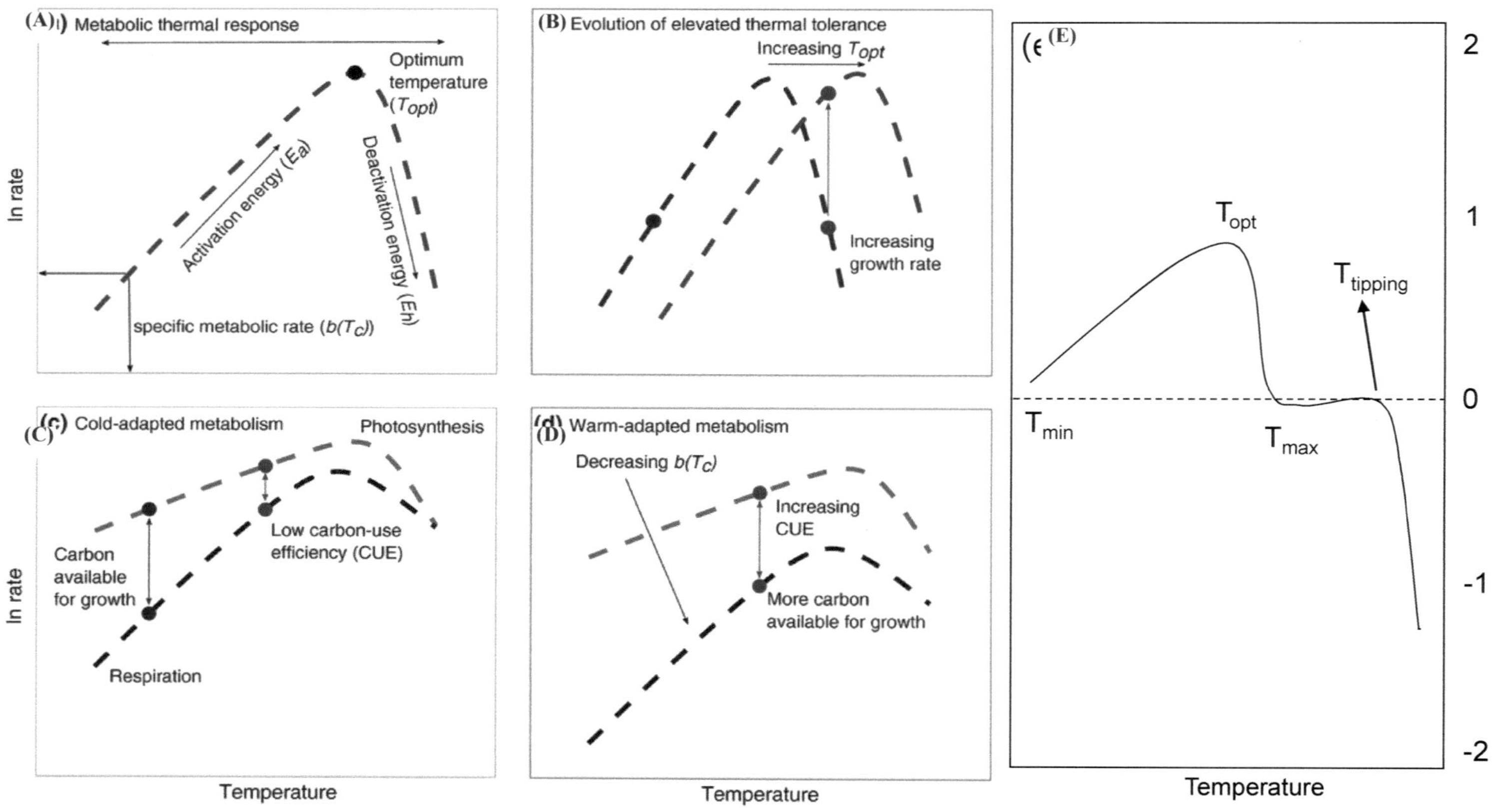

Figure 14.4. Effects of temperature on phytoplankton metabolism. (A) Metabolic rates, $b(T)$, increase as an exponential function of temperature to an 'optimum' temperature (T_{opt}), then decline rapidly. (B) Growth responses for a cold-adapted (point & line on the left) and warm-adapted (point & line on the right) phytoplankton. Evolution of elevated thermal tolerance entails increases in T_{opt} and growth rates at elevated temperatures. (C) Thermal responses for photosynthesis (top line, P) and respiration (bottom line, R) for a cold-adapted phytoplankton. The relative difference between P and R along the temperature gradient represents the carbon use efficiency (CUE = $1-R/P$). Because R is more sensitive to temperature than P, CUE declines (left to right vertical lines) with increasing temperature, limiting growth at high temperatures. As phytoplankton adapt to warmer temperature (D), selection should drive down specific rates of respiration, $R(T_c)$, more than those of photosynthesis, $P(T_c)$, increasing CUE and ensuring they are not limiting for growth. Reproduced from Padfield et al. (2015) under the Creative Commons Attribution License CC BY 4.0. (E) As temperature rises, growth rate increases, then decreases reaching 0. Beyond the tipping point, molecular machinery breaks down due to protein denaturation, giving negative growth.

10°C. Similarly, PCNA (Proliferating Cell Nuclear Antigen), necessary for DNA synthesis, showed denaturation and degradation at both elevated temperatures, more quickly at 15°C than at 10°C.

As temperature further increases, resources produced by photosynthesis are redistributed among respiration (basic metabolism), stress response (e.g., ROS scavenging), and damage repair. There is a tipping point beyond which molecular machinery breaks down due to irreversible protein denaturation, leading to cell death and population decline (Fig. 14.4).

A recent study investigated proteomic and metabolomic changes in the bloom-forming species *Prorocentrum shikokuense* (=*P. donghaiense*) between four temperatures, including the optimal 25°C, the supra-optimal 28°C, and the sub-optimal 19 and 22°C (Zhang et al. 2022). The authors observed decreases in Particulate Organic Carbon and Nitrogen (POC, PON) but increases in the POC/PON ratio as temperature increased, except at 28°C. These results are consistent with a higher cell division rate and photosynthetic carbon fixation rate at higher temperatures. At the supra-optimal temperature, however, cells increased the production of light harvesting, photoreaction, and protein homeostasis. The sub-optimal temperature induced upregulation of glutathione (antioxidant), transcription, and lipid biosynthesis, compensating for the decrease in translation efficiency and cell membrane fluidity. At 19°C, there was an increase in nitrate reduction and nitrogen flux towards asparagine, aspartic acid, and glutamine, accompanied by the accumulation of glutathione, glutarate-semialdehyde, and 5-KETE. Although not reported in the paper, the metabolomic dataset shows a remarkable increase in a sulfinpyrazone-like metabolite at 19°C relative to 25°C (13-fold) and 28°C (26-fold). Sulfinpyrazone (C23H20N2O3S) is a medicine for treating gout as it inhibits reabsorption (transport) of uric acid and enhances urinary excretion of uric acid (Margulies et al. 1980). In *P. shikokuense*, this sulfur- and nitrogen-containing compound may block intracellular cross-membrane trafficking of organic N molecules as part of the N-flux diversion at lower temperatures or simply accumulate as an intermediate metabolite at lower temperatures.

Temperature increase can change the dynamics of mutualism between a host and its symbiont. Symbiosis will be elaborated on in Chapter 19, but a brief discussion is fitting here. Rising temperatures, accompanied by abundant nutrients, may promote the photosynthesis, nitrogen assimilation, and cell division of photosymbionts. The increased cell proliferation, along with elevated respiration at higher temperatures, will lead to a decrease in photosynthate translocation to the coral host, rendering the photosymbionts functionally parasitic (Baker et al. 2018).

Besides the general metabolic processes influenced by temperature changes, one important metabolic component is the ribosome, the protein translation factory. Due to higher translation efficiency, higher temperatures lower the requirement of ribosomes in the cell, hence lower the content of rRNA and ribosomal proteins (Toseland et al. 2013). This might free up resources for growth, partially explaining higher growth rates at higher temperatures. Consequently, species adapted to cold temperatures are expected to synthesize a higher amount of ribosomal RNA and proteins through the upregulation of both transcription and translation machinery (Pearson et al. 2015). As a result, higher temperatures cause a significant increase in the N:P ratio of cellular organic matter (Toseland et al. 2013). This stoichiometric shift will have a stronger biogeochemical effect in polar ecosystems than in tropical ecosystems because resource allocation to biosynthesis (ribosome-mediated) is much higher in cold habitats. By the same token, as enzyme activities are lower at lower temperatures, it is no surprise that Rubisco synthesis in Antarctic phytoplankton blooms is upregulated relative to warmer water species (Young et al. 2015).

Physiological Plasticity and Adaptation to Warming

By reanalyzing global data, Chen (2015) found that the optimal, minimal, and maximal growth temperatures of marine phytoplankton decrease with increasing latitudes. This trend is less pronounced for freshwater lineages. Additionally, the thermal breadth of marine phytoplankton is narrower compared to freshwater phytoplankton, reflecting adaptation to more variable and extreme temperatures in freshwater ecosystems. Contrary to Jenzen's rule, thermal breadth does not increase with latitude. For most phytoplankton, minimal growth temperatures are lower than their environmental temperatures, and maximal growth temperatures exceed their habitat temperatures.

In marine phytoplankton, there is a strong phylogenetic signal in minimal growth temperature and thermal breadth. Cyanobacteria and dinoflagellates have higher minimal growth temperatures and narrower thermal breadths than diatoms. This aligns with the understanding that diatoms prefer colder, nutrient-rich environments, while dinoflagellates thrive in warmer, nutrient-poor environments. Cyanobacteria are prominent in tropical and oligotrophic areas for their role in nitrogen fixation.

Phytoplankton, with their short life cycles, have high potential to adapt to heat stress through genetic mutations. When temperatures exceed physiological limits, survival depends on selecting for beneficial mutations or emerging new mutations induced by heat stress. An experimental evolution study (Huertas et al. 2011) with 16 species of eukaryotic phytoplankton showed varied success in adapting to increased temperatures, indicating differences between taxa in the potential to evolve new genotypes under temperature selection. Another study found that the dinoflagellate *Alexandrium ostenfeldii* exhibited plastic responses rather than genetic adaptation to elevated temperatures (Jemey et al. 2019).

Experimental work on cyanobacteria revealed contrasting strategies between species to cope with long-term thermal regime shifts (Qu et al. 2022). *Trichodesmium* and *Crocosphaera*, filamentous and single-celled diazotrophic cyanobacteria, were grown at optimal (28°C), supraoptimal (32°C), and suboptimal (22°C) temperatures for several hundred generations. *Trichodesmium* exhibited physiological plasticity to temperature changes, while *Crocosphaera* showed genetic adaptation to selection temperatures. These different responses can have significant ecological implications. For instance, when exposed to further elevated temperatures (34°C), nitrogen fixation in both species initially declined. Subsequently, all *Trichodesmium* cell lines acclimated to this high temperature, while only the 32°C-selected cell line of *Crocosphaera* survived and fixed nitrogen at this temperature.

A recent study on the chlorophyte *Symbiochlorum hainanensis*, isolated from the coral *Porites lutea*, showed that this symbiotic chlorophyte is heat resistant, surviving up to 38°C. Transcriptomic analyses indicated that thiamine triphosphate (TTP) abundance and gene expression, along with antioxidase genes, increased at elevated temperatures (36°C and 38°C relative to 26°C), suggesting TTP's role in thermal tolerance (Xiao et al. 2023). While the exact role of TTP remains to be verified, it might act as an antioxidant to eliminate Reactive Oxygen Species (ROS) induced by heat stress, preventing oxidative damage to cells. Additionally, selenate reduction pathway genes were upregulated under heat stress, suggesting that selenoproteins might also contribute to thermal tolerance. The antioxidative enzyme glutathione peroxidase, the first discovered selenoprotein, contains selenium in selenocysteine. This 21st naturally occurring amino acid is encoded by the UGA codon, usually a stop codon. Incorporation of selenocysteine into proteins requires specialized machinery, including a selenocysteine tRNA, a secondary stem-loop structure named selenocysteine insertion sequence (SECIS), SECIS binding protein 2 (SBP2), and other protein factors, as demonstrated in the human system (Bellinger et al. 2009).

In summary, the adaptation of phytoplankton to temperature changes is influenced by their thermal tolerance limits, genetic makeup, and ability to undergo physiological and genetic changes. These adaptations vary significantly among different species and ecosystems, with marine phytoplankton showing distinct phylogenetic signals in their thermal responses. Understanding these mechanisms is crucial for predicting how phytoplankton populations will respond to global warming and its associated impacts on marine and freshwater ecosystems.

Heat Perception and Signaling

For algae as well as other organisms, an accurate assessment of rising ambient temperatures is crucial for the timely activation of molecular defense mechanisms before heat damage occurs.

However, little is known about how algal cells sense environmental heat and the signaling cascade responsible for transducing the heat stress signal, ultimately triggering gene expression. Fortunately, plant research, which has advanced significantly, offers valuable insights. Land plants experience greater temperature fluctuations than algae in aquatic ecosystems. Unable to escape unfavorable temperatures, land plants must anticipate damaging conditions early to express genes and accumulate Heat Shock Proteins (HSPs) for cellular defense against heat damage (Saidi et al. 2010).

The heat-induced gene network contributing to high-temperature response and thermotolerance is complex (Finka et al. 2011). Although much of the mechanism remains unclear, current data provide a general signaling pathway in plants (Fig. 14.5). As discussed in Chapter 12, sensing temperature and its changes may involve changes in cell membrane rigidity and transient receptor potentials. Specifically, a temperature increase induces membrane fluidity changes, activating Ca^{2+} channels and triggering a transient Ca^{2+} influx. Ca^{2+} activates Calmodulin (CaM) and promotes the phosphorylation of heat shock transcription factors (HSFs) via interaction with CaM-Binding protein Kinase (CBK) or Heat-Activated MAP Kinase (HAMK). Activated HSFs bind to HSP promoters, likely causing histone displacement, thus eliciting HSP expression and enhancing thermotolerance. Temperature increases and membrane state changes also elevate hydrogen peroxide (H_2O_2) levels, which might activate a similar pathway. Heat-induced Nitric Oxide (NO), cytoskeleton rearrangements, and protein denaturation also trigger HSF-dependent HSP expression.

Optimal growth temperatures vary among species, but beyond tolerance values, plants engage in a conserved response characterized by the upregulation of HSP genes (Finka et al. 2011). This response involves 0.5–1% of the total genome and is similar in animals, yeasts, and prokaryotes (Saidi et al. 2010). Notably, in the model diatoms *P. tricornutum* and *T. pseudonana*, HSFs are the

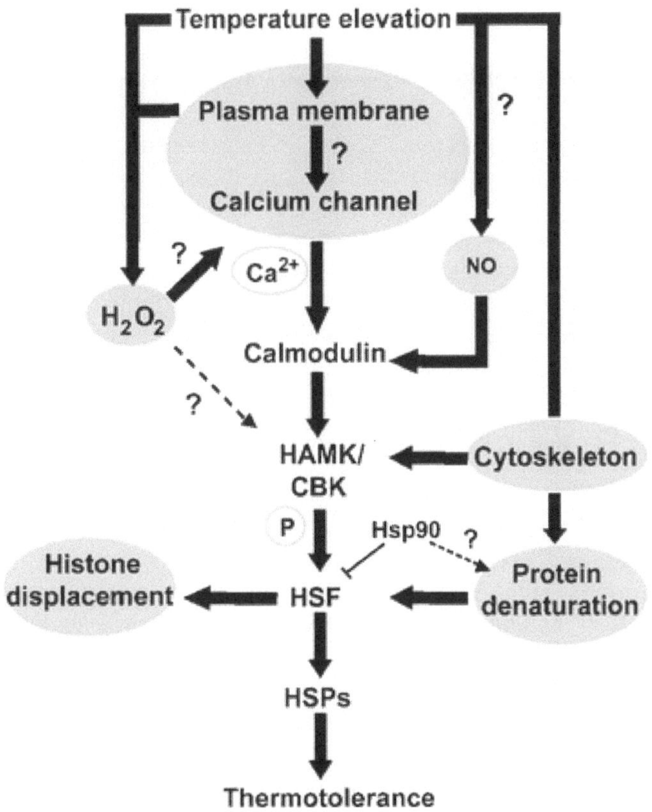

Figure 14.5. Schematic diagram of plant temperature signaling pathway involving heat-shock transcription factors (HSFs) and heat-shock proteins (HSPs). Reproduced with permission from Saidi et al. 2010.

most abundant class of transcription factors, reflecting an evolutionary adaptation to cope with thermal stress.

Recent studies on algae have provided additional insights into their heat stress responses. In the red alga *Neopyropia yezoensis* (= *Pyropia yezoensis*), both membrane-fluidization-dependent and -independent pathways are involved in heat-inducible gene expression (Khoa and Mikami 2022). The expression of genes encoding multiprotein-bridging factor 1 (NyMBF1) and HSP70s (NyHSP70-1 and NyHSP70-2) was induced by heat stress and membrane fluidization, while two homologs of the heat-inducible High Temperature Response 2 (HTR2) gene were induced by heat stress independently of membrane fluidization.

In the rhodophyte *Cyanidioschyzon merolae*, which inhabits acidic hot springs and resists hot temperatures up to 63°C, two small Heat Shock Protein (sHSP) genes play a crucial role in temperature sensing and survival (Kobayashi et al. 2014). The two sHSP genes are located side by side in an inverted repeat orientation on the same chromosome and sharing a promoter. Their expression is triggered by absolute temperature rather than temperature changes: below the threshold temperature (46–50°C), temperate rinses do not activate their expression. Once temperature exceeds the threshold, their expression is upregulated by over 1,000-fold in response to heat shock. Similar sHSP structural genes exist in the chlorophyte *Chlamydomonas reinhardtii*, but the threshold temperature to induce their expression (~ 36°C) differs from that for *C. merolae*.

These findings suggest that algae have evolved sophisticated mechanisms to sense and respond to temperature changes, which may differ from those observed in land plants. The ability to sense absolute temperatures and the presence of both membrane fluidization-dependent and independent signaling pathways highlight the sophistication of heat stress responses in algae. Further research is needed to fully elucidate these mechanisms and their potential applications in improving thermotolerance in both algae and land plants.

The Role of COP9 Signalosome in Temperature Signaling

The constitutive photomorphogenesis protein 9 (COP9) signalosome (CSN) is a conserved protein complex, typically consisting of eight subunits (CSN1 to CSN8) in higher eukaryotes such as plants and animals, but of fewer subunits in other eukaryotes such as yeasts (Qin et al. 2020). The CSN complex was originally identified in plants from a genetic screen for mutants that mimic light-induced photomorphogenic development when grown in the dark. Initially regarded as a regulator (repressor) of photomorphogenesis in *Arabidopsis* (Wei and Deng 1992), subsequent studies showed that CSN is also involved in many other cellular and developmental processes in various eukaryotes (Wei and Deng 2003). In accordance with its functional diversity, the architecture of the CSN complex varies remarkably with the organisms.

The primary function of CSN is to regulate protein degradation via the ubiquitin/26S proteasome pathway, by coordinating the activity of E3 ubiquitin ligase and the 26S proteasome in proteolysis (Qin et al. 2020). For instance, the CSN promotes deneddylation (removal of the ubiquitin-like conjugate Nedd8) of cullins, stabilizing them and recycling unstable, neddylated cullins into stable, unneddylated cullins, thus promoting cullin-organized E3 activity (Wu et al. 2005). Cullins are a family of structurally related proteins required for ubiquitin-dependent protein degradation and are named after their role in 'culling', sorting or 'selecting' cellular proteins for ubiquitin-mediated proteasomal degradation.

CSN5, also known as the Jun activating binding protein (JAB1), binds to $p27^{Kip1}$, an inhibitory regulator, and leads to cell cycle progression and the concomitant export of $p27^{Kip1}$ out of the nucleus and its degradation (Fig. 14.6). Binding of JAB1 to MIF allows the nuclear accumulation of $p27^{Kip1}$ and causes cell cycle arrest.

In *Arabidopsis* seedlings, the expression of cold-responsive genes in response to cold stress was delayed markedly in CSN knock-down transgenic plants, suggesting that CSN plays important roles in plant cold response (Schwechheimer et al. 2002). A proteomic study showed that after night-

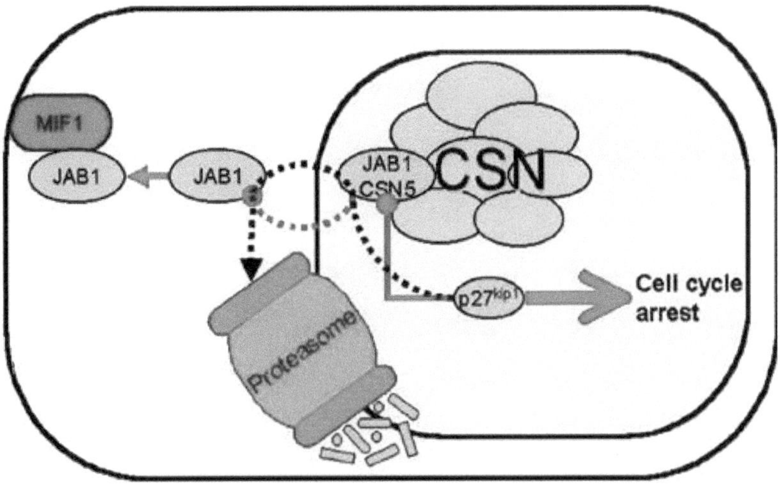

Figure 14.6. Model of JAB1/CSN5 function in p27^{Kip1} mediated cell cycle control. Thin lines ending with filled circles represent inhibitory interactions. Thin lines ending with closed arrow heads represent stimulatory interactions. Dotted lines show protein movement. Binding of p27^{Kip1} to JAB1 leads to cell cycle progression, and the concomitant export of p27^{Kip1} from the nucleus and its degradation. Binding of JAB1 to MIF allows the nuclear accumulation of p27^{Kip1} and cell cycle arrest. CSN, COP9 signalosome. Reproduced with permission from Chamovitz and Segal (2001) under the Creative Commons Public Domain Mark 1.0 license.

heat stress, the abundance of CSN3 increased significantly in heat-tolerant rice lines but decreased drastically in heat-sensitive rice lines at the early milky stage, suggesting that CSN may contribute to the heat tolerance of rice (Zhang et al. 2017). Similarly, a recent study on the diatom model *Phaeodactylum tricornutum* showed that CSN5- and CSN6-overexpressing transgenic cell lines exhibited higher temperature tolerance than the wild type, suggesting that CSN is involved in the regulation of temperature tolerance in this diatom (Qin et al. 2020).

CSN5 and CSN6 subunits are present in the genomes of various lineages of algae, including diatoms, dinoflagellates, chlorophytes, and rhodophytes (Figs. 14.7 and 14.8). These findings highlight the conservation of the COP9 signalosome components across diverse algal lineages, suggesting their fundamental importance in cellular processes. The presence of CSN5 and CSN6 in such a wide range of algal groups indicates that these proteins likely play crucial roles in protein degradation and cellular signaling pathways, similar to their functions in higher plants and animals. Interestingly, phylogenetic analyses show that CSN6 amino acid sequences cluster by taxonomic divisions, indicating that the evolution of this gene parallels that of algal species (Fig. 14.9). This pattern suggests that CSN6 has evolved alongside the diversification of algal species, potentially adapting to the specific needs of each lineage.

Genomic Adaptation to Temperature Niches

Genomic features that have been examined in the context of adaptation to habitat temperature ranges mainly include overall GC content, structural and functional stability of encoded proteins, and protein repertoire to prevent or mitigate temperature-induced damages. Few consensus features can be drawn from the algal genomes reported so far, but various unique characteristics are emerging.

Some algae are thermophilic. For instance, *Synechococcus lividus* can grow at temperatures as high as 73°C but stops growing when the temperature is lower than 54°C, with 60–63°C being its optimal temperature. Cells will die if kept at 30°C for 10 days (Meeks and Castenholz 1971, Lee 1999). Similarly, the unicellular rhodophytes *Cyanidioschyzon merolae* and *Galdieria sulphuraria* thrive in hot and acidic springs, with optimal growth temperatures at 42–50°C and 42°C, respectively.

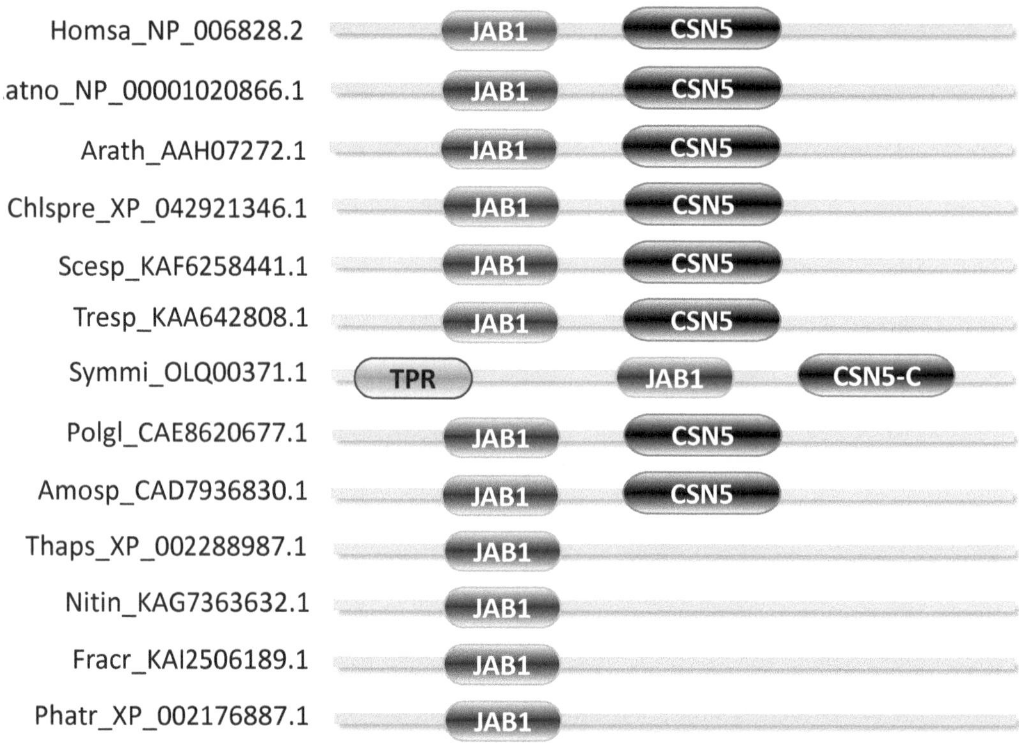

Figure 14.7. COP9 signalosome subunit 5 (CSN5) in dinoflagellates and diatoms in comparison to animals and plants. JAB1, Jun activating binding protein, which is another name of CSN5. TPR, Tetratricopeptide-like helical domain superfamily, also annotated as Intraflagellar transport protein 88 homolog. Homsa, *Homo sapiens*; Ratno, *Rattus norvegicus;* Arath, *Arabidopsis thaliania;* Chlre, *Chlamydomonas reinhardtii;* Scesp, *Scenedesmus* sp.; *Tresp, Trebouxia* sp.; Symmi, *Symbiodinium microadriaticum*; Polgl, *Polarella glacialis;* Amosp, *Amoebophrya* sp.; Thaps, *Thalassiosira pseudonana;* Nitin, *Nitzschina inconspicua;* Fracy, *Fragilaria crotonensis;* Phatr, *Phaeodactylum tricornutum*

A study analyzing 57 hot spring cyanobacterial Metagenome-Assembled Genomes (MAGs) along with mesophilic counterparts indicated that hot spring cyanobacteria had higher GC contents, smaller genomes, shorter and lower-molecular weight proteins, and more basic amino acids (histidine and arginine) and less acidic amino acids (aspartate and glutamate) than mesophilic cyanobacteria (Alcorta et al. 2020). However, the same is not observed in hot spring rhodophytes (Lee 1999). For example, GC contents are even lower in hot spring taxa shown above than intertidal marine red seaweeds (Table 6.1). Genome size reduction in hot spring cyanobacteria is not consistent with the notion that gene family expansion may provide versatility for environmental resource exploitation and hence adaptation (Chapter 4).

In the example of thermal adaptation described above, *Crocosphaera* cultures selected by the supraoptimal temperature 32°C performed better at high temperatures, indicating genetic adaptation (Qu et al. 2022). Genomic sequencing of both selected and non-selected (control) cultures indicated that the adaptive genome accumulated a significant number of single nucleotide mutations. These mutations impacted genes involved in important biological functions such as reverse transcriptase and bacterial plasmid partitioning machinery (ParMRC), both implicated in coping with environmental stress. This implies a possibility that those mutations have desensitized the strain.

A comparative genomic study revealed different patterns of transcription factor (TF) composition between thermophiles and phylogenetically related mesophilic reference cyanobacteria (Tang et al. 2024). The data indicate that thermophilic cyanobacteria contain more TFs putatively involved in various biological regulations, mainly for responses to ambient changes. For instance, the two

Figure 14.8. COP9 signalosome subunit 6 (CSN6) in dinoflagellates and diatoms in comparison to animals and plants. JMM/MPN, JAB1/MPN/MOV34 metalloenzyme domain. MitMem, maintenance of mitochondrial structure and function. Homsa, *Homo sapiens*; Lamsa, *Lamellibrachia satsuma* (Annelida); Dinoflagellates: Sympi, *Symbiodinium pilosum*; Symne, *S. necroappetens*; Polgl, *Polarella glacialis*. Diatoms: Thaps, *Thalassiosira pseudonana*; Nitin, *Nitzschia inconspicua*; Fracy, *Fragilariopsis cylindrus*; Phatr, *Phaeodactylum tricornutum*.

common response regulators, OmpR and NarL family, dominate the putative RRs identified in each genome. Dominant in the sigma factors identified were the Ecf and RpoE family, both of which were predicted to be extracytoplasmic. The RpoE potentially regulates genes protecting against photooxidative stress and may partially overlap with the heat-shock response, whereas the Ecf family may be involved in DNA repair response, cell wall stress, and amino acid starvation. Other potentially important TFs for thermophilic taxa are mainly involved in stress response to metal ions (ArsR) and Fe/Zn acquisition (Fur), adaptation and maintenance of homeostasis and cell protection (Rrf2). In addition, all thermophilic cyanobacterial genomes contain a SfsA transcription factor that is known to be involved in sugar fermentation, likely regulating energy generation. These mechanisms may facilitate survival of these thermophilic species in hot springs.

A study (Buerger et al. 2020) exposed 10 clonal strains of coral symbionts (Symbiodiniaceae) to elevated temperatures (31°C) for 4 years *in vitro*. As a result, all 10 heat-selected strains expanded their thermal tolerance *in vitro*. After reintroduction into coral host larvae, three of the 10 heat-selected endosymbionts increased the holobionts' bleaching tolerance. The tolerant symbionts exhibited higher constitutive expression of algal carbon fixation genes and coral heat tolerance genes.

In a study on the phaeophyte *Saccharina latissima* exposed to combinations of light intensities and temperatures, microarray analysis showed that 32% of genes responded to temperature variations, with high temperature eliciting stronger effects than low temperature (Heinrich et al. 2012). The gene expression changes included adjustments in primary metabolism and the induction of several ROS scavengers, as well as sophisticated regulation of heat shock proteins. In another study on the eelgrass *Zostera marina*, southern and northern populations were exposed to the same elevated temperature (Franssen et al. 2011). During high-temperature exposure, the transcriptomic

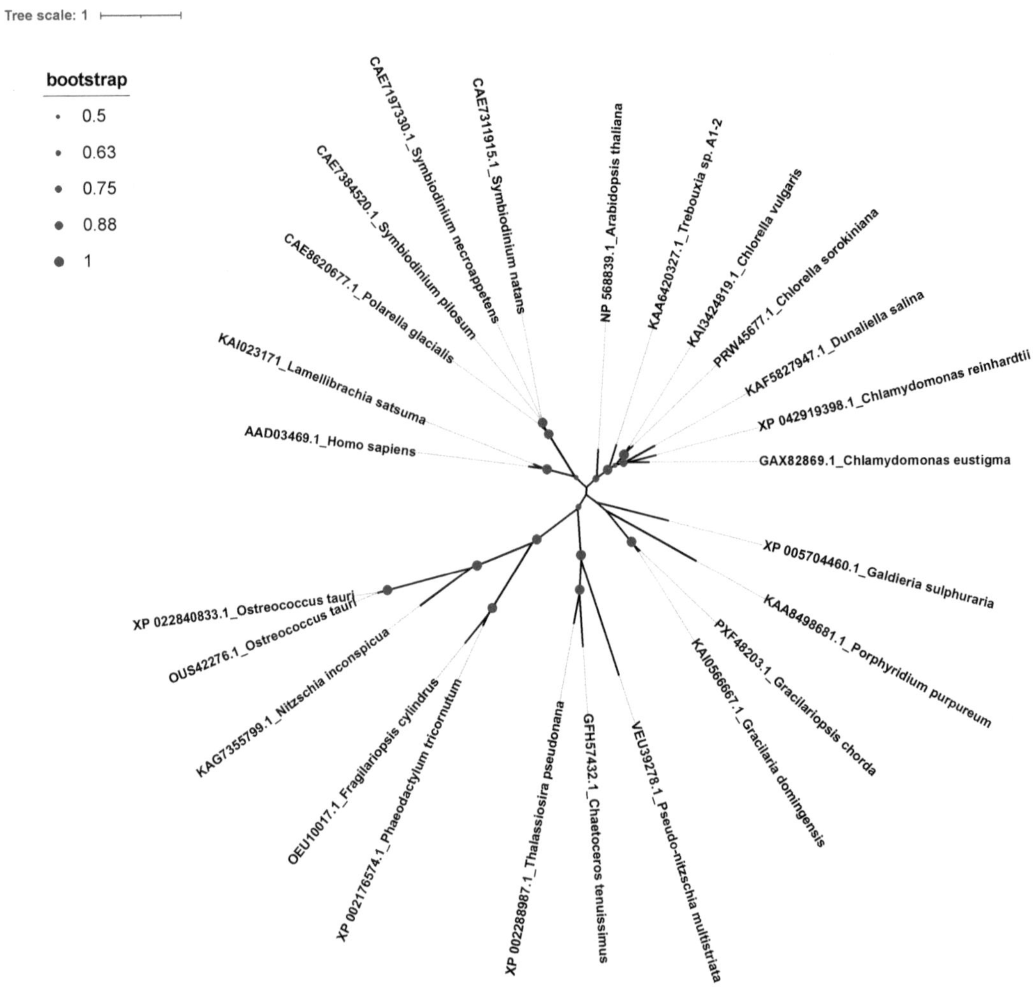

Figure 14.9. Phylogenetic relationships of COP9 signalosome subunit 6 (CSN6) from algae in comparison to animals and plants. Shown is a Maximum Likelihood tree of amino acid sequences based on LG+G model and 1000 bootstrap, values of which are indicated by the size of filled circles at nodes.

profiles of the two populations were similar; however, during recovery from heat stress, the southern population returned to a normal transcriptomic profile (similar to the control), while the northern population showed a very different profile, with signs of protein degradation. This suggests that the distinguishing characteristics between heat-adapted and non-adapted populations might lie in the recovery phase rather than the stress performance.

In the sequenced genome of *Fragilariopsis cylindrus*, a cold-adapted diatom originally isolated from near Antarctica (the Southern Ocean), nearly 25% of the 61-megabase diploid genome is composed of genetic loci with highly divergent alleles. These divergent alleles show differential expression across environmental conditions such as darkness, low iron, freezing, elevated temperature, and increased CO_2. Furthermore, the alleles with the highest ratios of nonsynonymous to synonymous nucleotide substitutions also exhibit the most pronounced condition-dependent expression patterns, suggesting that the allelic differentiation is a result of cold selection (Mock et al. 2017).

The Antarctic Sea ice chlorophytes *Chlamydomonas* sp. and *Tetrabaena solialis* both have an optimal growth temperature at 5°C. A comparative transcriptomic study indicates a substantial number of shared positively selected genes and significant molecular convergence in these two Antarctic algae (Zhang et al. 2019). There is notable evidence of positive selection and convergent substitutions in genes related to photosynthetic machinery, various antioxidant systems, and several key translation elements in these algae. These findings demonstrate that psychrophilic algae possess a more stable photosynthetic apparatus and multiple protective mechanisms.

Coping With Thermal Stress

As discussed above, many algae are adapted to live at extreme temperatures by human standards. These conditions cannot, strictly speaking, be considered stress. However, stress may occur when the temperature in the habitat changes, including daily sun and tide cycle, seasonality, and unusual climate events (e.g., storms, El Niño). Furthermore, the temperature in the microclimate in the alga's immediate vicinity may differ considerably from that of the wider surroundings. Additionally, low temperature *per se* is not necessarily damaging to an organism but the formation of ice crystals is lethal as it causes mechanical damage or changes concentrations of osmotically active solutes. At high temperatures, oxygen deficiency, which is in part because O_2 is less soluble in hot than in cold water, may be the proximal cause of stress, but large heat waves may also cause oxidative stress or protein denaturation.

Algae employ various genomic mechanisms to respond to heat and cold stresses, which involve expression regulation of temperature-responsive genes (Fig. 14.10). Overall, the molecular response varies among species, particularly to cold stress.

Heat Stress

The most common response to heat stress is the upregulation of Heat Shock Proteins (HSPs), particularly HSP70 and HSP97, which help protect proteins from denaturation and assist in protein folding under heat stress (Fig. 14.10A). Light harvesting complex protein genes appear to be the second most common responsive genes to heat stress, but in most cases are downregulated. Another response to thermal stress is the activation of antioxidant systems like superoxide dismutase (SOD) and increased Nitric Oxide (NO) production to combat oxidative stress caused by heat-induced Reactive Oxygen Species (ROS) accumulation. The expression of HSP genes is induced by heat shock transcription factors like HSFA1d. Additionally, glutathione (GSH) biosynthesis might be induced by heat stress as GSH acts as an antioxidant and plays a central role in mitigating heat stress effects. Membrane fluidization-dependent and independent pathways regulate the expression of certain heat-responsive genes like MBF1 (Multiprotein-Bridging Factor 1) and HSP70s, with membrane changes acting as a signal to elicit gene expression changes.

In cyanobacteria, *Synechocystis* is polyploid with the number of genome copies per cell varying from 58 to 218. Many genes such as hrcA and hik34 are related to this species' heat shock response. To understand the molecular basis of long-term temperature tolerance, *Synechocystis* sp. PCC6803 was subjected to random mutation and adaptive evolution to generate a mixture of strains with significantly increased thermal tolerance (Tillich et al. 2014). Comparison of the genomes of these mutants with that of the wild type revealed mutation of ATP-dependent clp protease ATPase subunit (clpC), polyribonucleotide nucleotidyltransferase (pnp), pyruvate kinase 2 (pyk2), group 3 RNA polymerase sigma factor (sigF), upstream of a putative peptidase (nlpD), uracil phosphoribosyltransferase (pyrR), chemotaxis protein required for thick pili and motility (pilJ), and the major class III adenylate cyclase (cya1), among others in the tolerant mutant strains.

In the rhodophyte *Neopyropia seriata*, heat stress was shown to induce the expression of two homologs of *high temperature response 2* (*HTR2*), named NyHTR2 and NyHTR2L, proteins that appear to be conserved only within the order Bangiales (Khoa and Mikami 2022). The membrane fluidizer Benzyl Alcohol (BA) did not induce the expression of these genes. However, the expression of multiprotein-bridging factor 1 (NyMBF1) and HSP70s (NyHSP70-1 and NYHSP70-2) was induced by heat stress and by BA. These observations corroborate that the heat stress response in *N. yezoensis* involves membrane fluidization-dependent and -independent pathways.

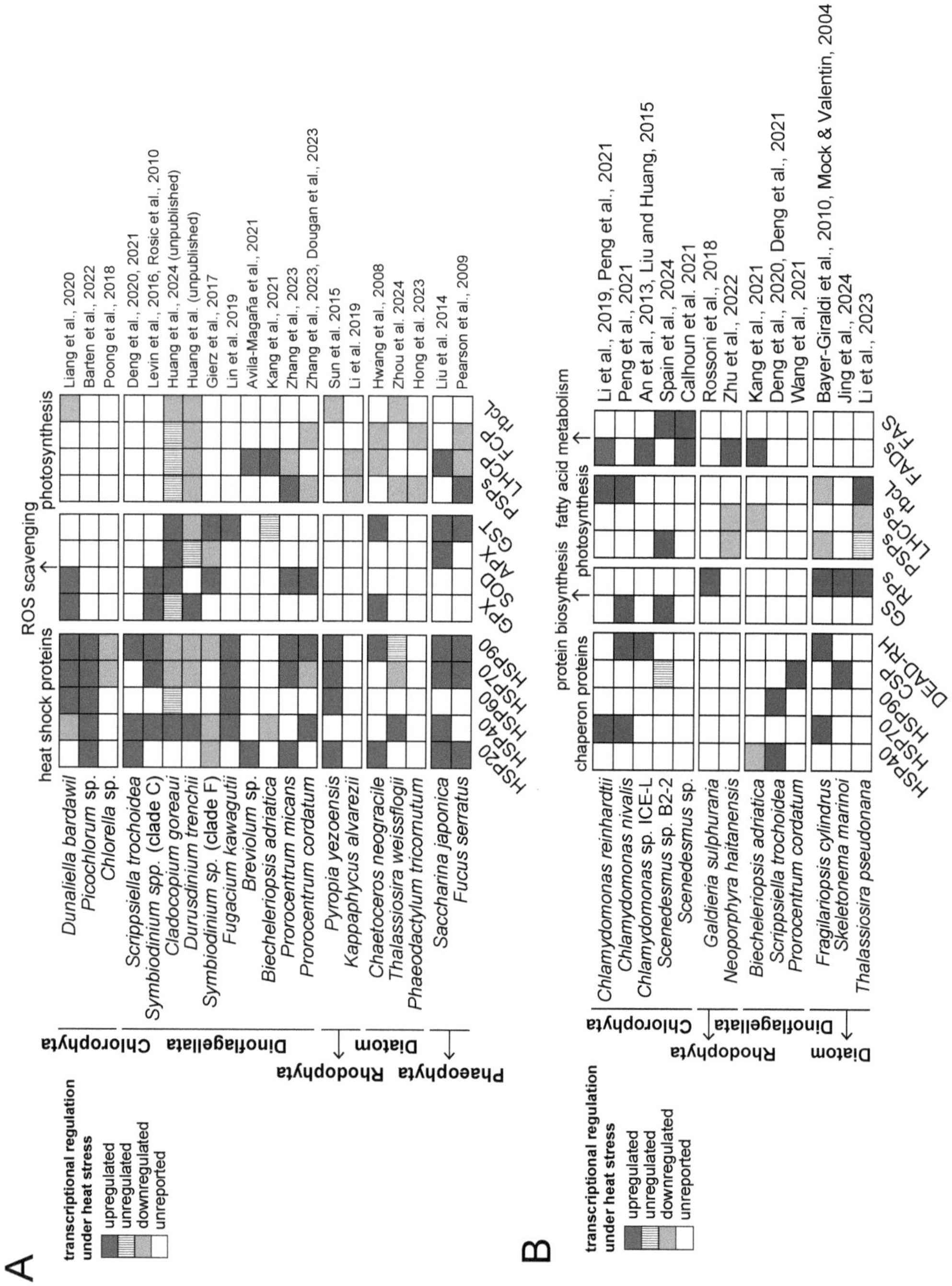

Figure 14.10. Common genes and their responses to heat stress (A) and cold stress (B) in algae.

As discussed in Chapter 9, dinoflagellates use complementary interplay between RNA editing and alternative splicing to regulate the expression and functional diversity of biomolecules, leading to reduction in photosynthesis, central metabolism, and protein synthesis in response to heat stress (Dougan et al. 2023).

Cold Stress

As shown in Fig. 14.10B, fatty acid synthesis genes appear to be the most common cold responsive genes. They are consistently upregulated under cold stress. Cold stress also often induces the accumulation of compatible solutes like proline, which acts as an osmoprotectant and stabilizes proteins and membranes under cold stress. It can also elicit the upregulation of HSFA1d, a Heat Shock Transcription Factor that also responds to cold stress by inducing cold-responsive genes. Increased GSH and HSP97 is another potential response to cold stress as these are core factors helping algae cope with both heat and cold stresses (Xing et al. 2022).

An interesting experimental evolution study provides insights about how algae cope with cold stress. The hot spring rhodophyte *G. sulphuraria* was subjected to continuous cold stress at 28°C, which is 14°C below its optimal growth temperature of 42°C, for over 100 generations (Rossoni and Weber 2019). After this prolonged cold exposure, the cold-adapted samples exhibited a growth rate approximately 30% faster than the initial population. Whole-genome sequencing identified 757 variants across 429 genes (6.1% of the transcriptome) involved in cell cycle regulation, gene regulation, signaling, morphogenesis, microtubule nucleation, and transmembrane transport. Genes functioning in the cell cycle accounted for ~ 34% of the enriched GOs under cold stress, including GO terms of "positive regulation of cell proliferation", "re-entry into mitotic cell cycle", "DNA replication, removal of RNA primer", "ATP-dependent 5'-3' DNA helicase activity", "attachment of mitotic spindle microtubules to kinetochore", "mitotic chromosome movement towards spindle pole", and "midbody", impacting every stage of the cell cycle. Regarding stress and signaling, functions of changed genes are enriched GO terms of information transformation such as "response to stress", "hyperosmotic response", and "ER overload response", terms of receptors (G-protein coupled, tyrosine kinases, and guanylate cyclases), and "stress-activated protein kinase signaling cascade", among others. A significant number of variants accumulated in CpG islands within the intergenic regions, suggesting epigenetic remodeling. From the dataset, 20 candidate genes and three putative *cis*-regulatory elements were identified, which were significantly affected by temperature. This study demonstrates that coping with cold involves reprogramming of an intricate gene network and epigenetic regulation may play an important role.

Production Of Dimethylsulfide and Methane as Potential Climate Feedback Mechanisms

Production of Dimethylsulfide (DMS)

DMS is a degradation product of dimethylsulfoniopropionate (DMSP), which is mainly produced by phytoplankton, as described for haptophytes in Chapter 9, as well as seaweed and plants (Yoch 2002). Recently, DMSP was also found to be produced by bacteria (Curson et al. 2017). DMS enters the atmosphere as aerosols and serves as the nuclei of cloud, promoting cloud formation, reducing radiation to earth's surface and potentially bringing a cooling effect. The main algal contributors of DMSP production include dinoflagellates, coccolithophores and other haptophytes, and to a lesser extent, diatoms, chrysophytes and prasinophytes (Keller 1989). Based on the type of the first reaction step, the biosynthesis pathway of DMSP can be divided into three types (Fig. 14.11), namely, the methyl transfer (MMT) pathway, the amino transfer (MAT) pathway, and the decarboxylation (MDC) pathway, all of which use methionine as the initial substrate (Carrión et al. 2023). The MMT pathway is used by plants, algae, and bacteria, MAT pathway is used by

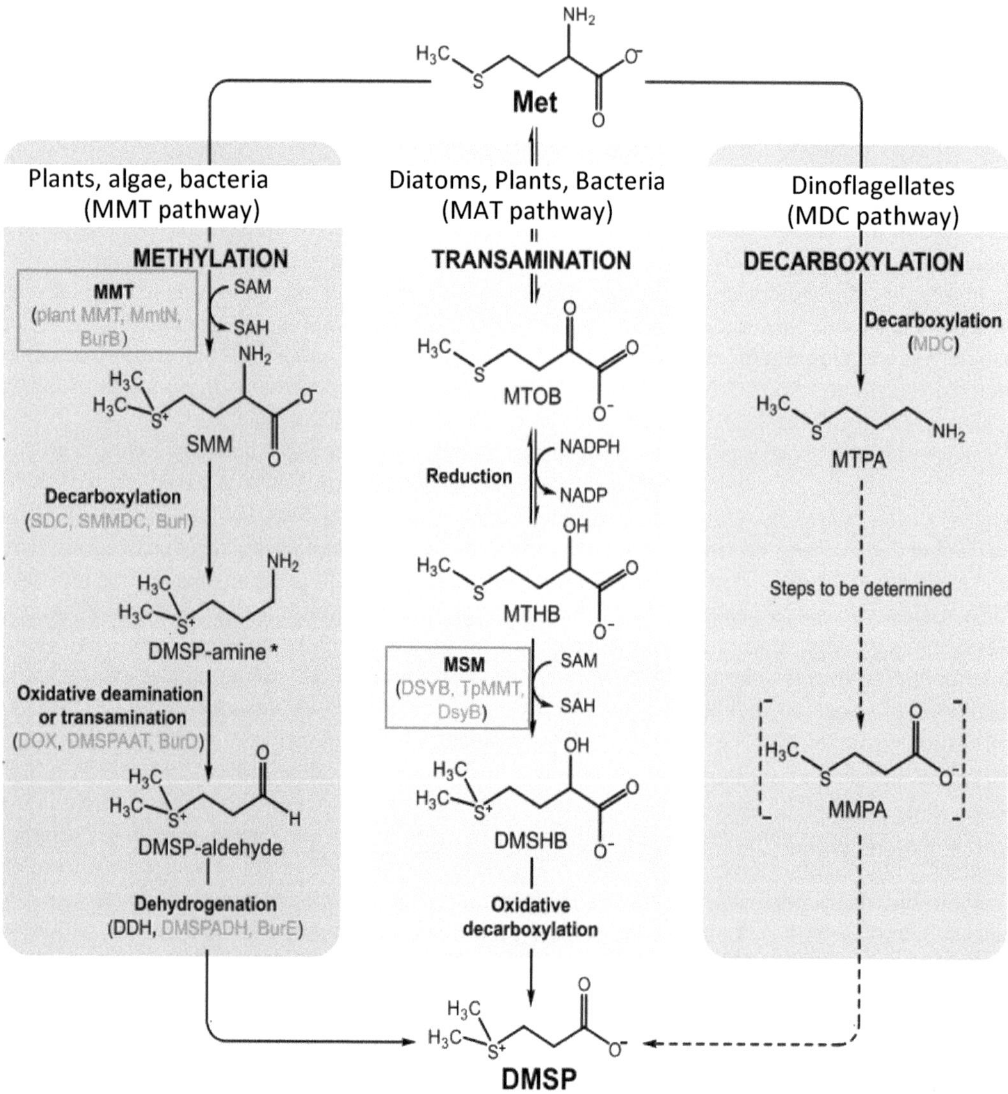

Figure 14.11. DMSP synthesis pathways: methionine methyl transfer pathway (MMT), methionine amino transfer pathway (MAT), and methionine decarboxylation pathway (MDC). Enzymes of eukaryotes are represented in blue, while proteins of prokaryotes are represented in pink. Met: L-methionine; MMT: Methionine S-methyltransferase reaction; MSM: Methionine S-methyltransferase reaction; SMM: S-methylmethionine; MTOB: 4-Methylthio-2-oxobutanoate; MTHB: 4-Methylthio-2-hydroxybutanoate; DMSHB: 4-Dimethylsulfonio-2-hydroxybutanoate; MTPA: 3-(Methylthio) propionic acid furan; SAM: S-Adenosyl methionine; SAH: S-Adenosyl homocysteine. Reproduced with permission from Carrión et al. (2023).

corals, algae, and bacteria, and the MDC pathway by dinoflagellates. For instance, in the saltmarsh plant *Spartina altermiflora*, DMSP synthesis takes this route of the MMT pathway: methionine → S-methylmethionine → DMSP-amine → DMSP-aldehyde → DMSP.

The first key enzyme in DMSP synthesis ever discovered was Dsyb in the MMT pathway in marine bacteria (Curson et al. 2017). Then, MMtN was found as a key gene in marine sediment bacteria for DMSP production through the MMT pathway. Later, *Burkholderia* was found to use the BurB gene to produce DMSP in the process of generating bacterial virulence factors (Trottmann et al. 2020). Recently, the protein structures and enzymatic characteristics of the key genes Dsyb and MmtN involved in bacterial DMSP synthesis were elucidated (Li et al. 2022, Peng et al. 2022).

In algae, a homolog of Dsyb was identified in the haptophyte *Prymnesium parvum* by bioinformatics and cytological analyses (Curson et al. 2018). Analysis of sequence similarity reveals that certain species of haptophytes and dinoflagellates possess genes homologous to DYSB, while others, such as *Thalassiosira* and *Pseudo-nitzschia* species, exclusively harbor homologous genes to TpMMT. Moreover, certain diatom species, like *Skeletonema* and *Chaetoceros*, possess both TpMMT and DYSB homologous genes (McParland et al. 2021). TpMMT represents a diatom-specific gene pivotal to DMSP synthesis (McParland et al. 2021, Carrión et al. 2023, Li et al. 2022). In *Fragilariopsis cylindrus*, the expression of the DSYB homologous gene was found to increase under conditions of elevated salinity and nitrogen limitation, coinciding with an elevation in DMSP content (Curson et al. 2018). Similarly, an upregulation of TpMMT gene expression was observed in *Pseudo-nitzschia* under salinity stress, corresponding to an increase in DMSP content (Kageyama et al. 2018). These findings underscore the crucial role of potential DMSP synthesis key genes in diatoms in response to stress. Studies of the DMSP biosynthesis gene expression have provided supportive evidence for the notion that algae are the primary DMSP producers in the ocean. A study showed that in coastal seawater samples from the North Pacific, the transcriptional abundance of algal DSYB homologous genes was twice that of bacterial DSYB genes. Similarly, analysis of the relative abundance of TpMMT and DSYB homologous genes in the *Tara Oceans* dataset revealed that eukaryotic algae possessing DSYB homologous genes are the main DMSP producers in the surface ocean (Li et al. 2022).

In the ocean, DMSP is reduced to dimethylsulfide (DMS) either by phytoplankton, the producers of DMSP, or bacteria. It involves DMSP lyase, which mediates the cleavage of DMSP into DMS and acrylate. While DMSP lyases were initially discovered in marine bacteria, DMSP lyase activity has been reported in eukaryotic phytoplankton at least since the 1990's. At least eight prokaryotic DMSP lyase genes have been identified to date (Li et al. 2021). Eukaryotic DMSP lyases or DMSP lyase-like (DL-L) genes were initially identified in the coccolithophore *Emiliania huxleyi* (Alcolombri et al. 2015) but are now known to exist also in dinoflagellates such as symbiotic species from the family Symbiodiniaceae and in animals such as scleractinian corals (Chiu and Shinzato 2022). Using the DMSP lyase (DL) domain to search through algal genomes and the *Tara Oceans* metatranscriptomic data, Shemi et al. (2023) found prevalence of the DL gene in algae in the global oceans. In particular, the data reveal differential expression of certain dinoflagellate DLs between the euphotic and mesopelagic zones, suggesting a functional specialization and an involvement in the metabolic plasticity of mixotrophic dinoflagellates.

Negative feedback on climate is expected of DMSP production because the major DMSP producer, species of dinoflagellates and haptophytes, are favored by warmer temperatures. However, experimental data have not been consistent in supporting this proposition. Spielmeyer and Pohnert (2012) discovered that both warming (from 14.5°C to 20.5°C) and ocean acidification can increase the DMSP content per cell in non-calcifying *Emiliania huxleyi*, while for calcifying *E. huxleyi*, warming decreases the DMSP content per cell. Subsequent studies have shown that ocean acidification, ocean warming (from 17°C to 21°C), and their combined effects all lead to an increase in DMSP content in *E. huxleyi* (Arnold et al. 2013). Recent studies have reported that ocean acidification decreases the production of DMSP in another species of coccolithophore, *Gephyrocapsa oceanica* (Larsen and Beardall 2021).

Methane

Counterintuitively, algae, the major sink of CO_2, can produce methane in some cases. Methane is more than ~ 80-fold greater Global Warming Potential (GWP) than CO_2 over the first 20 years after it reaches the atmosphere or ~ 28-fold global GWP of CO_2 over a 100-year period (IPCC2013). Although CO_2 has a longer-lasting effect, methane sets the pace for near-term warming. Methane used to be thought to only come from archaea, which convert CO_2 to CH_4 in anerobic environments.

In the past decade, increasing evidence suggests that eukaryotes, including algae, plants, animals, fungi, and likely humans, also produce CH_4 under oxic conditions, though at much lower rates (Bižić et al. 2020). Notably, nitrogen-fixing blue-green algae can produce CH_4 at substantial rates under light and dark and oxic and anoxic conditions (Bižić et al. 2020). Studies showed that one subgroup of nitrogenase, Fe-only nitrogenase, was responsible for the production of CH_4 from CO_2 in cyanobacteria, and 9% of the diverse nitrogen-fixing microorganisms from a range of environments encode Fe-only nitrogenase (Fixen et al. 2016, Zheng et al. 2018). This appears to be a 'sloppy' N_2-fixation reaction, but the biochemical mechanism underlying the production of CH_4 while fixing N_2 remains to be unraveled (Bižić et al. 2020). It is also unclear yet whether algal methane production is promoted or changed at all under global warming. As such, the nature and magnitude of climate feedback roles of both methane and DMS production by algae remain elusive.

15

CO_2 and Ocean Acidification

CO_2 and Ocean Acidification

Some algal habitats can be extremely acid, notably in geothermal and mining areas, whereas some other environments such as saline deserts can be strongly alkaline. Normally intermediate environments may experience fluctuations in pH due to circumstances such as influx of ammonia or intense photosynthesis in an already alkaline water. Due to anthropogenic emission of CO_2 from fossil fuel combustion, atmospheric CO_2 level has increased from ~ 280 ppm (more accurately, μatm) before industrialization to the current ~ 410 ppm and is projected to reach 750–1000 ppm at the end of the century. The effect of increasing atmospheric CO_2 is not limited to climate warming. Ocean absorbs ~ 30% of the human emitted CO_2. This 'extra' amount of CO_2 that enters the ocean perturbs the equilibrium of carbonate chemistry, as it reacts with water to form HCO_3^- and Hydrogen ions (H^+). The released H^+ acts to increase acidity and lower seawater pH, leading to the observed decreases of pH (Fig. 15.1), a phenomenon called Ocean Acidification (OA). Surface ocean pH is estimated to have dropped on global average by approximately 0.1 units from the preindustrial era to present (Fig. 15.1A), which is equivalent to ~ 30% increase in hydrogen ion concentration. If the current rate of CO_2 increase continues, another 0.2-unit decrease of pH is projected (Fig. 15.1B).

Accompanying the pH decline is the reduction of carbonate (CO_3^-) concentration and alkalinity, potentially reducing calcium carbonate precipitation in calcifying algae and other organisms. Besides, declining pH disrupts intracellular pH homeostasis, the restoration of which costs energy. The stress effects of OA on algae are not limited to the disruption of pH homeostasis, but have ancillary effects such as reduced iron bioavailability and exacerbated phosphorus deficiency (Lin 2023). Furthermore, global warming has caused decreases in solubility and hence concentration of O_2 in the ocean. These compounded changes in carbonate chemistry have profound impacts on organisms, but the effects are highly variable between groups of organisms, even within the algal community. Based on the chemistry, the effects can be classified into three types: direct effects of elevated CO_2 on algal photosynthesis, effects of elevated CO_2 on calcification, and effects of declined pH on algal growth.

High CO_2 Demand of Algal Photosynthesis and Evolution of Carbon Acquisition Strategies

CO_2 is essential for algal photosynthesis, and CO_2 concentrations in normal seawater (pH > 8.0) are markedly below the half-saturation level of Rubisco in algae. There are two reasons for the CO_2 inadequacy. First, at typical pH in the ocean (7.5–8.5), the majority of inorganic carbon (DIC) exists as bicarbonate (HCO_3^-) and only a small percentage as dissolved CO_2 (Text box 15.1). For instance,

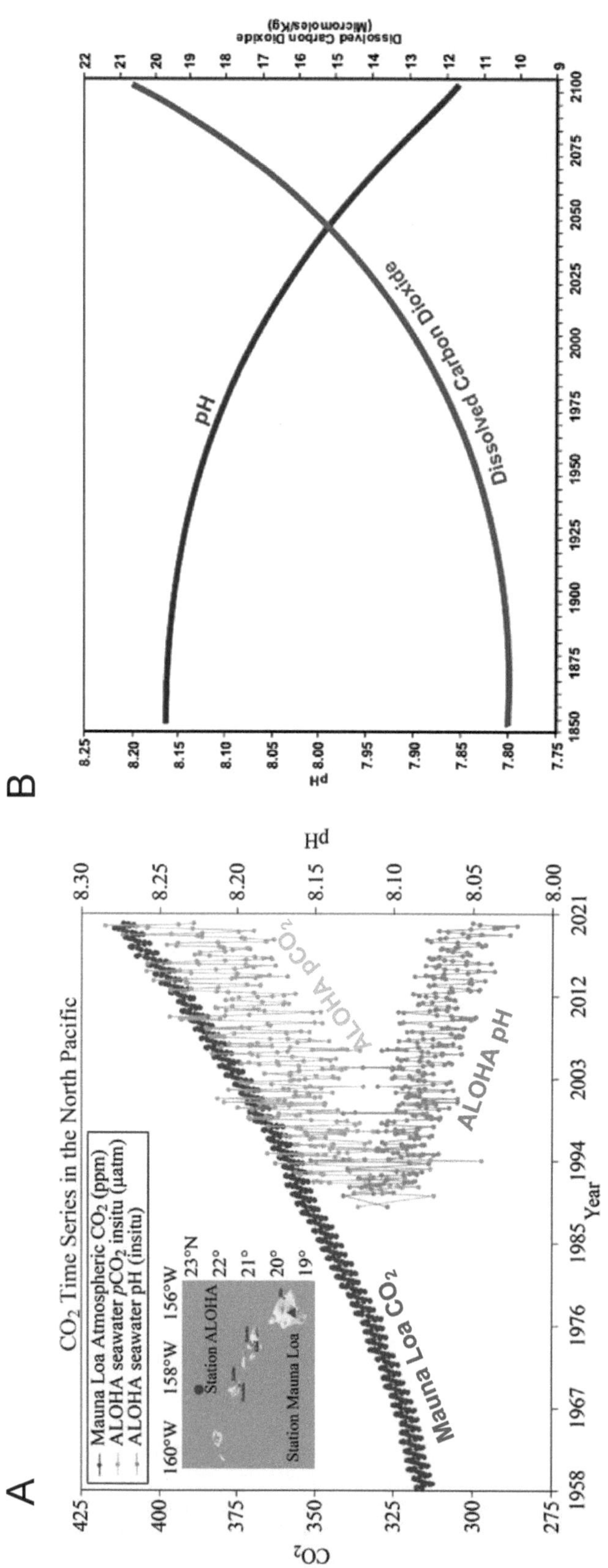

Figure 15.1. Trends of CO_2 increase and pH decrease in the ocean. (A) Atmospheric CO_2, seawater partial pressure of CO_2, and seawater pH as measured at Mauna Loa and Station ALOHA, respectively. From https://www.pmel.noaa.gov/co2/file/Hawaii+Carbon+Dioxide+Time-Series. (B) Historical and projected pH and dissolved CO_2 in the global ocean. Reproduced with permission from Feely et al. (2006).

Text box 15.1
In typical ocean surface (pH = 8.2)
CO_2 = ~ 10 – 15 μM (0.5%)
HCO_3^- = ~ 1955 μM (85%)
CO_3^{2-} = ~ 241 μM (10.5%)
Total DIC: ~ 2300 μM

at pH 8.2, of the total DIC (~ 2,300 μM), HCO_3^- accounts for 85% (~ 1,955 μM), dissolved CO_2 for 5% (10–15 μM), and CO_3^{2-} for 10.5% (~ 241 μM).

As has been discussed in Chapter 2, to compensate for the low affinity of Rubisco for CO_2, most algae have evolved the Carbon Concentrating Mechanisms (CCMs) so that cells can take up the abundant bicarbonate in the ocean as the source of CO_2 for the photosynthetic apparatus (Giordano et al. 2005). This includes biophysical and biochemical types of mechanisms. The biophysical CCM is composed of bicarbonate transporter to take up HCO_3^- and Carbonic Anhydrase (CA) to convert HCO_3^- to CO_2 and the reverse. The biochemical CCM includes the C4 pathway and the Crassulacean Acid Metabolism (CAM) pathway as known in plants, although CAM has not been proven to exist in algae.

Cyanobacteria possess five distinct inorganic carbon transport systems (Price 2011), including two types of Na^+-dependent bicarbonate transporters (BicA and SbtA), one traffic ATPase for bicarbonate uptake (BCT1) and two CO_2 uptake systems based on modified NADPH dehydrogenase complexes (NDH-1MS or NDH-I3 and NDH-1MS' or NDH-I4). NDH-1MS consists of NDHD3, NDHF3, and CUPA, whereas NDH-1MS' consists of NDHD4, NDHF4, and CUPB (Han et al. 2017). The five systems occur widely in cyanobacteria. The genetic diversity of inorganic carbon uptake system in eukaryotic algae is less understood.

As CCMs operation costs energy, these findings imply that increased CO_2 input into the ocean would relieve CO_2 limitation, lessen the demand for CCMs, and hence save energy, potentially promoting algal photosynthesis and growth rate. Indeed, some phytoplankton species exhibits enhanced growth rate under elevated CO_2. However, every species does not respond the same way, because high CO_2 causes ocean acidification, and acidic conditions cause stress in algae like in other organisms, as will be discussed in detail below.

Different pH Niches of Different Lineages of Algae

Lowered pH may disrupt acid-base balance and proton homeostasis inside algal cells, causing energy expenditure to cope with the stress. Experimental data have shown that different algal species have different physiological ranges of pH in which the species can grow (Fig. 15.2). Some species grow in low, or high, or intermediate narrow ranges of pH, whereas other species can grow in a wide range of pH, demonstrating physiologically plasticity. Calcifying species such as coccolithophorids (a subgroup of haptophytes) are particularly susceptible to low pH as acidic conditions promote $CaCO_3$ dissolution and impair its formation (see below for more detail). Their suitable pH range is narrower and mostly situated above pH7.5 (Fig. 15.2A). Effects of elevated CO_2-derived decline in pH (OA) are more complex, as they combine effects of raised CO_2 and lowered pH.

Variable Effects of OA on Algae: Potential Mechanisms and Ecological Consequences

Responses to OA differ between and even within groups of algae (Gao and Campbell 2014, Dutkiewicz et al. 2015). Conceivably, the effects of OA are determined by how much an algal species benefits from increased CO_2 for photosynthesis and/or is negatively affected by lowered pH. Dunkiewicz et al. (2015) showed that diazotrophic cyanobacteria, diatoms, dinoflagellates, and other large phytoplankton grew faster in high-CO_2 conditions, whereas coccolithophores,

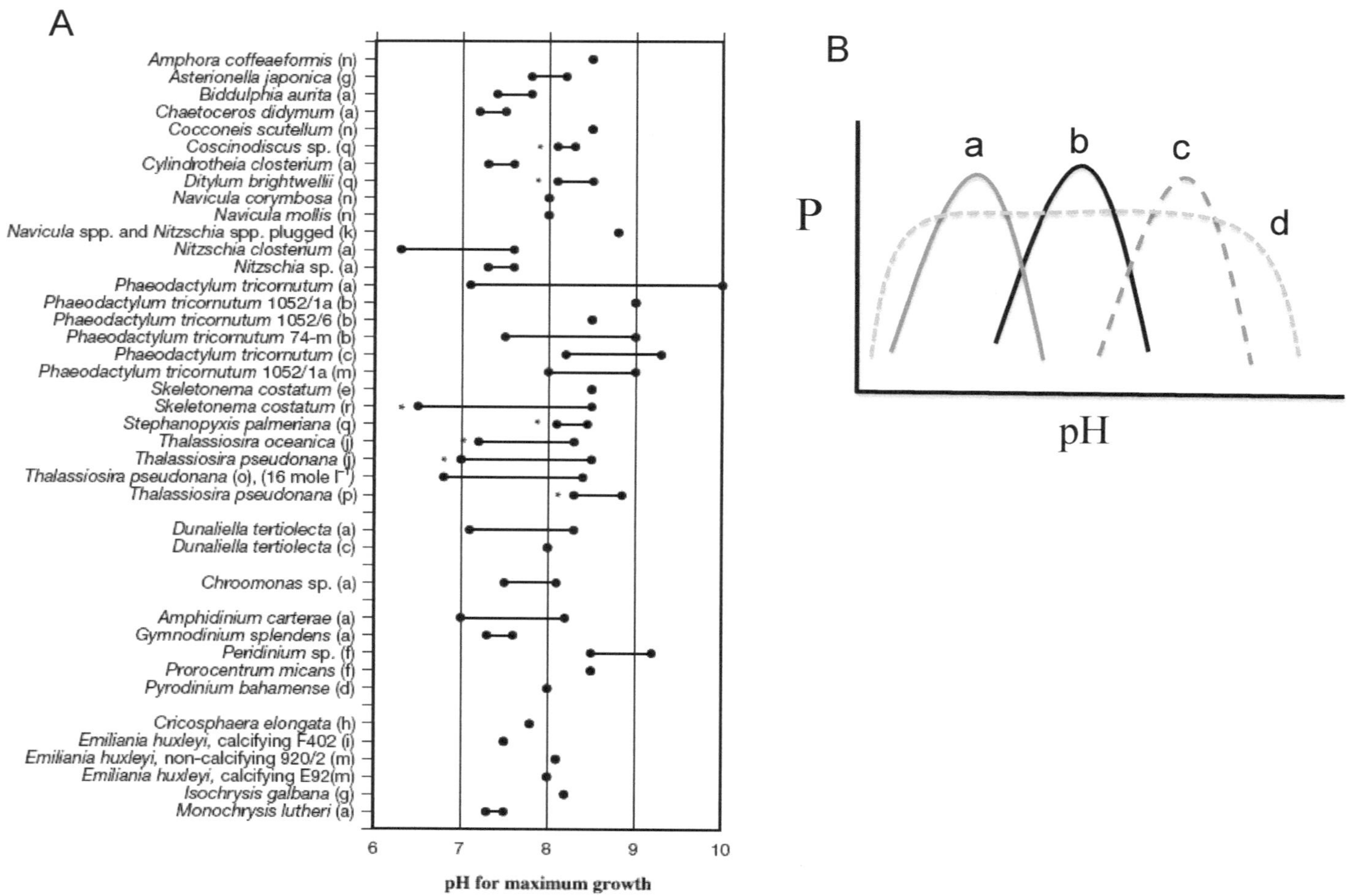

Figure 15.2. Different pH niches of algae. (A) Differential ranges of growth pH for different algal species. Reproduced with permission from Hinga (2002). (B) Schematic indicating different reaction norm of algal growth against pH conditions. Species a, b, and c grow in low, intermediate, and high narrow ranges of pH, respectively, whereas species d has a wide pH niche breadth, being physiologically plastic.

Synechococcus, and *Prochlorococcus* did not; yet there was remarkable variation in CO_2 response within these groups. Another study indicated that higher CO_2 (800 ppm) slowed down growth rate and increased chlorophyll content per cell in dinoflagellates (*Alexandrium monilatum, Prorocentrum cordatum* [formerly *P. minimum*]) but promoted growth and reduced chlorophyll content in the other taxa examined (the diatom *Chaetoceros affinis*, haptophytes *Chrysochromulina polylepis* and *Gephyrocapsa oceanica*, and the raphidophyte *Heterosigma akashiwo*). Transcriptomic analyses in that study indicated minimal change in CA gene expression, showing no evidence that elevated CO_2 relieves CCM-associated energy consumption. Besides, few genes were regulated in the shared trend across the species except for several functionally unknown genes. In general OA will not increase CO_2 abundance to a level that is adequate for Rubisco, and by reducing CCM demand may conserve only approximately 3–6% energy (Hopkinson et al. 2011).

Regardless of CCM-dependent or independent species, OA causes stress to algal cells as it perturbs intracellular pH homeostasis, which imposes energetic costs to maintain. Acidic condition causes alteration of protein structure and compromises enzyme activities. Proton Pump Rhodopsin (PPR), which occurs in many but not all phytoplankton species, can potentially help maintain pH homeostasis in PPR-containing species (Lin 2023). In contrast to the photosynthetic apparatus that requires hundreds of proteins, PPR functions alone and hence is a highly efficient energy harvesting system. Interestingly, PPR may not only provide energy in form of ATP to meet the energy demand for coping with acidic stress, but potentially can also directly pump excess protons from inside cells to maintain pH homeostasis. Evidence of the latter function exists in bacterioplankton PPR (Bunse et al. 2016) but that in algae has yet to be attained (Lin 2023).

Another significant effect of raised CO_2 on algae is the decreases of nutrient supply due to warming-induced water stratification depressing upwelling (or vertical mixing) that is responsible for bringing abundant nutrients from deep water into the euphotic zone. OA and nutrient limitation can interact while influencing algal physiology and growth. The interactive effects can be additive, synergistic or antagonistic. For instance, an ocean acidification–nitrogen limitation interaction was found to promote growth in the diatom *Chaetoceros didymus*, whereas ocean acidification alone showed no effect and nitrogen limitation depressed growth (Gao et al. 2019). In contrast, elevated pCO_2 partially mitigated the inhibitory effects of UV radiation (Gao et al. 2019) or nitrogen limitation (Valenzuela et al. 2018) on phytoplankton growth. Regarding phosphorus nutrient (P), phytoplankton have evolved adaptive strategies to cope with P deficiency, such as reducing phosphorus demand by substituting phospholipids with sulfur- or nitrogen-containing lipids (van Mooy et al. 2009, Shi et al. 2017b). More importantly, phytoplankton use alkaline phosphatase most commonly but also other hydrolases alternatively to scavenge Dissolved Organic Phosphorus (DOP) from the water column for phosphorus (Lin et al. 2016). While a basic pH is generally preferrable to alkaline phosphatase, its optimal pH varies between species (Fig. 15.3), with some species potentially benefiting from ocean acidification for scavenging DOP, as shown recently (Guo et al. 2023).

It can be inferred that OA can potentially intensify stress of P-deficiency and conversely, P limitation can aggravate the negative effects of OA. Indeed, synergistic effects of OA and P-deficiency have been demonstrated in the inhibitory effects on N_2 fixation in *Trichodesmium* (Zhang et al. 2022). It was previously shown that ocean acidification repressed nitrogen fixation in *Trichodesmium*, particularly under iron limitation (Shi et al. 2010). The major reason for the negative effect is, contrary to the intuition that acidic conditions usually favor metal solubility (Schwertmann 1991), ocean acidification changes the binding characteristics of some iron ligands, resulting in lowered bioavailability of dissolved iron. Zhang et al. (2022) postulated that ocean acidification exacerbates the negative effects of phosphate deficiency mainly because the cells under OA showed a high content of polyphosphate. Polyphosphate is a polymer of three to thousands of phosphates with high-energy phosphoanhydride bonds, and has been shown to primarily serve in storing luxuriously absorbed phosphate and responding to phosphate deficiency (Martin et al. 2014). However, its accumulation in acidified and P-stressed *Trichodesmium* cells is consistent with the potential of polyphosphate to be a pH (H^+) buffer previously proposed (Harold 1966).

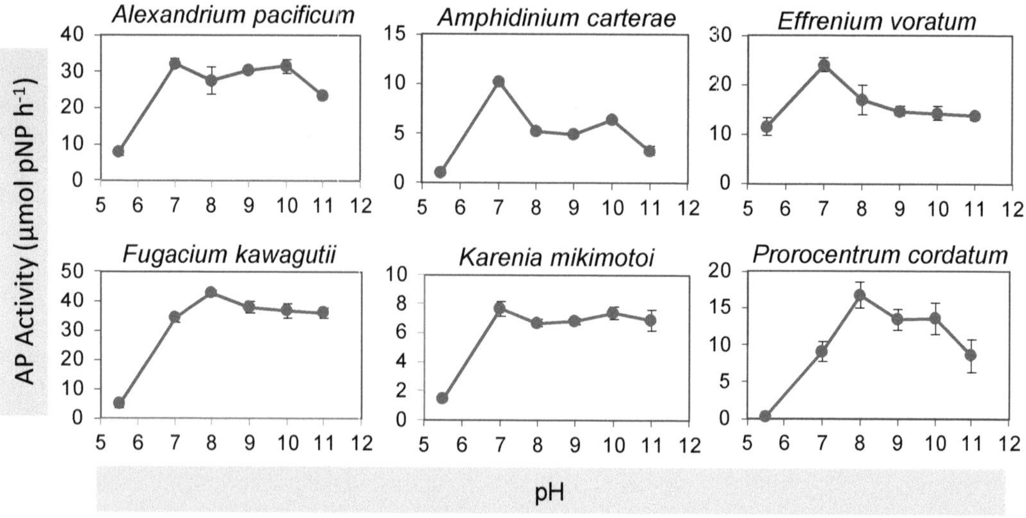

Figure 15.3. Differential pH effects on alkaline phosphatase activity (APA) among different algal species. Reproduced with permission from Guo et al. 2023.

Yet, polyphosphate accumulation raises phosphorus demand, increasing phosphate deficit for metabolism.

In some phytoplankton species, both ocean acidification and phosphorus stress can cause shifts of intracellular resources towards defense such as producing more toxins. For instance, phosphorus deficiency can induce toxin production in the dinoflagellate *Karlodinium veneficum* and this could be intensified by elevated CO_2 (Wang et al. 2019), another example of synergistic effects of OA and P limitation. Toxin production is one of the common strategies in phytoplankton to deter grazing, so there is potential for a strengthened defense to partially offset the population decline caused by slower growth under ocean acidification and/or phosphate limitation (Lin 2023).

Whereas calcifying species are generally more susceptible to high-CO_2 conditions than non-calcifying species, this trend is not uniform among species, and the form of $CaCO_3$ that species produce (calcite, aragonite) is not strongly linked to species sensitivity (Busch and McElhany 2017). Elevated CO_2 coupled with increased light exposure was found to synergistically reduce marine primary productivity, based on the experiments with the natural phytoplankton assemblages in the South China Sea (Gao et al. 2012).

For calcifying species, calcium carbonate ($CaCO_3$) is formed following the chemical equation,

$$Ca^{2+} + 2HCO_3^- = CaCO_3 + CO_2 + H_2O \qquad (1)$$

Higher CO_2 would drive the reaction toward the left of the equation, i.e., the dissolution of $CaCO_3$. Another way to understand this is, when more CO_2 is dissolved into the ocean, it reacts with water to produce H_2CO_3, which dissociates into HCO_3^- and H^+. The resulting more H^+ lowers pH, which promotes dissolution and inhibits formation of $CaCO_3$. Conversely, the formation of $CaCO_3$ releases CO_2, increasing the partial pressure of CO_2 in the seawater. Therefore, calcification of organisms does not immediately benefit carbon sequestration, as one may intuitively think, but rather exacerbate the CO_2 rise in the atmosphere (Frankignoulle et al. 1994, van Dam et al. 2021). As such, ocean acidification and $CaCO_3$ dissolution (which increases $CaCO_3$ saturation [Ω] in seawater) constitute negative feedback on atmospheric CO_2 (Riebesell et al. 2000), creating new equilibria at different CO_2 loadings. Recent geoengineering initiatives to load $CaCO_3$ materials into the coastal ocean aim to drive the equation toward left for CO_2 absorption.

The differential responses of different algal species to OA will have profound consequences to the algal communities. The OA stress will select for species that are physiologically plastic against pH changes or species that have higher fitness under lower pH. Therefore, OA can cause ecosystem shifts by altering competitive interactions between the species (Kroeker et al. 2013).

Given all these potential effects of OA and differential responses among different species to OA, the impacts of OA on the algal community most likely vary by species compositions and environmental conditions (ocean regions). Some studies found that OA promoted the productivity of the nutrient-rich, river-impacted coastal phytoplankton communities but depresses that of the oligotrophic high-salinity oceanic communities (Gao et al. 2021). Theoretically, the impacts of global warming and OA are most pronounced in the polar regions. For instance, temperature rise at the Arctic was nearly four times the global average rate since 1979 (Rantanen et al. 2022). More rapid OA was also found in the Arctic than other ocean areas (Qi et al. 2022). Models have predicted a decline of pH by 0.3–0.5 units from preindustrial levels at the end of the century, together with higher seasonal pH fluctuations particularly in coastal waters (Steinacher et al. 2009, Cai et al. 2011). Furthermore, the concurrent rapid ice melt causes reduction in alkalinity, exacerbating the impact of OA. Yet a study showed that the Arctic phytoplankton community was resilient to OA, showing no significant changes in productivity or species composition under OA treatment (Hoppe et al. 2018). Out of ten CO_2 treatment conditions used in this study, only one showed significant OA effects on primary productivity (at the low temperature of 1.8°C) and three exhibited shifts in species composition but without correlation to specific experimental conditions. The resilience of a phytoplankton community to OA may occur in three possible ways. One is physiological flexibility of all the species in the community, another is strain sorting in combination with high intraspecific diversity (strains adapted to different parts of the pH variability spectrum), and the other is shifts between functionally redundant species (Hoppe et al. 2018). In the Antarctic, a study (Beszteri et al. 2018) indicated that among the dominant species, the haptophyte *Phaeocystis antarctica* and the two species of diatoms *Fragilariopsis kergulensis* and *Chaetoceros debilis*, the haptophyte species appeared to be the most resistant to OA combined with high light. *C. debilis* exhibited various responses to OA, with upregulation of genes related to cell division and transcription under high light and reduced respiration, elevated expression of genes involved in light protection, transcription and translation, and reallocation of carbon to lipids under low light.

Caution is needed when generalizing results from laboratory studies to natural assemblages. One reason is as shown earlier that an assemblage may consist of species that are either plastic or ensembled of genetically differentiated strains or functionally redundant (Hoppe et al. 2018) such that community appears to be resilient to OA even though those species may show different responses to OA. The second reason is that after long-term exposure to OA, genetic adaptation may occur at different levels depending on the fitness trait being studied, and natural assemblages are selected not only by OA but also other factors such as competition or defense against grazing or viral attack (Schlüter et al. 2016).

Adaptation usually comes at a cost to fitness, so there is tradeoff. For instance, when the species adapts to high CO_2, the adaptive genotype will get a better fitness performance (e.g., growth rate) under high CO_2 than the non-adaptive genotype but will show a poorer performance at low CO_2 than the non-adaptive genotype.

Potential Genomic Mechanisms Underlying Adaptation to Ocean Acidification

As modeled in plants, mechanisms by which cells control cytosolic pH can be grouped into two major categories: one being the biophysical pH stat and the other the biochemical pH stat (Smith and Raven 1979). The biophysical pH stat involves the electrogenic transport of protons across the plasma membrane. In contrast, the biochemical pH stat works by changing cytosolic proton concentration via protonation and deprotonation of organic acids.

There are abundant lessons from bacterial research about how organisms respond to acidic environments because many bacteria live in acid environments (e.g., acidic soil, animal intestines). As thoroughly reviewed in Lund et al. (2014), various strategies are used by microorganisms to prevent intracellular pH from falling below lethal levels. These strategies, based on the degree of acidity they act under, are classified as 'Acid Tolerance Response' (ATR) to mildly acidic (up to pH 3.0) condition and 'Amino acid-dependent extreme acid resistance' (XAR) allowing unadapted cells to survive at levels too acidic (pH $\leq$ 2.5) to permit growth. These strategies include:

1. Preventing proton influx by modifying the cellular membrane;
2. Pumping protons out of the cell (via the F1F0-ATPase);
3. Biochemically binding protons (i.e., amino acid decarboxylation) or generate ammonia (i.e., amino acid deiminases, deaminases and urease);
4. Preventing or repairing acid-derived macromolecule damages.

These responses or protection mechanisms start with a transcriptional and translational response to a drop in pH, and if this response is prevented or reduced, growth and survival of these bacteria are impaired at the lowered pH. Species living in acidic environments or frequently exposed to acidic conditions have developed sophisticated mechanisms at the physiological and molecular levels to maintain intracellular pH homeostasis, thereby withstand and survive as well as adapt to acid stress (Guan and Liu 2020).

Although how protons move across membranes is not fully understood, direct measurement in *E. coli* shows that when the external medium is acidified, the pH in the periplasm falls very rapidly and remains to be similar to that outside the cell, suggesting that the outer membrane is not a significant barrier to proton movement and the periplasm has limited buffering capacity (Wilks and Slonczewski 2007, Lund et al. 2014). In contrast, the cytoplasm is more protected, due to its intrinsic buffering capacity and the relative impermeability of the cytoplasmic membrane to protons, and by inducible acid stress responses. Nevertheless, significant reduction in pH can occur in the cytoplasm. Proteins embedded in the membranes, particularly the periplasmic domains of inner membrane proteins will be exposed to the low pH of the periplasm if the external medium is acidified. The cellular consequence of internal pH decreases can be complex but includes lowered enzyme activity, acid-induced protein unfolding, membrane damage, and DNA damage (Lund et al. 2014).

In several acid-resistant species of Gram-negative enteric bacteria, specific periplasmic chaperones and proteases (e.g., HdeA, HdeB, DegP, SurA) play a major role in protecting periplasmic and membrane proteins against acidity-derived damages by preventing protein unfolding and facilitates renaturing of unfolded proteins (Hong et al. 2012). These proteins contribute to acid resistance in a range of different species (Lund et al. 2014). Acid stress also triggers cytoplasmic membrane protecting mechanisms against acid damage. For instance, Cyclopropane Fatty Acids (CFAs) show elevated levels in more acid-resistant cells of *E. coli* (Brown et al. 1997). Mutation of the CFA synthase gene (*cfa*) renders *E. coli* sensitive to low pH, and the sensitivity can be overcome partially by an exogenous supply of CFAs (Chang and Cronan 1999).

Decreasing proton permeability can also be achieved by the modification of the phospholipids in the internal membrane (Lund et al. 2014). In some bacteria, the resistance to acidic pH stems from converting unsaturated fatty acids (UFAs) into cyclopropane fatty acids (CFAs) by the addition of a methyl group (from S-adenosyl-methionine) to the double bond of UFA (Chang and Cronon 1999, Kim et al. 2005). In other microorganisms, such as oral bacteria, the acidic pH induces a shift in the membrane composition from short-chained saturated fatty acids to long-chained mono UFAs (Fozo et al. 2004).

F1F0-ATPase is a sophisticated molecular machine localized in the plasma membrane, which can either synthesize ATP (ATP synthase) with the energy-releasing proton influx, or conversely pump out protons using the energy provided by ATP hydrolysis (ATPase). The involvement of F1F0-ATPase in pH homoeostasis was first observed in *Streptococcus* spp., which do not have the respiratory chain and use this enzymatic complex to expel protons when the cytoplasmic pH decreases (Kobayashi et al. 1986). Subsequent studies have suggested a proton-pumping role of F1F0-ATPase in several other bacterial species (Lund et al. 2014).

The amino acid decarboxylases are cytosolic enzymes, mostly with pyridoxal 5′-phosphate (PLP), a derivative of vitamin B6, as a substrate. These enzymes, having an acidic pH optimum (< 6), perform a proton-consuming decarboxylation reaction on specific amino acids, such as glutamate, lysine, arginine and ornithine (Gale 1946). Each decarboxylase works with a cognate cell membrane-localized antiporter, which becomes active only when the extracellular pH drops below threshold levels, thereby providing a selective gate for entry of the amino acid substrates and exit of the decarboxylation products (Lund et al. 2014).

Deiminase and deaminase systems produce ammonia (NH_3), which combines with intracellular protons to become ammonium ion thereby raising pH. The arginine deiminase system consists of three enzymes, all being active at pH 3.1 or lower: the arginine deiminase, the ornithine transcarbamoylase, and the carbamate kinase. They catalyze the conversion of arginine into ornithine, NH_3, and carbon dioxide with the formation of one ATP per arginine (Cunin et al. 1986). Besides, glutaminase and adenosine deaminases also contribute to acid protection in *E. coli* (Sun et al. 2012, Lu et al. 2013), which use the relevant substrates to release NH_3 in the cytoplasm. In addition, urease catalyzes the hydrolysis of urea, yielding NH_3 and carbamate, which spontaneously decomposes into another molecule of NH_3 and CO_2, thereby achieving a relief from acid stress because NH_3 combines with intracellular protons to yield NH_4^+.

Exposure to acidic pH can also lead to the accumulation of damaged proteins in the cytoplasm, which may induce cytoplasmic (DnaK and GroEL) or periplasmic (HdeA and HdeB) chaperones. Furthermore, some microorganisms may use components of the Clp protease (caseinolytic protease that cleaves proteins such as casein and albumin) complex to remove damaged proteins and maintain protein homoeostasis (Lund et al. 2014). Low pH can also cause DNA damages such as strand breaks (Jeong et al. 2008). The key cellular component in *E. coli* that protects against this damage is the DNA-binding protein Dps. As a protective mechanism, low pH promotes the usually slow DNA depurination that can cause loss of genetic information, and hence induces DNA repair systems (Lund et al. 2014).

For algae, responses to ocean acidification have been studied mostly at the physiological and ecological levels, and molecular level understanding is much more limited and fragmentary than for microorganisms. Nevertheless, this research field has recently been growing rapidly. Similar to bacteria, algae typically exhibit transcriptional and translational responses to elevated CO_2 conditions, and the extent of such responses varies with species. A study found that the model diatom *T. pseudonana* rapidly responded to increasing CO_2 by differentially expressing gene clusters that regulate transcription and chromosome folding, and subsequently reduced transcription of photosynthesis and respiration gene clusters under steady-state elevated CO_2 (Hennon et al. 2015). These results indicate that CO_2 elevation first causes a shift in gene regulation and then a metabolic reconfiguration in this diatom. Furthermore, one of the CO_2-responsive gene clusters contain CCM and photorespiration genes that share a putative cAMP-responsive *cis*-regulatory sequence, implying that these genes are co-regulated in response to CO_2 changes, with cAMP as an intermediate messenger. In addition, inhibiting cAMP hydrolysis induced downregulation of CCM gene δ-CA3 in nutrient-replete diatom cultures, indicating an important role for cAMP in downregulating CCM and photorespiration genes under elevated CO_2. This provides insights into the signaling pathway that relay elevated CO_2 to CCM relaxation.

Algae living extreme acidic environments, such as the rhodophytes *Cyanidioschyzon merolae* and *Cyanidium caldarium*, also use similar strategies as acid-adapted bacteria to prevent breakdown of intracellular pH homeostasis. These acidophilic algae share common attributes such as cell surface barrier that is extremely impermeable to protons (H^+) and higher saturated fatty acid content in cell membrane resulting in decreased fluidity of the plasma membrane (Lee 1999). Conceivably, the molecular mechanisms for preventing proton influx must be different between acidophilic algae and bacterial counterparts. My BLAST analysis using bacterial acid resistant protein genes HdeA and HdeB as queries did not yield any significant matches in algal genomes housed in NCBI and SAGER (http://sampgr.org.cn/sequenceserver/) databases. However, significant hits were found in dinoflagellates for acid resistance-associated proteases-encoding genes DegP and SurA. DegP is an ATP-independent serine endopeptidases that is conserved in all domains of life, and in acidophilic bacteria it is activated after the environmental pH returns normal to degrade acid-damaged proteins. In plants, DegP has been reported as a thylakoid (luminal side) protein (Itzhaki et al. 1998), suggesting a potential role in the chloroplast instead of cytoplasm. SurA is a periplasmic molecular chaperone protein, which facilitates correct folding and maturation of outer membrane proteins in Gram-negative bacteria. The role of its homologs in algae or plants remains to be explored.

As discussed above, acid stress can alter protein structure and damage enzyme activity, and this can impose effects on the uptake and assimilation of nutrients, degradation of substrates, and synthesis of proteins and nucleic acids. In *Arabidopsis*, for instance, when grown at an alkaline condition (pH 7.5), its transport of N via NRT2-type nitrate and AMT1-type ammonium transporters was induced, conceivably to increase the cytosolic H^+ concentration due to co-transport of NO_3^- with H^+ and NH_4^+ tending to be dissociated upon entry into cells, respectively (Jain and Schmidt 2024). However, assimilation of NO_3^- will consume H^+. Some studies showed that low pH environments induced decreases in protein synthesis (Jan et al. 2001) and changes in protein phosphorylation (Koponen et al. 2012). For instance, acid-adapted bacterium *Propionibacterium freudenreichii* exposed to extreme low pH exhibited dramatic changes in morphology and decreases in viability, but acid-acclimated cells preserved integrity and viability, suggesting adaptability of the species to acid stress. However, acid-stressed cells were found to express general stress proteins such as the protein chaperonins GroEL andgroES, recR, repB, and Biotin Carboxyl Carrier Protein (BCCP). A few ASPs (Acid Stress Proteins) were significantly increased under acidic growth condition, but most of the cellular proteins were repressed (Jan et al. 2001). In the lactic acid bacterium *Lactobacillus rhamnosus*, acid-grown (pH 4.8) cells strongly upregulated gene encoding a surface antigen and expression of F0F1-ATP synthase genes, whereas the abundance of proteins participating in nucleotide biosynthesis and protein synthesis was significantly diminished, and proteins involved in central cellular pathways were phosphorylated, particularly the glycolytic enzymes (Koponen et al. 2012). Alteration in protein phosphorylation also has been reported for plants such as *Arabidopsis*, a pH-dependent phosphor-switch was observed for the ABC transporter PDR7, suggesting changes in activity and, possibly, substrate specificity.

Comparative genomic analyses can shed light on genomic adaptation to acid environments. A study compared the genomes of the neutrophilic chlorophyte model *Chlamydomonas reinhardtii* and its acidophilic relative *C. eustigma* (Hirooka et al. 2017). The result revealed several features in *C. eustigma* that probably contributes to the adaptation to an acidic environment. This species highly expresses genes encoding heat-shock proteins and plasma membrane H^+-ATPase. It also has lost fermentation pathways that acidify the cytosol and has acquired an energy shuttle and buffering system and arsenic detoxification genes through horizontal gene transfer. Furthermore, the arsenic detoxification genes have been multiplied in the genome. These features are shared by other acidophilic green algae as well as the above-mentioned acidophilic rhodophytes. The acquisition and expansion of arsenic detoxification genes suggest that arsenic solubility and hence toxic effects are enhanced by the acidic condition.

An interesting adaptive proteomic evolution was unveiled recently. Selection pressure seems to have driven the evolution of proteomes in acidophilic algae towards higher proportions of basic amino acids and hence higher pI values (isoelectric point) of the proteins. pI is the pH at which the protein has no net charge, or the sum of all changes from the constituent amino acids are zero. Out of the twenty standard amino acids, three (arginine, lysine, and histidine) are basic because they carry positive charge at neutral pH, and two (aspartic acid and glutamic acid) are acidic because they are negatively charged at neutral pH. This evolutionarily driven bias of amino acid composition is reflected in crucial genes. For instance, Alkaline Phosphatase (AP), the critical enzyme helping algae to exploit dissolved organic phosphate when dissolved inorganic phosphate is depleted, requires different pH for maximal activity in different species of dinoflagellates (Fig. 15.3). Meanwhile, the protein sequences of AP with low optimal pH have greater proportions of basic amino acid residues and higher PI whereas those with high optimal pH have lower proportions of basic amino acid residues and lower PI (Guo et al. 2023). Arginine and lysine are among the basic amino acids that are increased in the acid-adapted APs. This suggests that these dinoflagellates may employ a similar strategy as the above-mentioned bacteria, which use arginine deamination to produce ammonia to neutralize H^+.

Algal proteomes may respond to acidification by increasing basic amino acid composition. A study exposed the acid-tolerant chlorophyte siphonous alga *Halimeda opuntia* to elevated CO_2 levels and analyzed proteomic analyses of the alga compared to that under control CO_2 conditions. The results indicated that the alga changed its proteomes in such a way that proteins with isoelectric point (pI) of 7 exhibited little change in abundance, but there were many differentially expressed proteins with pI > 7 and pI < 7. The net result was an increase in average pI of the proteome in the elevated CO_2-acclimated algae (Table 15.1).

Table 15.1. Isoelectric point (PI) profile of differentially expressed proteins (DEPs) in *Halimeda* opuntia among three CO_2 treatments.

Treatment	Expression	pI > 7		pI = 7	pI < 7	
		DEP number	Average pI	DEP number	DEP number	Average pI
Medium CO_2 vs Control	Down	53	8.91	—	89	5.29
	Up	129	9.07	1	183	5.35
High CO_2 vs Control	Down	83	9.17	—	127	5.15
	Up	173	9.03	2	206	5.52

16

Energy Acquisition

Photosynthesis and Rhodopsin Proton Pump

Energy powers the biological processes to support growth and survival of organisms, thereby driving the global biogeochemical processes. Hence, energy metabolism is integrated with other metabolic processes such as nutrient uptake, cell division, chemotaxis, secretion of organic compounds, and efflux of waste metabolites and toxic compounds. A minimum amount of energy is required to bring the reactants to the transition state for a reaction to occur. The higher this minimum energy, called activation energy, the harder the reaction; enzymes (catalysts) serve to lower the activation energy and increase the rate of a chemical reaction. Energy in organisms is circulated in the chemical form ATP as the currency, and is generated and stored, and converted to kinetic energy through metabolism, which includes anabolic and catabolic pathways. Anabolic pathways consume energy to synthesize larger molecules, whereas catabolic pathways generate energy by breaking down larger molecules.

The energy-generating metabolism of stored molecules in the cell involves several pathways generally grouped into aerobic and anaerobic classes. About 95% of aerobic microorganisms generate ATP from Tricarboxylic Acid Cycle (TCA) and Oxidative Phosphorylation (OP), which occur in the inner membrane of the mitochondria of eukaryotes and on the inner part of the cell membrane of prokaryotes. This coupled TCA-OP metabolic chain starts with the input of acetyl CoA, which is derived from pyruvate produced by glycolysis. Prokaryotic organisms are very diverse in their ability to use metabolites for energy generation and to switch from one mode of energy generation to another. *Escherichia coli*, for instance, can use aerobic respiration with O_2 as the final electron acceptor but can also use alternative electron acceptors such as nitrate, fumarate, trimethylamine oxide (TMAO) and dimethylsulphoxide (DMSO). *E. coli* can also fine-tune the respiratory chain by using alternative cytochrome oxidases according to the O_2 concentration surrounding the cell. Some bacteria can perform anaerobic metabolism such as fermentation. Less is known about the diversity of energy-generating metabolic pathways in cyanobacteria, but their TCA cycle is non-canonical. The enzyme complex 2-oxogluterate dehydrogenase (OGDH) are absent in cyanobacteria and is bypassed by the functional complementation of glutamate decarboxylase, 4-aminobutyrate transaminase and succinate-semialdehyde dehydrogenase (Beck et al. 2012).

The ultimate source of energy for the biota is sunlight, which is introduced into the ecosystem by phototrophic organisms. Algae capture and convert sunlight energy into ATP to power metabolic processes or to be stored in molecules such as carbohydrates and lipids. The only known exceptions include anaerobic green sulfur bacteria living in dark hydrothermal vents by harnessing geothermal radiation (700–800 nm) for "photosynthesis" (Beatty et al. 2005), and some archaea that can consume and produce hydrogen as source of energy via the activity of hydrogenases (Greening et al. 2024 and references therein). The central component of ATP production is the proton (H+) translocation and the action of ATP synthase.

Conventionally, the only recognized pathway for sunlight energy to enter the marine ecosystem is photosynthesis by chlorophyll-based photosystems. This dogma changed in 2000 when a Proton-Pump Rhodopsin (PPR) was found to occur in marine bacteria living in sunlit surface ocean (Béjà et al. 2000). It was subsequently found that PPR also exists in various algal taxa and heterotrophic protists, redrawing the landscape of energy acquisition and adding a dimension to ecological niche in algae. Therefore, while all algae are dependent on the chlorophyll-based photosystem, some lineages also possess a second photoenergy capturing mechanism, like a dual engine in a vehicle. Between the two "engines", the chlorophyll-based system requires over 33 proteins and multiple metal ions, whereas the PPR system is based on one protein with one chromophore (all-*trans* retinal). As a much simpler machinery, PPR is presumably more efficient.

Light Distribution in the Ocean and Critical Depth

In the electromagnetic spectrum, the waveband from 400–700 nm (visible light) is known as Photosynthetically Active Radiation (PAR). Ultraviolet (UV) is damaging to cell components, while infra-red can only be absorbed by Chl *d* and *f*, which are found in a few cyanobacterial taxa.

Light attenuates exponentially from ocean surface to depths following $I_z = I_0 e^{-kZ}$ where I_0 and I_z are light intensities at surface and depth Z, respectively, k is attenuation coefficient. This is due to the absorption of light by water molecules and particles. The speed at which light disappears is not spectrally uniform. Short (UV and shorter) and long wavelengths are absorbed more quickly than the short visible light. This leaves only blue light in the bottom of the euphotic zone.

Light availability varies spatially and temporally. Tropical regions receive strong light, while high latitudes receive weaker light. There are also daily and seasonal fluctuations, with extreme seasonality in polar regions, where winters are constantly dark and summers constant light, and minimal seasonality in tropical regions, where light is fairly stable throughout the year.

The light field experienced by algae, particularly phytoplankton, is influenced by water column stability. For seaweeds and other benthic algae, the light availability in their shallow water habitats depends on water clarity, which decreases under turbulent conditions. For phytoplankton, cells are transported between ocean surface and depths where water mixing (turbulence) reaches. In tropical and subtropical regions, the water column is stable (stratified), whereas, in temperate and polar regions, strong vertical mixing occurs in winter, and the water is stable and stratified in summer.

In turbulent water, phytoplankton cells are transported between the surface and the bottom of the mixed layer. In the top layers, the energy produced by photosynthesis exceeds the energy consumed by metabolism, whereas in the bottom layers, there is an energy deficit, and the cells use the surplus from the top layers. Down to a certain depth, the depth-integrated photosynthesis is balanced by depth-integrated respiration, resulting in no net energy to support growth as phytoplankton cells travel between the surface and this depth. This depth is known as the critical depth (Z_c). If the mixed layer is shallower than Z_c, phytoplankton can grow. If the mixed layer is deeper than Z_c, phytoplankton population will decline.

This Sverdrup model (Sverdrup 1953) explains why blooms occur in the spring in the temperate region or summer in the polar region, and phytoplankton biomass is low in the winter. This model has received support from various studies but has also been challenged by others. Differences seem to lie mainly in the definition of the "mixed layer", which in Sverdrup's model means the layer over which phytoplankton cells are randomly translocated and spend nearly equal amounts of time at every depth. This layer is more accurately termed a "turbulent layer" (Franks 2015 and references therein). Even when a mixed layer, defined as a hydrographically uniform layer, is deeper than the critical depth, phytoplankton cells might stay well above the critical depth because a shallower turbulent layer might exist within the hydrographically uniform layer due to different mixing and turbulence time scales.

Proton Gradient and ATP Synthase

ATP synthesis is carried out by ATP synthase utilizing energy from a transmembrane proton gradient. Thus, a proton gradient across a membrane and ATP synthase are key to ATP production. The ATP synthase enzyme system is universal in living organisms, ancient, and unique in its mode of function. There are three types of ATPases: the A-, F-, and V-type. *In vitro*, all three types of ATP synthases are reversible: they can use the energy from ATP hydrolysis to move cations across ion-impermeable membranes, and they can use the energy stored in the transmembrane ion gradient to synthesize ATP (Nirody et al. 2020). *In vivo*, the V-type functions primarily as an ATPase.

The A and F types can operate in either direction in prokaryotes. In eukaryotic mitochondria and chloroplasts, F-type ATPases generally function as ATP synthases, with a notable exception in mammalian mitochondria, where F-type ATPases can pump ions to reenergize the membrane. A-type ATPase occurs in archaea. V-type ATPase is present in vacuoles, which are present in most eukaryotes. In secondary endosymbiosis algae such as diatoms, dinoflagellates, and coccolithophorids, a V-ATPase is localized around the chloroplast. This ATPase serves to create proton-enriched environment for enriching CO_2 and promote photosynthesis (Yee et al. 2023). It is posited that this V-ATPase is derived from the "food vacuole" in the process of plastid evolution and is co-opted for phagocytosis, symbiosis, and photosynthesis.

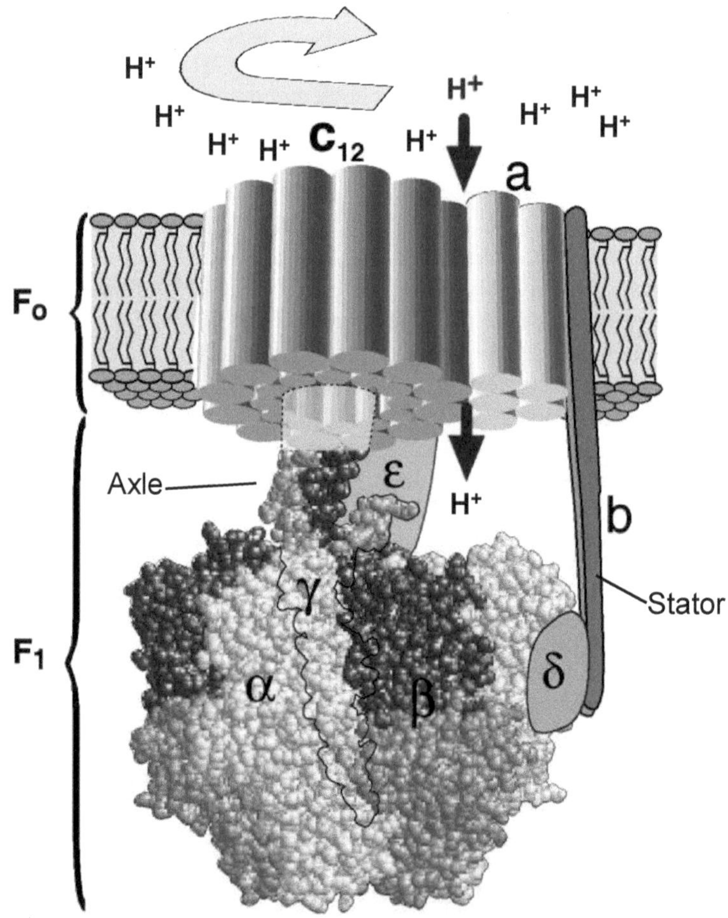

Figure 16.1. Model of F-ATPase architecture and function. Reproduced with permission from Wang and Oster (1998) with slight modifications.

The F-type ATPase is prevalent in all domains of life, where it serves as the core machinery of cellular energy production. It is located across the plasma membrane in prokaryotic cells and across the inner mitochondrial membrane in eukaryotes. Organisms capable of photosynthesis also have F-ATPase across the thylakoid membrane, which is situated in the chloroplast for eukaryotic lineages and in the cytoplasm in prokaryotic taxa.

The F-ATPase comprises F0 and F1. Both its synthetic and hydrolytic cycles are carried out through a rotary mechanism (Wang and Oster 1998). ATP synthase functions similarly to the turbines of hydroelectric plants that utilize the kinetic energy of water flowing through dams. In this analogy, a proton gradient is likened to dammed water, and the flow of protons down the gradient is like water driving the turbines of a hydroelectric engine. Like hydroelectric turbines, ATPase components rotate in response to the proton flow, and this rotational energy is then coupled to ATP synthesis.

The F-ATP synthase includes the Fo rotary motor complex embedded in the membrane, the F1 catalytic complex that synthesizes ATP, and a stator that connects them and prevents the rotation of the catalytic subunits. The central stalk (axle) is considered part of F1, and its rotation is coupled to that of the membrane rotor. The Fo rotor spins in response to proton (H^+) flow down a concentration gradient across the membrane. This rotation causes the central stalk (axle) to rotate, altering the conformation of components of the F1 base, driving the synthesis of ATP. The synthase can also act as an H^+ pump ATPase when its rotations are reversed by ATP hydrolysis. The F1 motor achieves high mechanical torque and nearly 100% efficiency by converting the free energy of ATP binding into elastic strain, which is then released through a coordinated kinetic and tightly coupled conformational mechanism to generate rotary torque (Wang and Oster 1998).

Chlorophyll-Dependent Photosynthesis

Photosynthesis occurs in the photosynthetic apparatus, which is not enveloped in cyanobacteria but is enveloped as chloroplasts in eukaryotic algae. The photosynthetic apparatus is complex and comprised of between 100 to several hundred proteins. Photosynthesis includes the light reaction that carries out energy harvesting and generation of ATP and NADPH and the dark reaction in which CO_2 fixation occurs through the Calvin-Benson-Bassham (CBB) cycle (Fig. 16.2). The light reaction involves two photosystems (PSI and PSII), each composed of a core complex and a peripheral light-harvesting complex known as antenna (LHCI and LHCII). With the aid of LHCII, PSII absorbs photoenergy and uses it to split H_2O, producing O_2 and extracting electrons, creating the parallel flow of electrons and protons towards PSI. PSI also absorbs photoenergy and boosts the energy level of the electron and proton flows, leading to the production of ATP and NADPH. Antennae, PSI, and PSII are all complexes of many subunits. The increasing accessibility of modern analytical technologies such as Cryo-electron microscopy (CryoEM) in recent years has been instrumental in unfolding the structure of these complexes at fine resolution.

As discussed in Chapter 2, the core subunits of both PSII and PSI are highly conserved, but the antenna vary substantially between groups of algae as they are adapted to distinct habitat light conditions (Bai et al. 2021). There are 11–15 subunits in PSI, ten of which are shared by all photosynthetic organisms, and 21–29 in PSII, 20 of which are shared by all photosynthetic organisms. Haptophytes have the fewest subunits, whereas the greatest number of subunits occur in chlorophytes (15) for PSI and cyanobacteria (30) for PSII. The efficiency of this process depends on how closely the protein binds to the pigments, which also influences the power of LHC to dissipate excess energy and provide photoprotection (Zhao et al. 2024). The capacity and efficiency of energy capture and conversion as well as photoprotection by the chlorophyll-based system vary between algal species.

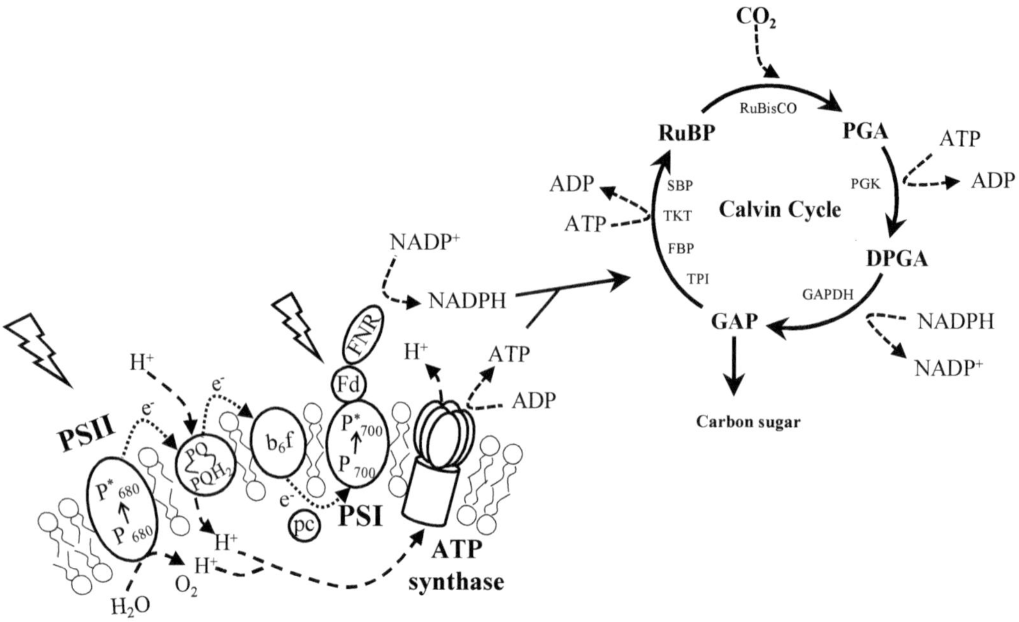

Figure 16.2. Schematic of photosynthesis. Light reaction occurs in the photosystem (PS) I and II composed of reaction centers and antennas. The ATP and NADPH produced in PSI are used to support carbon dioxide in the Calvin-Benson-Bassham (CBB) cycle.

Proton-Pumping Rhodopsin (PPR), A Chlorophyll-Independent Energy Acquisition System

Rhodopsins occur in all three domains of life and perform various functions and are widespread in the aquatic ecosystem. In microbes including microalgae and related protists rhodopsin is highly diverse in sequence but is consistently composed of protein (opsin) with seven transmembrane domains and a retinal as the chromophore. It is curious that seaweeds have not been reported to possess a rhodopsin. Rhodopsin as a photoreceptor and a sensory device is described in Chapter 12. The current chapter is dedicated to one subtype of rhodopsin, the proton-pumping rhodopsin (PPR).

PPR differs from other types of rhodopsins in its all-*trans* retinal, a proton Donor (D) and a receptor (E), residues that form the pocket to hold the retinal, and residues that tune absorption spectra (Fig. 16.3). The tuning residue, in most cases, is a Leucine or Methionine for absorbing green light or glutamine for absorbing blue light; however, variations could happen depending on the organism, and absorption of yellow spectrum has also been found (Pushkarvew et al. 2018). Upon activation by solar energy, the all-*trans* retinal changes its conformation to *cis*, and proton-donor residues are excited to shed protons, which are transferred from one side (e.g., cytoplasm) to the other (e.g., extracellular) side of the membrane. This creates a proton gradient that drives ATP production when protons flux back across the membrane (Fig. 16.4).

PPR was first found in the archaea *Halobacterium halobium* in the 1970s (Oesterhelt and Stoechenius 1971). It was later rediscovered in the SAR86 clade of bacteria in the early 2000's and later in many other bacteria in the surface ocean (Béjà et al. 2000, 2001). PPR occurs in 48% of small-sized plankton (< 0.8 μm) in the ocean's euphotic zone or in 13–70% of bacteria in the surface ocean (Ma et al. 2023). The possession of PPR enables the bacteria to grow or survive better in nutrient-poor environments under illumination (Gómez-Consarnau et al. 2010). A study along a Mediterranean-Atlantic transect estimated that prokaryotic PPR there captured as much photoenergy as chlorophyll *a*-based phototrophy, sufficient to support bacterial basal metabolism in the oligotrophic seas (Gómez-Consarnau et al. 2019).

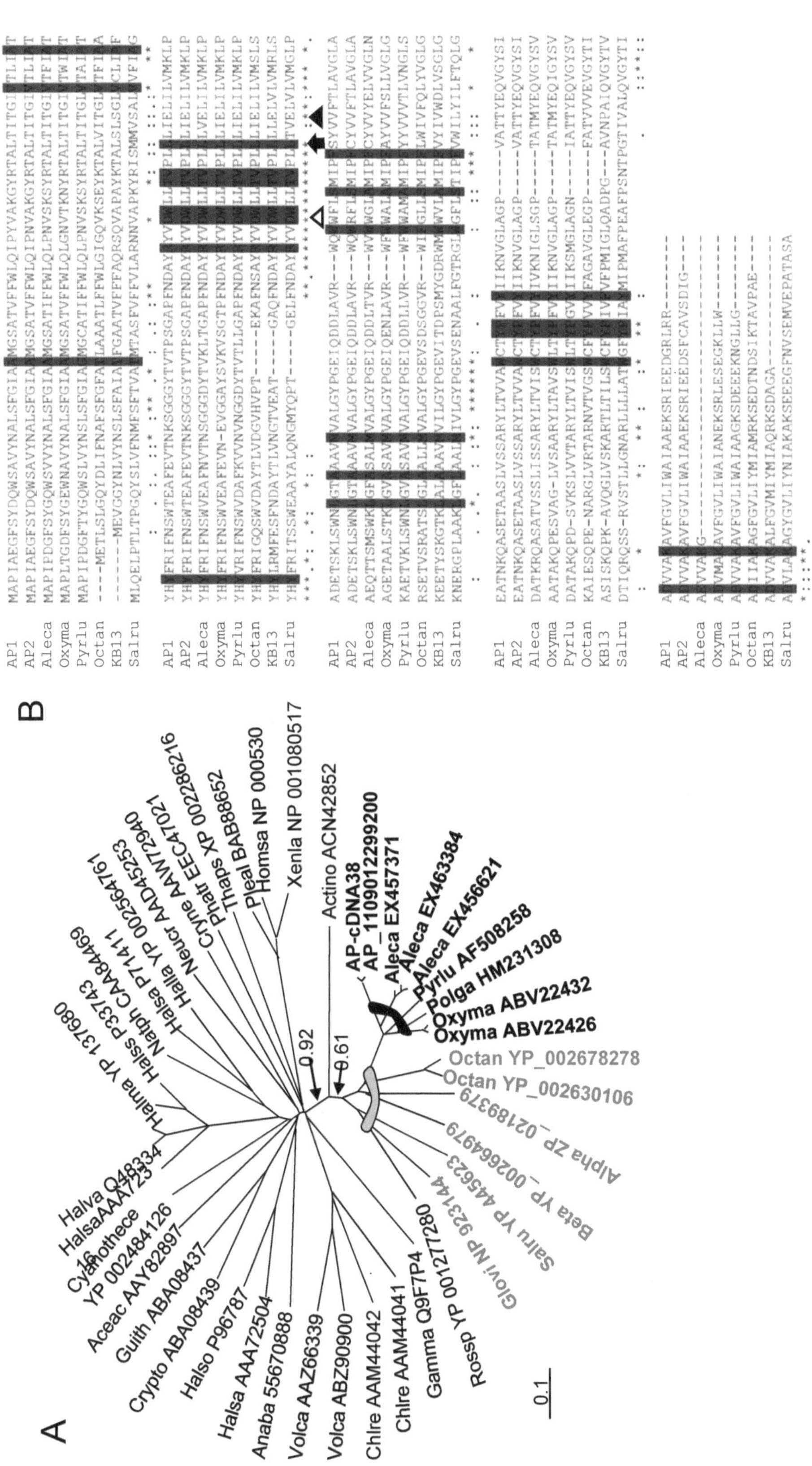

Figure 16.3. Phylogenetic tree (A) and alignment of rhodopsin sequences from dinoflagellates and bacteria (B). (A) Dinoflagellate sequences (bold-typed, grouped by dark curvy line) are clustered with proteorhodopsin sequences (grey). (B) The 22 residues in the retinal-binding pocket (shaded) are generally conserved but the proton donor residue (E; filled triangle) and acceptor (D; open triangle) are specific for ion pump rhodopsin. The spectrum-tuning residue is marked with black arrow, L or M for green and Q for blue absorption in most cases. Reproduced with permission from Lin et al. (2010).

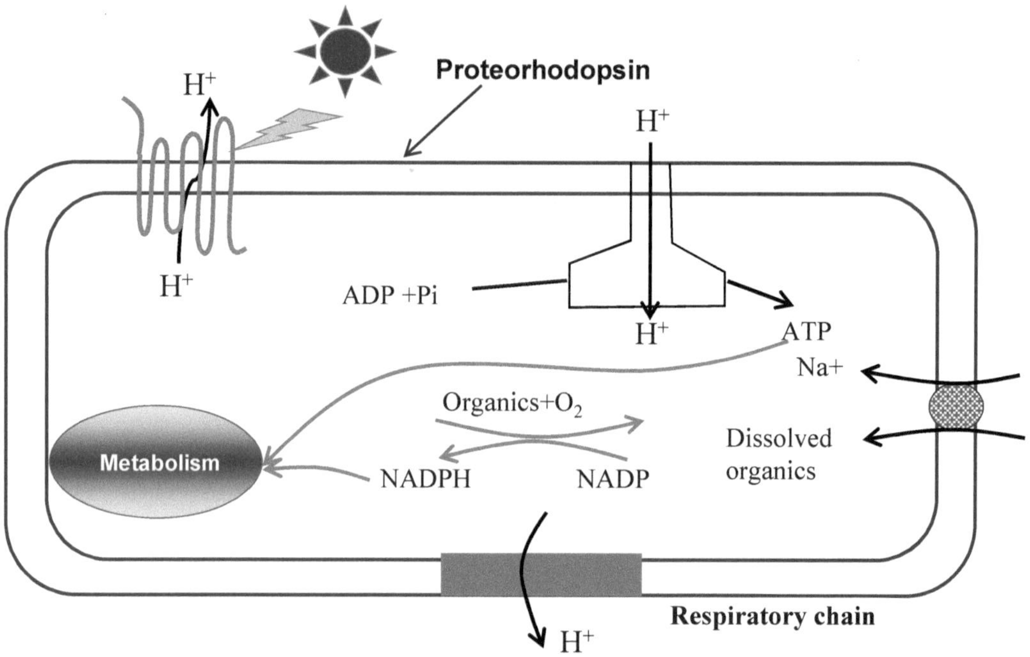

Figure 16.4. Schematic of photoenergy harvesting leading to ATP production by proton pump rhodopsin modeled on a bacteria cell.

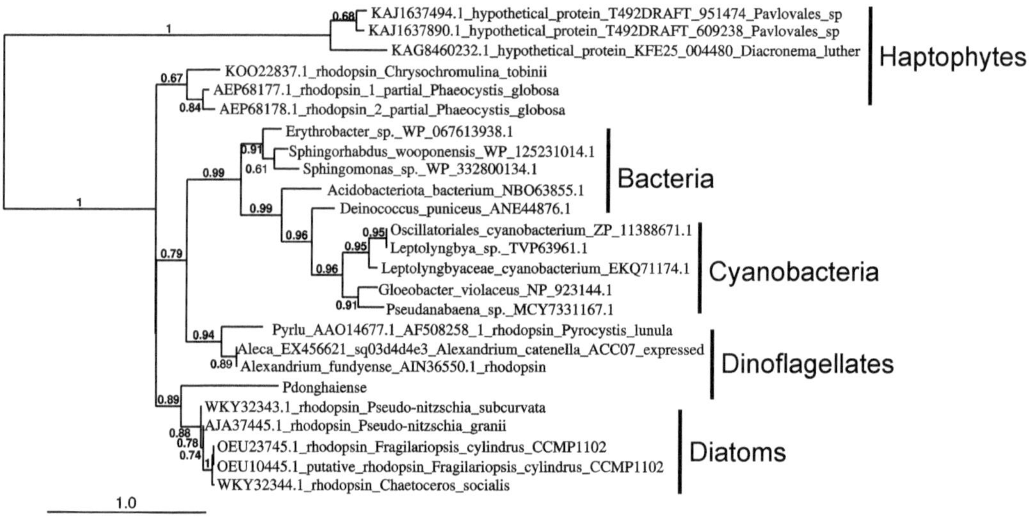

Figure 16.5. PPR in algae and related protists.

In photosynthetic organisms, PPR has been detected in cyanobacteria, chlorophytes dinoflagellates, diatoms, cryptophytes, haptophytes, among others (Ma et al. 2023; Fig. 16.5). Although it exists in green algae (e.g., *Acetabularia acetabulum* [Wada et al. 2011]), PPR has not been reported in land plants. In dinoflagellates, PPR in algae was apparently acquired from bacteria via horizontal transfer due to their high sequence similarities (Lin et al. 2010, Slamovits et al. 2011). In the heterotrophic dinoflagellate *Oxyrrhis marina*, PPR was found to be distributed in the peripheral region of the cell, and its proton-pumping activity was demonstrated (Janke et al. 2013, Rhiel et al. 2020, 2022, Westermann et al. 2023). As expected for photoreceptors and an ATP-producing proton pump, dinoflagellate PPR (dinoPPR)

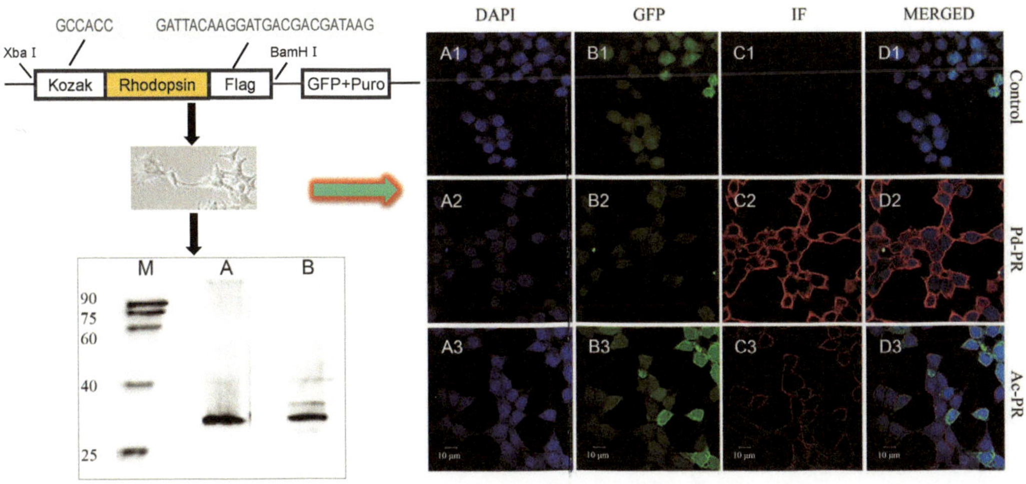

Figure 16.6. Expression and membrane localization of two dinoflagellate proton pump rhodopsin (PPR) genes in 293T cells. Left, construction of the expression plasmid (up), transformed cells (middle), and Western blot showing expression of PPR gene from *Prorocentrum shikokuense* (Pd-PR) (A) and that from Alexandrium pacificum (Ac-PR) (B). Right, Immunofluorescence microscopic images showing cell membrane localization of Pd-PR and Ac-PR. From A to D, images show blue fluorescence of DAPI (4,6-diamidino-2-phenylindole) staining of DNA (A), green fluorescence of GFP (B), red fluorescence of Flag (C), and merger for A, B, and C (D). D1, control group which was transfected with blank PB513b-1 vector. D2 293T cells transfected with PB513b-Pd-PR. D3 293T cells transfected with PB513b-Ac-PR. Reproduced with permission from Ma et al. (2020).

expressed heterologously in the HEK 293T cell line are localized in the cell membrane (Ma et al. 2020; Fig. 16.6). Some of this PPR was observed in internal structures of the heterotrophic dinoflagellate (Slamovits et al. 2011), likely attributable to food vacuoles (Westermann et al. 2023; Fig. 16.7). When expressed in the model diatom *P. tricornutum*, the PPR of the dinoflagellate *P. shikokuense* was found to localize in the peripheral region of the diatom cells (Fig. 16.7). The expression of dinoPPR appears to support the survival of starved heterotrophic dinoflagellate *O. marina* (Guo et al. 2014). Furthermore, in a recent study *O. marina* PPR expressed heterologously in *E. coli* was found to be localized on the cell membrane and to exhibit outward proton pumping activities (Kikuchi et al. 2021). In the photosynthetic species such as *Prorocentrum shikokuense* and *Breviolum minutum*, dinoPPR was shown to absorb blue to green light spectra and support dim-light and P-starved growth (Shi et al. 2015, 2017b, Ma et al. 2024). Strikingly, PPR in *P. shikokuense* was upregulated by ~ 48 folds during a bloom (Zhang et al. 2019), an exceptionally large dynamic range of gene transcript abundance in dinoflagellates. Its upregulation during bloom is consistent with previous reports that *P. shikokuense* tends to cause harmful bloom outbreaks during turbid and low-phosphate periods, supporting the postulation that PPR confers competitive advantages in this species to outgrow other phytoplankton species in light- and phosphate-limited environments.

In diatoms, PPR appears to garner photoenergy to support iron-deficient growth (Marchetti et al. 2015). The role of diatom PPR as proton pump was recently verified by measuring pH changes in the culture of *E. coli* heterologously expressing the PPR of the pennate diatom *Pseudo-nitzschia granii* named PngR (Yoshizawa et al. 2023). When expressed in the model diatom species *P. tricornutum*, PngR fused with eGFP (enhanced green fluorescence protein) was found to localize in the outmost envelop of the chloroplast, suggesting that the energy harnessed by this PPR might support carbon fixation (Yoshizawa et al. 2023). Similarly, significant positive correlations between expression of PPR and that of non-Calvin carbon fixation genes were documented in the subtropical South China Sea for several order of bacteria, suggesting the potential of PPR in those bacteria support their carbon fixation (Li et al. 2024). The function of PPR in other groups of algae remains to be investigated, and the mechanistic of PPR function in all algae is still largely unclear.

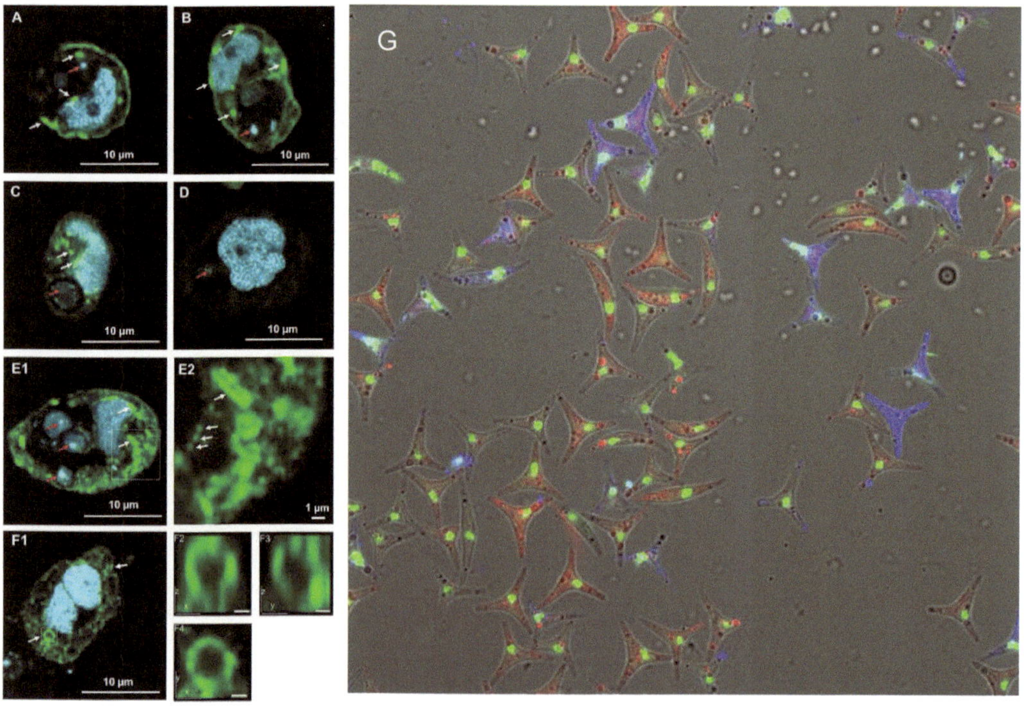

Figure 16.7. The subcellular localization of dinoflagellate PPR. (A-F) Immunofluorescence proton pump rhodopsin in the heterotrophic dinoflagellate *Oxyrrhis marina*. The localization of rhodopsin shown in green (Alexa Fluor 488) after primary detection with the affinity-purified antibody ADY17806 (A and B) and the affinity-purified antibody AEA49880 (C) show significant labeling of the cell periphery and cell internal structures, which resembled vacuoles, tiny vesicles, and rather compact structures, in comparison to the control without primary antibody (D). DAPI-stained DNA is shown in light blue. Reproduced with permission from Westermann et al. (2023) under the Creative Commons CC BY 4.0 license. G) *A. catenella* rhodopsin heterologously expressed in *Phaeodactylum tricornutum*. Shown is autofluorescence of the chloroplast (red) overlain with fluorescence of nuclear stain (green) and PPR (blue). PPR fluorescence comes from immunostaining of Flag contained in the PPR construct bound by anti-Flag, which was in turn recognized by the secondary antibody carrying a fluorochrome to emit blue fluorescence under UV excitation. Confocal image, courtesy of Dr. Minglei Ma.

Energy Niche Differentiation

Algal species with both the Chl *a*-based and the PPR-based photoenergy conversion machineries are apparently advantaged over those with only Chl *a*-based machinery. It is expected that the species with this "dual engine" will outcompete the co-occurring species operating with a "single engine". The competitive advantage has been demonstrated in dinoflagellates and diatoms, particularly in forming algal blooms and exhibiting superior growth in cultures under nutrient and light limitations (Marchetti et al. 2012, Shi et al. 2015, 2017b, Zhang et al. 2019, Yu et al. 2020). During a bloom of *Prorocentrum shikokuense* in a subtropical nearshore area, both the chlorophyll- and PPR- energy harvesting systems were expressed at very high levels (5,000-4,000 FPKM), suggesting a significant contribution of both to the species' rapid proliferation (Yu et al. 2020).

An evolutionary question arises: why haven't all algae acquire PPR? Why is the possession of PPR punctate on the tree algal life? Given that PPR requires only one protein and a chromophore and is theoretically much efficient than the Chl *a*-based photosystem, evolution seems to have taken a 'detour' by not replacing the Chl *a*-based system with PPR. It seems that PPR only serves as a supplemental energy source for algal cells. In oceanic bacteria, PPR can supply energy on par with bacteriochlorophyll (Gómez-Consarnau et al. 2019). However, eukaryotic PPR in the oceanic plankton appears less abundant than bacterial PPR, as observed in the South China Sea, where eukaryotic PPR was only about as abundant as bacterial PPR in certain cases (Ma et al. 2023). The relative contribution of PPR to the total energy supply for algae, and thus the magnitude of competitive advantage it confers remains to be studied. The fitness cost of operating PPR that may have limited the spread of this energy machinery in algae await to be unveiled.

17

Nutrient Acquisition and Regulation

Nutrient Cycle in the Marine Pelagic Food Chain

Nitrogen (N), Phosphorus (P), and trace metals like iron (Fe), Zinc (Zn), and Manganese (Mn) are essential for algal growth. Silicate (Si) is required for diatoms and some other algae. Nutrient sources include terrestrial inputs, atmospheric deposition, upwelling, and N_2 fixation (Fig. 17.1). Nutrients assimilated by algae transfer through the food chain, with particulate (POM) and Dissolved Organic Matter (DOM) released into the environment and recycled by bacteria. Some phytoplankton are mixotrophic, utilizing dissolved organic nutrients or ingesting bacteria/algae to compensate for nutrient shortages or in addition to nutrient uptake.

The availability of nutrients constrains the growth rate and yield of primary producers. A gradient of nutrient abundance exists from high abundance in coastal to low abundance in oceanic environments, creating a productivity gradient. In oceanic waters, upwelling in divergent zones supplies nutrients, creating productive areas in the generally oligotrophic open ocean, where nutrient supplies otherwise primarily rely on recycling. At steady state, C, N, Si, and P exist in a stoichiometric ratio of C:N:Si:P = 106:16:16:1 on average (Redfield 1934). Under this paradigm, deviations from this stoichiometry indicate nutrient deficiencies. For instance, an N:P ratio of 30 indicates P limitation, whereas a ratio of 10 suggests N limitation. However, optimal nutrient ratios vary between species due to differences in cellular biochemistry (e.g., relative quantities of P-rich ribosomal RNA, DNA, and phospholipids; N-rich proteins; and C-rich carbohydrates and lipids) (Geider and Roche 2002). Cellular C:N:P ratios also vary with environmental conditions. For example, under nutrient-replete conditions, prasinophytes have lower C:N ratios than diatoms due to higher protein content. In response to N deficiency, both groups increase their C:N ratio, but via different mechanisms (elevated carbohydrates in diatoms versus increased lipids in prasinophytes) (Finkel et al. 2016).

In oceanic waters, primary productivity (and N_2 fixation) is often co-limited by the deficiency of multiple nutrients such as P and Fe or Zn deficiency. Co-limitation is increasingly recognized (e.g., Browning and Moore 2023), revising Liebig's Law of the Minimum that growth is limited by the scarcest nutrient. Notably, N overloading from human activities in coastal oceans can cause P limitation (Karl 2014, Lin et al. 2016, Moon et al. 2021). N is considered a proximate limiting nutrient, while P is an ultimate limiting nutrient (Tyrrell 1999), as N has a biogenic renewable source (N_2 fixation), while P is sourced from slow rock weathering (Paytan and McLaughlin 2007). Human-induced N inputs extend the N-P imbalance from coastal zones to the global ocean (Peñuelas et al. 2013, Karl 2014).

Seaweed nutrition resembles microalgae, utilizing various N and P forms and maintaining species-specific C:N:P ratios (Roleda and Hurd 2019). Nutrient uptake involves passive, facilitated passive, and active transport. The molecular characteristics of nutrient uptake and regulation are less studied in seaweeds than in microalgae.

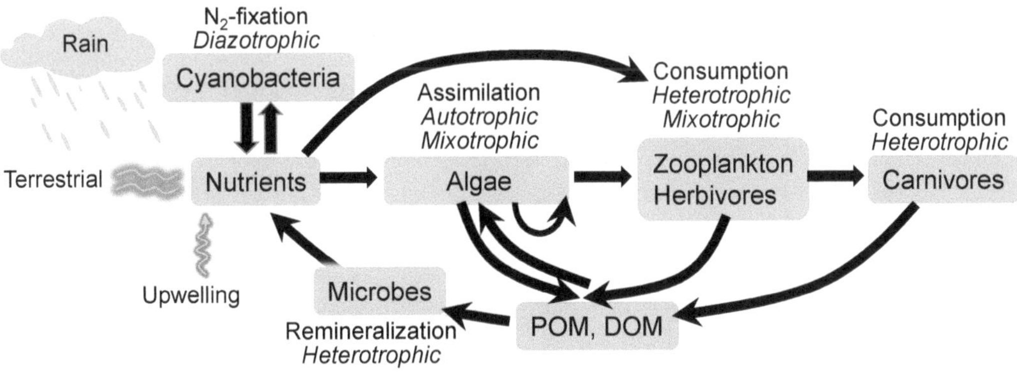

Figure 17.1. Nutrient cycle in marine food web. The sources of nutrients include terrestrial (river and groundwater), rain, upwelling, and biological N_2 fixation. After assimilated by algae, nutrients flux through the food chain, and in the process, parts of them are released into the environment as dissolved organic materials, which are decomposed by bacteria and recycled.

Uptake Kinetics and Regulating Mechanisms

Uptake rates of nutrients are determined by the ambient nutrient concentration, cell size and shape, and the density, binding affinity, and turnover rate of the transport proteins embedded in the cell's plasma membrane (Lin et al. 2016). Nutrient uptake kinetic can be described by the Michaelis-Mention function (Eq. 17.1; Fig. 17.2)

$$V = V_{max} \frac{Ks[S]}{Ks+[S]} \tag{17.1}$$

Where:

- V is the uptake rate
- V_{max} is the maximum uptake rate
- K_S is the half-saturation constant
- [S] is the substrate concentration

The maximum uptake rate (V_{max}) is determined by the density of transporters on the cell membrane (per cell volume or carbon weight) multiplied by the rate of intracellular transport and substrate release within the cell per transporter, as postulated for trace metals (Hudson and Morel 1993). A cell's surface can only accommodate a finite number of transporters, encompassing all the types of nutrients the species needs (Fig. 17.2). These include high-affinity and low-affinity transporters for the same nutrient (e.g., ammonium), symporters of two opposite-charge ions (e.g., Na^+ and NO_3^-), antiporters to import a nutrient while exporting an element with the same charge (e.g., H^+ and Zn^{2+}).

Ks provides a relative measure of the binding affinity of the uptake system for nutrient (Dugdale 1967). A high-affinity transporter has a low K_S while a low-affinity transporter has a high K_S. High-affinity nutrient transporters have stronger binding to the nutrient but are slower in releasing the nutrient once inside the cell and are suited to take up the nutrient at low concentrations (Lin et al. 2016). Low-affinity transporters, in contrast, have weaker binding to the nutrient but are faster in releasing the nutrient once inside the cells, and are suited for uptake when the nutrient is abundant. There is experimental evidence that a trade-off exists between the maximum nutrient uptake rate and affinity. For instance, a study documented substantial variations in N-nutrient utilization traits not only between, but also within, geographic populations of *Alexandrium ostenfeldii* (Brandenburg et al. 2018). An interesting pattern emerged from this study that strains with a higher maximum N-nutrient uptake rate exhibited a lower affinity.

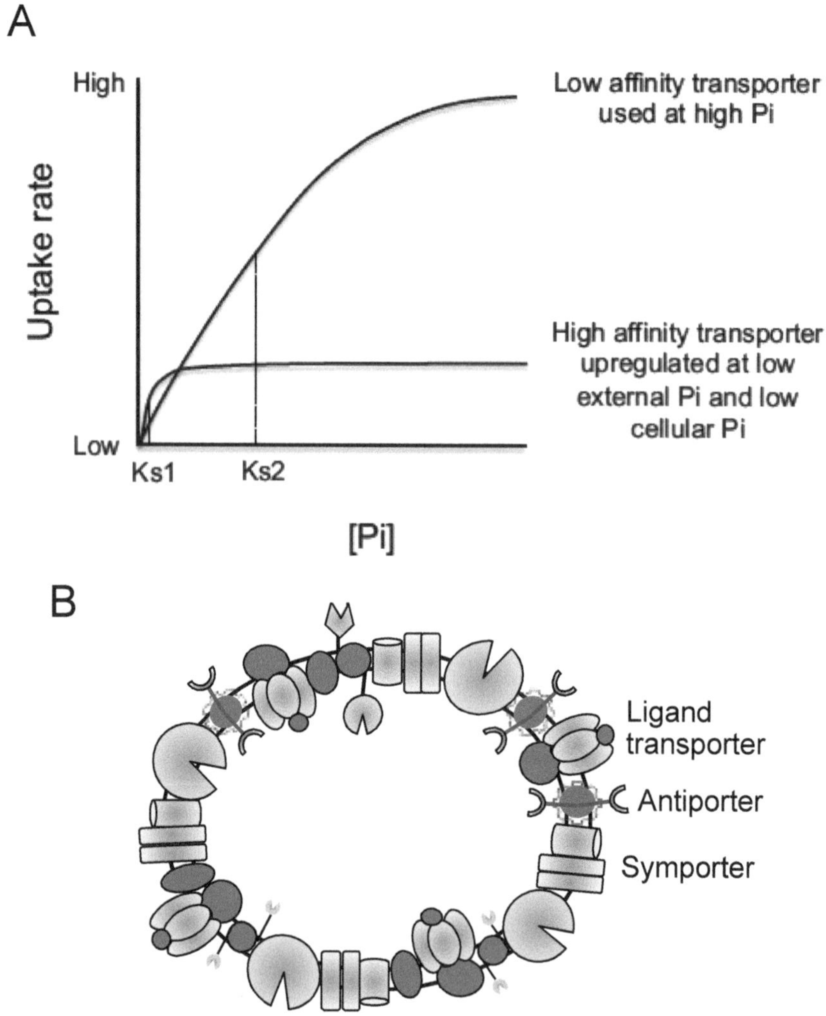

Figure 17.2. Kinetics of nutrient uptake and molecular basis. (A) Uptake kinetics by high affinity and low affinity transporter following the Monod model. (B) Finite number of transporters on cell surface, with example symporters, antiporters, and transporters of nutrients (e.g., iron) complexed by ligands labeled.

The relationship between algal growth rate and nutrient concentration takes two forms. One is the Monod model, which shows the same form as the relationship between uptake rate and nutrient concentration (Fig. 17.3).

$$\mu = \mu_{max} \frac{[N]}{K_N + [N]} \tag{17.2}$$

The other is the Droop model (Fig. 17.4), which posits that algal growth rate is dependent on intracellular nutrient concentration (Q) rather than environmental nutrient concentration and that the cell has a minimum nutrient quota (Q_0) only above which the algae can grow.

$$\mu = \mu_{max} (1 - \frac{Q_0}{Q}) \tag{17.3}$$

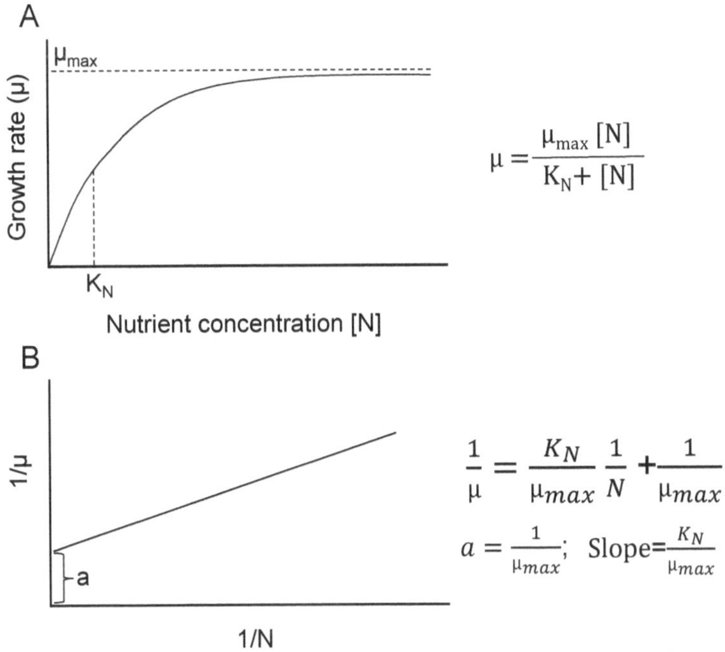

Figure 17.3. Monod model of growth rate versus nutrient concentration (A) and the approach to addressing the half-saturation coefficient (K_N) and maximum growth rate (μ_{max}) by reorganizing the Monod equation (B).

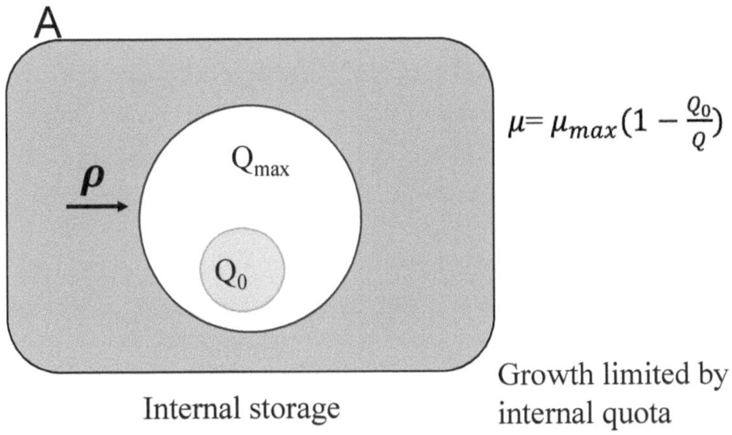

Figure 17.4. Schematic of the Droop model (A) and differential growth patterns of two algal species with different minimum quota and maximum growth rates (B). At low nutrient concentrations the species with low Q_0 and μ_{max} (species 1) will likely win, but when nutrient concentration is high, the species with high Q_0 and μ_{max} (species 2) will likely win.

General Effects of and Strategies to Cope with Nutrient Deficiency

Deficiency of different nutrients emanates common symptoms among algae including:

- A decrease in growth rate, often accompanied by cell cycle arrest;
- A reduction in pigmentation (in severe cases, chlorosis, or "bleaching") and a decrease in the cellular concentration of reaction centers;
- A reduction in Fv/Fm, increase in accessory pigment/Chl *a* ratio and C:Chl *a* ratio;

- An increase in cellular quota of C, decrease in cellular content of the nutrient element and nutrient element/C ratio, and change in biochemical composition (e.g., content of RNA, protein, lipid, etc);
- A reduction in maximal "final yield" of the population or standing stock (i.e., Liebig limitation);
- A shift in species dominance (due to different k_N for different species);
- Autolysis of cells;
- Induction of toxin production in toxin-producing species: e.g., P limitation promotes production of domoic acid in diatoms, saxitoxins in dinoflagellates; iron and Si limitation promote production of domoic acid.

Phenotypic effects of nutrient deficiencies include cell cycle arrest (typically at the G1 phase), stalling population growth, cell size enlargement (in P limitation) or reduction (in N limitation), and downregulation of photosynthesis and carbon concentrating mechanisms (Dagenasis-Bellefeuille and Morse 2013, Lin et al. 2016, Chen et al. 2018). Deficiencies in major nutrients like N, P, or Si cause reprogramming of central metabolic pathways such as glycolysis, the tricarboxylic acid cycle, carbohydrate metabolism, lipid metabolism, and the Ornithine-Urea Cycle (OUC).

Despite common responses, there are marked differences in how algal species respond to nutrient deficiencies. For instance, N- or P-deprived *Amphidinium* cells arrest in G1, whereas *Pfiesteria piscicida* cells stall at both G1 and G2/M phases (Olson and Chisholm 1986, Lin et al. 2004, Li et al. 2015). Under N deficiency, algal photosynthesis continues, often at lower rates, leading to carbon accumulation in the form of starch and/or lipids. N deficiency caused a two-fold higher accumulation of triacylglycerol in dinoflagellates (Fuentes-Grunewald et al. 2012), while in the heterotrophic dinoflagellate *Crypthecodinium cohnii*, N limitation did not significantly affect lipid production (Pleissner and Eriksen 2012).

Algae have evolved strategies to cope with nutrient deficiencies, including:

- Enhancing uptake of the limited nutrient by increasing transporter affinity or producing specific uptake molecules (e.g., siderophores for Fe in cyanobacteria);
- Using alternative forms of nutrients (e.g., switching from NH_4^+ to NO_3^-);
- Inducing enzymes to facilitate the use of organic nutrients (e.g., alkaline phosphatase for phosphate, peptidase, amino acid oxidase, urease, and urea amidolyase for N);
- Vertical migration to access nutrients below the nutricline (e.g., dinoflagellates, large diatoms);
- Induction of substitution molecules (e.g., flavodoxin replacing ferredoxin, plastocyanin replacing cytochrome c6, nitro- or sulfo-lipids replacing phospholipids) to reduce cellular demand for the limited nutrient (e.g., Fe, P);
- Enhanced intracellular recycling of nutrients (e.g., via autophagy);
- Changing trophic modes (e.g., shifting from autotrophic to mixotrophic or from free-living to symbiotic).

At the molecular level, these adaptive measures involve upregulation of nutrient acquisition capability, reprogramming of metabolic pathways to minimize nutrient requirements, and enhancing intracellular recycling. Upregulation of acquisition capability includes increased expression of nutrient transporters and assimilation genes, as well as enzymes that help scavenge alternative forms of the deficient nutrient (e.g., dissolved organic nitrogen and phosphorus).

Reducing cellular demand for the nutrient is typically achieved by replacing functional molecules with equivalents that contain less or none of the deficient nutrient. For instance, algae and microorganisms can shift from (4Fe-4S) to (2Fe-2S) clusters associated with ferredoxin (Fd) and upregulate flavodoxin (Fld) expression while downregulating Fd expression under Fe limitation (Lodeyro et al. 2012, Gao et al. 2021, van Creveld et al. 2023). In the photosystem electron transfer chain, cytochrome (Cyt) c6 can be substituted by plastocyanin, an iron-free, copper-containing molecule (Gao et al. 2021). Under P stress, algal cells often produce sulfolipids and/or nitrolipids to replace phospholipids (van Mooy et al. 2009). Cells can also accelerate intracellular recycling of nutrients through catabolic activities, such as ATP metabolism to recycle P (Shi et al. 2017) and autophagy to recycle cellular constituents (Kajikawa and Fukuzawa 2020, Li et al. 2024).

Nitrogen (N)

N nutrients typically include ammonium (NH_4^+), nitrate (NO_3^-), nitrite (NO_2^-), and dissolved organic N-containing compounds (DON) such as urea (H_2NCONH_2), nucleotides, amino acids, and cyanates (OCN^-). In these diverse forms, N has different valence states. Inorganic N can result from nitrification or denitrification, the latter including dissimilatory and assimilatory reactions. N is the only nutrient with a biogenic source, as diazotrophic blue-green algae and bacteria can fix N_2 into NH_4^+. N_2 can diffuse into cells, whereas the uptake of environmental N-compounds relies on transporters. In marine and freshwater environments, besides N_2 fixation, NH_4^+ also comes from organism metabolism, qualifying NH_4^+ as a regenerated or remineralized nutrient. The oxidized forms, NO_3^-/NO_2^-, usually enter the ocean from terrestrial sources (e.g., river discharge), thus qualifying as new nutrients. Urea, cyanate, and other DON are metabolic products of organisms. However, ammonium and urea are widely industrially produced as fertilizers, providing ample terrestrial supply for ocean and freshwater ecosystems. These external supplies, along with biological N_2-fixation, constitute the new N-nutrient.

N-Nutrient Uptake and Assimilation

Transporters for NH_4^+, NO_3^-/NO_2^-, and urea have been well documented for algae (Fig. 17.5). Involved in the uptake of other DON are ABC cassette transporters, but these are less characterized. Multiple transporters may exist for each nutrient. For instance, there are high-affinity and low-affinity transporters for ammonium and nitrate. For urea uptake, plants have two transport systems: one being a sodium solute symporter that co-transports urea with protons at high affinity, and the other being members of the tonoplast intrinsic protein (a type of aquaporin) that transport urea in a channel-like manner (Kojima et al. 2007).

Once transported into the cell, nitrate is reduced in the cytoplasm to nitrite by a cytosolic NADH-dependent Nitrate Reductase (NR). Nitrite is then transported into the chloroplast, where it is further reduced to ammonium by a ferredoxin-dependent nitrite reductase (Fd-NiR). Ammonium is incorporated into glutamate by Glutamine Synthetase (GS), producing glutamine. Finally, glutamine reacts with 2-oxoglutarate under the action of glutamine-oxoglutarate aminotransferase (GOGAT), also known as glutamate synthase, to produce two molecules of glutamate (Bowler et al. 2010).

Diatoms possess a plastid-localized GSII of red algal origin, believed to be responsible for the assimilation of ammonium produced by nitrate reduction (Robertson et al. 1999). Accordingly, the transcript of GSII is more abundant in diatom cells assimilating nitrate than in those assimilating ammonium. Additionally, diatoms have an NAD(P)H-dependent GOGAT (NAD[P]H-GOGAT) and a ferredoxin-dependent counterpart (Fd-GOGAT), believed to be plastid-localized. Diatoms also harbor a cytosolic NAD(P)H-dependent nitrite reductase (NAD[P]H-NiR) homologous to nirB of bacteria and fungi, and a mitochondrial GSIII, neither of which occur in green algae or plants, suggesting a bacterial origin of this gene.

Diatoms are known to take up higher amounts of nitrate when it is abundant in the environment (luxury uptake) and store it inside cells for subsequent photosynthetic assimilation or surviving dark and anoxic conditions (e.g., on sediment) through dissimilatory nitrate reduction to ammonium (Stief et al. 2022). Nitrate concentrations inside diatom cells can exceed ambient nitrate concentrations by ~ 100–4000 folds.

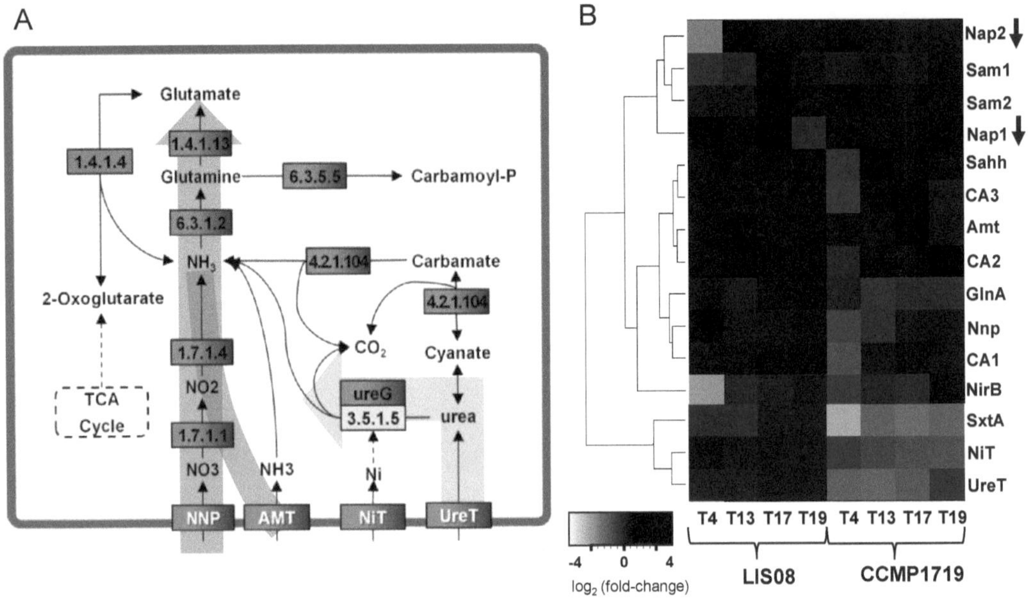

Figure 17.5. N-nutrient uptake and assimilation pathways (A) and genes responding to N-limitation (B) in the dinoflagellate *Alexandrium catenella*. (A) Genes for transporters of nitrate/nitrite (NNP), ammonium (AMT), nickel (NIT), and urea (UreT) and for enzymes involved in N assimilation were expressed at higher levels in the bloom (shading on the right of the boxes) than in lab cultures (shaking on the, left of the boxes). (B) Transcriptional responses of genes related to N nutrition, carbon concentrating mechanism (CA), and toxin synthesis (sxtA). Gene names are shown on the right of the heatmap. Nap, nitrogen-associated protein; Sams, S-adenosylmethionine synthetase; Sahh, S-adenosylhomocysteine hydrolase; CA, carbonic anhydras; Amt, ammonium transporter; GlnA, glutamine synthetase; Nnp, nitrate/nitrite transporter; NirB, nitrite reductase (NAD(P)H) large subunit; SxtA, saxitoxin A; NiT, nickel transporter; UreT, urea transporter. Reproduced with permission from Zhuang et al. (2015).

It is generally held that the above-outlined diatom model of N-nutrient acquisition applies to algae in general (Bowler et al. 2010). However, unique additional mechanisms may occur in other lineages. For instance, as mentioned above and elaborated in Chapter 8, dinoflagellates can utilize cyanate and use polyguanine as N storage. Most, if not all, algae can utilize urea.

Ornithine-Urea Cycle (OUC)

Nitrogen may be metabolized through the Ornithine-Urea Cycle (OUC) in algae. The OUC in algae involves five enzymes: Carbamoyl Phosphate Synthetase (CPS), ornithine transcarbamylase (OTC), argininosuccinate synthase (ASuS), argininosuccinate lyase (ASL), and arginase (ARG). OUC in *T. pseudonana* (Armbrust et al. 2004) was the first documented outside metazoans. It is now known to serve as a distribution and repackaging hub to mobilize N into different metabolic pathways in diatoms (Allen et al. 2011; see Chapter 7). Many other algae also possess the complete gene set for OUC, including stramenopiles, haptophytes, cryptophytes, rhodophytes, and dinoflagellates, suggesting OUC is common in eukaryotic algae (Horák et al. 2020). For dinoflagellates, OUC genes have been identified in *Alexandrium tamarense* (Dagenasis-Bellefeuille and Morse 2013), *Gambierdiscus caribaeus* (Price et al. 2016), among others. OUC also exists in vascular plants, where it plays a significant role in ammonium detoxification (Esteban et al. 2016). However, OUC in vascular plants is considered incomplete because they do not have Carbamoyl Phosphate Synthase-type I (CPSI), the enzyme that synthesizes urea directly from NH_4^+; instead, they use CPSII (EC 6.3.5.5.) to synthesize urea from glutamine (Urra et al. 2022).

Response of Uptake Mechanisms to N-nutrient Variability

A consensus emerges about how the N-nutrient uptake and assimilation steps respond to fluctuating nutrient availability. As shown in Fig. 17.6, when N-nutrient is limited, the transporters of various forms of N-nutrients (NH_4^+, NO_3^-/NO_2^-, urea, DON) are upregulated in most species, with a few exceptions mostly in dinoflagellates. Among dinoflagellates, the free-living Symbiodiniaceae *Effrenium voratum* downregulated both transporters and assimilation-related genes under N limitation, whereas *Scrippsiella trochoidea* and *Karenia mikimotoi* have some transporters upregulated and others downregulated. Genes for urea and nitrate utilization are upregulated in most species except the diatoms *T. pseudonana* and *P. tricornutum* under N limitation. For ammonium assimilation, Glutamine Synthetase (GS), GSIII, glutamate synthase (GOGAT), NADPH-dependent GOGAT, and ferredoxin-dependent GOGAT are all consistently upregulated with few exceptions under N limitation. However, GSII and glutamate dehydrogenase are upregulated in about half of the species and downregulated in the other half under N limitation, potentially indicating lineage-specific responses.

Response of Central Metabolic Pathways to N-Nutrient Variability

N-nutrient conditions influence the central metabolism of algal cells. Consensus from omics data across different algal lineages reveals common responses, though variability exists between lineages in some metabolic pathways (Fig. 17.7). The strongest consensus, based on both transcriptomics and proteomics, is the upregulation of glycolysis and the TCA cycle, and the downregulation of photosynthesis under N limitation. Exceptions in these pathways may be due to species-specific distinctions or different experimental conditions or sampling methods. Oxidative phosphorylation, the pentose phosphate pathway, and BCAA (Branched-Chain Amino Acid) degradation are consistently upregulated, while the glyoxylate pathway and lipid degradation are consistently downregulated under N limitation, though data coverage is limited. BCAA degradation can provide acetyl-CoA and pyruvate for fatty acid biosynthesis. The most variable response occurs in lipid biosynthesis, which is upregulated under N limitation in some species and downregulated in others. For instance, the dinoflagellate *Prorocentrum shikokuense* shows opposite responses in the transcriptome (upregulated) and proteome (downregulated).

Based on transcriptomic studies, N-deprivation leads to the upregulation of glycolysis and the tricarboxylic acid cycle (TCA) in the diatoms *Thalassiosira pseudonana* and *Phaeodactylum tricornutum* (Hockin et al. 2012, Alipanah et al. 2015). Additionally, N deficiency in *P. tricornutum* causes shifts towards lipid accumulation through reconfiguration of intermediate metabolism rather than upregulation of fatty acid and lipid biosynthesis (Levitan et al. 2014, Longworth et al. 2016). In a proteomic study on *T. pseudonana*, Chen et al. (2018) confirmed the promoting effect of N deficiency on glycolysis and TCA, and showed that 23 ribosomal proteins were downregulated under N deficiency, compared to 89 downregulated proteins under P deficiency and 11 (mostly from the chloroplast ribosome) upregulated under Si deficiency.

Dinoflagellates exhibit similar responses to diatoms. In *Alexandrium catenella* (formerly *A. fundyense* or *A. tamarense* Complex Group I), glycolysis and the TCA cycle were upregulated at lower nitrate concentrations (0.4 mM vs. 0.9 mM) and downregulated at elevated nitrate concentrations (1.3 mM vs. 0.9 mM) (Saldivia et al. 2023). Upregulated glycolytic enzymes included glyceraldehyde 3-phosphate dehydrogenase (GAPDH), enolase (ENO), pyruvate kinase (PK), aldolase (ALD), and phosphoglycerate kinase (PGK). Since glycolysis generates pyruvate, which feeds into the TCA via acetyl-CoA, the upregulation of glycolysis is expected to promote the TCA. Similar metabolic responses were reported in another typical (peridinin type) thecate dinoflagellate *P. shikokuense*; additionally, the phagotrophic pathway was upregulated (Li et al. 2021). In *Karenia mikimotoi*, athecate species with tertiary plastids, genes involved in glycolysis,

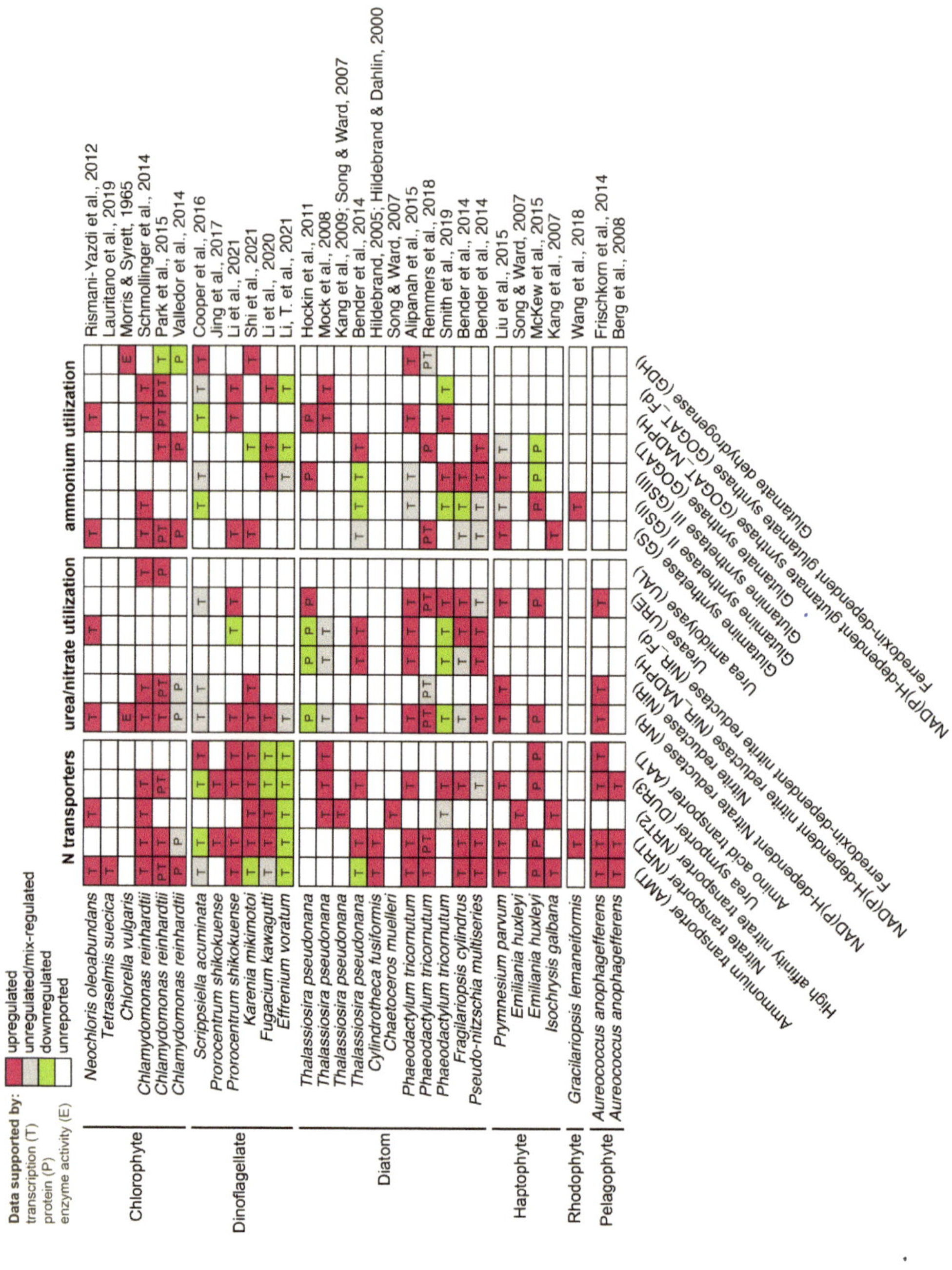

Figure 17.6. Responses of N uptake and assimilation genes to N limitation.

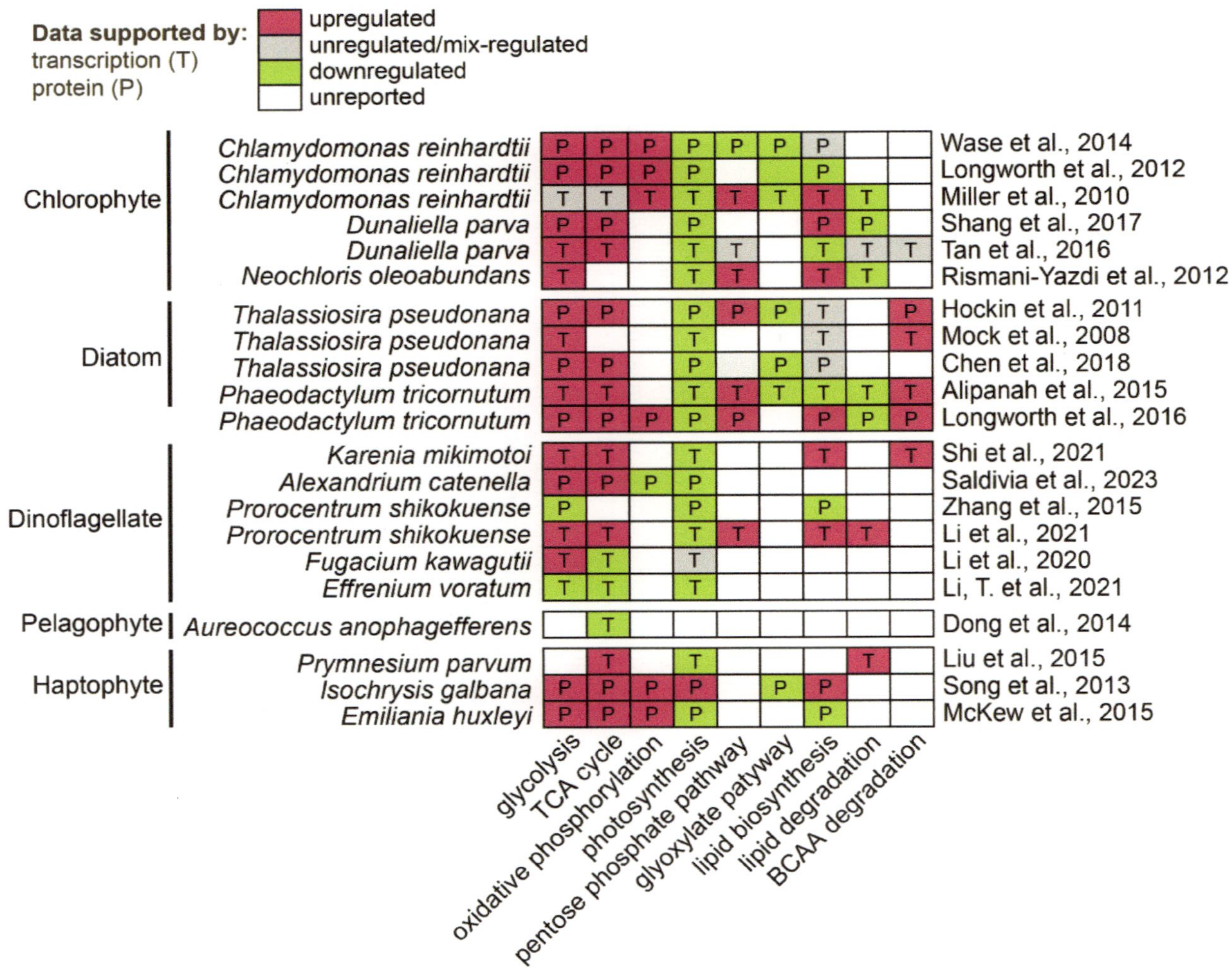

Figure 17.7. Responses of central metabolism to N limitation.

the TCA cycle, fatty acid metabolism, and lipid accumulation were significantly upregulated under N-nutrient limitation (Shi et al. 2021).

Urea Utilization

Abundant Sources of Urea

Urea is a significant part of the ornithine-urea cycle (Allen et al. 2011). Internal urea can be released into the environment through excretion or decomposition of dead organisms. Additional sources of external urea come from agriculture (man-made urea), aquaculture, and wastewater discharge from human activities (Glibert et al. 2006, Bouwman et al. 2013). Consequently, urea can reach up to 150 μM and account for over 50% of the total dissolved N in some aquatic systems (Bogard et al. 2012). Concentrations of 25–50 μM are common in coastal oceans (Glibert et al. 2006, 2016). Increasing environmental urea can differentially stimulate the growth of some phytoplankton, promoting shifts to harmful species. For example, dinoflagellate dominance in phytoplankton communities increased with the percentage of urea contributing to total N uptake in Moreton Bay and the Florida Shelf (Glibert et al. 2006). There is mounting evidence that some phytoplankton species prefer urea over DIN, such as *Lingulodinium polyedrum* off the Baja Peninsula, Mexico (Kudela and Cochlan 2000), *Prorocentrum cordatum* (formerly *P. minimum*) in the Chesapeake Bay, USA (Fan et al. 2003), and *Alexandrium pacificum* in the Thau Lagoon, France (formerly *A. catenella*; Collos et al. 2004).

Two Enzyme Systems for Urea Utilization

Environmental urea is taken up by plants through passive or active transport systems involving aquaporin or urea/amide channel proteins and the DUR3 protein (Kojima et al. 2007). DUR3 is a high-affinity urea transporter. As a nitrogen source, urea must be broken down to ammonium after uptake through urea transporters, mediated by either urease (URE) or Urea Aamidolyase (UA) (Mobley and Hausinger 1989, Hausinger 2004). URE is widespread in algae and is an amidohydrolase with nickel as the cofactor. In contrast, UA is better known in bacteria and fungi and exists in multi-domain or multi-protein complexes. This enzyme system includes a urea transporter (DUR3A, B, C), urea carboxylase (UC or DUR1), and allophanate hydrolase (AH or DUR2) domains, with the latter two catalyzing ATP- and biotin-dependent urea carboxylation (consuming HCO_3^-) to produce allophanate and deamination of allophanate into ammonium and CO_2, respectively. DUR1 and DUR2 are encoded by separate genes in some organisms (Zhao et al. 2018), such as chlorophytes, in which they are often located separately, and bacteria, in which they are next to each other. In fungi, DUR1 and DUR2 are fused. Recently, UA was identified in 10 phyla of algae, including Cyanophyta, Rhodophyta, Chlorophyta, Dinophyta, Haptophyta, Bacillariophyta, Cryptophyta, Eustigmatophyta, Phaeophyta, and Prasinodermophyta (Liang et al. 2024).

Genes encoding urea transporters and the two enzymes are commonly upregulated under N-limited conditions (Li et al. 2021). Urea transporters are also upregulated under urea-replete conditions relative to DIN-replete conditions. The two urea-utilizing enzymes respond differently to nutrient conditions. URE gene expression in the dinoflagellate *Prorocentrum shikokuense* (formerly *P. donghaiense*) increased markedly under N-limited conditions but did not significantly increase under urea-replete conditions (Jing et al. 2017). In contrast, UA was significantly upregulated in both N-deficient and urea-replete conditions (Navarathna et al. 2011). A recent study using CRISPR/Cas9-based gene knockout technology revealed that DUR2 is essential for urea utilization, whereas DUR3 seems to have functional counterparts in *C. reinhardtii*, as its knockout only decreased, but did not abolish, urea uptake (Liang et al. 2024).

Phosphorus (P)

P is a major nutrient, second only to Nitrogen (N), essential to the growth or algae and other organisms. P deficiency causes bone diseases such as rickets in children and osteomalacia in adults, while in plants, it leads to purple casts and wilting of leaves. P is a central component of nucleic acids (both DNA and RNA), which account for 42–78% (on average ~ 57%) of total non-storage cellular P in plants, green algae, and cyanobacteria (Raven 2013). Phospholipids, key component of cell membranes, hold 15–30 % of total cellular P in the plant model *Arabidopsis thaliana* (Pant et al. 2015) or 5–32% of non-storage P in plants and algae (Raven et al. 2013). Evidently, nucleic acids and membrane phospholipids represent the bulk of cellular P in algae (Dyhrman 2016).

P also plays a central role in the production of chemical energy (ATP) and of reducing equivalents (NADH, NADPH) during photosynthesis and respiration, which are essential for carbon fixation and cell metabolism (Falkowski and Raven 2007). Polyphosphate is synthesized and stored in the algal cells to serve as a storage of P and energy, and potentially as a buffering agent of intracellular pH (Zhang et al. 2022). Phosphorous also regulates the activity and function of many proteins and metabolic processes through phosphorylation and dephosphorylation, modulating signaling pathways in cells (e.g., through adenosine monophosphate [AMP] or inositol trisphosphate [IP3]) (Cooper 2000).

Unlike nitrogen that occurs in several valence states, P in compounds has only valence states of 5 or 3 (Paytan and McLaughlin 2007). The dominant form of P nutrient is inorganic orthophosphate (DIP, PO_4^{3-}), although phosphine (PH_3) and phosphide (P^{3-}) are also present in the ocean. The inorganic polymer of phosphate, polyphosphate (polyP) contains high-energy bonds, thus providing both P nutrient and energy. Dissolved Organic Phosphate (DOP) compounds such as Glucose-6-Phosphate (G6P), adenosine triphosphate (ATP), glycerol-phosphate, phospholipids, phosphoproteins are alternative but also important sources of P nutrient due to their relatively high abundance in the euphotic zone (Duhamel et al. 2021). DOP as P-nutrient is particularly important in supporting primary production in areas where DIP is limited, which is much more prevalent than previously thought (Ustick et al. 2021). It is estimated that DOP supports up to 30% of the primary productivity in the North Atlantic Ocean (Mather et al. 2008).

Differential Mechanisms of P Nutrient Acquisition

Uptake of DIP relies on phosphate transporters, including low-affinity and high-affinity ones (Fig. 17.8). Once inside the cell, phosphate is incorporated into ATP to start its participation in the cell's metabolism. DOP utilization either starts with extracellular hydrolysis followed by uptake of the released phosphate by DIP transporters or transport of DOP molecules by specialized transporters such as ABC cassette followed by intracellular hydrolysis or possible direct participation in metabolic processes (Fig. 17.8).

The ability to use different forms of P nutrients and underlying mechanisms regulating the utilization are differentiated between different phytoplankton taxa. Studies have shown that G6P and ATP can as efficiently support growth of diatoms and dinoflagellates as DIP (Zhang et al. 2017, Luo et al. 2017, Zhang et al. 2020). The dinoflagellate *Karenia mikimotoi* utilizes ATP via extracellular hydrolysis involving 5'-nucelotidase as well as direct uptake of ADP/AMP (Luo et al. 2017) whereas *Prorocentrum shikokuense* (formerly *P. donghaiense*) uses ATP by direct uptake (Li et al. 2015). In the diatom *Sketonema costatum*, ATP supplied as the sole source of P is utilized via extracellular hydrolysis, potentially involving NAD+ diphosphatase rather than AP or 5'-nucelotidase (Zhang et al. 2020). *Phaeodactylum tricornutum* also uses different mechanisms to utilize ATP and G6P, through extracellular hydrolysis (possibly by 5'-nucelotidase due to upregulated gene expression) and direct uptake followed by intracellular hydrolysis (possibly by phoD as its expressed is upregulated)

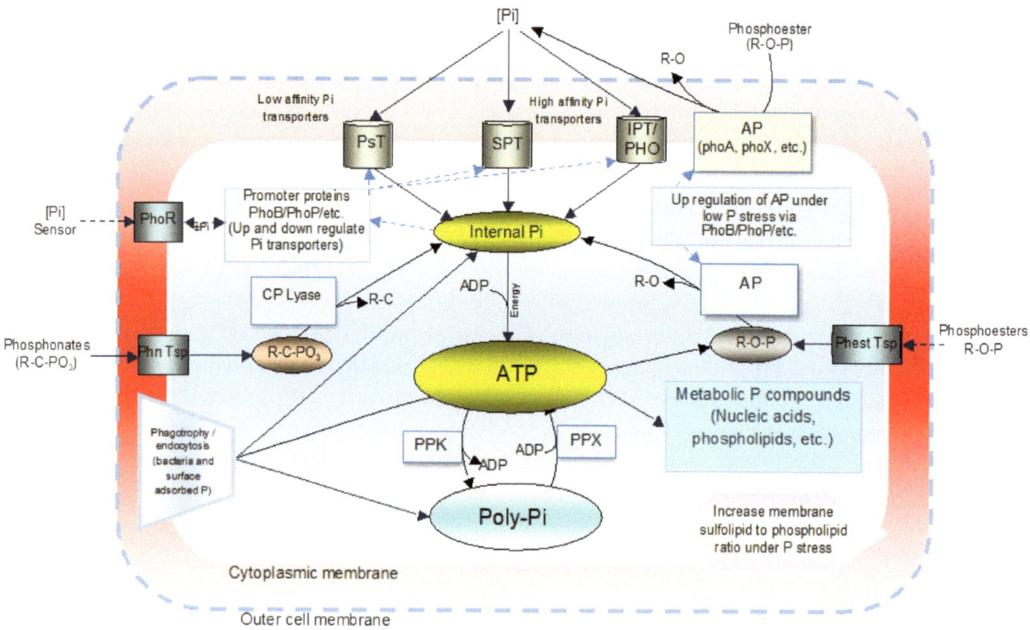

Figure 17.8. Uptake and assimilation of inorganic and organic phosphorus nutrients. Reproduced with permission from Lin et al. (2016).

(Zhang et al. 2023). These findings indicate that the mechanism by which phytoplankton utilize DOP is species- and substrate-dependent (Fig. 17.9). Due to the functional differentiation among taxa, a phytoplankton community with high species diversity can emanate high P use efficiency (Lomas et al. 2014).

Although G6P and ATP are both as effective P nutrients as DIP to support the growth of dinoflagellates examined so far, in general the ability and efficiency of DOP utilization are species- and substrate-dependent. In a mixed-species culture, the fast-growing diatom *Skeletonema*

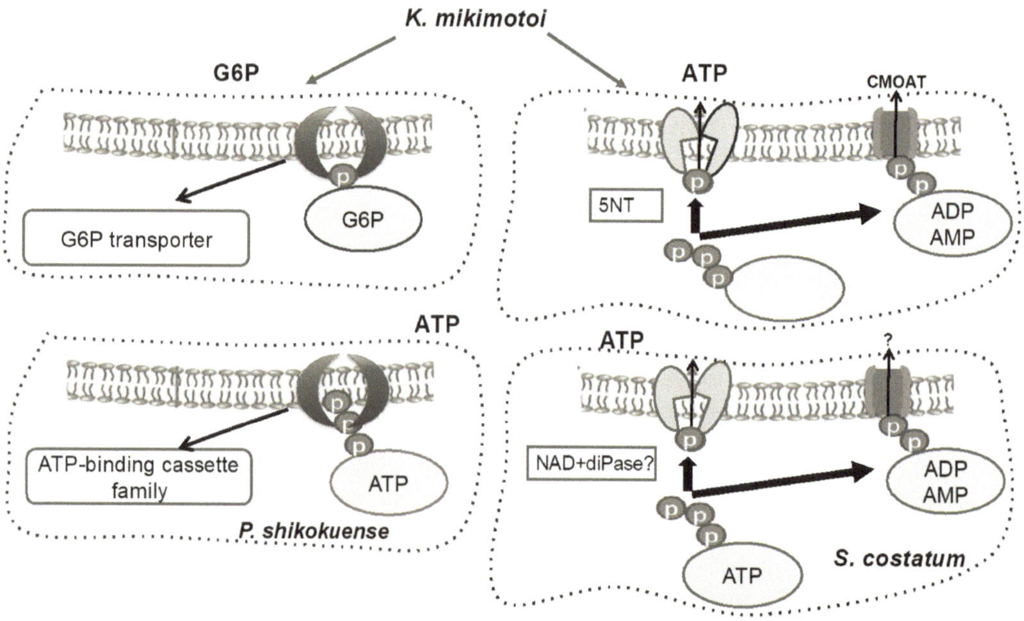

Figure 17.9. Variable mechanisms underlying DOP utilization depending on the type of DOP and algal species.

costatum (sensu lato) always outgrows the dinoflagellates *Prorocentrum shikokuense* (*formerly P. donghaiense*) and *Alexandrium pacificum* (formerly *A. catenella*) regardless of DIP (KH_2PO_4) concentrations or DOP type. *P. shikokuense* grows the best on lecithin, with a 2-4 fold higher growth rate than on KH_2PO_4. *A. pacificum*, in contrast, does not grow as well on DOP as the other two species (Ou et al. 2015). The diatom *P. tricornutum* can utilize glycerol-3-phosphate and phytate but is unable to utilize triethyl phosphate (Zhang et al. 2022).

Alkaline Phosphatase: P-Stress Indicator and Functional Differentiation

Alkaline Phosphatase (AP) is the most common enzyme implicated in DOP utilization. It hydrolyzes phosphomonoesters or phosphodiesters. The prototypic AP (PhoA), discovered in *Escherichia coli*, contains Zn and Mg as dual metal cofactors (2Zn-Mg), whereas other types of AP may contain Fe and Ca, e.g., PhoX containing 2Fe-3Ca (Yong et al. 2014) and PhoD (phosphodiesterase) containing Fe-2Ca (Rodriguez et al. 2014). Algae use PhoA, a similar but distinct variant of archetype PhoA (atypical PhoA or $PhoA^{aty}$), as well as PhoX and PhoD (Lin et al. 2013, Lin et al. 2015). Interestingly, CRISPR (Clustered Regularly Interspaced Short Palindromic Repeats), the genetic feature now widely used for genome editing, was initially discovered in *E. coli* during research on variants of the PhoA gene (Ishino et al. 1987).

Differential Mechanisms of DOP Utilization and AP as Indicator of P Stress

Recent studies show that DOP utilization does not always involve AP. For example, the dinoflagellate *P. shikokuense* grows on ATP as the sole source of P without hydrolyzing it extracellularly, suggesting that ATP is transported into cells (Li et al. 2015). An ATP-binding cassette might be involved in ATP uptake. In contrast, *Karenia mikimotoi* growing on ATP as the sole source of P hydrolyzes ATP and releases phosphate into the medium, but with no AP activity detected from the cells (Luo et al. 2017). Instead, a 5'-nucleotidase is a potential enzyme responsible for ATP hydrolysis due to its elevated expression in ATP cultures compared to DIP cultures. In addition, *K. mikimotoi* utilizes the hydrolysis products ADP/AMP as well, potentially via a canalicular Multispecific Organic Anion Transporter (cMOAT). Similar mechanisms of DOP utilization were found in Symbiodiniaceae. When grown on glycerol phosphate as the sole source of P, *Cladocopium goreaui* hydrolyzed the DOP extracellularly. However, no extracellular hydrolysis was detected with phytate, and upregulation of endocytosis genes was observed, suggesting uptake of phytate via endocytosis (Li et al. 2023). In the diatom *Skeletonema costatum*, utilization of ATP involves extracellular hydrolysis, possibly by NAD+ diphosphatase (NAD+ diPase) instead of AP (Fig. 17.8; Zhang et al. 2020). Examining previous studies indicates that AP activity is always detected and increases when algae are under P-nutrient stress, suggesting that AP is more an indicator of P stress than an indicator of DOP utilization.

High Divergence of AP Gene Sequence and Differentiation of AP Function

AP sequences (amino acid) in eukaryotic algae are highly divergent (Lin et al. 2015, Zhang et al. 2022). $PhoA^{aty}$ seems to use Ca^{2+} as the major cofactor, while its activity is also promoted by Mg^{2+} and Mn^{2+} but not Zn^{2+} (Lin et al. 2019). AP subcellular localization can be highly differentiated too, ranging from intracellular, extracellular (cell surface or secreted), to diffuse, implying different pathways of DOP utilization (extracellular hydrolysis, import followed by intracellular hydrolysis, or both) (Lin et al. 2012). Such localization and functional differentiation can happen to the multiple isoforms of AP within a species. Using CRISPR/Cas9 genome editing technique, Zhang et al. (2022) demonstrated that PhoA and PhoD in *P. tricornutum* are extracellular and intracellular AP, respectively, and PhoA is responsible for the utilization of G3P, whereas Phytate is transported into the cell and hydrolyzed by intracellular APs such as PhoD. This example demonstrates that diverse AP isoforms or isoenzymes co-existing within a species have likely undergone functional differentiation.

DOP Transporters

Transporters for DOP are poorly understood overall. For ATP and its degradation products (ADP, AMP), a homolog of cMOAT was identified in the dinoflagellate *Karenia mikimotoi* (Luo et al. 2017). cMOAT is one of the ATP-binding cassette transporters, which in humans serves as multidrug resistance anion exporter. Similarly, ATP-binding cassette transporter is the suspected transporter of glucose-6-phosphate in *K. mikimotoi* (Zhang et al. 2017). Some phytoplankton can grow on glyphosate, the globally used herbicide as it effectively inhibits the Shikimate pathway that is responsible for the synthesis of aromatic amino acids (Wang et al. 2016). Chemically, glyphosate is a phosphonate characterized by a C-P bond. In bacteria capable of metabolizing glyphosate, there is a gene operon encoding a C-P lyase system responsible for the transport and breakdown of phosphonates (phnCDEFGHIJKLMNOP). PhnC/K/L homologs were recently detected in the haptophyte *Isochrysis galbana*, which is able to grow on glyphosate as the sole source of P (Wang et al. 2024).

Polyphosphate

Inorganic polyphosphate (polyp) is defined as polymers of three to thousands of phosphate residues linked by high-energy phosphoanhydride bonds. PolyP occurs widely across the tree of life, in both bacteria and eukaryotes. In mammals, polyP occurs in low abundances but plays an important regulatory role in various physiological functions such as blood clotting, bone mineralization, vascular homeostasis and energy metabolism (Borghi and Saiardi 2023). There is mounting evidence that PolyP plays highly diverse roles beyond serving as P and energy storage. Although it could be localized to the cell wall, mitochondria, nucleus, and endoplasmic reticulum, polyP mainly accumulates in acidocalcisomes and acidic vacuoles. Its synthesis in blue-green algae as well as bacteria and amoebae involves polyP kinase (PPK). In most unicellular eukaryotes, its synthesis involves polyP polymerase and occurs in the Vacuole Transport Chaperon (VTC), a membrane-associated protein complex (Hothorn et al. 2009, Borghi and Saiardi 2023). Despite the different synthesis machinery, both are implemented by the transfer of the terminal phosphate of ATP to the growing polyP chain, forming the phosphoanhydride bonds and releasing ADP. In yeast, the VTC complex has three polypeptide components, VTC1, VTC4, and VTC2/VTC3. VTC4 is the catalytic subunit of the complex and has a SPX domain at the N-terminus, a polyP polymerase domain in the central region, and a VTC domain in the C-terminus. The SPX domain binds inositol pyrophosphate and activates polyP polymerase activity. VTC4 is widely distributed in microbes but absent in mammals and the plant clade Viridiplantae except for the Chlamydomonadales order of microalgae, suggesting the ancestral polyp synthesis machinery was secondarily lost in evolution and replaced by a different mechanism (Borghi and Saiardi 2023).

It has been shown that different organisms require polyP to cope with various stress conditions such as high temperature, oxidative stress, osmotic shock, survival during stationary phase, heavy metal accumulation, and nutrient deprivation. Recent studies reported that many nucleolar proteins undergo a unique post-translational modification called lysine-polyphosphorylation. In this modification, polyP chains are covalently attached to lysine residues, a novel mechanism for protein regulation and function in ribosome biogenesis (Azevedo et al. 2015, Bentley-DeSousa et al. 2018). Furthermore, there seems to be a high complexity of interactions of polyP with other elements such as phospho-inositol that can potentially influences nucleolar function and synthesis of polyP itself (Borghi and Saiardi 2023).

For phytoplankton, polyP used to be thought of as a phosphate storage when there is copious supply from the environment, which would be hydrolyzed by enzymes such as polyphosphatase (PPX) to release phosphate for metabolic use when phosphate in the environment is depleted (Solórzano and Strickland 1968, Sicko-Goad and Jensen 1976, Kornberg et al. 1999, Diaz et al. 2008, Li and Dittrich 2019). However, this long held not on was challenged by recent studies, which showed that polyP was syn sized under P deficiency. Furthermore, in the bacterium *Ruegeria pomeroyi*, polyP seemed to be hydrolyzed by alkaline phosphatases PhoD and PhoX and a 5'-nucleotidase (Adams et al. 2022).

In the diatom *Thalassiosira pseudonana*, two VTC family genes were identified, and the transcription of VTC4 and the cellular content of polyP were both found to increase under P limitation (Dyhrman et al. 2012). Similarly, plankton in natural environments show higher levels of polyP in areas where phosphate is scarce (Martin et al. 2014). In the chlorophyte model *Chlamydomonas reinhardtii*, a study conducting a suite of physiological and functional genetic analyses showed that polyP synthesis enabled the algal cells to acclimate to Sulfur (S) deprivation (Sanz-Luque et al. 2020). Initial data linked sulfur deprivation to polyP accumulation. However, through comparative analyses between VTC4 mutants and wild type using normal growth conditions and treatments with electron transport or proton gradient inhibitors, the authors demonstrated that it was the synthesis, not the accumulation, of polyP that facilitated the acclimation to S deprivation. This was because the direct cause of S-deprivation acclimation was the electron flow in the chloroplast and/or mitochondria and ATP homeostasis, which was regulated by polyP synthesis. The results of the study also indicate that the synthesis of polyP is essential not for the initiation, but for the sustenance, of the acclimation to S deprivation. In addition, the study revealed that polyP synthesis in *C. reinhardtii* also serves to regulate acclimation to N deprivation.

As suggested earlier for bacteria (Harold 1966), another role of polyP might be to buffer intracellular pH variation when external pH fluctuates. In the diazotrophic blue-green alga *Trichodesmium erythraeum* IMS101, polyP abundance was found to increase under acidified (by elevated CO_2) and P-limited conditions, suggesting that polyP might be accumulated to buffer intracellular acidification (Zhang et al. 2022). Much more research is needed to unveil the potentially diverse functions of polyP in algae.

SPX-PHR-PSI Regulatory Cascade

In land plants, Phosphorus (P) homeostasis is regulated by a sophisticated molecular mechanism involving SPX (SYG1/PHO81/XPR1) domain-containing proteins and the phosphate starvation response protein (PHR). SPX acts as an upstream negative regulator, while PHR is a positive regulator of P uptake (Secco et al. 2012, Puga et al. 2014, Wang et al. 2014). This SPX-PSR cascade regulates the expression of P Starvation-Induced (PSI) proteins. SPX domains can either stand alone or combine with other functional domains, such as SPX-EXS, SPX-MFS, and SPX-RING.

SPX-EXS (ERD1/XPR1/SYG1) is crucial for the acquisition, translocation, and allocation of phosphate (Wege et al. 2016). SPX-MFS (Major Facilitator Superfamily) represents a new group of vacuolar P transporters in plants (Liu et al. 2015). SPX-RING, which contains both N-terminal SPX and C-terminal RING (Really Interesting New Gene) domains, is known for N limitation adaptation but also regulates P transporters, suggesting its role in N/P stoichiometric homeostasis (Lin et al. 2013, Peng et al. 2007, Kant et al. 2011). The SPX-PSR cascade is ancient, as it appears to be shared with algae.

In the chlorophyte *C. reinhardtii*, SPX-VTC (Vacuolar Transporter Chaperone) and SPX-SLC (permease solute carrier 13), previously identified in yeast, regulate the influx and efflux of vacuolar phosphate (Sanz-Luque et al. 2020, Wang et al. 2021). Recently, SPX and PHR were also identified in the diatom *P. tricornutum*. Using CRISPR/Cas9 gene knockout, SPX in this diatom was shown to be a negative regulator of PSI, including phosphate transporters and alkaline phosphatases, mediated by PHR (also known as PSR1 in algae) (Kumar Sharma et al. 2020, Zhang et al. 2021).

Crosstalk Between N and P Signaling and N/P Homeostasis

Because nutrient supply in the soil is highly variable, prompting land plants to rely on a coordinated network that integrates Nitrogen (N) and Phosphorus (P) signaling to control uptake and maintain N:P stoichiometric homeostasis (Helliwell 2023). Algae face similar nutrient variations and deficiencies in aquatic environments. Despite extensive research on physiological and molecular responses to nutrient changes and stress, understanding whether phytoplankton have an integrative N-P signaling pathway like plants has been limited. A transcriptomic study of *Symbiodinium* isolates under heat stress and P-limitation revealed downregulation of N uptake and assimilation genes, including nitrate transporter (NRT), nitrate reductase (NR), nitrite reductase (Nir), glutamine synthetase (GS), and glutamine-oxoglutarate aminotransferase or glutamate synthase (GOGAT) (Li et al. 2024). This suggests a potential that *Symbiodinium* has a mechanism to regulate the N:P stoichiometric homeostasis.

A study investigating the regime shift from a diatom-dominant phytoplankton community to a dinoflagellate (*Prorocentrum shikokuense*, a.k.a. *P. donghaiense*) showed high diatom trypsin expression before the bloom, coinciding a rapid decrease in P concentration (Zhang et al. 2019). Trypsin, a well-studied enzyme primarily known as a digestive enzyme in animals and a biotechnological tool in protein research, was discovered by Claude Bernard in 1846 and isolated by Kühne in 1867. Over the following centuries, trypsin has been found to respond to various environmental factors, such as temperature, light, and growth hormone, and to influence plasma insulin and growth rate in animals (Rungruangsak-Torrissen et al. 2006).

Research has found trypsin genes across all major lineages of algae (You et al. 2022, 2023). The number of trypsin genes in a species does not correlate with genome size. For example, the diatom *Thalassiosira psedonana* has 19 trypsin genes in a 30-Mbp genome, while *P. tricornutum* has 10 trypsin genes in its 27-Mbp genome. Conversely, the dinoflagellate *Fugacium kawagutii* has only two trypsin genes in a 1.2 Gbp genome. It was speculated that trypsins in diatoms might support silicate frustule formation by providing amines for polyamine synthesis (Zhang et al. 2019), necessary for silica morphogenesis (Sumper and Kröger 2004). However, correlation analyses indicated that diatom trypsins are highly responsive to N and P nutrient conditions, unlike trypsins in *F. kawagutii*, which showed little response (You et al. 2023). The lack of transcriptional response in dinoflagellates may be due to the scarcity of transcriptional regulation in these organisms.

Using CRISPR/Cas9 and overexpression methods, You et al. (2022) indicated that trypsin co-regulates N and P nutrient uptake and serves to maintain N:P stoichiometric homeostasis in *P. tricornutum*. As discussed in Chapter 12, a Ca^{2+} signaling pathway is involved in P and N nutrition regulation in *P. tricornutum* (Helliwell et al. 2021). It is unclear whether this pathway integrates with the trypsin mechanism, or whether it is upstream or downstream of trypsin. Current observations suggest trypsin directly regulates N and P, while the Ca^{2+} pathway primarily regulates P, likely serving downstream of trypsin.

Vitamins

B-type Vitamins

Vitamins are cofactors of enzymes in various biosynthesis pathways. Vitamins A, B family, C, and E, all abundant in algae, also serve as antioxidants for humans, algae, and plants (Arora and Philippidis 2023). Vitamin B12 (cobalamin; VB12) primarily acts as a cofactor of methionine synthase (Croft et al. 2005). Algae either lack or do not express the genes required to synthesize VitB12, making over 50% of algal species dependent on an external source of VB12, hence VB12-auxotrophic, due to their VB12-dependent methionine synthase (metH) (Croft et al. 2005). The remaining algal species use VB12-independent methionine synthase (metE), bypassing the need for VB12.

Vitamin B1 (thiamine; VB1) is a coenzyme in metabolic pathways, particularly in energy production and central metabolism, including carbon assimilation and respiration. VB1 is also crucial for plant stress responses and has several non-coenzyme roles (Fitzpatrick and Chapman 2020). VB1 functions through its derivatives, primarily thiamine pyrophosphate (TPP) and thiamine triphosphate (TTP), with TTP potentially contributing to high-temperature tolerance (Xiao et al. 2023). Vitamin B7 (biotin; VB7) is involved in metabolic processes related to fats, carbohydrates, and amino acids (Penberthy et al. 2020).

Dinoflagellates are one of the most vitamin-dependent algal lineages, with 91% requiring an external VB12 supply (Tang et al. 2010). Dinoflagellates can deplete VB12 from coastal waters within hours to days (Sañudo-Wilhelmy et al. 2012). A study examined 58 species of dinoflagellates showed that all possess metH, while 40 have metE-like genes that lack the N-terminus of typical functional metE. Expression of the metH gene responds to vitamin conditions in the growth medium, whereas metE-like genes do not, explaining the VB12-auxotrophy in most dinoflagellates (Lin et al. 2022). Additionally, about 40% of dinoflagellates are auxotrophic for VB1 and VB7.

Among diatoms, some species possess metE, while others rely on metH. A study on diatoms showed that *Fragilariopsis cylindrus* and *F. kerguelensis* grew equally well with or without B12, whereas *Pseudo-nitzschia granii*, *Chaetoceros* sp., and *Skeletonema* sp. could not grow without B12. The presence of the metE gene enabled *Fragilariopsis cylindrus* to grow without B12. In contrast, *Pseudo-nitzschia granii* lacks a functional metE gene (Ellis et al. 2017).

Trace Metals

Trace metals such as iron, zinc, manganese, cobalt, and copper are essential in electron transfer systems and metalloenzymes in algal cells (Table 17.1). These metals play crucial roles in acquiring and assimilating nitrogen and phosphorus nutrients. For instance, Zn, Fe, or Co are part of the dual-metal cofactors of alkaline phosphatases, which enables algae to utilize DOP. Nickel is the cofactor of urease, enabling cells to utilize urea.

The demand and distribution of dissolved metals vary significantly among species and ocean regions, generally ranked as $Fe \approx Zn > Mn \approx Ni \approx Cu >> Co \approx Cd$ (Twining and Baines 2013). High concentrations of these metals, often due to industrial pollution, can be toxic (Blaby-Haas and Merchant 2013; see Chapter 13).

The uptake and storage of trace metals in cells depend on their binding to organic complexes, influencing uptake rates and intracellular buffering. Stronger binding results in less free metal in the cell. The binding strengths follow the Irving-Williams series: $Mn^{2+} < Fe^{2+} < Co^{2+} < Ni^{2+} < Cu^{2+} > Zn^{2+}$ (Irving and Williams 1953). The dissociation coefficients follow $K_{Mn} < K_{Zn} < K_{Cd} < K_{Cu}$ (Sunda and Huntsman 1983, 2000).

For some metals like Zn and Cd, cellular content increases proportionally with free metal ion concentrations. For others like Fe, Mn, and Cu, cells regulate their contents even at low levels, likely by downregulating transporters. Cellular contents often approach an asymptote at concentrations above those needed for maximal growth, though Fe content can continue increasing, indicating luxury uptake. Metal availability can also influence the uptake of other metals due to competition for transport channels (Sunda and Huntsman 1998a,b). Laboratory data suggest marine phytoplankton have a range of metal quotas, generally ranked as $Fe > Zn > Mn \approx Cu > Co \approx Cd$, with quotas spanning one to two orders of magnitude (Twining and Baines 2013).

Trace metal abundance and bioavailability depend on supply and environmental redox conditions. Coastal waters generally have higher metal concentrations due to terrestrial input, while open oceans are more scarce. Some metals, like iron, are more soluble in reduced environments, whereas others, like copper, are more soluble in oxygenic environments. The shift from anoxygenic to oxygenic eras decreased iron availability and increased copper availability, driving the evolution of algal dominance of the ocean from green-type to red-type algae (Chapter 2). Genomic analyses show evolutionary trends in metal-binding proteins align with algal lineages, with higher percentages of Fe-, Mn-, and Zn-binding proteins correlating with the half-saturation coefficients (Ks) of these metals (Zhang et al. 2022). Secondary endosymbiosis-derived lineages have lower percentages of

Table 17.1. Common metalloproteins present within marine phytoplankton. Updated from Twining and Baines (2013).

Metal	Protein(s)	Function(s)
Fe	Cytochromes	Electron transport in photosynthesis and respiration
	Other Fe-S proteins	Electron transport in photosynthesis and respiration
	Nitrate and nitrite reductase	Conversion of nitrate to ammonia
	Chelatase	Porphyrin and phycobiliprotein synthesis
	Nitrogenasen	N fixation
	Catalase	Conversion of hydrogen peroxide to water
	Peroxidase	Reduction of reactive oxygen species
	Superoxide dismutas	Disproportionation of superoxide to hydrogen peroxide and O_2
	Alkaline phosphatase	Hydrolysis of phosphate esters
Zn	Carbonic anhydrase (α, β, γ)	Hydration and dehydration of carbon dioxide
	Alkaline phosphatase	Hydrolysis of phosphate esters
	RNA polymerase	Nucleic acid replication and transcription
	tRNA synthetase	Synthesis of tRNA
	Reverse transcriptase	Synthesis of single-stranded DNA from RNA
	Carboxypeptidase	Hydrolysis of peptide bonds
	Superoxide dismutase	Disproportionation of superoxide to hydrogen peroxide and O_2
Mn	O2-evolving enzyme	Oxidation of water during photosynthesis
	Superoxide dismutase	Disproportionation of superoxide to hydrogen peroxide and O_2
	Arginase	Hydrolysis of arginine to ornithine and urea
	Phosphotransferases	Phosphorylation reactions
Ni	Urease	Hydrolysis of urea
	Superoxide dismutase	Disproportionation of superoxide to hydrogen peroxide and O_2
Cu	Plastocyannin	Photosynthesis electron transport
	Cytochrome oxidase	Mitochondrial electron transport
	Ascorbate oxidase	Ascorbic acid oxidation and reduction
	Superoxide dismutase	Disproportionation of superoxide to hydrogen peroxide and O_2
	Multicopper ferroxidase	High-affinity transmembrane Fe transport
Co	Vitamin B12	C and H transfer reactions
	Carbonic anhydrase (α, β, γ)	Hydration and dehydration of carbon dioxide
Cd	Carbonic anhydrase (ζ)	Hydration and dehydration of carbon dioxide
Mo	Nitrate reductase	Conversion of nitrate to ammonia
	Nitrogenase	N fixation

metalloproteins and metal requirements than primary endosymbiosis-derived lineages (Zhang et al. 2022).

In the modern ocean, trace metal concentrations are spatially heterogeneous due to varying physical and chemical conditions. Up to 40% of the ocean, notably the Southern Ocean, equatorial Pacific Ocean, and north Pacific Ocean, is iron-limited, categorizing these areas as High-Nutrient, Low-Carbon (HNLC). Trace metal content in natural algal species is highly variable taxonomically, ecotypically, spatially, and temporally. The recent use of Synchrotron X-ray Fluorescence (SXRF) has enabled estimation of trace metal content in natural phytoplankton populations. Fe content ranges from 1 to 4 mmol (mol P)$^{-1}$, Zn content from 2 to 5 mmol (mol P)$^{-1}$, with rare cases up to 15 mmol (mol P)$^{-1}$, while Mn, Ni, and Co are below 1 mmol (mol P)$^{-1}$ (Twining and Baines 2013).

Iron

Iron is one of the most important trace elements for living organisms. In algae, iron is required in the respiratory chain, photosynthesis, various enzymes (e.g., SOD, alkaline phosphatase), and nitrogen fixation and metabolism. In photosynthesis alone, iron is required for the PSII complex at two iron atoms per unit, the Cytb6f complex at five iron atoms per monomer, Cytc6 at one iron atom per monomer, the PSI complex at 12 iron atoms per unit, and ferredoxin at two iron atoms. Partially shared and largely distinct Fe uptake mechanisms are used by prokaryotic and eukaryotic marine algae (Sutak et al. 2020, Gao et al. 2021).

The present-day surface ocean is oxygen-rich and slightly alkaline (global average pH ~ 8.2). This environment promotes the oxidation of iron into its Ferric state (Fe^{3+}), resulting in the predominant form of iron being Fe^{3+} bound to organic material or colloid particles of oxyhydroxides (Gledhill and Buck 2012). The organic ligands complexed with iron include both small, well-defined ligands such as siderophores and diverse macromolecular complexes. Unchelated iron is present in hydrolyzed forms, $Fe(OH)_x^{(3-x)+}$, with the neutral tri-hydroxy species, $Fe(OH)_3$, having extremely low solubility (Sutak et al. 2020).

Uptake

Prokaryotes use three mechanisms for Fe uptake:

1. Transport of iron-loaded siderophores across the cell outer membrane via a TonB-Dependent Transporter (TBDT) and then into the cells by an ABC transporter. Cyanobacteria do not excrete siderophores but can use siderophores produced by other organisms.
2. Transport of inorganic ferric species through the outer membrane by TBDT or by porins and then into the cells by the ABC transporter system FutABC.
3. Reduction of ferric iron in the periplasmic space by the Alternative Respiratory Terminal Oxidase (ARTO), followed by ferrous ion transport by divalent metal transporters (FeoB or ZIP) or by reoxidation and uptake by the high-affinity permease Ftr1.

In eukaryotes, there are also three main models proposed so far for iron transport into the cells (Sutak et al. 2020). First, iron is transported via endocytosis after carbonate-dependent binding of Fe(III)' (inorganic soluble ferric species) to phytotransferrin (pTF) at the cell surface. This mechanism allows cells to take up iron from very low iron concentration, with the rate of uptake depending on the concentration of unchelated Fe species and thus limited thermodynamically. pTF is also known as Iron Starvation-Induced Protein 2A (ISIP2A) in diatoms (e.g., gene under NCBI accession number EEC48748 for *P. tricornutum* and that under ANP23381 for *Pseudo-nitzschia granii*). ISIP2A is a phytotransferrin capable of mediating high-affinity Fe(III) uptake. The functional domain PF07692, initially identified in the Fe-Assimilation (FEA) proteins FEA1 and FEA2 of the green alga *Chlamydomonas reinhardtii* (Allen et al. 2007), was further characterized in *Dunaliella salina* (Paz et al. 2007). In *C. reinhardtii*, FEA1 and FEA2 are prominently expressed under low-iron conditions, facilitating high-affinity iron uptake, likely in its ferrous form (Narayanan et al. 2011). The ecological importance of ISIP2A is evident from its expression across diverse marine lineages, including diatoms, dinoflagellates, and haptophytes, as shown by metatranscriptomic profiles from Antarctica and Monterey Bay (Sutak et al. 2020).

Iron release from phytotransferrin requires endocytosis, while efficient iron binding to phytotransferrin depends on CO_3^-. The transcriptional expression of the endocytosis pathway in diatoms in a subtropical ocean suggests their potential to uptake iron via endocytosis (Li et al. 2024). Additionally, Fe' uptake rates positively correlate with CO_3^- concentration, implying that ocean acidification, which reduces CO_3^- levels, could diminish phytotransferrin-mediated iron acquisition (McQuaid et al. 2018). This aligns with findings that media acidification reduces iron uptake rates by diatoms and coccolithophores (Shi et al. 2010).

Second, iron uptake involves the extracellular reduction of all Fe species, including Fe'. This strategy allows cells to absorb iron from various ferric species. Transcriptomic analysis of *P. tricornutum* revealed that iron starvation induces the expression of ferric reductase genes FRE1-FRE4, with FRE2 located near an iron-responsive gene encoding a putative ferrichrome-binding protein (FBP1), similar to the bacterial FhuD protein involved in ferrichrome transport (Allen et al. 2008).

Third, Fe chelated by hydroxamate siderophores is transported by endocytosis, dependent on ISIP1, after binding to the FBP1 protein, with iron released from siderophores via FRE2-dependent reduction. A proposed fourth pathway in eukaryotic algae, seen in species like the dinoflagellate *Alexandrium catenella*, involves using TonB-dependent transporters for siderophore-chelated Fe uptake (Sanchez et al. 2018).

Marine microorganisms produce siderophores containing iron-binding moieties similar to terrestrial counterparts: catecholates, hydroxamates, and α-hydroxycarboxylates. Eukaryotic algae generally do not produce siderophores but can utilize bacterium-generated ones. For example, the chlorophyte *Chlorella sorokiniana* relies on bacterial siderophores for low-abundance iron uptake (Rawat et al. 2022). Marine diatoms like *Thalassiosira pseudonana* and *P. tricornutum* use a ferric-siderophore complex reducing system, dissociating ferric complexes at the cell surface for ferrous iron uptake (Armbrust et al. 2004, Kustka et al. 2007).

Pseudo-nitzschia diatoms secrete Domoic Acid (DA), a neurotoxin similar to the photosiderophore mugineic acid produced by Graminea plants. DA chelates Fe and Cu, enhancing Fe solubilization from sediments in coastal regions and promoting Fe uptake and growth of *Pseudo-nitzschia* (Rue and Bruland 2001, Wells et al. 2005). The underlying mechanism remains unclear.

Strategies to Cope with Iron Deficiency

Algae optimize iron uptake by employing high-affinity or iron stress-induced mechanisms and reducing iron demand under low-Fe conditions. A common strategy is to replace iron-containing enzymes or proteins with iron-free equivalents. For instance, the ratio of flavodoxin (Fld, iron-free) to ferredoxin (Fd, iron-sulfur protein) increases under iron-limited conditions in diatoms and dinoflagellates (Erdner et al. 1999). Similarly, the Cu-containing plastocyanin (PC) versus the Fe-dependent cytochrome c6 (Cytc6), fructose-bisphosphate aldolase type 1 versus type 2 in gluconeogenesis, and Mn/Fe-SOD versus Cu/Zn- or Ni-SOD modulate iron needs based on availability.

Algae have evolved differential iron/copper nutrition strategies: Group 1 algae contain both Cytc6 and PC, group 2 only have Cytc6, and group 3 only have PC (Blaby-Haas and Merchant 2013). Terrestrial green algae like *Chlamydomonas* belong to group 1, expressing PC but replacing it with Cytc6 under Cu-deficiency. Marine green algae and land plants (group 3) have lost Cytc6 but possess Cu/Zn SOD. The rhodophyte *Cyanidioschyzon merolae* (group 2) lacks PC or Cu/Zn SOD, consistent with its copper-scarce habitat. Chromalveolates are represented in all three groups, with groups 1 and 2 primarily in coastal waters and group 3 in the open ocean (Blaby-Haas and Merchant 2013).

Heterologous expression of chlorophyte plastocyanin in *P. tricornutum* improves growth under Fe deficiency (Castell et al. 2020). Research shows 50–70% of intracellular copper in diatoms and green algae is allocated to plastocyanin (Merchant et al. 2006, Peers et al. 2006). Copper is crucial for high-affinity iron uptake (Askwith et al. 1994). In Cu-limited diatoms, both cellular iron quotas and uptake rates decrease, and more copper is needed for growth when iron is limited (Peers et al. 2005, Annett et al. 2008).

In *Chlamydomonas reinhardtii*, two pathways detoxify superoxide during Fe-deficiency stress (Page et al. 2012). One pathway upregulates MSD3, leading to a plastid-specific MnSOD, while the other preferentially retains plastid FeSOD over other Fe proteins, indicating prioritized Fe allocation. FeSOD is maintained through *de novo* resynthesis, suggesting a salvage mechanism for intracellular Fe recycling and reallocation.

The evolution of Fld has spurred debates. It is found in all major algal taxa, including those arising from secondary endosymbiosis events like dinoflagellates, haptophytes, diatoms, and cryptomonads (Zurbriggen et al. 2007, Erdner et al. 1999). However, streptophytes, including embryophytes and charophytes, lack Fld (Zurbriggen et al. 2007). This distribution suggests that the gene was lost during the transition from green algae to terrestrial plants (Lodeyro et al. 2012).

In blue-green algae, an iron stress-induced protein (isiA) occurs in the photosystem as a chlorophyll-binding protein, initially found to be similar to the PSII core protein CP43 (Laudenback and Straus 1988). Subsequent research revealed that isiA connects to PSI, forming an octadecameric ring within which the PSI trimer sits (Fig. 17.10), facilitating efficient energy transfer to PSI (Bibby et al. 2001). Cryo-EM analysis confirmed the octadecameric ring structure and isiA-PSI association (Akita et al. 2020). Time-resolved fluorescence analysis demonstrated clear energy transfer from isiA to PSI. Many cyanobacteria possess multiple isiA genes, all expressed under iron deficiency, highlighting its importance in coping with iron stress (Shen et al. 2016, Nagao et al. 2021). The isiA-chlorophyll binding reduces light energy harvesting, a protective mechanism to prevent photodamage under iron deficiency (Ma et al. 2017).

Death-Specific Protein (DSP)

A protein initially identified in the diatom *S. costatum* (Chung et al. 2005) was later found in other diatoms and the haptophyte *Emiliania huxleyi* (Chung et al. 2008, Thamatrakoln et al. 2013). The mRNA level of this gene increased with genomic DNA fragmentation in aging cell populations of *S. costatum*, suggesting its association with cell death, hence named Death-Specific Protein (DSP) (Chung et al. 2005).

In *S. costatum* (ScDSP-1), DSP was upregulated under dim light or blockage of PSII electron transport by 3-(3,4-dichlorophenyl)-1,1-dimethylurea (DCMU) or 2,5-dibromo-3-methyl-6-isopropyl-p-benzoquinone (DBMIB). Treatment with the Nitric Oxide (NO) donor diethylamine nitric oxide induced ScDSP-1 expression, which was inhibited by the NO scavenger hemoglobin. Pretreatment with a cell-penetrating NO scavenger reduced ScDSP-1 mRNA expression elicited by DCMU and DBMIB. In *T. weissflogii* and *P. tricornutum*, diatom-derived aldehyde products induced calcium-dependent NO synthase-like activity, suggesting NO and calcium ions as secondary messengers in the PCD regulatory pathway (Vardi et al. 2006). These findings suggest that the upregulation of ScDSP-1 expression in the stationary phase is mediated by NO signaling (Chung et al. 2008).

Thamatrakoln et al. (2013) isolated two DSP genes in *Thalassiosira pseudonana* (TpDSP1 and TpDSP2) with characteristics similar to ScDSP-1 and ScDSP-2, such as a membrane-spanning region and a pair of Ca^{2+}-binding EF-hand motifs. They suggested the DSP family is most similar to proteins associated with Ca^{2+}-dependent signal transduction cascades. Using homologous overexpression of TpDSP1, they found this protein localized in the chloroplast, enhancing growth under Fe limitation at subsaturated light by modulating photosynthetic electron transport, although growth was impaired at high irradiance levels.

DSP seems absent in green algae. However, using ScDSP-1 as a query, BLAST analysis revealed DSP-like sequences in some dinoflagellates (e.g., *Polarella glacialis*, *Karlodinium veneficum*) and other haptophytes. These results suggest DSP is prevalent in the red-type lineages of algae (Chung et al. 2008). The sequences from different lineages share conserved domains of Ca^{2+} binding and EF-hand motifs. The function of DSP in these lineages and the signaling cascade remains to be elucidated.

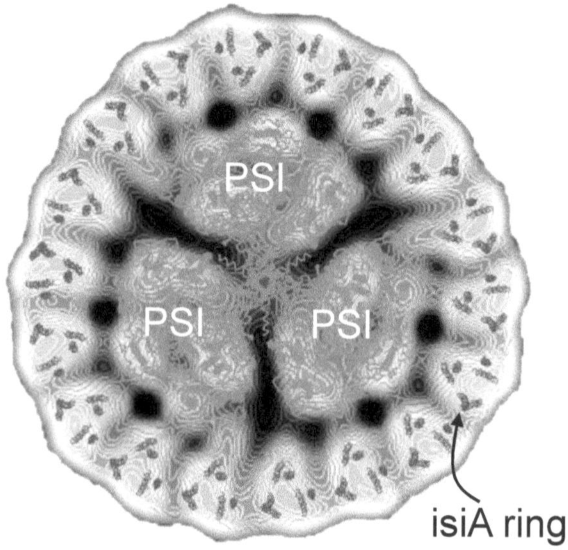

Figure 17.10. Octadecameric isiA ring structure in connection to PSI trimer. The image indicates the presence of a PSI trimer within the center of an 18-member ring of isiA. Reproduced with permission from Bibby et al. (2001).

Copper

Copper (Cu) is a micronutrient essential for optimal metabolism in algae, affecting growth, photosynthesis, and electron flow (Peers and Price 2006). Cu is used in both photosynthetic and respiratory electron transport chains, Zn/Cu-SOD, and may be involved in transmembrane uptake of Fe (Askwith et al. 1994, Peers et al. 2005, Twining and Baines 2013).

The copper uptake pathway involves organically bound copper (Cu-L) undergoing dissociation at the cell surface, yielding Cu^{2+} through equilibrium or Cu^{+} via reduction (Fig. 17.11). Cu^{+} is transported across the cell membrane by high-affinity CTR cuprous transporters, while Cu^{2+} is taken up by low-affinity Divalent Metal Transporters (DMT). The expression of CTR transporters and reductases (FRE) is regulated by environmental copper levels. Interestingly, both deficiency and excess of Cu negatively impact electron flow, photosynthesis, and growth (Rocha et al. 2021).

In the oceanic centric diatom *Thalassiosira oceanica*, Cu limitation induces upregulation of ferredoxin and plastidial Fd-dependent isoenzymes involved in nitrogen assimilation, as well as enzymes involved in glutathione synthesis. This suggests an integration of nitrogen uptake and metabolism with photosynthesis and oxidative stress resistance. There is also evidence of chloroplast-mitochondrial crosstalk to channel excess electrons and/or ATP between the two organelles, and a malate shunt connecting metabolic processes in the two compartments (Hoppmann et al. 2022).

In contrast to copper deficiency, excess copper can alter the production of proteins, which are suggested to be the main metal-binding sites, fatty acids, lipids (changing their saturation), carbohydrates, and pigments such as Chl *a* (Rocha et al. 2021). In the diatom *Thalassiosira pseudonana*, excess copper stress induces the production of a cell surface protein of 100 kDa (Davis and Palenik 2008). This protein (p100) is comprised almost entirely of a novel repetitive cysteine-rich domain (TpRCR), containing conserved aromatic, acidic, and potential metal-binding amino acids.

P100 belongs to a large family of modular proteins that consist of a variable number of TpRCR domain repeats. Its expression dynamics in the cell cycle are similar to those of the diatom cell wall proteins silaffins, suggesting a role for TpRCR in silicate frustule formation. A BLASTP search using this TpCRC sequence (ABS19631.1) as a query against various algal genomes (http://sampgr.org.cn/) indicates that this protein also occurs in *T. antarctica, T. gravida, T. miniscula, T. punctigera, T. weissflogii, Skeletonema grethea, S. japonicum, S. marinoi, S. costatum, S. dohrnii,*

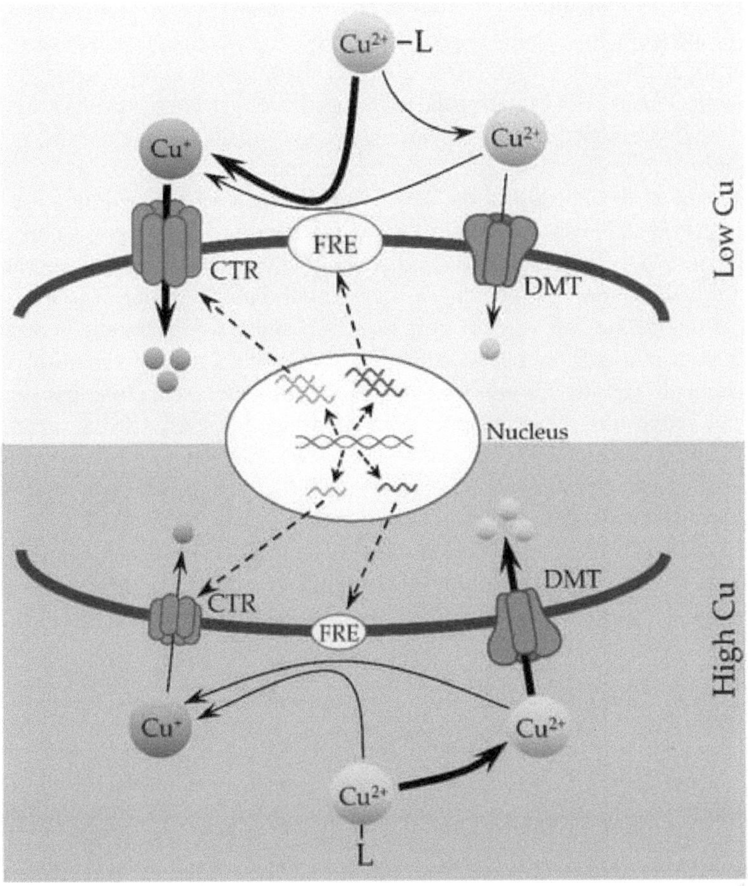

Figure 17.11. Model of copper uptake by a marine diatom in low and high Cu environments. Organically complexed Cu (Cu-L) is dissociated to Cu^{2+} by equilibrium or to Cu^+ by reduction at the cell surface. Cu^+ is transported across the cell membrane by high-affinity CTR cuprous transporters and Cu2+ is transported into cells by low-affinity divalent metal transporters (DMT). Expression of CTR transporters and reductases (FRE) are regulated by environmental Cu concentrations. Reproduced with permission from Kong 2022 under the Creative Commons CC BY 4.0 license.

and *S. menzelii*. However, no significant matches were found in the pennate diatom *Phaeodactylum tricornutum*, the green alga *Ulva prolifera*, or various cryptophytes and dinoflagellates in the database.

Zinc

Zinc is a cofactor for various enzymes, including superoxide dismutase, carboxypeptidase, carbonic anhydrase, alkaline phosphatase, and various dehydrogenases, and is a necessary mineral nutrient for algal and plant growth (Coleman 1998). Zinc deficiency in plants leads to a significant reduction in electron transport and phosphorylation. Although zinc is a micronutrient, it is often found in high concentrations in algal biomass due to accumulation and sequestration in polyphosphate bodies in cell sectors. The 'zinc quota' in a cell (minimum amount required for a cell to thrive) varies from 10^5 atoms in *E. coli* to 10^7 atoms in yeast and 10^8 atoms in mammalian cell types. When normalized by cell volume, the range of zinc quota narrows down to 0.1–0.5 mM (Eide 2006).

In bacteria, zinc is primarily taken up by the ZnuABC transporter and the ZIP family transporter. In eukaryotes, known Zn transporters belong mainly to two solute carrier (SLC) families: the ZIP (SLC39) and CDF/ZnT (SLC30) proteins. The ZIP family, which widely occurs, transports Zn and/or other metal ion substrates from the extracellular space or organellar lumen into the cytoplasm.

The CDF (cation diffusion facilitator), known as ZnT in mammals, transport Zn from the cytoplasm into the lumen of intracellular organelles or to the outside of the cell, opposite to the ZIP transporters (Eide 2006). In the plant *A. thaliana*, IRT1, a ZIP member, has a histidine-rich intracellular loop between TM3-4 that senses Zn^{2+}. Mutating the loop in pathogenic bacteria results in Zn accumulation, as indicated by the fluorescence Zn^{2+} dye FluoZin3 (Pang et al. 2023). The binding of Zn to the loop promotes ubiquitination, endocytosis, and vacuolar degradation.

When dissolved Zn is depleted in the ocean, diatoms and other algae may use Cd and Co to functionally replace Zn. This has been demonstrated in carbonic anhydrase, a core element of the Carbon Concentrating Mechanism (CCM) in diatoms (Kellogg et al. 2022). Laboratory data show that many phytoplankton species can achieve high growth rates by replacing Zn with Cd or Co in cultures. Field data indicate that vertical profiles of Co and Cd concentrations resemble those of major and other trace nutrients, and the consumption of Co and Cd approximates that of Zn (Morel et al. 2020). Although Mg and Zn are the best-known cofactors of alkaline phosphatase, there is growing evidence suggesting that algae, including dinoflagellates, use Ca^{2+} as one of the cofactors (Lin et al. 2019).

Using global proteomic analysis, Kellogg et al. (2022) identified two proteins, ZCRP-A/B (Zn/Co Responsive Protein A/B), among four diatom species in Zn/Co-limited environments. Reverse genetic analysis and homology data suggest that ZCRP-A serves as a Zn/Co chaperone, whereas ZCRP-B is a membrane-bound transport complex component. These findings demonstrate that multiple adaptive responses to Zn scarcity occur in marine diatoms in low Zn regions of the Pacific Ocean.

Manganese (Mn)

Manganese (Mn) acts either as an enzyme cofactor or as a metal with catalytic activity. Mn is an essential cofactor for the Oxygen-Evolving Complex (OEC) of the photosynthetic machinery, catalyzing the water-splitting reaction in photosystem II (PSII). Mn is also a cofactor of the Fe/Mn-type SOD. Compared to iron and zinc, Mn is relatively abundant and has been thought not to limit algal growth. However, recent studies demonstrate that it is limited in the Southern Ocean (Browning et al. 2021).

Several transporter gene families have been implicated in Mn transport in plants (Socha and Guerinot 2014), including NRAMP (Natural Resistance-Associated Macrophage protein), YSL (Yellow Stripe-Like), ZIP (Zinc-Regulated Transporter/Iron-Regulated Transporter [ZRT/IRT1]-related protein), CAX (cation exchanger), CCX (calcium cation exchangers), CDF/MTP (Cation Diffusion Facilitator/Metal Tolerance Protein), P-type ATPases, and VIT (vacuolar iron transporter).

Interestingly, most proteins that transport Mn across membranes are not specific for the metal. For instance, plant Mn^{2+} transporters can transport, to varying extents, other divalent cations such as Fe^{2+}, Zn^{2+}, Cu^{2+}, Cd^{2+}, Ca^{2+}, Co^{2+}, and Ni^{2+}. It appears that transporter proteins contain specific binding domains for different metal ions. For example, in the Mn/Fe transporter AtMTP8, Fe transport activity could be inhibited by introducing point mutations in different Fe-binding domains without affecting its Mn^{2+} transport capability (Chu et al. 2017).

Nutrient Co-limitation in Algae

There are different types of co-limitation (Arrigo 2005). The simplest case is when two or more nutrients are too low for cellular uptake. In this type of co-limitation, algal growth is stimulated by the addition of all the limiting nutrients (Moore et al. 2013). Examples include the Fe and P co-limitation of diazotrophs (Held et al. 2020), N and P co-limitation of phytoplankton in the Baltic Sea (Seppala et al. 1999), Si and P co-limitation near the Pearl River estuary (Yin et al. 2000), and N and Fe co-limitation in the southern Northwest Pacific (The Philippine Sea) (Browning et al. 2022). In the Drake Passage, Fe and Mn were found to co-limit two different dominant groups of phytoplankton, *Fragilariopsis* and picoeukaryotes, whereas other algal groups exhibited Fe limitation (Balaguer et al. 2022).

A more complex case of co-limitation arises when algal growth is stimulated by the addition of one of the co-limiting nutrients. Two scenarios exist in this case. The first involves biochemical interaction (biochemical co-limitation), in which the addition of one limiting nutrient may facilitate the uptake or assimilation of another. For instance, algal growth can be co-limited by the deficiency of phosphate (PO_4^{3-}) and zinc (Zn^{2+}). In this case, the addition of Zn^{2+} promotes the expression of alkaline phosphatase, for which zinc is an essential cofactor, enabling phytoplankton to scavenge DOP for P nutrient to support growth. The second scenario, referred to as community co-limitation, occurs when adding one nutrient stimulates the growth of one subgroup of the community, while adding a different nutrient increases the growth of another. In Fe-replete oligotrophic waters where N_2-fixing blue-green algae are abundant and N and P are scarce, the addition of NO_3^- promotes non-N_2 fixers capable of utilizing DOP with the aid of AP. The addition of P stimulates species such as *Trichodesmium* spp., which support their high N requirement through N_2 fixation (Arrigo 2005). These differential responses to the deficiency or supply of multiple limiting nutrients can be explained by the differential uptake efficiencies (K_N) of different species of algae discussed earlier in this chapter.

Autophagy

Autophagy is a major catabolic pathway by which eukaryotic cells degrade and recycle unnecessary or damaged cytoplasmic materials in a membrane-trafficking pathway to maintain cellular homeostasis. During this degradative process, cytosolic components including proteins, membranes, and entire organelles are engulfed within a double-membrane vesicle (named autophagosome) and delivered to the vacuole (Nakatogawa et al. 2009). Autophagy is conserved in algae (Pérez-Pérez and Crespo 2014), which involves the TORC1 signaling pathway in the activation regulation of autophagy, although important parts of the regulatory cascade are still uncharacterized (Fig. 17.12).

Studies have shown that nutrient deficiency upregulates the expression of autophagy genes (Shemi et al. 2016). For instance, trace metal deficiencies in the dinoflagellate *Fugacium kawagutii* induce upregulation of autophagy (Li et al. 2020). Similarly, phosphate depletion in laboratory cultures and during a natural bloom of the dinoflagellate *Prorocentrum shikokuense* was accompanied by elevated expression of autophagy pathway genes (Zhang et al. 2019, Li et al. 2024). Limitation of carbon (Avin-Wittenberg et al. 2015), nitrogen (Masclaux-Daubresse et al. 2017), and sulfur (Wawrzyńska et al. 2015, Prigent et al 2024) have all been shown to induce autophagy.

Autophagy exists in two types, selective and non-selective. Plants possess various forms of selective autophagy including chlorophagy, reticulophagy, mitophagy, proteaphagy, ribophagy, among many others (degradation of chloroplasts, endoplasmic reticulum, mitochondria, proteasomes, ribosomes, and others) (Chen et al. 2019). In the dinoflagellate *P. shikokuense*, P deficiency was found to induce the expression of nine autophagy (ATG) genes known in other organisms to function in the three stages of autophagy, and to decrease the expression of mitochondrial metabolism genes, results suggestive of mitophagy (Li et al. 2024).

Mixotrophy

Many microalgae are mixotrophic, performing photosynthesis when nutrients are available and switching to phagotrophy when nutrients are depleted. Phagotrophy operates by phagocytosis, a mechanism where the organism ingests particles > 0.45 μm in diameter through endocytosis (Uribe-Querol and Rosales 2020). This involves the formation of membrane-bound vesicles called phagosomes, which fuse with lysosomes to form phagolysosomes where prey cells are digested by enzymes. This process resembles the phagocytic response of innate immune cells like macrophages

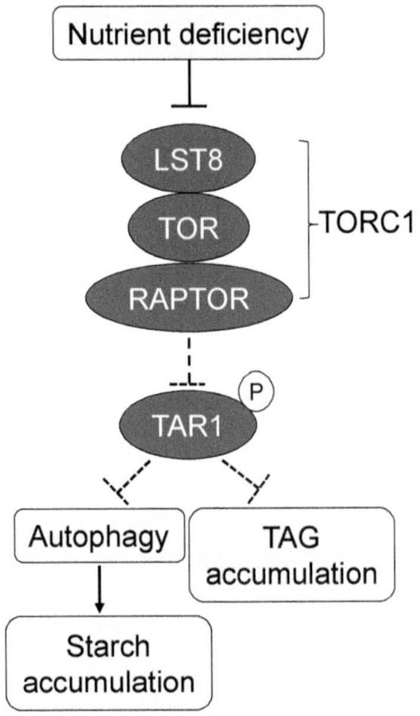

Figure 17.12. Hypothetical regulatory cascade of autophagy in response to nutrient deficiency. Dashed lines depict uncharacterized parts of the cascade. Based on Kajikawa and Fukuzawa (2020).

(Underhill and Ozinsky 2002). Phagotrophy is an ancestral trait that facilitated the emergence and spread of chloroplasts into eukaryotes and is crucial for intracellular phytosymbioses.

Phagomixotrophy is phylogenetically and ecologically widespread, especially in oligotrophic regions. This suggests that optimizing physiological processes for resource competition and minimizing losses (e.g., through predation) is a favorable strategy, subject to complex selective processes generating diverse mixotrophic strategies in extant phytoplankton (Raven et al. 2009). The regulation of the switch between photosynthesis and phagotrophy in algae is not fully understood, but there is growing information about the molecular machinery of phagotrophy, primarily documented in model organisms, which aids explorations in algae.

Phagotrophy involves four major steps: endocytosis (KO04144), formation of phagosomes (KO04145), fusion of phagosomes with lysosomes (KO04142), and autophagy. The genes involved in this process are shown in Table 17.2. A study on two polar algae, the prasinophytes *Micromonas polaris* and *Pyramimonas tychotreta*, found that low nutrient conditions induced upregulation of 153 genes related to phagotrophic ingestion, 37 of which were in both species, with predicted functions in endocytosis, phagosome formation, lysosome production, and autophagy regulation (McKie-Krisberg et al. 2018). This study also found transcripts containing Ankyrin domains, which are suggested to play a role in phagocytosis.

In the phagotrophic haptophyte *Prymnesium parvum*, feeding on bacteria and ciliates induced upregulation of genes involved in fatty acid oxidation and the TCA cycle. Meanwhile, genes related to inorganic nitrogen uptake and transformation and iron uptake were downregulated when prey was present, suggesting nitrogen and iron were obtained from prey ingestion (Liu et al. 2015). Similarly, in the dinoflagellate *P. shikokuense*, nitrogen limitation triggered significant upregulation of phagotrophy pathway genes (Li et al. 2021). Upregulation of phagotrophy genes was also reported for natural blooms of *P. shikokuense* (Yu et al. 2020) and other harmful algal bloom species like *Heterosigma akashiwo* (Ji et al. 2018) as bloom formation depleted nitrogen nutrients in the environment. Phagotrophic pathways were also active in the continental shelf and slope areas in a subtropical marginal sea (South China Sea) (Li et al. 2024). Metatranscriptomic data showed that phagotrophy gene expression was positively correlated with Calvin carbon fixation

Table 17.2. Genes involved in Phagotrophy. Reproduced with permission from McKie-Krisberg et al. (2018) under the Creative Commons CC BY 4.0 license.

KO	Gene symbol	Description	EC
		(A) ENDOCYTOSIS	
K00889	PIP5K	1-phosphatidylinositol-4phophate 5-kinse	2.7.1.68
K12492	ARFGAP1	ADP-ribosylation factor GTPase-activating protein 1	
K12200	PDCD6IP, ALIX, RIM20	Programmed cell death 6-interacting protein	-
K11824	AP2A	AP-2 complex subunit alpha	-
K12489	ACAP	Arf-GAP with coiled-coil ANK repeat and PH domain-containing protein	-
K18442	ARFGEF, BIG	Brefelclin A-inhibited guanine nucleotide-exchange protein	-
K12483	EHD1	EH domain-containing protein 1	-
K07901	RAB8A, MEL	Ras-related protein Rab-8A	-
K07889	RAB5C	Ras-related protein Rab-5C	-
K12194	CHMP4, SNF7, VPS32	Charged multivesicular body protein 4	
K18646	RTSC, SPG8	WASH complex subunit strumpellin	-
K12198	CHMP5, VPS60	Charged multivesicular body protein 5	
K12562	AMPH	Amphiphysin	-
K04646	CLTC	Clathrin heavy chain	-
K12471	EPN	Epsin	-
K13649	FOLR	Folate receptor	-
K18443	GBF1	Golgi-specific brefeldin A-resistance guanine nucleotide exchange factor 1	-
K03283	HSPA1_8	Heat shock 70kDa protein 1/8	-
K17917	SNX1_2	Sorting nexin-1/2	-
K12486	SMAP	Stromal membrane-associated protein	-
		(B) PHAGOSOME	
K03661	ATPeV0B ATP6F	V-type H+-transporting ATPase 21kDa protelipid subunit	-
K10414	DYNC2H DNCH2	dynein heavy chain 2 cytosolic	-
K00921	PIKFYVE FAB1	1-phosphatidylinositol-3-phosphate 5-kinase	2.7.1.150
K08054	CANX	Calnexin	-
K08057	CALR	Calreticulin	-
K01365	CTSL	Cathepsin L	3.4.22.15
K10956	SEC61A	Protein tranport protein SEC61 subunit alpha	-
K07374	TUBA	Tubulin alpha	-
		(C) LYSOZONE	
K02154	ATPeV0A ATP6N	V-type H+-trasporting ATPase subunit a	-
K12391	AP1G1	AP-1 complex subunit gamma-1	-
K12396	AP3D1	AP-3 complex subunit delta-1	-
K12400	AP4E1	AP-4 complex subunit epsilon-1	-
K02146	ATPeV0D ATP6D	V-type H+-tranporting ATPase subunit d	-
K03661	ATPeVOB ATP6F	V-type H+-transporting ATPase 21kDa proteolipid subunit	-
K02155	ATPeV0C ATP6L	V-type H+-transporting ATPase 16kDa proteolipid subunit	-
K05656	ABCB9 TAPL	ATP-binding cassette subfamily B (MDR/TAP) member 9	-

Table 17.2 contd. ...

...Table 17.2 contd.

KO	Gene symbol	Description	EC
K04646	CLTC	Clathrin heavy chain	-
K01074	PPT	Palmitoyl-protein thioesterase	3.1.2.22
K02144	ATPeV1H	V-type H+-transporting ATPase subunit H	-
(D) REGULATION OF AUTOPHAGY			
K08269	ULK1_2_3, ATG1	Serine/treonine-protein kinase ULK/ATG 1	2.7.11.1
K07198	PRKAA, AMPK	5'-AMP-activated protein kinase catalytic alpha subunit	2.7.11.11
K08333	PIK3RA, VPS15	phosphoinostiide-3-kinase regulatory subunit 4	2.7.11.1
K08334	BECN1, VPS30, ATG6	Beclin 1	-
K00914	PIK3C3, VPS34	Phosphatidylinositol 3-kinase	2.7.1.137
K08339	ATG5	Autophagy-related protein 5	-
K08342	ATG4	Cysteine protease ATG4	3.4.22.-

KO, KEGG Orthology number; EC, Commission nomenclature number produced by the IUBMB/IUPAC Biochemical Nomenclature Committee.

genes in some lineages (diatoms, chlorophytes, and haptophytes) but negatively correlated in others (dinoflagellates). These contrasting correlations suggest that phagotrophy may play differential roles, likely promoting photosynthetic carbon fixation by acquiring limited nutrients and complementing photosynthetic carbon fixation when environmental nutrients are depleted.

The discussed obligate phototrophs that use phagotrophy to supplement deficient nutrients belong to the 'Constitutive Mixotrophs' (CMs) class in the recently proposed plankton functional group classification scheme (Mitra et al. 2016). On the other end of the mixotroph spectrum are obligate heterotrophs that use ingested and retained specific algae species to perform photosynthesis ('non-constitutive mixotrophs with specific prey' or SNCM). Some of these mixotrophs ingest whole prey cells, while others funnel the intracellular contents of prey through a peduncle. The bioluminescent symbiont-containing dinoflagellate *Noctiluca scintillans* represents one of the SNCM mixotrophs. This green type *N. scintillans* has a grazing-driven positive nutrient feedback loop that integrates its heterotrophy with photosynthesis by the endosymbiont (*Protoeuglena noctilucae*). This *Noctiluca* ecotype, restricted to tropical marine areas, feeds on diatoms and other phytoplankton, releasing ammonium and phosphate, which stimulate the growth of prey algae. The metabolism-derived and environmentally acquired nutrients also fuel the growth of the endosymbiont *P. noctilucae*. Comparative metatranscriptomic analyses for the symbiont-carrying *N. scintillans* and isolated cultured *P. noctilucae* revealed that *N. scintillans* controls the endosymbiont, promoting its photosynthesis but suppressing its cell division, showcasing a remarkable trophic strategy (Luo et al. 2022).

Eutrophication

Eutrophication occurs when the environment becomes enriched with nutrients, increasing the amount of algal growth in estuaries and coastal waters. Excessive algal growth defines eutrophication, which has significant ecological consequences. A system is diagnosed as eutrophic with Chl-*a* concentrations greater than 2.21 mg L^{-1}, or oligotrophic with the concentrations less than 0.1 mg L^{-1} (Karydis and Kitsiou 2019). Importantly, eutrophication is characterized not only by increases in nutrient loading, but also changes in stoichiometry of different nutrients. In unperturbed ecosystems, freshwater systems are usually phosphorus-deficient whereas oceans are nitrogen-deficient. Thus, Nitrogen (N) overloading is the major driver of eutrophication in coastal seas, whereas phosphorus input, often from legacy deposit in sediment, is more important in freshwater and estuarine systems. N overloading can cause relative phosphorus deficiency in the impacted water. This P deficiency has been reported in estuaries such as Changjiang Estuary and Pearl Estuary in China as well as Gulf of Mexico and Long Island Sound in the US (Dai et al. 2023, Devlin and Brodie 2023).

The ultimate source of nutrient overloading is the production of fertilizers to increase crop yields for the rapidly growing human population, particularly nitrogen fertilizers. The Haber-Bosch process, which produces ammonium from atmospheric N_2 and hydrogen under high pressure and temperature, has revolutionized agriculture but also led to aquatic eutrophication.

Eutrophication leads to multiple environmental problems, including harmful algal blooms, hypoxia, and seawater acidification. The problems arise from specific algal species' responses to increased nutrient concentrations. For a particular species, growth rate (μ) increases with nutrient concentration until the rate reaches its maximum (μ_{max}), following the Monod model (Eq. 17.2). However, the Monod model's applicability is questioned due to the luxury uptake of nutrients and storage for later growth, leading to a temporal uncoupling between cell division rates and environmental nutrient concentrations (Sommer 1991). In such cases, the Droop model is more appropriate (Eq. 17.3), where a cell must have the minimum reserve of the nutrient in the cell, the so-called minimum quota (Q_0), to grow and proliferate.

Based on both models, eutrophication can alter dominant species in the algal community due to the different K_N or Q_0 and μ_{max} values among species. This change can affect herbivore grazing on the algal assemblage. Often, the "newly emerging" dominant species are unsuitable as food for zooplankton or herbivorous animals, leading to excess algal biomass and HAB outbreaks. This scenario is exemplified by *Aureococcus anophaggerens*, whose resistance to grazing led to the collapse of scallop industry in Long Island, New York. This excess biomass will eventually die and is decomposed by microbes, consuming oxygen, and releasing CO_2, creating hypoxic and acidic conditions.

18

Defense Against Grazing and Microbial Insults

The ability to defend against adversaries is crucial for an algal population to survive and grow in the natural marine environment, where threats may exist in multitude. These adversaries include both abiotic and biotic factors. While the abiotic insults encompass various environmental stressors, the biotic adversaries consist of grazers and microbes. Defense necessitates a reallocation of the internal resources and a reprogramming of the metabolic landscape to produce a chemical arsenal, undergo morphological alterations, or change reproductive strategies. The pressure imposed by these stressors and attacks drives adaptive evolution in algae, enabling them to reduce the costs associated with this crucial resource investment and optimize their overall fitness (Pansch et al. 2009).

Defense Against Environmental Stress

As algae constantly face environmental stressors, cells must allocate resources toward preventing or recovering from their damaging effects. These stressors include UV radiation, thermal stress, nutrient deficiency, and pollution, among others. The mechanisms for coping with UV-induced DNA damage, thermal stress, oxidative stress, and nutrient stress have been discussed in Chapter 13.

Pollution by heavy metals imposes stress on algae even though many of them are required in trace amounts for photosynthesis and growth. The most common ill effect is the production of Reactive Oxygen Species (ROS). For instance, in response to copper toxicity, *Prorocentrum cordatum* (formerly *P. minimum*) exhibited increased ROS production and a transcriptional shift (detected via microarray analyses), with 515 genes (~ 8.6%) being differentially expressed (Guo et al. 2016). These Differentially Expressed Genes (DEGs) were enriched in KEGG pathways related to ribosomal function, RNA transport, carbon metabolism, biosynthesis of amino acids, and photosystem maintenance, among other processes, suggesting photosystem damage, photosynthesis impairment, and oxidative stress. In the phaeophyte *Laminaria digitata*, exposure to copper within 24 hours elicited significant increase in lipid peroxidation and the release of unsaturated Fatty Acids (FAs) (Ritter et al. 2008).

Some of the increasing manufactured organic compounds can inhibit growth of many algae. One example is glyphosate, which disrupts the shikimate pathway in susceptible photosynthetic organisms. It inhibits growth in many algae while exhibiting growth-promoting or no effects on others (Wang et al. 2016). However, this chapter will primarily focus on defense against biotic insults.

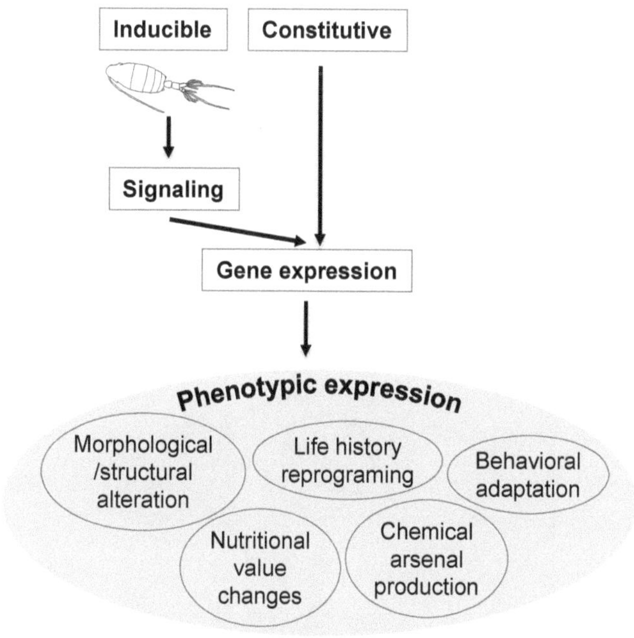

Figure 18.1. Algal defense strategies against grazing.

Defense Against Grazing

Grazing is one of the most important biotic factors controlling algal biomass. Grazers of phytoplankton include copepods, ciliates and other heterotrophic protists, herbivores, filter-feeders, while those of seaweeds are mainly invertebrate foragers such as amphipods, isopods, polychaetes, gastropods, crabs, sea urchins, fishes, Siganidae, reptiles, green turtles, sirenians, and birds (Iken 2012, Lürling 2021). As demonstrated for phytoplankton, but likely applicable to seaweeds, the major defense strategies include avoiding ingestion and avoiding digestion (Lürling 2021). These strategies manifest in highly diverse forms, which can be classified into five categories (Fig. 18.1): (1) behavioral adaptation, (2) morphological and structural alteration, (3) life history reprogramming, (4) changes in nutritional value, and (5) production of chemical deterrents. Some of the defense strategies are constitutive, likely a result of adaptive evolution, whereas others are induced by grazer presence or grazing activity.

Algae often employ multiple strategies for defense and switch between them in response to different stressors. For example, *Microcystis* species are known to use microcystins for defense (Ger et al. 2016). However, a transcriptomic study on *Microcystis* exposed to *Daphnia* showed that genes encoding the synthesis of microcystins (mycABCDEFG) and other cyanobacterial toxins were not upregulated; instead, genes encoding heat shock protein and gas vesicles were upregulated, potentially to enhance colony formation as a means to deter grazing (Harke et al. 2016).

Behavioral Adaptation

Flagellated algae can migrate vertically diurnally to avoid encounters with grazers. In their own defense, many zooplankton species engage in Diel Vertical Migration (DVM) to optimize their survival and growth. Typically, these organisms ascend to shallower depths during nighttime hours to forage in prey-abundant surface waters, and descend to deeper, darker layers of the ocean during daytime to reduce their visibility to visual predators (Bandara et al. 2021). Flagellated algae migrate in the opposite direction to these zooplankters: they dive to deeper, nutrient-rich waters at nighttime to scavenge nutrients and ascend to surface during daytime to photosynthesize. The DVM behaviors

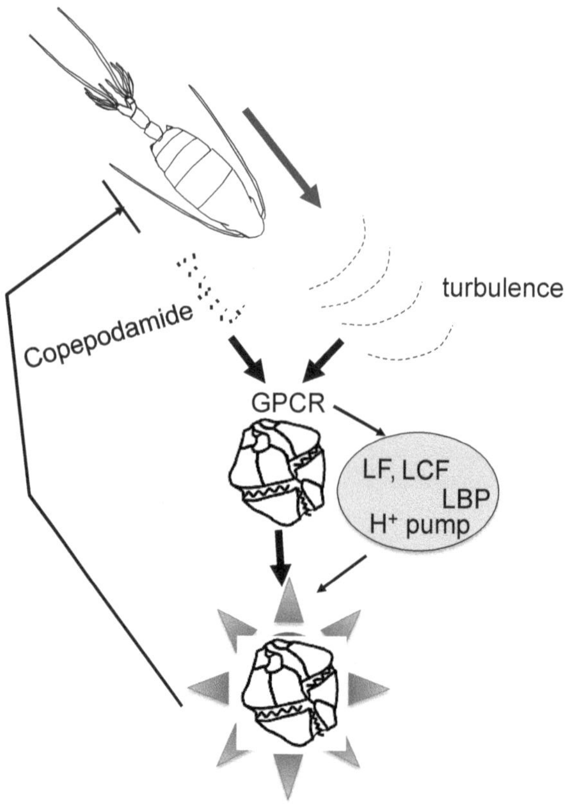

Figure 18.2. Schematic of dinoflagellate bioluminescence induced by chemical or mechanical cues grom grazer to deter grazing. GPCR, G protein-coupled receptor; LF, luciferin; LCF, luciferase; LBP, luciferin-binding protein. Perpendicular line at the end of an arrow indicates inhibitory regulation.

of both the algae and zooplankton are under circadian clock control and entrained by periodically oscillating cues like the light-dark cycle. Evidently, the flagellated algae are phototactic whereas the zooplankton are counter-phototactic. Thus, photoreceptors such as cryptochrome may play an important role in the diel rhythm of DVM (Bandara et al. 2021).

Some dinoflagellates use bioluminescence to discourage grazers, apparently through a feedback signaling system (Fig. 18.2). For instance, bioluminescence of *Lingulodinium polyedrum* deters grazing by the copepod *Acartia tonsa*, changing the dinoflagellate from a preferred prey when not bioluminescent to totally rejected alga when bioluminescent (Prevett et al. 2019). Furthermore, chemical cues of copepod's presence, copepodamite, and possibly turbulence created when copepods are approaching, stimulate bioluminescence of this dinoflagellate. The "grazer approaching→mechanical cue created→stimulating bioluminescence→discouraging grazer" pathway remains to be experimentally verified.

Morphological and Structural Alteration

Morphological alteration is another strategy used by algae to deter grazing. Ingestion can be compromised if phytoplankton grow to sizes or shapes beyond the ingestion capacity of most grazers or develop thick silicate cell walls that provide protection against grazers that crush cells prior to ingestion. For example, the colonial haptophyte *Phaeocystis globosa*, when sensing a predator's chemical signal, significantly increases the size of the colony to hinder ingestion by zooplankton (Tang 2003). Diatoms and some other lineages of phytoplankton modulate the length of their multi-cell chains in response to the presence of the grazer or grazing activity, physically increasing the difficulty of grazing by the grazer (Bergkvist et al. 2012). In other cases, constitutive anti-ingestion defense is always operational, such as the physical properties of diatom frustules to resist grazing by animals.

Some seaweeds have developed structural adaptations to deter grazing. They may have tough or fibrous tissues, thorns, or spines that make them less palatable or difficult to consume. Some species have flexible fronds that allow them to bend and reduce damage caused by grazers. Some microalgae thicken their cell walls or embed cells in mucous to avoid digestion and survive gut passage in grazers.

Life History Reprogramming

Some algae employ reproductive strategies that help ensure their survival despite grazing. Some seaweeds grow very fast, allowing them to recover quickly from grazing damage. Dinoflagellates undergo sexual reproduction and produce cysts in response to environmental stress, whereas diatoms and many other lineages of algae can produce resting cells. These specialized forms of cells are resistant to grazing and digestion. Some seaweeds produce abundant spores or resistant structures to enable them to disperse and colonize new areas.

Changes in Nutritional Values

Becoming less nutritional and/or less palatable may achieve either or both ingestion and digestion deterrence. This often involve a metabolic shift. The shift may change the biochemical composition, manifesting in decreased nutritional molecules such as proteins and fatty acids and increased inhibitory or grazing-deterring metabolites. Some seaweeds produce secondary metabolites like polyphenols, terpenoids, and alkaloids that discourage herbivory. Phytoplankton also produce various grazer-deterring compounds.

Chemical Defense Strategy

The classical example of a chemical anti-grazing strategy is the production of trypsin inhibitors in land plants to inhibit digestion by herbivores (Koiwa et al. 1997). Functional equivalents in algae include, depending on algal lineages, the production of Reactive Oxygen Species (ROS), oxylipinds that mainly include polyunsaturated aldehydes (PUAs) and non-volatile oxylipins (a.k.a Linear Ozxygenated Fatty Acids, LOFAs), or other metabolites (e.g., dimethylsulfide or DMS) that inhibit the development or feeding activity of grazers. Microalgae produce a large suite of toxins, including cyanotoxins (CTXs), Paralytic Shellfish poisoning Toxins (PSTs), Diarrhetic Shellfish poisoning toxins (DSTs), neurotoxic shellfish poisoning toxins (NSTs), azaspiracid shellfish poisoning Toxins (AZTs), Ciguatera Fish poisoning Toxins (CFTs), Amnesic Shellfish poisoning Toxins (ASTs), palytoxin poisoning toxins (PTTs), cytotoxins (CTT), and hemolysins (HML). Additionally, certain cyanobacteria (e.g., *Trichodesmium*), diatoms, and dinoflagellates produce β-N-methylamino-L-alanine (MBAA), a non-proteinogenic neurotoxin. While most of these algal toxins are poisonous to mammals and vertebrates (including to humans) through food chain transfer and accumulation, some of them have been shown to deter grazing (Turner and Tester 1997).

Algal Anti-Grazing Toxins

Anti-grazing toxins are typically induced by grazing pressure, but many of the algal toxins listed above may be induced by nutrient conditions or temperature. Research on the anti-grazing mechanisms in algae is generally limited, and taxon sampling is inadequate for drawing general conclusions. However, common responses to herbivory can be identified in the studied rhodophyte (Collén et al. 2012) and phaeophyte seaweeds (Flöthe et al. 2014), including regulation of amino acid abundance, fatty acid profile, and lipid oxidized derivatives

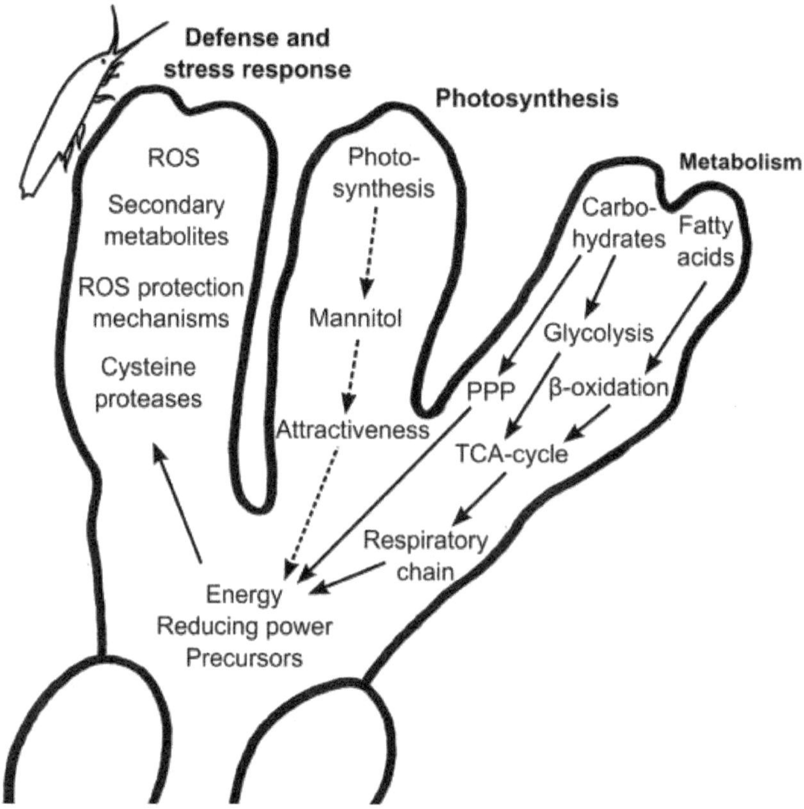

Figure 18.3. Model showing grazing-induced changes in photosynthesis, carbohydrate and lipid metabolism, and defences in *Fucus vesiculosus*. Grazing by *Idotea baltica* results in down-regulation of photosynthesis that would lead to a reduced availability of energy, reducing power and precursors (dashed arrows). Meanwhile, grazing results in the activation of glycolysis, pentose phosphate pathway (PPP) and β-oxidation, leading to the accumulation of energy, reducing equivalents and precursors that may then be used to induce defense mechanisms, such as reactive oxygen species (ROS), secondary metabolites and cysteine proteases (solid arrows). Reproduced with permission from Flöthe et al. 2014.

(especially oxylipins), with accompanying transcriptomic modifications (Fig. 18.3). Compared to abiotic stress, grazing seems to elicit smaller transcriptomic responses in algae. In the brown alga *Fucus vesiculosus*, 1.7% of total genes were found to be differentially expressed in response to grazing by *Littorina obtusata* (Flöthe et al. 2014), whereas in *L. digitata* and *Lessonia spicata*, only 0.8% of total genes were induced upon grazing by herbivores (Ritter et al. 2017). Among the mechanisms represented by the differentially expressed genes, the production of inhibitory compounds is a notable common feature in algae (Howe and Schimiller 2002).

Cyanobacteria can produce cyanopeptolins (CPs), a six-residue peptide that has inhibitory activities against trypsin and chymotrypsin, the common digestive enzyme in zooplankton (Gademann et al. 2010). CPs are synthesized by non-ribosomal peptide synthase complexes. *Trichodesmium* is known to release different secondary metabolites, such as the lipophilic chlorinated trichotoxin, the palytoxin causing clupeotoxism in humans via fish, and the neurotoxin ß-N-methylamino-L-alanine (BMAA), some of which may serve as grazer deterrents (reviewed by Bergman et al. 2013).

In eukaryotic phytoplankton, it has been reported that presence of copepod grazers in *Alexandrium* cultures can induce synthesis of saxitoxin in the dinoflagellates (Selander et al. 2006, Bergkvist et al. 2008, Wohlrab et al. 2010), in addition to avoiding behaviors (Selander et al. 2011). Similarly, in diatoms, grazing pressure induces transcriptomic and metabolomic responses

(Amato et al. 2018) as well as chain length changes discussed earlier (Bergkvist et al. 2012). In *Pseudo-nitzschia seriata*, one of the few diatom species that produce neurotoxin, was also found to increase production of the toxin when the cells were exposed to grazers (Harðardóttir et al. 2019). The grazer deterring role of toxin production has also been shown for the haptophyte *Chrysochromulina polylepsis*, the toxic strain of which was grazed upon by *Oxyrrhis marina* at only 27% of the rate of grazing on a non-toxic strain (John et al. 2002).

Bacterial Toxins Produced by Symbiotic Bacteria

Some algae can employ symbiotic bacteria to produce toxins for its own protection from grazing. *Bryopsis* sp., for instance, harbors "*Candidatus* Endobryopsis kahalalidefaciens" as an endosymbiont, which produces a library of complex molecules (the kahalalides) for chemical defense of the host (Zan et al. 2019). Interestingly, these compounds are subsequently hijacked by the mollusk *Elysia rufescens* and are also utilized for defense. In adapting to the symbiotic lifestyle, the kahalalide-producing bacteria have lost essential pathways that are present in free-living bacteria. This shift grants control to the *Bryopsis* host, effectively allowing it to operate a microbial factory for defense.

Counter Anti-Grazing and A Potential "Arms Race"

As remarkable example of co-evolution between rivals, toxic cyanobacteria and their grazer *Daphnia* in the freshwater ecosystem engage in an arms race. In response to grazing by *Daphnia*, which uses trypsin and chymotrypsin as major digestive enzymes, some cyanobacteria have evolved the ability to produce the trypsin inhibitor cyanopeptolins (CPs), as mentioned earlier, in addition to toxins such as microcystins. To overcome the toxic effect of CPs, *Daphnia* has developed tolerance by selecting for CPs-tolerant chymotrypsin haplotypes (Schwarzenberger et al. 2020). Whether some cyanobacteria further evolved new CPs to undermine the CPs-tolerant chymotrypsin will be interesting to study.

Oxylipins

Among the variety of toxins induced by grazing, oxylipin is arguably the most common in seaweeds and microalgae as well as land plants and animals. It not only deters grazing, but also defends against wounds, pathogen infection, metal toxicity, desiccation, and other kinds of stress (Barbosa et al. 2016, Linares-Maurizi et al. 2023). Oxylipins are oxidative degradation PolyUnsaturated FAs (PUFAs) by radical and/or enzymatic conversion (Frankel 1980). Among microalgae, chlorophytes, diatoms, and dinoflagellates have been found to produce oxylipins for anti-grazing defense (De Los Reyes et al. 2014, Johnson et al. 2020). Structures of some PUFAs and corresponding oxylipins are shown in Fig. 18.4 as an example. The high diversity of oxylipins in macroalgae differs in the position of oxygenation by lipoxygenases (LOXs) and downstream reactions (Barbosa et al. 2016). However, in many instances, the abundance of PUFA substrates does not closely match their utilization in oxylipin biosynthetic pathways. For example, phaeophytes have relatively low levels of C18 PUFA, yet they frequently use this substrate in LOX-initiated biosynthetic pathways. Chlorophyta, like vascular plants, predominantly oxidize C18 substrates. Rhodophyta, on the other hand, extensively utilize both C18 and C20 PUFA for oxylipin production.

From an adaptive evolutionary perspective, it is striking how PUFAs are co-opted to serve multiple roles. They have high nutritional values, can serve as a sink of radicals, and when receiving radicals change to anti-grazing compounds and other regulatory molecules. The other oxylipin molecules include Jasmonate, a plant hormone that regulates wound healing, defense against pathogens, inhibition of root growth and seed germination, and stimulation of senescence, among other developmental processes in land plants.

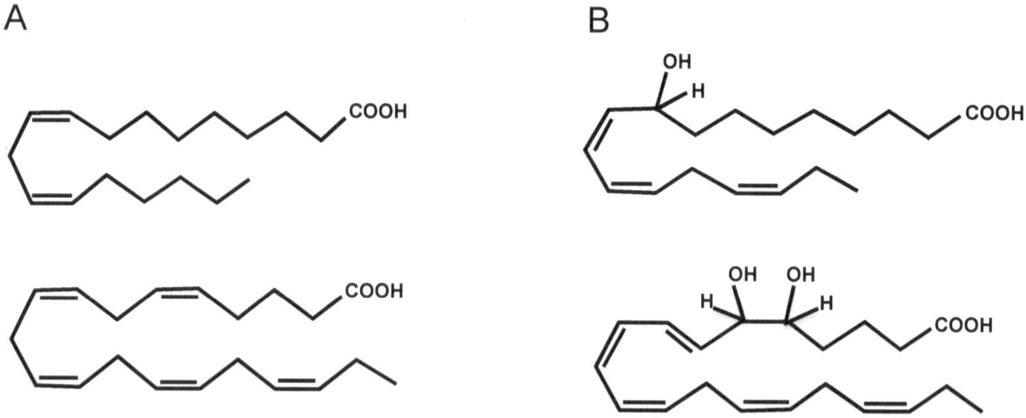

Figure 18.4. Schematic of representative substrate fatty acids (A) and oxylipins (B).

Types of Oxylipins and Biosynthesis Pathways

There have been extensive metabolite studies on algal (mainly seaweed) oxylipins for industrial purposes, which have resulted in the identification of numerous compounds. The efforts have also helped sketch a diversity of biosynthesis pathways for oxylipins, both in plants and algae (Fig. 18.5). In land plants, 18:2 or 18:3 serves as the substrate, which is oxidized by α-dioxygenase (αDOX) to produce 2-hydroperoxy FA aldehydes or lipoxygenase (LOX) to produce 9- or 13-hydroperoxides. 9- or 13-hydroperoxides are then further transformed into various types of oxylipins with the aid of different enzymes such as reductase, divinylether synthase, peroxygenase, hydroperoxide lyase, LOX, epoxyalcohol synthase, and allene oxide synthase (Browse 2009). In algae, substrates and enzymes involved in oxylipin synthesis vary with species, being better understood in rhodophytes and phaeophytes than in chlorophytes (Barbosa et al. 2016). The enzymes involved (Table 18.1) consist of LOX with specificity of oxygenation on various carbon positions on the substrate and hydroperoxide lyase (HPL).

The biosynthesis pathway of oxylipins in chlorophytes is similar to that in land plants (Bernart et al. 1993). Chlorophytes are generally rich in C18 PUFA and hence are expected to produce C18 oxylipins. Species in this phylum use 9- or 13-LOX, but also HPL, which results in the production of various short-chain carbohydrates, aldehydes, and alcohols (Andreou et al. 2009). For instance, *Ulva lactuca* expresses a calcium-stimulated LOX, which can cleave different PUFA substrates into various hydroxy-fatty acids, including 9-hydroxyoctadecadienoic acid (9-HODE), 13-HODE, 9-hydroxyoctadecatrienoic acid (9-HOTrE), 12-hydroxyeicosatetraenoic acid (HETE), 15-HETE, and many others. *Cladophora columbiana* also produces abundant oxylipins, some of which may arise from HPL cleavage of a 9-hydroperoxyoctadecatrienoic acid (9-HpOTrE) precursor, itself formed by the action of a linoleate 9-LOX (Barbosa et al. 2016).

Red and brown seaweeds produce oxylipins via both the animal (eicosanoid) and the plant (octadecanoid) pathways (Howe and Schilmiller 2002, Bouarab et al. 2004, Flöthe et al. 2014, Ritter et al. 2017). Rhodophytes transform mainly C20 FAs, involving a 12-lipoxygenase to introduce oxygen to C12 (Jiang and Gerwick 1991). However, *Gracilariopsis lemaneiformis* synthesizes 13 different oxylipin molecules from C18 and C20 FAs by LOX activity at C8, C9, and C12 positions (Browse 2009). It was found that the chemical defense of *Gracilaria* spp. against grazing by *Echinolittorina peruviana* was stimulated by tissue disruption and proceeded via the release of arachidonic acid derived oxylipins (Rempt et al. 2012).

Phaeophytes oxidize C18 and C20 FAs to produce highly cyclic oxylipins employing a lipoxygeanse (Proteau and Gerwick 1992). For instance, *Ecklonia stolonifera* was found to produce C18 tricyclic compounds ecklonialactones A (50) and B (51) with antifeedant activity against the

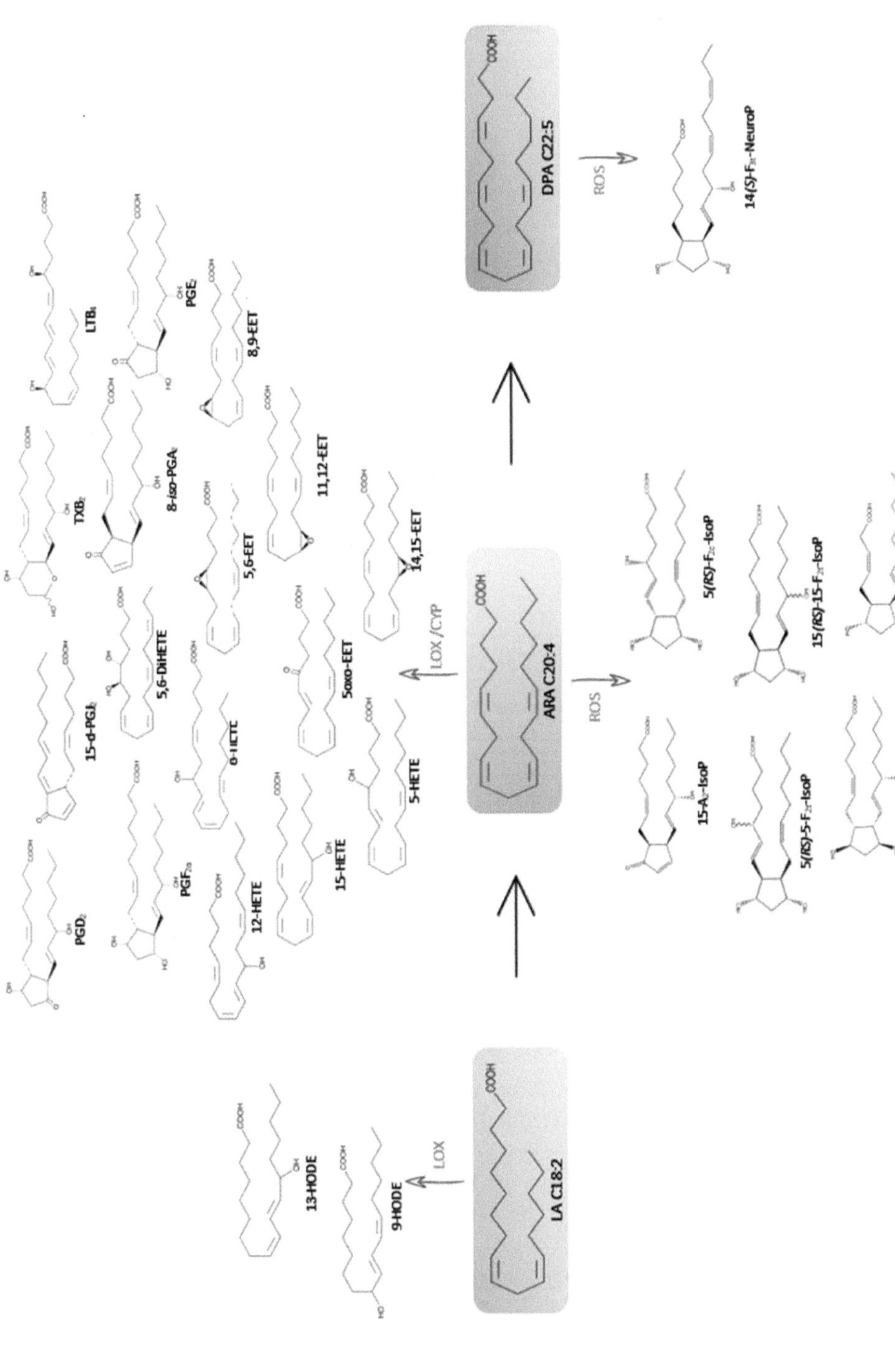

Figure 18.5. Biosynthesis pathways of oxylipins in microalgae. Reproduced from Linares-Maurizi et al. 2023 under the Creative Common CC BY License.

Table 18.1. Enzymes involved in the biosynthesis of oxylipins in algae.

Phylum	Species	Enzyme system*	Reference (GenBank accession number)
Rhodophyta	Species in the phylum in general	12-LOX, arachidonate 5R-, 8R, 15S-LOX, linoleate 9S- and 13S-LOX	Jiang and Gerwick 1991, Gerwick and Singh 2002, Bouarab et al. 2004
	Gracilariopsis lemaneiformis	8-LOX, 9-LOX, 12-LOX	Browse 2009
	Porphyra purpurea	LOX	Liu and Reith 1994
	Pyropia haitanensis	Two LOX isoforms	Chen et al. 2015, Zhu et al. 2015
	Sarcodiotheca gaudichaudii	8-LOX	
	Agardhiella subulata	8-LOX	
Phaeophyta	Species in the phylum in general	LOX, HPL	Andreou et al. 2009
	Ecklonia stolonifera	LOX	Kurata et al. 1989
	Laminaria angustata	12-LOX, 15-LOX	Boonprab et al. 2004
Chlorophyta	Species in the phylum in general	9-LOX, 13-LOX, HPL, and various downstream enzymes	Andreou et al. 2009
	Cladophora columbiana	HPL	
	Ulva conglobata	9-LOX, 11-LOX, HPL	Akakabe et al. 2002, 2003
	Cladophora columbiana	9-LOX, HPL	Gerwick et al. 1993
	Monoraphidium sp.	LOX	(XP_013898442.1)
	Chlamydomonas reinhardti	LOX	(XP_042918114.1)
Cyanobacteria	*Nostoc*	LOX	(WP_010994078.1)
	Anabaena catenula	LOX	(WP_190907571.1)
Diatoms	*Speices in the phylum in general*	LOX, COX, CYP	
	Skeletonema marinoi	LOX	(KAK1734479.1)
	Pseudo-nitzschia arenysensis	LOX	(QWC64745.1)
Dinoflagellates	*Durusdinium trenchii*	LOX	(CAK9069906.1)
	Durusdinium trenchii	ALOX15	(CAK9107062.1)
	Prorocentrum shikokuense	LOX	(Unigene31179_All.p1)
Haptophytes	*Emiliania huxleyi*	LOX	(20130905|17507_1)

* LOX, lipoxygenase, numbers indicating carbon position as the site of oxygenation; HPL, hydroperoxide lyase

abalone *Haliotis discus hannai* (Kurata et al. 1989). Subsequently, additional ecklonialactones (C–F (52–55)) and other related classes of cyclical oxylipins were found in various species (Barbosa et al. 2016).

Microalgae also produce oxylipins to deter grazing. Like in seaweeds, microalgae produce oxylipins through non-enzymatic (e.g., ROS) and enzymatic reactions, with the latter relying on LOX, cyclooxygenase (COX) and cytochrome (CY) P450 (Fig. 18.5). Omega-3 and Omega-6, two abundant fatty acids in microalgae, are the major substrates (Linares-Maurizi et al. 2023). Among microalgae, diatoms are one of the most important groups that actively synthesize oxylipins from membrane-bound PUFAs (Fontana et al. 2007, Russo et al. 2020). Diatom oxylipins (aldehydes) can deter grazing by copepods through impairing copepod egg hatching success and embryo and larval development (Poulet et al. 1995, Blasio and Balzano 2021). This discovery completely changed the classical notion that diatoms are dominant and preferred prey for copepods. In natural blooms of diatoms (*Skeletonema costatum* and *Pseudo-nitzschia delicatissima*), copepod egg hatching rate was as low as 12% compared to 90% in post-bloom conditions (Miralto et al. 1999). The detrimental effect of oxylipins provides potential alternative explanation for why diatoms can escape grazing and form blooms compared to the classical notion that cold early-spring temperature retards copepod development causing a mismatch between time windows of copepod feeding and phytoplankton growth. Diatom oxylipins can also deter micrograzers such as ciliates and heterotrophic dinoflagellates (Lavrentyev et al. 2015, Johnson et al. 2020). Oxylipins are not detectable in intact cells, but they are typically released after cells are damaged by grazing activities (Jüttner 2001).

Oxidative Burst

Like oxylipin formation, oxidative bursts are also a common stress response in animals, plants, red algae and brown algae. An oxidative burst is a rapid, transient, production of huge amounts of reactive oxygen species (ROS), which is one of the earliest observable aspects of a plant's defense strategy. These organisms activate a class of enzymes, such as NADPH oxidase, and cause the overexpression of related proteins (e.g., Phenylalanine AmmoniaLyase or PAL) involved in defense after the invasion of pathogens (Weinberger 2007). The up-regulated expression promotes the immune effect of algae on pathogenic bacteria to a certain extent (Bouarab et al. 2004).

Marine macroalgae can release ROS upon wounding and grazing. A study on the phaeophyte *Ascoseira mirabilis* indicated that wounding and the presence of hydrogen peroxide (H_2O_2) or non-H_2O_2 ROS had a direct anti-herbivore effect on the amphipod grazer *Gondogeneia antarctica* (McDowell et al. 2014b). Similar anti-grazing effect of strong oxidants induced by wounding was also found in the rhodophyte *Palmaria decipiens* (McDowell et al. 2014a). These findings suggest that a "grazing (wound) → ROS → deterring grazing" response pathway operates in brown and red algae.

Microalgae also use ROS for defense against grazing. The toxic dinoflagellate *Alexandrium catenella* not only produces PST, but also ROS, which inhibits grazing by the ciliate *Tiarina fusus* and heterotrophic dinoflagellate *Polykrikos kofoidii* (Flores et al. 2012). In my laboratory, we found that the intensity of the anti-grazing effect depends on the cell abundance ratio between *A. catenella* and the grazer *Oxyrrhis marina* (Fig. 18.6).

Defense Against Microbial Insults

Microbial attack from bacteria and viruses is another type of significant biotic insults for algae. The interaction between phytoplankton and microbes is complex, including mutualistic and antagonistic. Mutualistic relationships are detailed in Chapters 20 and 21. One of the antagonistic relationships is competition for nutrients that are limited in the environment (Liu et al. 2012, Le Chevanton et al. 2016). The other, which is the major focus here, is virulent attacks of the microbes, to which phytoplankton respond by producing chemical arsenals such as peroxides to wage oxidative burst or antimicrobial peptides (AMPs) to kill the microbes.

Bacterial Insults and Defensin

For exploring the potential of biological control of harmful algal blooms, bacteria inhibiting the growth of or killing their host algae have attracted research attention (Coyne et al. 2022). Virulent attacks by microbes are prevalent, with effects ranging from inhibiting algal growth to causing algal cell death. Algicidal interactions can be direct or indirect. In the indirect mode, algicidal bacteria secrete dissolved chemicals that can kill algae, whereas in the direct mode the algicidal bacteria cells effectively antagonize the alga through direct contact. A direct contact mode may also result from the involvement of a compound that reaches effective concentrations only in the phycosphere. Growth-inhibitory effects of phycosphere bacteria have been well documented (e.g., Tang et al. 2022).

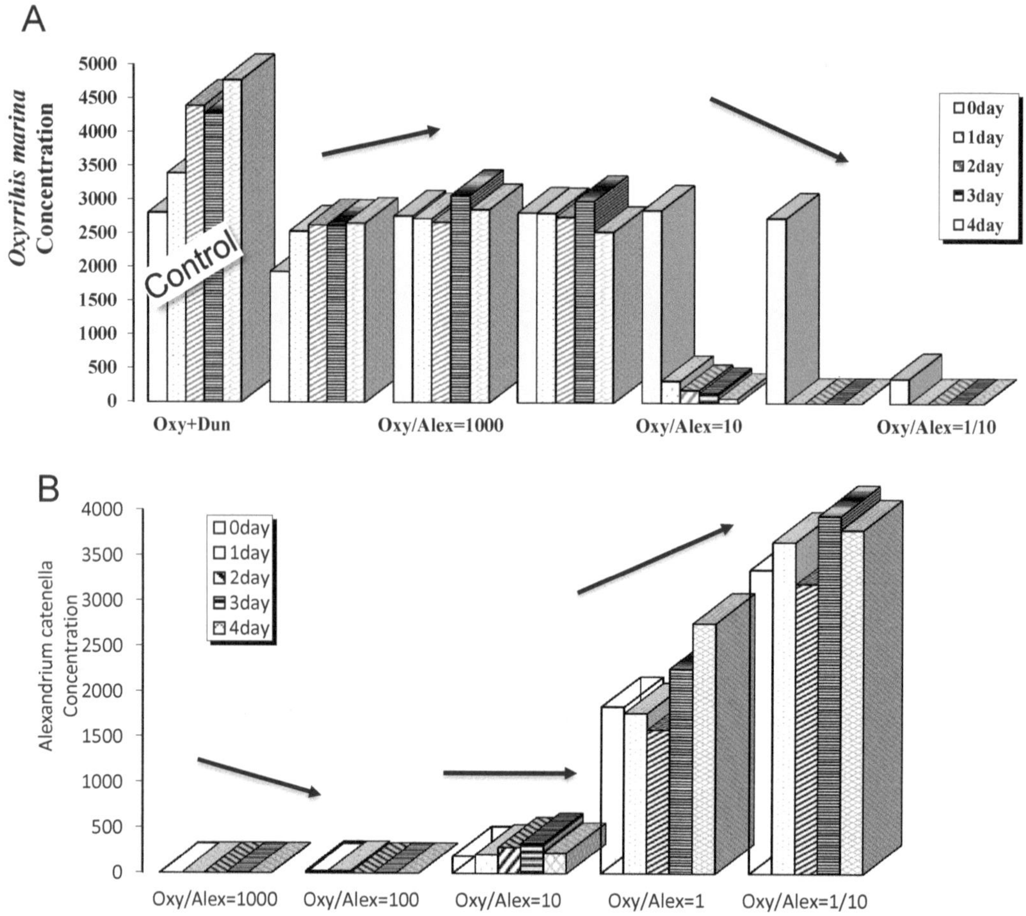

Figure 18.6. Effect of anti-grazing defense depending on grazer-alga abundance ratio. (A) Growth trend of grazer (*Oxyrrhis marina*) in control (*Dunaliella tertiolecta* [Dun] as food) and different grazer (Oxy)/alga (Alex) ratio treatments. (B) Growth trend of *Alexandrium catenella* in different grazer/alga ratio treatments.

Diverse Bacterial Adversaries

Algicidal bacteria are phylogenetically diverse and polyphyletic, although recent studies revealed a high proportion of algicidal interactions involving bacteria from the phylum Proteobacteria, particularly from the class γ-Proteobacteria. An algicidal bacterium from the genus of *Alteromonas* was isolated from a bloom of the dinoflagellate *Prorocentrum shikokuense* (=*P. donghaiense*) and found to likely produce beta-glucosidase and target the megacytic growth zone of the dinoflagellate cell, causing its death (Shi et al. 2018). *P. shikokuense* can also be killed by prodigiosin, a secondary metabolite of various microorganisms including *Seratia*, *Vibrio*, and *Streptomyces* (Chen et al. 2021). The same alga is vulnerable to *Sulfitobacter porphyrae*, a bacterium originally identified from *Porphyra* (Fukui et al. 2014), which kills the dinoflagellate by producing soluble compounds and destroying its cell membrane (Zhang et al. 2020). *Sulfitobacter* strain D7 is algicidal to *Emiliania huxleyi* by utilizing DMSP produced by the haptophyte and releasing high amount of DMS (Barak-Gavish et al. 2018). A new algicidal compound, 3,3',5,5'-tetrabromo-2,2'-biphenyldiol (4-BP), was identified and found to be prevalent geographically and taxonomically (Zhang et al. 2023). Phytoplankton respond to microbial attacks by producing chemical arsenals such as peroxides to wage oxidative burst or antimicrobial peptides (AMPs) to kill the infecting microbes.

Antimicrobial Peptides (AMPs)

AMPs, mainly defensins and cathelicidins, are small polypeptides (< 100 amino acids) with a conserved signature of cysteines that have antimicrobial activity at physiological concentrations under conditions prevailing in the organisms (Ganz 2003). Initially discovered in insects and mammals, defensins are now known to occur widely, from vertebrates, invertebrates, plants, and molds (Antcheva et al. 2013), and mainly target bacteria and fungi but can also combat viruses (Daher et al. 1986, Zhang et al. 2002). In plants, defensins, mostly acting against fungal pathogens, form a well-conserved three-dimensional structure consisting of a cysteine-stabilized α/β (CSαβ) motif, which forms one α-helix and three anti-parallel β-sheets. The amino acid sequence has six to eight cysteine residues, the majority of which forms three to four disulfide bridges in the sequence of Cys1-Cys8, Cys2-Cys5, Cys3-Cys6, and Cys4-Cys7 (Lay and Anderson 2005, Lacerda et al. 2014).

Defensin genes have a tripartite structure (Fig. 18.7A). The roles of defensins include antimicrobial activity and digestion inhibition, among others. The antimicrobial activity stems from forming pores into invader's cell membrane or interacting with phospholipids, both of which lead to change of ion or peptide permeability (Lacerda et al. 2014), leading to microbial membrane breakdown (Fig. 18.7B). Despite the conserved cysteine residues, the general amino acid sequence and quaternary structure vary widely among different defensins, which might confer specificity to target microbes (Ganz 2003). Recent studies have revealed evidence of defensins in algae.

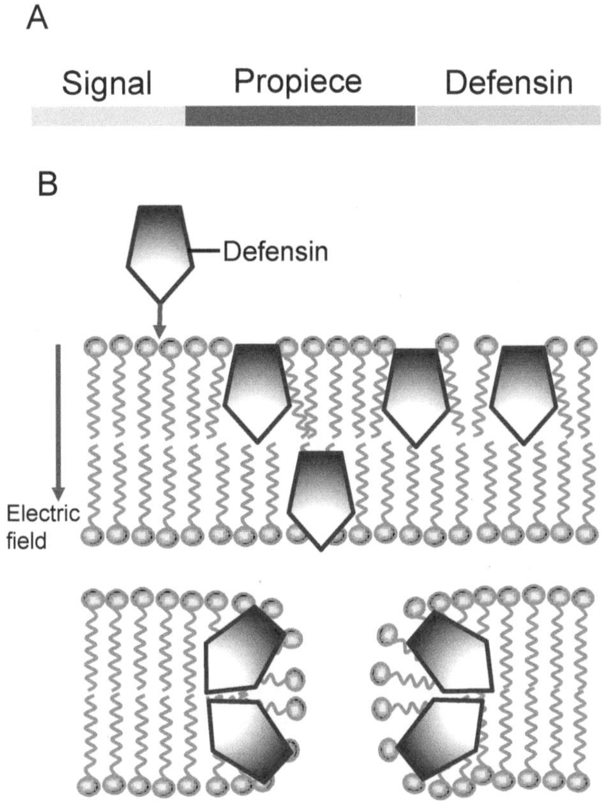

Figure 18.7. Schematic of the structure (A) and the carpet–wormhole model of action of defensins (B). (A) Defensin precursor polypeptide structure. (B) Most defensins (shown as pentagon) are amphipathic molecules that have clusters of positively charged amino-acid side chains (dark) and hydrophobic amino-acid side chains (light). They interact with microbial membranes, In the top panel, electrostatic attraction and the transmembrane bioelectric field pull the peptide molecules towards and into the membrane. As peptide molecules accumulate in a 'carpet', the membrane is strained and the peptides transition into another arrangement (shown in the lower panel) that lowers the strain but results in the formation of membrane 'wormholes' or pores.

Table 18.2. Putative defensin coding genes of *Prorocentrum shikokuense* before (T0) and during a bloom (T1, T2, T3). In bold-type are genes also expressed in another bloom of the same species.

Gene ID	PFAM ID	PFAM description	T0_FPKM	T1_FPKM	T2_FPKM	T3_FPKM
comp103852_c0	PF01361/ PF06954	Tautomerase enzyme// Resistin	0	107.79	122.10	179.56
comp104043_c0	PF00323	**Mammalian defensin**	0	78.80	79.32	93.48
comp111955_c0	PF01097	Arthropod defensin	0	60.94	82.00	48.48
comp113060_c0	PF01097	Arthropod defensin	0	50.78	77.36	62.91
comp86684_c0	PF08018	Frog antimicrobial peptide	0	50.42	52.74	25.73
comp114487_c0	PF08102	Scorpion antimicrobial peptide (ATP5H)	0	44.54	32.71	21.78
comp91075_c0	PF08026	Bee antimicrobial peptide	0	41.58	31.32	43.21
comp118715_c0	PF07359	**Liver-expressed antimicrobial peptide 2 precursor**	0	35.41	34.91	23.54
comp96942_c0	PF06451	**Moricin**	0	27.21	31.97	20.00
comp69037_c0	PF00321	**Plant thionin**	0	48.39	57.72	68.05
comp106855_c0	PF08107	**Pleurocidin family**	0	76.11	57.50	82.53
comp98089_c4	PF0095	Synaptobrevin	0	38.58	33.54	36.41

In two studies, metatranscriptomic analyses of harmful algal blooms caused by the dinoflagellate *Prorocentrum shikokuense* revealed active expression of genes annotated as defensins previously known as antimicrobial peptides in plants, arthropods, and other organisms (Table 18.2; Zhang et al. 2019, Yu et al. 2020). Among these, synaptobrevin has been implicated in innate immunity in nematodes and plants (Waheed et al. 2021). Cyclotide has been reported as an antibiotic of microbial origin (Craik et al. 1999). Gallidermin is a type of lanthionine-containing polypeptide antibiotic (Kellner et al. 1988). Several of these, including pleurocidin, mammalian defensin, Liver-Expressed Antimicrobial Peptide (LEAP) 2 precursor, moricin, and plant thionin, were actively expressed in both bloom events. Pleurocidin is known to create pores in the membrane of bacteria, leading to cell lysis and hence acts innate host defense (Saint et al. 2002). LEAPs are a family of antimicrobial peptides serving as part of the host innate immune system in various animals, including fish (Liu et al. 2023) and have not been reported in plants or algae. Moricin is a 42-amino acid peptide initially discovered in the hemolymph of the silkworm *Bombyx mori* (Hara and Yamakawa 1995). Thionins belong to a rapidly growing family of biologically active peptides in the plant kingdom (Höng et al. 2021). They are small (~ 5 kDa), cysteine-rich peptides with toxic and antimicrobial properties. Structural simulation for one of these peptides showed a secondary structure similar to human defensin (one alpha helix and two beta sheets; Fig. 18.8). The upregulation of these defense in genes during the bloom relative to pre-bloom conditions indicated the potential importance of microbial defense in the bloom development of this dinoflagellate. However, the prevalence and exact functions of these putative defensins in dinoflagellates remain to be experimentally demonstrated, and the regulation of their encoding genes needs to be elucidated.

Viral Attacks and Antiviral Defense

Viruses are the most abundant biological entities in the ocean, with abundance ranging from 10^6 to 10^8 mL^{-1} and a total biomass of ~ 0.03 gigaton (Gt) carbon, playing crucial roles in the complex biogeochemical processes (Lara et al. 2017). Marine viruses display myriad morphologies, including helical, icosahedral, elongated, or lemon-like forms, and can be enveloped in a lipid membrane, with or without a tail, measuring between 20 nm to 1.2 μm (Vincent and Vardi 2023). They also occur in

Figure 18.8. Structure simulation for a putative defensin identified in *P. shikokuense* using AlphaFold3 (TM-score 0.36). The dinoflagellate defensin structure (blue) is overlain by the structure of human defensin (yellow). Only the conserved region is shown, which includes an alpha helix and two beta sheets.

various genetic conformations, including single-stranded (ss) RNA, ssDNA, or double-stranded (ds) DNA, and possess genomes up to 2.4 Mb.

Diverse virus resistance mechanisms have been reported in microalgae. At the physiological level, these include variations in susceptibility in the host cell, blockade of intracellular virus genome replication, bacterially mediated virus resistance, and host cell physiological barriers; at the molecular level, recent genomic and transcriptomic studies have revealed virus resistance systems at higher resolutions (Hongo et al. 2021 and references therein).

In the cyanobacteria *Prochlorococcus*, variabilities in the genomic islands containing viral-attachment genes facilitate host-virus coexistence (Avrani et al. 2011). For the prasinophyte *Ostrecoccus tauri*, the variable size of the hypervariable region in chromosome 19 among strains is proposed to be associated with the alga's susceptibility to infection by its dsDNA virus, OtV (Yau et al. 2016). Furthermore, in many host-virus relationships, whole- or partial-viral genome sequences in the host genome can act as a viral resistance factor (Bertsch et al. 2009, Bondy-Denomy et al. 2016).

Diatom viruses were initially discovered in 2004, and since then research data have grown rapidly. These viruses are broadly classified into two groups based on genome conformation: single-stranded DNA (ssDNA) virus (Bacilladnavirus) or ssRNA virus (Bacillarnavirus). Morphologically, they are icosahedral, < 40 nm in diameter, and have no tail. Viral infection inflicts fatal damages on the host in axenic culture, but in nature, diatom host populations maintain their sizes despite the attack of the viruses, suggesting effective virus resistance strategies (Tomaru et al. 2015).

In one of the smallest diatoms *Chaetoceros tenuissimus*, infectious viruses can coexist during blooms. Recent genome sequencing for the species revealed a gene fragment of a replication-associated gene from the infectious ssDNA virus integrated into the 41 Mb haploid genome assembly. This gene fragment was transcriptionally active and conserved in nine other geographic populations of *C. tenuissimus*. Furthermore, this viral gene fragment is flanked by retrotransposon genes with reverse transcriptase function. These data suggest that this virus was likely integrated by a retrotransposon during viral infection (Hongo et al. 2021).

Even for the same algal species, strategies to defend viral impacts can differ with ecosystems or bloom events. For instance, *P. shikokuense* formed blooms in April-May period of 2014, about 500 km apart, one in Changjiang Estuary area in the East China Sea (ECS) and the other off Baicheng Beach (BB, off Xiamen University campus) in Xiamen. The bloom in ECS was observed to persist during the one-week cruise ship monitoring whereas the bloom in BB disappeared the next day. Metatranscriptomic analyses indicated that the algae-infecting viral communities had very similar viral taxon composition, both dominantly by Mimiviridae and Phycodnaviridae (Wang et al. 2023). However, the interaction between the viruses and *P. shikokuense* was quite different between the two bloom outbreaks. In the prolonged bloom in ECS, *P. shikokuense* appeared to harbor metabolically active viruses and maintained active photosynthesis and other cellular activities, a scenario of co-thriving. In contrast, the BB bloom did not show evidence of host-virus co-thriving.

The recent identification of a putative homolog of CRISPR-associated protein 9 (Cas9) gene in a dinoflagellate (Li et al. 2024) suggests that this antiviral defense mechanism may operate in algae. Further explorations to more systematically search for its prevalence and experimentally verify its function are required.

Cost of Defense

For land plants, while the benefit of producing trypsin inhibitors is obvious, a substantial cost is apparent (Zavala et al. 2004), because the defending organisms need to allocate resource to produce the weaponry. This is also true in algae (Pansch et al. 2009, Pančić and Kiorboe 2018). The fitness cost of producing chemical defense has recently been demonstrated in a dinoflagellate, *Alexandrium catenella*, that produces Paralytic Shellfish poisoning Toxins (PST) to deter grazing (Park and Dam 2021). It was found that PST production in *A. catenella* was induced by grazing pressure and chemical cues of the grazer, and toxin production led to a reduction of growth rate. The slowdown of proliferation stemming from the metabolic cost of producing PST and cell loss by grazing seem to balance the natural growth potential, resulting in a steady state population level (Fig. 18.9).

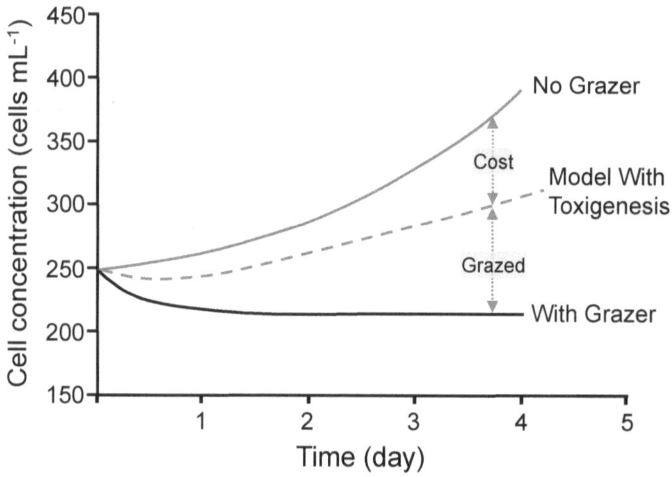

Figure 18.9. Fitness cost of toxin production in *Alexandrium catenella*. The combination of fitness costs of toxin production and cell loss due to grazing led to no net growth in *A. catenella*. Based on Park and Dam 2021.

Signaling in Defense

Stress signal (e.g., chemical cue) usually binds to a receptor on the cell surface, which activates a signal transduction pathway and eventually elicits gene expression changes and physiological response (Chapter 12). Signaling in plant defense against insects has been modeled on tomatoes, which is reviewed in Browse (2009). Basically, local signals, such as oligouronides, and systemic signals, such as the peptide hormone system, interact with receptors to initiate a series of cellular responses. Oligouronides are polymeric substances consisting of uronic acid units with glycosidic linkages that are commonly combined with monosaccharides. Peptide hormone systemin is an 18-amino acid peptide produced upon wounding of tomato leaves. These responses include the opening of plasma membrane ion channels, increases in cytoplasmic [Ca^{2+}], and the activation of a MAP kinase. Additionally, a phospholipase is activated, which releases 18:3 (or OPDA) from membrane lipids, leading to the synthesis of Jasmonic Acid (JA). After this initial rapid response (within 2–10 minutes), JA signaling activates many genes involved in these processes. While other signaling mechanisms also play a role, mutant analysis has shown that JA synthesis and JA signaling are crucial for effective defense (Browse 2009).

The sensing and downstream signaling pathway in algae is in general poorly understood. Presence (of grazers), contact (by microbes), or action (wounding of seaweed by grazers) must produce a cue or signal to interact with a specific receptor. Some exciting advances have occurred in recent years about chemical cues of grazer presence. Selander and colleagues have reported that copepods leave trail of polar lipids in ambient seawater, named copepodamides (Fig. 18.10). Copepodamides trigger toxin production and bioluminescence in toxic dinoflagellates (*Alexandrium*) and defense

A

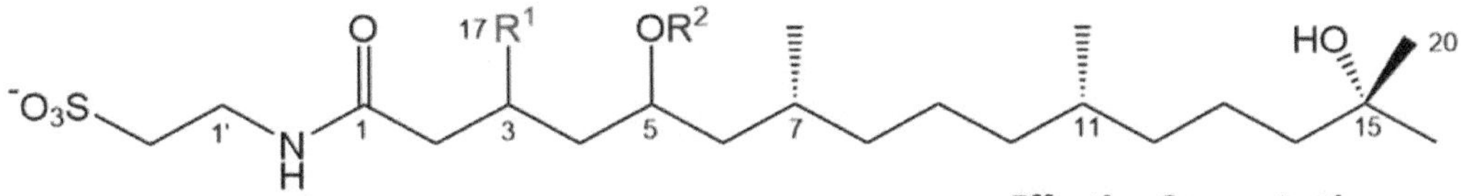

Copepodamide	MW (Da)	R^1	R^2	Effective Concentration for Toxin Doubling (nM)
A	761.5	$-CH_3$	Docosahexaenoic acid (DHA)	0.062 ± 0.006
B	735.5	$-CH_3$	Eicosapentaenoic acid (EPA)	0.023 ± 0.002
C	709.5	$-CH_3$	Stearidonic acid (SDA)	0.14 ± 0.03
D	759.5	$=CH_2$	Docosahexaenoic acid (DHA)	0.64 ± 0.11
E	733.5	$=CH_2$	Eicosapentaenoic acid (EPA)	1.7 ± 0.2
F	707.5	$=CH_2$	Stearidonic acid (SDA)	1.2 ± 0.1
G	451.3	$-CH_3$	-H	18 ±1
H	449.3	$=CH_2$	-H	57±1

B

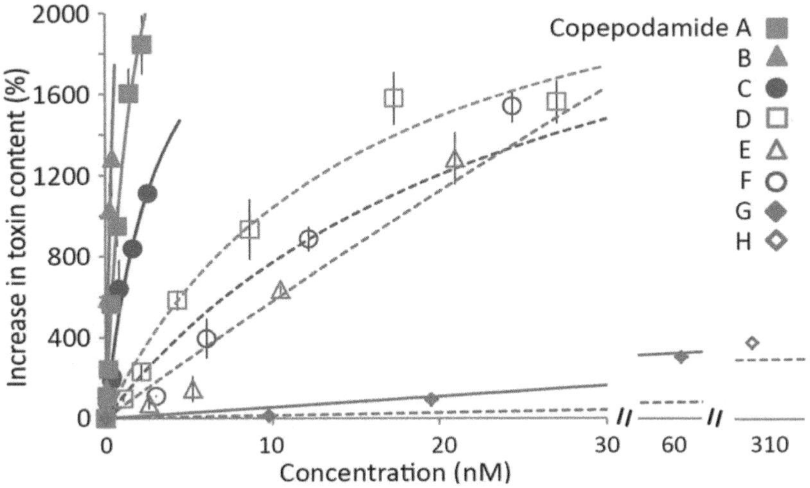

Figure 18.10. Copepod-derived chemical cue and dose response in inducing toxin production in algae. (A) Chemical structure and diversity of the chemical cue copepodamide. (B) Increases in copepod presence (copepodamide concentrations) induces elevated toxin content in *Alexandrium minutum*. Reproduced with permission from Selander et al. 2015.

traits in diatoms (*Skeletonema*, *Pseudo-nitzschia*) in a dosage-dependent manner (Selander et al. 2015, 2019). The copepod cue induces synthesis of saxitoxin (Selander et al. 2006, 2015, Bergkvist et al. 2008, Wohlrab et al. 2010) and avoiding behaviors in *Alexandrium* (Selander et al. 2011). Similarly, in diatoms, grazing pressure induces behavioral (Bergkvist et al. 2012) and transcriptomic as well as metabolomic responses (Amato et al. 2018). In *Pseudo-nitzschia seriata*, one of the few diatom species that produce neurotoxin, was also found to increase production of the toxin when the cells were exposed to grazers (Harðardóttir et al. 2019). These findings suggest that there are at least two and likely more signaling pathways in these dinoflagellates and diatoms that transmit the grazer cue to activate gene expression or protein activation thereby triggering chemical or behavioral responses.

In a study on the diatom *P. tricornutum* as prey for *Oxyrrhis marina* (Johnson et al. 2020), exposure of the diatom to *O. marina* resulted in Nitric Oxide (NO) production by *P. tricornutum* and pronounced change in its dissolved oxylipin profile. Experimentally elevating NO levels also led to an increase in oxylipin production and decrease in overall grazing rates. Under lower levels of NO, *O. marina* preferentially grazed on *P. tricornutum*. This suggests that presence of the micrograzer activated the oxylipin anti-grazing mechanism through NO signaling. In the brown alga *Saccharina japonica*, a conserved peptide in the N-terminal part of bacterial flagellin, flg22, can also induce the oxidative burst in the gametophytic stage, suggesting its role as Microbe-Associated Molecular Pattern (MAMP) elicitor in algae (Wang et al. 2013).

In anti-microbial defense, bacterial lipopolysaccharides can be the trigger of early events of defense reactions in the kelp *Laminaria digitata*. A study showed that the challenge of *L. digitata* sporophytes with bacterial lipopolysaccharides resulted in an oxidative burst and rapid release of free saturated and unsaturated FAs, with concomitant accumulation of oxylipins such as 13-hydroxyoctadecatrienoic and 15-hydroxyeicosapentaenoic acids (Küpper et al. 2006). Much work remains to be done to unravel the receptors and signaling pathways in algal defense against grazing and microbial attacks.

19

Life Cycle, Sex, and the Cell Cycle

Whereas cyanobacteria only perform asexual reproduction, the life cycle of eukaryotic algae consists minimally of a sexual phase and an asexual phase. The life cycle in seaweed is more complex due to the multiple morphological and developmental stages (spores, sporophyte and gametes, gametophyte) in each of the phases and variations between lineages (Fig. 19.1). Furthermore, parthenogenesis and endoreduplication occur in some algae (e.g., phaeophytes; Bothwell et al., 2010), which bypass syngamy. In addition, sometimes vegetative cells can fuse during wound healing or morphogenesis (Borg et al. 2023).

In microalgae, asexual reproduction is the primary mode of population growth, while sexual reproduction is typically triggered by stress conditions as a survival strategy. As is true in many protists, sex results in the transformation of the cells into a resting stage (e.g., cysts in dinoflagellates) for dormancy under harsh environmental conditions. However, like seaweeds, microalgal life cycles may vary between lineages, ranging from haplontic, diplontic, to haplo-diplontic life cycles represented by dinoflagellates, diatoms, and coccolithophorids, respectively (Fig. 19.2). Despite the absence of sex, cyanobacteria also have two-stage life cycles, which are known in filamentous taxa, characterized by the production of akinetes to survive adverse environments, like dinoflagellate cysts or resting cells of other algal lineages (see Chapter 4 for more information).

For microalgae, sex is not easy to observe directly and often relies on molecular tools to detect. In the laboratory, asexual reproduction is the predominant mode of population growth since clonal cultures are not conducive to sexual reproduction, which typically requires a mix of different mating types or genotypes (strains). In the field, as an algal population often comprises multiple genotypes, the probability of sex is higher. In some dinoflagellate species (e.g., *Noctiluca scintillans*), sexual reproduction is very important and efficient in causing blooms (Luo et al. 2022). Whether the same is true for photosynthetic species of dinoflagellates or other groups of algae is an open question, but there is evidence that sex promotes cell proliferation and bloom development (Lin et al. 2022).

Asexual reproduction proceeds with the active progression and completion of the cell cycle, which in typical eukaryotes consists of four discrete phases (Fig. 19.3): the DNA duplication (Synthesis) phase (S), the Mitosis phase (M), and the two growth (or gap) phases between them (G1 for the interval from M to S and G2 for that from S to M). Under some conditions (e.g., environmental stress), cells leave the active cycle and enter G0, a quiescent stage (e.g., dormancy). Typically, mitosis is immediately followed by cytokinesns, producing two daughter cells. However, exceptions occur where nucleus division results in four or eight nuclei leading to the production of four or eight daughter cells. This has been well documented in green microalgae (Zachleder et al. 2016) and dinoflagellates (directly observed in Parrow and Burkholder 2003 or

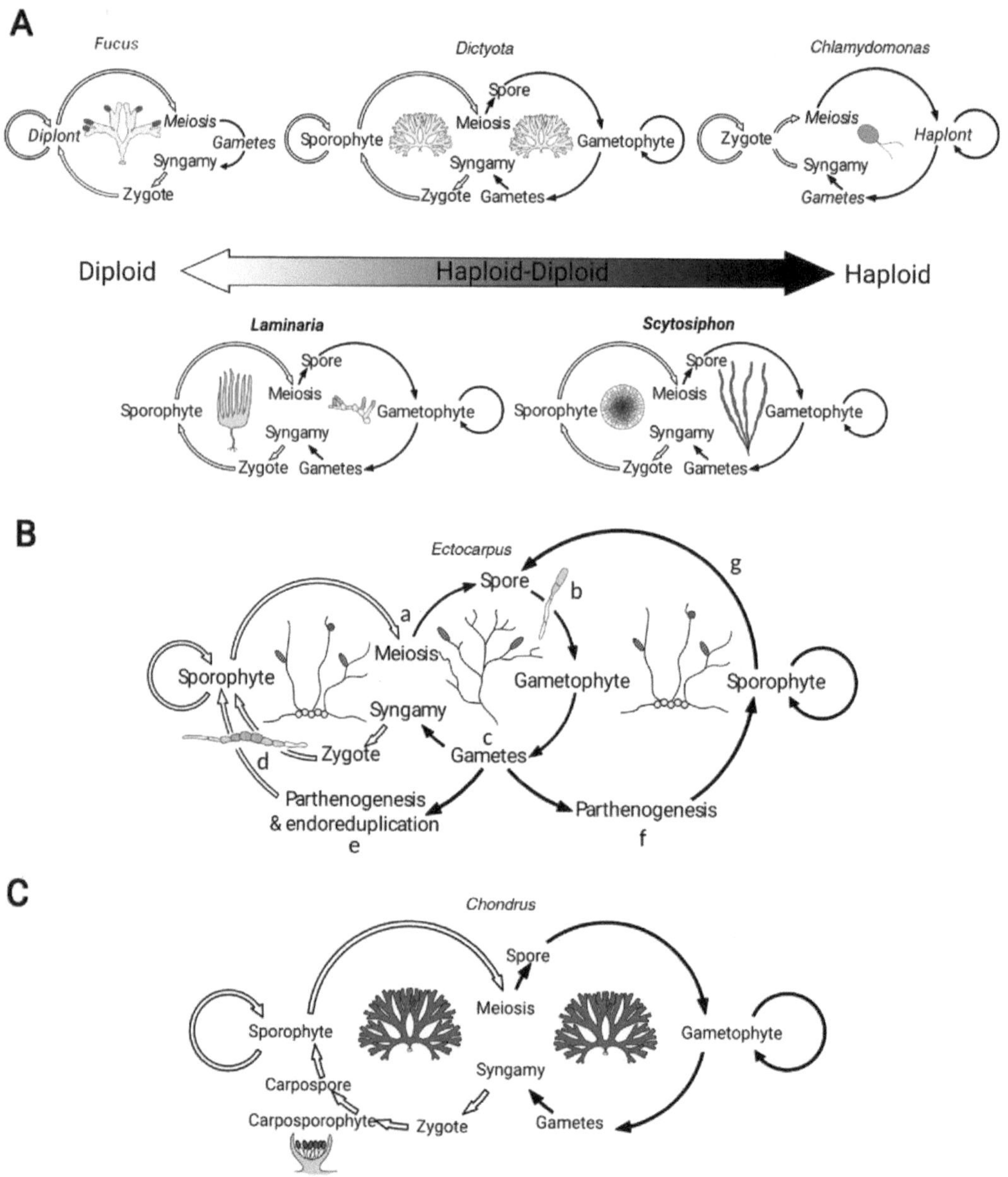

Figure 19.1. Schematic of algal life cycles. Diploid and haploid stages are marked by white and black arrows, respectively. (A) Brown algal life cycle. Fucales are characterized by a diplontic life cycle. Many unicellular algae (*Chlamydomonas*) are characterized by a haplontic life cycle where the formation of a zygote is immediately followed by a meiotic division. (B) Simplified diplohaplontic life cycle of *Ectocarpus siliculosus*. (C) Life cycle of the red alga *Chondrus* showing three distinct stages: the gametophyte, the sporophyte and the carposporophyte, which develops parasitically on the gametophyte after fertilization. Loops connecting a generation with itself denote asexual reproduction mediated by vegetative reproduction (e.g. fragmentation, propagule formation) (*Fucus, Dictyota*), the formation of asexual spores (mitospores) (*Dictyota, Ectocarpus, Laminaria*), mitosis (*Chlamydomonas*) or parthenogenetic development of unfertilized (female) gametes (*Scytosiphon, Laminaria, Ectocarpus*). Adapted with permission from Liu et al. (2017) under the Creative Commons CC BY 4.0 license.

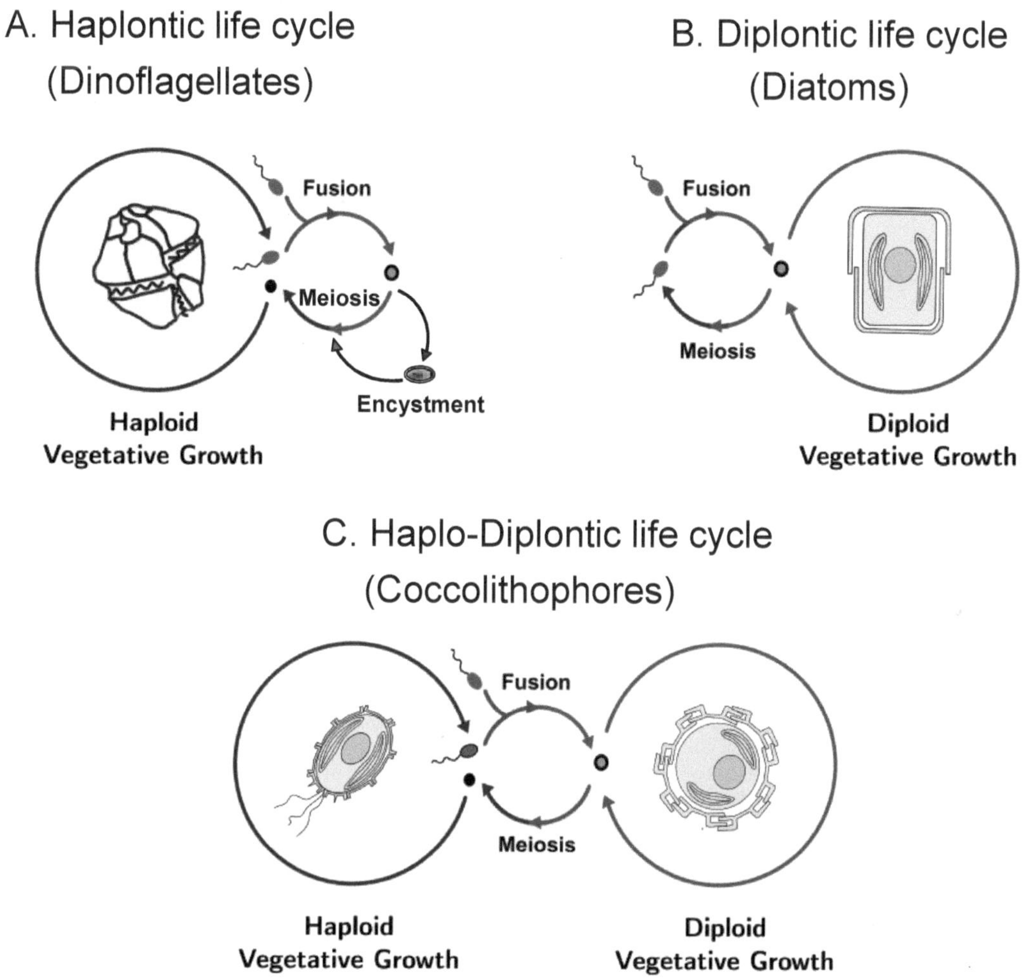

Figure 19.2. Schematic of life cycle of microalgae. (A) Simplified haplontic life cycle represented by dinoflagellates, in which sexual reproduction either leads to encystment to endure adverse environments or immediately to meiosis to promote population growth. (B) Diplontic life cycle represented by diatoms, in which sexual reproduction is triggered by cell size reaching minimum. (C) Haplo-Diplontic life cycle represented by coccolithophores, in which the haploid cells are less calcified than the diploid cells. Modified with permission from de Vries et al. (2021) under the Creative Commons CC BY 4.0 license.

via DNA analysis in Lin et al. 2022). The multi-division reproduction is thought to represent a "continuum" instead of segregated (with gaps) S-M transitions in filamentous chlorophytes (Zachleder et al. 2016). For *Pseudopfiesteria shumwayae* (formerly *Pfiesteria shumwayae*), this multi-division reproduction was described as primary, secondary, and tertiary reproductive cysts (Parrow and Burkholder 2003). Whether this involves sequential nuclear divisions by meiosis and mitosis, akin to gamete production in diploid sexual organisms, is worth exploring in future research. The effects of environmental factors on the growth rate of an algal population are coupled with the regulation of the cell cycle. The timing of cell cycle progression is controlled by an endogenous circadian clock and is entrained to the light dark cycle.

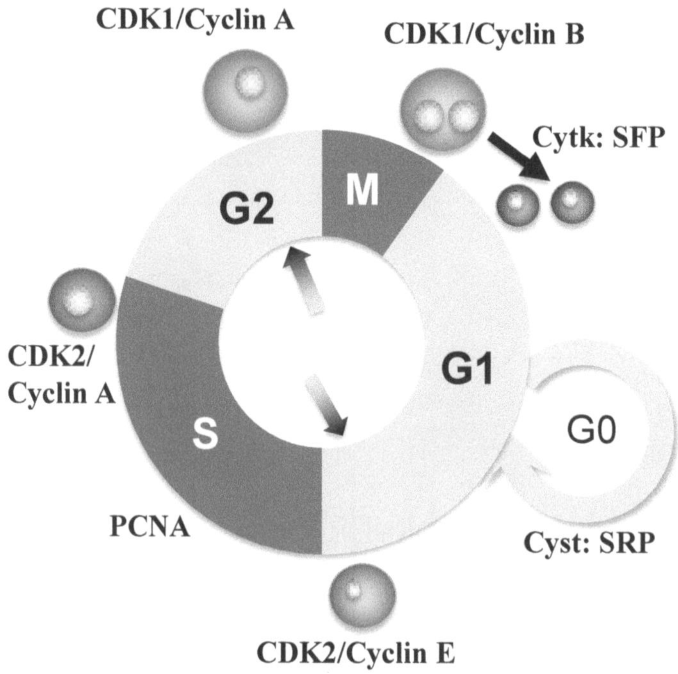

Figure 19.3. Schematic of algal cell division cycle (CDC) and core regulatory genes. The algal CDC resembles typical eukaryotic CDC, consisting of four discrete phases: DNA replication (S), mitosis (M), and the two gaps (G) between them. When cells leave the active cycle they enter the G0 loop as quiescent stage (with G1 DNA status). The progression from one phase to the other is strictly regulated by cell cycle protein complexes composed of stage-specific cyclins and their cognate cyclin-dependent kinases (CDKs). PCNA, proliferating cell nuclear antigen; Cytk, cytokinesis; SFP, septum formation protein; SRP, sporulation related protein. Arrows within the circle indicate checkpoints.

Regulators of the Life Cycle

Developmental processes must be precisely coordinated with life cycle progression. For multicellular lineages with haploid-diploid life cycles, distinct developmental programs for sporophytes and gametophytes must be appropriately deployed at specific points within a single life cycle (Arun et al. 2019). There must be a molecular switch that regulates the transitions between sexual and asexual stages, and molecular regulators orchestrating both processes. Interestingly, some of these molecular mechanisms are common between macro- and micro-algae.

TALE-HD System

In the unicellular green alga *Chlamydomonas*, vegetative cells are haploid with 17 small chromosomes. Under N-nutrient starvation, vegetative cells develop into haploid gametes of plus (mt^+) and minus (mt^-) mating types (Fig. 19.4). Mt^+ and mt^- gametes are identical in appearance, thus isogamous, but express different Three Amino acid Loop Extension (TALE) Family Homeodomain (HD) Transcription Factors (TFs), Gsm1 and Gsp1 (Lee et al. 2008). Upon gamete fusion, these proteins heterodimerize and move to the nucleus to regulate the diploid phase of the life cycle. Gsm1 (gamete specific minus 1) belongs to the KNOX (knotted 1-like homeobox) class, while Gsp1 (gamete specific plus 1) is part of the BEL class. In the multicellular moss *Physcomitrella patens*, deletion of KNOX genes MKN1 and MKN6 blocks sporophyte development, converting it into a diploid gametophyte (Sakakibara et al. 2013). The BEL class gene BELL1 (BEL1-like protein 1) is crucial for sporophyte development, and its ectopic expression in gametophytes induces sporophyte formation (Horst et al. 2016). KNOX and BEL proteins act as master regulators of the sporophyte

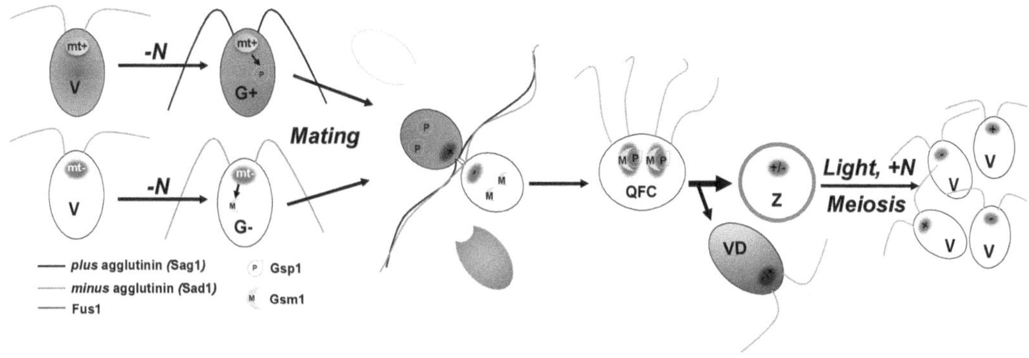

Figure 19.4. Life cycle of *Chlamydomonas reinhardtii*. Haploid vegetative (V) cells of two mating types (mt^+ and mt^-) divide mitotically. Under N-nutrient limitation, they differentiate into gametes (G^+ and G^-), expressing mating type-specific gametic traits. When gametes are mixed, the plus (Sag1) and minus (Sad1) agglutinins displayed on their flagellar surfaces mediate the initial adhesion reaction, causing a rise in intracellular cAMP which triggers gamete cell wall release and mating-structure activation; the Fus1 protein on the plus mating structure interacts with partner protein(s) on the minus mating structure to trigger cell fusion and the formation of the binucleate quadriflagellated cell (QFC). Homeoproteins Gsp1 and Gsm1, pre-synthesized in plus and minus gametes, respectively, interact to activate transcription of zygote-specific genes. Nuclei fuse, flagella are resorbed, and a thick cell wall is assembled around the zygote (Z). After a period of dormancy and when N-nutrient is resupplied, zygotes undergo meiosis to release four haploid meiotic products that resume vegetative growth. Occasional QFCs forego the meiotic pathway and instead resume vegetative growth as +/− vegetative diploids (VD). Based on Goodenough et al. 2007 and Lee et al. 2008.

program, coupling it with life cycle progression through KNOX/BEL heterodimers (Horst et al. 2016). Homologs of the genes can be found in various chlorophyte and prasinophyte genomes (Table 19.1).

In the phaeophyte filamentous alga *Ectocarpus*, a mutation at the OUROBOROS (ORO) locus causes sporophyte generation to be converted into gametophyte (Coelho et al. 2011). Furthermore, mutation of a second locus, SAMSARA, also results in conversion of the sporophyte generation into a gametophyte (Arun et al. 2019). Intriguingly, both OUROBOROS and SAMSARA encode TALE-HD transcription factors, and the two proteins associate to form a heterodimer. Consistent with these phenotypes, ORO and SAM transcripts were found to be most abundant in gametes (Arun et al. 2019). The findings from these studies on different phyla and cellularity indicate in concert that TALE-HD TF is a common life cycle regulatory system controlling sporophyte developmental programs with deep evolutionary origins (Arun et al. 2019).

Cues Triggering Sex

Besides cues such as low temperatures and nutrient depletion as well as cell size miniaturization, induction of sex in diatoms involves pheromones, which ensure synchronization and support the mate-finding process of the motile cells (Sato et al. 2011, Montresor et al. 2016, Basu et al. 2017). Sato et al. (2011) showed that the araphid pennate diatom *Pseudostaurosira trainorii* produces motile male gametes, which moved with the extrusion and retraction of microtubule-based "threads". These threads resemble flagella detected in the male gamete-producing *L. danicus* cells, but with distinct differences in their absence of tubule structure of flagella. This is reminiscent of the flagellum development-related genes found in *Leptocylindrus danicus* and some other centric diatom species (Nanjappa et al. 2017). Whether *P. trainorii* has and uses the same genes to generate the threads remains to be studied. More importantly, Nanjappa et al. (2017) also presented evidence for sex pheromones that reciprocally stimulate sexualization of compatible clones and orient motility of the male gametes after an initial random movement. In *S. robusta*, diproline acts as an attraction pheromone, whereas molecule SIP^+, secreted by mt^+ cells, triggers both cell cycle arrest and diproline production in mt^- cells (Moeys et al. 2016).

Table 19.1. Gamete-specific genes in algae.

Phylum	Species	KNOX class	BELL class	NCBI accession number (Genome coordinate)
Chlorophyta	*Chlamydomonas reinhardtii*	gsm1	gsp1	ABJ15867.1, AF108140_1
	Chlamydomonas incerta	gsm1	gsp2	ABJ15868.1, AAW82031.1
	Chlamydomonas schloesseri	gsm1		KAG2430025.1
	Volvox reticuliferus	gsm1		BCL66242.1
Prasinophyta	*Pterosperma* sp. CCMP1384	KNOX-like		MMETSP1438-20131217|24340_1
	Micromonas sp. CCMP2099	KNOX-like		Micromonas-sp-CCMP2099-20130918|7634_1
	Micromonas pusilla CCMP1545	KNOX-like		XP_003060802.1
	Micromonas pusilla RCC1614	KNOX-like		MMETSP1402-20130829|26360_1
	Prasinophyceae RCC1871	KNOX-like		MMETSP1456-20131121|40534_1
	Pyramimonas CCMP2087	KNOX-like		MMETSP1445-20131203|114053_1
	Pycnococcus sp. CCM1998	KNOX-like		MMETSP1085-20130531|1672_1
Phaeophyta	*Ectocarpus* sp. JMC-2016a	ORO	SAM	ANN44522.1, ANN44523.1
	Ectocarpus sp. CCAP1310/34	ORO		CAB1112438.1
	Ectocarpus siliculosus		SAM	CBJ31259.1

Experimental manipulation of *Pseudo-nitzschia multistriata* has provided evidence that the chemical cues that trigger sex require no physical contact between cells of opposite mating types. In a comparative genomic study, two strains of *P. multistriata* were placed in separate compartments of an apparatus that allowed free exchange of the medium but no contact between the cells. These "Communicating but Not Contacting (CNC)" strains in the co-culture underwent cell cycle arrest in the G1 phase, whereas both strains cultured in isolation continued to progress through the cell cycle (Basu et al. 2017). In a cross study, both MTs exhibited similar cell cycle arrest for most cells and about 20% of cells undergoing meiosis (Scalco et al. 2016). Genes associated with sexual reproduction were regulated in the CNC culture relative to the control, markedly more so in the mt^- (596 genes) than the mt^+ (182 genes) mating types. Among these regulated genes, cohesion complex genes (SMC1, SMC3, SCC3, RAD21) and meiosis-related genes (RAD51-A1, RAD51-C, SMC5, PSD5, and Smc-containing proteins) were upregulated, whereas cell cycle genes such as diatom-specific cyclins dsCYC5 and dsCYC4 were downregulated. Signaling pathway genes were also upregulated. Notably, expression of a soluble guanylate cyclase was elevated by 6.3-fold in the CNC cells, consistent with the bifunctional guanylyl cyclase/phosphodiesterase (GC/PDE) in *S. robusta* cells in response to the SIP^+ pheromone, although the gene involved in the two species (25% identical) were not orthologs. In addition, G Protein-Coupled Receptors (GPCRs) were upregulated, as was the inositol phospholipid signaling pathway.

Dinoflagellates are a good model for microalgal life cycle because their haplontic life cycle manifests in the annual blooms of many species. Our knowledge about their life cycle has evolved dramatically, from the simple linear pathway, vegetative cells/asexual reproduction → gametes/sexual reproduction → Encyst/dormancy → Germination → Vegetative cells, to a reticulated pathway. The major complication has come from newer observations that sometimes cysts are formed without sex and conversely, sexual reproduction does not result in encystment; furthermore, there are different types of cysts: resting cysts, reproductive cysts, and pellicle cysts (Parrow and Burkholder 2003, Bravo and Figueroa 2014, and references therein). Accordingly, in this process cells have DNA contents of 1C, 2C, 4C, and possibly more, where C stands for copy of genome (Fig. 19.5; Figueroa et al. 2015, 2021, Lin et al. 2022, Kalinina et al. 2023).

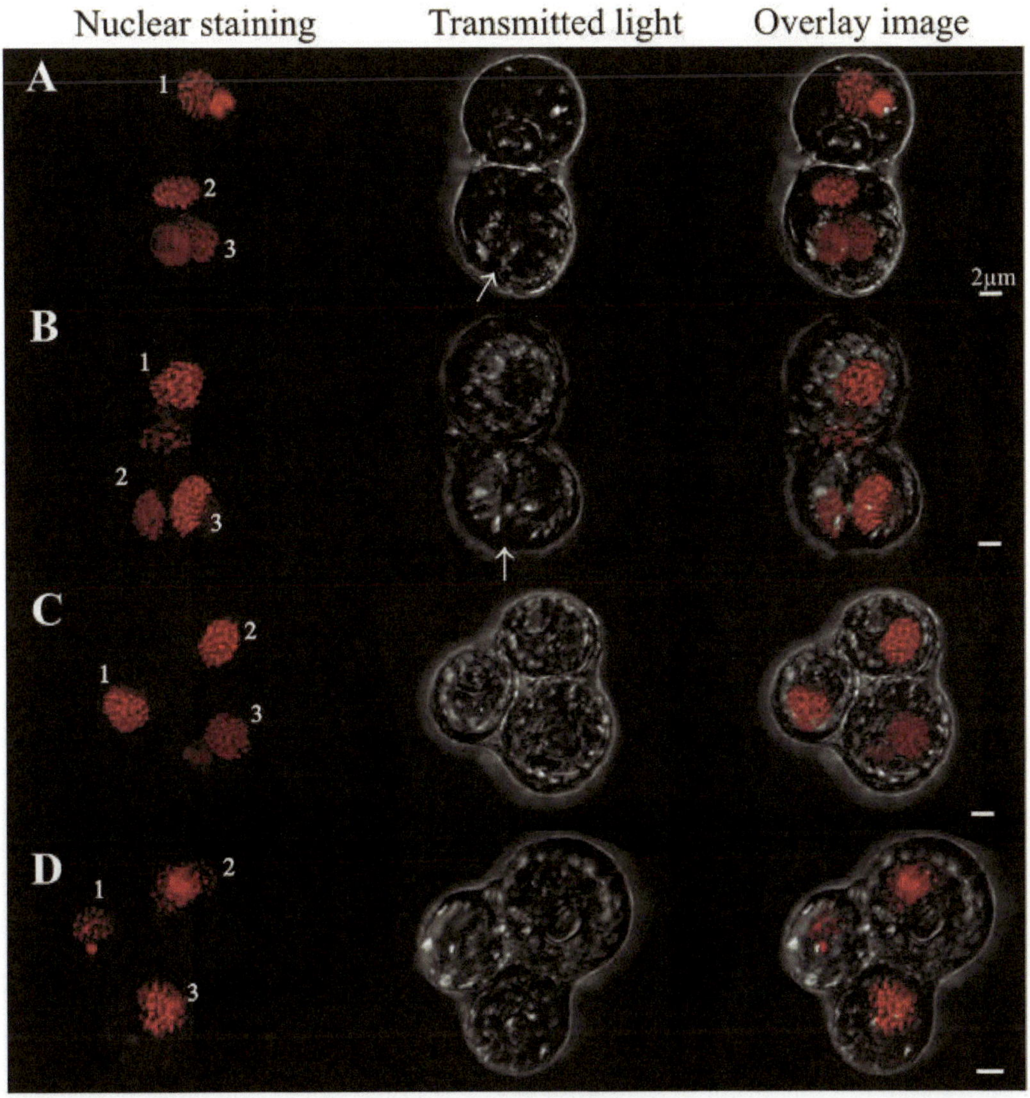

Figure 19.5. Confocal images of Symbiodiniaceae cells sorted in the (>2C–4C) DNA content gate that have three nuclei and different numbers of lobes in their external morphology (lobes clearly visible in the transmitted light column). (A, B) cells with two lobes, one showing a dividing furrow (arrow); (C, D) three lobes. Reproduced with Figueroa et al. (2021) under the Creative Commons Attribution 4.0 International License.

Figueroa et al. (2015) found that in *Alexandrium minutum*, sex led to the production of planozygotes, most of which produced four daughter cells through a round of meiosis and mitosis, while far fewer planozygotes progressed into resting cysts, which only happened under severe nutrient limitation (Figueroa et al. 2015). In addition, some algae exhibit "budding" behavior under normal growth conditions (Fig. 19.6). The "reticulation" of the life cycle raises a question of whether induced cyst formation is always sexual. Many species can be coaxed to form cysts in the laboratory using cold, salinity stress, unusual light or pH conditions, and nutrient (N, P) deficiency. It is curious how these diverse types of stress signals are perceived and transduced convergently to trigger cyst formation. In addition, whether some of these cases are asexual is not entirely clear. Similarly, the stress-triggered signaling pathway is also unclear in other groups of microalgae.

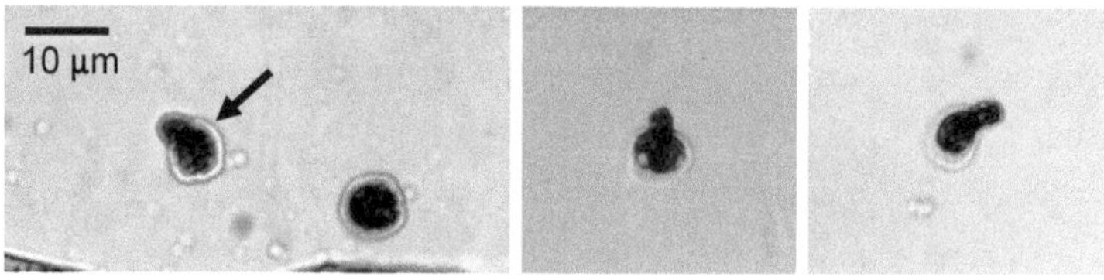

Figure 19.6. Budding of a *Prorocentrum shikokuense* cells.

Sexual Reproduction and Molecular Toolkit

Sexual reproduction is widespread among eukaryotes and has a deep evolutionary origin, with meiosis likely millions of years old, serving as a mechanism for chromosome recombination and generating genetic diversity (Vesteg and Krajcovic 2011, Goodenough and Heitman 2014, Speijer et al. 2015). In algae, sexual reproduction is believed to occur in most macrophytes and microphytes except cyanobacteria (Lahr et al. 2011, Agrawal 2012, Fučíková et al. 2015). Sex in algae serves a variety of roles beyond maintaining genetic diversity. For diatoms, sexual reproduction also serves to restore cell size, although the latter function may not be as necessary or frequent as previously thought (Bulankova et al. 2022). Dinoflagellates engage in sexual reproduction to form cysts that help them survive environmental challenges (Bravo and Figueroa 2014) or as recently postulated, potentially to boost population growth (Lin et al. 2022). To explore evolutionary drivers of sexuality, several hypotheses have been proposed, invoking recombination, nutrition, and repair (review in Liu et al. 2017).

One of these hypotheses, put forth by Cavalier-Smith (2002), argues that the doubling of cell volume resulting from syngamy of two gametes significantly increase the survival of the organisms during periods of food or nutrient shortage, while smaller, haploid cells resulting from meiosis are advantaged under resource-replete conditions. This seems to be consistent with the life cycle of dinoflagellates in which syngamy of two haploid cells produce zygotes under resource-depleted or environmental stress conditions to develop cysts or hyperactive meiosis under resource-replete conditions to develop blooms (Bravo and Figueroa 2014, Luo et al. 2022, Lin et al. 2022). The hypothesis also fits the cell size-restoring role of sex in diatoms.

It is challenging to directly observe sexual reproduction in microalgae because gametes, zygotes or chemical signals associated with sex are difficult to document, particularly in the natural environment. As evidence, information on the sexual cycle of diatoms is available for only about 200 out of the > 10,000 known species (Montresor et al. 2016). Sexual reproduction in coral symbionts has escaped observations until recently. Therefore, gene markers specific to sexual reproduction constitute a critical toolkit for the identification of sexual species.

Meiosis Genes

Meiosis is a hallmark of sexual reproduction. The complete gene set involved in meiosis is in general not entirely understood but is known to vary among species. Some genes considered essential to sexuality are absent in certain sexual organisms (Reedy et al. 2009). Despite the variability, a set of meiosis genes has been developed as the "core" meiosis genes (Schurko and Logsdon 2008). This 12-gene set includes SPO11, HOP1, HOP2, MND1, REC8, DMC1, MSH4, and MSH5 as meiosis-specific genes with RAD21, RAD51, MSH2, and MSH6 as meiosis-associated genes (Table 19.2). SPO11 is one of the meiosis core genes in most sexual eukaryotes, and its presence in the genome is indicative of the species' sexual practice. Its function is to create DNA double-strand breaks and triggers meiotic recombination (Cao et al. 1990, Keeney et al. 1997, Malik et al. 2008, Schurko and Logsdon 2008). The importance of SPO11 is consistent with its prevalence and sequence conservation (Fig. 19.7A). However, SPO11 is absent in *Dictyostelium discoidium* and closely related slime molds that have meiotic sexual cycles (Goodenough and Heitman 2014).

Table 19.2. Meiosis genes identified so far in algae (*) compared to core (bold type) and common eukaryotic meiosis gene repertoire.

Protein	Function
DMC1*	Meiosis-specific homolog of Rad51, has similar function but promotes interhomolog recombination.
EXO1*	Double-strand breaks processing.
HAP2*	Plasmogamy.
HOP1	Protein that binds DSBs and oligomerizes early during meiotic prophase, and forms axial and lateral elements of the synaptonemal comple.
HOP2*	With Mind1, ensures accurate and efficient homology searching, downstream of Rad51 and Dmc1, during pachytene stage of meiotic prophase.
MC3*	Structural maintenance of chromosomes.
MEI2*	The mei2 gene of *S. pombe* is a positive regulator of meiosis and encodes an RNA binding protein required for premeiotic DNA synthesis and entry into meiosis I.
MER11*	Double Strand Breaks processing.
MER3	Meiosis-specific DEAD-box helicase that promotes Hollidav junction resolution with crosover interference together with ZMM proteins including Msh4 and Msh5.
MLH1*	Mismatch repair of dinucleotide and trinucleotide sequences, interacts with Msh2, forms heterodimers with Mh2, Mh3 and Pms1.
MLH2	Forms a heterodimer with Mlh1. interacts with Msh2/3 or Msh2/6 for removal of cisplatin adducts.
MLH3	Forms a heterodimer with Mlh1. interacts with Msh23 or Msh2/6 for frame shift repair in mitosis or meiosis, or with Msh4/5 to promotmeiotic crossovers.
MND1*	With Hop2, functions after meiotic DSB formation, and required for stable heteroduplex DNA formation; stimulation of DMC1 activity.
MNS1*	Control of meiotic division.
MRE11*	3'-5'dsDNA exonuclease and ssDNA endonuclease, trims back broken DNA ends and harpins; involved in meiotic recombination.
MSH2*	Forms a heterodimer with Msh3 or Msh6.
MSH4*	Forms a heterodimer with Msh5, interacts with Mlh1/Mlh3 heterodimer. Directs Holiday junction resolution towards crossover with interference.
MSH5*	Forms a heterodimer with Msh4, interacts with Mlh1/Mh3 heterodimer. Directs Holiday junction resolution towards crossover with interference.
MSH6*	Forms a heterodimer with Msh2, binds base-base mismatches.
MUG157*	Meiotically upregulated gene.
MUS81*	Crossing over II.
MutS2*	Endonuclease.
PDS5	Important for maintenance of sister chromatid cohesion in late prophase.
PMS1*	Mismatch repair. Interacts with Msh2/3 or Msh2/6 as a heterodimer with Mlh1.
RAD1 (MEI9)	Forms a heterodimer with Rad10 (Ercc1). 5' > 3' endonuclease, essential for nucleotide excision repair. Required for meiotic crossing over, normal meiotic chromosome disjunction, to repair mismatches in heteroduplex DNA and to resolve reciprocally exchanged recombination intermediates in *Drosophila*.
RAD21 (SCC1)*	Holds smc1 and smc3 heads together by binding N-terminal domain to Smc3 and C-terminal domain to smc1, thus holding sister chromatids together during mitosis and meiosis.
RAD50*	Dimer, holds broken DNA ends together while Mre11 trims. ATPase, has DNA-binding activity.
RAD51*	Forms helical filaments on single-stranded and double-stranded DNA and catalyzes homologous DNA pairing and strand exchange (Intrahomologous recombination).
RAD52	Binds to the ssDNA ends of DsBs and initiates DsB repair by homologous recombination. Stimulates Rad51-mediated strand invasion binteraction with Rad51 and RPA, and promotes single strand annealing (SSA).

Table 19.2 contd. ...

...*Table 19.2 contd.*

Protein	Function
RAD54*	Stimulation of RAD51 activity.
REC8	Meiotic homolog of Rad21. Holds smc1 and Smc3 heads together by binding N-terminal domain to Smc3 and C-terminal domain to smc, thus holding sister chromatids together during meiosis.
SCC3	Necessary for sister chromatid cohesion, and required for DsB repair. Interacts with smc1, smc3 and Rec8/Rad21 in holding cohesiring together.
SMC1*	Forms a heterodimer with Smc3 to form core sister chromatid cohesin subunits, with ring shape around sister chromatids.
SMC2*	Forms a heterodimer with Smc4 to form core condensin subunits, ring shape, essential for chromosome assembly and segregation.
SMC3*	Forms a heterodimer with Smc1 to form core sister chromatid cohesin subunits, with ring shape around sister chromatids.
SMC4*	Forms a heterodimer with Smc2 to form core condensin subunits, ring shape, essential for chromosome assembly and segregation.
SMC5	Forms a heterodimer with Smc6 (Rad18) and is involved in DNA repair and checkpoint responses.
SMC6 (RAD18)	Binds ssDNA, has important role in postreplication DNA repair. Forms a heterodimer with Smc5 and is involved in DNA repair & checkpoint responses.
SPO11*	Transesterase, creates DNA double-strand breaks (DSB) in homologous chromosomes.

No consensus could be drawn for algal sex detection toolkit, in part because our knowledge about the complete set of meiosis-specific gene repertoire is very limited for most algal lineages. As such, the list of meiosis genes for algae has continued to grow as research advances. In dinoflagellates, from genomes and transcriptomes, a range of genes (31-51) with homology to meiosis genes of model organisms have been identified (Chi et al. 2014, Lin et al. 2015, Morse 2019). These include most of the 12 core meiosis genes previously developed for eukaryotes (Table 19.2) but miss some of these some others that yeast possesses. For instance, securin, which normally serve in preventing the transition from metaphase to anaphase by inhibiting the protease separin, was not found; besides, some components of the cohesion complex with the function to hold homologous chromosomes together are absent (Morse 2019). The absence of cohesion complex elements is consistent with more recent findings suggesting that dinoflagellates may have lost the canonical synaptonemal complex (Shah et al. 2020). However, dinoflagellates possess additional genes that are not in the core meiosis gene repertoire, such as meiosis protein 2 (Mei2), Meiotically Up-regulated Gene (MUG) 71, 89, 157, and 190, meiosis expressed gene 1 protein (MEG1), and several others (Table S8 in Lin et al. 2015). These genes were initially documented in yeast but are now known to exist in dinoflagellate genomes. Homologs of some of these genes in *Prorocentrum shikokuense* exhibit extremely high expression (e.g., *mei2*) during bloom events (e.g., Zhang et al. 2019, Yu et al. 2020). Mei2-encoding genes occur widely, including vascular plants, all major lineages of algae, and other protists (Fig. 19.7B).

Gamete Fusion Gene: Generative Cell-Specific Protein 1 (GSP1)

Gamete fusion is as essential as meiosis to sexual reproduction. One of the key proteins regulating gamete fusion is the generative cell-specific protein 1 (GSP1). Due to the importance, this protein is highly conserved from microalgae to seaweeds (Fig. 19.8). While the specific mechanisms may vary between species, the conservation of GSP1 and its related proteins suggests a common underlying strategy for regulating the crucial transition from gametes to zygote in sexual reproduction.

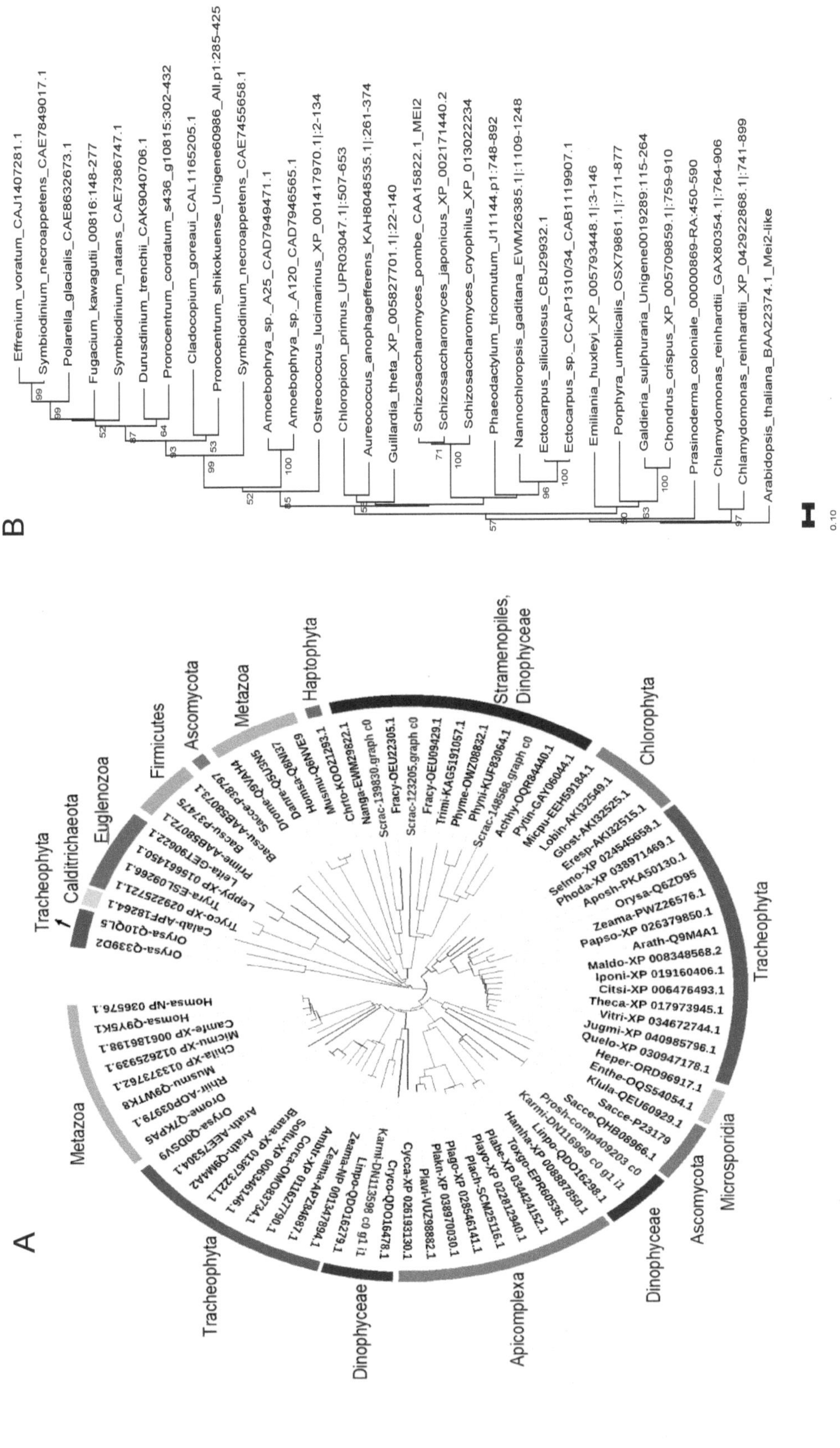

Figure 19.7. Highly conserved meiosis genes in algae. A) SPO11. B) Mei2.

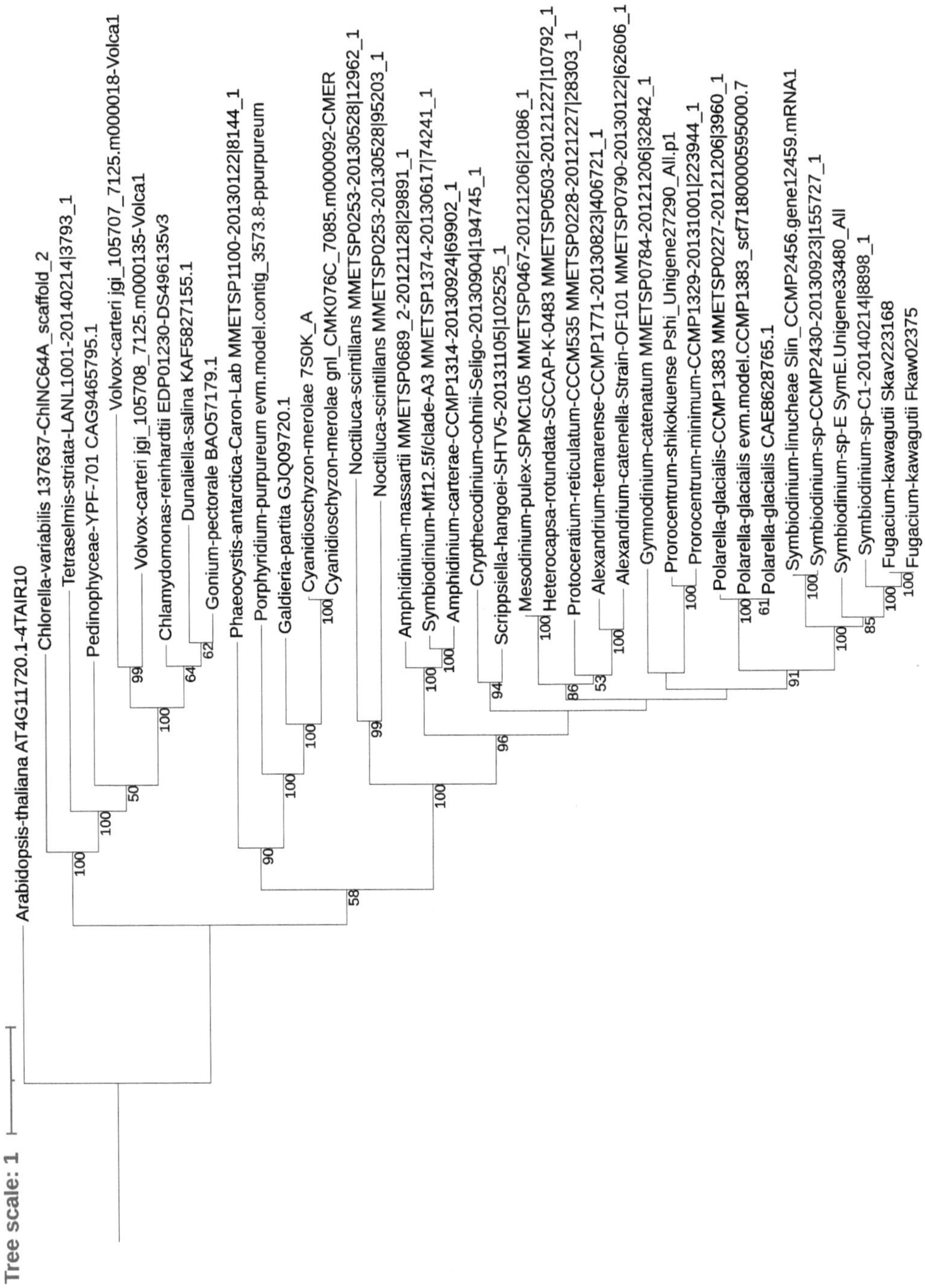

Figure 19.8. Prevalence and high similarity of generative cell-specific protein 1 (GSP1) across major lineages of algae.

Other Sex Genes

Flagellar genes and their expression are a hallmark of sexual reproduction in diatoms. This is because flagella have been lost in the vegetative stage of the diatom life cycle but are still present in male gametes of centric diatoms (Nanjippa et al. 2017). Based on the genomes of *T. pseudonana* and *P. tricornutum*, it has been proposed that pennate diatoms have completely lost the flagellum or cilium structure complexes (van Dam et al. 2013). From the transcriptome of *Leptocyclindrus danicus*, however, Nanjippa and colleagues (2017) identified 22 transcripts related to flagellum development and confirmed synchronous high expression of six of these during male gamete formation. Interestingly, these genes are mostly absent in the closely related species *L. aporus*, which is not known to perform sexual reproduction. Among the 22 transcripts were Intra Flagellar Transport (IFT) subcomplex B and IFT-A, the latter previously thought not to occur in diatoms. The flagellum-related genes were also found in the transcriptomes of several other centric species, indicating the general conservativeness of these genes and sexual reproduction. In particular, IFT172 and IFT88 are conserved across protist species.

Sexuality-Asexuality Switch

The sexual-asexual switch in dinoflagellates and other microalgae, if it exists, has not been identified. Some potential regulators have been explored, however. Mei2, an RNA-binding protein encoded by the *mei2* gene (Shimoda et al. 1987), along with a long non-coding RNA (meiRNA), is crucial for meiosis in *Schizosaccharomyces pombe* (Watanabe and Yamamoto 1994, Yamamoto 1996). Mei2 triggers meiosis by binding meiRNA at U-rich motifs, leading to its accumulation in a nuclear structure called the "mei2-dot", which plays a pivotal role in the mitosis-meiosis switch (Fig. 19.9; Jeffares et al. 2004, Shen et al. 2022). meiRNA acts as a "sponge" RNA to sequester the meiosis inhibitor Mmi1 protein. Mei2 also binds Mmi1's mRNA and the transcription factor Rep2 (Mukherjee et al. 2018). The Mei2-meiRNA complex controls both the protein and mRNA of Mmi1 and is hence considered a "double sponge" (Mukherjee et al. 2018). Mei2 homologs in land plants play roles in meiosis and vegetative growth (Kaur et al. 2006). As shown above, the *mei2* gene is widespread and highly conserved in algae. Recent studies have revealed the ubiquity of the *mei2* gene and its high expression level in dinoflagellate blooms (Zhang et al. 2019, Yu et al. 2020). However, how Mei2 functions in algae, whether it partners with meiRNA and Mmi1, remains to be studied.

The KNOX/BELL system that occurs in chlorophytes and phaeophytes could not be detected in other phyla from my analysis using BLAST. The Chromalveolata group may use different mechanisms to regulate the diploid/haploid life stage development, or their TALE-HD system is too divergent from the chlorophyte and phaeophyte system to be identified using blast analysis. In fact, the gsm1/gsp1 and oro/sam sequences are very different pairs in primary sequence that BLAST analysis produces no hit with each other. Given the lack of information in algae, some evolutionary perspective from animals may lend some insight. While sexuality is believed to be the basal trait of animals, punctate asexuality (e.g., parthenogenesis) must be derived from the secondary loss of sexuality. A study comparing transcriptomes of sexual with asexual populations of the insect *Timema* (Parker et al. 2019) revealed no asexuality-specific genes and limited gene-level convergence (< 1%), but showed marked convergent changes in gene expression (up to 8%). The study indicates that a key change crucial to asexual reproduction—the production of unreduced eggs capable of developing without fertilization—likely involves reprograming of meiotic spindle regulation. Therefore, differential expression of existing genes, with the involvement of epigenetic regulation, may be as important to look for as absence or presence of specific genes.

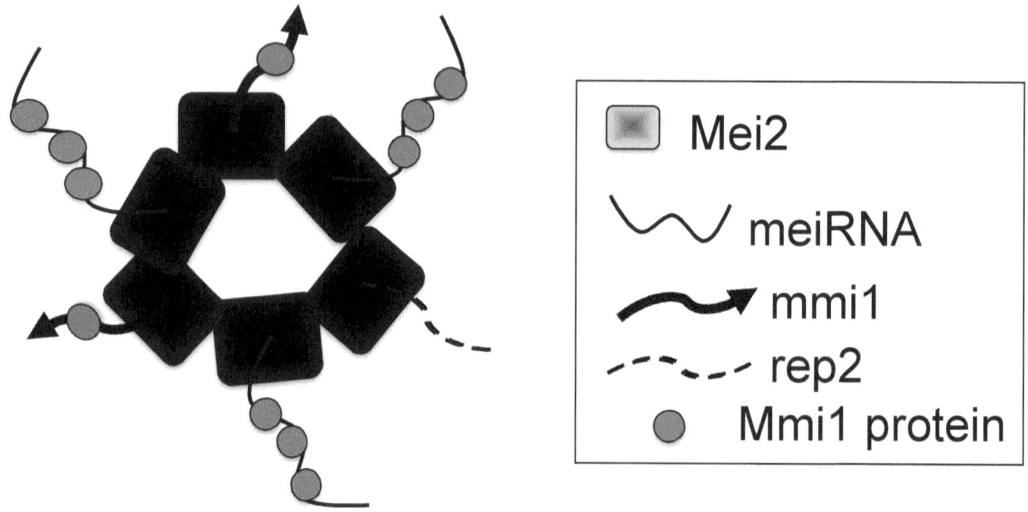

Figure 19.9. Schematic of how MEI2 functions. Mei2 binds the long non-coding RNA meiRNA at U-rich motifs, which acts like a "sponge" RNA to capture the meiotic inhibitor protein Mmi1. Mei2 also binds the mRNA of Mmi1 and the transcription factor rep2. The interactive complex activation or inactivation of the meiosis machinery. Mei2 contains three RNA recognition motifs (RRM1-3), of which RRM3 is highly conserved across lineages. Based on Jeffares et al. (2004) and Mukherjee et al. (2018).

Cell Cycle Machinery

Based on the best studied cell cycle models in yeast, humans, vascular plant, the architecture of the cell cycle regulatory machinery is highly conserved in eukaryotes, although the specific constituent proteins vary between organisms. As shown in Fig. 19.3, there are two major checkpoints in cell cycle regulation, the G1/S and G2/M transitions, which serve as surveillance systems that ensure the order, integrity, and accuracy of key events within the cell cycle. At these checkpoints, cells are committed to progressing towards mitosis, with no possibility of "return". If any error occurs, the checkpoints halt cell cycle progression and, in severe cases, trigger apoptosis-like cell death. Therefore, the checkpoints are crucial for cells to maintain genetic integrity and cellular homeostasis.

The core molecular machinery of the cell cycle (Fig. 19.3), which is shared by algae and other eukaryotes, is composed of cyclins and Cyclin-Dependent Kinases (CDKs). CDKs are serine/threonine protein kinases that phosphorylate key substrates to facilitate DNA synthesis and mitotic progression (Gould et al. 1991). Although the catalytic subunits of CDKs are present in excess and expressed constitutively, they lack activity until bound by their corresponding cyclin subunits, whose levels oscillate throughout the cell cycle due to a tight control through synthesis and ubiquitin-dependent proteolysis (Evans et al. 1983, Glotzer et al. 1991, Murray and Hunt 1993). Whereas yeast has one single CDK, which is the originally discovered p34cdc2 in *Schizosaccharomyces pomb* or CDC28 in *Saccharomyces cerevisiae*, now both known as CDK1, other eukaryotes have multiple CDKs. Cyclin binding allows inactive CDKs to assume an active configuration. Additionally, CDK activity is subject to negative regulation by the binding of small inhibitory proteins, known as CKIs, or by inhibitory tyrosine phosphorylation, which prevents phosphate transfer to substrates.

Along with the Cyclin/CDK core is Proliferating Cell Nuclear Antigen (PCNA). PCNA is typically not included in the cell cycle machinery because it is not only involved in the S phase of the cell cycle, but also is required for DNA repair. PCNA is indispensable for the timely and faithful replication of the genome, hence extremely important in the cell cycle. It functions as a DNA sliding clamp, forming a ring-shaped homo-trimer that encircles double-stranded DNA and acts as a scaffold to recruit proteins involved in DNA replication, repair, and recombination.

As such, PCNA enhances the processivitiy of the DNA polymerases delta and epsilon and allows for rapid DNA synthesis. Here the three core components will be discussed.

PCNA

PCNA is a highly conserved nuclear protein of about 34 kDa. The protein was first discovered in the 1970's in the serum of patients with systemic lupus erythematosus (Miyachi et al. 1978). It has since been found in plants and all other eukaryotes, archaea and found to be structurally similar to DNA pol III β subunit in bacteria (Strzalka and Ziemienowicz 2011). In algae, it was first detected in phytoplankton in attempts to find molecular markers of growth rate for use to estimate *in situ* growth rate of phytoplankton (Lin et al. 1994). As demonstrated in the chlorophyte *Dunaliella tertiolecta,* PCNA in phytoplankton is localized in the nucleus and shows higher abundances in the S phase (Lin et al. 1995). Similar studies revealed presence of PCNA in dinoflagellates (Leveson and Wong 1999, Liu et al. 2005). The gene sequence, at the amino acid level, is about 70% or more similar between algae and humans or rice as well as between species within algae (Cheng et al. 1997, Lin and Carpenter 1998, Lin and Corstjiens 2002, Zhang et al. 2003), indicating a strong purifying selection and implying the pivotal importance of this protein.

From genomes and transcriptomes, PCNA can be easily identified. Interestingly, usually PCNA gene occurs in one or two copies in the genome, in genomes exceeding 4 Gbp, the copy number increases exponentially with genome size, up to hundreds of copies in large dinoflagellate genomes (Hou et al. 2019). PCNA amino acid sequences are highly conserved, a striking feature emanating from a phylogenetic tree or simulated structure (Fig. 19.10).

Cyclin/CDK System

CDKs are typically constitutively expressed whereas cyclins are synthesized when they are needed and degraded when their mission is completed (Glotzer et al. 1991, Murray and Hunt 1993, Yamano et al. 1998). CDKs are inactive until binding to a specific cyclin. In the G2 to M transition (Fig. 19.11), for instance, the binding of cyclin B triggers triple-phosphorylation at active site threonine and tyrosine residues involving Wee1 (a tyrosine kinase) and CDK-activating kinase (CAK). This initial step brings the CDK into an activation-ready state but is held inactive (hence Wee1 is considered an inhibitory kinase). Later, two of the three phosphorylated sites were dephosphorylated by phosphatase CDC25, producing an active CDK/CYC complex. When mitosis is completed, Anaphase-Promoting Complex (APC) drives the exit from M, which involves polyubiquitination and degradation of CYCB. Throughout the cell cycle, different cyclins act with respective partner CDKs and regulate the progression from one stage to the next.

Different nomenclature systems have been used for the cyclin/CDK systems in different groups of model organisms, and a clarification is warranted. In yeast, identification of the cyclin-dependent kinases follows the Cell Division Cycle protein (CDC) tradition for the cyclin-dependent kinase, e.g., CDC2/CDC28, and Clns (G1/S-specific) and Clbs (B-type or mitotic) for cyclins. Plant CDKs are named CDKA, CDKB, CDKC, CDKD, CDKE, CDKF, CDKG, and CKL (CDK-like) and cyclins termed CYCA, CYCB, and so on. The equivalent naming system in animals goes with numbers (CDK1, CDK2, etc.) for CDKs and CCNs for cyclins. Functionally, CDK1/CDK2 = CDKA, CDKC = CDK9, CDKD = CDK7, CDKE = CDK8, CDKG = CDK10/CDK11, whereas CDKB and CDKF are plant exclusive (Carneiro et al. 2021). The equivalent terms are sometimes used interchangeably for non-model organisms. In this book, plant CDK (with alphabets) and animal cyclin (CYC with numbers) naming systems will be followed in general, and attempts will be made to ensure clarity.

In the budding yeast *Saccharomyces cerevisiae,* the cell cycle has one checkpoint, known as START. After passing START, cells begin to replicate their DNA, duplicate the spindle pole body (the yeast equivalent of centrosomes), and initiate the growth of a new daughter cell as a bud. In this

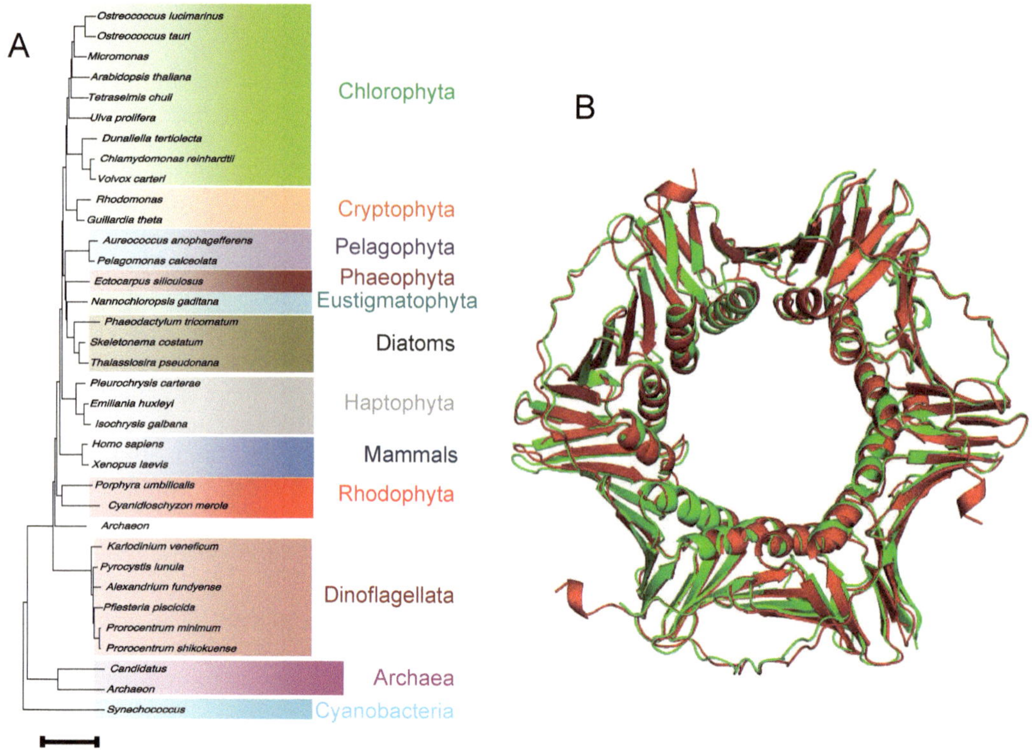

Figure 19.10. High similarity of PCNA across kingdoms. (A) Phylogenetic tree based on amino acid sequences of PCNA. (B) Simulated homotrimer structure of PCNA from Alexandrium catenella (green) overlain on the human homolog (red). Simulation was carried out using AlphaFold3 at the AlphaFold Server (https://golgi.sandbox.google.com), and the result showed very strong reliability (TM-score: 0.97, RMSD: 0.95Å).

process, cyclins bind to and activate the constitutively expressed CDK1. CDK1 interacts with nine cyclins expressed in different phases of the cell cycle, three G1 cyclins: Cln1, Cln2, Cln3, and six B-type cyclins: Clb1, Clb2, Clb3, Clb4, Clb5, Clb6. Except for Cln3, cyclins function in pairs, with paralogs Cln1/2, Clb1/2, Clb3/4, and Clb5/6 having roughly equivalent functions (Adler et al. 2022).

The cell cycle machinery in vascular plants is a bit more complex, with two checkpoints (G1/S and G2/M). *Arabidopsis thaliana* contains over 70 cell cycle regulators, including CDKs and cyclins (Carneiro et al. 2021). Its genome comprises eight groups of genes encoding CDKs and more than 50 genes encoding various types of cyclins. These cyclins are classified into groups CYCA, CYCB, CYCC, CYCD, CYCH, CYCL, CYCT, CYCU, SDS, and CYCJ (Shimotohno et al. 2021). Despite the large number of CYCs, the key cyclins that directly control the cycle are the CYCA and CYCB, which are mainly expressed in the G2/M transition, and CYCD, which regulates the G1/S progression. CYCD-CDKA phosphorylates the Retinoblastoma-Related (RBR) protein at the end of G1, releasing the E2F transcription factor and Dimerization Partner (DP) heterodimer to promote transcription of key G1/S transition genes. CDK-activating kinases (CAKs), composed of CDKD and CDKF in plants, phosphorylate and activate CDKA/B. CDKF functions independently, while CDKD requires CYCH. In G2, CDKA/B interact with CYCA/B/D to phosphorylate and activate Activator Myb (Act-MYB), inducing G2/M-specific genes, including mitotic cyclins. CDKA/B-CYCA/B/D also phosphorylate and destabilize Repressor Mybs (Rep-MYBs) to prevent their binding to the mitosis-specific activator (MSA) element, crucial for the G2/M transition.

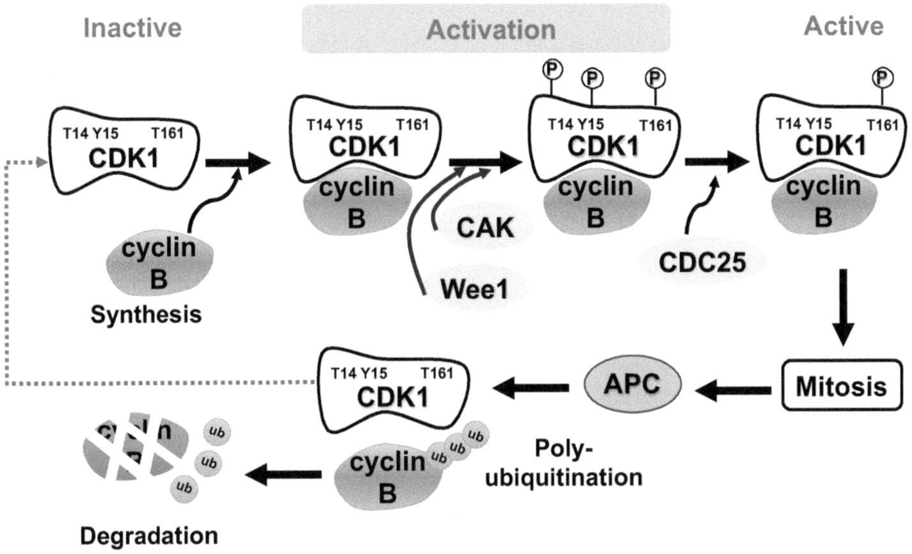

Figure 19.11. Schematic of CDK activation and cyclin synthesis and degradation in mitosis. In CDK1, threonine (T) 161 and 14 as well as tyrosine (Y) 15 are targets of phosphorylation and dephosphorylation triggered by binding of cyclin B. Wee1 is tyrosine kinase whereas CAK is a CDK-activating kinase. CDC25 is a phosphatase that removes the repressing phosphates from T14 and Y15, producing active CDK/CYC complex (mitosis-promoting factor) to promote mitosis. Once mitosis is completed, anaphase-promoting complex (APC) drives the transition from M to anaphase, which involves polyubiquitination and degradation of CYCB.

Cyclin-Dependent Kinases (CDKs)

CDKs can be distinguished by distinct cyclin-binding domains, which are generally conserved across lineages of organisms. The major types include the following (Carneiro et al. 2021):

CDKA (CDK1/CDK2): PSTAIRE
CDKB (plant only): PPTALRE or PPTTLRE
CDKC (CDK9): PITAIRE
CDKD/E/F (CDK7/-/CDK8): SPTAIRE
CDKG (CDK10/CDK11): PLTSLRE
CKL: (V,I,L)(K,R)FMAREI

Among these, CDKA is a homolog to yeast CDC2/CDC28 and mammalian CDK1 and CDK2, all with the PSTAIRE domain. It is the most ubiquitous CDK in eukaryotes and function in both the G1-to-S and G2-to-M transitions (Nurse 1990, Ferreira et al. 1991). CDKD and CDKF are CDK-activating kinases (CAKs), involved in activating other CDKs (Lu et al. 2017).

Like the algal PCNA studies, the identification of algal CYC/CDK began with immunodetection, followed by PCR-based gene cloning, when DNA sequencing was more of a privilege than commonplace. Using antibodies developed from human p34cdc2, a protein with expected molecular mass was detected in the chlorophyte *Dunaliella tertiolecta* (Lin et al. 1996), suggesting that CDC2 (CDK1) is conserved between chlorophyte algae and mammals. Furthermore, its expression appeared to be higher in the exponential phase than in the late stationary phase. Immunofluorescence showed p34cdc2-specific antibody stained the fewest cells during the S phase. The p34cdc2-like gene in *D. tertiolecta* was isolated and sequenced but was found to contain a unique PSTTLRE cyclin-binding domain (Lin and Carpenter 1999). Activation of CDC2 through phosphorylation was also detected in the dinoflagellate *Gambierdiscus toxicus* (van Dolah et al. 1995), following the immunodetection of a putative p34cdc2 protein in the heterotrophic dinoflagellate *Crypthecodinium cohnii* (Rodriguez et al. 1993).

From NGS data, many more CDK sequences could be identified in algae; however, it is challenging to discriminate one type of CDK from another based on the primary sequence. First, identification of the signature cyclin-binding domain (C-dom) is not straightforward, because algal CDKs do not necessarily follow the pattern of model organisms, as in the case of CDKA in *D. tertiolecta*. Second, phylogenetic tree analysis neither clusters CDKs by the type of C-dom nor always agrees with original annotation of the genes. In dinoflagellate CDKs as an example, BLAST analyses with plant CDK sequences as a query revealed homologs for most of the plant CDKs in dinoflagellates as a group, but some species miss some of these genes (Fig. 19.12). CDKA is most abundant, and all contains the PSTAIRE domain. CDKB (canonically with a PPTALRE or PPTTLRE domain) has been regarded as an exclusive plant CDK, but a sequence from *Symbiodinium* with the

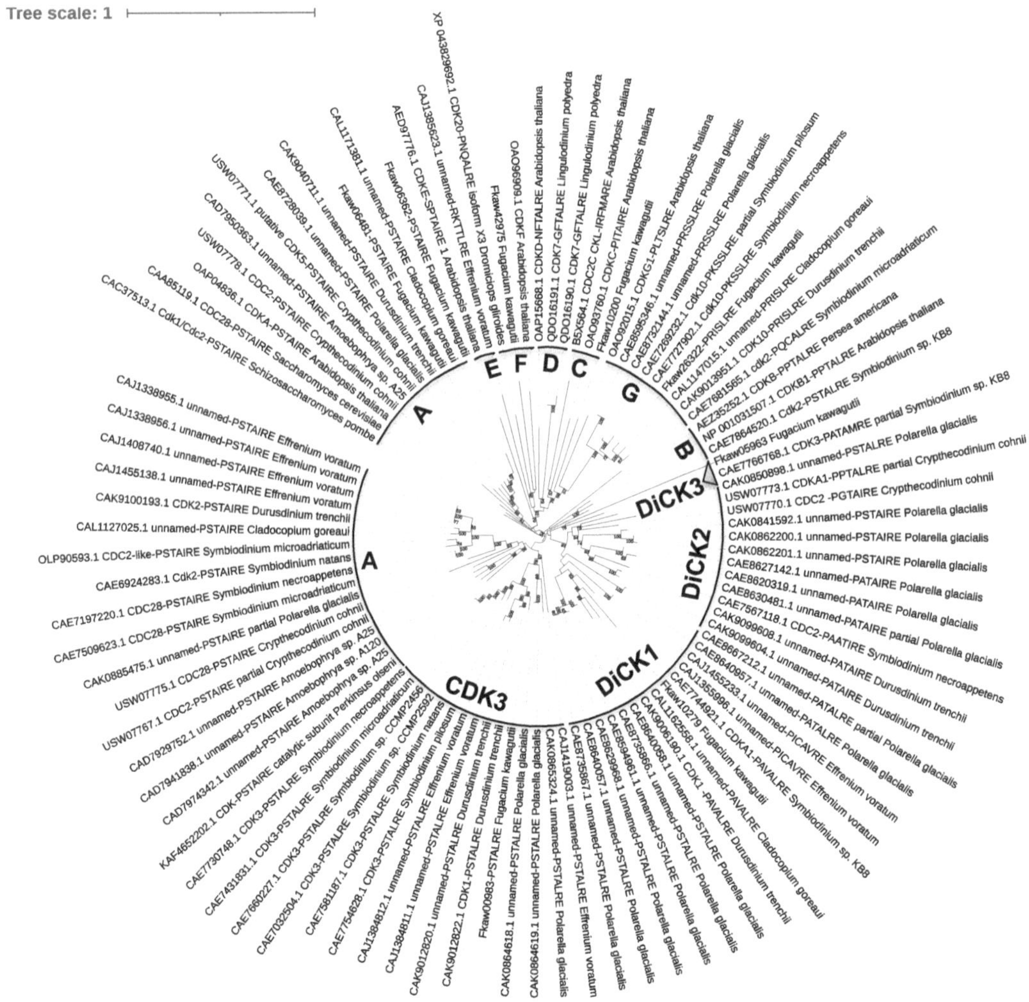

Figure 19.12. Phylogenetic analysis of CDK classification in dinoflagellates. Classification from the current analysis is shown inside the circle, which may not agree with names in the gene IDs from original data source. Letters refer to plant types of CDKs. CDK3, an animal type. DiCKs are putatively unique dinoflagellate CDKs.

PSTALRE domain is clustered with CDKB of *Arabidopsis* and *Persea* (both with the PPTALRE domain) in the phylogenetic tree, suggesting that a homolog may exist in dinoflagellates. However, a sequence from *C. cohnii* contains PPTTLRE domain but is not phylogenetically affiliated with CDKB. This and many other dinoflagellate sequences form an independent group (DiCk2), one of the two groups that appear to be dinoflagellate-specific (named DiCK1 and DiCK2 herein).

Both the chlorophyte *Chlamydomonas reinhardtii* and prasinophyte *Ostreococcus tauri* contain only one CDKB gene that resembles the CDKB1 genes of land plants (Inzé and De Veylder 2006). CDKB in *O. tauri* contains a PSTALRE domain, halfway between the PSTAIRE CDKA and PPTALRE/PPTTLRE CDKB domains.

Cyclins

Cyclins are variable, like CDKs, in the primary sequence within and between species, but they have a similar tertiary structure called the cyclin fold, with two domains of five alpha helices each. The first five helices correspond to a conserved region known as "cyclin box", while the second domain, named the Destruction box (D-box), is located upstream of cyclin box and is less conserved. Cyclin box, about 100 amino acids in length (Fig. 19.13), is the target for CDK binding and hence essential for CDK activation (Horton and Templeton 1997). The D-box motif is the target for cyclin polyubiquitination and degradation. For instance, cyclin B, associated with mitosis, pairs with CDK1 to regulate the G2 to M phase transition. Its level rises during mitosis and falls during the rest of the cell cycle, with degradation targeted by the destruction box (D-box) motif. Additionally, cyclin B has a signature motif (H/Q)(K/R/Q)(F/L) within the cyclin box (Zhuang et al. 2013 and references therein).

Using antibodies developed from human cyclin B1, a protein with expected molecular mass were detected in the chlorophyte *Dunaliella tertiolecta* (Lin et al. 1996), suggesting that cyclin B1 is conserved between chlorophyte algae and mammalians. Furthermore, the expression of this protein appeared to be higher in the exponential phase than in the late stationary phase. Western blot analysis detected a higher cyclin B1-like protein abundance approximately during G2/M period.

Using a conserved cyclin box sequence as the query in BLAST analysis against algal genomic and transcriptomic databases, various forms of cyclins were identified. To group these cyclins based on sequence similarity, a Neighbor-Joining phylogenetic tree was built. Defined types of cyclins in yeasts and *Arabidopsis* were included in the tree to guide the identification of separate clades. Despite the effort, no clear identification of the clades to canonical types of cyclins could be achieved, because cyclins from the model organisms, *Saccharomyces cerevisiae* and *Arabidopsis thaliana* tend to cluster together (in the yCYC/pCYC/dino clade), and algal homologs form four separate clades (Fig. 19.14). Among algae, dinoflagellates form three exclusive clades, suggesting that these are unique types of cyclins in dinoflagellates. One of these three, DinoCYCB1, contains many that has previously been annotated as G2/mitotic-specific cyclin, but also contains *Arabidopsis* SDS-like cyclin. The second, DinoCYC2-1, contains G1/S-specific cyclin previously identified in *Prorocentrum shikokuense* (Shi et al. 2017a) and other species, cyclin Pho80 from the yeast *S. cerevisiae*, as well as CYC3 from the malaria alveolate *Plasmodium*. The G1/S-specific cyclin in *P. shikokuense* was significantly downregulated under P-limited growth conditions (Shi et al. 2017), whereas Pho80 in yeast interacts with CDK Pho85p and regulates responses to P limitation as well as other nutrient conditions (Huang et al. 2007). The other dinoflagellate-specific clade, DinoCYC2-2, represents diverse dinoflagellates including *Noctiluca scintillans*, peridinin-type lineages, fucoxanthin-type lineages, and "dinotom" species, but no model organisms. It is unclear which type of cyclin this clade belongs to. The fourth clade contains *Arabidopsis* B-type cyclins and cover all major lineages of algae including some dinoflagellates from the groups *Amoebophrya* and *Gymnodinium*.

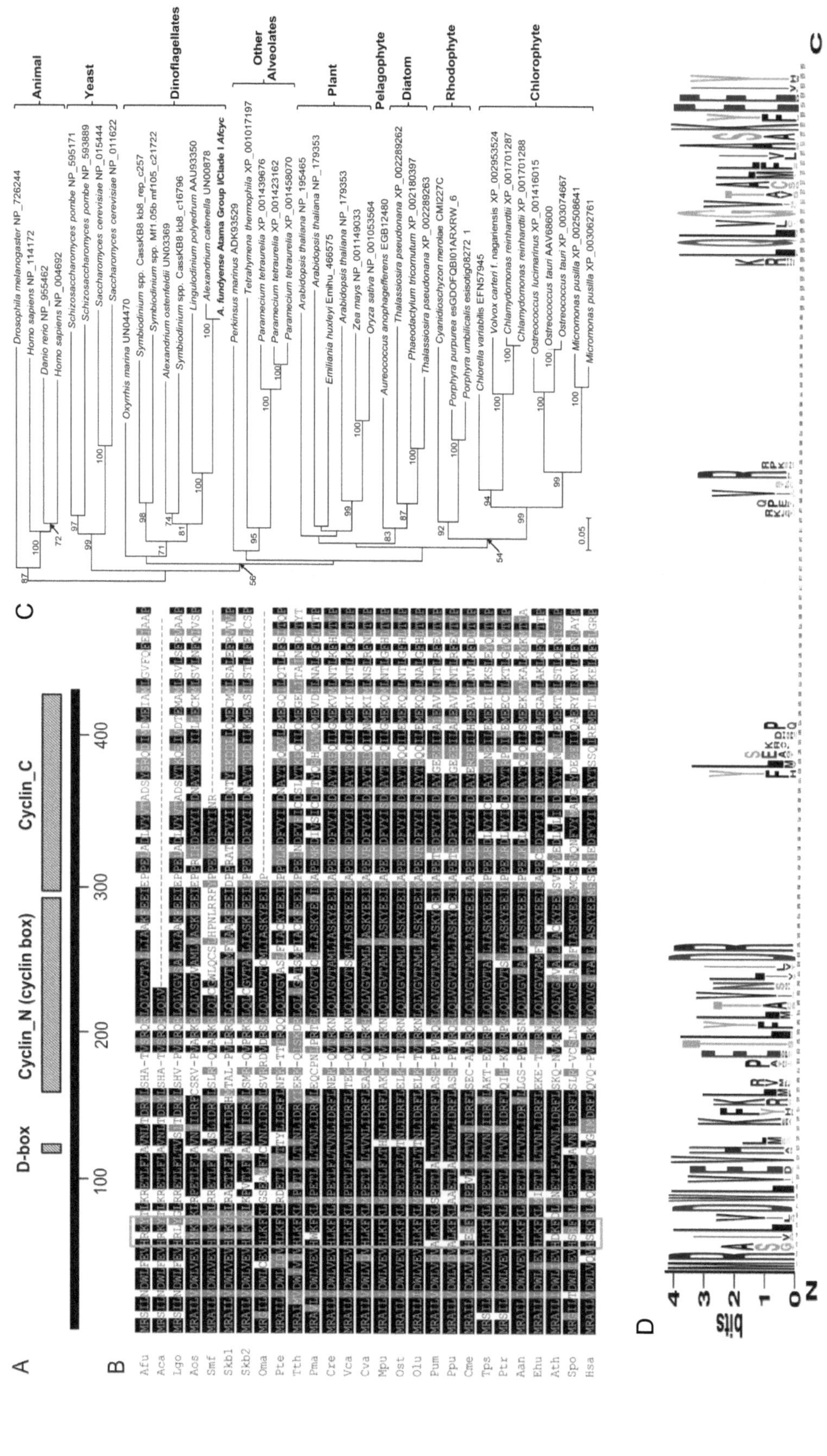

Figure 19.13 contd. ...

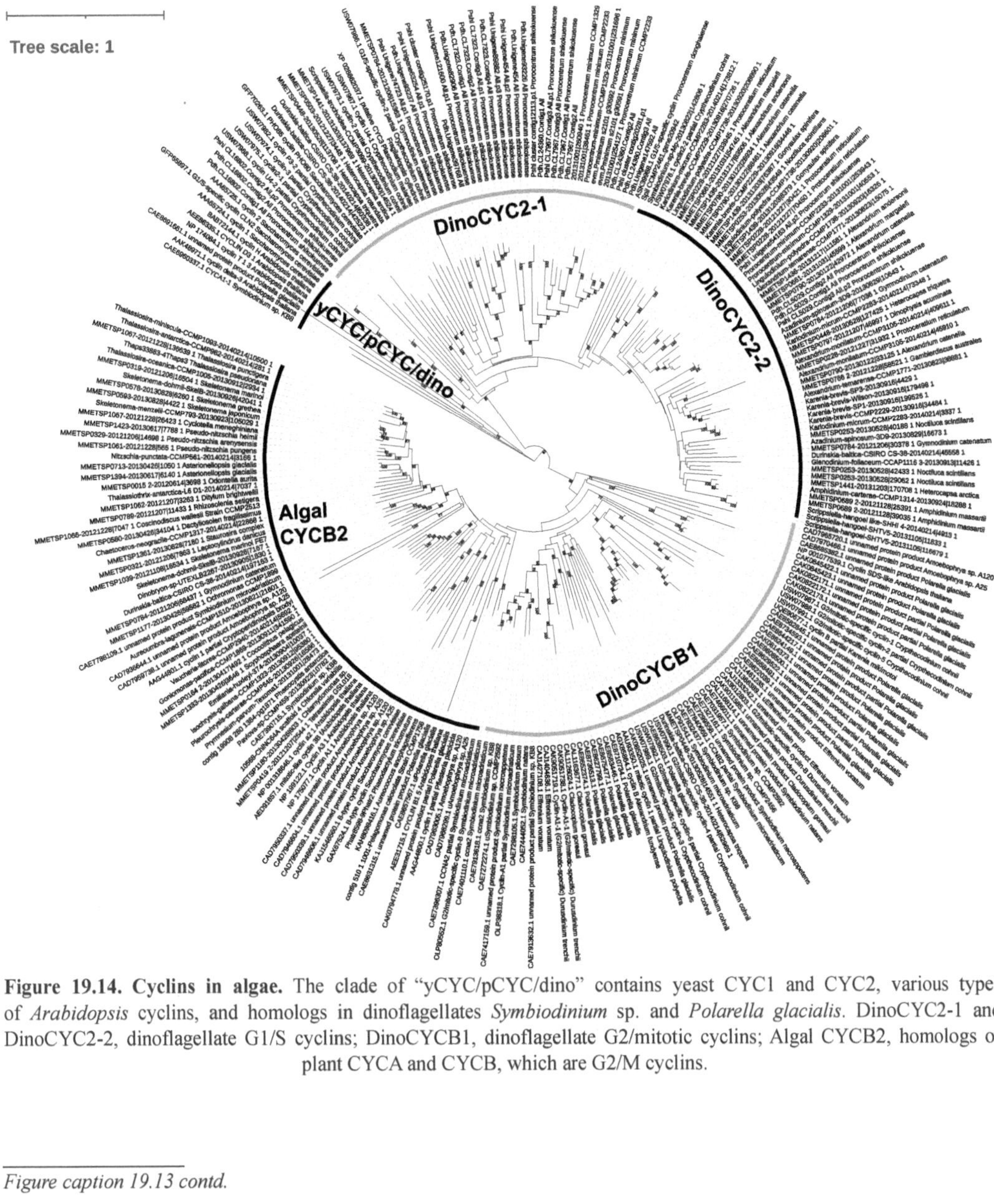

Figure 19.14. Cyclins in algae. The clade of "yCYC/pCYC/dino" contains yeast CYC1 and CYC2, various types of *Arabidopsis* cyclins, and homologs in dinoflagellates *Symbiodinium* sp. and *Polarella glacialis*. DinoCYC2-1 and DinoCYC2-2, dinoflagellate G1/S cyclins; DinoCYCB1, dinoflagellate G2/mitotic cyclins; Algal CYCB2, homologs of plant CYCA and CYCB, which are G2/M cyclins.

Figure caption 19.13 contd.

Figure 19.13. Prevalence and high similarity of mitotic cyclin (CycB) and cyclin box. (A) Schematic of domain and motif organization. Two cyclin domains and destruction box (D-box) are shown in hatched boxes. (B) Sequence alignment of cyclin N-terminal domain (cyclin box) in model organisms and algae. Identical amino acid residues are black shaded, similar residues are gray shaded and cyclin B signature motif is boxed. Species names are as follows: Afu, *Alexandrium fundyense* (now renamed as *A. catenella*; Aca, *Alexandrium catenella* (now renamed as *A. pacificum*); Lgo, *Lingulodinium polyedrum*; Aos, *Alexandrium ostenfeldii*; Smf, *Symbiodinium* spp. Mf1.05b; Skb, *Symbiodinium* spp. CassKB8; Oma, *Oxyrrhis marina*; Pte, *Paramecium tetraurelia*; Tth, *Tetrahymena thermophile*; Pma, *Perkinsus marinus*; Cre, *Chlamydomonas reinhardtii*; Vca, *Volvox carteri* f. *nagariensis*; Cva, *Chlorella variabilis*; Mpu, *Micromonas pusilla*; Ota, *Ostreococcus tauri*; Olu, *Ostreococcus lucimarinus*; Pum, *Porphyra umbilicalis*; Ppu, *Porphyra purpurea*; Cme, *Cyanidioschyzon merolae*; Tps, *Thalassiosira pseudonana*; Ptr, *Phaeodactylum tricornutum*; Aan, *Aureococcus anophagefferens*; Ehu, *Emiliania huxleyi*; Ath, *Arabidopsis thaliana*; Spo, *Schizosaccharomyces pombe*; Has, *Homo sapiens*. (C) Neighbor Joining tree inferred from cyclin B protein sequences. Cyclin B from animals were used to root the tree. Only the bootstrap values higher than 50% are shown at the nodes. (D) WEBLOGO of aligned 139 algal cyclin box sequences showing highly conserved positions and variable regions. (A-C) Based on Zhuang et al. 2013.

CDC25

Cdc25 phosphatases belong to the Rhodanese Homology Domain superfamily and play a crucial role in cell cycle progression by activating cell division kinases. These phosphatases dephosphorylate phosphotyrosine and phosphothreonine residues, thereby activating their Cdk/cyclin substrates (Fig. 19.11). Specifically, Cdc25A regulates the entry into the S phase, while Cdc25B is essential for the G2/M phase transition of the cell cycle. A BLAST search of the genomes of *P. tricornutum* and *T. pseudonana* did not find CDC25 (Huysman et al. 2010). Similarly, no CDC25 has been found in the land plants *Arabidopsis* and *Oryza sativa*, in which the CDC25-mediated regulatory mechanisms might have been replaced by a mechanism relying on the plant-specific CDKB. The failure to find CDC25 based on sequence similarity could be due to large divergence. Alternatively, it is possible that CDC25 has been replaced by CDKA2, because its CDK hallmark domain is PSTALRE, which, as in the case of CDKB of *O. tauri*, is halfway between CDKA and CDKB. This appears to occur in dinoflagellates as well, as some of their CDKs in many species carry the PSTALRE domain (Fig. 19.12). A BLAST search of algal genomes revealed a CDC25-like protein in the cryptophyte *Guillardia theta* strain CCMP2712 (XP_005826366.1), but none in dinoflagellates. Further analysis using the *G. theta* CDC25-like protein as a query yielded hits from other cryptophytes, chlorarachniophytes, fungi, and animals, with E values from –24 to –50, but no matches to any other algae. Given the importance of CDC25's function, how it evolved or was replaced in most algae warrants further investigation.

WEE1

WEE1 is a nuclear protein, a tyrosine kinase belonging to the Ser/Thr family of protein kinases. This protein catalyzes the inhibitory tyrosine phosphorylation of CDC2/cyclin B kinase, helping to coordinate the transition between DNA replication and mitosis by protecting the nucleus from cytoplasmically activated CDC2 kinase. WEE1 is present in all algae examined. A BLAST analysis using WEE1 from *Arabidopsis* and Humans as queries produced significant hits (with e values <-20) from all major groups of algae (Fig. 19.15), indicating that WEE1 is prevalent and conserved. In the Maximum Likelihood tree inferred from the hits along with the query sequences, WEE1 sequences from green algae and land plants forms a monophyletic clade. Various lineages of algae, including rhodophytes, prasinophytes, diatoms, raphidophytes, phaeophytes, haptophytes, chlorarachniophytes, pelagophytes, cryptophytes, and an euglenid species, are represented in this clade, along with human WEE1. In addition, there is a major clade composed exclusively of dinoflagellates. This phylogenetic pattern suggests strong lineage-specific WEE1 gene divergence for chlorophytes/streptophytes and dinoflagellates.

Nucleus-Plastid Crosstalk to Synchronize Their Divisions

In green algae, nuclear mitosis is synchronized with plastid division through signaling mediated by Mg-Protoporphyrin IX (Mg-protoIX), a molecule used by chloroplasts to signal the nucleus to modulate nuclear gene expression (retrograde signaling) (Kobayashi et al. 2009). Recent studies on pennate diatoms have revealed the timing of chloroplast division. In diatoms, the number of chloroplasts per cell varies with species, with centric diatoms being generally polyplastidic whereas pennate diatoms or mono-, di-, tetra- or oligoplastidic (Mann 1996). In mono-, di- and tetraplastidic diatoms, chloroplast ontogeny is tightly coupled with the cell cycle, with division generally occurring just before or after cytokinesis. In *Seminavis robusta* (Gillard et al. 2008) and *P. tricornutum*, chloroplast division occurs between the onset of S-phase and the completion of mitosis (Huysman et al. 2010). When *S. robusta* cells were chemically blocked from entering the S phase, its chloroplast failed to duplicate. Furthermore, when treated with the cytostatic Sex-Inducing Pheromone (SIP), which arrests the cell cycle in G1 phase, key chloroplast division gene DRP5B and FtsZ were downregulated (Bilcke et al. 2021a). Examining various groups of

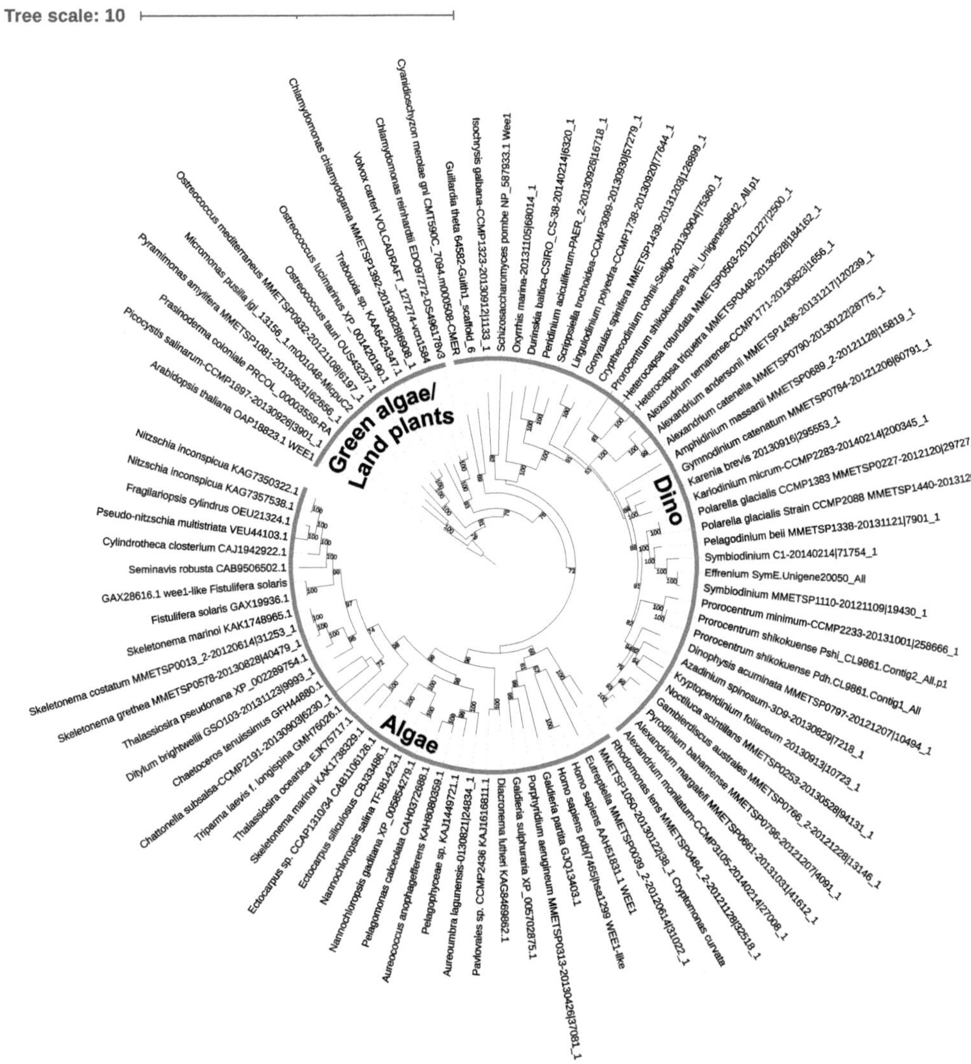

Figure 19.15. Prevalence of WEE1 in algae. Shown is a Maximum Likelihood tree based on the best evolutionary model identified for this dataset (VT+F+R6). Green, chlorophytes and *Arabidopsis*; Dino, dinoflagellates; Algae, non-green and non-dinoflagellate algae as well as humans.

algae further suggests that a G2/M checkpoint exists to ensure chloroplast division is complete before cells enter mitosis (Sumiya et al. 2016). In addition, other organelles such as mitochondria show synchronous rhythmicity in division with the chloroplast and the cell, with mitochondrial and plastid FtsZ genes being expressed synchronously (Bilcke et al. 2021b).

Adaptive Cell Cycle Regulation Under Environmental Stress

In plants under unfavorable conditions, CDK inhibitors like WEE (Wee1-like protein kinase), SIM (SIAMESE), and KRP (Kip-related proteins) sense external signals and inhibit CDK activity to arrest the cell cycle (Carneiro et al. 2021). DNA damage activates Ataxia Telangiectasia Mutated/Ataxia Telangiectasia and Rad3 Related (ATM/ATR), which phosphorylate and activate Suppressor of Gamma Response 1 (SOG1), triggering DNA Damage Response (DDR). SOG1 induces the expression of ANAC044/085 and CDK inhibitors (CKIs), indirectly repressing Act-MYB.

ANAC044/085 stabilize Rep-MYBs, repressing G2/M-specific genes and arresting the cell cycle at G2. The elevated expression of CKIs alone is insufficient to arrest the cell cycle at G2 in DDR, and the simultaneous repression of G2/M-specific genes is required (Carneiro et al. 2021).

Environmental Stress and Endoreduplication

Environmental stress induces endoreduplication in some algae, an alternative form of the cell cycle (known as endocycle) where DNA replication occurs repetitively without mitosis or cytokinesis, increasing DNA content and ploidy (Lang and Schnittger 2020). One of the underlying mechanisms to trigger the endocycle is transcriptional repression of mitotic CDK and cyclin genes and cell cycle arrest at G2 resulting from the accumulation of Rep-Myb transcription factors that represses the expression of G2/M-specific genes (Carneiro et al. 2021).

A second mechanism is the proteasomal degradation of mitotic cyclins, particularly CYCA2;3 that associates with CDKB1;1, by the APC/C (De Veylder et al. 2011). The Anaphase-Promoting Complex/Cyclosome (APC/C) is a multi-subunit ubiquitin ligase responsible for targeting mitotic regulators for degradation during mitotic exit. At the end of mitosis, APC/C is activated through phosphorylation and its association with the WD-40 protein Cdc20/Fizzy (Yudkovsky et al. 2000).

Another important mechanism contributing to endocycle onset is the suppression of CDK activities by CKIs such as SIM and SIM-related proteins (SMRs). In land plants, *Arabidopsis* NAC domain containing protein (ANAC) 044 and 085 are induced by heat, cold, and salinity stresses as well as DNA damage, suggesting that ANAC044/ANAC085 is an important component of stress signaling, serving by stabilizing Rep-MYBs to trigger G2 arrest (Takahashi et al. 2019).

Environmental Stresses and Cell Cycle Arrest

Heat stress. In algae, environmental stresses can cause cell cycle arrest in G1 or G2 and in rare cases S phase or induce dormancy in the form of cysts or spore-like cells. Cell cycle arrest induced by hyperthermal stress (39°C) in *C. reinhardtii* can occur with high mitotic cyclin-dependent kinase (CDK) activity, indicating that the arrest is not due to a lack of CDK (Zachleder et al. 2019). If verified to be true in other chlorophytes, this is indicative of differences in the molecular mechanism of cell cycle arrest between green algae and land plants.

N-nutrient limitation. Nitrogen deficiency typically causes G1 phase extension or arrest. This is true in *C. reinhardtii*, which could even leave the active cycle and enter G0 phase (Takeuchi and Benning 2019). *C. reinhardtii* has a nuclear protein named COMPROMISED HYDROLYSIS OF TRIACYLGLYCEROL 7 (CHT7), initially identified from a mutant library with the phenotype of being slow or unable to degrade TAG when N is resupplied after N deprivation (Tsai et al. 2014). Degradation of TAG allows the cell to reutilize the carbon and energy stored in that molecule, as its loss-of-function Δ*cht7* was a cell cycle inhibitory regulator, functionally opposite of CDKs. Most of the N-deprived Δ*cht7* cells were unresponsive to N-refeeding, leading to the overall delayed TAG degradation, and some of these cells had multiple nuclei and a swollen cell appearance and showed derepressed expression of cell cycle genes following N deprivation. The Δ*cht7* was later shown to cause misregulation of 43 genes, suggesting that it may be an upstream regulator of the cell cycle (Lin et al. 2022). The CHT7 protein belongs to the Lin-54 family (IPR028307) and contains a CHC (CXC-Hinge-CXC) domain with two cysteine-rich CXC motifs. Lin-54 is part of the DREAM complex (DP, RB-like, E2F, and MuvB, where MuvB stands for multi-vulval class B), which in humans represses cell cycle-dependent genes in quiescent cells and is involved in the cell cycle-dependent activation of G2/M genes. Studies indicate that CHC proteins in plants and animals are crucial for cell cycle regulation, forming complexes with transcription factors and directly binding to DNA via their CHC domain. CHT7 homologs can be found in vascular plants, green algae (chlorophytes and prasinophytes), diatoms, haptophytes, cryptophytes, pelagophytes,

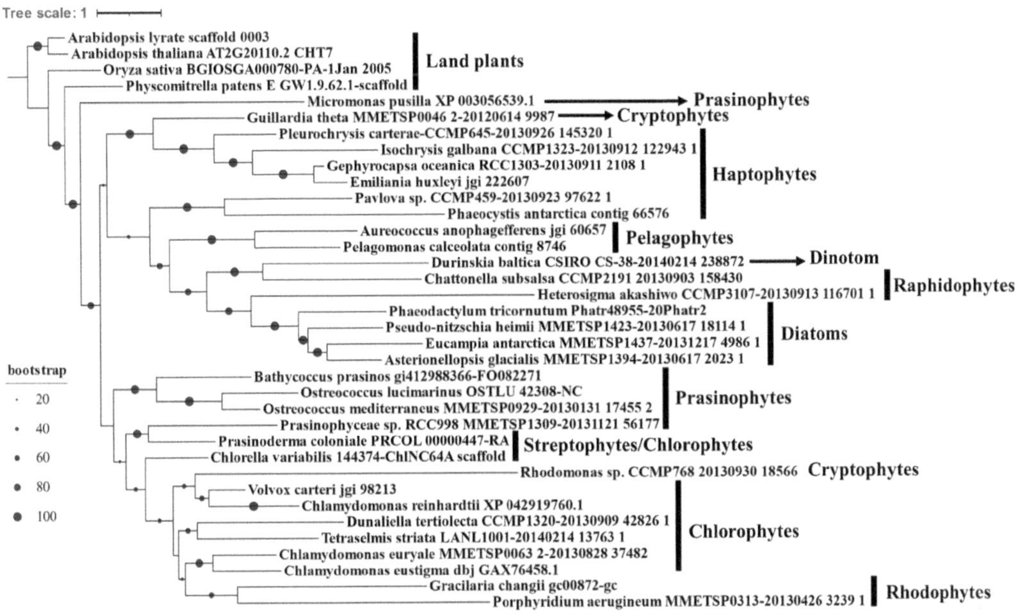

Figure 19.16. Presence of CHT7-like proteins in a wide range of algae except dinoflagellates. CHT7 is member of the Lin-64 family, which is involved in cell cycle regulation in general and has been shown to play a role in cell cycle regulation in response to N-starvation in *Chlamydomonas reinhardtii.*

and raphidophytes (Fig. 19.16), but not detected in dinoflagellates. Whether CHT7 is a regulator of cell cycle response to N-starvation in most algal lineages remains to be studied.

N limitation can trigger life cycle transition from vegetative cells to resting cysts in some dinoflagellates. In *Scrippsiella acuminata*, cyst induction was observed under N limitation, but not under P limitation, and the expression levels of cyclin B and CDK1 were downregulated (Yue et al. 2022).

P-nutrient limitation. While N limitation usually causes reduction of cell size, P limitation causes cell size enlargement (Li et al. 2015, 2016). Cell cycle analysis has revealed that P-nutrient limitation induces cell cycle arrest in the G1 phase (Li et al. 2015, 2016), as does N limitation. The G1 arrest was accompanied by the downregulated expression of G1/S-specific cyclin in the dinoflagellate *Prorocentrum shikokuense* (Shi et al. 2017a). Although G1 arrest appears to be a common cell cycle response to P limitation, the molecular mechanism may vary with species. In *Alexandrium pacificum* (formerly *A. catenella*), a study based on subtraction hybridization analyses revealed the occurrence of a homolog of Fizzy/cdc20, and its upregulation in P-depleted cultures where cells were arrested in G1 (Zhang et al. 2014). Meanwhile cell cycle related calcium-dependent protein kinase genes were downregulated. As CDC20/Fizzy needs to be phosphorylated (inhibitory) and dephosphorylated (activating) to trigger mitotic CYCB degradation in model organisms, how the upregulation of cdc20/fizzy acts to induce or reinforce G1 arrest in dinoflagellates is unclear.

Light limitation. Light limitation can arrest algal cell cycle in G1 or G2 phases, depending on the species (Vaulot et al. 1986). For coccolithophore *Hymenomonas carterae*, cells transferred from light to prolonged darkness were arrested at G1 phase. For the diatom *Thalassiosira weissflogii* (now renamed *Conticribra weissflogii*), dark treatment stalled the cell cycle at G1 and G2 phases. For dinoflagellates such as *Alexandrium catenella* (=*A. fundyense*) and *Prorocentrum shikokuense*, placing cultures in darkness for 48 hours resulted in cell cycle arrest in the G1 phase, which has been used as a tool to synchronize the cell population (Shi et al. 2013, Zhuang et al. 2013).

Circadian Rhythm and Light/Dark Entrainment

Cell cycle in the macroalgal chlorophyte *Ulva compressa* is diurnally gated at the G1 phase in pace with light dark cycles. The light-dark cycle entrains the cell cycle "clock" and causes the cell population to synchronize with respect to cell cycle progression.

Cyanobacteria can grow with more than one doubling per day, but meanwhile exhibit circadian rhythms of cell division and gene expression. For instance, *Synechococcus* PCC7942 can divide at rates as rapid as once every 10 h, but with a circadian periodicity of low or nil cell division (Mori et al. 1996). This suggests that *Synechococcus* cells can simultaneously sustain two timing circuits that express different periods.

Intriguingly, in filamentous algae, the cell cycle progression is not always synchronous among the cells in the same filament. For instance, in *Microspora* sp., some cells in the chain can have one nucleus (apparently in G1 phase) while others in the same chain have two nuclei (post-mitosis but pre-cytokinese) (Zachleder 2016).

Cell cycle behavior in algae has been best studied in chlorophytes. Chlorophytes such as *Euglena gracilis* and *Chlamydomonas reinhardtii* have been used as models for research on synchronous behavior and light-dark entrainment of the cell cycle (Edmunds 1966, Spudich and Sager 1980). Research by Edmunds, a professor at Stony Brook University, has shown that algal cell cycles are under the control of a clock, which interacts with the circadian rhythm clock (Edmunds and Adams 1981). This interaction ensures that cellular metabolism and cell division are tightly coordinated to optimize the utilization of energy for cell growth and proliferation. Genes for photosynthesis are highly expressed in the G1 stage during the day to ensure rapid cell growth using light energy. Conversely, genes for cell division are expressed at the end of the day, post cell size-determined commitment to mitosis. The peak expression of genes for DNA replication, nuclear division, and chloroplast fission at this stage indicates rapid cell division cycles. Post-division, cells enter a brief G0 (quiescent) period in the dark, downregulating cell division and metabolism genes, while genes for flagellar assembly are actively expressed. The daughter cells emerge at dawn to start a new life cycle (Zones et al. 2015).

Application of Cell Cycle Markers in Ecological Studies

Cell cycle stage-specific genes have the potential to serve as a marker for ecological studies. This application requires that the expression of the gene can be detected and quantified in a species-specific manner. For instance, the *in situ* growth rate for individual species of phytoplankton in the natural aquatic systems is a crucial parameter for ecological studies but very difficult to estimate. A cell cycle-based approach has been developed (Carpenter and Chang 1988), and can be made more widely applicable if an easily detectable cell cycle marker is established (Lin et al. 1997). PCNA has been used to assess cell cycle progression and characterize growth status of a natural population of the giant diatom *Ethmodiscus rex* (Lin and Carpenter 1995). An antibody was developed for the cyclin box derived from the harmful algal bloom pelagophyte alga *Aureococcus anophagefferns* (Lin et al. 2000). The characterization of cyclin B and the establishment of a positive correlation between its expression and growth rate in *Alexandrium catenella* has allowed its use in studying how grazers induce toxin production and growth depression (Park and Dam 2021). CHT7, a cell cycle regulator responsive to N-nutrient deprivation, has the potential to serve as a marker indicating growth status in nutrient-variable environments. One drawback that has limited the applications of these cell cycle genes is that their expression changes in magnitude rather than between presence and absence. This issue needs to be addressed for the cell cycle marker to be widely useful. One potential solution is to use ratios of two or more genes. More research investment in this area is required.

20

Genomics of Algal Symbioses

Symbiosis is an ancestral trait of life forms as it was responsible for the birth of algae and eukaryotes (Sagan 1967, Margulis 1970). Chloroplasts are believed to have emerged from serial endosymbioses, as discussed in Chapter 2. Briefly, these include primary endosymbiosis, secondary endosymbiosis, and tertiary endosymbiosis. Primary endosymbiosis gave rise to Archaeplastida, secondary endosymbiosis drove the radiation of the green and red algae into their highly diverse descendants, and the tertiary replacement resulted from the symbiosis of a secondary endosymbiosis-derived alga with a eukaryotic heterotrophic host.

The ocean hosts a diverse array of mutualistic relationships between algae and various organisms (Table 20.1). These include symbioses where algae serve as symbionts in partnerships with other algae, heterotrophic protists, or animals, as well as symbioses in which algae act as hosts for bacteria. In heterotrophic protist-alga symbioses, the integration of the symbiont into the host spans a gradient. This ranges from a Morphologically and Functionally Intact Alga (MFIA) serving as a permanent endosymbiont or a transient one, to the retention of multiple organelles or only the plastid as transient "symbionts" (kleptoplastids). These diverse heterotrophic protist-alga symbioses represent the different types of mixotrophy discussed in Chapter 17.

Alga-Alga Symbioses

Some photosynthetic algae harbor another photosynthetic alga as endosymbionts. In this relationship, the alga-cyanobacterium (blue-alga) is widespread, and the mutualism is obvious. Other alga-alga relationships (dinoflagellate-diatom) are more functionally cryptic.

Alga-Cyanobacterium Symbioses

Cyanobacteria have developed broad symbiotic relationships with various lineages of algae, seagrasses, and other organisms (Table 20.2). The most striking example is N_2-fixing (diazotrophic) cyanobacteria that form symbiotic association (the symbiotic cyanobacteria are commonly named cyanobionts) with algae, plants, and various other organisms (Carpenter 2002, Rai et al. 2000, Foster and O'Mullan 2008, Foster et al. 2011). The best-known alga-cyanobiont symbiosis involves diatoms as the host (known as diatom-diazotroph associations or DDA), which was documented as early as the beginning of the 20th century (see Foster and O'Mullan 2008). This symbiosis most frequently involves the diatoms *Rhizosolenia*, *Hemiaulus*, *Guinardia*, and *Bacteriastrum* and the cyanobacterium *Richelia intracellularis* or coccoid unicellular species. *R. intracellularis* belongs to the order Nostocales and forms short filaments with a terminal heterocyst. There appears to be notable genetic divergence between *R. intracellularis* symbionts in different host genera (Janson

Table 20.1. Mutualism between diverse groups of organisms and algae.

Host	Endosymbiont	Style	Distribution
Algae (e.g., diatoms)	N_2-fixing cyanobacteria	Endosymbiotic	Ocean, estuaries
Algae (e.g., dinoflagellates)	Algae	Endosymbiotic	Ocean
Heterotrophic protists	Algae	Endosymbiotic	Ocean, freshwater
Invertebrates (e.g., corals)	Algae	Endosymbiotic	Ocean, estuaries
Vertebrates	Algae	Endosymbiotic	Ocean, freshwater

Table 20.2. Documented diazotrophic cyanobacteria symbioses with various organisms.

Host	Endosymbiont	Style	Distribution	References
		Diatoms		
Bacteriastrum	*Richelia intracellularis*	Endosymbiotic	Marine	Rai et al. 2001
Chaetoceros sp.	*Richelia intracellularis*	Epiphytic	Pacific Ocean	Janson et al. 1999
Chaetoceros spp.	*Calothrix rhizosoleniae*	Epiphytic	ALOHA (subtropical North Pacific Ocean)	Foster et al. 2010
Chaetoceros sp.	*Richelia intracellularis*	Epiphytic	South Pacific Ocean	Janson et al. 2000
Clacodium frauenfeldianum	*Cyanothece* like	Endosymbiotic	Tropical Atlantic and Pacific Oceans	Carpenter and Janson 2000
Coscinodiscus sp.	Anabaena	Endosymbiotic	Marine	Taylor 1982
Coscinodiscus sp.	Similar to *Synechocystis* sp.	Endosymbiotic	near Zanzibar	Carpenter 2002
Guinardia cyclindrus	*Richelia intracellularis*	Endosymbiotic	Marine	Villareal 1992
Hemiaulus hauckii	*Richelia intracellularis*	Endosymbiotic	Pacific Ocean	Janson et al. 1999
Hemiaulus membranaceus	*Richelia intracellularis*	Endosymbiotic	Pacific and Atlantic Oceans	Janson et al. 2000
Rhizosolenia clevei	*Richelia intracellularis*	Endosymbiotic	South Pacific Ocean	Janson et al. 1999
Rhopalodia gibba	*Cyanothece* like	Endosymbiotic	Freshwater	Floener and Bothe, 1980, Prechtl et al. 2004
Roperia tessellata	*Nostoc*	Endosymbiotic	Marine	Villareal 1992
		Haptophytes		
Various species	UCYN-A	Endosymbiotic	North Pacific Ocean	Thompson et al. 2012
		Dinoflagellates		
Various species	*Synechococcus, Prochlorococcus*	Symbiotic	Oceans	Foster et al. 2006a, Foster and Zehr 2019
		Ciliates		
Tintinnids	*Synechococcus*	Symbiotic	Oceans	Foster et al. 2006a, b
		Radiolarians		
Various species	*Synechococcus*	Symbiotic	Oceans	Foster et al. 2006a, b
		Seaweeds and seagrasses		
Various species	Various species	Symbiotic	Oceans	Mutalipassi et al. 2021
Other organisms	Various species	Symbiotic	Oceans	Mutalipassi et al. 2021

et al. 1999). Besides *Richelia*, *Calothrix* also forms symbiotic relationship with diatoms such as *Chaetoceros* sp. (Fig. 20.1; Foster and Zehr 2006, Foster et al. 2011).

The pennate diatom *Rhopalodia gibba* was noted to contain intracellular cyanobionts as early as in the 1960's (Floener and Bothe 1980) and the cyanobiont was later found to be closely related to *Cyanothece* (Prechtl et al. 2004). Similarly, DDA between Rhopalodiaceae diatoms and N_2-fixing cyanobionts also occurs in the ocean. *Epithemia pelagica* and *E. catenate* were detected repeatedly from the subtropical North Pacific Ocean and detected throughout the global ocean (Schvarcz et al. 2022). The cyanobionts in these diatom species lack fluorescent photopigments but possess *nif*H genes similar to those of free-living unicellular cyanobacteria. These DDA holobionts, which appear to be present throughout the global tropical oceans, show the highest N_2-fixing levels on daytime and lowest at nighttime. A simulation study (Fig. 20.2) indicated that the host diatom (e.g., *Hemiaulus*) provides 25% of its fixed organic carbon to the holobiont whereas the diazotrophic cyanobiont (e.g., *Richelia*) provides over 80% of its fixed N to the host in return; N transfer to the host stimulates N_2 fixation in the cyanobionts by 5.5 folds (Inomura et al. 2020). The model estimates that ~ 90% of photosynthesis by the *Hemiaulus–Richelia* holobiont is performed by *Hemiaulus* (the diatom). The largest part of the fixed C (825: 67% for diatom and 15% for the *Richelia* trichome) is used for biosynthesis (including respiration that provides energy for biosynthesis), and the remaining 18% is used for supplying energy and electron for N_2 fixation by the cyanobiont.

R. intracellularis can reside in the periplasmic region of the host cell (between the cytoplasmic membrane and frustule), as in *Rhizosolenia clevei*, or inside the cytoplasm of the host, as in *Hemiaulus hauckii*. Genome sequencing of these two *Richelia* symbionts showed distinct differences, with the internal *Richelia* genome being far reduced containing incomplete sequences and lacking genes in several nitrogen assimilatory pathways, among others (Hilton et al. 2013, Nieves-Morión et al. 2020). In the *H. hauckii-R. intracellularis* symbiosis, the cyanobacterium is not enclosed by a host membrane, evidencing its being well integrated as an organelle of the host.

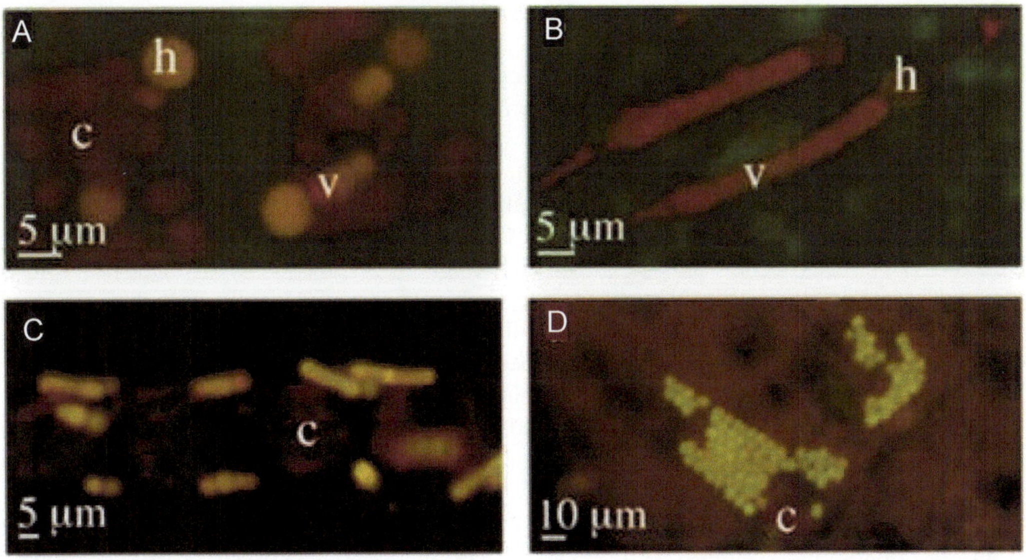

Figure 20.1. Epifluorescent micrographs of diatom-diazotrophic associations (DDAs) common in the ocean. (A) Two *Hemiaulus membranaceus* diatoms with two *Richelia intracellularis* associated to each diatom. The chlorophyll a within the chloroplast (c) of the diatom fluoresces red, whereas the pigments in the vegetative (v) cells and the terminal heterocyst (h) of *Richelia* fluoresce yellow–orange. (B) The apical end of a *Rhizosolenia clevei* diatom with two associated trichomes of *R. intracellularis*. (C) A chain of *Chaetoceros* spp. diatoms with *Calothrix rhizosoleniae* attached to the spines (not visible). (D) A chain of *Climacodium frauenfeldianum* diatoms associated with yellow-fluorescing unicellular cyanobacteria (cyanobionts). Reproduced with permission from Foster et al. (2011) under the Creative Commons Attribution-NonCommercial-No Derivative Works 3.0 Unported License and Dr. Foster's courtesy.

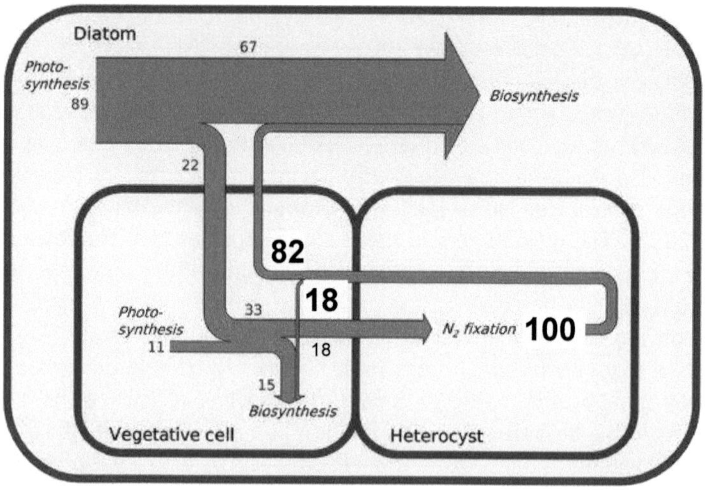

Figure 20.2. Simulated C (blue arrows) and N (brown arrows) exchanges between the diatoms (*Hemiaulus*) and trichomes (*Richelia*). The width of the arrows represents the relative magnitude of fluxes in mass. Small-font numbers represent percentages of C fluxes (here, 100 indicates total sources and sinks of C in the symbiosis; e.g., 89 on the top left means 89% of photosynthesis in the symbiosis is done by a diatom) and large-font numbers are percentages of N (separate from those of C). Reproduced with permission from Inomura et al. 2020 under the Creative Commons CC BY 4.0 license.

The cyanobiont appears to have developed a specialized biochemistry exemplified by the presence of specific glycolipids and the lack of central metabolic enzymes such as glutamine oxoglutarate aminotransferase. Ultrastructural analysis (Flores et al. 2022) indicates that the endosymbiont has also developed specialized structures such as a characteristic heterocyst envelope and close association with host mitochondria in adaptation to symbiosis and avoiding oxygen poisoning of nitrogenase.

The algae-cyanobiont mutualism is not limited to diatoms. Some haptophyte species have been found to harbor the N_2-fixing cyanobacterium UCYN-A (Thompson et al. 2012). This was discovered by sorting raw seawater samples from California coast and Station ALOHA without preserving or filtering them that could dislodge the cyanobacteria cells from the picoeukaryotic hosts. 18S rRNA sequencing indicated that the symbiotic partner was closely related to the calcareous haptophytes *Braarudosphaera bigelowii* and *Chrysochromulina parkeae*. $^{15}N_2$- and ^{13}C-bicarbonate labeling and halogenated *in situ* hybridization anometer-scale Secondary Ion Mass Spectrometry (HISH-SIMS) analysis revealed the association of the UCYN-A cells with the haptophyte cells, in which UCYN-A provided N to the host and carbon fluxed the other direction. Genome sequence data obtained from the isolated cells showed that the cyanobiont had no TCA cycle, PSII, and lack some biosynthetic pathways, indicating genome streamlining in adaptation to the symbiotic lifestyle. This uncultivated cyanobacterium was hence described as *Candidatus Atelocyanobacterium thalassa*, literally meaning incomplete marine cyanobacterium.

UCYN-A is a geographically widespread and has multiple sublineages (UCYN-A1, -A2, -A3, -A4), which are potentially distinct ecotypes (Zehr et al. 2017). Earlier genome sequencing for flow cytometrically sorted cells showed that UCYN-A is most similar to the unicellular cyanobacterium *Cyanothece* sp. ATCC51142, *Cyanothece* sp. CCY0110 or *Crocosphaera watsonii* WH8501 (Zehr et al. 2008). In *B. bigelowii*, UCYN-A symbionts are located in between the two chloroplasts of the host (Fig. 20.3; Cornejo-Castillo et al. 2024), suggesting a potential interaction with photosynthesis of the host. Interestingly, the biovolume of the UCYN-A symbiont in *B. bigelowii* exhibits a power function of the biovolume of the host (Fig. 20.3; Cornejo-Castillo et al. 2024). The nitrogenase gene arrangement and composition in this species are very similar to those of *Cyanothece* sp. ATCC51142 and of the *R. gibba* symbiont.

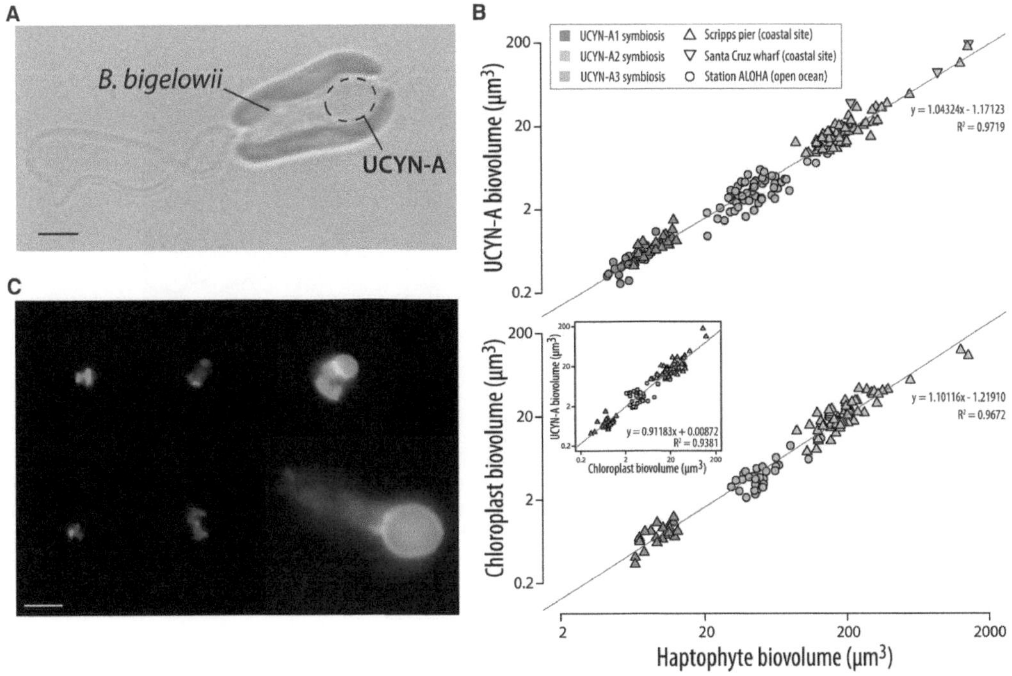

Figure 20.3. Localization of the cyanobacterial symbiont (UCYN-A) in the haptophyte (*Braarudosphaera bigelowii*) cell and positive correlation between the biovolume of the symbiont and the host. (A) Light microscopy image of the showing the position of the symbiont in between the two chloroplasts of the host. (B) Upper panel: UCYN-A volume (sublineages UCYN-A1, -A2, and -A3; see inset) as a function of associated haptophyte cell volume; lower panel: chloroplast volume as a function of associated haptophyte cell volume; inset in lower panel: UCYN-A volume vs. chloroplast volume of associated host cells. (C) Composite of catalized reporter deposition fluorescence *in situ* hybridization (CARD-FISH) images showing the range of sizes of the different UCYN-A sublineages (left images: UCYN-A1 sublineage; middle images: UCYN-3 sublineage; right images: UCYN-A2 sublineage) and their symbiotic haptophyte partners. Scale bar, 5 μm. Reproduced with permission from Cornejo-Castillo et al. 2024.

Besides, some cyanobacteria are endosymbionts of dinoflagellates (Table 20.2, Fig. 20.4). For instance, *Sinophysis canaliculate* is a benthic dinoflagellate and it harbors cyanobionts (Escalera et al. 2011). *Amphisolenia* spp. and species of *Ornithocercus*, *Histioneis*, and *Citharistes* also host cyanobionts (Lucas 2012). Using single-cell sequencing, Nakayama et al. (2019) found cyanobacteria in the dinoflagellate *Ornithocercus magnificus*. Besides, seaweeds such as *Codium decorticatum* and seagrasses such as *Thalassia testudinum* and *Cymodocea rotundata* harbor cyanobionts (Mutalipassi et al. 2021).

Sargassum distribution seems to be expanding in the recent decade as a consequence of global climate and environmental changes (Wang et al. 2019), and *Sargassum* harbors epiphytic diazotrophs. Wang and colleagues measured N_2 fixation rates from the whole fronds of *Sargassum* for 24 times during the period of 2015–2021 from the western edge of the Gulf Stream off Cape Hatteras, North Carolina. They found that the rate of N_2 fixation by these diazotrophs ranged 0–37.77 μmol N g^{-1} h^{-1} with a mean of 4.156 μmol N g^{-1} h^{-1}, which by extrapolation scales to a range of 0–30,916 μmol N m^{-2} d^{-1} with a mean of 3,697 μmol N m^{-2} d^{-1}. These estimated rates are substantial and should not be neglected. The taxonomic identity of the diazotroph, the nature of the relationship, and the molecular mechanisms regulating the relationship are yet unclear.

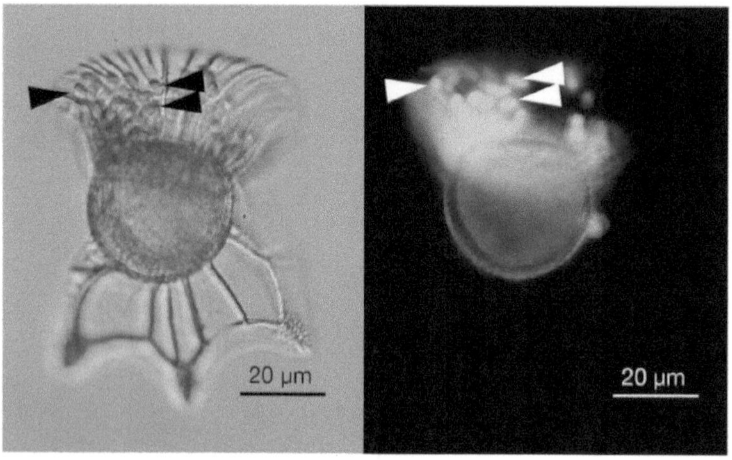

Figure 20.4. Microscopy images of an *Ornithocercus magnificus* dinoflagellate with its cyanobacterial symbiont. (Left) Bright-field image of *O. magnificus* from the same sampling area where the individual used in the present study was collected, showing brown cyanobacterial symbionts (arrowheads) residing inside the symbiotic chamber. (Right) Epifluorescence image of the same individual at the same angle of view as that of the left image. Arrowheads in both images point to the same cyanobacterial cells. Reproduced with permission from Nakayama et al. (2019).

Dinoflagellate-Alga Symbioses

A range of dinoflagellate species harbors other algae as endosymbionts. Because most of these cases involve a heterotrophic dinoflagellate as the host, the symbiotic relationship is often deemed as an intermediate stage in the evolution of plastids and studied to examine the endosymbiosis theories of plastid evolution. The *Kryptoperidinium-Nitzschia* (diatom) relation probably represents the earliest stage of plastid evolution because the symbiotic holobiont retains and operates the nucleus, mitochondria, and chloroplast of the endosymbiotic diatom as well as the nucleus and mitochondria of the host (Fig. 20.5). This would be a tertiary plastid acquisition in progress as the endosymbiotic diatom represents a secondary plastid lineage. Additionally, dinoflagellates in the genus *Durbinskia* harbor "mature" chloroplasts that are also diatom-originated, which are closely related to the chloroplasts of *Nitzschia palea*. As revealed by Crowell et al. (2018), the chloroplast genomes of *N. palea* and *D. baltica* have identical gene content, synteny, and a 92.7% pairwise sequence similarity, with most differences occurring in intergenic regions. The *N. palea* mitochondrial genome and *D. baltica*'s endosymbiont mitochondrial genome also have identical gene content and order, with a sequence similarity of 90.7%. Genome-based phylogenies demonstrated that *D. baltica* is more similar to *N. palea* than to any other diatom sequence currently available.

Algae as Symbionts

Heterotrophic Protist-Alga Symbioses

Protist-alga association span a wide spectrum, ranging from heterotrophic Protist Harboring Algal Symbionts (HPAS) on one end to Heterotrophic Protist retaining prey Alga Plastids (kleptoplastids) (HPAP) on the other, with intermediates that retain parts of algal cells (e.g., multi-organelles). The relationships of ciliates, foraminifera (forams), and heterotrophic dinoflagellates such as *Noctiluca scintillans* with algae are some important examples of HPAS. Ciliates and forams, both unicellular organisms, have symbiotic relationships with various lineages of algae such as symbiodiniacean dinoflagellates and green algae.

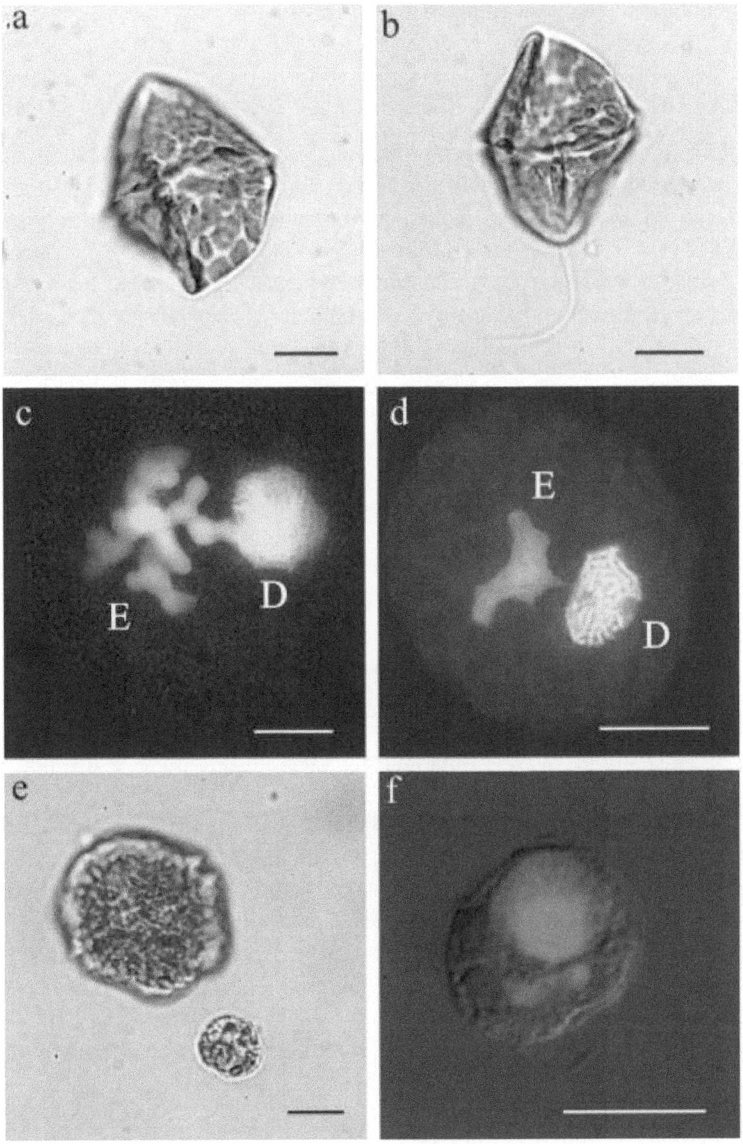

Figure 20.5. *Kryptoperidinium foliaceum* (strain Baiona B9) with diatom symbionts. External appearance (a, b) and nuclear morphology (c, d) of vegetative stages, D is showing the dinokaryon and E the endosymbiont nucleus; (e) comparison of sizes among culture cells; (f) nuclear morphology of the smaller cell shown in (e). Scale bars: 10 µm. Reproduced with permission from Figueroa et al. 2009.

Paramecium-Chlorella Symbiosis

A *Paramecium bursaria* cell carries several hundred *Chlorella* cells in its cytosol. Studies examining the effect of the symbiotic relationship on the symbiont showed that the symbiotic strain *Chlorella* sp. utilized monosaccharides such as glucose and fructose, not polysaccharides, for growth (Kamako et al. 2005) and produced high amounts of maltose (Brechignac and Wolf 1994). Besides, axenic isolated symbiont cultures could utilize NH_4^+ and amino acids, but not nitrate or nitrite, which supported normal growth of free-living *Chlorella* species such as *C. vulgaris* and *C. kessleri*. While nitrate stimulated only nitrite reductase (NiR) activity and not Nitrate Reductase (NR) activity for the symbiotic *Chlorella* sp., it did both for the two free-living species. Finally, nitrate utilization of the Japanese symbiotic strain differed from that of European and American symbionts (Kamako et al. 2005).

Foraminifera-Alga Symbiosis

Foraminifera (forams) carrying endosymbiotic algae are abundant in shallow tropical and semitropical seas (Lee 2006). Phylogenetically diverse forams (sorites and others) are the hosts for a wide variety of endosymbiotic algae (diatoms, dinoflagellates, unicellular chlorophytes, unicellular rhodophytes and/or cyanobacteria) or their plastids. Two factors predispose symbiosis in the group: (1) their overall organization into compartments, which keeps different cellular activities separate (for example, digestion is kept apart from symbionts) and (2) asexual reproduction insures vertical transmission of symbionts. The relationship between hosts and symbionts in diatom-carrying forams is not selective; a single host species can accommodate many different types of pennate diatoms, numbering in the dozens. *Nitzschia frustulum* var. *symbiotica* is the most common of the diatom symbionts, being found in about 30% of the associations. Sometimes one endosymbiotic diatom can be replaced by another.

However, forams in general are selective for what algae they assimilate. Some forams with algal symbionts such as *Heterostigina depressa* can grow in the light in the absence of obvious feeding on algae whereas others cannot grow without feeding on prey even in the light, indicating various degree of dependency on symbiosis. Chlorophyte species serving as symbionts of forams are mainly from the genus *Chlamydomonas* (*C. hedleyi*, *C. provasolii*, *C. eugametos*). *Porphyridium purpurum* from Rhodophyta, red cyanobacteria, many species from Haptophyta (e.g., *Pleurochrysis scherfelii*), and various species from the dinoflagellate family Symbiodiniaceae are frequently found endosymbionts of various species of forams.

Based on *Chlamydomonas hedleyi* isolated from *Archaias angulatus* as the model, it was found that the alga grew best on urea (20 μM) as the N source in the medium but at half the growth rate on NH_4^+ (2 μM) or NO_3^- (20 μM) as N sources. Besides, when urea was the N source, the optimum concentration of PO_4^{3-} was 0.1 μM, but was higher (I μM) when NO_3^- was the N source. No requirements for vitamins were demonstrated; however, a supplement of thiamine boosted growth (Lee et al. 1974). Similarly, *C. provasolii* isolated from *Cyclorbiculina compressa* also showed growth enhancement by a supply of vitamins (biotin and thiamine), high growth rates on NO_3^- or NH_4^+ (both ~ 200 μM) as the N source, and maximum growth in media with high PO_4^{3-} (100 μM) (Lee et al. 1979). In contrast, no obvious nutritional benefit was observed for endosymbiotic diatoms isolated from diatom-bearing hosts from the oligotrophic waters of the Gulf of Eilat, and this alga could grow on NO_3^- as the N source but optimal concentration varied from 2 μM to 2 mM depending on the clones (Lee et al. 1980).

Algae in foraminifera cells were also influenced by external nutrient supplies. High nutrient abundances overstimulate growth of the algae such that the forams lose control of the endosymbiont. In experiments in which the foraminifera were incubated in chemostats, *A. lobifera* species and *Marginopora kudakajimensis* withdrew nitrate and phosphate from the medium (Lee et al. 1991). Enrichment of the media in the chemostats with either nitrogen, phosphate or both to levels above 8 μM NaH_2P0_4 and 1.5 mM $NaNO_3$ led to algal overgrowth and eventual death of the hosts. The overpopulation problem resembles what is believed to happen in corals (Hallock 2000). Generally, feeding on algae seems to be necessary for most symbiotic forams (Lee 2006). Using ^{14}C isotope as tracer, most studies showed that the photosynthate produced by the algae and translocated to foram hosts were glycerol, although lipid, floridoside, and polyglucan also have been found (Lee 2006). Bleaching of forams can occur under high light, UV exposure, and high temperatures.

In immunochemical studies comparing multiple symbiotic species and non-symbiotic species of diatoms, a 104 kDa glycoprotein (CSSA, Cornmon Symbiont Surface Antigen) was found on the surfaces of all the symbiotic species tested and was absent from the non-symbiotic species tested (Chai and Lee 1999, 2000). Blocking this CSSA with antibody caused a loss of the ability of the diatom to bypass digestion and be drawn into the test to become an endosymbiont within the forams. Further immunocytochemical and structural analyses revealed abundant receptors for the CSSA on the pseudopodia making initial contact with the diatoms and on the primary organic lining of the test. These provide evidence that the initial recognition between the host foraminifera and the potential symbiotic diatoms is mediated by a cell signaling system involving molecules on the surfaces of diatoms and the pseudopods of the foraminifera. One of the characters that predisposes foraminifera to be hosts for algal symbionts is that their digestion begins in the granuloreticular network just after the pseudopods contact their prey (Faber and Lee 1991). Shortly following contact, the symbiotic diatom is phagocytosed and once in a phagosome, the alga escapes initial digestion and its surrounding membrane is converted to one of a symbiosome. It is then drawn into a foraminiferal test and is spatially separated from most digestive activity. Larger foraminifera are so morphologically modified from their ancestors that it is most likely that the evolutionary changes were driven by adaptation to symbiosis (Lee and Hallock 1987). The greenhouse nature of the transparent tests and the canals and pores of many larger foraminifera are seen as adaptations to symbiosis (Lee 2006). In addition, the CSSA is produced by the diatom even after it has lost its normal cell envelope, and it seems necessary to maintain the association even after symbiosis has been established (Chai and Lee 2000).

Mesodinium-Cryptophyte and Dinophysis-Cryptophyte Symbioses

The *Mesodinium rubrum*-cryptomonad (*Teleaulax amphioxeia*) relationship was initially found to be symbiotic (Hansen and Fenchel 2006), but later shown to be a kleptokaryotic, kleptoplastidic, and kleptomitochondriatic relationship (Johnson et al. 2007). However, at least in the Long Island Sound population, it is a host-symbiont relationship because the cytoplasmic membrane appears to be functionally intact (although structurally deformed), and its associated nutrient transporters are actively expressed (Qiu et al. 2016). Importantly, metatranscriptomic analyses indicated that in this symbiosis, ammonium was the most actively utilized nitrogen nutrient, involving ammonium transporters of both the host and the symbiont.

Another genus of ciliate, *Strombidinium*, also has kleptoplastidic relationship with the cryptophyte (McManus et al. 2018). A single-cell transcriptomic analysis on *Strombidium* cf. *basimorphum* showed that this species is also able to retain the chloroplast and transcriptionally active genetic materials from the cryptophyte prey (Maselli et al. 2021).

Dinophysis spp. represent another intriguing type of protist-alga symbiosis. It appears to be an intermediate stage of plastid development. As discussed in Chapter 8, at least some *Dinophysis* species are heterotrophic protists *per se*. However, they cannot ingest algal prey directly, but rely on *Mesodinium* to serve as the "feeder". They then retain the chloroplast of the prey algae (cryptophytes or haptophytes) for transient function to photosynthesize. The kleptoplastids quickly lose their functions and need to be replenished continually, apparently because the *Dinophysis* hosts do not have all the genes needed to maintain and regulate the chloroplast (Wisecaver and Hackett 2010).

Invertebrate-Alga Symbioses

Invertebrates that harbor algal symbionts include corals, giant clams and other mollusks, jellyfish, polycystinea, and acantharia. Due to the high diversity of such symbioses, a comprehensive overview is beyond the scope of this section. Therefore, only a few conspicuous examples will be discussed here. Arguably, coral-Symbiodiniaceae mutualism is the most ecologically important and well-studied symbiosis occurring in the ocean.

Coral-Alga Symbiosis

Coral reefs are among the world's most biodiverse and productive ecosystems. Despite covering less than 0.5% of the global ocean, they are home to over 25% of ocean biodiversity and support 30% of the world's marine fish species. Coral reefs provide essential products and services through fisheries and tourism. While most coral reefs are found in warm, shallow waters, over 600 species of deep-sea corals thrive at temperatures of 6-14°C (Cairns et al. 1999). Scleractinians like *Desmophyllum*, *Goniocorella*, *Lophelia*, *Oculina*, and *Solenosmilia* can form reefs (Stanley and Cairns 1988).

Shallow reef corals rely on the photosynthesis of endosymbiotic dinoflagellates from the Symbiodiniaceae family (LaJeunesse et al. 2018). Some deep-sea corals also harbor Symbiodiniaceae symbionts, such as *Cladocopium* spp. in Hawaiian black corals (Wagner et al. 2011) and *Durusdinium* in Red Sea mesophotic corals (Terraneo et al. 2023). Due to the lack of light, deep-sea coral symbionts likely live in a heterotrophic mode (Dimond and Carrington 2008, Wagner et al. 2011).

For tropical and subtropical corals, optimal water temperatures range from 23° to 29°C, limiting their distribution to within 30° of the equator. In this mutualistic relationship, corals provide shelter and nutrients (CO_2, ammonium, phosphates, and lipids), while Symbiodiniaceae dinoflagellates supply photosynthates (primarily glycerol and glucose) and essential organic compounds (e.g., amino acids) that the host cannot synthesize (Davy et al. 2012). These photosynthates account for 90% of the coral host's energy needs (Muscatine 1990).

Establishing this symbiotic relationship begins with recognition and infection, followed by material exchanges and potentially gene expression cross-regulation. While not fully understood, research in this area has expanded significantly, largely due to advancements in omics and molecular genetic tools. Most advances in coral-Symbiodiniaceae mutualism have come from studies on isolated Symbiodiniaceae. Some research uses the easily cultured anemone *Exaiptasia* as a coral model. Recently, efforts to study natural coral systems have increased, thanks to the feasibility of sample collection and metatranscriptomic or metaproteomic analyses.

Mollusks-Alga Symbiosis

The best-known examples of mollusks with symbiotic algae include certain marine bivalves and a gastropod lineage. Giant clams (subfamily Tridacninae) and heart cockles (subfamily Fraginae) harbor symbiodiniacean dinoflagellates as endosymbionts. While giant clams exhibit relatively conservative shell forms, photosymbiotic fragines display more anatomical variability, adapting to their lifestyles with features such as flattened shells, leafy mantle extensions, and lens-like microstructural structures (Li et al. 2020).

Herbivorous sacoglossan sea slugs are oligophagous or monophagous, feeding suctorially on siphonaceous, coenocytic algae by puncturing cell walls and extracting the contents (Middlebrooks et al. 2014). They retain the chloroplasts of the prey algae they ingest, storing them inside specialized cells of the digestive tubules. Strictly speaking, these sea slugs are kleptoplastidic rather than symbiotic. Thylakoid protein maintenance, chloroplast gene expression, and photosynthesis can continue for months without the algal nucleus (Mujer et al. 1996, Green et al. 2000). For instance, *Elysia chlorotica* retains the chloroplasts of *Vaucheria litorea*, which it ingests and uses as photosynthetic factories (Kawaguti and Yamasu 1965, Rumpho et al. 2000). *E. chlorotica* holds one of the longest kleptoplast retention durations among sea slugs, lasting approximately 9–10 months, by efficiently incorporating and maintaining functional plastids from *V. litorea*, which possesses inherently robust chloroplasts (Green et al. 2000, Chan et al. 2018). Remarkably, some algal nuclear genes have been transferred to the genome of *E. chlorotica*, suggesting that the host has gained some control over chloroplast function (Rumpho et al. 2008). Two other sea slugs, *E. timida* and *Plakobranchus ocellatus*, do not seem to have acquired such control (Wägele et al. 2011).

E. crispata (also known as the lettuce sea slug) is found along the northern coast of South America, throughout the Caribbean Sea, and up to the Florida Keys (Cartaxana et al. 2023). It can retain chloroplasts for 3–4 months after feeding. In the wild, they feed on *Caulerpa mexicana*, *C. verticillata*, or *Penicillus capitatus*, depending on their habitat location (Middlebrooks et al. 2014). Consequently, the algal sources of sequestered chloroplasts in *E. crispata* vary by location. These prey and chloroplast source algae are often not the dominant species in the sea slug's habitats (Middlebrooks et al. 2014). The longevity of the retained plastids depends on the source algal species (Cartaxana et al. 2023).

Vertebrate-Alga Symbiosis

Salamanders are the only vertebrates known to form symbiosis with algae inside their tissues (Rumpho et al. 2011). Salamander eggs contain algae and are laid in ponds where algae are present. If laid in autoclaved water without algae, the eggs have no algae (Bishop 2022). Over time, the eggs accumulate algae, which become distributed throughout the embryo during development. Nutrients regulate the expression of symbiosis-related genes. For instance, provided L-glutamate, N transporter genes are turned off. When P is depleted, the DIP transporter gene is turned off. Inside the animal, algae conduct fermentation and express metabolites not found in free-living algae. These metabolites may vary between tissues or regions of the salamander. Algae may attach to sperm to hitchhike into the egg.

Algae As Hosts

Alga-Bacterium Symbioses

Bacteria have been observed inside many types of algae (Maire et al. 2021 and references therein), indicating the prevalent existence of intracellular interactions. In these symbiotic relationships, bacteria play many roles, including as sources of vitamin B12 (Croft et al. 2005, Kazamia et al. 2012, Xie et al. 2013). Euglenoid algae are commonly known to possess intracellular bacteria (Surek and Melkonian 1983), even localized within the nucleus (Leedale 1969). An *Eutreptiella* species isolated from Long Island Sound was also found to contain endosymbiotic bacteria closely related to *Rickettsia* in addition to ectobiotic bacteria (Kuo and Lin 2013). Experimental evidence from this study suggests that the ectobiotic bacteria serve as a source of vitamins, but the role of *Rickettsia*-like bacteria inside euglenids remains to be unraveled.

A recent metagenomic study revealed that a bacterium from the order Rhizobiales, known for symbiosis with legumes, lived symbiotically inside a pennate diatom (*Haslea* sp.) and fixed N_2 in the tropical North Atlantic (Tschitschko et al. 2024). Named *Candidatus Tectiglobus diatomicola*, this alphaproteobacterium harbored a nitrogenase gene cluster (*nif*) similar to that of a *Vibrio*-related bacterium (gammaproteobacterium). The mismatch of the bacterium identity and *nif* gene cluster suggests horizontal gene transfer. The localization of *Ca. T. diatomicola* inside the diatom host was demonstrated using fluorescence *in situ* hybridization with a 16S rRNA gene probe. Further N_2-fixation measurements suggested that the level of N_2 fixation of the *Ca. T. diatomicola-Haslea* symbiosis is comparable to that of diazotrophic cyanobacteria. By mining global ocean metagenomic databases, the authors found that this newly recognized bacterium, and potentially its close relative *Ca. T. profundi*, are widespread in the ocean and may represent a missing source of fixed nitrogen in the global ocean nitrogen budget. Interestingly, *Rhizobium* bacteria have also been found in the culture of the dinoflagellate *Gambierdiscus balechii*, and a *nifH*-like gene was identified, but the sequence shared a higher similarity to bchX, rendering it unlikely to fix N_2 (Wu et al. 2022).

Molecular Bases of Phytosymbioses Gleaned from Genomes

The field of aquatic phytosymbiosis research has advanced rapidly in the last decade thanks to technological improvements. However, the molecular mechanisms underlying the initiation, maintenance, and breakdown of symbiosis remain poorly understood. This section aims to glean insights from existing omic studies to encourage further research in these areas.

One approach to understanding how organisms establish symbiotic relationships is through comparative genomic analyses of symbiotic versus non-symbiotic organisms. The initiation of symbiosis can be likened to the onset of pathogen infection (Weis et al. 2008), involving microbial invasion into the target organism. A comparative genomic analysis involving symbiodiniacean algae, other algae, the land plant *A. thaliana*, and the parasite *Plasmodium falciparum* (Lin et al. 2015) identified genes unique to *Plasmodium* and Symbiodiniaceae, such as those related to cell recognition (e.g., merozoite surface protein 1) or unknown function (uncharacterized protein PF11_0213, Protein dpy-19 homolog 1). Merozoite surface proteins in apicomplexans are crucial for red blood cell invasion (Beeson et al. 2016). The study also revealed genes unique to Symbiodiniaceae, including nodulation proteins U, nolO, and Z, host cell factor 2, and Fibronectin type II domain-containing protein 3C1. In addition, comparative genomic studies have highlighted the importance of protein secretion systems and horizontal gene transfer in symbioses between bacteria and other organisms (Sandkvist 2001, Cianciotto and White 2017, Epstein and Tiffin 2021).

The review by Davy et al. (2012) provides a mechanistic framework for examining the critical steps of symbiosis in corals and other systems. Symbiont recruitment by hosts may occur through mechanisms similar to microbial infection or food ingestion via endocytosis in animal models. As hosts typically select specific symbionts, the decision to retain, expel, or digest symbiont cells likely involves an immune system-mediated winnowing step. Maintaining the symbiotic relationship requires sustained immune suppression, while the breakdown may result from immune suppression failure and symbiont rejection. Based on this concept, symbiosis undergoes six phases: (i) initial host-symbiont contact, (ii) symbiont engulfment, (iii) dynamic intracellular sorting of symbionts, (iv) symbiont proliferation within host tissues, (v) dynamic stability, and (vi) dysfunction and breakdown. From a molecular perspective, these phases can be categorized into three major stages: initiation, maintenance, and breakdown.

Initiation of Symbiosis

Host Recognition

The first critical step in symbiosis establishment is the recognition of the symbiont cells by the host organism through the "matching" of the Microbe-Associated Molecular Pattern (MAMP) on the symbiont cells with the MAMP recognition receptor (PRR) of the host (Fig. 20.6). MAMP molecules include lectins, complement receptor 3, and scavenger receptors (Davy et al. 2012). Lectins are carbohydrate-binding proteins that specifically bind the sugar groups of other molecules. Glycans on the symbiont cell surface are critical MAMPs recognized by the host through lectins (Wood-Charlson et al. 2006). The mannose branch length of glycans, which varies among symbiodiniacean dinoflagellates (Lin et al. 2015), may contribute to symbiont specificity.

In *Pocillopora damicornis*, galectin-1 (GLT-1) binds the cell surface of all five different Symbiodiniaceae species examined, but with different binding affinities and spatial pattern (Wang et al. 2021). For *Symbiodinium microadriaticum* and *Breviolum minutum*, binding sites are discrete loci on the cell surface, but for *Fugacium kawagutii* and *Cladocopium goreaui*, binding is widespread on the whole cell surface. While GLT-1 exhibits the potential to cause agglutination in *F. kawagutii*, there is evident preferential binding to *C. goreaui*. Furthermore, the binding affinity is influenced by temperature (Wang et al. 2021). PdGLT-1 had high binding affinity to *C. goreaui* and *D. trenchii*

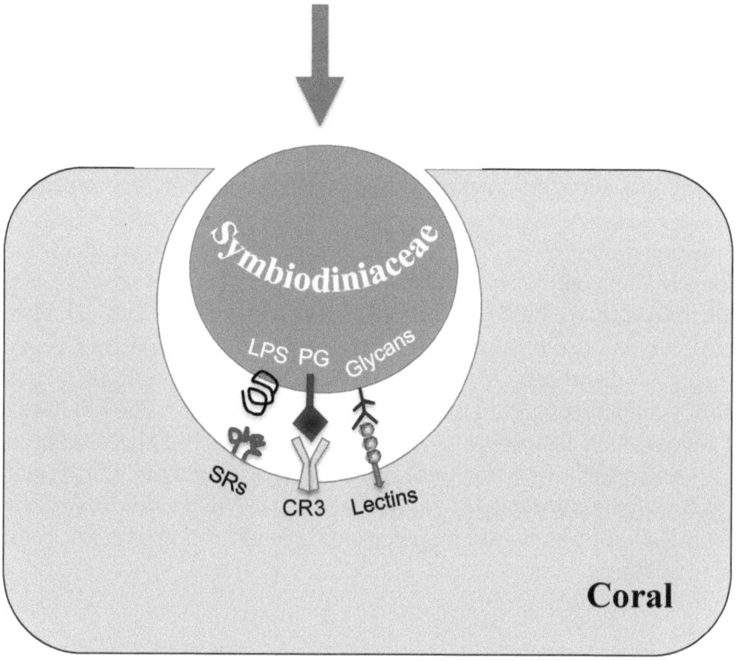

Figure 20.6. Schematic of host-symbiont recognition mechanisms.

even under thermal stress (30°C). These results indicate that corals with high expression of PdGLT-1 might have a higher chance of survival in the gradually warming ocean.

Experimental evidence supports the importance of glycan characteristics in symbiosis development. In *Fungia scutaria* larvae, successful recruitment of symbiodiniacean symbionts required intact cell surfaces, and enzymatic modification of the algal cell surface reduced infection rates (Wood-Charlson et al. 2006). In *Acropora tenuis*, exposure to ITS-1 type C1 symbionts resulted in 20–70% higher infection success on the coral larvae and five-fold higher Symbiodiniaceae abundance than ITS-1 type D, with enzyme treatment of C1 cells to modify cell surface 40–100 kDa glycome promoting retention (Bay et al. 2011). This suggests that cell surface recognition modulates post-infection sorting for suitable endosymbionts.

Using *Exaiptasia diaphana* as a coral model, Tortorelli et al. (2022) examined glycan-lectin interactions in host-symbiont recognition with three symbiodiniacean species. They found differences in cell surface glycomes among the algal species, and trypsin treatment affected recruitment rates, indicating a crucial role of protein-based moieties in symbiont recognition. D-galactose residues were important for cnidarian-dinoflagellate symbiosis, and L-fucose, D-xylose, and D-galacturonic acid were potentially important in early mutualism steps.

Notably, trypsin treatment had species-specific effects on infection success, reducing *Breviolum minutum* uptake but promoting *Cladopodium goreaui* uptake by *E. diaphana* (Tortorelli et al. 2022). Similar differential effects were reported in *Fungia scutaria* and *Acropora tenuis* (Wood-Charlson et al. 2006, Bay et al. 2011).

Investigating the role of photosynthesis in symbiont infection, Jinkerson et al. (2022) found that infection of sea anemone (*Aiptasia*), coral (*Acropora*), and jellyfish (*Cassiopea*) occurred even without photosynthesis. Wild-type *B. minutum* was maintained within *Aiptasia* after six months in darkness. This "photosynthesis-independence" contrasts with studies showing that photosynthate presence triggers infection in *Fugia scutaria* (Hagedorn et al. 2015). The Arrested Phagosome Hypothesis suggests that photosynthate release by symbionts mimics prey digestion, allowing ingested algae to evade the host's immune system (Malcom and April 2012). The disparity in findings may be due to interspecific differences.

Infection

The molecular mechanisms underpinning symbiont entry into the host are among the least understood aspects of symbiosis biology. Endocytosis may play a role, as it is an enriched GO term (GO:0006897) in symbiodiniacean genomes, such as that of *Fugacium kawagutii* (Lin et al. 2015). However, endocytosis is also involved in phagotrophy, and Symbiodiniaceae have been shown to ingest food, making it challenging to attribute the endocytosis machinery specifically to symbiont infection.

Circumsporozoite protein (CSP) is a secreted membrane protein that signals for the invasion of hosts by sporozoites of apicomplexan parasites. A BLAST analysis revealed the presence of a CSP-like protein gene in various symbiodiniacean taxa, including clades A, B, C, D, E, and F (Fig. 20.7). A similar protein gene was also identified in the non-symbiodiacean relative *Polarella glacialis*, but with very high sequence divergence. The *F. kawagutii* genome contains 16 CSP-coding genes and an additional forkhead domain-containing CSP-like gene (Fkaw34466). The forkhead domain belongs to a family of transcription factors. *In silico* analysis predicts that this Forkhead/DinoCSP may regulate gene sets responsible for vesicle-mediated transport, potentially influencing the secretion of CSP-like proteins. In the chlorophyte symbiont of corals, *Symbiochloris hainanensis*, a CSP-like protein was identified and shown to be upregulated by acidification alone or combined with heat stress (Gong et al. 2020).

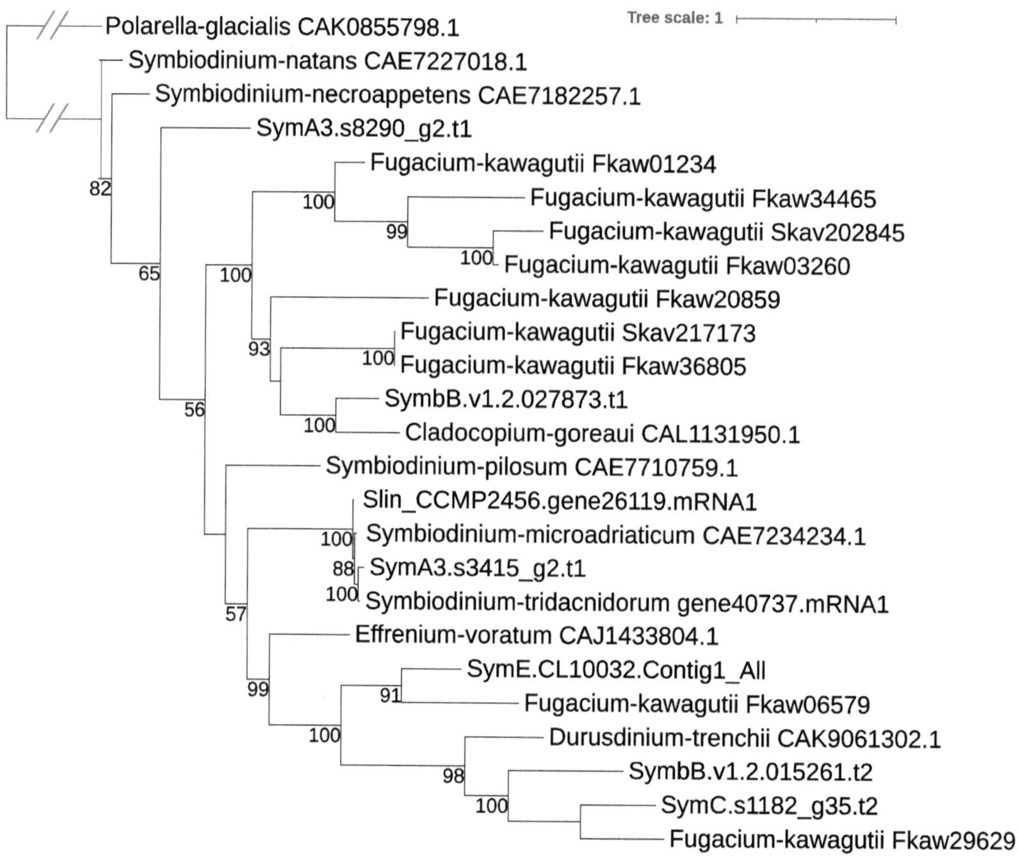

Figure 20.7. Circumsporozoite surface protein (CSP) like proteins in Symbiodiniaceae. BLAST hit in *Polarella glacialis* is weak (with very long branch) but presented as outgroup.

The secretion system has been shown to be important for symbiosis involving microbes. A comparative genomic study on light organ bacteria of fishes revealed that these symbiotic bacteria commonly possess the 12-protein type II secretion (T2S) system (Gould and Henderson 2023). Six of these proteins (gspC, gspD, gspJ, gspK, gspL, and gspM) differ among bacterial species, suggesting divergence among co-existing symbionts. Protein secretion is a major virulence mechanism in bacterial infection, and T2S is one such system encoded by at least 12 genes, potentially up to 15, named from A to O (Sandkvist 2001). In symbiotic gram-negative bacteria, the T2S system secretes proteins into the host environment, likely exporting virulence factors and playing a role in bacteria-eukaryote mutualism (Evans et al. 2008, Cianciotto and White 2017).

Host Immune Suppression or Evasion

A recent study using *Aiptasia* as the model revealed that the establishment of cnidaria-Symbiodiniaceae mutualism was mediated by symbionts evading vomocytosis (spitting out) by host cell through immune suppression (Jacobovitz et al. 2021). The work indicated that uptake of microalgae was non-selective, but non-symbiotic species were expelled through vomocytosis, whereas the symbiotic cells induced host cell innate immune suppression and form a lysosome-associated membrane protein 1-positive niche.

Macrophage Migration Inhibitory Factors (MIF) are important cytokines and chemokines for vertebrate immune system, involved in various diseases and pathological disorders (Calandra and Roger 2003). MIFs are multifunctional proteins and can control major cell functions such as migration, proliferation, and p53-mediated apoptosis (Jaouannet et al. 2020). Cytokines play a crucial role in innate immunity and cell signaling in mammals and plants. They are released upon binding of Pathogen-Associated Molecular Patterns (PAMPs) to Toll-Like Receptors (TLRs) or Pattern Recognition Receptors (PRRs) (Calandra and Roger 2003, Choi et al. 2010, Merk et al. 2012, Panstruga et al. 2015, Gruner et al. 2021). From the genomes of Symbiodiniaceae and MMETSP dataset, my blast analysis detected MIF homolog genes in Symbiodiniaceae species as well as other dinoflagellates. Pfam search of the sequences confirmed that they contained MIF functional domain (Lin 2024). The role of symbiodiniacean MIF homologs, specifically in suppressing immune rejection by coral hosts, requires further investigation.

Maintenance of Symbiosis

Nutrient Exchanges

Transport of nutrients and photosynthate translocation is crucial to the maintenance of mutualism. Comparative genome analyses on *Fugacium kawagutii*, *Breviolum minutum*, and *Symbiodinium microadriaticum* with non-symbiotic algae revealed significant expansion of gene families involved in uptake of nutrients including N, C, P, vitamins, and metal ions (Lin et al. 2015, Aranda et al. 2016). As shown in Fig. 20.8, these include transporters for uptake of various forms of N and P and a host of vitamins and metal ions as well as transporters for translocation of sugar and carbohydrates. For instance, ammonium transporters are encoded by 38 genes in *F. kawagutii* and 25 genes in *B. minutum* but only 0–13 genes in diatoms as well as green and red algae (Table S21 in Lin et al. 2015).

As shown in Chapter 3 (Fig. 3.5), retrotransposition contributed to the gene family expansion as two major bouts of this genomic innovation coincided with emergence of coral-Symbiodiniaceae mutualism and radiation of Symbiodiniaceae, respectively, and impacted nutrient transport and photosynthesis, respectively (Song et al. 2017). For *B. minutum*, a transcriptomic study compared freshly isolated the population from *Exaiptasia diaphana* with a free-living population (Maor-Landaw et al. 2020). Many of the of 380 DEGs were related to photosynthesis and morphologies, and many transporters were upregulated, including transporters of ammonium, aromatic amino acids, organic cations, solute (carrier family 22), monocarboxylate, zinc, GABA, and glycine and

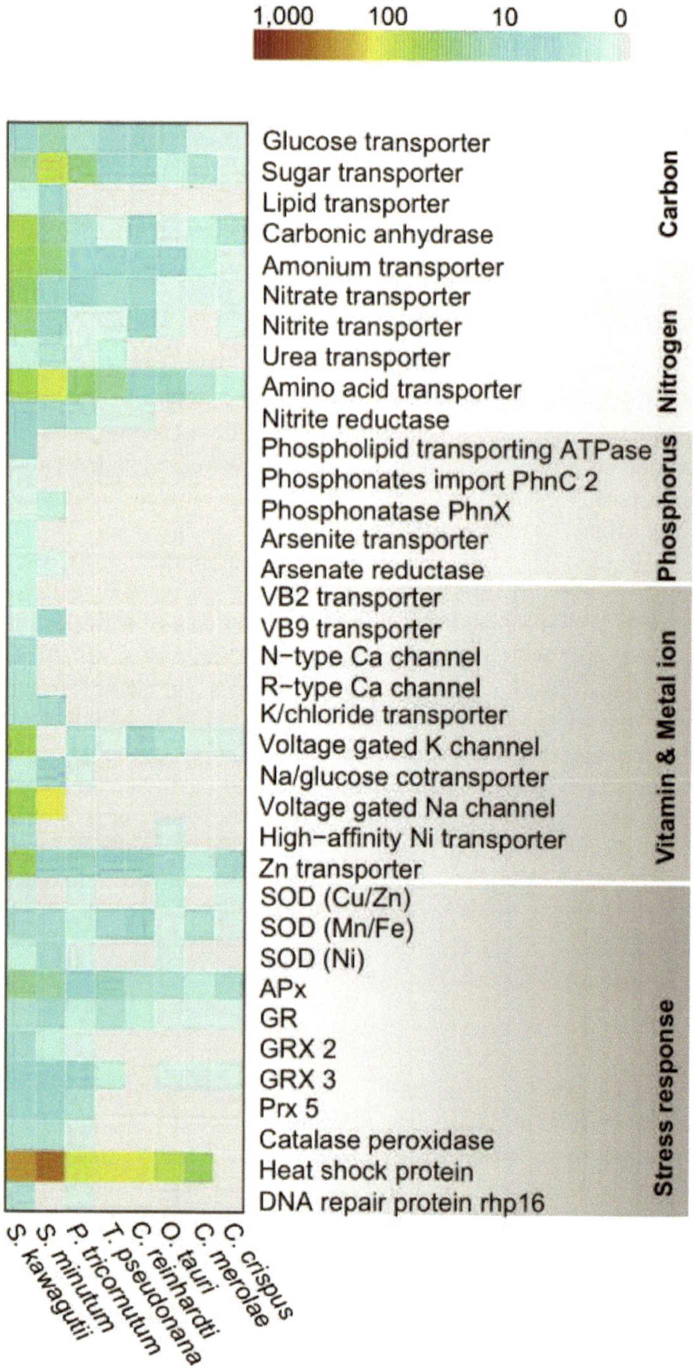

Figure 20.8. Remarkable expansion of gene families involved in nutrient exchanges and anti-stress protection in symbiodiniacean dinoflagellates compared to other lineages of algae. Reproduced with permission from Lin et al. (2015).

Calcium-transporting ATPase, and one of the potential glucose exporters identified, SWEET1, a member of the Sugars Will Eventually be Exported Transporters (SWEETS).

Ammonium: Ancient Nutrient "Trade" Currency in Symbiosis. Although symbiosis involves complex interactions between the host and the symbiont, sharing of nutrients is a common thread.

Nitrogen (N) nutrients, especially ammonium and amino acids, are the best recognized currency of "trade" between the symbiotic partners. Evolutionarily, ammonium uptake and assimilation prevailed all algae, including cyanobacteria and eukaryotic algae from primary endosymbiosis and secondary endosymbiosis. As discussed in Chapter 4, some cyanobacteria lack nitrate transporters and nitrate reductase, but have ammonium transporters in all examined genomes.

The classical example is the symbiosis of legumes and other flowering plants with root nodular bacteria *Rhizobium*, which fixes N_2 into NH_4^+ and provides the host plant with necessary nitrogen. The oceanic counterpart is the symbioses between diazotrophic cyanobacteria and eukaryotic algae such as diatoms and haptophytes. Heterocystous cyanobacteria *Richelia* spp. live as endosymbionts within centric and pennate diatoms (Villareal 1992, Schvarcz et al. 2022), thriving in oligotrophic oceanic waters (Pierella Karlusich et al. 2021).

The mutualistic relationship between the bryophyte *Anthoceros punctatus* and *Nostoc punctiforme* requires genes involved in nitrogen control and heterocyst differentiation initiation (Wong and Meeks 2002). A strain of the *N. punctiforme* with a mutation in the global nitrogen regulator ntcA could not infect the hornwort, but when complemented with one or more copies of ntcA, the mutant strain restored its ability to infect *A. punctatus* at a similar frequency as the wild type.

The use of NH_4 as the trade currency extends beyond the alga-diazotroph partners. In the *Mesodinium rubrum* temporary symbiotic relationship with the cryptophyte *Teleaulax amphioxeia*, NH_4^+ transporter was the most actively expressed nutrient transporter during a red-tide bloom (Qiu et al. 2016). In *Dinophysis acuminata*, NH_4^+ was the most effective nutrient among N compounds (NH_4^+, nitrate, urea) influencing growth, with the ammonium-grown cultures showing the highest uptake rate (Hattenrath-Lehmann et al. 2021).

Another example is the green population of *Noctiluca scintillans*, which harbors endosymbiont cells from the chlorophyte group *Protoeuglena noctilucae*. Although this bioluminescent and heterotrophic dinoflagellate is globally distributed and usually causes red-tide blooms, the green population occurs primarily in the tropical oceans. During green-tide blooms, *Noctiluca* not only provides ammonium to support the growth of *P. noctilucae*, but also releases copious amounts of ammonium and phosphate to stimulate prey algal growth in the ambient water (Luo et al. 2022).

In coral-Symbiodiniaceae symbiosis, the photosymbionts have evolved an expanded gene family of NH_4^+ transporters, with *F. kawagutii* and *B. minutum* having significantly more NH_4^+ transporter genes compared to other algae (Table S21 in Lin et al. 2015). Nutrient transporters may be regulated by microRNA mechanisms (Fig. 20.9).

Glycerol Versus Glucose: The Form of Photosynthate Translocated to the Host. The photosynthates of symbiotic algae are transferred to the host mainly in the form of glycerol or glucose, depending on the species of the host and the symbiont. Studies from the 1960s to the 1990s indicated that a "host factor" enhances the release of photosynthates from the symbionts, with glycerol being the main product released (Muscatine 1967, Trench 1971, Trench et al. 1981, Hinde 1983, Sutton and Hoegh-Guldberg 1990). However, many studies report that the transferred photosynthates are mainly glucose in intact cnidarians. For example, Burriesci et al. (2012) found that *Aiptasia* line H1 with clade B Symbiodiniaceae as symbionts primarily transferred glucose.

A study using ^{13}C- and ^{14}C-bicarbonate to trace the allocation of photosynthetically fixed carbon in the giant clam *Tridacna crocea* (Ishikura et al. 1999) showed that in the presence of clam mantle tissue homogenate (MH), Symbiodiniaceae cells isolated from the clam released up to 5.6% of the total fixed ^{14}C, mostly as glycerol, whereas the major intracellular photosynthate was glucose. In intact *T. crocea* incubated in seawater containing ^{14}C-bicarbonate, 46–80% of the fixed ^{14}C was translocated from the symbionts to the host. Again, isolated Symbiodiniaceae incubated with MH released mostly glycerol, whereas those in the intact mantle released glucose. In a study on *Fungia scutaria* (Hagedorn et al. 2015), it was found that symbiodiniacean symbionts freshly isolated from the coral constantly released trehalose, but released glycerol only in the presence of coral tissue.

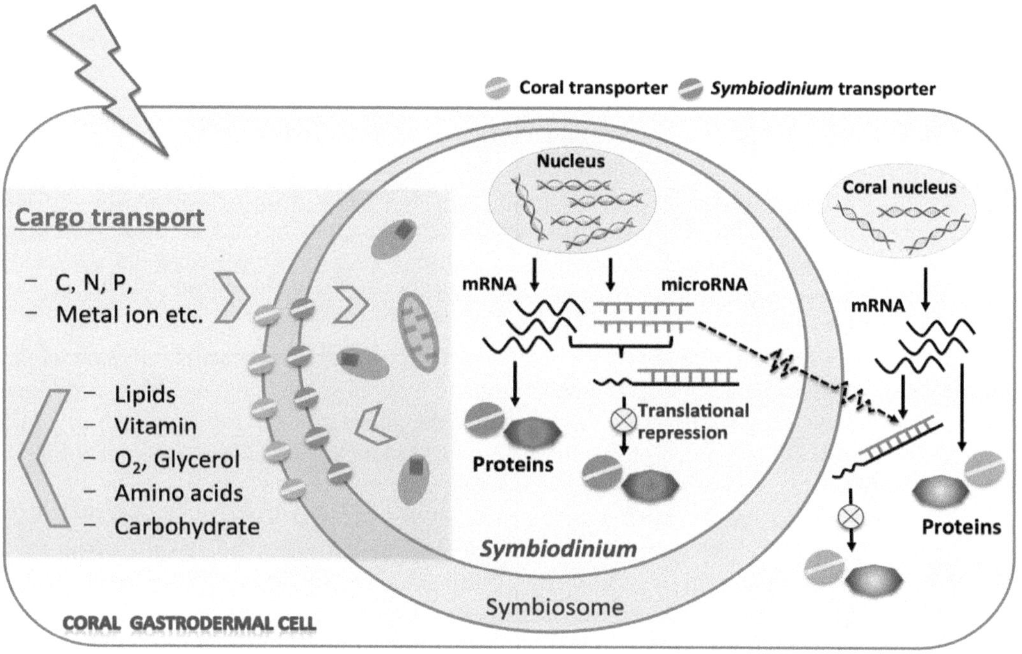

Figure 20.9. Potential molecular mechanism of symbiosis through miRNA.

The translocation of these two carbohydrates relies on different transporters. Glucose transporters include importers and exporters. SWEETs are exporters, with most plants having roughly 20 homologs, some localized in the plasma membrane (SweetTrac1) and others in the vacuole (SweetTrac2) (Gwon et al. 2023). SWEETs have been identified in *Breviolum minutum*, with three out of seven transcripts upregulated when the dinoflagellate was supplemented with glucose (Xiang et al. 2018). In a transcriptomic study comparing the *in hospite* with the free-living *B. minutum*, 10 BmSWEET transcripts were identified, but only one, BmSweet1, was upregulated *in hospite* (Maor-Landaw et al. 2020). These findings suggest that symbiodiniacean symbionts may use SWEETs for both sugar uptake from the environment (especially in the free-living state) and sugar translocation to hosts in the symbiotic state. Given that some symbiodiniacean species can conduct phagotrophic nutrition (Jeong et al. 2012), sugar import is not unexpected.

Glycerol can be transported through glycerol transporters, known as aquaglyceroporins (GLPs), which belong to the Membrane Intrinsic Protein (MIP) family. MIPs are channels across cell membranes that allow the transport of small solutes and water in both plants and animals (Laloux et al. 2018). In mammalian cells, aquaporins (AQP) 3, 6, 7, 8, 9, 10, and 11 transport glycerol, whereas in plants, TIPs, NIPs, and XIPs are glycerol transporters (Laloux et al. 2018). The green alga *Chlamydomonas reinhardtii* possesses a MIP homolog (CrMIP1), which serves as a glycerol exporter (Anderca et al. 2004). A GLP homolog was identified in *Fugacium kawagutii* (Lin et al. 2015), but its function remains to be experimentally demonstrated. Whether GLPs are prevalent in coral symbionts also needs investigation.

The form of photosynthate provided by symbionts to coral hosts depends on the transporters in both partners and other influencing factors. For instance, *F. kawagutii* has both aquaglyceroporin and glucose transporters, but if the coral host associated with this symbiodiniacean dinoflagellate has only glucose transporters, only glucose can be provided (Lin et al. 2015). This biochemical complementarity likely plays a crucial role in determining the species specificity of symbiont recruitment. Photosynthate translocation may be under the control of the miRNA regulatory system (Fig. 20.9).

Symbiont Population Control by Host

A balanced allocation of photosynthates to meet the needs of both the algal endosymbiont's growth and the host's requirements is pivotal for maintaining symbiosis. When the endosymbiont population overgrows, the reserve of photosynthates available to the host decreases (Wooldridge 2013). This "selfish" behavior of the endosymbionts, often resulting from eutrophication, is detrimental to the host's health, leading to coral bleaching (Cunning and Baker 2013, DeCarlo et al. 2020, Wooldridge 2020). Hosts have evolved mechanisms to control their symbiont populations.

One major mechanism involves controlling nutrient delivery to the symbionts. Corals, for instance, possess and express nitrogen assimilation enzymes to compete with the endosymbionts for nutrient uptake and assimilation. Under normal (nutrient-poor) conditions, host corals express Nitrogen (N) nutrient transporters to take up N from the environment and presumably deliver it to the endosymbionts. However, the corals also express glutamine synthetase and Glutamine Oxoglutarate AminoTransferase (GOGAT, also known as glutamate synthase) to assimilate N, thereby reducing its availability to their symbionts. This control mechanism has been directly demonstrated in the mutualistic relationship between the sea anemone (*Exaiptasia pallida*) and Symbiodiniaceae (*Breviolum minutum*) (Xiang et al. 2020).

Another potential mechanism is for the host to stimulate photosynthesis but repress the cell proliferation of the endosymbionts, even under nutrient-rich conditions. This was recently demonstrated in the symbiotic relationship between the dinoflagellate *Noctiluca scintillans* and prasinophyte *Protoeuglena noctilucae* (Luo et al. 2022). Comparing metatranscriptomic data generated from a bloom of this green *Noctiluca* and transcriptomic data from an isolated *Protoeuglena noctilucae* culture indicated that the expression of photosynthesis genes was remarkably elevated *in hospite* compared to its free-living form. However, cell cycle genes of the symbiont were expressed at much lower levels *in hospite* than in the free-living form. These results suggest that the host exploits the symbiont for energy by stimulating its photosynthesis while repressing its population growth.

Photoprotection and Anti-Stress

In coral-Symbiodiniaceae mutualism, both partners contribute to photoprotection. Interestingly, the biosynthesis of "sunscreen," the Mycosporine-like Amino Acids (MAAs), is either "commissioned" to the symbionts or the coral host. Symbiodiniacean dinoflagellates from ancestral lineages (e.g., *Symbiodinium*, *Durusdinium*) possess the four core genes necessary for MAA synthesis. In other cases, the gene cluster responsible for MAA biosynthesis appears to have been transferred to the coral host, as evidenced by the clustering of coral MAA biosynthesis genes with homologs from dinoflagellates (Lin et al. 2015). The gene cluster appears to have been lost also in *Cladocopium* and *Breviolum* (Shoguchi et al. 2018, 2021).

Compared with non-symbiotic algae, symbiodiniacean species exhibit notable expansions of gene families involved in the synthesis of antioxidants such as superoxide dismutase (SOD), ascorbate peroxidase (APx), glutaredoxin (GRX), peroxiredoxin (PrxR), and catalase (CA) (Lin et al. 2015). These expansions are likely adaptations to the symbiotic lifestyle, which demands heightened photoprotection and oxidative stress management.

Symbiodiniaceae in symbiosis with corals may be more sensitive to environmental stress than their free-living counterparts. Comparative transcriptomic studies support this notion. Mohamed et al. (2019) analyzed *Cladocopium goreaui* responses to the switch from a free-living to a symbiotic lifestyle and found downregulation of genes related to stress response, immune activities, protein synthesis, and N-glycan biosynthesis. The downregulation of stress response genes suggests a less stressful environment inside the coral or reliance on host anti-stress capacity.

In another study, Bellantuono et al. (2019) compared the transcriptomes of *Durusdinium trenchii in hospite* and free-living states under normal and thermal stress conditions. They found distinct gene expression patterns, with lower general transcriptional activity under stable conditions *in hospite* but increased activity under thermal stress. In the free-living state, *D. trenchii* exhibited upregulation in cell signaling pathways and functions related to environmental response, as well as transcripts for mitosis, meiosis, and motility. However, *in hospite* cells showed increased transcriptional activity for photosynthesis, carbohydrate transport, chromatin modifications, and a disrupted circadian clock. Hyperthermal stress caused significant changes in transcriptional activity *in hospite*. These findings suggest that the host provides a stable environment for its algal symbionts under normal conditions, but the environment becomes more stressful for the algae during thermal stress.

These studies underscore the intricate balance of photoprotection and stress response in coral-Symbiodiniaceae mutualism, highlighting the roles of both partners in maintaining this delicate symbiotic relationship.

Breakdown of Symbiosis

Mutualism between corals and their dinoflagellate endosymbionts can break down if the exchange of nutrients is disrupted or if one partner produces harmful metabolites. Nutritional exchanges are susceptible to environmental stressors such as heat waves, ocean acidification, and nutrient pollution. Among these, temperature stress is a primary cause of coral bleaching (Hughes et al. 2017). Thermal stress, combined with strong irradiance, damages the photosystem, inhibits

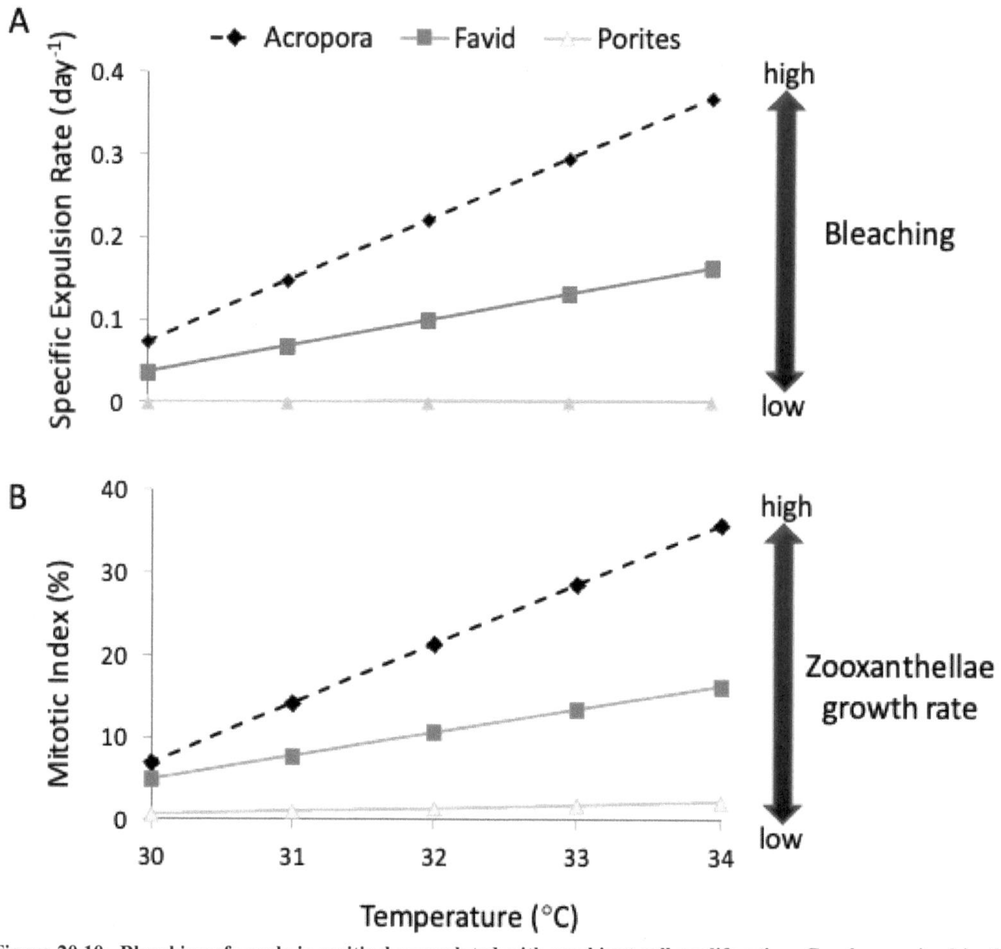

Figure 20.10. Bleaching of corals is positively correlated with symbiont cell proliferation. Corals examined in this study included *Acropora hyacinthus, Favites complanata*, and *Porites solida*. (A) Zooxanthellae expulsion rates (per day) for the 3 coral species. (B) Corresponding intracellular MI of the zooxanthellae within the 3 coral species. Reproduced with permission from Wooldrige (2013) under the Creative Commons CC BY 3.0 license.

photosynthesis, and induces the production and accumulation of reactive oxygen species (ROS), harming cell membranes and functions (Lesser 2010). The exact role of ROS—whether from the symbionts or the coral host—in triggering symbiont expulsion is not well understood.

As alluded to earlier (Wooldridge 2013), endosymbiont overgrowth, induced by higher temperatures and nutrient enrichment, could disrupt the mutualistic relationship (Fig. 20.10). Overpopulation of symbionts can lead to CO_2 limitation and increased ROS production, aside from decreased photosynthate translocation. To control symbiont populations, coral hosts regulate nutrient supply. For instance, corals possess nutrient assimilation enzymes like Glutamine Synthetase (GS) to compete with algal GS for ammonium. This control mechanism aligns with the hypothesis that coral bleaching occurs only when warming is combined with nutrient overloading (DeCarlo et al. 2020, Donovan et al. 2020). For example, in the Red Sea, a severe bleaching event occurred in 2015 due to high nutrient concentrations from stronger upwelling, although the temperature was not the warmest (DeCarlo et al. 2020).

Parasitic Behavior Under Environmental Stress

Some strains of Symbiodiniaceae (e.g., clade A) can act parasitically under certain conditions (Stat et al. 2008), and thermal stress can drive this shift from mutualism to parasitism (Sachs and Wilcox 2006). For instance, in *Orbicella faveolata*, higher temperatures with sufficient nutrient supply led to increased carbon fixation and nitrogen assimilation by the symbionts. However, the symbionts did not translocate proportionately more organic carbon to the host, instead using it for cell proliferation, reducing the holobiont's net primary production by 60% (Baker et al. 2018). Again, this relates to the partition of photosynthates between the host and the symbiont.

Using a non-targeted metabolomics approach, Williams et al. (2021) analyzed metabolite responses of *Montipora capitata* and *Pocillopora acuta* holobionts to heat stress. They found a variety of co-regulated dipeptides upregulated in heat-stressed holobionts, suggesting deep evolutionary origins and involvement in symbiosis.

Metabolic Changes and Thermal Tolerance

The trigger of bleaching may come from metabolites produced under stress. A multi-chemical omics study on *Durusdinium trenchii*, a thermoresistant species, exposed to heat stress (33°C), revealed changes in 53 analytes. Marked increases were noted in glutathione, xanthine, guanine, and tryptophan, some of which are important in stress response, while decreases were observed in nutritious compounds such as glucose, citric acid, glutamic acid, aminobutyric acid, and galacturonic acid (Matthews et al. 2024). Furthermore, indole-3-Acetic Acid (IAA) and tryptophan increased, raising a suspicion of potential harmful effect as a tryptophan-bacteria-IAA-algal proliferation link has been shown for diatoms (Amin et al. 2015). Heat stress also led to the release of more Biogenic Volatile Organic Compounds (BVOCs), including halogenated compounds, hydrocarbons, sulfur-containing compounds, and myo-inositol. As their increases appear to be unique to *D. trenchii*, they might contribute to this species' thermal tolerance (Matthews et al. 2024).

Symbionts significantly impact the thermal tolerance of the coral holobiont. Bleaching is proposed as a way to reshuffle symbionts, replacing thermally susceptible types with resistant ones (Baker 2001). Symbiodiniacean dinoflagellates can genetically adapt to temperature elevation within months (Huertas et al. 2011). Heat-selected symbionts have been shown to increase coral bleaching tolerance. Buerger et al. (2020) exposed 10 clonal strains of coral symbionts to elevated temperatures (31°C) for four years. After reintroduction into coral host larvae, three of the 10 heat-selected endosymbionts increased the holobionts' bleaching tolerance.

These findings illustrate the complex interplay between coral hosts and their symbionts in response to environmental stressors, highlighting the importance of nutrient regulation and metabolic adjustments in maintaining mutualistic relationships and enhancing thermal tolerance.

Molecular Basis of Kleptoplastidy

Kleptoplastidy, the process of stealing and retaining plastids, has evolved in diverse marine eukaryotes, including ciliates, dinoflagellates, foraminifera, saccoglossan gastropods (sea slugs), and flatworms (Cruz and Cartaxana 2022). These 'stolen' chloroplasts originate from various algal lineages, including cryptophytes, haptophytes, chlorophytes, and diatoms, as discussed in Chapter 2.

A critical aspect of kleptoplastidy is how the host regulates the function of the kleptoplastid. In algae, the majority of chloroplast genes are encoded in the nuclear genome. Key proteins essential for photosynthesis, such as photosystem subunits and light-harvesting antennae, are synthesized in the cytoplasm and then imported into the chloroplast (Nowack and Weber 2018). Kleptoplast longevity has been hypothesized to depend on the Horizontal Gene Transfer (HGT) of algal nuclear-encoded genes to the nuclear genome of kleptoplastic host, allowing the subsequent targeting of encoded proteins to the kleptoplast (Rumpho et al. 2011).

In *Dinophysis acuminata*, which holds cryptophyte-originated plastids, transcriptomic studies have indicated that the dinoflagellate nuclear genome encodes three genes involved in photosynthesis, including a photosystem II subunit (psbU), plastid ferredoxin (petF), and a gene encoding an auxiliary light harvesting protein (LI818) (Wisecaver and Hackett 2010). Using DinoSL as the selecting primer to enrich dinoflagellate transcripts, an independent transcriptomic study (Hattenrath-Lehmann et al. 2021) provided data verifying the dinoflagellate origin of these genes and uncovering additional genes such as Oxygen Evolving Enhancer I (OEE1) gene.

A study on a heterotrophic Kareniaceae species from the Antarctic Ross Sea suggests that host control of kleptoplastids has stemmed from various stages of the dinoflagellate evolutionary history (Hehenberger et al. 2019). This dinoflagellate retains a relict peridinin plastid with genes encoding enzymes for the isoprenoid, heme, and iron-sulfur cluster biosynthesis pathways. Their dinoflagellate origin is verified by the presence of DinoSL at the 5' end of the transcripts. This dinoflagellate also harbors genes of haptophyte origin, as do Kareniaceae dinoflagellates with tertiary plastids of haptophyte origin. More strikingly, some of these haptophyte-originated genes appear to be derived from a different haptophyte lineage than where the kleptoplastid is originated.

Kleptoplastidic foraminifera appear to acquire chloroplasts only from diatoms (Leutenegger 1984, Bernhard and Bowser 1999). Foram's kleptoplastidy occurs in diverse environments, ranging from deep oceans to transitional waters, from oxic to anoxic conditions, and from photic to aphotic habitats (Bernhard and Bowser 1999, Pińko et al. 2023). Some species retain kleptoplastids for only a few days, necessitating constant replenishment of their chloroplast reservoir, while others can keep kleptoplasts functional for several months (Lopez 1979, Grzymski et al. 2002, Jauffrais et al. 2016). Whether these kleptoplastic forams harbor various numbers of photosynthesis related genes to keep their kleptoplastids functional is unclear.

The prevalence of kleptoplastidy points to a highly diverse functionality of kleptoplastids, extending the host's physiological flexibility. Aside from providing photosynthates, the kleptoplastids may aid in maintaining high intracellular pH of the host (Tsuchiya et al. 2020). In addition to photosynthetic activity in photic environments, kleptoplastids in aphotic habitats have been shown to be active in sulfur and nitrogen uptake (Lopez 1979, Grzymski et al. 2002, Jauffrais et al. 2016). This functional diversity allows kleptoplastidic organisms to thrive in a wide range of environmental conditions.

21

Genomics of Algae-Microbes Interactions

Diverse Types of Alga-Microbe Interactions

Microbes (bacteria, archaea, viruses) are the most abundant life forms in the ocean. Bacteria and archaea metabolize organic matter produced by algae. In addition, bacteria form various functional relationships with algae. Viruses infect algae in lytic and nonlytic fashions. To avoid redundancy, the *microbes* here refer to non-diazotrophic bacteria, and symbiosis between algae and bacteria treated in Chapter 20 will not be repeated here. Fungi are less studied in terms of their relationships with algae but because they are prevalent and some helpful insights are emerging, they deserve discussion. The interaction between algae and microbes encompasses both physically and functionally intimate relationships on one end and physically distant ecological linkages on the other.

Intimate relationships include infection, attachment to the surface of algae or reside in the phycosphere (epiphytic). In no-contact ecological interactions, algae photosynthetically produce dissolved organic carbon available to bacteria, while bacteria remineralize sulfur, nitrogen and phosphorous to support growth of algae. Besides, heterotrophic bacteria can supply B vitamins as organic cofactors for algae or produce siderophores to bind iron, rendering iron available to algae. Furthermore, the functional interactions between algae and microbes can occur through metabolites for both the intimately associated and co-occurring partners. For instance, bacteria may produce metabolites that stimulate or inhibit the growth of algae. Many algal taxa are known to be infected by parasites or viruses, resulting in death or cell lysis. However, recent studies show that some viral infection is nonlytic and appears to be co-adapted with the host for extended infection. Current understanding of the regulation of the algae-microbes interactions will be discussed below.

Algae-Parasites Relationships

There exist two types of parasitism involving algae. Parasitoids were initially defined as insects whose larvae infect (inside or attached on the surface of) a host and eventually kill the host, typically before the host produces offspring. The definition has been expanded to include any eukaryote that resides in or on the surface of its host as eggs or larvae, develops, and eventually kills the host. In contrast, a canonical parasite lives in or on the host at the expense of the host's fitness but does not always kill it (Godfray 1994). Here, both groups are collectively called parasites or parasitic organisms for convenience unless specified otherwise.

Documented parasites of algae are mainly from the phylogenetic groups Amoeba, Fungus, Stramenopiles, Perkinsids, and Dinoflagellata (Table 21.1; Carney and Lane 2014). They infect

Table 21.1. Parasites of algae.

Group	Species	Targets	References
Amoebae	*Leptophyrs vorax*	*Closterium* sp.	Hess et al. 2012
Amoebae	*Vampyrella* sp.	Various algae	Hess et al. 2012
Aphelid/Fungus	*Amoeboaphelidium protococcarum*	*Semedesmus* sp.	Latcher et al. 2013, Gromov and Mamkaeva 1970
Chytrid/Fungus	*Paraphysoderma sedebokerensis*	*Haematococcus pluvalis*	Hoffman et al. 2008, Gutman et al. 2009
Chytrid/Fungus	*Rhizophydium algavorum*	Various algae	Gromov et al. 1999
Chytrid/Fungus	*Chytromyces* sp. and *Zygorhizidium* sp.	*Asterionella formosa* and various other diatoms	Canter and Jaworski 1979, Beakes et al. 1988, Bruning 1991, Grami et al. 2011, Kagami et al. 2011
Chytrid/Fungus	*Entoplyctis apiculata*	*Chlamydomonas* sp.	Shin et al. 2001
Chytrid/Fungus	*Phlyctidium scenedesmi*	*Secenedesmus* sp.	Fott 1967, Ilkov 1975
Chytrid/Fungus	*Dinomyces*	Dinoflagellates	Lepelletier et al. 2014
Chytrid/Fungus	*Paradinomyces triforaminorum*	*Kryptoperidinium folicaceum*	Reñé et al. 2022
Fungus	*Aspergillus pseudoglaucus*	*Prorocentrum lima*	Berry et al. 2022
Labyrinthulids/ Stramenopiles	*Labyrinthula Cienk*	Green algae, Cyanobacteria	Raghukumar 1986, 1987
Oomycete/ Stramonopiles	*Ectrogella* sp.	*Pseudo-nitzschia*	Hanic et al. 2009
Oomycete/ Stramenopiles	*Lagenisma coscinodisci*	*Coscinodiscus centralis*	Gotelli 1971
Oomycete/ Stramenopiles	*Sirolpidium bryopsidis*	Green algae	Buaya et al., 2021
Perkinsids	*Parvilucifera*	Dinoflagellates	Norén et al. 1999
Dinoflagellates	*Amoebophrya* sp.	Dinoflagellates	Coats and Park 2002, Guillou et al. 2008, Chambouvet et al. 2011
Dinoflagellates	*Amoebophrya ceratii*	Dinoflagellates	Chambouvet et al. 2008

cyanobacteria, green algae (chlorophytes and charophytes), diatoms, dinoflagellates, and other algal lineages.

Species from the perkinsid genus *Parvilucifera* infect a wide range of dinoflagellates (Norén et al. 1999). There are currently four recognized species in this genus, *P. infectans* being the type, *P. corolla*, *P. multicavata*, and *P. rostrata*. The life cycle of the generalist parasitoids starts with nanoflagellate infective stages (trophocyte) that invade dinoflagellate host cells, which lose motility and sink, followed by sporocyte, which produces and releases hundreds of zoospores (Garces et al. 2013, Jephcott et al. 2016). The infection of *P. rostrata* on two bloom-forming dinoflagellates, *Alexandrium minutum* and *Heterocapsa triquestra*, was shown to exhibit different modes of temperature response (Schmitt et al. 2022). For *A. minutum*, the infectivity of *P. rostrata* was higher at higher temperatures (higher infections at 20–22°C), the same trend as the dinoflagellate's growth rate; for *H. triquestra*, the infectivity showed a unimodal trend (higher infections at 19°C), also showing the same trend as the host's growth rate (maximal growth at 15–18°C).

As indicated in Chapter 8, two basal groups of dinoflagellates are parasitic. Of these, species from the order Syndiniales, particularly the genus *Amoebophrya* from the Amoebophryaceae, are common parasites of many dinoflagellate species (Coats and Park 2002). *Amoebophrya* is highly diverse and widely distributed, often dominating plankton assemblages. Its life cycle consists of an infective dispersal stage (dinospore), an intracellular growth stage (trophont), and an extracellular reproductive stage (vermiform) (Cachon and Cachon 1987). *Amoebophrya* infection is specific, with a genotype infecting a species of dinoflagellate; as such, HAB species-targeting genotypes can control Harmful Algal Blooms (Chambouvet et al. 2008). Some host dinoflagellates have evolved defense mechanisms against *Amoebophrya*. Target dinoflagellates can be induced by *Amoebophrya* infection to produce metabolites that suppress or kill the parasites (Long et al. 2021).

In adapting to the parasitic lifestyle, *Amoebophrya* genomes are highly reduced, with cryptic genomes (John et al. 2019, Kayal and Smith 2021). Some insights can be drawn from transcriptomic analyses into what is involved in the initiation and maturation of infection (Table 21.2). Farhat et al. (2018) reported transcriptomic changes in strains AT120 and A20 of *A. ceratii* during infection on *Scrippsiella acuminata*, whereas Lu et al. (2014) documented changes in strain AT5.2 during infection on *Alexandrium tamarense*. As summarized in Table 21.2, marked differences were observed between the two studies, suggesting host- and strain-specific speed or order of metabolic shifting. For instance, signal transduction pathway genes were upregulated in strain AT5.2 during the initial infection stage, while downregulated in strains A120 and A20 in that stage before they were upregulated in the late infection stage. Nevertheless, some common features can be identified. For instance, cytoskeleton genes were upregulated during early infection (parasite penetration) in all strains. In addition, cell division was promoted in the late infection stage in both studies. Farhat et al. (2018) found very few homologs of the genes critical for the parasitic lifestyle of apicomplexans as highly expressed in these *Amoebophrya* strains, suggesting unique infection mechanisms in these parasitic dinoflagellates.

Table 21.2. Gene expression changes in *Amoephrya* in the infection process.

	Strain AT5.2		Strains A120 & A25	
	Upregulated	**Downregulated**	**Upregulated**	**Downregulated**
Target recognition	Lectins: ricin-type beta-trefoil, C-type; SOD		Cytoskeleton, dynein, kinesin	
Initial penetration	Cytoskeletal organization; signal transduction; HSPs	TCA cycle, electron transport chain, ATP synthase complex	Translation and transcription: initiation factors, spliceosome and ribosome biogenesis; folding, sorting and degradation: exosome; cytoskeleton, transproters	Signaling, transport
Late infection & maturation	Cell division, reconstruction, and proliferation; stress response		Transport and catabolism (dynein, kinesin, intraflagellar transport, inorganic pyrophosphatase, lysosome), cellular community, cell growth and death (kinases, cyclins), signal transduction (MAPK, RAP1, RAs), energy metabolism (F-type, V-type, and H+-type ATPase subunits); trombospondin 1 (TSP1) and von-Willebrand factor type A (vWF)	
References	Lu et al. 2014		Farhat et al. 2018	

Algae-Fungi Mutualism and Parasitism

Mutualistic relationships between free-living algae and fungi are widespread in nature and are thought to have facilitated the evolution of land plants and lichens. In most of the known algal-fungal mutualisms, including lichens, algal cells reside externally to fungal cells. However, in a recent study, the stramenopile alga *Nannochloropsis oceanica* was found to be internalized within the hyphae of the fungus *Mortierella elongata* (Du et al. 2019). In this relationship, there is a carbon flow from the alga to the fungus and nitrogen flow in the opposite direction.

Various Chytridiomycete fungi are parasites of algae, including green algae, diatoms, and dinoflagellates (Table 21.1). Recently, a new genus with a new species (*Paradinomyces triforaminorum*) was established and found to specifically infect blooms of the dinotom (dinoflagellate containing a diatom symbiont) *Kryptoperidinium foliaceum* (Reñé et al. 2022).

The common filamentous fungus *Aspergillus pseudoglaucus* was recently found to co-exist and interact with the toxigenic epiphytic dinoflagellate *Prorocentrum lima* (Berry et al. 2022). This fungus was found to have physical contact with the dinoflagellate, with a biofilm of *P. lima* formed within and around the *A. pseudoglaucus* colonies, and fungal mycellum attached to the apex of the dinoflagellate cell. More importantly, a chemical interaction was detected, as the supernatant of the dinoflagellate-fungus co-culture contained secreted compounds. The clear effect of the presence of *A. pseudoglaucus* was the upregulation of the dinoflagellate toxins okadaic acid and dinophysistoxin 1, implicating a role of this fungus in inducing toxin production by *P. lima*.

Alga-Bacterium Interactions

Algae-bacteria associations are prevalent and profoundly impact algal growth, metabolism, and morphogenesis. Algae-cyanobacteria associations have been discussed in Chapter 20 and are excluded from the current chapter. The algae-bacteria interactions encompass those involving epiphytic bacteria that live on the surface of algae (phycosphere), endosymbiotic bacteria that live inside algal cells, and sympatric bacteria that interact with algae via chemical cues without physical contact.

Epiphytic Bacteria

The microscale mucus region surrounding an algal cell, rich in organic matter and hence a microcosm for bacteria, is named phycosphere (Bell and Mitchell 1972, Bell et al. 1974, Sapp et al. 2007). Dwelling inside the phycosphere, the bacteria utilize the Organic Matter (OM) and synthesize a plethora of metabolites that define their relationships with the host, encompassing mutualism, commensalism, antagonism, parasitism, and competition (Seymour et al. 2017). Others may provide the breakdown products of the algal extracellular OM or vitamins they synthesize as a source of nutrients for their host (Kuo and Lin 2013, Zachar and Boza 2022).

The species composition of phycospheric microbiomes varies with algal species (Sapp et al. 2007, Baker et al. 2023). This variability is likely due to the distinct physical or chemical conditions in the phycosphere of each algal species rather than the phylogenetic position of the alga (Baker et al. 2023). Nevertheless, the most dominant bacteria seem to concentrate in several phyla. In marine microalgal phycosphere, Alphaproteobacteria (*Roseobacter*), Gammaproteobacteria (e.g., *Alteromonas* spp.), and Bacteroidota (e.g., *Flavobacterium-Sphingobacterium*) are common (Sapp et al. 2007). From all three phyla of seaweeds, common dominant phycospheric bacteria have also been identified. These include Proteobacteria, Bacteroidota, Verrucomicrobiota, Planctomycetota, Firmicutes, Patescribacteria, and Cyanobacteria (Tujula et al. 2010, Burke et al. 2011a, Lachnit et al. 2011, Lemay et al. 2020).

Lu et al. (2023) sampled phycospheric bacteria from *Ulva*, *Saccharina*, *Grateloupia*, and *Gelidium* as representatives of all three phyla of macroalgae in four seasons and sequenced 16S rRNA gene amplicons. They detected Alphaproteobacteria, Gammaproteobacteria, Firmicutes, Bacteroidota, and Actinobacteriota, and from these identified 14 core bacterial genera, defined as macroalgae-common bacteria, most dominantly Flavobacteriaceae and Rhodobacteraceae. Furthermore, these authors isolated 4,426 bacterial strains and by sequencing select strains yielded 820 non-redundant draft genomes. The authors found that the assembly of phycospheric microbiomes was shaped by their functional profiles rather than species identities (Burke et al. 2011a), as functionally redundant taxa can replace each other (Tujula et al. 2010, Burke et al. 2011b).

Among the common phycospheric bacteria, species from Bacteroidota are known for their ability to degrade algal polysaccharides such as alginate, laminarin, and carrageenan via genes from the Polysaccharide Utilization Loci (PULs). In Lu et al. (2023), 61.8% of the carbohydrate-active enzymes (CAZymes) from the phycospheric microbiome were attributed to Bacteroidota. From the sequenced genomes of the large number of phycospheric bacteria, Lu et al. (2023) predicted 4,451 PULs and 8,810 Biosynthetic Gene Clusters (BGCs). While the PULs are critical in OM degradation, the highly diverse biosynthetic pathways produce molecules including terpenes (28.3%), bacteriocins (12.3%), Non-Ribosomal Peptides (NRPS, 10.5%), and NRPS-like clusters (8.0%), homoserine lactones (7.8%), type III polyketide synthases (PKS; 7.5%), type I PKS (5.9%), and beta-lactones (5.4%). The largest BGCs were from a *Streptomyces* species retrieved from the rhodophyte *Gelidium* sp., which encoded over 22 PKS and NRPS modules. Ninety-three of the top 100 genomes with the highest number of BGCs were from phycospheric bacteria, and ten of the top 20 BGC-rich genomes were from phycospheric Bacteroidota. This suggests that the potential for secondary metabolite production in this phylum may have been underestimated. Bacteroidota exhibited high proportions of BGCs for terpene and NRPS biosynthesis.

Baker et al. (2023) tested 368 bacterial strains from pond microalgae for their effects on their hosts. They found that positive and neutral effects on host growth were more prevalent than negative effects. The proportion of each effect type varied among host species. Bacteria had a greater impact on algal carrying capacity than on growth rate, indicating that nutrient remineralization and the metabolism of toxic byproducts might be the dominant mechanisms involved.

Complex interactions occur not only between the alga and bacteria but also between the bacteria within the assemblage. A study on the phycospheric bacteria of *Senedesmus quadricauda* revealed dominant species from the genera *Variovorax* (Betaproteobacteria), *Porphyrobacter* (Alphaproteobacteria), and *Dyadobacter* (Bacteroidota) (Astafyeva et al. 2022). *Dyadobacter* was shown to be key to the growth and fitness of *S. quadricauda* and highly adapted to living in the phycosphere. It produces and releases a large variety of polysaccharide-modifying enzymes accompanied by the high expression level of the type IX secretion system (T9SS) in algal co-cultures. The co-cultured bacteria also actively expressed Quorum-Quenching proteins (QQ) and biosynthesis of vitamins B1, B2, B5, B6, and B9, and Leucine-Rich Repeat (LRR) proteins and enzymes that are involved in bacterial ROS tolerance. Complementary to *Dyadobacter*, *Variovorax* expresses the genes for the biosynthesis of vitamins B2, B5, B6, B7, B9, and B12, and *Porphyrobacter* is specialized in the production of vitamins B2 and B6. Thus, the shared currency between partners are vitamins, microalgae growth-promoting substances, protection against ROS, and dissolved organic carbon.

Intriguingly, epibiotic bacteria from lineages such as *Cytophaga* (Bacteroidota) and *Roseobacter* (Alphaproteobacteria) play a critical role in the morphogenesis of Ulvales seaweeds (Wichard 2015). In the case of *Ulva mutabilis*, *Roseobacter* sp. seems to produce an effect resembling that of cytokines and promotes cell division, whereas *Cytophaga* sp. acts like auxin promoting rhizoid and cell wall formation. Strikingly, when both bacterial species were provided to axenic *U. mutabilis* "callus morphotype" cultures, the wild-type morphology was restored. When only one of the bacteria was provided, the culture displayed either dark green germlings with cell walls covered by bubble-like structures (with *Roseobacter*) or short thalli with partly degenerated cell walls (with *Cytophaga*). However, the chemical identity of the compounds (growth factors, hormones, or else) and the signaling or regulatory cascade involved in the algal-bacterial interaction is mysterious.

Functionally Linked Bacteria

Free-living bacteria too can have mutualistic or antagonistic interactions with sympatric algae. While antagonistic effects of bacteria have been elaborated in Chapter 18, the focus here is mutualistic or commensal relationships. An intriguing interaction was reported for the psychrotolerant bacterium, *Pseudomonas pelagia*, and the Antarctic green alga *Pyramimonas gelidicola* (Koh et al. 2013).

P. gelidicola was found to upregulate the expression of antifreeze proteins in response to cold conditions, while *P. pelagia* exhibited no antifreeze activity despite possessing genes for cold shock DEAD box protein A and cold shock protein, which are involved in cold adaptation. The results suggest that the psychrotolerant bacterium may exploit the green alga for antifreeze proteins.

More complex interactions occur between bacteria and algae. Bacteria may metabolize specific type of organic matter released by algae and produce signaling molecules or nutritional compounds that stimulate the growth of the interacting algae. For instance, Amin et al. (2015) reported that *Roseobacter* and *Sulfitobacter* sp. SA11 use tryptophan from the diatom *Pseudo-nitzschia multiseries* and produces Indole-3-AceticAacid (IAA), which acts as an inter-kingdom signaling molecule and stimulates the growth of the diatom. IAA, a crucial plant auxin, influences development through mechanisms such as altering cell orientation, organ development, fertility, and cell elongation.

IAA's role in metabolically linking phytoplankton and microbes is significant (Amin et al. 2015). Plants, microalgae, bacteria, and fungi produce IAA, an ability that appears to have evolved independently. IAA regulates the physiological responses and gene expression in these microorganisms. In plants, IAA regulates various aspects of growth and development. Microbial IAA plays an important role in plant-microbe interactions (Spaepen and Vanderleyden 2011). Bacteria and plants possess highly similar IAA biosynthesis pathways, including tryptophan (Trp)-dependent and Trp-independent pathways, although some intermediates differ. In plants, four Trp-dependent pathways have been proposed: indole-3-acetamide (IAM), indole-3-pyruvic acid (IPA), tryptamine (TRA), and indole-3-acetaldoxime pathways (Mano and Nemoto 2012).

IAA-mediated algae-bacteria interactions occur beyond diatoms. A similar relationship was reported for the chlorophyte *Chlorella sorokiniana* and the bacterium *Azospirillum brasilense*, where the bacterium uses tryptophan and thiamine released by the alga and produces IAA, influencing the host (Palacios et al. 2016). Recently, a study on two common symbiodiniacean endosymbionts of corals, *Symbiodinium microadriaticum* and *Breviolum minutum*, revealed a similar algae-bacteria interaction (Matthews et al. 2023). Treating these dinoflagellates with detergent (Trixon-X100) and antibiotic (Ab+Tx) significantly reduced the total number of bacterial cells per Symbiodiniaceae by 1.19 times in *S. microadriaticum* and 2.60 times in *B. minutum*, altering the microbial profile of both symbiodiniacean species. The suppressed bacteria included *Labrenzia* (*Roseibium*), *Marinobacter*, and *Muricauda*, members of Symbiodiniaceae core microbiome (Lawson et al. 2018, Maire et al. 2021). Consequently, a significance reduction was observed in the growth of both dinoflagellates and in metabolites, including IAA and TCA cycle intermediates. Further experiments traced the origin of IAA to the core bacterial species and showed its effect on the growth of these dinoflagellates. These results, along with a follow-up study (Matthews et al. 2024), establish the interactive loop of Symbiodiniaceae-metabolites (tryptophan)-bacteria-metabolites (IAA)-Symbiodiniaceae growth.

Algal-Viral Interactions

The role of viruses in infecting and degrading algal cells, which can control the abundance of phytoplankton, is detailed in Chapter 18. This role is biogeochemically significant as it shortcuts the elemental cycle and accelerates nutrient regeneration (Suttle 2007). It is estimated that viruses release 145 Gt carbon per year in the global tropical and subtropical oceans (Lara et al. 2017). However, this chapter focuses on an additional dimension of algae-virus interactions, involving genetic exchanges and functional coadaptation.

Alga-To-Virus Gene Transfer

Many viruses encode Auxiliary Metabolic Genes (AMGs) and express them during infection to redirect host's metabolic pathways towards cellular processes that optimize viral production. This also expands the metabolic capabilities of the infected cells. For instance, some viruses carry genes for photosynthesis, sphingolipid biosynthesis, and photoreception. Many of these AMGs have resulted from horizontal gene transfer between the viruses and their hosts, which can go either from host to virus or vice versa. Genomic analyses have shown that marine viruses, either phages (viruses infecting bacteria) or eukaryote viruses, harbor host-derived genes specifying various functions, even complete biochemical pathways (Monier et al. 2009). This allows viruses to manipulate the host via metabolic reprogramming during infection. In extreme cases, such interaction can alter the central carbon metabolism and pigment biosynthesis of the host, as demonstrated in cyanobacteria (Dammeyer et al. 2008, Thompson et al. 2011), or reprogram lipid biosynthesis of the host, as shown in some eukaryotic algae (Rosenwasser et al. 2014, Malitsky et al. 2016, Ziv et al. 2016). A virus that infects the prasinophyte *Ostreococcus tauri* carries a host-derived nitrogen transporter gene, which is transcribed during the infection cycle. Functional complementation experiments using a yeast mutant indicated that this viral-stolen gene is functional for transporting ammonium, methylammoium, and potentially multiple other nitrogen compounds (Monier et al. 2017). Infection thus reshapes the genomic and transcriptomic content of host cells into that of a new entity, the "virocell", the genetic information combining host and virus genomes (Vincent and Vardi 2023).

The largest giant viruses encode thousands of genes, including those previously thought to be restricted to cellular organisms, such as major components of the translation machinery, lipid metabolism, cellular structures (actin, fibers), and nutrient uptake systems (N, P, S, Fe), thus blurring the classical definitions of virion and host cell (Vincent and Vardi 2023). The giant virus that infects a predatory choanoflagellate encodes 900 proteins, many related to cellular metabolism and organic matter degradation (Needham et al. 2019). This viral genome contains three proton pump rhodopsin genes, which are better known in bacterial and various species of microalgae.

Virus-To-Alga Gene Transfer

One of the most striking examples of HGT from viruses to algae is the Dinoflagellate Viral Nuclear Protein (DVNP). Sequence and phylogenetic analyses indicate that DVNP occurs only in dinoflagellates outside viruses known to infect various algae such as the phaeophyte *Ectocarpus* and the prasinophyte *Micromonas* (Gornik et al. 2012, Irwin et al. 2018). DVNP does not occur in the closest relative of the Dinophyta phylum, *Perkinsus marinus*, which uses canonical histones to organize DNA with nucleosomes. Conversely, DVNP universally exists in dinoflagellates, from ancestral to later-diverging lineages, which do not have nucleosomes, indicating the coincidence of the evolutionary event with the disappearance of nucleosomes (Cornik et al. 2012). However, core histone genes (H2A, H2B, H3, and H4) are still present and expressed in dinoflagellates (Lin et al. 2010, Roy et al. 2014). More phylogenetic studies suggest that the acquisition of DVNP might be due to the loss of histone H1, and DVNP in dinoflagellates diverged rapidly, with one subset evolving to resemble histone H1 (Wang et al. 2024). Furthermore, the divergence of DVNP was accompanied, and presumably influenced, by the acquisition of the bacteria-originated histone-like protein (a.k.a major basic nuclear protein), which also shows similarity to histone H1.

There are many more documented events of virus-to-dinoflagellate gene transfer. Recent comparative genomics research revealed multiple waves of viral gene transfer to the *Symbiodinium* genus of symbiodiniacean dinoflagellates, with 59 high-confidence viral-derived HGT genes identified (Benites et al. 2022). These involved genes from both DNA and RNA viruses, including those from the phylum Nucleocytoviricota, the negative-strand RNA virus group Mononegavirales, and Unclassified RNA viruses. These transferred genes have a wide range of functions, as demonstrated by their functional annotation and GO term affiliations: DNA/RNA polymerase superfamily, RNA-directed RNA polymerase, viral RNA helicase, ATP binding, viral RNA genome replication, mRNA capping. The wide taxonomic and functional distribution of the HGT genes in just one genus of dinoflagellates indicates rampant virus-dinoflagellate interactions. In these interactions, the viral genes are integrated into the *Symbiodinium* genome by hitchhiking on the host's retrotransposition (a.k.a mRNA recycling) machinery (Song et al. 2017), either from newly infecting viruses or proviruses (previously invading, hidden in host cells) (Benites et al. 2022).

Host-Virus Co-Adaptation

It has generally been believed that viruses cause cell death, but recent research has revealed that alga-virus interactions are dynamic and the host and the viruses may co-adapt and co-thrive. For example, the haptophyte *Phaeocystis pouchetii* can coexist with its virus PpV-01 for 40 days after infection. Similarly, the green alga *Chlorella variabilis* was reported to maintain a stable interaction with its virus PBCV-1 for 15 days after being infected for 30 days. The prasinophyte *Ostreococcus* can live in symbiosis with its infecting double-stranded DNA (dsDNA) virus OmV2. The virus of the coccolithophorid *Emiliania huxleyi* can adopt a chronic infection strategy in the early stage, allowing it to multiply rapidly without causing cell lysis, a condition called pseudolysogenic state. Similar strategies seem to operate in Harmful Algal Bloom (HAB)-causing dinoflagellates.

Wang et al. (2023) described the viral composition and metabolic landscape in two HAB events of the dinoflagellate *Prorocentrum shikokuense*. Research data shows that *P. shikokuense* in both blooms was infected by Mimiviridae and Phycodnaviridae viruses. Contrary to the expectation that active viruses would depress the formation of blooms or bring blooms to an end, these viruses are extremely transcriptionally active during the blooms that last for weeks. The data show that the blooms and chronic infections occurred in parallel. Moreover, based on the viral transcriptomes isolated from the metatranscriptomic dataset, these viruses seemed to maintain a stable infection by producing Single-Nucleotide Polymorphisms (SNPs) and continuing to exploit host cell metabolic machinery. This suggests co-adaptation between viruses and the HAB-causing dinoflagellate. This novel insight will help to fully understand and model the role of viruses in regulating blooms of dinoflagellates and other algae.

References

Abassi, S., H. Wang, B.S. Park, J.W. Park and J.S. Ki. 2017. A novel cyclophilin B gene in the red tide dinoflagellate *Cochlodinium polykrikoides*: Molecular characterizations and transcriptional responses to environmental stresses. Biomed. Res. Int. 2017(1): 4101580. DOI:10.1155/2017/4101580.

Adams, D.R., D. Ron and P.A. Kiely. 2011. RACK1, a multifaceted scaffolding protein: structure and function. Cell Commun. Signal. 9: 22.

Adams, J.C., R. Steffen, C.-W. Chou, S. Duhamel and J.M. Diaz. 2022. Dissolved organic phosphorus utilization by the marine bacterium *Ruegeria pomeroyi* DSS-3 reveals chain length-dependent polyphosphate degradation. Environ. Microbiol. 24(5): 2259–2269.

Adler, S.O., T.W. Spiesser, F. Uschner, U. Münzner, J. Hahn, M. Krantz et al. 2022. A yeast cycle model integrating stress, signaling, and physiology. FEMS Yeast Res. 22(1): foac026.

Agre, S. Sasaki and M.K. Chrispeels. 1993. Aquaporins: a family of water channel proteins. Am. J. Physiol. 265: (Renal Fluid Electrolyte Physiol. 34): F461.

Ahmad, M., P. Galland, T. Ritz, R. Wiltschko and W. Wiltschko. 2007. Magnetic intensity affects cryptochrome-dependent responses in *Arabidopsis thaliana*. Planta 225: 615–624.

Ajonina-Ekoti, I., M.A. Kurosinski, A.E. Younis, D. Ndjonka, M.K. Tanyi, M. Achukwi et al. 2013. Comparative analysis of macrophage migration inhibitory factors (MIFs) from the parasitic nematode *Onchocerca volvulus* and the free-living nematode *Caenorhabditis elegans*. Parasitol. Res. 112: 3335–3346. DOI:10.1007/s00436-013-3513-1.

Akakabe, Y., K. Matsui and T. Kajiwara. 2002. Enantioselective formation of (R)-9-HPODE and (R)-9-HPOTrE in marine green alga *Ulva conglobata*. Bioorg. Med. Chem. 10: 3171–3173.

Akakabe, Y., K. Matsui and T. Kajiwara. 2003. 2,4-Decadienals are produced via (R)-11-HPITE from arachidonic acid in marine green alga *Ulva conglobata*. Bioorg. Med. Chem. 11: 3607–3609.

Akita, F., R. Nagao, K. Kato, Y. Nakajima, M. Yokono, Y. Ueno et al. 2020. Structure of a cyanobacterial photosystem I surrounded by octadecameric IsiA antenna proteins. Commun. Biol. 3: 232.

Alcolombri, U., S. Ben-Dor, E. Feldmesser, Y. Levin, D.S. Tawfik and A. Vardi. 2015. Identification of the algal dimethyl sulfide-releasing enzyme: a missing link in the marine fulfur cycle. Science 348(6242): 1466–1469.

Alcorta, J., T. Alarcórta, O. Salgado and B. Diez. 2020. Taxonomic novelty and distinctive genomic features of hot spring cyanobacteria. Front. Genet. 11: 568223.

Aldon, D., M. Mbengue, C. Mazars and J.-P. Galaud. 2018. Calcium signalling in plant biotic interactions. Int. J. Mol. Sci. 19(3): 665.

Alipanah, L., J. Rohloff, P. Winge, A.M. Bones and T. Brembu. 2015. Whole-cell response to nitrogen deprivation in the diatom *Phaeodactylum tricornutum*. J. Exp. Botan. 66(20): 6281–6296.

Allaf, M.M. and H. Peerhossaini. 2022. Cyanobacteria: model microorganisms and beyond. Microorganisms 10(4): 696.

Allagulova, C.R., A.R. Libyanova and A.M. Avalbaev. 2023. Multiple ways of nitric oxide production in plants and its functional activity under abiotic stress conditions. Int. J. Mol. Sci. 24(14): 11637.

Allen, A.E., C.L. Dupont, M. Oborník, A. Horák, A. Nunes-Nesi, J.P. McCrow et al. 2011. Evolution and metabolic significance of the urea cycle in photosynthetic diatoms. Nature 473: 203–207.

Allen, A.E., J. LaRoche, U. Maheswari, M. Lommer, N. Schauer, P.J. Lopez et al. 2008. Whole-cell response of the pennate diatom *Phaeodactylum tricornutum* to iron starvation. Proc. Natl. Acad. Sci. USA 105: 10438–10443.

Allen. J.R., M. Roberts, A.R. III Loeblich and L.C. Klotz. 1975. Characterization of the DNA from the dinoflagellate *Crypthecodinium cohnii* and implications for nuclear organization. Cell 6(2): 161–169.

Alley, R.B, T. Bernsten, N.L. Bindoff, Z. Chen, A. Chidthaisong, P. Friedlingstein et al. 2007. Summary for policymakers. *In:* Solomon, S., D. Qin, M. Manning, Z. Chen, M. Marquis, K.B. Averty et al. [eds.]. Climate Change 2007: The Physical Science Basis. Contribution of Workshop Group I to the fourth assessment report of the Intergovernmental Panel on Climate Change. Cambridge University Press, Cambridge and New York.

Allorent, G. and D. Petroutsos. 2017. Photoreceptor-dependent regulation of photoprotection. Curr. Opin. Plant Biol. 37: 102–108.

Alscher, R.G., N. Erturk and L.S. Heath. 2002. Role of superoxide dismutases (SODs) in controlling oxidative stress in plants. J. Exp. Bot. 53(372): 1331–1341.

Alva, V, and A.N. Lupas. 2019. Histones predate the split between bacteria and archaea. Bioinformatics 35(14): 2349–2353.

Alvarez-Carreño, C., A. Becerra and A. Lazcano. 2016. Molecular evolution of the oxygen-binding hemerythrin domain. PLoS ONE 11: e0157904.

Alverson, A.J. 2007. Strong purifying selection in the silicon transporters of marine and freshwater diatoms. Limnol. Oceanogr. 52: 1420–1429.

Amato, A., V. Sabatino, G.M. Nylund, J. Bergkvist, S. Basu, M.X. Andersson et al. 2018. Grazer-induced transcriptomic and metabolomic response of the chain-forming diatom *Skeletonema marinoi.* ISME J. 12: 1594–1604.

Amin, S.A., D.H. Green, M.C. Hart, F.C. Küpper, W.G. Sunda and C.J. Carrano. 2009. Photolysis of iron-siderophore chelates promotes bacterial-algal mutualism. Proc. Natl. Acad. Sci. USA 106: 17071–17076.

Amin, S.A., L.R. Hmelo, H.M. van Tol, B.P. Durham, L.T. Carlson, K.R. Heal et al. 2015. Interaction and signaling between a cosmopolitan phytoplankton and associated bacteria. Nature 522(7554): 98–101.

Anderca, M.I., S. Suga, T. Furuichi, K. Shimogawara, M. Maeshima and S. Muto. 2004. Functional identification of the glycerol transport activity of *Chlamydomonas reinhardtii* CrMIP1. Plant Cell Physiol. 45(9): 1313–1319.

Andersen, R.A., J.L. Kim, I. Tittley and H.S. Yoon. 2015. *Chrysotila dentata* comb. nov., *Chrysotila roscoffensis* comb. nov. and *Chrysocapsa wetherbeei* sp. nov. Phycologia 54(3): 321–322.

Anderson, A., A. Laohavisit, I.K. Blaby, P. Bombelli, C.J. Howe, S.S. Merchant et al. 2016. Exploiting algal NADPH oxidase for biophotovoltaic energy. Plant Biotechnol. J. 14(1): 22–28.

Andresen, I.J., R.J.S. Orr, K. Shalchian-Tabrizi and J. Brate. 2021. Compartmentalization of mRNAs in the giant, unicellular green alga *Acetabularia acetabulum*. Algal Res. 59: 102440.

Andrew, S.M., C.M. Moreno, K. Plumb, B. Hassanzadeh, L. Gomez-Consarnau, S.N. Smith et al. 2023. Widespread use of proton-pumping rhodopsin in Antarctic phytoplankton. Proc. Natl. Acad. Sci. USA 120(39): e2307638120.

Andreou, A., F. Brodhun and I. Feussner. 2009. Biosynthesis of oxylipins in non-mammals. Prog. Lipid Res. 48: 148–170.

Annenkov, V.V., R. Gordon, S.N. Zelinskiy and E.N. Danilovtseva. 2020. The probable mechanism for silicon capture by diatom algae: assimilation of polycarbonic acids with diatoms-is endocytosis a key stage in building of siliceous frustules? J. Phycol. 56: 1729–1737.

Annett, A.L., S. Lapi, T.J. Ruth and M.T. Maldonado. 2008. The effects of Cu and Fe availability on the growth and Cu:C ratios of marine diatoms. Limnol. Oceanogr. 53: 2451–2461.

Annunziata, R., A. Ritter, A.E. Fortunato, A. Manzotti, S. Cheminant-Navarro, N. Agier et al. 2019. bHLH-PAS protein RITMO1 regulates diel biological rhythms in the marine diatom *Phaeodactylum tricornutum*. Proc. Natl. Acad. Sci. USA 116: 13137–13142.

Antcheva, N., F. Guida and A. Tossi. 2013. Chapter 18. Defensins. *In:* Kastin, A.J. [ed.]. Handbook of Biologically Active Peptides. p.101.

Aranda, M., Y. Li, Y.J. Liew, S. Baumgarten, O. Simakov, M.C. Wilson et al. 2016. Genomes of coral dinoflagellate symbionts highlight evolutionary adaptations conducive to a symbiotic lifestyle. Sci. Rep. 6: 39734. DOI:10.1038/srep39734.

Archibald, J.M. 2009. The puzzle of plastid evolution. Curr. Biol. 19: R81–R88.

Archibald, J.M. and P.J. Keeling. 2002. Recycled plastids: a 'green movement' in eukaryotic evolution. Trends Genet. 18(11): 577–584.

Arimoto, A., K. Nishitsuji, Y. Higa, N. Arakaki, K. Hisata, C. Shinzato et al. 2019. A siphonous macroalgal genome suggests convergent functions of homeobox genes in algae and land plants. DNA Res. 26(2): 183–192.

Armbrust, E.V., J.A. Berges, C. Bowler, B.R. Green, D. Martinez, N.H. Putnam et al. 2004. The genome of the diatom *Thalassiosira pseudonana*: ecology, evolution, and metabolism. Science 306: 79–86.

Arnold, H.E., P. Kerrison and M. Steinke. 2013. Interacting effects of ocean acidification and warming on growth and DMS-production in the haptophyte coccolithophore *Emiliania huxleyi*. Glob. Chang. Biol. 19: 1007–1016.

Arora, N. and G.P. Philippidis. 2023. The prospects of algae-derived vitamins and their precursors for sustainable cosmeceuticals. Processes 11(2): 587.

Arribas-Hernández, L., S. Bressendorf, M.H. Hansen, C. Poulsen, S. Erdmann and P. Brodersen. 2018. An m^6A-YTH module controls developmental timing and morphogenesis in *Arabidopsis*. Plant Cell 30: 952–967.

Arrigo, K.R. 2005. Marine microorganisms and global nutrient cycles. Nature 437: 349–355.

Arun, A., S.M. Coelho, A.F. Peters, S. Bourdareau, L. Peres, D. Scornet et al. 2019. Convergent recruitment of TALE homeodomain life cycle regulators to direct sporophyte development in land plants and brown algae. eLife 8: e43101.

Ashby, M.K. and C.W. Mullineaux. 1999. Cyanobacterial ycf27 gene products regulate energy transfer from phycobilisomes to photosystems I and II. FEMS Microbiol. Lett. 181(2): 253–260.

Ashby, M.K. and J. Houmard. 2006. Cyanobacterial two-component proteins: structure, diversity, distribution, and evolution. Microbiol. Mol. Biol. Rev. 70(2): 472–509.

Askwith, C., D. Eide, A. Van Ho, P.S. Bernard, L. Li, S. Davis-Kaplan et al. 1994. The FET3 gene of *S. cerevisiae* encodes a multicopper oxidase required for ferrous iron uptake. Cell 76: 403–410.

Astafyeva, Y., M. Gurschke, M. Qi, L. Bergmann, D. Indenbirken, I. de Grahl et al. 2022. Microalgae and bacteria interaction-evidence for division of diligence in the alga microbiota. Microbiol. Spectr. 10(4): e0063322.

Aubert, C., P. Mathis, A.P.M. Eker and K. Brettel. 1999. Brettel Intraprotein electron transfer between tyrosine and tryptophan in DNA photolyase from *Anacystis nidulans*. Proc. Natl. Acad. Sci. USA 96: 5423–5427.

Augustijn, K.D., R. Kleemann, J. Thompson, T. Kooistra, C.E. Crawford, S.E. Reece et al. 2007. Functional characterization of the *Plasmodium falciparum* and *P. berghei* homologues of macrophage migration inhibitory factor. Infect. Immun. 75: 00902–06. DOI:10.1128/iai.00902-06.

Avin-Wittenberg, T., K. Bajdzienko, G. Wittenberg, S. Alseekh, T. Tohge, R. Bock et al. 2015. Global analysis of the role of autophagy in cellular metabolism and energy homeostasis in *Arabidopsis* seedlings under carbon starvation. Plant Cell 27(2): 306–322.

Avrani, S., O. Wurtzel, I. Sharon, R. Sorek and D. Lindell. 2011. Genomic island variability facilitates *Prochlorococcus*-virus coexistence. Nature 474: 604–608.

Aylett, C.H.S., E. Sauer, S. Imseng, D. Boehringer, M.N. Hall, N. Ban et al. 2016. Architecture of human mTOR complex 1. Science 351(6268): 48–52.

Azevedo, C., T. Livermore and A. Saiardi. 2015. Protein polyphosphorylation of lysine residues by inorganic polyphosphate. Mol. Cell 58: 71–82.

Bachvaroff, T.R., E. Williams, R. Jagus and A.R. Place. 2015. A noncryptic noncanonical multi-module PKS/NRPS found in dinoflagellates. In: Proceedings of the 16th International Conference on Harmful Algae. Wellington: Cawthron Institute, Nelson, New Zealand.

Bachvaroff, T.R. 2019. A precedented nuclear genetic code with all three termination codons reassigned as sense codons in the syndinean *Amoebophrya* sp. ex *Karlodinium veneficum*. PLoS ONE 14(2): e0212912.

Backer, L.C. and D.J. McGillicuddy, Jr. 2006. Harmful algal blooms at the interface between coastal oceanography and human health. Oceanography 19(2): 94–106.

Badger, M.R., D. Hanseon and G.D. Price. 2002. Evolution and diversity of CO_2 concentrating mechanisms in cyanobacteria. Funct. Plant Biol. 29(3): 161–173.

Bai, T., L. Gao, M. Xu and L. Tian. 2021. Structural diversity of photosystem I and its light-harvesting system in eukaryotic algae and plants. Front. Plant Sci. 12: 781035.

Baker, A.C. 2001. Ecosystems: Reef corals bleach to survive change. Nature 411 (6839): 765–766.

Baker, D., J. Lauer, A. Ortega, S.L. Jackrel and V.J. Denef. 2023. Effects of phycosphere bacteria on their algal host are host species-specific and not phylogenetically conserved. Microorganisms 11(1): 62.

Baker, D.M., C.J. Freeman, J.C.Y. Wong, M.L. Fogel and N. Knowlton, 2018. Climate change promotes parasitism in a coral symbiosis. JSME J. 12: 921–930.

Baker, K.G. and R.J. Geider. 2021. Phytoplankton mortality in a changing thermal seascape. Glob. Chang. Biol. 27(20): 5253–5261.

Balaguer, J., F. Koch, C. Hassler and S. Trimborn. 2022. Iron and manganese co-limit the growth of two phytoplankton groups dominant at two locations of the Drake Passage. Commun. Biol. 5: 207.

Balskus, E.P. and C.T. Walsh. 2010. The genetic and molecular basis for sunscreen biosynthesis in cyanobacteria. Science 329: 1653–1656.

Bandara, K., Ø. Varpe, L. Wijewardene, V. Tverberg and K. Elane. 2021. Two hundred years of zooplankton vertical migration research. Biol. Rev. 96(4): 1547–1589.

Banerjee, A., E. Herman and T. Kottke. 2016. Aureochrome 1a LOV domain dimer from *Phaeodactylum tricornutum*. Structure 24(2): 213–215.

Barak-Gavish, N., M.J. Fada, C. Ku, P.A. Lee, G.R. DiTullio, S. Malitsky et al. 2018. Bacterial virulence against an oceanic bloom-forming phytoplankter is mediated by algal DMSP. Sci. Adv. 4: eaau5716.

Barbosa, A., P. Valentão and P.B. Andrade. 2016. Biologically active oxylipins from enzymatic nonenzymatic routes in macroalgae. Mar. Drugs 14(1): 23.

Barki, F., M.M. Sandin and M. Jamy. 2021. Diversity and ecology of protists revealed by metabarcoding. Curr. Biol. 31(19): R1267–R1280.

Barlow, J.T., S.R. Bogatyrev and R.F. Ismagilov. 2020. A quantitative sequencing framework for absolute abundance measurements of musosal and luminal microbial communities. Nat. Commun. 11: 2590.

Bartolo, A.G., G. Zammit, A.F. Peters and F.C. Kupper. 2020. The current state of DNA barcoding of macroalgae in the Mediterranean Sea: presently lacking but urgently required. Bot. Mar. 63(3): 253–272.

Bassler, B.L. and R. Losick. 2006. Bacterially speaking. Cell 125(2): 237–246.

Basu, S., S. Patil, D. Mapleson, M.T. Russo, L. Vitale and C. Fevola. 2017. Finding a partner in the ocean: molecular and evolutionary bases of the response to sexual cues in a planktonic diatom. New Phytol. 215(1): 140–156.

Baumgarten, S., T. Bayer, M. Aranda, Y.J. Liew, A. Carr, G. Micklem et al. 2013. Integrating microRNA and mRNA expression profiling in *Symbiodinium microadriaticum*, a dinoflagellate symbiont of reef-building corals. BMC Genomics 14: 704.

Baurain, D., H. Brinkmann, J. Petersen, N. Rodriguez-Ezpeleta, A. Stechmann, V. Demoulin et al. 2010. Phylogenomic evidence for separate acquisition of plastids in cryptophytes, haptophytes, and stramenopiles. Mol. Biol. Evol. 27(7): 1698–1709.

Bay, L.K., V.R. Cumbo, D. Abrego, J.T. Kool, T.D. Ainsworth and B.L. Willis. 2011. Infection dynamics vary between *Symbiodinium* types and cell surface treatments during establishment of endosymbiosis with coral larvae. Diversity 3(3): 356–374.

Beatty, J.T., J. Overmann, M.T. Lince, A.K. Manske, A.S. Lang, R.E. Blankenship et al. 2005. An obligately photosynthetic bacterial anaerobe from a deep-sea hydrothermal vent. Proc. Natl. Acad. Sci. USA 102(26): 9306–9310.

Beauchemin, M. and D. Morse. 2017. A proteomic portrait of dinoflagellate chromatin reveals abundant RNA-binding proteins. Chromosoma 127: 1–15.

Beauchemin, M., S. Roy, S. Pelletier, A. Averback, F. Lanthier and D. Morse. 2016. Characterization of two dinoflagellate cold shock domain proteins. mSphere 1: e00034-15.

Beauchemin, M., R. Sougata, P. Daoust, S. Daneais-Bellefeuille, T. Bartomeu, L. Letourneau et al. 2012. Dinoflagellate tandem array gene transcripts are highly conserved and not polycistronic. Proc. Natl. Acad. Sci. USA 109 (39): 15797–15798.

Beakes, G.W., H.M. Canter and G.H.M. Jaworski. 1988. Zoospore ultrastructure of *Zygorhizidium affluens* and *Z. planktonicum*, two chytrids parasitizing the diatom *Asterionella formosa*. Can. J. Bot. 66: 1054–1067.

Beaulieu, J.M., I.J. Leitch, S. Patel, A. Pendharkar and C.A. Knight. 2008. Genome size is a strong predictor of cell size and stomatal density in angiosperms. New Phytol. 179(4): 975–986.

Beck, C., H. Knoop, I.M. Axmann and R. Steuer. 2012. The diversity of cyanobacterial metabolism: genome analysis of multiple phototrophic microorganisms. BMC Genomics 13: 56.

Becker, B., K. Hoef-Emden and M. Melkonian. 2008. Chlamydial genes shed light on the evolution of photoautotrophic eukaryotes. BMC Ecol. Evol. 8: 203.

Beedessee, G., K. Hisata, M.C. Roy, F.M. Van Dolah, N. Satoh and E. Shoguchi. 2019. Diversified secondary metabolite biosynthesis gene repertoire revealed in symbiotic dinoflagellates. Sci. Rep. 9: 1204. DOI:10.1038/s41598-018-37792-0.

Beedessee, G., T. Kubota, A. Arimoto, K. Nishitsuji, R.F. Waller, K. Hisata et al. 2020. Integrated omics unveil the secondary metabolic landscape of a basal dinoflagellate. BMC Biol. 18: 139. DOI:10.1186/s12915-020-00873-6.

Beeson, J.G., D.R. Drew, M.J. Boyle, G. Feng, F.J.I. Fowkes and J.S. Richards. 2016. Merozoite surface proteins in red blood cell invasion, immunity and vaccines against malaria. FEMS Microbiol. Rev. 40(3): 343–372.

Behrenfeld, M.J., R.T. O'Malley, D.A. Siegel, C.R. McClain, J.L. Sarmiento, G.C. Feldman et al. 2006. Climate-driven trends in contemporary ocean productivity. Nature 444(7120): 752–755.

Béjà, O., L. Aravind, E.V. Koonin, M.T. Suzuki, A. Hadd, L.P. Nguyen et al. 2000. Bacterial rhodopsin: evidence for a new type of phototrophy in the sea. Science 289: 1902–1906.

Béjà, O., E.N. Spudich, J.L. Spudich, M. Leclerc and E.F. DeLong. 2001. Proteorhodopsin photography in the ocean. Nature 411: 786–789.

Bekker, A., H.D. Holland, P.L. Wang, D. Rumble III, H.J. Stein, J.L. Hahhah et al. 2004. Dating the rise of atmospheric oxygen. Nature 427: 117–120.

Bell, N., J.J. Lee and M.L. Summers. 2017. Characterization and *in vivo* regulon determination of an ECF sigma factor and its cognate anti-sigma factor in *Nostoc punctiforme*. Mol. Microbiol. 104(1): 179–194.

Bell, W.H. and R. Mitchell. 1972. Chemotactic and growth response of marine bacteria to algal extracellular products. Biol. Bull. 143: 265–277.

Bell, W.H., J.M. Lang and R. Mitchell. 1974. Selective simulation of marine bacteria by algal extracellular products. Limnol. Oceanogr. 19: 833–839.

Bellantuono, A.J., K.E. Dougan, C. Granados-Cifuentes and M. Rodriguez-Lanetty. 2019. Free-living and symbiotic lifestyles of a thermotolerant coral endosymbiont display profoundly distinct transcriptomes under both stable and heat stress conditions. Mol. Ecol. 28(24): 5265–5281.

Bellinger, F.P., A.V. Raman, M.A. Reeves and M.J. Berry. 2009. Regulation and function of selenoproteins in human disease. Biochem. J. 422: 11–22.

Benedetti, F., M. Vogt, U.H. Elizondo, D. Righetti, N.E. Zimmermann and N. Gruber. 2021. Major restructuring of marine plankton assemblages under global warming. Nat. Commun. 12: 5226.

Benga, G., O. Popescu, V. Borza, V.I. Pop, A. Muresan, I. Mocsy et al. 1986. Water permeability in human erythrocytes: identification of membrane proteins involved in water transport. Eur. J. Cell Biol. 41: 252–262.

Benites, L.F., T.G. Stephens and D. Bhattacharya. 2022. Multiple waves of viral invasions in symbiodiniaceae algal genomes. Virus Evol. 8(2): 1–15.

Bentley-DeSousa, A., C. Holinier, H. Moteshareie, Y.C. Tseng, S. Kajjo, C. Nwosu et al. 2018. A screen for candidate targets of lysine polyphosphorylation uncovers a conserved network implicated in ribosome biogenesis. Cell Rep. 22: 3427–3439.

Berdieva, M.A., I.A. Pozdnyakov, V.O. Kalinina, S.O. Skarlato. 2021. Putative meiotic toolkit in the dinoflagellate *Prorocentrum cordatum*: Additional evidence for sexual process from transcriptome. J. Euk. Microbiol. 68: e12845.

Berera, R., I.H. van Stokkum, S. d'Haene, J.T. Kennis, R. van Grondelle and J.P. Dekker. 2009. A mechanism of energy dissipation in cyanobacteria. Biophys J. 96: 2261–2267.

Berges, J.A. and P.G. Falkowski. 1998. Physiological stress and cell death in marine phytoplankton: induction of proteases in response to nitrogen or light limitation. Limnol. Oceanogr. 43(1): 129–135.

Bergkvist, J., E. Selander and H. Pavia. 2008. Induction of toxin production in dinoflagellates: the grazer makes a difference. Oecologia 156(1): 147–154.

Bergkvist, J., P. Thor, H.H. Jakobsen, S.A. Wängberg and E. Selander. 2012. Grazer-induced chain length plasticity reduces grazing risk in a marine diatom. Limnol. Oceanogr. 57(1): 318-324.

Bergman, B., G.S andh, S. Lin, J. Larsson and E.J. Carpenter. 2013. *Tichodesmium* - a widespread marine cyanobacterium with unusual nitrogen fixation properties. FEMS Microbiol. Rev. 37: 286–302. DOI:10.1111/j.1574-6976.2012.00352.x.

Berman-Frank, I., P. Lundgren, Y.B. Chen, H. Kupper, Z. Kolber, B. Bergman et al. 2001. Segregation of nitrogen fixation and oxygenic photosynthesis in the marine cyanobacterium *Trichodesmium*. Science 294:1534–1537.

Bernard, G., C.X. Chan, Y.B. Chan, X.Y. Chua, Y. Cong, J.M. Hogan et al. 2019. Alignment-free inference of hierarchical and reticulate phylogenomic relationships. Brief Bioinform. 20(2): 426–435.

Bernard, G., P. Greenfield, M.A. Ragan and C.X. Chan. 2018. *k*-mer similarity, networks of microbial genomes, and taxonomic rank. mSystems. 3(6): e00257-18.

Bernhard, J.M., S.S. Bowser. 1999. Benthic foraminifera of dysoxic sediments: Chloroplast sequestration and functional morphology. Earth Sci. Rev. 46: 149–165.

Bernart, M.W., G.G. Whatley and W.H. Gerwick. 1993. Unprecedented oxylipins from the marine green alga *Acrosiphonia coalita*. J. Nat. Prod. 56: 245–259.

Berney, C. and J. Pawlowski. 2006. A molecular time-scale for eukaryote evolution recalibrated with continuous microfossil record. Proc. Roy. Soc. Lond. B. 273: 1867–1872.

Bertsch, C., M. Beuve, V.V. Jolja, M. Wirth, F. Pelsy, E. Herrbach et al. 2009. Retention of the virus-derived sequences in the nuclear genome of grapevine as a potential pathway to virus resistance. Biol. Direct. 4: 21.

Berry, O., E. Briand, A. Bagot, M. Chaigné, L. Meslet-Cladière, J. Wang et al. 2022. Deciphering interactions between the marine dinoflagellate *Prorocentrum lima* and the fungus *Asperigillus pseudoglaucus*. Appl. Microbiol. Internat. 25(2): 250–267.

Bertomeu, T., J.W. Hastings and D. Morse. 2003. Vectorial labeling of dinoflagellate cell surface proteins. J. Phycol. 39(6): 1254-1260.

Berube, P.M., S.J. Biller, T. Hackl, S.L. Hogle, B.M. Satinsky, J.W. Becker. 2018. Single cell genomes of *Prochlorococcus*, *Synechococcus*, and sympatric microbes from diverse marine environments. Sci. Data 5: 180154.

Beszteri, S., S. Thomas, V. Benes, L. Harms and S. Trimborn. 2018. The response of three Southern Ocean phytoplankton species to ocean acidification and light availability: a transcriptomic study. Protist 169: 958–975.

Bhattacharya, D., H.S. Yoon and J.D. Hackett. 2004. Photosynthetic eukaryotes unite: Endosymbiosis connects the dots. Bioessays 26: 50–60.

Bhattacharya, D., J.M. Archibald, A.P.M. Weber and A. Reyes-Prieto. 2007. How do endosymbionts become organelles? Understanding early events in plastid evolution. BioEssays 29: 1239–1246.

Bhattacharya, D., D.C. Price, C.X. Chan, H. Qiu, N. Rose, S. Ball et al. 2013. Genome of the red alga *Porphyridium purpureum*. Nat. Comm. 4: 1941.

Bhaud, Y., J.-M. Salmon and M.-O. Soyer-Gobillard. 1991. The complex cell cycle of the dinoflagellate protoctist *Crypthecodinium cohnii* as studied *in vivo* and by cytofluorimetry. J. Cell Sci. 100: 675–682.

Bian, Y., H. Lin, Y. Qi, Z. Fang and D. Xu. 2022. PIFs- and COP1-HY5-mediated temperature signaling in higher plants. Stress Biol. 2: 35.

Bibby, T.S., J. Nield and J. Barber. 2001. Iron deficiency induces the formation of an antenna ring around trimeric photosystem I in cyanobacteria. Nature 412: 743–745.

Bibikova, M., M. Golic, K.G. Golic and D. Carroll. 2002. Targeted chromosomal cleavage and mutagenesis in *Drosophila* using zinc-finger nucleases. Genetics 161(3): 1169–1175.

Bicudo, C.E.de M. and M. Menezes. 2016. Phylogeny and classification of Euglenophyceae: A brief review. Front. Ecol. Evol. 4: 17.

Biddle, K.D. 2016. Programmed cell death in unicellular phytoplankton. Curr. Biol. 26(13): R594–R607.

Bilcke, G., K. Van den Berge, S. De Decker, E. Bonneure, N. Poulsen, P. Bulankova, et al. 2021a. Mating type specific transcriptomic response to sex inducing pheromone in the pennate diatom *Seminavis robusta*. ISME J 15: 562–576.

Bilcke, G., C.M. Osuna-Cruz, M.S. Silva, N. Poulsen, S. D'hondt, P. Bulankova et al. 2021b. Diurnal transcript profiling of the diatom *Seminavis robusta* reveals adaptations to a benthic lifestyle. Plant J. 107: 315–336.

Bishop, C.D. 2022. Pond water is the principal source of algae that form a symbiosis with embryos of the spotted salamander *Ambystoma maculatum*: an experimental test. Symbiosis 87(2): 107–120.

Bižić, M., T. Klintzsch, D. Ionescu, M.Y. Hindiyeh, M. Günthel, A.M. Muro-Pastor et al. 2020. Aquatic and terrestrial cyanobacteria produce methane. Sci. Adv. 6(3): eaax5343.

Blaby-Haas, C.R. and S.S. Merchant. 2013. The ins and outs of algal metal transport. Biochim. Biophys. Acta 1823(9): 1531–1552.

Blair, J.E., P. Shah and S.B. Hedges. 2005. Evolutionary sequence analysis of complete eukaryote genomes. BMC Bioinformatics 6: 53.

Blanc, G., G. Duncan, I. Agarkova, M. Borodovsky, J. Gurnon, A. Kuo et al. 2010. The *Chlorella variavbilis* NC64A genome reveals adaptation to photosymbiosis, coevolution with viruses, and cryptic sex. Plant Cell 22: 2943–2955.

Blank, C.E. and P. Sánchez-Baracaldo. 2010. Timing of morphological and ecological innovations in the cyanobacteria–a key to understanding the rise in atmospheric oxygen. Geobiology 8: 1–23.

Blankenship, R.E. 2010. Early evolution of photosynthesis. Plant Physiol. 154(2): 154(2): 434–438.

Blasio, M. and S. Balzano. 2021. Fatty acids derivatives from eukaryotic microalgae, pathways and potential applications. Front. Microbiol. 12: 718933.

Blifernez, O., L. Wobbe, K. Niehaus and O. Kruse. 2011. Protein arginine methylation modulates light-harvesting antenna translation in *Chlamydomonas reinhardtii*. Plant J. 65(1): 119–130.

Blommaert, J. 2020. Genome size evolution: towards new model systems for old questions. Proc. R. Soc. B 287: 20201441.

Blumenthal, T. 2004. Operons in eukaryotes. Brief. Functional Genomics 3(3): 199–211.

Boccaletto, P., F. Stefaniak, A. Ray, A. Cappannini, S. Mukherjee, E. Purta et al. 2022. MODOMICS: A Database of RNA Modification Pathways. 2021 Update. Nucleic Acids Res. 50: D231–D235.

Bodyl, A., J.W. Stiller and P. Mackiewicz. 2009. Chromalveolate plastids: Direct descent or multiple endosymbioses? Trend. Ecol. Evol. 24: 119–121.

Bogan, S.N. and V.Y. Soojin. 2024. Potential role of DNA methylation as a driver of plastic responses to the environment across cells, organisms, and populations. Genome Biol. Evol. 16(2): evae022.

Bogard, M.J., D.B. Donald, K. Finlay and P.R. Leavitt 2012. Distribution and regulation of urea in lakes of central North America. Freshw. Biol. 57: 1277–1292.

Bolhuis, H., I. Severin, V. Confurius-Guns, U.I.A. Wollenzien and L.J. Stal. 2010. Horizontal transfer of the nitrogen fixation gene cluster in the cyanobacterium microcoleus chthonoplastes. ISME J. 4: 121–130.

Bondy-Denomy, J., J. Qian, E.R. Westra, A. Buckling, D.S. Guttman, A.R. Davidson et al. 2016. Prophages mediate defense against phage infection through diverse mechanisms. ISME J. 10: 2854–2866.

Bonni, A., A. Brunet, A.E. West, S.R. Datta, M.A. Takasu and M.E. Greenberg. 1999. Cell survival promoted by the Ras-MAPK signaling pathway by transcription-dependent and -independent mechanisms. Science 286: 1358–1362.

Boonprab, K., K. Matsui, Y. Akakabe, N. Yotsukura and T. Kajiwara. 2004. Arachidonic acid conversion by lipoxygenase in the brown alga, *Laminaria angustata*. Kasetsart J. Nat. Sci. 38: 72–77.

Boraas, M.E., D.B. Seale and J.E. Boxhorn 1998. Phagotrophy by a flagellate selects for colonial prey: a possible origin of multicellularity. Evol. Ecol. 12: 153–164.

Borg, M., S.A. Krueger-Hadfield, C. Destombe, J. Collén, A. Lipinska and S.M. Coelho. 2023. Red macroalgae in the genomic era. New Phytol. 240(2): 471–488.

Borghi, F. and A. Saiardi. 2023. Evolutionary perspective on mammalian inorganic polyphosphate (polyp) biology. Biochem. Soc. Trans. BST20230483. DOI: 10.1042/BST20230483.

Borthakur, P.B., C.C. Orozco, S.S. Young-Robbins, R. Haselkorn and S.M. Callahan. 2005. Inactivation of patS and hetN causes lethal levels of heterocyst differentiation in the filamentous cyanobacterium *Anabaena* sp. PCC 7120. Mol. Microbiol. 57: 111–123.

Bothwell, J.H., H. Marie, A.F. Peters, J.M. Cock and S.M. Coelho. 2010. Role of endoreduplication and apomeiosis during parthenogenetic reproduction in the model brown alga. New Phytol. 188(1): 111–121.

Bouarab, K., F. Adas, E. Gaquerel, B. Kloareg, J.P. Salaün and P. Potin. 2004. The innate immunity of a marine red alga involves oxylipins from both the eicosanoid and octadecanoid pathways. Plant Physiol. 135(3): 1838–1848.

Bouchard, J.N. and H. Yamasaki. 2009. Implication of nitric oxide in the heat-stress-induced cell death of the symbiotic alga *Symbiodinium microadriaticum*. Mar. Biol. 156(11): 2209–2220.

Boulias, K. and E.L. Greer. 2023. Biological roles of adenine methylation in RNA. Nat. Rev. Genet. 24: 143–160.

Bourdareau, S., L. Tirichine, B. Lombard, D. Loew, D. Scornet, Y. Wu et al. 2021. Histone modifications during the life cycle of the brown algal *Ectocarpus*. Genome Biol. 22: 12.

Bouwman, A.F., A. Beusen, P.M. Glibert, C. Overbeek, M. Pawlowski, J. Herrera et al. 2013. Mariculture: Significant and expanding cause of coastal nutrient enrichment. Environ. Res. Lett. 8: 044026.

Bowazolo, C., B. Song, S. Dorion, M. Beauchemin, S. Chevrier, J. Rivoal and D. Morse. 2022. Orchestrated translation specializes dinoflagellate metabolism three times per day. Proc. Natl. Acad. Sci. USA 119(30): e2122335119.

Bowler, C., A.E. Allen, J.H. Badger, J. Grimwood, K. Jabbari, A. Kuo et al. 2008. The *Phaeodactylum* genome reveals the evolutionary history of diatom genomes. Nature 456: 239–244.

Bowler C, A. Vardi and A.E. Allen. 2010. Oceanographic and biogeochemical insights fromdiatom genomes. Annu. Rev. Mar. Sci. 2: 333–365.

Bramucci, A.R. and R.J. Case. 2019. Phaeobacter inhibens induces apoptosis-like programmed cell death in calcifying *Emiliania huxleyi*. Sci. Rep. 9: 5215.

Brandenburg, K.M., S. Wohlrab, U. John, A. Kremp, J. Jerney, B. Krock et al. 2018. Intraspecific trait variation and trade-offs within and across populations of a toxic dinoflagellate. Ecol. Lett. 21: 1561–1571.

Brawley, S.H., N.A. Blouin, E. Ficko-Blean, G.L. Wheeler, M. Lohr, H.V. Goodson et al. 2017. Insights into the red algae and eukaryotic evolution from the genome of *Porphyra umbilicalis* (Bangiophyceae, Rhodophyta). Proc. Natl. Acad. Sci. USA 114(31): E6361–E6370.

Brechignac, F. and L. Wolf. 1994. SYMBIOSE system for microgravity bioregnerative support of experiments. Adv. Space. Res. 14: 79–88.

Brisbin, M.M. and S. Mitarai. 2019. Differential gene expression supports a resource-intensive, defensive role for colony production in the bloom-forming haptophyte, *Phaeocystis globosa*. J. Eukaryot. Microbiol. 66(5): 788–801.

Brocchieri, L., E. Conway de Macario and A.J. Macario. 2008. Hsp70 genes in the human genome: conservation and differentiation patterns predict a wide array of overlapping and specialized functions. BMC Evol. Biol. 8: 19.

Brocks, J.J., A.J.M. Jarrett, E. Sirantoine, C. Hallmann, Y. Hoshino and T. Liyanage. 2017. The rise of algae in Cryogenian oceans and the emergence of animals. Nature 548: 578–581.

Bravo, I. and R.I. Figueroa. 2014. Towards ecological understanding of dinoflagellate cyst functions. Microorganisms 2(1): 11–32.

Brown, A.F.M., Q. Dortch, F.M. Van Dolah, T.A. Leighfield, W. Morrison, A.E. Thessen et al. 2006. Effect of salinity on the distribution, growth, and toxicity of *Karenia* spp. Harmful Algae 5(2): 199–212.

Brown, J.L., T. Ross, T.A. McMeekin and P.D. Nichols. 1997. Acid habituation of *Escherichia coli* and the potential role of cyclopropane fatty acids in low pH tolerance. Int. J. Food Microbiol. 37: 163–173.

Browning, T.J., E.P. Achterberg, A. Engel and E. Mawji. 2021. Manganese co-limitation of phytoplankton growth and major nutrient drawdown in the Southern Ocean. Nat. Commun. 12: 884.

Browning, T.J., X. Liu, R. Zhang, Z. Wen, J. Liu, Y. Zhou et al. 2022. Nutrient co-limitation in the subtropical Northwest Pacific. Limnol. Oceanogr. 7(1): 52–61.

Browning, T.J. and C.M. Moore. 2023. Global analysis of ocean phytoplankton nutrient limitation reveals high prevalence of co-limitation. Nat. Commun. 14: 5014.

Brownlee, C., K.E. Helliwell, Y. Meeda, D. McLachlan, E.A. Murphy and G.L. Wheeler. 2022. Regulation and integration of membrane transport in marine diatoms. Sem. Cell Dev. Biol. Doi: 10.1016/j.semcdb.2022.03.006.

Browse, J. 2009. Oxidation of membrane lipids and functions of oxylipins. pp. 389–405. *In:* Wada, H. and N. Murata [eds.]. Lipids in Photosynthesis: Essential and Regulatory Functions. Springer Science + Business Media B. V.

Brückner, A., C. Polge, N. Lentze, D. Auerbach and U. Schlattner. 2009. Yeast two-hybrid, a powerful tool for systems biology. Int. J. Mol. Sci. 10(6): 2763–2788.

Brunelle, S.A., E.S. Hazard, E.E. Sotka and F.M. van Dolah. 2007. Characterization of a dinoflagellate cryptochrome blue-light receptor with a possible role in circadian control of the cell cycle. J. Phycol. 43(3): 509–518.

Bruning, K. 1991. Infection of the diatom Asterionella by a chytrid. I. Effects of light on reproduction and infectivity of the parasite. J. Plank. Res. 13: 103–117.

Brzezinski, M.A., R.J. Olson and S.W. Chisholm. 1990. Silicon availability and cell-cycle progression in marine diatoms. Mar. Ecol. Progr. Ser. 67: 83–96.

Brunson, J.K., S.M.K. McKinnie, J.R. Chekan, J.P. McCrow, Z.D. Miles, E.M. Bertrand et al. 2018. Biosynthesis of the neurotoxin domoic acid in a bloom-forming diatom. Science 361: 1356–1358.

Buaya, A.T., B. Scholz and M. Thines. 2021. *Sirolpidium bryopsidis*, a parasite of green algae, is probably conspecific with *Pontisma lagenidioides*, a parasite of red algae. Fungal Syst. Evol. 7: 223–231.

Buerger, P., C. Alvarez-Roa, C.W. Coppin, S.L. Pearce, L.J. Chakravarti, J.G. Oakeshott et al. 2020. Heat-evolved microalgal symbionts increase coral bleaching tolerance. Sci. Adv. 6: eaba2498.

Bulankova, P., G. Bilcke, W. Vyverman and L. De Veylder. 2022. Cellular hallmarks and regulation of the diatom cell cycle. *In:* Falciatore, A. and T. Mock [eds.]. The Molecular Life of Diatoms. Springer, Cham. https://doi.org/10.1007/978-3-030-92499-7_9.

Bunnik, E.M. and K.G. Le Roch. 2013. An introduction to functional genomics and systems biology. Adv. Wound Care (New Rochelle) 2(9): 490–498.

Bunse, C., D. Lundin, C.M.G. Karlsson, N. Akram, M. Vila-Costa, J. Palovaara et al. 2016. Response of marine bacterioplankton pH homeostasis gene expression to elevated CO_2. Nat. Clim. Change 6: 483–487.

Burke, C., P. Steinberg, D. Rusch, S. Kjelleberg and T. Thomas. 2011a. Bacterial community assembly based on functional genes rather than species. Proc. Natl. Acad. Sci. USA 108: 14288–14293.

Burke, C., T. Thomas, M. Lewis, P. Steinberg and S. Kjelleberg. 2011b. Composition, uniqueness and variability of the epiphytic bacterial community of the green alga *Ulva australis*. ISME J. 5: 590–600.

Burki, F., M.M. Sandin and J. Mahwash. 2021. Diversity and ecology of protists revealed by metabarcoding. Curr. Biol. 31(19): R1267–R1280.

Burri, B.J., M.R. La Frano and C. Zhu. 2016 Absorption, metabolism, and functions of β-cryptoxanthin. Nutr. Rev. 74(2): 69–82.

Burriesci, M.S., T.K. Raab and J.R. Pringle. 2012. Evidence that glucose is the major transferred metabolite in dinoflagellate–cnidarian symbiosis. J. Exp. Biol. 215(19): 3467–3477.

Busch, A. and S. Hess. 2021. Sunscreen mucilage: a photoprotective adaptation found in terrestrial green algae (Zygnematophyceae). Eur. J. Phycol. 57(1): 107–124.

Busch, A., J.V. Gerbracht, K. Davies, U. Hoecker and S. Hess. 2024. Comparative transcriptomics elucidates the cellular responses of an aeroterrestrial zygnematophyte to UV radiation. J. Exp. Bot. DOI: 10.1093/jxb/erae131.

Busch, D.S. and P. McElhany. 2017. Using mineralogy and higher-level taxonomy as indicators of species sensitivity to pH: a case-study of Puget Sound. Elementa 5: 53.

Cachon, J. and M. Cachon. 1987. Parasitic dinoflagellates. pp. 571–610. *In*: Taylor, F.J.R. [ed.]. The Biology of Dinoflagellates. Blackwell Science, Oxford.

Cai, W.-J., X. Hu, W.-J. Huang, M.C. Murrell, J.C. Lehrter, S.E. Lohrenz et al. 2011. Acidification of subsurface coastal waters enhanced by eutrophication. Nat. Geosci. 4: 766–770.

Cain, T.J. and A.T. Smith. 2021. Ferric iron reductases and their contribution to unicellular ferrous iron uptake. J. Inorg. Biochem. 218: 111407.

Cairns, S.K. 1999. Species richness of recent Scleractinia. Atoll Res. Bull. 459: 1–46.

Calandra, T. and T. Roger. 2003. Macrophage migration inhibitory factor: A regulator of innate immunity. Nat. Rev. Immunol. 3(10): 791–800. DOI:10.1038/nri1200.

Calatrava, V., T.G. Stephens, A. Gabr, D. Baya, D. Bhattacharya and A.R. Grossman. 2022. Retrotransposition facilitated the establishment of a primary plastid in the thecate amoeba *Paulinella*. Proc. Natl. Acad. Sci. USA 119(23): e2121241119.

Caldeira, K. and M.E. Wickett. 2005. Ocean model predictions of chemistry changes from carbon dioxide emissions to the atmosphere and ocean. J. Geophys. Res. 110: C09S04.10.1029/2004JC002671.

Calhoun, S., B. Kamel, T.A.S. Bell, C.P.S. Kruse, R. Riley, V. Singan et al. 2022. Multi-omics profiling of the cold tolerant *Monoraphidium minutum* 26B-AM in response to abiotic stress. Algal Res. 66: 102794.

Caló, G., M.A. De Marco, G.L. Salerno and G.M.A. Martínez-Noël. 2022. TOR signaling in the green picoalga *Ostreococcus tauri*. Plant Sci. 323: 111390.

Cao, L., E. Alani and N. Kleckner. 1990. A pathway for generation and processing of double-strand breaks during meiotic recombination in *S. cerevisiae*. Cell 61(6): 1089–1101.

Carlson, M.C.G., F. Ribalet, I. Maidanik, B.P. Durham, Y. Hulata, S. Ferrón et al. 2022. Viruses affect picocyanobacterial abundance and biogeography in the North Pacific Ocean. Nat. Microbiol. 7: 579–580.

Calvaruso, C., A. Rokka, E.-M. Aro and C. Büchel. 2020. Specific Lhc proteins are bound PSI or PSII supercomplexes in the diatom *Thalassiosira pseudonana*. Plant Physiol. 183(1): 67–79.

Campbell, E.L., K.D. Hagen, M.F. Cohen, M.L. Summers and J.C. Meeks. 1996. The devR gene product is characteristic of receivers of two-component regulatory systems and is essential for heterocyst development in the filamentous cyanobacterium *Nostoc* sp. strain ATCC 29133. J Bacteriol. 178(7): 2037–2043.

Canapa, A., M. Barucca, M.A. Biscotti, M. Forconi and E. Olmo. 2016. Transposons, genome size, and evolutionary insights in animals. Cytogenet. Genome Res. 147: 217–239.

Canter, H.M. and G. Jaworski. 1979. The occurrence of a hypersensitive reaction in the planktonic diatom *Asterionella formosa* Hassall parasitized by the chytrid *Rhizophydium planktonicum* Canter emend., in culture. New phytol. 82: 187–206.

Caputi, L., Q. Carradec, D. Eveillard, A. Kirilovsky, E. Pelletier, J.J.P. Karlusich, et al. 2019. Community-level responses to iron availability in open ocean plankton ecosystems. Global Biogeochem. Cycles 33: 391–419.

Cardona, T., P. Sanchez-Baracaldo, A.W. Rutherford and A.W. Larkum. 2019. Early Archean origin of Photosystem II. Geobiol. 17(2): 127–150.

Carneiro, A.K., P. da Fonseca Montessoro, A.F. Fusaro, B.G. Araújo and A.S. Hemerly. 2021. Plant CDKs-Driving the cell cycle through climate change. Plants 10(9): 1804.

Carney, L.T. and T.W. Lane. 2014. Parasites in algae mass culture. Front. Microbiol. 5: 278.

Carpenter, E.J. 2002. Marine cyanobacterial symbioses. Biol. Environ.: Proc. R. Irish Acad. 102B1: 15–18.

Carpenter, E.J. and J. Chang. 1988. Species-specific phytoplankton growth rates via diel DNA synthesis cycles. I. Concept of the method. Mar. Ecol. Prog. Ser. 43(1/2): 105–111.

Carradec, Q., E. Pelletier, C. Da Silva, A. Alberti, Y. Seeleuthner, R. Blanc-Mathieu et al. 2018. A global ocean atlas of eukaryotic genes. Nat. Commun. 9: 373.

Carreto, J., M. Carignan and N. Montoya. 2001. Comparative studies on mycosporine-like amino acids, paralytic shellfish toxins and pigment profiles of the toxic dinoflagellates *Alexandrium tamarense*, *A. catenella* and *A. minutum*. Mar. Ecol. Prog. Ser. 223: 49–60.

Carrión, O., X.-Y. Zhu, B.T. Williams, J. Wang, X.-H. Zhang and J.D. Todd. 2023. Molecular discoveries in microbial DMSP synthesis. Adv. Microb. Physiol. 83: 59–116.

Carroll, D. 2017. Genome editing: Past, present, and future. Yale J. Biol. Med. 90(4): 653–659.

Carroll, S.B. 2000. Endless forms: the evolution of gene regulation and morphological diversity. Cell 101: 577–580.

Cartaxana, P., L. Morelli, E. Cassin, V. Havurinne, M. Cabral and S. Cruz. 2023. Prey species and abundance affect growth and photosynthetic performance of the polyphagous sea slug Elysia crispate. Roy. Soc. Open Sci. 10: 230810.

Casanova-Ferrer, S. Ares and J. Muñoz-García. 2022. Terminal heterocyst differentiation in the *Anabaena* patA mutant as a result of posttranscriptional modifications and molecular leakage. PLoS Comput. Biol. 18(8): e1010359.

Castell, C., P. Bernal-Bayard, J.M. Ortega, M. Roncel, M. Hervás and J.A. Navarro. 2020. The heterologous expression of a plastocyanin in the diatom *Phaeodactylum tricornutum* improves cell growth under iron-deficient conditions. Physiol. Plant. 171(2): 277–290.

Cattolico, R.A., J.C. Boothroy and S.P. Gibbs. 1976. Synchronous growth and plastid replication in the naturally wall-less alga *Olisthodiscus luteus*. Plant Physiol. 57(4): 497–503.

Cavalier-Smith, T. 1998. A revised six-kingdom system of life. Biol. Rev. 73: 203–266.

Cavalier-Smith, T. 1999. Principles of protein and lipid targeting in secondary symbiogenesis: Euglenoid, dinoflagellate, and sporzoan plastid origins and eukaryote family tree. J. Euk. Microbiol. 46: 347–366.

Cavalier-Smith, T. 2002. Origins of the machinery of recombination and sex. Heredity 88: 125–141.

Cavalier-Smith, T. 2003. Genomic reduction and evolution of novel genetic membranes and protein-targeting machinery in eukaryote-eukaryote chimaeras (meta-algae). Phil. Trans. Roy. Soc. London Ser. B. 358: 109–133.

Cavalier-Smith, T. 2005. Economy, speed and size matter: evolutionary forces driving nuclear genome miniaturization and expansion. Ann. Bot. 95(1): 147–175.

Cavalier-Smith, T. 2006. Cell evolution and earth history: stasis and revolution. Phils. Trans. R. Soc. Lond. B. Biol. Sci. 361(1470): 969–1006.

Cedar, H. and Y. Bergman. 2009. Linking DNA methylation and histone modification: patterns and paradigms. Nat. Rev. Genet. 10(5): 295–304.

Celen, A.B. and U. Sahin. 2020. Sumoylation on its 25th anniversary: mechanisms, pathology, and emerging concepts. FEBS J. 287(15): 3110–3140.

Celis-Plá, P.S.M., F. Rodríguez-Rojas, L. Méndez, F. Moenne, P.T. Muñoz, M.G. Lobos et al. 2019. MAPK pathway under chronic copper excess in green macroalgae (Chlorophyta): influence on metal exclusion/extrusion mechanisms and photosynthesis. Int. J. Mar. Sci. 20: 4547.

Cenci, U., S.J. Sibbald, B.A. Curtis, R. Kamikawa, L. Eme, D. Moog et al. 2018. Nuclear genome sequence of the plastid-lacking cryptomonad *Goniomonas avonlea* provides insights into the evolution of secondary plastids. BMC Biol. 16: 137.

Chai, J. and J.J. Lee. 1999. Initial recognition of endosymbiotic diatoms by the larger foraminifer *Amphistegina lobifera*. Symbiosis 26: 39–53.

Chai, J. and J.J. Lee. 2000. Recognition, establishment and maintenance of diatom endosymbioses in foraminifera. *In*: Lee, J.J. and P.H. Muller [eds.]. Advances in the Biology of Foraminifera. Micropaleontology 46 (supplement I): 182–195.

Chambouvet, A., C. Alves-de-Souza, V. Cueff, D. Marie, S. Karpov and L. Guillou. 2011. Interplay between the parasite *Amoebophrya* sp. (Alveolata) and the cyst formation of the red tide dinoflagellate *Scrippsiella trochoidea*. Protist. 162: 637–649.

Chambouvet, A., P. Morin, D. Marie and L. Guillou. 2008. Control of toxic marine dinoflagellate blooms by serial parasitic killers. Science 322(5905): 1254–1257.

Chamizo-Ampudia, A., E. Sanz-Luque, Á. Llamas, F. Ocaña-Calahorro, V. Mariscal, A. Carreras et al. 2016. A dual system formed by the ARC and NR molybdoenzymes mediates nitrite-dependent NO production in *Chlamydomonas*. Plant, Cell Environ. 39(10): 2097–2107.

Chamovitz, D.A. and D. Segal. 2001. JAB1/CSN5 and the COP9 signalosome. A complex situation. EMBO Rep. 2(2): 96–101.

Chan, C.X., P. Vaysberg, D.C. Price, K.N. Pelletreau, M.E. Rumpho and D. Bhattacharya. 2018. Active host response to algal symbionts in the sea slug *Elysia chlorotica*. Mol. Biol. Evol. 35: 1706–1711.

Chan, Y.H. and J.T. Wong. 2007. Concentration-dependent organization of DNA by the dinoflagellate histone-like protein HCc3. Nuc. acid. Res. 35: 2573–2583.

Chang, Y.Y. and J.E. Cronan, Jr. 1999. Membrane cyclopropane fatty acid content is a major factor in acid resistance of *Escherichia coli*. Mol. Microbiol. 33: 249–259.

Charlson, R.J., J. Lovelock, M.O. Andreae and S.G. Warren. 1987. Oceanic phytoplankton, atmospheric sulphur, cloud albedo and climate. Nature 326: 655–661.

Charrier, B., M.H. Abreu, R. Araujo, A. Bruhn, J.C. Coates, O. De Clerck et al. 2017. Furthering knowledge of seaweed growth and development to facilitate sustainable aquaculture. New Phytol. 216: 967–975.

Chazan, A., A. Rozenberg, K. Mannen, T. Nagata, R. Tahan, S. Yaish et al. 2022. Diverse heliorhodopsins detected via functional metagenomics in freshwater Actinobacteria, Chloroflexi and Archaea. Environ. Microbiol. 24(1): 110–121.

Chen, A.K., M.I. Latz, P. Sobolewski and J.A. Frangos. 2007. Evidence for the role of G-proteins in flow stimulation of dinoflagellate bioluminescence. Am. J. Physiol. Regl. Integr. Comps. Physiol. 292: R2020–R2027.

Chen, B. 2015. Patterns of thermal limits of phytoplankton. J. Plankton Res. 37(2): 285–292.

Chen, D., M. Lyu, X. Kou, J. Li, Z. Yang, L. Gao et al. 2022. Integration of light and temperature sensing by liquid-liquid phase separation of phytochrome B. Mol. Cell 82(16): 3015–3029.

Chen, H., Z. Zhu, J. Chen, R. Yang, Q. Luo, J. Xu, H. Shan and X. Yan. 2015. A multifunctional lipoxygenase from *Pyropia haitanensis*—The cloned and functioned complex eukaryotic algae oxylipin pathway enzyme. Algal Res. 12: 316–327.

Chen, J., L. Lu, S.M.C. Robb, M. Collin, Y. Okumoto, J. Stajich and S.R. Wessler. 2020. Genomic diversity generated by a transposable element burst in a rice recombinant inbred population. Proc. Natl. Acad. Sci. USA 117: 26288–26297.

Chen, K., X. Jiao, A. Ashton, A.D. Rocco, T.G. Pestell, Y. Sun et al. 2020. The membrane-associated form of cyclin D1 enhances cellular invasion. Oncogenesis 9: 83.

Chen, M. 2019. Chapter four-Chlorophylls *d* and *f*: Synthesis, occurrence, light-harvesting, and pigment organization in chlorophyll-binding protein complexes. Adv. Bot. Res. 90: 121–139.

Chen, M.-Y., W.-K. Teng, L. Zhao, C.-X. Hu, Y.-K. Zhou, B.-P. Han et al. 2021. Comparative genomics reveals insights into cyanobacterial evolution and habitat adaptation. ISME J. 15: 211–227.

Chen, N., Q. Xu, J. Zhu, H. Song, L. He, S. Liu et al. 2024. Chromosome-scale genome assembly reveals insights into the evolution and ecology of the harmful algal bloom species *Phaeocystis globosa* Scherffel. iScience DOI: 10.1016/j.isci.2024.110575.

Chen, Q., D. Shinozaki, J. Luo, M. Pottier, M. Havé, A. Marmagne et al. 2019. Autophagy and nutrients management in plants. Cells 8(11): 1426.

Chen, X.-H., Y.-Y. Li, H. Zhang, J.-L. Liu, Z.-X. Xie, L. Lin et al. 2018. Quantitative proteomics reveals common and specific responses of a marine diatom *Thalassiosira pseudonana* to different macronutrient deficiencies. Front. Microbiol. 9: 2761.

Chen, Y., G. Luo, S. Chen, D. Zhang, W. Xie, Z. Wang et al. 2021. The potential of prodigiosin for control of *Prorocentrum donghaiense* blooms: algicidal properties and acute toxicity to other marine organisms at various trophic levels. Ecotoxicol. Environ. Safety 228: 112913.

Chen, Y., K.E. Dougan, D. Bhattacharya and C.X. Chan. 2024. Nuclear genomes of dinoflagellates reveal evolutionary conserved pattern of RNA editing relative to stress response. Front. Protistol. 2: 1320917.

Chen, Z., Y. Dong, S. Duan, J. He, H. Qin, C. Bian, et al. 2024. A chromosome-level genome assembly for the paramylon-producing microlagal *Euglena gracilis*. Sci. Data 11: 780.

Cheng, D., V. Vemulapalli and M.T. Bedford. 2012. Chapter four-methods applied to the study of protein arginine methylation. Method. Enzymol. 512: 71–92.

Cheng, H., C. Bowler, X. Xing, V. Bulone, Z. Shao and D. Duan. 2021. Full-length transcriptome of *Thalassiosira weissflogii* as a reference resource and mining of chitin-related genes. Mar. Drugs 19(7): 392.

Cheng, L., K.E. Trenberth, N. Gruber, J.P. Abraham, J.T. Fasullo, G. Li et al. 2020. Improved estimates of changes in upper ocean salinity and the hydrological cycle. J. Climate 33(23): 10357–10381.

Cheng, L.-C., S.-P.L. Hwang and J. Chang. 1997. Gene sequence and expression of an analog of proliferating cell nuclear antigen (PCNA) in the algal *Tetraselmis chui* and detection of the encoded protein with anti-rat PCNA monoclonal antibody. Appl. Environ. Microbiol. 63(10): 4010–4014.

Cheng, S., W. Xian, Y. Fu, B. Marin, J. Keller, T. Wu et al. 2019. Genomes of subaerial Zygnematophytes provide insights into land plant evolution. Cell 179: 11057–1067.

Cheng, Y., O. Zak, P. Aisen, S.C. Harrison and T. Walz. 2004. Structure of the human transferrin receptor-transferrin complex. Cell 116: 565–576.

Chi, J., M.W. Parrow and M. Dunthorn. 2014. Cryptic sex in *Symbiodinium* (Alveolata, Dinoflagellata) is supported by an inventory of meiotic genes. J. Eukaryot. Microbiol. 61: 322–327.

Chisholm, S.W., S.L. Frankel, R. Goericke, R.J. Olson, B. Palenik, J.B. Waterbury et al. 1992. *Prochlorococcus-marinus* nov. gen. nov. sp.: An oxy-phototrophic marine prokaryote containing divinyl chlorophyll *a* and *b*. Arch Microbiol. 157: 297–300.

Chisholm, S.W., R.J. Olson, E.R. Zettler, R. Goericke and J.B. Waterbury. 1988. A novel free-living prochlorophyte abundant in the oceanic euphotic zone. Nature 334: 340–343.

Cho, K., M. Ueno, Y. Liang, D. Kim and T. Oda. 2022. Generation of reactive oxygen species (ROS) by harmful algal bloom (HAB)-forming phytoplankton and their potential impact on surrounding living organisms. Antioxidans 11(2): 206.

Choi, Y.-E., J.-K. Rhee, H.-S. Kim, J.-W. Ahn, H. Hwang and J.-W. Yang. 2015. Chemical genetics approach reveals importance of cAMP and MAP kinase signaling to lipid and carotenoid biosynthesis in microalgae. J. Microbiol. Biotechnol. 25(5): 637–647.

Chiu, Y.-L. and C. Shinzato. 2022. Evolutionary history of DMSP lyase-like genes in animals and their possible involvement in evolution of the scleractinian coral genus, *Acropora*. Front. Mar. Sci. 9: 889866.

Chu, H.-H., S. Car, A.L. Socha, M.N. Hindt, T. Punshon and M.L. Guerinot. 2017. The *Arabidopsis* MTP8 transporter determines the localization of manganese and iron in seeds. Sci. Rep. 7: 11024.

Chung, C.C., S.P.L. Hwang and J. Chang. 2005. Cooccurrence of ScDSP gene expression, cell death, and DNA fragmentation in a marine diatom *Skeletonema costatum*. App. Environ. Microbiol. 71(12): 8744–8751.

Chung, C.C., S.-P.L .Hwang and J. Chang. 2008. Nitric oxide as a signaling factor to upregulate the death-specific protein in a marine diatom, *Skeletonema costatum*, during blockage of electron flow in photosynthesis. Appl. Environ. Microbiol. 74(21): 6521–6527.

Cianciotto, N.P. and R.C. White. 2017. Expanding role of type II secretion in bacterial pathogenesis and beyond. Infect Immun. 85: e00014–17.

Clapham D. 2003. TRP channels as cellular sensors. Nature 426: 517–524.

Clement, R., S. Lignon, P. Mansuelle, E. Jensen, M. Pophillat, R. Lebrun et al. 2017. Responses of the marine diatom *Thalassiosira pseudonana* to changes in CO_2 concentration: a proteomic approach. Sci Rep. 7: 42333.

Clement, R., L. Dimnet, S.C. Maberly and B. Gontero. 2016. The nature of the CO_2-concentrating mechanisms in a marine diatom, *Thalassiosira pseudonana*. New Phytol 209: 1417–1427.

Clode, P.L., M. Saunders, G. Maker, M. Ludwig and C.A. Atkins. 2009. Uric acid deposits in symbiotic marine algae. Plant Cell Environ. 32(2): 170–177. DOI: 10.1111/j.1365-3040.2008.01909.x.

Coale, T.H., M. Moosburner, A. Horák, M. Obornik, K.A. Barbeau and A.E. Allen. 2019. Reduction-dependent siderophore assimilation in a model pennate diatom. Proc. Natl. Acad. Sci. USA 116(47): 23609–23617.

Coates, C.J. and H. Decker. 2017. Immunological properties of oxygen-transport proteins: hemoglobin, hemocyanin and hemerythrin. Cell Mol. Life Sci. 74(2): 293–317.

Coats, D.W. and M.G. Park. 2002. Parasitism of photosynthetic dinoflagellates by three strains of Amoebophrya (Dinophyta): parasite survival, infectivity, generation time, and host specificity. J. Phycol. 38: 520–528.

Cock, J.M., V. Vanoosthuyse and T. Gaude. 2002. Receptor kinase signalling in plants and animals: distinct molecular systems with mechanistic similarities. Curr. Opin. Cell Biol. 14: 230–236.

Cock, J.M., L. Sterck, P. Rouze, D. Scornet, A.E. Allen, G. Amoutzias et al. 2010. The *Ectocarpus* genome and the independent evolution of multicellularity in brown algae. Nature 465: 617–621.

Coesel, S., M. Mangogna, T. Ishikawa, M. Heijde, A. Rogato, G. Finazzi et al. 2009. Diatom PtCPF1 is a new cryptochrome/photolyase family member with DNA repair and transcription regulation activity. EMBO Rep. 10(6): 655–661.

Coesel, S.N., B.P. Durham, R.D. Groussman, S.K. Hu, D.A. Caron, R.L. Morales et al. 2021. Diel transcriptional oscillations of light-sensitive regulatory elements in open-ocean eukaryotic plankton communities. Proc. Natl. Acad. Sci. USA 118(6): e2011038118.

Cohen, N.R., W. Gong, D.M. Moran, M.R. Mcllvin, M.A. Saito and A. Marchetti. 2018. Transcriptomic and proteomic responses of the oceanic diatom *Pseudo-nitzschia granii* to iron limitation. Environ. Microbiol. 20: 3109–3126.

Coleman, J.E. 1998. Zinc enzymes. Curr. Opin. Chem. Biol. 2(2): 222–234.

Collén, J., B. Porcel, W. Carre, S.G. Ball, C. Chaparro, T. Tonon et al. 2013. Genome structure and metabolic features in the red seaweed *Chondrus crispus* shed light on evolution of the Archaeplastida. Proc. Natl. Acad. Sci. USA 110(13): 5247–5252.

Collén, J., I. Guisle-Marsollier, J.J. Léger and C. Boyen. 2007. Response of the transcriptome of the intertidal red seaweed *Chondrus crispus* to controlled and natural stresses. New Phytol. 176(1): 45–55.

Collens, A.B. and L.A. Katz. 2021. Opinion: genetic conflict with mobile elements drives eukaryotic genome evolution, and perhaps also eukaryogenesis. J. Hered. 112: 140–144.

Collier, J.L. and A.R. Grossman. 1994. A small polypeptide triggers complete degradation of light-harvesting phycobiliproteins in nutrient-deprived cyanobacteria. EMBO J. 13(5): 1039–1047.

Collos, Y., C. Gagne, M. Laabir and A. Vaquer. 2004. Nitrogenous nutrition of *Alexandrium catenella* (Dinophyceae) in cultures and in Thau Lagoon, Southern France. J. Phycol. 40: 96–103.

Conde, A., M.M. Chaves and H. Gerós. 2011. Membrane transport, sensing and signaling in plant adaptation to environmental stress. Plant Cell Physiol. 52: 1583–1602.

Condon, C., D. Liveris, C. Squires, I. Schwartz and C.L. Squires. 1995. rRNA operon multiplicity in *Escherichia coli* and the physiological implications of rrn inactivation. J. Bacteriol. 177(14): 4152–4156.

Connolly, J.A., M.J. Oliver, J.M. Beaulieu, C. Knight, L. Tomanek and M.A. Moline. 2008. Correlated evolution of genome size and cell volume in diatoms (Backllariophyceae). J. Phycol. 44(1): 124–131.

Cooney, E.C., N. Okamoto, A. Cho, E. Hehenberger, T.A. Richards, A.E. Santoro et al. 2020. Single-cell transcriptomics of *Abedinium* reveals a new early-branching dinoflagellate lineage. Genome Biol. Ecol. 12: 2417–2428.

Cooper, G.M. 2000. The Cell: A Molecular Approach, 2nd edition. Sinauer Associates, Boston, MA. 689 pp.

Cornejo-Castillo, F.M., A.M. Cabello, G. Salazar, P. Sánchez-Baracaldo, G. Lima-Mendez, P. Hingamp et al. 2016. Cyanobacterial symbionts diverged in the late Cretaceous towards lineage-specific nitrogen fixation factories in single-celled phytoplankton. Nat. Commun. 7: 11071.

Cornejo-Castillo, F.M., K. Inomura, J.P. Zehr and M.J. Follows. 2024. Metabolic trade-offs constrain the cell size ratio in a nitrogen-fixing symbiosis. Cell 187(7): 1762–1768.

Corrales-Guerrero, L., V. Mariscal, D.J. Nürnberg, J. Elhai, C.W. Mullineaux, E. Flores et al. 2014. Subcellular localization and clues for the function of the HetN factor influencing heterocyst distribution in *Anabaena* sp. strain PCC 7120. J. Bacteriol. 196: 3452–3460.

Correa, A.M.S., R.M. Welsh and R.L. Vega Thurber. 2013. Unique nucleocytoplasmic dsDNA and +ssRNA viruses are associated with the dinoflagellate endosymbionts of corals. ISME J. 7: 13–27.

Couso, I., M.E. Perez-Perez, M.M. Ford, E. Martinez-Force, L.M. Hicks, J.G. Umen et al. 2020. Phosphorus availability regulates TORC1 signaling via LST8 in *Chlamydomonas*. Plant Cell 32: 69–80.

Cox, C.D., N. Bavi and B. Martinac. 2019. Biophysical principles of ion-channel-mediated mechanosensory transduction. Cell Rep. 29(1): 1–12.

Coyne, K.J., Y. Wang and G. Johnson. 2022. Algicidal bacteria: a review of current knowledge and applications to control harmful algal blooms. Front. Microbiol. 13: 871177.

Craik, D.J., N.L. Daly, T. Bond and C. Waine. 1999. Plant cyclotides: a unique family of cyclic and knotted proteins that defines the cyclic cystine knot structural motif. J. Mol. Biol. 294 (5): 1327–1336.

Crespo, J.L., S. Díaz-Troya and F.J. Florencio. 2005. Inhibition of target of rapamycin signaling by rapamycin in the unicellular green alga *Chlamydomonas reinhardtii*. Plant Physiol. 139: 1736–1749.

Croft, M.T., A.D. Lawrence, E. Raux-Deery, M.J. Warren and A.G. Smith. 2005. Algae acquire vitamin B_{12} through a symbiotic relationship with bacteria. Nature 438: 90–93.

Crow, K.D. and G.P. Wagner. 2005. What is the role of genome duplication in the evolution of complexity and diversity? Mol. Biol. Evol. 23: 887–892.

Crowell, R., J. Nienow and A.B. Cahoon. The complete chloroplast and mitochondrial genomes of the diatom *Nitzschia palea* (Bacillariophyceae) demonstrate high sequence similarity to the endosymbiont organelles of the dinotom *Durinskia baltica*. J. Phycol. 55(2): 352–364.

Cruz, S. and P. Cartaxana. 2022. Kleptoplasty: getting away with stolen chloroplasts. PLoS Biol. 20(11): e3001857.

Cuadrado, Á., A. De Bustos and R.I. Figueroa. 2019. Chromosomal markers in the genus *Karenia*: towards an understanding of the evolution of the chromosomes, life cycle patterns and phylogenetic relationships in dinofagellates. Sci. Rep. 9: 3072.

Cuadrado, Á., R.I. Figueroa, M. Sixto, I. Bravo and A. De Bustos. 2022. First record of the spatial organization of the nucleosome-less chromatin of dinoflagellates: The nanrandom distribution of microsatellites and bipolar arrangement of telomeres in the nucleus of *Gambierdiscus australes* (Dinophyceae). J. Phycol. 58: 297–307.

Cunin, R., N. Glansdorff, A. Pierard and V. Stalon. 1986. Biosynthesis and metabolism of arginine in bacteria. Microbiol. Rev. 50: 314–352.

Cunning, R. and A.C. Baker. 2013. Excess algal symbionts increase the susceptibility of reef corals to bleaching. Nat. Clim. Chang. 3: 259–262.

Curson, A.R.J., J. Liu, A. Bermejo Martínez, R.T. Green, Y. Chan, O. Carrión et al. 2017. Dimethylsulfoniopropionate biosynthesis in marine bacteria and identification of the key gene in this process. Nat. Microbiol. 2(5): 17009.

Curson, A.R.J., B.T. Williams, B.J. Pinchbeck, L.P. Sims, A.B. Martinez, P.P.L. Rivera et al. 2018. DSYB catalyses the key step of dimethylsulfoniopropionate biosynthesis in many phytoplankton. Nat. Microbiol. 3: 430–439.

Curtis, B.A., G. Tanifuji, F. Burki, A. Gruber, M. Irimia, S. Maruyama et al. 2012. Algal genomes reveal evolutionary mosaicism and the fate of nucleomorphs. Nature 492: 59–65.

Dagenasis-Bellefeuille, S. and D. Morse. 2013. Putting the N in dinoflagellates. Front. Microbiol. 4: 369.

Daher, K.A., M.E. Selsted and R.I. Lehrer. 1986. Direct inactivation of viruses by human graulocyte defensins. J. Virol. 60: 1068–1074.

Dai, M., Y. Zhao, F. Chai, M. Chen, N. Chen, Y. Chen et al. 2023. Persistent eutrophication and hypoxia in the coastal ocean. Coastal Futures 1: e19.

Dale, B. 2023. Paleontological evidence for dinoflagellates and ciliates as early eukaryotes. J. Mar. Sci. Eng. 11(3): 533. DOI: 10.3390/jmse11030533.

Dammeyer, T., S.C. Bagby, M.B. Sullivan, S.W. Chisholm and N. Frankenberg-Dinkel. 2008. Efficient phage-mediated pigment biosynthesis in oceanic cyanobacteria. Curr. Biol. 18: 442–448.

Dana, H., G.M. Chalbatani, H. Mahmoodzadeh, R. Karimloo, O. Rezaiean, A. Moradzadeh et al. 2017. Molecular mechanisms and biological functions of siRNA. Int. J. Biomed. Sci. 13(2): 48–57.

Dang, Y. and B.R. Green. 2009. Substitutional editing of Heterocapsa triquetra chloroplast transcripts and a folding model for its divergent chloroplast 16S rRNA. Gene 442: 73–80. doi: 10.1016/j.gene.2009.04.006.

Daufresne, M., K. Lengfellnera and U. Sommer. 2009. Global warming benefits the small aquatic ecosystems. Proc. Natl. Acad. Sci. USA 106(31): 12788–12793.

Davis, A.K. and B. Palenik. 2008. Characterization of a modular, cell-surface protein and identification of a new gene family in the diatom *Thalassiosira pseudonana*. Protist 159: 195-207. Microbiol. Mol. Biol. Rev. 76(2): 229–261.

Davison, J.R., R. Rajwani, G. Zhao and C.A. Bewley. 2023. The genome of antibiotic-producing colonies of the Pelagophyte alga *Chrysophaeum taylorii* reveals a diverse and non-canonical capacity for secondary metabolism. Sci. Rep. 13: 11944.

Davy, S.K., D. Allemand and V.M. Weis. 2012. Cell biology of cnidarian-dinoflagellate symbiosis.

Dayhoff, M.O., R. Schwartz, B. Orcutt and M.O. Dayhoff. 1972. A model of evolutionary change in proteins. Atlas of protein sequence and structure. 5 (suppl. 3): 345-352. Washington (DC) National Biomedical Research Foundation.

De Clerck, O., S.-M. Kao, K.A. Bogaert, J. Blomme, F. Foflonker, M. Kwantes et al. 2018. Insights into the evolution of multicellularity from the sea lettuce genome. Curr. Biol. 28(18): 2921–2933.

De Los Reyes, C., J. Ávila-Román, M.J. Ortega, A. de la Jara, S. García-Mauriño, V. Motilva et al. 2014. Oxylipins from the microalgae *Chlamydomonas debaryana* and *Nannochloropsis gaditana* and their activity as TNF-α Inhibitors. Phytochemistry 102: 152–161.

De Mendoza, A., A. Bonnet, D.B. Cargas-Landin, N. Ji, H. Li, F. Yang et al. 2018. Recurrent acquisition of cytosine methyltransferases into eukaryotic retrotransposons. Nat. Comm. 9: 1341.

De Vargas, C., M.P. Aubry, I.A.N. Probert and J. Young. 2007. Origin and evolution of coccolithophores: from coastal hunters to oceanic farmers. pp. 251–285. *In*: Falkowski, P. and A.H. Knoll [eds.]. Evolution of primary producers in the sea. Academic Press, London, UK.

De Veylder, L., J.C. Larkin and A. Schnittger. 2011. Molecular control and function of endoreplication in development and physiology. Trends Plant Sci. 16: 624–634.

De Vries, J., F. Monteiro, G. Wheeler, A. Poulton, J. Godrijan, F. Cerino et al. 2021. Haplo-diplontic life cycle expands coccolithophore niche. Biogeosciences 18: 1161–1184.

De Vries, S., J.M.R. Fürst-Jansen, I. Irisarri, A.D. Ashok, T. Ischebeck, K. Feussner et al. 2021. The evolution of the phenylpropanoid pathway entailed pronounced radiations and divergences of enzyme families. Plant J. 107(4): 975–1002.

DeCarlo, T.M., L. Gajdzik, J. Ellis, D.J. Coker, M.B. Roberts, N.M. Hammerman et al. 2020. Nutrient-supplying ocean currents modulate coral bleaching susceptibility. Sci. Adv. 6(34): eabc5493.

Del Cortona, A., F. Liliaert, K.A. Bogaert, M. Turmel, C. Boedeker, J. Janouskovec et al. 2017. The plastid genome in Cladophorales green algae is encoded by hairpin chromosomes. Curr. Biol. 27: 3771–3782.

Dell'Aquila, G., M.I. Ferrante, M. Gherardi, M.C. Lagomarsino, M.R. d'Alcala, D. Ludicone et al. 2017. Nutrient consumption and chain tuning in diatoms exposed to storm-like turbulence. Sci. Rep. 7: 1828.

Delmont, T.O., Gaia, M., Hinsinger, D.D., Frémont, P., Vanni, C., Fernandez-Guerra, A. et al. 2022. Functional repertoire convergence of distantly related eukaryotic plankton lineages abundant in the sunlit ocean. Cell Genomics 2(5): 100123.

Delwiche, C.F. 2007. The origin and evolution of dinoflagellates. pp. 191–205. *In*: Falkowski, P. and A.H. Knoll [eds.]. Evolution of primary producers in the sea. Academic Press, London, UK.

Deng, Y., Z. Hu, L. Shang, Q. Peng and Y.Z. Tang. 2017. Transcriptomic analyses of *Scrippsiella trochoidea* reveals processes regulating encystment and dormancy in the life cycle of a dinoflagellate, with a particular attention to the role of abscisic acid. Front. Microbiol. 8: 2450.

Deng, Y., Z. Hu, Z. Chai and Y.Z. Tang. 2019. Cloning and partial characterization of a cold shock domain-containing protein gene from the dinoflagellate *Scrippsiella trochoidea*. J. Euk. Microbiol. 66(3): 393–403.

Deng, Y., F. Li, Z. Hu, C. Yue and Y.Z. Tang. 2021. The implication inferred from the expression of small heat-shock protein genes in dinoflagellate resting cysts buried in marine sediment. Diversity 13: 471.

Deng Y., J. Yao, X. Wang, H. Guo and D. Duan. 2012. Transcriptome sequencing and comparative analysis of *Saccharina japonica* (Laminariales, Phaeophyceae) under blue light induction. PLoS ONE 7: e39704.

Deppisch, P., C. Helfrich-Förster and P.R. Senthilan. 2022. The gain and loss of cryptochrome/photolyase family members during evolution. Genes (Basel) 13(9): 1613.

Derelle, E., C. Ferraz, S. Rombauts, P. Rouzé, A.Z. Worden, S. Robbens et al. 2006. Genome analysis of the smallest free-living eukaryote *Ostreococcus tauri* unveils many unique features. Proc. Natl. Acad. Sci. USA 103(31): 11647–11652.

Derks, A., K. Schaven and D. Bruce. 2015. Diverse mechanisms for photoprotection in photosynthesis. Dynamic regulation of photosystem II excitation in response to rapid environmental change. Biochim. Biophys. Acta 1847: 468–485.

Deschamps, P., I. Haferkamp, D. Dauvillée, S. Haebel, M. Steup, A. Buléon et al. 2006. Nature of the periplastidial pathway of starch synthesis in the cryptophyte *Guillardia theta*. Euk. Cell 5(6): 954–963.

Deschamps, P. and D. Moreira. 2012. Reevaluating the green contribution to diatom genomes. Genome Biol. Evol. 4(7): 683–688.

Devlin, M. and J. Brodie. 2023. Nutrients and Eutrophication. *In:* Reichelt-Brushett, A. [eds.]. Marine Pollution – Monitoring, Management and Mitigation. Springer Textbooks in Earth Sciences, Geography and Environment. Springer, Cham. DOI: 10.1007/978-3-031-10127-4_4.

Dexheimer, P.J. and L. Cochella. 2020. MicroRNAs: from mechanism to organism. Front. Cell Dev. Biol. 8: 409. DOI: 10.3389/fcell.2020.00409.

Dextro, R., E. Delbaje, V. Geraldes, E. Pinto, P.F. Long and M.F. Fiore. 2023. Exploring the relationship between biosynthetic gene clusters and constitutive production of mycosporine-like amino acids in Brazilian cyanobacteria. Molecules 28(3): 1420.

Diaz, J.M. and S. Plummer. 2018. Production of extracellular reactive oxygen species by phytoplankton: past and future directions. J. Plankton Res. 40(6): 655–666.

Diaz, J.M., E. Ingall, C. Benitez-Nelson, D. Paterson, M.D. De Jonge, I. McNulty et al. 2008. Marine polyphosphate: A key player in geologic phosphorus sequestration. Science 320(5876): 652–655.

Di Dato, V., F. Musacchia, G. Petrosino, S. Patil, M. Montresor, R. Sanges et al. 2015. Transcriptome sequencing of three *Pseudo-nitzschia* species reveals comparable gene sets and the presence of nitric oxide synthase genes in diatoms. Sci. Rep. 5: 12329.

Diesel, J., G. Molano, G.J. Montecinos, K. Feweese, S. Calhoun, A. Kuo et al. 2023. A scaffolded and annotated reference genome of giant kelp (*Macrocystis pyrifera*). BMC Genomics 24: 543.

Dimier, C., S. Giovanni, T. Ferdinando and C. Brunet. 2009. Comparative ecophysiology of the xanthophyll cycle in six marine phytoplanktonic species. Protist 163(3): 397–411.

Dimond, J. and E. Carrington. 2008. Symbiosis regulation in a facultatively symbiotic temperate coral: zooxanthellae division and expulsion. Coral Reefs 27: 601–604.

Dipolo, R. and L. Beaugé. 2006. Sodium/Calcium exchanger: influence of metabolic regulation on ion carrier interactions. Physiol. Rev. 86(1): 155–203.

Dobrogojski, J., M. Adamiec and R. Lucinski. 2020. The chloroplast genome: a review. Acta Physiologiae Plantarum 42: 98.

Domozych, D.S. and C.E. Domozych. 2014. Multicellularity in green algae: upsizing in a walled complex. Front. Plant Sci. 5: 649.

Doney, S.C., D.S. Busch, S.R. Cooley and K.J. Kroeker. 2020. The impacts of ocean acidification on marine ecosystems and reliant human communities. Ann. Rev. Environ. Resour. 45: 83–112.

Dong, X., C. Zhang, Y. Peng, H.-X. Zhang, L.-D. Shi, G. Wei et al. 2022. Phylogenetically and catabolically diverse diazotrophs reside in deep-sea cold seep sediments. Nat. Commun. 13: 4885.

Donovan, M.K., T.C. Adam, A.A. Shantz, K.E. Speare, K.S. Munsterman, M.M. Rice et al. 2020. Nitrogen pollution interacts with heat stress to increase coral bleaching across the seascape. Proc. Natl. Acad. Sci. USA 117(10): 5351–5357.

Dooijes, D., I. Chaves, R. Kieft, A. Kirks-Mulder, W. Martin and P. Borst. 2000. Base J originally found in Kinetoplastida is also a minor constituent of nuclear DNA of *Euglena gracilis*. Nuclear. Acid. Res. 28(16): 3017–3021.

Dorion, S., J.C. Ouellet and J. Rivoal. 2021. Glutathione metabolism in plants under stress: beyond reactive oxygen species detoxification. Metabolites 11(9): 641.

Dorrell, R.G. and C.J. Howe. 2012. Functional remodeling of RNA processing in replacement chloroplasts by pathways retained from their predecessors. Proc. Natl. Acad. Sci. 109(46): 18879–18884.

Dorrell, R.G. and C. Howe 2015. Integration of plastids with their hosts: lessons learned from dinoflagellates. Proc. Natl. Acad. Sci. 112(33): 10247–10254.

Dorrel, R.G., R.E.R. Nisbet, A.C. Barbrook, S.J.L. Rowden and C.J. Howe. 2019. Integrated genomic and transcriptomic analysis of the peridinin dinoflagellate *Amphidinium carterae* plastid. Protist 170(4): 358–373.

Dorrell, R.G., A. Villain, B. Perez-Lamarque, G.A. de Kerdrel, G. McCallum, A.K. Watson et al. 2021. Phylogenomic fingerprinting of tempo and functions of horizontal gene transfer within ochrophytes. Proc. Natl. Acad. Sci. USA 118(4): e2009974118.

Dougan, K.E., Z.-L. Deng, L. Wöhlbrand, C. Reuse, B. Bunk, Y. Chen et al. 2023. Multi-omics analysis reveals the molecular response to heat stress in a "red tide" dinoflagellate. Genome Biol. 24: 265.

Dreyer, I., K. Li, J. Riedelsberger, R. Hedrich, K.R. Konrad and E. Michard. 2022. Transporter networks can serve plant cells as nutrient sensors and mimic transceptor-like behavior. iScience 25(4): 104078.

Droop, M.R. 1973. Some thoughts on nutrient limitation in algae. J. Phycol. 9: 264–272.

Droop, M.R. 1983. 25 years of algal growth kinetics. Botanica Marina 26: 99–112.

Du, H., C. Yang, G. Ding, L. Shi and F. Xu. 2017. Genome-wide identification and characterization of SPX domain-containing members and their responses to phosphate deficiency in *Brassica napus*. Front Plant Sci. 8: 35.

Du, Z.-Y., K. Zienkiewicz, N.V. Pol, N.E. Ostrom, C. Benning and G.M. Bonito. 2019. Algal-fungal symbiosis leads to photosynthetic mycelium. eLife 8: e47815.

Dufresne, A., M. Salanoubat, F. Partensky, F. Artiguenave, I.M. Axmann, V. Barbe et al. 2003. Genome sequence of the cyanobacterium *Prochlorococcus marinus* SS120, a nearly minimal oxyphototrophic genome. Proc. Natl. Acad. Sci. USA 100(17): 10020–10025.

Dufresne, A., M. Ostrowski, D.J. Scanlan, L. Garczarek, S. Mazard, B.P. Palenik et al. 2008. Unraveling the genomic mosaic of a ubiquitous genus of marine cyanobacteria. Genome Biol. 9(5): R90.

Dugdale, R.C. 1967. Nutrient limitation in sea: dynamics, identification, and significance. Limnol. Oceanogr. 12: 685–695.

Duhamel, S., J.M. Diaz, J.C. Adams, K. Djaoudi, V. Steck and E.M. Waggoner. 2021. Phosphorus as an integral component of global marine biogeochemistry. Nat. Geosci. 14: 359–368.

Durak, G.M., C. Brownlee and G.L. Wheeler. 2017. The role of the cytoskeleton in biomineralisation in haptophyte algae. Sci. Rep. 7: 15409.

Durak, G.M., A.R. Taylor, C.E. Walker, I. Probert, de Vargas, C., S. Audic et al. 2016. A role for diatom-like silicon transporters in calcifying coccolithophores. Nat. Commun. 7: 10543.

Durkin, C.A., J.A. Koester, S.J. Bender and E.V. Armbrust. 2016. The evolution of silicon transporters in diatoms. J. Phycol. 52(5): 716–731.

Durkin, C.A., T. Mock and E.V. Armbrust. 2009. Chitin in diatoms and its association with the cell wall. Eukaryot. Cell 8(7): 1038–1050.

Dutkiewicz, S., J.J. Morris, M.J. Follows, J. Scott, O. Levitan, S.T. Dyhrman et al. 2015. Impact of ocean acidification on the structure of future phytoplankton communities. Nat. Clim. Chang. 5: 1002–1006.

Dyhrman, S.T. 2016. Nutrient and their acquisition: phosphorus physiology in microalgae. pp. 155–183. *In*: Borowitzka, M.A, J. Beardall and J.A. Raven [eds.]. The Physiology of Microalgae, Developments in Applied Phycology 6, Springer.

Dyhrman, S.T., P.D. Chappell, S.T. Haley, J.W. Moffett, E.D. Orchard, J.B. Waterbury et al. 2006. Phosphonate utilization by the globally important marine diazotroph *Trichodesmium*. 439(7072): 68–71.

Dyhrman, S.T., B.D. Jenkins, T.A. Rynearson, M.A. Saito, M.L. Mercier, H. Alexander et al. 2012. The transcriptome and proteome of the diatom *Thalassiosira pseudona* reveal a diverse phosphorus stress respose. PLoS ONE 7(3): e33768.

Ebenezer, T.E., M. Zoltner, A. Burrell, A. Nenarokova, A.M.G.N. Vanclová, B. Prasad et al. 2019. Transcriptome, proteome and draft genome of *Euglena gracilis*. BMC Biol. 17: 11.

Edmunds, I.N. 1966. Studies on synchronously dividing cultures of *Euglena gracilis* klebs (strain z) .3. Circadian components of cell division. J. Cell. Physiol. 67: 35.

Edmunds, L.N. Jr. and K.J. Adams. 1981. Clocked cell cycle clocks. Science 211(4486): 1002–1013.

Edvardsen, B., E.S. Egge and D. Vaulot. 2016. Diversity and distribution of haptophytes revealed by environmental sequencing and metabarcoding-a review. Perspect. Phycol. 3: 77–91.

Ehrlich, E., U.H. Thygesen and T. Kiørboe. 2022. Evolution of toxins as a public good in phytoplankton. Proc. Biol. Sci. 289(1977): 20220393.

Eide, D.J. 2006. Zinc transporters and the cellular trafficking of zinc. Biochim. Biophys. Acta-Mol. Cell Res. 1763(7): 711–722.

Ellegren, H. 2004. Microsatellites: simple sequences with complex evolution. Nat. Rev. Genet. 5: 435–445.

Ellis, K.A., N.R. Cohen, C. Moreno and A. Marchetti. 2017. Cobalamin-independent methionine synthase distribution and influence on vitamin B12 growth requirements in marine diatoms. Protist 168(1): 32–47.

Elser, J.J., R.W. Sterner, E. Gorokhova, W.F. Fagan, T.A. Markow, J.B. Cotner et al. 2000. Biolgoical stoichiometry from genes to ecosystems. Ecol. Lett. 3(6): 540–550.

Endo, H., Y. Hanawa, H. Araie, I. Suzuki and Y. Shiraiwa. 2018. Overexpression of *Tisochrysis lutea* Akd1 identifies a key cold-induced alkenone desaturase enzyme. Sci. Rep. 8: 11230.

Engelken, J., H. Brinkmann and I. Adamska. 2010. Taxonomic distribution and origins of the extended LHC (light-harvesting complex) antenna protein superfamily. BMC Evol. Biol. 10: 233.

Epstein, B. and P. Tiffin. 2021. Comparative genomics reveals high rates ofhorizontal transfer and strong purifying selection on rhizobial symbiosis genes. Proc. Biol. Sci. 288(1942): 20201804.

Erdner, D.L. and D.M. Anderson. 2006. Global transcriptional profiling of the toxic dinoflagellate *Alexandrium fundyense* using massively parallel signature sequencing. BMC Genomics 7: 88.

Erdner, D.L., N.M. Price, G.J. Doucette, M.L. Peleato and D.M. Anderson. 1999. Characterization of ferredoxin and flavodoxin as markers of iron limitation in marine phytoplankton. Mar. Ecol. Prog. Ser. 184: 43–53.

Ermilova, E. 2020. Cold stress response: an overview in Chlamydomonas. Front. Plant Sci. 11: 569437.

Errera, R.M., A. Bourdelais, M. Drennan, E. Dodd, D. Henrichs and L.J.T. Campbell. 2010. Variation in brevetoxin and brevenal content among clonal cultures of *Karenia brevis* may influcence bloom toxicity. Toxicon 55(2-3): 195–203.

Escalera, L., B. Reguera, K. Takishita, S. Yoshimatsu, K. Koike and K. Koike. 2011. Cyanobacterial endosymbionts in the benthic dinoflagellate *Sinophysis canaliculata* (Dinophysiales, Dinophyceae). Protist 162: 304–314.

Esteban, R., I. Ariz, C. Cruz and J.F. Moran. 2016. Review: Mechanisms of ammonium toxicity and the quest for tolerance. Plant Science 248: 92–101.

Evans, F.F., S. Egan and S. Kjelleberg. 2008. Ecology of type II secretion in marine gammaproteobacteria. Environ. Microbiol. 10: 1101–1107.

Evans, T., E.T. Rosenthal, J. Youngbloom, d. Distel and T. Hunt. 1983. Cyclin: a protein specified by maternal mRNA in sea urchin eggs that is destroyed at each cleavage division. Cell 33: 389–396.

Ewe, D., M. Tachibana, S. Kikutani, A. Gruber, C.R. Bártulos, G. Konert et al. 2018. The intracellular distribution of inorganic carbon fixing enzymes does not support the presence of a C4 pathway in the diatom *Phaeodactylum tricornutum*. Photosynth. Res. 137: 263–280.

Faber, W.W. and J.J. Lee. 1991. Histochemical evidence for digestion in *Heterostegina depressa* and *Operculina ammonoides* (Foraminifera). Endocytobiol. Cell Res. 8: 53–59.

Faktorová, D., R.E.R. Nisbet, J.A.F. Robledo, E. Casacuberta, L. Sudek, A.E. Allen et al. 2020. Genetic tool development in marine protists: Emerging model organisms for experimental cell biology. Nat. Methods 17: 481–494.

Falciatore, A., M.R. d'Alcalà, P. Croot and C. Bowler. 2000. Perception of environmental signals by a marine diatom. Science 288: 2363–2366.

Falkowski, P.G. 1997. Photosynthesis: The paradox of carbon dioxide efflux. Curr. Biol. 7: R637–639.

Falkowski, P.G. and J.A. Raven 1997. Aquatic Photosynthesis. Princeton University Press, Princeton, New Jersey, USA.

Fan, C., P.M. Glibert and J.M. Burkholder 2003. Characterization of the nitrogen uptake kineticsof *Prorocentrum minimum* in natural blooms and laboratory cultures. Harmful Algae 2: 283–299.

Fan, W., W. Guo, J.L. Van Etten and J.P. Mower. 2017. Multiple origins of endosymbionts in Chlorellaceae with no reductive effects on the plastid or mitochondrial genomes. Sci. Rep. 7: 10101.

Farhat, S., I. Florent, B. Noel, E. Kayal, C.D. Silva, E. Bigeard et al. 2018. Comparative time-scale gene expression analysis highlights the infection processes of two amoebophrya strains. Front. Microbiol. 9: 2251.

Farhat, S., P. Le, E. Kayal, B. Noel, E. Bigeard, E. Corre et al. 2021. Rapid protein evolution, organellar reductions, and invasive intronic elements in marine aerobic parasite dinoflagellate *Amoebophrya* spp. BMC Biol. 19: 1.

Featherston, J., Y. Arakaki, E.R. Hanschen, P.J. Ferris, R.E. Michod, B.J.S.C. Olson et al. 2017. The 4-celled *Tetrabaena socialis* nuclear genome reveals the essential components for genetic control of cell number at the origin of multicellularity in the volvocine lineage. Mol. Biol. Evol. 35(4): 855–870.

Feely, R. A., C.L.S. Sabine and V.J. Fabry. 2006. Carbon Dioxide and Our Ocean Legacy. http://www.pmel.noaa.gov/pubs/PDF/feel2899/feel2899.pdf.

Feinstein, T.N., R.P. Traslavina, M.-Y. Sun and S. Lin. 2002. Effects of light on photosynthesis, grazing, and population dynamics of the heterotrophic dinoflagellate *Pfiesteria piscicida* (Dinophyceae). J. Phycol. 38(4): 659–669.

Fensome, R.A., R.J. Taylor, G. Norris, W.A. Sarjeant, D.I. Wharton and G.L. Williams. 1993. A classification of living and fossil dinoflagellates. Micropaleontol. Spec. Publ. 7. Hanover PA: Sheridan Press.

Fernández, P.A., J.D. Gaitán-Espitia, P.P. Leal, M. Schmid, A.T. Revill and C.L. Hurd. 2020. Nitrogen sufficiency enhances thermal tolerance in habitat-ofrming kelp: implications for acclimation under thermal stress. Sci. Rep. 10: 3186.

Ferrari, M., A. Muto, L. Bruno and R. Cozza. 2023. DNA methylation in algae and its impact on abiotic stress responses. Plants (Basel) 12(2): 241.

Ferreira, P.C.G., A.S. Hemerly, R. Villarroel, M. Van Montagu and D. Inze. 1991. The *Arabidopsis* functional homolog of the p34cdc2 protein kinase. Plant Cell 3: 531–540.

Ficko-Blean, E., C. Hervé and G. Michel. 2015. Sweet and sour sugars from the sea: the biosynthesis and remodeling of sulfated cell wall 750 polysaccharides from marine macroalgae. Perspect. Phycol. 2: 51–64.

Figueroa, R.I., I. Bravo, S. Fraga, E. Garcés and G. Llaveria. 2009. The life history and cell cycle of *Kryptoperidinium foliaceum*, a dinoflagellate with two eukaryotic nuclei. Protist 160: 285–300.

Figueroa, R.I., C. Dapena, I. Bravo and A. Cuadrado. 2015. The hidden sexuality of *Alexandrium minutum*: an example of overlooked sex in dinofalgellates. PLoS ONE 10(11): e0142667.

Figueroa, R.I., L.I. Howe-Kerr and A.M.S. Correa. 2021. Direct evidence of sex and a hypothesis about meiosis in Symbiodiniaceae. Sci. Rep. 11: 18838.

Finka, A., R.U.H. Mattoo and P. Goloubinoff. 2011. Meta-analysis of heat- and chemically upregulated chaperone genes in plant and human cells. Cell Stress Chaperones 16(1): 15–31.

Finkel, Z.V., M.J. Follows, J.D. Liefer, C.M. Brown, I. Benner and A.J. Irwin. 2016. Phylogenetic diversity in the macromolecular composition of microalgae. PLoS ONE 11: e0155977.

Finking, R. and M.A. Marahiel. 2004. Biosynthesis of nonribosomal peptides. Ann. Rev. Microbiol. 58: 453–488.

Finzi-Hart, J.A., J. Pett-Ridge, P.K. Weber, R. Popa, S.K. Fallon, T. Gunderson et al. 2009. Dixation and fate of C and N in the cyanobacterium *Trichodesmium* using nanometer-scale secondary ion mass spectrometry. Proc. Natl. Acad. Sci. USA 106(15): 6345–6350.

Fitzpatrick, T.B. and L.M. Chapman. 2020. The importance of thamine (vitamin B1) in plant health: from crop yield to biofortification. J. Biol. Chem. 295(34): 12002–12013.

Fixen, K.R., Y. Zheng, D.F. Harris, S. Shaw, Z.-Y. Yang, D.R. Dean et al. 2016. Light-driven carbon dioxide reduction to methane by nitrogenase in a photosynthetic bacterium. Proc. Natl. Acad. Sci. USA 113(36): 10163–10167.

Flamholz, A.I., N. Prywes, U. Moran, D. Davidi, Y.M. Bar-On, L.M. Oltrogge et al. 2019 Revisiting trade-offs between Rubisco kinetic parameters. Biochemistry 58: 3365–3376.

Floener, L. and H. Bothe. 1980. Nitorgen fixation in *Rhopalodia gibba*, a diatom containing blue-greenish inclusions symbiotically. pp. 541–552. *In:* Schwemmler, W. and H.E.A. Schenk [eds.]. Endocytobiology, endosymbiosis and cell biology. Walter de Gruyter & Co., Berlin.

Flores, E., D.K. Romanovicz, M. Nieves-Morión, R.A. Foster and T.A. Villareal. 2022. Adaptation to an intracellular lifestyle by a nitrogen-fixing heterocyst-forming cyanobacterial endosymbiont of a diatom. Front. Microbiol. 13: 799362.

Flores, H.S., G.H. Wikfors and H.G. Dam. 2012. Reactive oxygen species are linked to the toxicity of the dinoflagellate *Alexandrium* spp. to protists. Aquat. Microb. Ecol. 66(2): 199–209.

Flores-Molina, M.R., D. Thomas, C. Lovazzano, A. Núñez, J. Zapata, M. Kumar et al. 2014. Desiccation stress in intertidal seaweeds: effects on morphology, antioxidant response and photosynthetic performance. Aquat. Bot. 113: 90–99.

Flöthe, C.R., M. Molis and U. John. 2014. Induced resistance to periwinkle grazing in the brown seaweed *Fucus vesiculosus* (Phaeophyceae): molecular insights and seaweed-mediated effects on herbivore interactions. J. Phycol. 50: 564–576.

Flöthe, C.R., M. Molis, I. Kruse, F. Weinberger and U. John. 2014. Herbivore-induced defence response in the brown seaweed *Fucus vesiculosus* (Phaeophyceae): temporal pattern and gene expression. Eur. J. Phycol. 49(3).

Fojtová, M., J.T. Wong, M. Dvořáčková, K.T. Yan, E. Sýkorová, J. Fajkus. 2010. Telomere maintenance in liquid crystalline chromosomes of dinoflagellates. Chromosoma 119: 485–493.

Fontana, A., G. d'Ippolito, A. Cutignano, A. Miralto, A. Ianora, G. Romano and G. Cimino. 2007. Chemistry of oxylipin pathways in marine diatoms. Pure Appl. Chem. 79: 481–490.

Footitt, S., I. Douterelo-Soler, H. Clay and W.E. Finch-Savage. 2011. Dormancy cycling in *Arabidopsis* seeds is controlled by seasonally distinct hormone-signaling pathways. Proc. Natl. Acad. Sci. 108(50): 20236–20241.

Forterre, P. 2013. The virocell concept and environmental microbiology. ISME J. 7: 233–236.

Foster, R.A. and G.K. O'Mullan. 2008. Nitrogen-fixing and nitrifying symbioses in the marine environment. pp. 1197–1218. *In:* Capone, D.G., D.A. Bronk, M.R. Mulholland and E.J. Carpenter [eds.]. Nitrogen in the Marine Environment (2nd edition). Academic Press

Foster, R.A. and J.P. Zehr. 2006. Characterization of diatom-cyanobacteria symbioses on the basis of nifH, hetR and 16S rRNA sequences. Environ. Microbiol. 8(11): 1913–1925.

Foster, R.A. and J.P. Zehr. 2019. Diversity, genomics, and distribution of phytoplankton-cyanobacterium single-cell symbiotic associations. Annu. Rev. Microbiol. 73: 435–456.

Foster, R.A., N.L. Goebel and J.P. Zehr. 2010. Isolation of *Calothrix rhizosoleniae* (Cyanobacteria) strain SC01 from *Chaetoceros* (Bacillariophyta) spp. diatoms of the subtropical North Pacific Ocean. J. Phycol. 46(5): 1028–1037.

Foster R.A., J.L. Collier and E.J. Carpenter. 2006a. Reverse transcription PCR amplification of cyanobacterial symbiont 16S rRNA sequences from single non-photosynthetic eukaryotic marine planktonic host cells. J. Phycol. 42: 243–250.

Foster, R.A., E.J. Carpenter and B. Bergman. 2006b. Unicellular cyanobionts in open ocean dinoflagellates, radiolarians, and tintinnids: ultrastructural characterization and immunolocalization of phycoerythrin and nitrogenase. J. Phycol. 42(2): 453–463.

Fučíková, K., M. Pažoutová and F. Rindi. 2015. Meiotic genes and sexual reproduction in the green algal class Trebouxiophyceae (Chlorophyta). J. Phycol. 51: 419–430.

Foster, R.A., M.M.M. Kuypers, T. Vagner, R.W. Paerl, N. Musat and J.P. Zehr. 2011. Nitrogen fixation and transfer in open ocean diatom-cyanobacterial symbioses. ISME J. 5(9): 1484–1493.

Fott, B. 1967. Phlyctidium scenedesmi spec. nova, a new chytrid destroying mass cultures of algae. Z. Allg. Mikrobiol. 7: 97–102.

Foyer, C.H., M. Lelandais and K.K. Kunert. 1994. Photooxidative stress in plants. Physiol. Plant. 92: 696–717.

Fozo, E.M. and R.G. Quivey Jr. 2004. The fabM gene product of *Streptococcus mutans* is responsible for the synthesis of monounsaturated fatty acids and is necessary for survival at low pH. J. Bacteriol. 186: 4152–4158.

Frada, M., I. Probert, M.J. Allen, W.H. Wilson and C. de Vargas. 2008. The "Cheshire Cat" escape strategy of the coccolithophore *Emiliania huxleyi* in response to viral infection. Proc. Natl. Acad. Sci. USA 105(41): 15944–15949.

Frankel, E. 1980. Lipid oxidation. Progr. Lipid Res. 19: 1–22.

Frankignoulle, M., C. Canon and J.-P. Gattuso. 1994. Marine calcification as a source of carbon dioxide: positive feedback of increasing atmospheric CO_2. Limnol. Oceanogr. 39(2): 458–462.

Franks, F., S.F. Mathias and R.H.M. Hatley. 1990. Water, temperature and life. Philos. Trans. R. Soc. Lond. B 326: 517–533.

Franks, P.J.S. 2015. Has Sverdrup's critical depth hypothesis been tested? Mixed layers vs. turbulent layers. ICES J. Mar. Sci. 72(6): 1897–1907.

Franssen, S.U., J. Gu, N. Bergmann, G. Winters, U.C. Klostermerier, P. Rosenstiel et al. 2011. Transcriptomic resilience to global warming in the seagrass *Zostera marina*, a marine foundation species. Proc. Natl. Acad. Sci. USA 108: 19276–19281.

Frischkorn, K.R., M.J. Harke, C.J. Gobler and S.T. Dyhrman. 2014. De novo assembly of *Aureococcus anophagefferens* transcriptomes reveals divers responses to the low nutrient and low light conditions present during blooms. Front. Microbiol. 5: 375.

Frommolt, R., S. Werner, H. Paulsen, R. Goss, C. Wilhelm, S. Zauner et al. 2008. Ancient recruitment by chromists of green algal genes encoding enzymes for carotenoid biosynthesis. Mol. Biol. Evol. 25(12): 2653–2667.

Fu, H.-Y. and M.-W. Wang. 2023. Ascorbate peroxidase plays an important role in photoacclimation in the extremophilic red alga *Cyanodiococcus yangmingshanensis*. Front. Plant Sci. 14: 1176985.

Fu, H.-Y., S.-L. Liu and Y.-R. Chiang. 2020. Biosynthesis of ascorbic acid as a glucose-induced photoprotective process in the extremophilic red alga *Galdieria partita*. Front. Microbiol. 10: 3005.

Füßl, M., A.-C., König, J. Eirich, M. Hartl, L. Kleinknecht, A.-V. Bohne et al. 2021. Dynamic light- and acetate-dependent regulation of the proteome and lysine acetylome of *Chlamydomonas*. Plant J. 109: 261–277.

Fučíková, K., F. Leliaert, E.D. Cooper, P. Škaloud, S. D'Hondt, O. De Clerck et al. 2014. New phylogenetic hypotheses for the core Chlorophyta based on chloroplast sequence data. Front. Ecol. Evol. 2: 63.

Fuentes-Grünewald, C., E. Garcés, E. Alacid, N. Sampedro, S. Rossi and J. Camp. 2012. Improvement of lipid production in the marine strains *Alexandrium minutum* and *Heterosigma akashiwo* by utilizing abiotic parameters. J. Ind. Microbiol. Biotechnol. 39(1): 207–216.

Fuhrman, J. 2003. Genome sequences from the sea. Nature 424(6952): 1001–1002.

Fukui, Y., M. Abe, M. Kobayashi, Y. Shimada, H. Saito, H. Oidawa, et al. 2014. *Sulfitobacter porphyrae* sp. nov., isolated from the red alga *Porphyra yezoensis*. Int. J. Syst. Evol. Microbiol. 64(Pt 2): 438–443.

Fulton J.M., H.F. Fredricks, K.D. Bidle, A. Vardi, B.J. Kendrick, G.R. DiTullio et al. 2014. Novel molecular determinants of viral susceptibility and resistance in the lipidome of *Emiliania huxleyi*. Environ. Microbiol. 16: 1137–1149.

Fusté, N.P., R. Fernández-Hernáadez, T. Cemeli, C. Mirantes, N. Pedraza, M. Rafel et al. 2016. Nat. Commun. 7: 11581.

Gabr, A., T.G. Stephens and D. Bhattacharya. 2022. Loss of key endosymbiont genes may facilitate early host control of the chromatophore in *Paulinella*. iScience 25: 104974.

Gademann, K., C. Portmann, J.F. Blom, M. Zeder and F. Jüttner. 2010. Multiple toxin production in the cyanobacterium *Microcystis*: isolation of the toxic protease inhibitor cyanopeptolin 1020. J. Nat. Prod. 73: 980–984.

Galdieri, L., T. Zhang, D. Rogerson, R. Lleshi and A. Vancura. 2014. Protein acetylation and acetyl coenzyme A metabolism in budding yeast. Eukaryot. Cell. 13(12): 1472–1483.

Gale, E.F. 1946. The bacterial amino acid decarboxylases. Adv. Enzymol. 6: 1–31.

Galino-González, C. Mhiri, M.K. Deyholos and M.-A. Grandbastien. 2017. LTR-retrotransposons in plants: Engines of evolution. Gene 626: 14–25.

Gallaher, S. D., R.J. Craig, I. Ganesan, S.O. Purvine, S.R. McCorkle, J. Grimwood et al. 2021. Widespread polycistronic gene expression in green algae. Proc. Natl. Acad. Sci. USA 118: e2017714118. Doi: 10.1073/pnas.2017714118.

Ganini, D., H.C. Hollnagel, P. Colepicolo and M.P. Barros. 2013. Hydrogen peroxide and nitric oxide trigger redox-related cyst formation in cultures of the dinoflagellate *Lingulodinium polyedrum*. Harmful Algae 27: 121–129.

Ganz, T. 2003. Defensins: Antimicrobial peptides of innate immunity. Nat. Rev. Immunol. 3: 710–720.

Gao, K. and D.A. Campbell. 2014. Photoshysiological response of marine diatoms to elevated CO_2 and decreased pH: a review. Funct. Plant. Biol. 41: 449–459.

Gao, K., J. Beardall, D.-P. Häder, J.M. Hall-Spencer, G. Gao and D. Hutchins. 2019. Effects of ocean acidification on marine photosynthetic organisms under the concurrent influences of warming, UV radiation, and deoxygenation. Front. Mar. Sci. 6: 322.

Gao, K., J. Xu, G. Gao, Y. Li, D.A. Hutchins, B. Huang, et al. 2012. Rising CO_2 and increased light exposure synergistically reduce marine primary productivity. Nat. Clim. Change 2: 519–523.

Gao, S., S. Shen, G. Wang, J. Niu, A. Lin and G. Pan. 2011. PSI-driven cyclic electron flow allows intertidal macro-algae *Ulva* sp. (Chlorophyta) to survive in desiccated conditions. Plant Cell Physiol. 52(5): 885–893.

Gao, S., W. Yang, X. Li, L. Zhou, Z. Liu, S. Wu et al. 2024. Cryptochrome PtCPF1 regulates high temperature acclimation of marine diatoms through coordination of iron and phosphorus uptake. ISME J. 18(1): wrad019.

Gao, S., L. Zhou, W. Yang, L. Wang, X. Liu, Y. Gong et al. 2022. Overexpression of a novel gene (Pt2015) endows the commercial diatom *Phaeodactylum tricornutum* high lipid content and grazing resistance. Biotechnol. Biofuels Bioprod. 15: 131.

Gao, T., Z. Mo, L. Tang, X. Yu, G. Du and Y. Mao. 2022. Heat shock protein 20 gene superfamilies in red algae: evolutionary and functional diversities. Front. Plant Sci. 13: 817852.

Gao, X., C. Bowler and E. Kazamia 2021. Iron metabolism strategies in diatoms. J. Exp. Bot. 72: 2165-2180.

Garces, E., E. Alacid, A. Reñé, K. Petrou and R. Simo. 2013. Host-released dimethylsulphide activates the dinoflagellate parasitoid *Parvilucifera sinerae*. ISME J 7: 1065–1068.

Garcia-Cuetos, L., Ø. Moestrup, P.J. Hansen and N. Daugbjerg. 2010. The toxic dinoflagellate *Dinophysis acuminata* harbors permanent chloroplasts of cryptomonad origin, not kleptochloroplasts. Harmful Algae 9: 25–38.

García-Portela, M., B. Reguera, M.R. d'Alcalà, F. Rodríguez and M. Montresor. 2019. Effects of small-scale turbulence on two species of *Dinophysis*. Harmful Algae 89: 101654.

Garg, R. and I. Maldener. 2021. The formation of spore-like akinetes: a survival strategy of filamentous cyanobacteria. Microb. Physiol. 31(3): 296–305.

Gast, R., D. Moran, D. Dennett and D. Caron. 2007. Kleptoplasty in an Antarctic dinoflagellate: caught in evolutionary transition? Environ. Microbiol. 9(1): 39–45.

Gawad, C., W. Koh and S.R. Quake. 2016. Single-cell genome sequencing: current state of the science. Nat. Rev. Genetics 17: 175–188.

Geffroy, S., M.M. Lechat, M. Le Gac, G.A. Rovillon, D. Marie, E. Bigeard et al. 2021. From the sxtA4 gene to saxitoxin production: what controls the variability among *Alexandrium minutum* and *Alexandrium pacificum* strains? Front Microbiol. 12: 613199.

Geider, R.J. and J. La Roche. 2002. Redfield revisited: variability of C:N:P in marine microalgae and its biochemical basis. Eur. J. Phycol. 37: 1–17.

Ger, K.A., E.J. Faassen, M.G. Pennino and M. Lürling. 2016. Effect of the toxin (microcystin) content of *Microcystis* on copepod grazing. Harmful Algae 52: 34–45.

Gerard, V.A. 1997. The role of nitrogen nutrition in high-temperature tolerance of the kelp, *Laminaria saccharina* (Chromophyta). J. Phycol. 33(5): 800–810.

Gerwick, W.H. and I.P. Singh. 2002. Structural diversity of marine oxylipins. pp. 249–275. *In*: Gardner, H.W., T.M. Kuo [eds]. Lipid Biotechnology. Marcel and Dekker; New York, NY, USA.

Gerwick, W.H. P.J. Proteau, D.G. Nagle, M.L. Wise, Z.D. Jiang, M.W. Bernart et al. 1993. Biologically active oxylipins from seaweeds. Hydrobiologia. 260: 653–665.

Ghurye, J., A. Rhie, B.P. Walenz, A. Schmitt, S. Selvaraj, M. Pop et al. 2019. Integrating Hi-C links with assembly graphs for chromosome-scale assembly. PLoS Comput. Biol. 15(8): e1007273.

Gilbert, W.V., T.A. Bell and C. Schaening. 2016. Messenger RNA modifications: form, distribution, and function. Science 352(6292): 1408–1412.

Gillard, J., V. Devos, M.J.J. Huysman, L. De Veylder, S. D'Hondt, C. Martens et al. 2008. Physiological and transcriptomic evidence for a close coupling between chloroplast ontogeny and cell cycle progression in the pennate diatom *Seminavis robusta*. Plant Physiol. 148: 1394–1411.

Gillooly, J.F., A. Hein and R. Damiani. 2015. Nuclear DNA content varies with cell size across human cell types. Cold Spring Harbor Perspectives in Biology 7: a019091.

Gillott, M.A. and S.P. Gibbs. 1980. The cryptomaonad nucleomorph: its ultrastructure and evolutionary significance. J. Phycol. 16(4): 558-568.

Giordano, M., A. Norici, M. Forssen, M. Eriksson and J.A. Raven. 2003. An anaplerotic role for mitochondrial carbonic anhydrase in *Chlamydomonas reinhardtii*. Plant Physiol. 132: 2126–2134.

Giordano, M., J. Beardall and J.A. Raven. 2005. CO_2 concentrating mechanism in algae: mechanisms, environmental modulation, and evolution. Ann. Rev. Plant Biol. 56: 99–131.

Giovannoni, S.J., J. Cameron Thrash and B. Temperton. 2014. Implications of streamlining theory for microbial ecology. ISME J. 8: 1553–1565.

Gittings, J.A., D.E. Raitsos, G. Krokos and I. Hoteit. 2018. Impacts of warming on phytoplankton abundance and phenology in a typical tropical marine ecosystem. Sci. Rep. 8: 2240.

Gledhill, M. and K.N. Buck. 2012. The organic complexation of iron in the marine environment: a review. Front. Microbiol. 3: 69.

Glibert, P.M. 2016. Margalef revisited: a new phytoplankton mandala incorporating twelve dimensions including nutritional physiology. Harmful Algae 55: 25–30.

Glibert, P.M., J. Harrison, C. Heil, and S. Seitzinger. 2006. Escalating worldwide use of urea-a global change contributing to coastal eutrophication. Biogeochemistry 77: 441–463.

Glibert, P.M., F.P. Wilkerson, R.C. Dugdale, J.A. Raven, C.L. Dupont, P.R. Leavitt et al. 2016. Pluses and minuses of ammonium and nitrate uptake and assimilation by phytoplankton and implications for productivity and community composition, with emphasis on nitrogen-enriched conditions. Limnol. Oceanogr. 61(1): 165–197.

Glotzer, M., A.W. Murray and M.W. Kischner. 1991. Cyclin is degraded by the ubiquitin pathway. Nature 349: 132–137.

Gobler, C.J., D.L. Berry, S.T. Dyhrman, S.W. Wilhelm, A. Salamov, A.V. Lobanov et al. 2011. Niche of harmful alga *Aureococcus anophagefferens* revealed through ecogenomics. Proc. Natl. Acad. Sci. USA 108(11): 4352–4357.

Gobler, C.J., A.V. Lobanov, Y.-Z. Tang, A.A. Turanov, Y. Zhang, M. Doblin et al. 2013. The central role of selenium in the biochemistry and ecology of the harmful pelagophyte, *Aureococcus anophagefferens*. ISME J. 7: 1333–1343.

Godde, J.S., S. Baichoo, Z. Mungloo-Dilmohamud and Y. Jaufeerally-Fakim. 2018. Comparison of genomic islands in cyanobacteria: evidence of bacteriophage-mediated horizontal gene transfer from eukaryotes. Microbiol. Res. 211: 31–46.

Godfray, H.C.J. 1994. Parasitoids: behavioral and evolutionary ecology. Princeton University Press, Princeton, NJ, USA. 473 pp.

Goff, L.J. and A.W. Coleman. 1990. Red algal plasmids. Curr. Genet. 18: 557–565.

Goiris, K., K. Muylaert, S. Voorspoels, B. Noten, D. De Paepe, G.J.E. Baart et al. 2014. Detection of flavonoids in microalgae from different evolutionary lineages. J. Phycol. 50: 483–492.

Gómez, F. 2012. A quantitative review of the lifestyle, habitat and trophic diversity of dinoflagellates (Dinoflagellata, Alveolata). System. Biodiv. 10(3): 267–275.

Gómez-Consarnau, L., N. Akram, K. Lindell, A. Pedersen, R. Neutze, D.L. Milton et al. 2010. Proteorhodopsin phototrophy promotes survival of marine bacteria during starvation. PLoS Biol. 8: e1000358.

Gómez-Consarnau, L, J.A. Raven, N.M. Levine, L.S. Cutter, D. Wang, B. Seegers et al. 2019. Microbial rhodopsins are major contributors to the solar energy captured in the sea. Sci Adv. 5: eaaw8855.

Gomez-Garcia, M.R., M. Davison, M. Blain-Hartnung, A.R. Grossman and D. Bhaya. 2011. Alternative pathways for phosphonate metabolism in thermophilic cyanobacteria from microbial mats. ISME J. 5(1): 141–149.

Gomez-Osuna, A., V. Calatrava, A. Galvan, E. Fernandez and A. Llamas. 2020. Identification of the MAPK cascade and its relationship with nitrogen metabolism in the green alga *Chlamydomonas reinhardtii*. Int. J. Mol. Sci. 21: 3417.

Goncalves, E.C., J. Koh, N. Zhu, M.-J. Yoo, S. Chen, T. Matsuo et al. 2016. Nitrogen starvation-induced accumulation of triacylglycerol in the green algae: evidence for a role for ROC40, a transcription factor involved in circadian rhythm. Plant J. 85(6): 743–757.

Gong, S., X. Jin, Y. Xiao and Z. Li. 2020. Ocean acidification and warming lead to increased growth and altered chloroplast morphology in the thermos-tolerant alga *Symbiochlorum hainanensis*. Front. Plant Sci. 11: 585202.

Gong, B., X. Liang, Y. Li, Q. Xiao, P. Yang and Y. Wu. 2018. Expression and purification of cytochrome P450 55B1 from *Chlamydomonas reinhardtii* and its application in nitric ocide biosensing. Appl. Biochem. Biotechnol. 184: 102–112.

Gonzalez-Ballester D., S.V. Pollock, W. Pootakham and A.R. Grossman. 2008. The central role of a SNRK2 kinase in sulfur deprivation responses. Plant Physiol. 147: 216–227.

González-Pech, R., T.G. Stephens, Y. Chen, A.R. Mohamed, Y. Cheng, S. Shah et al. 2021. Comparison of 15 dinoflagellate genomes reveals extensive sequence and structural divergence in family Symbiodiniaceae and genus *Symbiodinium*. BMC Biol. 19: 73.

Goodenough, U. and J. Heitman. 2014. Origins of eukaryotic sexual reproduction. Cold. Spring. Harb. Perspect. Biol. 6, a016154.

Goodenough, U., H. Lin and J.-H. Lee. 2007. Sex determination in *Chlamydomonas*. Semin. Cell Dev. Biol. 18: 350–361.

Goreau, T.F. 1963. Calcium carbonate deposition by coralline algae and coral in relation to their roles as reef builders. Ann. N.Y. Acad. Sci. 109: 127–167.

Gornik, S.G., Febrimarsa, A.M. Cassin, J.J. MacRae, A. Ramaprasad, Z. Rchiad et al. 2015. Endosymbiosis undone by stepwise elimination of the plastid in a parasitic dinoflagellate. Proc. Natl. Acad. Sci. USA 112(18): 5767–5772.

Gornik, S.G., K.L. Ford, T.D. Mulhem, A. Bacic, G.I. McFadden and R.F. Waller. 2012. Loss of nucleosomal DNA condensation coincides with appearance of a novel nuclear protein in dinoflagellates. Curr. Biol. 22(24): 2303–2312.

Gornik, S.G., I. Hu, I. Lassadi and R.F. Waller. 2019. The biochemistry and evolution of the dinoflagellate nucleus. Microorganisms 7(8): 245.

Gornik, S.G., I. Maegele, E.A. Hambleton, P.A. Voss and R. Waller. 2022. Nuclear transformation of a dinoflagellate symbiont of corals. Front. Mar. Sci. 9: 1035413.

Goss, R. and T. Jakob. 2010. Regulation and function of xanthophyll cycle-dependent photoprotection in algae. Photosynthesis Res. 106: 103–122.

Gotelli, D. 1971. *Lagenisma coscinodisci*, a parasite of the marine diatom *Coscinodiscus*, occurring in the Puget Sound, Washington. Mycologia. 63: 171–174.

Gould, A.L. and J.B. Henderson. 2023. Comparative genomics of symbiotic *Photobacterium* using highly contiguous genome assemblies from long read sequences. Microb. Genom. 9(12): 001161.

Gould, K.L., S. Moreno, D.J. Owen, S. Sazer and P. Nurse. 1991. Phosphorylation at Thr167 is required for *Schizosaccharomyces pomb* p34cdc2 function. EMBO J. 10: 3297–3309.

Gould, S.B., W.-H. Tham, A.F. Cowman, G.I. McFadden and R.F. Waller. 2008. Alveolins, a new family of cortical proteins that define the protist infrakingdom Alveolata. Mol. Biol. Evol. 25(6): 1219–1230.

Govorunova, E.G., K.-H. Jang, O.A. Sineshchekov and J.L. Spudich. 2004. *Chlamydomonas* sensory rhodopsins A and B: cellular content and role in photophobic responses. Biophys. J. 86(4): 2342–2349.

Graham, L.E. and L.W. Wilcox. 2000. Algae. 640 pp. Prentice Hall.

Graham, L.E., J.M. Graham, L.W. Wilcox and M.E. Cook. 2022. Algae (fourth edition). 853 pp. LJLM Press.

Gray, M.W. 2012. Evolutionary origin of RNA editing. Biochemistry 51(26): 5235–5242.

Grami, B., S. Rasconi, N. Niquil, M. Jobard, B. Saint-Béat and T. Sime-Ngando. 2011. Functional effects of parasites on food web properties during the spring diatom bloom in Lake Pavin: a linear inverse modeling analysis. PLoS ONE 6:e23273.

Grébert, T., L. Garczard, V. Daubin, F. Humily, D. Marie, M. Ratin et al. 2022. Diversity and evolution of pigment types in marine *Synechococcus* cyanobacteria. Genome Biol. Evol. 14(4): evac035.

Green, B.J., W.-Y. Li, J.R. Manhart, T.C. Fox, E.J. Summer, R.A. Kennedy et al. 2000. Mollusc-algal chloroplast endosymbiosis. Photosynthesis, thylakoid protein maintenance, and chloroplast gene expression continue for many months in the absence of the algal nucleus. Plant physiol. 124: 331–342.

Green, D.H., C. Rad-Menéndez, C. Campbell and E.S. Kilias. 2023. The genome sequence of *Pycnococcus provasolii* (CCAP190/2) (Guillard, 1991). Wellcome Open Res. 8: 520.

Greenbaum, E., J.W. Lee, C.V. Tevault, S.L. Blaninship and L.J. Mets. 1995. CO_2 fixation and photoevolution of H_2 and O_2 in a mutant of *Chlamydomonas* lacking photosystem I. Nature 376: 438–441.

Greening, C., P.R. Cabotaje, L.E. Valentine Alvarado, P.M. Leung, H. Land, T. Rodrigues-Oliveira et al. 2024. Minimal and hybrid hydrogenases are active from archaea. Cell 187: 1–16.

Gregory, T.R. 2001. Coincidence, coevolution, or causation? DNA content, cell size, and the C-value enigma. Biol. Rev. Camb. Philos. Soc. 76(1): 65–101.

Gregory, T.R. 2005. Synergy between sequence and size in large-scale genomics. Nat. Rev. Genet. 6(9): 699–708.

Gregson, A.J., J.C. Green and B.S.C. Leadbeater. 1993. Structure and physiology of the haptonema in *Chrysochromulina* (Prymunesiophyceae). II. Mechanisms of haptonemal coiling and the regeneration process. J. Phycol. 29: 686–700.

Griese, M., C. Lange and J. Soppa. 2011. Ploidy in cyanobacteria. FEMS Microbiol. Lett. 323(2): 124–131.

Grime, J.P. 1989. The stress debate: symptom of impending synthesis? Biol. J. 37(1-2): 3–17.

Gromov, B.V. and K.A. Mamkaeva. 1970. The fine structure of *Amoeboaphelidium protococcarum*—an endoparasite of green alga *Scenedesmus*. Arch. Hydrobiol. 67: 452–459.

Gromov, B.V., A.V. Plujusch and K.A. Mamkaeva. 1999. Morphology and possible host range of *Rhizophydium algavorum* so. nov. (Chytridiales)—an obligate parasite of algae. Protistology 1: 62–65.

Gruner, K., F. Leissing, D. Sinitski, H. Thieron, C. Axstmann, K. Baufgarten et al. 2021. Chemokine-like MDL proteins modulate flowering time and innate immunity in plants. J. Biol. Chem. 296: 100611.

Grzymski, J., O.M. Schofield, P.G. Falkowski and J.M. Bernhard. 2002. The function of plastids in the deep-sea benthic foraminifer, *Nonionella stella*. Limnol. Oceanogr. 47: 1569–1580.

Gu, Y.-Z., J.B. Hogenesch and C.A. Bradfield. 2000. The PAS superfamily: sensors of environmental and developmental signals. Ann. Pharmacol. Toxicol. 40: 519–561.

Gu, Y.Z., J.B. Hogenesch and C.A. Bradfield. 2000. The PAS superfamily: sensors of environmental and developmental signals. Annu. Rev. Pharmacol. Toxicol. 40: 519–561.

Guan, N. and L. Liu 2020. Microbial response to acid stress: mechanisms and applications. Appl. Microbiol. Biotechnol. 104(1): 51–65.

Guérin, N., M. Ciccarella, E. Flamant, P. Frémont, S. Mangenot, B. Istace et al. 2022. Genomic adaptation of the picoeukaryote *Pelagomonas calceolate* to iron-poor oceans revealed by a chromosome-scale genome sequence. Commun. Biol. 5(1): 983.

Guillard, R.R.L., M.D. Keller, C.J. O'Kelly and G.L. Floyd. 1991. *Pycnococcus provasolii* gen. et sp. nov., a coccoid prasinoxanthin-containing phytoplankter from the western North Atlantic and Gulf of Mexico. J. Phycol. 27(1): 39–47.

Guillebault, D., S.S. Sasorth, E. Derelle, J.-M. Wurtz, J.-C. Lozano, S. Bingham et al. 2002. A new class of transcription initiation factors, intermediate between TATA box-binding proteins (TBPs) and TBP-like factors (TLFs), is present in the marine unicellular organism, the dinoflagellate *Crypthecodinium cohnii*. J. Biol. Chem. 277(43): 40881–40886.

Guillén-García, A., S.E.R. Gibson, C.J.C. Jordan, V.K. Ramaswamy, V.L. Linthwaite, E.H.C. Bromley et al. 2022. Allophycocyanin A is a carbon dioxide receptor in the cyanobacterial phycobilisome. Nat. Commun. 13: 5289.

Guillou, L., M. Viprey, A. Chambouvet, R.M. Welsh, A.R. Kirkham, R. Massana et al. 2008. Widespread occurrence and genetic diversity of marine parasitoids belonging to Syndiniales (Alveolata). Environ. Microbiol. 10(12): 3349–3365.

Guiry, M.D. and G.M. Guiry. 2017. AlgaeBase. World-wide electronic publication, National University of Ireland, Galway. https://www.algaebase.org; searched on 08 July 2024.

Guo, C., L. Li, S. Lin and X. Lin. 2023. Species-dependent effects of seawater acidification on alkaline phosphatase activity in dinoflagellates. J. Phycol. 59(6): 1347–1352.

Guo, R., W.-A. Lim and J.-S. Ki. 2016. Genome-wide analysis of transcription and photosynthesis inhibition in the harmful dinoflagellate *Prorocentrum minimum* in response to the biocide copper sulfate. Harmful Algae 57(Pt A): 27–38.

Guo, Z., H. Zhang and S. Lin. 2014. Light-promoted rhodopsin expression and starvation survival in the marine dinoflagellate *Oxyrrhis marina*. PLoS ONE 9: e114941.

Gutman, J., A. Zarka and S. Boussiba. 2009. The host range of *Paraphysoderma sedebokerensis*, a chytrid that infects *Haematococcus pluvialis*. Eur. J. Phycol. 44: 509–514.

Gwon, S., J. Park, A.K.M.M. Hugue and L.S. Cheung. 2023. The *Arabidopsis* SWEET1 and SWEET2 uniporters recognize similar substrates while differing in subcellular localization. J. Biol. Chem. 299(12): 105389.

Hackett, J.D., D.M. Anderson, D.L. Erdner and D. Bhattacharya. 2004. Dinoflagellates: A remarkable evolutionary experiment. Am. J. Bot. 91: 1523–1534.

Hackett, J.D., L. Maranda, H.S. Yoon and D. Bhattacharya. 2003. Phylogenetic evidence for the cryptophyte origin of the plastid of *Dinophysis* (Dinophysiales, Dinophyceae). J. Phycol. 39(2): 440–448.

Hadhav, D.B., Y. Sriramkumar and S. Roy. 2022. The enigmatic clock of dinoflagellates, is it unique? Front. Microbiol. 13: 1004074.

Hagedorn, M., V. Carter, N. Zuchowicz, M. Phillips, C. Penfield, B. Shamenek et al. 2015. Trehalose is a chemical attractant in the establishment of coral symbiosis. PLoS ONE 10(1): e0117087.

Hahn, M.W. and G.A. Wray. 2002. The g-value paradox. Evol. Dev. 4(2): 73–75.

Haimovich-Dayan, M., N. Garfinkel, D. Ewe, Y. Marcus, A. Gruber, H. Wagner et al. 2013. The role of C_4 metabolism in the marine diatom *Phaeodactylum tricornutum*. New Phytol. 197(1): 177–185.

Hallock, P. 2000. Symbiont-bearing foraminifera: harbingers of global change? Micropaleontol. 46(S1): 95–104.

Han, X., N. Sun, N. Zu and H. Mi. 2017. Co-ordination of NDH and Cup proteins in CO_2 uptake in cyanobacterium *Synechocystis* sp. PCC6803. J. Exp. Bot. 68(14): 3869–3877.

Hancock, J.T. 2020. Nitric oxide signaling in plants. Plants (Basel) 9: 1550.

Haney, A.M., J.E. Sanfilippo, L. Garczarek, F. Partensky and D.M. Kehoe. 2022. Multiple photolyases protect the marine cyanobacterium *Synechococcus* from ultraviolet radiation. mBio 13(4): e01511–22.

Hanic, L.A., S. Sekimoto and S.S. Bates. 2009. Oomycete and chytrid infections of the marine diatom *Pseudo-nitzschia pungens* (Bacillariophyceae) from Prince Edward Island, Canada. Botany. 87: 1096–1105.

Hanschen E.R., T.N. Marriage, P.J. Ferris, T. Hamaji, A. Toyoda, A. Fujiyama et al. 2016. The *Gonium pectorale* genome demonstrates co-option of cell cycle regulation during the evolution of multicellularity. Nat. Commun. 7: 11370.

Hansen, J.P. and T. Fenchel. 2006. The bloom-forming ciliate *Mesodinium rubrum* harbours a single permanent endosymbiont. Mar. Biol. Res. 2: 169–177.

Hara, S. and M. Yamakawa. 1995. Moricin, a novel type of antibacterial peptide isolated from the silkworm, *Bombyx mori*. J. Biol. Chem. 270(50): 29923–29927.

Harada, A., S. Ohtsuka and T. Horiguchi. 2007. Species of the parasitic genus *Duboscquella* are members of the enigmatic marine alveolate group I. Protist 158: 337–347.

Haraguchi, T., K. Ito, Z. Duan, S. Rula, K. Takahashi, Y. Shibuya et al. 2018. Functional diversity of class XI Myosins in *Arabidopsis thaliana*. Plant Cell Physiol. 59(11): 2268–2277.

Harigaya, Y., H. Tanaka, S. Yamanaka, K. Tanaka, Y. Watanabe, C. Tsutsumi et al. 2006. Selective elimination of messenger RNA prevents an incidence of untimely meiosis. Nature 442 (7098): 45–50.

Harke, M.J., J.G. Jankowiak and B.K. Morrell. 2016. Transcriptomic responses in the bloom forming cyanobacterium *Microcystis* induced during exposure to zooplankton. Appl. Environ. Microbiol. 83(5): e02832–16.

Harðardóttir, S., S. Wohlrab, D.M. Hhort, B. Krock, T.G. Nielsen, UJohn and N. Lundhlom. 2019. Transcriptomic responses to grazing reveal the metabolic pathway leading to the biosynthesis of domoic acid and highlight different defense strategies in diatoms. BMC Mol. Biol. 20: 7.

Harold, F.M. 1966. Inorganic polyphosphates in biology: structure, metabolism, and function. Bacteriol. Rev. 30: 772–794.

Harper, J.T. and P.J. Keeling. 2003. Nucleus-encoded, plastid-targeted glyceraldehyde-3-dehydrogenase (GAPDH) indicates a single origin for chromalveolate plastids. Mol. Biol Evol. 20: 1730–1735.

Hartz, A.J., B.F. Sherr and E.B. Sherr. 2011. Photoresponse in the heterotrophic marine dinoflagellate *Oxyrrhis marina*. J. Euk. Microbiol. 58(2): 171–177.

Haslbeck, M. and E. Vierling. 2015. A first line of stress defense: small heat shock proteins and their function in protein homeostasis. J. Mol. Biol. 427: 1537–1548.

Haslbeck, M, S. Weinkauf and J. Buchnerm. 2019. Small heat shock proteins: Simplicity meets complexity. J. Biol. Chem. 294: 2121–2132.

Hattendorf, D.A., A. Anna, A. Gangar, P.J. Brennwald and W.I. Weis. 2007. Structure of the yeast polarity protein Sro7 reveals a SNARE regulatory mechanism. Nature 446 (7135): 567–571.

Hattenrath-Lehmann, T., D. Nanjaappa, H. Zhang, L. Yu, J.A. Goleski, S. Lin et al. 2021. Transcriptomic and isotopic data reveal central role of ammonium in facilitating the growth of the mixotrophic dinoflagellate *Dinophysis acuminata*. Harmful Algae 104: 102031.

Hausinger, R.P. 2004. Metabolic versatility of prokaryotes for urea decomposition. J. Bacteriol. 186(9): 2520–2522.

Hayami, N., Y. Sakai, M. Kimura, T. Saito, M. Tokizawa, S. Luchi et al. 2015. The responses of Arabidopsis early light-induced protein2 to ultraviolet B, high light, and cold stress are regulated by a transcriptional regulatory unit composed of two elements. Plant Physiol. 169(1): 840–855.

Hayden, H.S., J. Blomster, C.A. Maggs, P.C. Silva, M.J. Stanhope and J.R. Waaland. 2003. Linnaeus was right all along: *Ulva* and *Enteromorpha* are not distinct genera. Eur. J. Phycol. 38(3): 277–294.

He, J., Y. Huang, L. Li, S. Lin, M. Ma, Y. Wang et al. 2024. Novel plastid genome characteristics in *Fugacium kawagutii* and the trend of accelerated evolution of plastid proteins in dinoflagellates. Genome Biol. Evol. 16(1): evad237.

Hebert, P.D.N., A. Cywinska, S.L. Bali and J.R. deWaard. 2003. Biological identifications through DNA barcodes. Proc. R. Soc. B: Biol. Sci. 270(1512): 313–321.

Hehenberger, E., R.J. Gast and P.J. Keeling. 2019. A kleptoplastidic dinoflagellate and the tipping point between transient and fully integrated plastid endosymbiosis. Proc. Natl. Acad. Sci. USA 116(36): 17934–17942.

Heinrich, S., K. Valentin, S. Frickenhaus, U. John and C. Wiencke. 2012. Transcriptomic analysis of acclimation to temperature and light stress in *Saccharina latissima* (Phaeophyceae). PLoS ONE 7(8): e44342.

Held, N.A., E.A. Webb, M.M. McIlvin, D.A. Hutchins, N.R. Cohen, D.M. Moran et al. 2020. Co-occurrence of Fe and P stress in natural populations of the marine diazotroph *Trichodesmium*. Biogeosciences 17(9): 2537–2551.

Helliwell, K.E. 2023. Emerging trends in nitrogen and phosphorus signaling in photosynthetic eukaryotes. Trends Plant Sci. 28(3): 344–358.

Helliwell, K.E., A. Chrachri, J.A. Koester, S. Wharam, F. Verret, A.R. Taylor et al. 2019. Alternative mechanisms for fast Na^+/Ca^{2+} signaling in eukaryotes via a novel class of single-domain voltage-gated channels. Curr. Biol. 29: 1503–1511.

Helliwell, K.E., E.L. Harrison, J.A. Christie-Oleza, A.P. Rees, F.H. Kleiner, T. Gaikwad et al. 2021. A novel Ca^{2+} signaling pathway coordinates environmental phosporus sensing and nitrogen metabolism in marine diatoms. Curr. Biol. 31(5): 978–989.

Hennon, G.M.M., J. Ashworth, R.D. Groussman, C. Berthiaume, R.L. Morales, N.S. Baliga et al. 2015. Diatom acclimation to elevated CO2 via cAMP signalling and coordinated gene expression. Nat. Clim. Chang. 5: 761–765.

Hennon, G.M.M., O.M. Williamson, M.D.J. Limón, S.T. Haley and S.T. Dyhrman. 2019. Non-linear physiology and gene expression responses of harmful alga *Heterosigma akashiwo* to rising CO_2. Protist 170(1): 38–51.

Hernández-Prieto, M.A., R. Hiller and M. Chen. 2022. Chlorophyll f can replace chlorophyll a in the soluble antenna of dinoflagellates. Photosynth. Res. 152(1): 13–22.

Herzog, M., M.-O. Soyer. 1983. The native structure of dinoflagellate chromosomes and their stabilization by Ca^{2+} and Mg^{2+} cations. Eur. J. Cell Biol. 30: 33–41.

Herrero, A and E. Flores. 2019. Genetic responses to carbon and nitrogen availability in *Anabaena*. Enviro Microbiol. 21(1): 1–17.

Herrero, A. and J. Stavans. 2016. The multicellular nature of filamentous heterocyst-forming cyanobacteia. FEMS Microbiol. Rev. 40(6): 831–854.

Herrero A, A.M. Muro-Pastor, A. Valladares and E. Flores. 2004. Cellular differentiation and the NtcA transcription factor in filamentous cyanobacteria. FEMS Microbiol. Rev. 28(4): 469–487.

Herron, M.D., J.M. Borin, J.C. Boswell, J. Walker, I.-C.K. Chen, C.A. Knox et al. 2019. De novo origins of multicellularity in response to predation. Sci. Rep. 9: 2328.

Herrero, A., S. Picossi and E. Flores. 2013. Gene Expression during Heterocyst Differentiation. pp. 281–329. *In*: Chauvat, F. and Cassier-Chauvat, C. [eds.]. Genomics of Cyanobacteria. 65 of Advances in Botanical Research. Academic Press.

Hershberg, R. and D.A. Petrov. 2008. Selection on codon bias. Annu. Rev. Genet. 42: 287–299.

Hertweck, C. 2009. The biosynthetic logic of polyketide diversity. Angew. Chem. Int. Ed. 48(26): 4688–4716. DOI:10.1002/anie.200900414.

Hess, S., N. Sausen and M. Melkonian. 2012. Shedding light on vampires: the phylogeny of Vampyrellid amoebae revisited. PLoS ONE 7: e31165.

Higgs, P.G. and W. Ran. 2008. Coevolution of codon usage and tRNA genes leads to alternative stable states of biased codon usage. Mol. Biol. Evol. 25: 2279–2291.

Hildebrand, M., B.E. Volcani, W. Gassman and J.I. Schroeder. 1997. A gene family of silicon transporters. Nature 385: 688–689.

Hildebrand, M., L.G. Frigeri and A.K. Davis. 2007. Synchronized growth of *Thalassiosira pseudonana* (Bacillariophyceae) provides novel insights into cell-wall synthesis processes in relation to the cell cycle. J. Phycol. 43(4): 730–740.

Hiller, R.G., P.M. Wrench and F.P. Sharples. 1995. The light-harvesting chlorophyll a-c-binding protein of dinoflagellates: a putative polyprotein. FEBS Letters. 363(1-2): 175–178.

Hilton, J.A., R.A. Foster, H.J. Tripp, B.J. Carter, J.P. Zehr and T.A. Villareal. 2013. Genomic deletions disrupt nitrogen metabolism pathways of a cyanobacterial diatom symbiont. Nat. Commun. 4:1767. doi: 10.1038/ncomms2748.

Himmel N.J. and D.N. Cox. 2020. Transient receptor potential channels: Current perspectives on evolution, structure, function and nomenclature. Proc. R. Soc. B Biol. Sci. 287: 20201309.

Hinde, R. 1983. Host release factors in symbioses between algae and invertebrates. *In*: Schenk, H.E.A. [ed.]. Intracellular Space as Oligogenetic Ecosystem. Proceedings. Endocytobiology II: 709–726. De Gruyter, Boston, USA.

Hinga, K.R. 2002. Effects of pH on coastal marine phytoplankton. Mar. Ecol. Prog. Ser. 238: 281–300.

Hippmann, A.A., N. Schuback, K.-M. Moon, J.P. McCrow, A.E. Allen, L.F. Foster et al. 2022. Proteomic analysis of metabolic pathways supports chloroplast-mitochondria cross-talk in a Cu-limited diatom. Plant Direct 6: e376.

Hirooka, S., Y. Hirose, Y. Kanesaki, S. Higuchi, T. Fujiwara, R. Onuma et al. 2017. Acidophilic green algal genome provides insights into adaptation to an acidic environment. Proc. Natl. Acad. Sci. USA 114(39): E8304–E8313.

Ho, C.L., W.K. Lee and E.L. Lim 2018. Unraveling the nuclear and chloroplast genomes of an agar producing red macroalga, *Gracilaria changii* (Rhodophyta, Gracilariales). Genomics 110: 124–133.

Hocker, A., S.P. Laursen, P. Radford, J. Tyson, C. Lambert, K.M. Stevens et al. 2023. Histones with an unconventional DNA-binding mode in vitro are major chromatin constituents in the bacterium *Bdellovibrio bacteriovorus*. Nat. Microbiol. 8: 2006–2019.

Hochmal, A.K., S. Schulze, K. Trompelt and M. Hippler. 2015. Calcium-dependent regulation of photosynthesis. Biochim. Biophys. Acta 1847: 993–1003.

Hockin, N.L., T. Mock, F. Mulholland, S. Kopriva and G. Malin. 2012. The response of diatom central carbon metabolism to nitrogen starvation is different from that of green algae and higher plants. Plant Physiol. 158(1): 299–312.

Hoegh-Guldberg, O. and J.F. Bruno. 2010. The impact of climate change on the world's marine ecosystems. Science 328(5985): 1523–1528.

Hoffman, Y., C. Aflalo, A. Zarka, J. Gutman, T.Y. James and S. Boussiba. 2008. Isolation and characterization of a novel chytrid species (phylum Blastocladiomycota), parasitic on the green alga *Haematococus*. Mycol. Res. 112: 70–81.

Hofmann, E., P.M. Wrench, F.P. Sharples, R.G. Hiller, W. Welte and K. Diederichs. 1996. Structural basis of light harvesting by carotenoids: Peridinin-chlorophyll-protein from *Amphidinium carterae*. Science 272: 1788–1791.

Hohmann, S. 2002. Osmotic stress signaling and osmoadaptation in yeasts. Microbiol. Mol. Biol. Rev. 66: 300–372.

Hohmann-Marriott and R.E. Blankenship. 2011. Evolution of photosynthesis. Annu. Rev. Plant Biol. 62: 515–548.

Holland, H.D. 2006. The oxygenation of the atmosphere and oceans. Philos. Trans. R. Soc. London B: Biol. Sci. 361: 903–915.

Höng, K., T. Austerlitz, T. Bohlmann and H. Bohlmann. 2021. The thionin family of antimicrobial peptides. PLoS ONE 16(7): e0254549.

Hong, W. Y.E. Wu, X. Fu and Z. Chang. 2012. Chaperone-dependent mechanisms for acid resistance in enteric bacteria. Trends Microbiol. 20: 328–335.

Hongo, Y., A. Yabuki, K. Fujikura and S. Nagai. 2019. Genes functioned in kleptoplastids of *Dinophysis* are derived from haptophytes rather than from cryptophytes. Sci. Rep. 9: 9009.

Horton, L.E. and D.J. Templeton. 1997. The cyclin box and C-terminus of cyclins A and E specify CDK activation and substrate specificity. Oncogene 14(4): 491–498.

Hongo, Y., K. Kimura, Y. Takaki, Y. Yoshida, S. Baba, C. Kobayashi et al. 2021. The genome of the diatom *Chaetoceros tenuissimus* carries an ancient integrated fragment of an extant virus. Sci. Rep. 11: 22877

Honjo, T. 1992. Harmful red tides of Heterosigma akashiwo. pp. 29–32. *In*: Svrjeced, R.S. [ed.]. Control of Disease in Aquaculture. NOAA Technical Report NMFS111.

Hopes, A., V. Nekrasov, S. Kamoun and T. Mock. 2016. Editing of the urease gene by CRISPR-Cas in the diatom *Thalassiosira pseudonana*. Plant Methods 12: 49.

Hopkins, J.F., D.F. Spencer, S. Laoissiere, J.A.D. Neilson, R.J.M. Eveleigh, D.G. Durnford et al. 2012. Proteomics reveals plastid- and periplastid-targeted proteins in the chlorarachniophyte alga *Bigelowiella natans*. Genome Biol. Evol. 4(12): 1391–1406.

Hopkinson, B.M., C.L. Dupont, A.E. Allen and F.M.M. Morel. 2011. Efficiency of the CO_2-concentrating mechanism of diatoms. Proc. Natl. Acad. Sci. USA 108: 3830–3837.

Hoppe, C.J.M., K.K.E. Wolf, N. Schuback, P.D. Tortell and B. Rost. 2018. Compensation of ocean acidification effects in Arctic phytoplankton assemblages. Nat. Clim. Change 8: 529–533.

Hoppmann, A.A., N. Schuback, K.-M. Moon, J.P. McCrow, A.E. Allen, L.F. Foster et al. 2022. Proteomic analysis of metabolic pathways supports chloroplast-mitochondria cross-talk in a Cu-limited diatom. Plant Direct. 6(1): e376.

Horák, A., A.E. Allen and M. Obornik. 2020. Common origin of ornithine-urea cycle in opisthokonts and stramenopiles. Sci. Rep. 10: 16687.

Hori, K., J. Okamoto, Y. Tanji and H. Unno. 2003. Formation, sedimentation and germination properties of *Anabaena* akinetes. Biochem. Engin. J. 14(1): 67–73.

Horst, N.A., A. Katz, I. Pereman, E.L. Decker, N. Ohad and R. Reski. 2016. A single homeobox gene triggers phase transition, embryogenesis and asexual reproduction. Nat. Plants 2: 15209.

Hothorn, M., H. Neumann, E.D. Lenherr, M. Wehner, V. Rybin, P.O. Hassa et al. 2009. Catalytic core of a membrane-associated eukaryotic polyphosphate polymerase. Science 324: 513–516.

Hotz, M., N.H. Thayer, D.G. Hendrickson, E. Schinski, J. Xu and D.E. Gottschling. 2022. rDNA array length is a major determinant of replicative lifespan in budding yeast. Proc. Natl. Acad. Sci. USA 119(5): e2119593119.

Hou, Y. and S. Lin. 2009. Distinct gene number-genome size relationships for eukaryotes and non-eukaryotes: gene content estimation for dinoflagellate genomes. PLoS ONE 4(9): e6978. doi 10.1371/journal.pone.0006978.

Hou, Y., N. Ji, H. Zhang, X. Shi, H. Han and S. Lin. 2019. Genome size-dependent PCNA gene copy number in dinoflagellates and molecular evidence of retroposition as a major evolutionary mechanism. J. Phycol. 55(1): 37–46.

Howe, G.A. and A.L. Schilmiller. 2002. Oxylipin metabolism in response to stress. Curr. Opin. Plant Biol. 5: 230–236.

Huang, D., H. Friesen and B. Andrews. 2007. Pho85, a multifunctional cyclin-dependent protein kinase in budding yeast. Mol. Microbiol. 66(2): 303–314.

Hu, C., J. Yang, Z. Qi, H. Wu, B. Wang, F. Zou et al. 2020. Heat shock proteins: biological functions, pathological roles, and therapeutic opportunities. MedComm 3(3): e161.

Hu, H. and J.L. Blanchard. 2009. Environmental sequence data from the Sargasso Sea reveal that the characteristics of genome reduction in *Prochlorococcus* are not a harbinger for an escalation in genetic drift. Mol. Biol. Evol. 26(1): 5–13.

Hu, L., N. Li, Z. Zhang, X. Meng, Q. Dong, C. Xu et al. 2020. CG hypomethylation leads to complex changes in DNA methylation and transpositional burst of diverse transposable elements in callus cultures of rice. Plant J. 101: 188–203.

Hu, E.-Z., X.-R. Lan, Z.-L. Liu, J. Gao and D.-K. Niu. 2022. A positive correlation between GC content and growth temperature in prokaryotes. BMC Genom. 23: 110.

Hu, Q., H. Miyashita, I. Iwasaki, N. Kurano, S. Miyachi, M. Iwaki et al. 1998. A photosystem I reaction center driven by chlorophyll d in oxygenic photosynthesis. Proc. Natl. Acad. Sci. USA 95(22): 13319–13323.

Huang, G.T., S.-L. Ma, L.-P. Bai, L. Zhang, H. Ma, P. Jia et al. 2012. Signal transduction during cold, salt, and drought stresses in plants. Mol. Biol. Rep. 9: 969–987.

Huang, T., F. Hu, Y. Pan, C. Li and H. Hu. 2023. Pyruvate orthophosphate dikinase is required for the acclimation to high bicarbonate concentrations in *Phaeodactylum tricornutum*. Algal Res. 72: 103131.

Hudson, R.J.M. and F.M.M. Morel. 1993. Trace metal transport by marine microorganisms: implications of metal coordination kinetics. Deep-Sea Res. 40: 129–150.

Huertas, I.E., M. Rouco, V. López-Rodas and E. Costas. 2011. Warming will affect phytoplankton differently: evidence through a mechanistic approach. Proc. Biol. Sci. 278(1724): 3534–3543.

Hughes, T.P., J.T. Kerry, M. Álvarez-Noriega, J.G. Álvarez-Romero, K.D. Anderson, A.J. Baird et al. 2017. Global warming and recurrent mass bleaching of corals. Nature 543: 373–377.

Hügler, M. and S.M. Sievert. 2011. Beyond the Calvin cycle: autotrophic carbon fixation in the ocean. Ann. Rev. Mar. Sci. 3: 261–289.

Hulatt, C.J., R.H. Wijffels and M.C. Posewitz. 2021. Genome Biol. Evol. 13(8): evab178.

Hutchinson, G.E. 1961. The paradox of the plankton. American Naturalist 95: 137–145.

Huysman, M.J.J., C. Martens, K. Vandepoele, J. Gillard, E. Rayko, M. Heijde et al. 2010. Genome-wide analysis of the diatom cell cycle unveils a novel type of cyclins involved in environmental signaling. Genome Biol. 11(2): R17.

Ianora, A., A. Miralto, S.A. Poulet, Y. Carotenuto, I. Buttino, G. Romano et al. 2004. Aldehyde suppression of copepod recruitment in blooms of a ubiquitous planktonic diatom. Nature 429: 403–407.

Iken, K. 2012. Grazers on benthic seaweeds. *In*: Wiencke, C. and K. Bischol [eds.]. Seaweed Biology, Ecological Studies 219. pp. 157–175.

Ilagan, R.P., J.F. Koscielecki, R.G. Hiller, F.P. Sharples, G.N. Gibson, R.R. Birge et al. 2006. Femtosecond time-resolved absorption spectroscopy of main-form and high-salt peridinin-chlorophyll a-proteins at low temperatures. Biochemistry 45: 14052–14063.

Im, S.H., B. Lepetit, N. Mosesso, S. Shrestha, L. Weiss, M. Nymark et al. 2024. Identification of promoter targets by Aureochrome 1a in the diatom *Phaeodacylum tricornutum*. J. Exp. Bot. 75(7): 1834–1851.

Imhoff, J.F., T. Rahn, S. Künzel and S.C. Neulinger. 2018. Photosynthesis is widely distributed among proteobacteria as demonstrated by the phylogeny of PufLM reaction center proteins. Front. Microbiol. 8: 2679.

Imamura, S., Y. Kawase, I. Kobayashi, T. Sone, A. Era, S.Y. Miyagishima, et al. 2015. Target of rapamycin (TOR) plays a critical role in triacylglycerol accumulation in microalgae. Plant Mol. Biol. 89: 309-318.

Imanian, B. and P.J. Keeling. 2007. The dinoflagellate *Durinskia baltica* and *Kryptoperidinium foliaceum* retain functionally overlapping mitochondria from two evolutionarily distinct lineages. BMC Evol. Biol. 7, 172.

Imanian, B., J.-F. Pombert and P.J. Keeling. 2010. The complete plastid genomes of the two 'Dinotoms' *Durinskia baltica* and *Kryptoperidinium foliaceum*. PLoS ONE 5(5): e10711.

Inoki, K., Ho Ouyang, Y. Li and K.-L. Guan. 2005. Signaling by target of rapamycin proteins in cell growth control. Microbiol. Mol. Biol. Rev. 69(1): 79–100.

Inomura, K., C.L. Follett, T. Masuda, M. Eichner, O. Prášil and C. Deutsch. 2020. Carbon transfer from the host diatom enables fast growth and high rate of N_2 fixation by symbiotic heterocystous cyanobacteria. Plants 9(2): 192.

Inoue, K., S. Ito, Y. Kato, Y. Nomura, M. Shibata, T. Uchihashi et al. 2016. A natural light-driven inward proton pump. Nat. Commun. 7: 13415.

Inzé, D. and L. De Veylder. 2006. Cell cycle regulation in plant development. Ann. Rev. Genet. 40: 77–105.

IPCC. 2013. Intergovernmental Panel on Climate Change, Ed., Climate Change 2013 - The Physical Science Basis (Cambridge Univ. Press, Cambridge, 2014).

Iranzo, J., A.E. Lobkovsky, Y.I. Wolf and E.V. Koonin. 2014. Virus-host arms race at the joint origin of multicellularity and programmed cell death. Cell Cycle 13: 3083–3088.

Irving, J. and R.J.P. Williams. 1953. The stability of transition-metal complexes. J. Chem. Soc. 0: 3192–3210.

Irwin, N.A.T., B.J.E. Martin, B.P. Young, M.J.G. Browne, A. Flaus, C.J.R. Loewen et al. 2018. Viral proteins as a potential driver of histone depletion in dinoflagellates. Nat. Commun. 9(1): 1535.

Irwin, N.A.T., A.A. Pittis, T.A. Richards and P.J. Keeling. 2022. Systematic evaluation of horizontal gene transfer between eukaryotes and viruses. Nat. Microbiol. 7: 327–336.

Ishida, K. and B.R. Green. 2002. Second- and third-hand chloroplasts in dinoflagellates: phylogeny of oxygen-evolving enhance 1 (PsbO) protein reveals replacement of a nuclear-encoded plastid gene by that of a haptophyte tertiary endosymbiont. Proc. Natl. Acad. Sci. USA 99: 9294–9299.

Ishikawa, M., F. Takahashi, H. Nozaki, C. Nagasato, T. Motomura and H. Kataoka. 2009. Distribution and phylogeny of the blue light receptors aureochromes in eukaryotes. Planta 230(3): 543–552.

Ishikura, M, K. Adachi and T. Maruyama. 1999. Zooxanthellae release glucose in the tissue of a giant clam, *Tridacna crocea*. Mar. Biol. 133: 665–673.

Ishino, Y., H. Shinagawa, K. Makino, M. Amemura and A. Nakata. 1987. Nucleotide sequence of the iap gene, responsible for alkaline phosphatase isozyme conversion in *Escherichia coli*, and identification of the gene product. J. Bacteriol. 169(12): 5429–5433.

Islas-Flores, T., E. Galan-Vasquez and M. Villanueva. 2021. Screening a spliced leader-based *Symbiodininium microadriaticum* reveals a hemerythrin-like protein as a putative SmicRACKI ligand. Microorganisms 9: 791.

Ito, H. and A. Tanaka. 2011. Evolution of a dininyl chlorophyll-based photosystem in *Prochlorococcus*. Proc. Natl. Acad. Sci. USA 108(44): 18014–18019.

Ito, H. M. Yokono, R. Tanaka and A. Tanaka. 2008. Identification of a novel vinyl reductase gene essential for the biosynthesis of monoviynl chlorophyll in *Synechocystis* sp. PCC6803. J. Biol. Chem. 283: 9002–9011.

Itzhaki, H., L. Naveh, M. Lindahl, M. Cook and Z. Adam. 1998. Identification and characterization of DegP, a serine protease associated with the luminal sides of the thylakoid membrane. J. Biol. Chem. 273(12): 7094–9098.

Iwasaki, H., S.B. Williams, Y. Kitayama, M. Ishiura, S.S. Golden and T. Kondo. 2000. A KaiC-interacting sensory histidine kinase SasA, necessary to sustain robust circadian oscillation in cyanobacteria. Cell 101(2): 223–233.

Jackson, C.J., J.E. Norman, M.N. Schnare, M.W. Gray, P.J. Keeling and R.F. Waller. 2007. Broad genomic and transcriptional analysis reveals a highly derived genome in dinoflagellate mitochondria. BMC Biol. 5: 41.

Jackson, C.J., S.G. Gornik and R.F. Waller. 2012. The mitochondrial genome and transcriptome of the basal dinoflagellate *Hematodinium* sp.: character evolution within the highly derived mitochondrial genomes of dinoflagellates. Genome Biol. Evol. 4(1): 59–72.

Jackson, C.J., S.G. Gornik and R.F. Waller. 2013. A tertiary plastid gains RNA editing in its new host. Mol. Biol. Evol. 30(4): 788–792.

Jacobovitz, M.R., S. Rupp, P.A. Voss, I. Maegele, S.G. Gornik and A. Guse. 2021. Dinoflagellate symbionts escape vomocytosis by host cell immune suppression. Nat. Microbiol. 6: 769–782.

Jadhav, D.B., Y. Sriramkumar and S. Roy. 2022. The enigmatic clock of dinoflagellates, is it unique? Front. Microbiol. 13: 1004074.

Jaeckisch, N., I. Yang, S. Wohlrab, G. Glöckner, J. Kroymann, H. Vogel et al. 2011. Comparative genomic and transcriptomic characterization of the toxigenic marine dinoflagellate *Alexandrium ostenfeldii*. PLoS ONE 6(12): e28012.

Jagodzik, P., M. Tajdel-Zielinska, A. Ciesla, M. Marczak and A. Ludwikow. 2018. Mitogen-activated protein kinase cascades in plant hormone signaling. Front. Plant Sci. 9: 1387.

Jain, D. and W. Schmidt. 2024. Protein phosphorylation orchestrates acclimations of *Arabidopsis* plants to environmental pH. Mol. Cell. Proteomics 23(1): 100685.

Jain, S., G. Prajapat, M. Abrar, L. Ledwani, A. Singh and A. Agrawal. 2017. Cyanobacteria as efficient producers of mycosporine-like amino acids. J. Basic Microbiol. 57: 715–727.

Jalal, M., N.K. Shala, M.W. Wojewodzic, T. Andersen and D.O. Hessen. 2014, Multigenerational genomic responses to dietary phosphorus and temperature in *Daphnia*. Genome 57: 439–448.

Jan, G., P. Leverrier, V. Pichereau and P. Boyaval. 2001. Changes in protein synthesis and morphology during acid adaptation of *Propionibacterium freudenreichii*. App. Environ. Microbiol. 67(5): 2029–2036.

Janke, C., F. Scholz, J. Becker-Baldus, C. Glaubitz, P.G. Wood, E. Bamberg et al. 2013. Photocycle and vectorial proton transfer in a rhodopsin from the eukaryote *Oxyrrhis marina*. Biochemistry 52(16): 2750–2763.

Jaouannet, M., A.-S. Pavaux, S. Pagnotta, O. Pierre, C. Michelet, S. Marro et al. 2020. Atypical membrane-anchored cytokine MIF in a marine dinoflagellate. Microorganisms 8: 1263.

Janouškovec, J., G.S. Gavelis, F. Burki, D. Dinh, T.R. Bachvaroff, S.G. Gornik et al. 2017. Major transitions in dinoflagellae evolution unveiled by phylotranscriptomics. Proc. Natl., Acad. Sci. USA 114(2): E171–E180.

Janson, S. 2004. Molecular evidence that plastids in the toxin-producing dinoflagellate genus *Dinophysis* originate from the free-living cryptophyte *Teleaulax amphioxeia*. Environ. Microbiol. 6: 1102–1106.

Janson, S., J. Wouters, B. Gergman and E.J. Carpenter. 1999. Host specificity in the *Richelia*-diatom symbiosis revealed by hetR gene sequence analysis. Environ. Microbiol. 1(5): 431–438.

Jantschke, A., I. Pinkas, A. Hirsch, N. Elad, A. Schertel, L. Addadi et al. 2019. Anhydrous β-guanine crystals in a marine dinoflagellate: Structure and suggested function. J. Struct. Biol. 207(1): 12–20. DOI:10.1016/j.jsb.2019.04.009.

Jauffrais, T., C. LeKieffre, M. Schweizer, E. Geslin, E. Metzger, J.M. Bernhard et al. 2018. Kleptoplastidic benthic foraminifera from aphotic habitats: insights into assimilation of inorganic C, N, and S studied with sub-cellular resolution. Environ. Microbiol. 21(1): 125–141.

Jauffrais, T., B. Jesus, E. Metzger, J.-L. Mouget, F. Jorissen and E. Geslin. 2016. Effect of light on photosynthetic efficiency of sequestered chloroplasts in intertidal benthic foraminifera (*Haynesina germanica* and *Ammonia tepida*). Biogeosciences 13: 2715–2726.

Jeandroz, S., D. Wipf, D.J. Stuehr, L. Lamattina, M. Melkonian, Z. Tian et al. 2016. Occurrence, structure, and evolution of nitric oxide synthase-like proteins in the plant kingdom. Science Signaling 9: 9(417): re2.

Jeffares, D., M.J. Phillips, S.A. Moore and B. Veit. 2004. A description of the Mei-2 like protein family: structure, phylogenetic distribution and biological context. Dev. Gene. Evol. 214(3): 149–158.

Jeffrey, S.W., H.S. MacTavish, W.C. Dunlap, M. Vesk and K. Groenewound. 1999. Occurrence of UVA- and UVB-absorbing compounds in 152 species (206 strains) of marine microalgae. Mar. Ecol. Prog. Ser. 189: 35–51.

Jemey, J., S. Suikkanen, E. Lindehoff and A. Kremp. 2019. Future temperature and salinity do not exert selection pressure on cyst germination of a toxic phytoplankton species. Ecol. Evol. 9: 4443–4451.

Jeong, H.J., Y.D. Yoo, J.S. Kim, K.A. Seong, N.S. Kang and T.H. Kim. 2010. Growth, feeding, and ecological roles of the mixotrophic and heterotrophic dinoflagellates in marine planktonic food webs. Ocean Sci. J. 45: 65–91. DOI:10.1007/s12601-010-0007-2.

Jeong, H.J., Y.D. Yoo, N.S. Kang, A.S. Lim, K.A. Seong, S.Y. Lee et al. 2012. Heterotrophic feeding as a newly identified survival strategy of the dinoflagellate *Symbiodinium*. Proc. Natl. Acad. Sci. USA 109(31): 12604-12609.

Jeong, K.C. K.F. Hung, D.J. Baumler, J.J. Byrd and C.W. Kaspar. 2008. Acid stress damage of DNA is prevented by Dps binding in *Escherichia coli* O157:H7. BMC Microbiol. 8: 181.

Jephcott, T.G., C. Alves-De-Souza, F.H. Gleason, F.F. va Ogtrop, T. Sime-Ngando, S.A. Karpov et al. 2016. Ecological impacts of parasitic chytrids, syndiniales and perkinsids on populations of marine photosynthetic dinoflagellates. Fungal Ecol. 19: 47–58.

Ji, N., L. Li, L. Lin and S. Lin. 2017. Identification and expression analysis of blue light receptor aureochrome in the harmful algal *Heterosigma kaashiwo* (Raphidophyceae). Harmful Algae 61: 71–79.

Ji, N., L. Lin, L. Li. L. Yu, Y. Zhang, H. Luo et al. 2018. Metatranscriptome analysis reveals environmental and diel regulation of a *Heterosigma akashiwo* (Raphidophyceae) bloom. Environ. Microbiol. 20(3): 1078–1094.

Ji, N., X. Yin, Y. Chen, Y. Chen, M. Xu, J. Huang et al. 2023. Full-length transcriptome analysis of the ichthyotoxic harmful alga *Heterosigma akashiwo* (Raphidophyceae) using single-molecule real-time sequencing. Microorganisms 11(2): 389.

Jiao, C., I. Sørensen, X. Sun, H. Sun, H. Behar, S. Alseekh et al. 2020. The *Penium margaritaceum* genome: hallmarks of the origins of land plants. Cell 181: 1097–1111.

Jin, P., Y. Ji, Q. Huang, P. Li, J. Pan, H. Lu et al. 2022a. A reduction in metabolism explains the tradeoffs associated with the long-term adaptation of phytoplankton to high CO_2 concentrations. New Phytol. 233(5): 2155–2167.

Jin, P., J. Wan, Y. Zhou, K. Gao, J. Beardall, J. Lin et al. 2022b. Increased genetic diversity loss and genetic differentiation in a model marine diatom adapted to ocean warming compared to high CO_2. ISME J. 16: 2587–2598.

Jiang, Y., M. Shi and L. Shi. 2019. Molecular underpinning for microbial extracellular electron transfer during biogeochemical cycling of earth elements. Sci. China: Life Sci. 62: doi: 10.1007/s11427-018-9464-3.

Jiang, Y., T. Cao, Y. Yang, H. Zhang, J. Zhang and X. Li. 2023. A chlorophyll c synthase widely co-opted by phytoplankton. Science 382: 92–98.

Jiang, Z.D. and W.H. Gerwick. 1991. Eicosanoids and other hydroxylated fatty acids from the marine alga *Gracilariopsis lemaneiformis*. Phytochemistry 30: 1187–1190.

Jing, X., S. Lin, H. Zhang, C. Koerting and Z. Yu. 2017. Utilization of urea and expression profiles of related genes in the dinoflagellate *Prorocentrum donghaiense*. PLoS ONE 12: 1–15.

Jinkerson, R.E., J.A. Russo, C.R. Newkirk, A.L. Kirk, R.J. Chi, M.Q. Martindale et al. 2022. Cnidarian-Symbiodiniaceae symbiosis establishment is independent of photosynthesis. Curr. Biol. 32: 2402–2415.

John, J., R.A. Fensome and L.K. Medlin. 2010. The application of a molecular clock based on molecular sequences and the fossil record to explain biogeographic distributions within the *Alexandrium tamarense* "species complex" (Dinophyceae). Mol. Biol. Evol. 20: 1015–1027.

John, U., U. Tillmann and L.K. Medlin. 2002. A comparative approach to study inhibition of grazing and lipid composition of a toxic and non-toxic clone of *Chrysochromulina polylepis* (Prymnesiophyceae). Harmful Algae 1: 45–57.

John, U., Y. Lu, S. Wohlrab, M. Groth, J. Janouškovec, G.S. Kohli et al. 2019. An aerobic eukaryotic parasite with functional mitochondria that likely lacks a mitochondrial genome. Sci. Adv. 5(4): eaav1110.

Johns, C.T., K.G. Bondoc-Naumovitz, A. Matthew, P.G. Matson, M.D. Iglesias-Rodriguez, H.L. Fuchs et al. 2023. Absorptive exchange of coccolith biominerals facilitates viral infection. Sci. Sdv. 9(3): eadc8728.

Johnson, M.D., D. Oldach, C.F. Delwiche and D.K. Stoecker. 2007. Retention of transcriptionally active cryptophyte nuclei by the ciliate *Myrionecta rubra*. Nature 445: 426–428.

Johnson, M.D., B.R. Edwards, D.J. Beaudoin, B.A.S. Van Mooy and A. Vardi. 2020. Nitric oxide mediates oxylipin production and grazing defense in diatoms. Environ. Microbiol. 22(2): 629–645.

Jones, D.T., W.R. Taylor and J.M. Thornton. 1992. The rapid generation of mutation data matrices from protein sequences. Comput. Appl. Biosci. 8: 275–282.

Jones, G.D., E.P. Williams, A.R. Place, R. Jagus and T.R. Bachvaroff. 2015. The aleolate translation initiation factor 4E family reveals a custom toolkit for translational control in core dinoflagellates. BMC Ecol. Evol. 15: 14.

Joung, J.K. and J.D. Sander. 2013. TALENs: a widely applicable technology for targeted genome editing. Nat. Rev. Mol. Cell Biol. 14(1): 49–55.

Jung, H., H.A. Seong and H. Ha. 2008. Direct interaction between NM23-H1 and macrophage migration inhibitory factor (MIF) is critical for alleviation of MIF-mediated suppression of p53 activity. J. Biol. Chem. 283(47): 32669–32679. DOI:10.1074/jbc.M804504200.

Jung, H., H.-A. Seong and H. Ha. 2008. Critical role of cysteine residue 81 of macrophage migration inhibitory factor (MIF) in MIF-induced inhibition of p53 activity. J. Biol. Chem. 283(29): 20383–20396.

Jüttner, F. 2001. Liberation of 5,8,11,14,17-eicosapentaenoic acid and other polyunsaturated fatty acids from lipids as a grazer defense reaction in epilithic diatom biofilms. J. Phycol. 37: 744–755.

Kagami, M., Y. Amano and I. Ishii. 2011. Community structure of planktonic fungi and the impact of parasitic chytrids on phytoplankton in Lake Inba, Japan. Microb. Ecol. 63: 358–368.

Kageyama, H., Y. Tanaka, A. Shibata, R. Waditee-Sirisattha and T. Takabe 2018. Dimethylsulfoniopropionate biosynthesis in a diatom *Thalassiosira pseudonana*: Identification of a gene encoding MTHB-methyltransferase. Arch. Biochem. Biophys. 645: 100–106.

Kajikawa, M. and H. Fukuzawa. 2020. Algal autophagy is necessary for the regulation of carbon metabolism under nutrient deficiency. Front. Plant Sci. 11: 36.

Kalapos, B., M. Hlavová, T.V. Nádai, G. Galiba, K. Bišová and R. Dóczi. 2019. Early evolution of the mitogen-activated protein kinase family in the plant kingdom. Sci. Rep. 1: 4094.

Kalinina, V., M. Berdieva, N. Aksenov and S. Skarlato. 2023. Phosphorus deficiency induces sexual reproduction in the dinoflagellate *Prorocentrum cordatum*. Sci. Rep. 13: 14191.

Kalvelage, J., L. Wöhlbrank, R.-A. Schoon, F.-M. Zink, C. Correll, J. Senkler et al. 2023. The enigmatic nucleus of the marine dinoflagellate *Prorocentrum cordatum*. mPhere 8(4): e0003823. DOI: 10.1128/msphere.00038-23.

Kamako, S.-i., R. Hoshina, S. Ueno and N. Imamura. 2005. Establishment of axenic endosymbiotic strains of Japanese *Paramecium bursaria* and the utilization of carbohydrate and nitrogen compounds by the isolated algae. Eur. J. Protistol. 41: 193–202.

Kamikawa, R., G. Tanifuji, M. Kawachi, H. Miyashita, T. Hashimoto and Y. Inagaki. 2015. Plastid genome-based phylogeny pinpointed the origin of the green-colored plastid in the dinoflagellate *Lepidodinium chlorophorum*. Genome Biol. Evol. 7(4): 1133–1140.

Kamikawa, R., D. Moog, S. Zauner, G. Tanifuji, K.-I. Ishida, H. Muyashita et al. 2017. A non-photosynthetic diatom reveals early steps of reductive evolution in plastids. Mol. Biol. Evol. 34(9): 2355–2366.

Kamikawa R., T. Mochizuki, M. Sakamoto, Y. Tanizawa, T. Nakayama, R. Onuma et al. 2022. Genome evolution of a nonparasitic secondary heterotroph, the diatom *Nitzschia putrida*. Sci. Adv. 8: eabi5075.

Kanaya, S., Y. Yamada, Y. Kudo and T. Ikemura. 1999. Studies of codon usage and tRNA genes of 18 unicellular organisms and quantification of *Bacillus subtilis* tRNAs: gene expression level and species-specific diversity of codon usage based on multivariate analysis. Gene 238(1): 143–155.

Kanehisa, M., M. Furumichi, Y. Sato, M. Kawashima and M. Ishiguro-Watanabe. 2023. KEGG for taxonomy-based analysis of pathways and genomes. Nucleic Acids Res. 51(D1): D587–D592.

Kang, Y., M.J. Karke, D.L. Berry, J.L. Collier, S.W. Wilhelm, S.T. Fyhrman et al. 2021. Transcriptomic responses of four pelagophytes to nutrient (N, P) and light stress. Front. Mar. Sci. 8: 636699. Doi: 10.3389/fmars.2021.636699.

Kant, S., M. Peng and S.J. Rothstein. 2011. Genetic regulation by NLA and microRNA827 for maintaining nitrate-dependent phosphate homeostasis in *Arabidopsis*. PLOS Genet. 7(3): e1002021.

Kaplan-Levy, R.N., O. Hadas, M.L. Summers, J. Rücker and A. Sukenik. 2010. Akinetes: Dormant cells of cyanobacteria. Top Curr. Genet. 21: 5–27.

Karentz, D., J.E. Cleaver and D.L. Mitchell. 1991. Cell survival characteristics and molecular responses of Antarctic phytoplankton to ultraviolet-B radiation. J. Phycol. 27: 326–341.

Karl, D.M. 2014. Microbially mediated transformations of phosphorus in the sea: new views of an old cycle. Ann. Rev. Mar. Sci. 6: 279–337.

Karnkowska, A., N. Yubuki, A. Yamaguchi, Y. Kashiyama, T. Suzaki, P.J. Keeling et al. 2023. Euglenozoan kleptoplasty illuminates the early evolution of photoendosymbiosis. Proc. Natl. Acad. Sci. USA 120 (12): e2220100120.

Karsten, U., T. Sawall, D. Hanelt, K. Bischof, F.L. Figueroa, A. Flores-Moya et al. 1998. An inventory of UV-absorbing mycosporine-like amino acids in macroalgae from polar to warm-temperate regions. Bot. Mar. 41: 443–454.

Karydis, M. and K. Kitsiou. 2019. Marine eutrophication: a global perspective. CRC Press. pp. 206.

Kato, H., R. Tokutsu, H. Kubota-Kawai, R.N. Burton-Smith, S. Kawaguti and T. Yamasu. 1965. Electron microscopy on the symbiosis between an elysioid gastropod and chloroplasts from a green alga. Biol. J. Okayama Univ. II: 57–64.

Kato, S., O. Misumi, S. Maruyama, H. Nozaki, Y. Tsujimoto-Inui, M. Takusagawa et al. 2023. Genomic analysis of an ultrasmall freshwater green alga, *Medakamo hakoo*. Commun. Biol. 6: 89.

Kato, H., R. Tokutsu, H. Kubota-Kawai, R.N. Burton-Smith, E. Kim and J. Minagawa. 2020. Characterization of a giant PSI supercomplex in the symbiotic dinoflagellate Symbiodiniaceae. Plant Physiol. 183(4): 1725–1734.

Kaushal, S.S., G.E. Likens, M.L. Pace, R.M. Utz, S. Haq, J. Gorman et al. 2018. Freshwater salinization syndrome on a continental scale. Proc. Natl. Acad. Sci. USA 115(4): E574–E583.

Kaur, J., J. Sebastian and I. Siddiqi. 2006. The *Arabidopsis*-mei2-like genes play a role in meiosis and vegetative growth in *Arabidopsis*. Plant Cell 18(3): 545–559.

Kawachi, M., I. Inouye, O. Maeda and M. Chihara. 1991. The haptonema as a food-capturing device: observations on *Chrysochromulina hirta* (Prymnesiophyceae). Phycologia 30(6): 563–573.

Kawaguti, S. and T. Yamasu. 1965. Electron microscopy on the symbiosis between an elysioid gastropod and chloroplasts of a green alga. Biol. J. Okayama University 11: 57–65.

Kay, H., E. Grünewald, H.K. Feord, S. Gil, S.Y. Peak-Chew, A. Stangherlin et al. 2021. Deep-coverage spatiotemporal proteome of the picoeukaryote *Ostreococcus tauri* reveals differential effects of environmental and endogenous 24-hour rhythms. Commun. Biol. 4: 1147.

Kayal, E. and D.R. Smith. 2021. Is the dinoflagellate *Amoebophrya* really missing an mtDNA? Mol. Biol. Evol. 38(6): 2493–2496.

Kazamia, E., H. Czesnick, T.T. Van Nguyen, M.T. Croft, E. Sherwood, S. Sasso et al. 2012. Mutualistic interactions between vitamin B12-dependent algae and heterotrophic bacteria exhibit regulation. Environ. Microbiol. 14: 1466–1476.

Kazamia, E., R. Sutak, J. Paz-Yepes, R.G. Dorrell, F.R.J. Vieira, J. Mach et al. 2018. Endocytosis-mediated siderophore uptake as a strategy for Fe acquisition in diatoms. Sci. Adv. 4: eaar4536.

Kearns, L.P. and D.C. Sigee. 1980. The occurrence of period IV elements in dinoflagellate chromatin: an X-ray microanalytical study. J. Cell Sci. 46: 113–127.

Keeling, P.J. 2009. Chromalveolates and the evolution of plastids by secondary endosymbiosis. J. Euk. Microbiol. 56: 1–8.

Keeling, P.J. 2010. The endosymbiotic origin, diversification and fate of plastids. Philos. Trans. R. Soc. Lond. B. Biol. Sci. 365(1541): 729–748.

Keeling, P.J. 2013. The number, speed, and impact of plastid endosymbioses in eukaryotic evolution. Ann. Rev. Plant Biol. 64: 583–607.

Keeney, S., C.N. Giroux and N. Kleckner. 1997. Meiosis-specific DNA double-strand breaks are catalyzed by Spo11, a member of a widely conserved protein family. Cell 88: 375–384.

Keller, M.D. 1989. Dimethyl sulfide production and marine phytoplanktpn: the importance of species composition and cell size. Biol. Oceanogr. 6: 375–382.

Kellis, M., N. Patterson, M. Wndrizzi, B. Birren and E.S. Lander. 2003. Sequencing and comparison of yeast species to identify genes and regulatory elements. Nature 423: 241–254.

Kellmann, R., A. Stüken, R.J. Orr, H.M. Svendsen and K.S. Jakobsen. 2010. Biosynthesis and molecular genetics of polyketides in marine dinoflagellates. Mar. Drugs 8(4): 1011–1048. DOI:10.3390/md8041011.

Kellner, R., G. Jung, T. Horner, H. Zahner, N. Schell, K.D. Entian et al. 1988. Gallidermin: a new lanthionine-containing polypeptide antibiotic. Eur. J. Biochem. 177 (1): 53–59.

Kellogg, R.M., M.A. Moosburner, N.R. Cohen, N.J. Hawco, M.R. Mcllvin, D.M. Moran et al. 2022. Adaptive responses of marine diatoms to zinc scarcity and ecological implications. Nat. Commun. 13: 1995.

Kerswell, A.P. 2006. Global biodiversity patterns of benthic marine algae. Ecology 87(10): 2479–2488.

Kettler, G.C., A.C. Martiny, K. Huang, J. Zucker, M.L. Coleman, S. Rodrigue et al. 2007. Patterns and implications of gene gain and loss in the evolution of *Prochlorococcus*. PLoS Genet. 3(12): e231.

Khalili, N. and G. Usup. 2015. Ultrastructure and molecular analysis of the harmful armored dinoflagellate *Alexandrium minutum*. J. Adv. Microbiol. DOI: 10.1166/jamr.2015.1251.

Khan, S. and J.M. Scholey. 2018. Assembly, functions and evolution of archaella, flagella and cilia. Curr. Biol. 28(6): R278–R292.

Khoa, H.V. and K. Mikami. 2022. Membrane-fluidization-dependent and -independent pathways are involved in heat stress-inducible gene expression in the marine red alga *Neopyropia yezoensis*. Cells 11(9): 1486.

Khudyakov, I.Y. and J.W. Golden. 2004. Different functions of HetR, a master regulator of heterocyst differentiation in *Anabaena* sp. PCC 7120, can be separated by mutation. Proc. Natl. Acad. Sci. USA 101: 16040–16045.

Kikuchi, M., K. Kojima, S. Nakao, S. Yoshizawa, S. Kawanishi, A. Shibukawa et al. 2021. Functional expression of the eukaryotic proton pump rhodopsin OmR2 in *Escherichia coli* and its photochemical characterization. Sci. Rep. 11: 14765.

Kikutani, S., K. Nakajima, C. Nagasato, Y. Tsuji, A. Miyatake and Y. Matsuda. 2016. Thylakoid luminal θ-carbonic anhydrase critical for growth and photosynthesis in the marine diatom *Phaeodactylum tricornutum*. Proc. Natl. Acad. Sci. USA 113: 9828–9833.

Kim, B.H., S. Kim, H.G. Kim, J. Lee, I.S. Lee and Y.K. Park. 2005. The formation of cyclopropane fatty acids in *Salmonella enterica* serovar Typhimurium. Microbiology 151: 209–218.

Kim, M.C., W.S. Chung, D.-J. Yun and M.J. Cho. 2009. Calcium and calmodulin-mediated regulation of gene expression in plants. Mol. Plant 2(1): 13–21.

Kim, D., K. Yamaguchi and T. Oda. 2006. Nitric oxide synthase-like enzyme mediated nitric oxide generation by harmful red tide phytoplankton, *Chattonella marina*. J. Plankton Res. 28(6): 613–620.

Kim, D., Y.S. Kang, Y. Lee, K. Yamaguchi, K. Matsuoka, K.-W. Lee et al. 2008. Detection of nitric oxide (NO) in marine phytoplankters. J. Biosci. Bioeng. 105(4): 414–417.

Kim, P., M. Kaur, H.-I. Jang and Y.-I. Kim. 2020. The circadian clock-A molecular tool for survival in cyanobacteria. Life 10(12): 365.

Kimber, M.S. 2014. Carboxysomal carbonic anhydrase. Subcell Biochem. 75: 89–103.

Kimura, M. 1968. Evolutionary rate at the molecular level. Nature 217 (5129): 624–626.

Kimura, M. 1983. The Neutral Theory of Molecular Evolution. Cambridge University Press. 367pp.

Kirk, D.L. 2005. A twelve-step program for evolving multicellularity and a division of labor. Bioessays 27(3): 299–310.

Klappenbach, J.A., J.M. Dunbar and T.M. Schmidt. 2000. rRNA operon copy number reflects ecological strategies of bacteria. Appl. Environ. Microbiol. 66(4): 1328–1333.

Klinger, C.M., L. Paoli, R.J. Newby, M.Y.-W. Wang, H.D. Carroll, J.D. Leblond et al. 2018. Plastid transcript editing across dinoflagellate lineages shows lineage-specific application but conserved trends. Genome Biol. Evol. 10(4): 1019–1038.

Kloareg, B., Y. Badis, J.M. Cock and G. Michel. 2021. Role and evolution of the extracellular matrix in the acquisition of complex multicellularity in eukaryotes: a macroalgal perspective. Genes 12: 1059.

Knight, H., A.J. Trewavas and M.R. Knight. 1996. Cold calcium signaling in *Arabidopsis* involves two cellular pools and a change in calcium signature after acclimation. Plant Cell 8: 489–503.

Knoll, A.H., R.E. Summons, J.R. Waldbauer, and J.E. Zum-Berge. 2007. The geological succession of primary producers in the oceans. pp. 133–163. *In*: Falkowski, P. and A.H. Knoll. [eds.]. Evolution of primary producers in the sea. Academic Press, London, UK.

Kobayashi, H., T. Suzuki and T. Unemoto. 1986. *Streptococcal* cytoplasmic pH is regulated by changes in amount and activity of a proton-translocating ATPase. J. Biol. Chem. 261: 627–630.

Kobayashi, Y., N. Harada, Y. Nishimura, T. Saito, M. Nakamura, T. Fujiwara et al. 2014. Algae sense exact temperatures: small heat shock proteins are expressed at the survival threshold temperature in *Cyanioschyzon merolae* and *Chlamydomonas reinhardtii*. Genome Biol. Evol. 6(10): 2731–2740.

Kobayashi, Y., Y. Kanesaki, A. Tanaka, H. Kurosawa, T. Kurosawa and K. Tanaka. 2009. Tetrapyrrole signal as a cell-cycle coordinator from organelle to nuclear DNA replication in plant cells. Proc. Natl. Acad. Sci. USA 106(3): 803–807.

Koga, M., R. Zwaal, K. Guan, L. Avery and Y. Ohshima. 2000. A *Caenorhabditis elegans* MAP kinase kinase, MEK-1, is involved in stress responses. EMBO J. 19: 5148–5156.

Koh, H.Y., W. Jung, H. Do, S.G. Lee, J.H. Lee and H.J. Kim. 2013. Draft genome sequence of *Pseudomonas pelagia* CL-AP6, a pcychrotolerant bacterium isolated from culture of Antarctic green alga *Pyramimonas gelidicola*. mSphere 1(5): e00699–13.

Kohli, G.S., U. John, F.M. Van Dolah and S.A. Murray. 2016. Evolutionary distinctiveness of fatty acid and polyketide synthesis in eukaryotes. ISME J. 10(8): 1877–1890.

Koike, K., H. Sekiguchi, A. Kobiyama, K. Takishita, M. Kawachi, K. Koike et al. 2005. A Novel type of kleptoplastidy in *Dinophysis* (Dinophyceae): Presence of haptophyte-type plastid in *Dinophysis mitra*. Protist 156: 225–237.

Koiwa, H., R.A. Bressan and P.M. Hasegawa. 1997. Regulation of protease inhibitors and plant defense. Trends Plant Sci. 2(10): 379–384.

Kojima, S., A. Bohner and N. von Wirén. 2007. Molecular mechanisms of urea transport in plants. J. Membrane Biol. 212: 83–91.

Koksharova, O. and N.A. Safronova. 2022. Non-proteinogenic amino acid β-N-Methylamino-L-Alanine (BMAA): bioactivity and ecological significance. Toxins (Basel) 14(8): 539.

Kolodny, Y., H. Zer, M. Propper, S. Yochelis, Y. Paltiel and N. Karen. 2021. Marine cyanobacteria tune energy transfer efficiency in their light-harvesting antennae by modifying pigment coupling. FEBS J. 288: 980–994.

Kong, L. 2022. Copper requirement and acquisition by marine microalgae. Microorganisms 10(9): 1853.

König, S., M. Juhas, S. Jäger, T. Kottke and C. Büchel. 2017. The cryptochrome-photolyase protein family in diatoms. J. Plant Physiol. 217: 15–19.

Kono, T., S. Mehrotra, C. Endo, N. Kizu, M. Matusda, H. Kimura et al. 2017. A RuBisCO-mediated carbon metabolic pathway in methanogenic archaea. Nat. Commun. 8: 14007.

Kooistra, W.H.C.F. and L.K. Medlin. 1996. Evolution of the diatoms (Bacillariophyta). IV. A reconstruction of their age from small subunit rRNA coding regions and fossil record. Mol. Phylgenet. Evol. 6: 391–407.

Kooistra, W.H.C.F., R. Gersonde, L.K. Medlin and D.G. Mann. 2007. The origin and evolutionof the diatoms: Their adaptation to a planktonic existence. pp. 207–249. *In*: Falkowski, P.G and A.H. Knoll [eds.]. Evolution of Primary Producers in the Sea. Elsevier, Boston, USA.

Kooistra, W.H.C.F., D. Sarno, S. Balzano, H. Gu, R.A. Andersen and A. Zingone. 2008. Global diversity and biogeography of *Skeletonema* species (Bacillariophyta). Protist 159(2): 177–193.

Koponen, J., K. Laakso, K. Koskenniemi, M. Kankainen, K. Savijoki, T.A. Nyman et al. 2012. Effect of acid stress on protein expression and phosphorylation in *Lactobacillus rhamnosus* GG. J. Proteomics 75(4): 1357–1374.

Kopp, C., M. Pernice, I. Domart-Coulon, C. Djediat, J.E. Spangenberg, D.T. Alexander et al. 2013. Highly dynamic cellular-level response of symbiotic coral to a sudden increase in environmental nitrogen. MBio. 4(3): 10–128. DOI: 10.1128/mbio.00052-13.

Kottmeier, D.M., A. Chrachri, G. Langer, K.E. Helliwell, G.L. Wheeler and C. Brownlee. 2022. Proc. Natl. Acad. USA 119(19): e2118009119.

Kozminski, K.G., K.A. Johnson, P. Forscher and J.L. Rosenbaum. 1993. A motility in the eukaryotic flagellum unrelated to flagellar beating. Proc. Natl. Acad. Sci. USA 90: 5519–5523.

Krauspe, V., M. Fahrner, P. Spät, C. Steglich, N. Frankenberg-Dinkel, B. Maček et al. 2020. Discovery of a small protein factor involved in the coordinated degradation of phycobilisomes in cyanobacteria. Proc. Natl. Acad. Sci. USA 118(5): e2012277118.

Krochko, J.E., G.D. Abrams, M.K. Loewen, S.R. Abrams and A.J. Cutler. 1998. (+)-Abscisic acid 8'-hydroxylase is a cytochrome P450 monooxygenase. Plant Physiol. 118(3): 849–860.

Kroeker, J.K., F. Micheli and M.C. Gambi. 2013. Ocean acidification causes ecosystem shifts via altered competitive interactions. Nat. Clim. Change 3(2): 156–159.

Kropf, D.L. 1997. Induction of polarity in fucoid zygotes. Plant Cell 9: 1011–1020.

Krouk, G. 2017. Nitrate signaling: calcium bridges the nitrate gap. Nat. Plants. 3: 1–2.

Kuczynska, P., M. Jemiola-Rzeminska, B. Nowicka, A. Jakubowska, W. Strzalka, K. Burda et al. 2020. The xanthophyll cycle in diatom *Phaeodactylum tricornutum* in response to light stress. Plant Physiol. Biochem. 152: 125–137.

Kudela, R.M. and W.P. Cochlan. 2000. Nitrogen and carbon uptake kinetics and the influence of irradiance for a red tide bloom off southern California. Aquat. Microb. Ecol. 21: 31–47.

Kühl, M., E. Trampe, M. Mosshammer, M. Johnson, W.W.D. Larkum, N.-U. Frigaard et al. 2020. Substantial near-infrared radiation-driven photosynthesis of chlorophyll *f*-containing Cyanobacteria in a natural habitat. Elife. 9: e50871.

Kühne, W. 1867. Ueber die Verdauung der Eiweisstoffe durchden Pankreassaft. Virchows Arch. Pathol. Anat. Physiol. Klin. Med. 39: 130–174.

Kulkarni, S.D., F. Zhou, N.D. Sen, H. Zhang, A.G. Hinnebusch and J.R. Lorsch. 2019. Temperature-dependent regulation of upstream open reading frame translation in *S. cerevisiae*. BMC Biol. 17: 101.

Kumar Sharma, A., A. Mühlroth, J. Jouhet, E. Marechal, L. Alipanah, E. Kissen et al. 2020. The Myb-like transcription factor phosphorus starvation response (PtPSR) controls conditional P acquisition and remodeling in marine microalgae. New Phytol. 225: 2380–2395.

Kuo, R.C. and S. Lin. 2013. Ectobiotic and endobiotic bacteria associated with *Eutreptialla* sp. isolated from Long Island Sound. Protist 164(1): 60–74.

Kuo, R.C., H. Zhang, Y. Zhuang, L. Hannick and S. Lin. 2013. Transcriptomic study reveals widespread spliced leader trans-splicing, short 5'-UTRs and potential complex carbon fixation mechanisms in the euglenoid alga *Eutreptiella* sp. PLoS ONE 8(4): e60826.

Küpper, F.C., E. Gaquerel, E. Boneberg, S. Morath, J. Salaün and P. Potin. 2006. Early events in the perception of lipopolysaccharides in the brown alga Laminaria digitata include an oxidative burst and activation of fatty acid oxidation cascades. J. Exp. Bot. 57: 1991–1999.

Kurata, K., K. Taniguchi, K. Shiraishi, N. Hayama, I. Tanaka and M. Suzuki. 1989. Ecklonialactone-A and B, two unusual metabolites from the brown alga *Ecklonia stolonifera* Okamura. Chem. Lett. 18(2): 267–270.

Kustka, A.B., A.E. Allen and F.M.M. Morel. 2007. Sequence analysis and transcriptional regulation of iron acquisition genes in two marine diatoms. J. Phycol. 43(4): 715–729.

Kustka, A.B., A.J. Milligan, H.Y. Zheng, A.M. New, C. Gates, K.D. Bidle et al. 2014. Low CO_2 results in a rearrangement of carbon metabolism to support C_4 photosynthetic carbon assimilation in *Thalassiosira pseudonana*. New Phytol. 204(3): 507–520.

Kwok, A.C.M. and J.T.Y. Wong. 2010. The activity of a wall-bound cellulase is required for and is coupled to cell cycle progression in the dinoflagellate *Crypthecodinium cohnii*. Plant Cell 22(4): 1281–1298.

Lacerda, A.F., E.A.R. Vasconcelos, P.B. Palegrini and M.F. Grossi de Sa. 2014. Antifungal defensins and their role in plant defense. Front. Microbiol. 5: 116.

La Claire II, J.W., G.C. Zuccarello and S. Tong. 1997. Abundant plasmid-like DNA in various members of the orders Siphonocladales and Cladophorales (Chlorophyta). J. Phycol. 33: 830–837.

La Claire II, J.W. and J. Wang. 2000. Localization of plasmid-like DNA in giant-celled marine green algae. Protoplasma 213: 157–164.

Lachnit, T., D. Meske, M. Wahl, T. Harder and R. Schmitz. 2011. Epibacterial community patterns on marine macroalgae are host-specific but temporally variable. Environ Microbiol. 13: 655–665.

Lahaye, R., M. van der Bank, D. Bogarin and V. Savolainen. 2008. DNA barcoding the flora of biodiversity hotspots. Proc. Natl. Acad. Sci. USA 105(8): 2923–2928.

Lahr, D.J.G., L.W. Palfrey, E.A.D. Mitchell, L.A. Katz and E. Lara. 2011. The chastity of amoebae: re-evaluating evidence for sex in amoeboid organisms. Proc. Roy. Eco. B. 278(1715): 2081–2090.

LaJeunesse, T.C., G. Lambert, K.A. Andersen, M.A. Coffroth and D.W. Galbraith. 2005. *Symbiodinium* (Pyrrhophyta) genome sizes (DNA content) are smallest among dinoflagellates. J. Phycol. 41(4): 880–886.

LaJeunesse, T.C., J.E. Parkinson, P.W. Gabrielson, H.J. Jeong, J.D. Reimer, C.R. Voolstra et al. 2018. Systematic revision of Symbiodiniaceae highlights the antiquity and diversity of coral endosymbionts. Cur. Biol. 28(16): 2570–2580.

Laloux, T., B. Junqueira, L.C. Maistriaux, J. Ahmed, A. Jurkiewicz and F. Chaumont. 2018. Plant and mammal aquaporins: same but different. Int. J. Mol. Sci. 19(2): 521.

Lampe, R.H., E.L. Mann, N.R. Cohen, C.P. Till, K. Thamatrakoln, M.A. Brzezinski et al. 2018. Different iron storage strategies among bloom-forming diatoms. Proc. Natl. Acad. Sci. USA 115(52): E12275-E12284.

Lang, B.F. and A.M. Nedelcu. 2012. Plastid genomes of algae. pp. 59–87. *In*: Bock, R. and V. Knoop [eds.]. Genomics of Chloroplasts and Mitochondria. Springer Nature, Switzerland AG.

Lang, G., L.J. De Nooijer and I. Oetjen. 2010. On the role of the cytoskeleton in coccolith morphogenesis: the effect of cytoskeleton inhibitors. J. Phycol. 46: 1252–1256.

Lang, D. and S.A. Rensing. 2015. The evolution of transcriptional regulation in the Viridiplantae and its correlation with morphological complexity. *In:* Ruiz-Trillo, I. and A. Nedelcu [eds.]. Evolutionary Transitions to Multicellular Life. Advances in Marine Genomics, Vol 2. Springer, Dordrecht. https://doi.org/10.1007/978-94-017-9642-2_15.

Lang, L. and A. Schnittger. 2020. Endoreplication—a means to an end in cell growth and stress response. Curr. Opin. Plant Biol. 54: 85-92.

Langer, G., L.J. de Nooijer and K. Oetjen. 2010. On the role of the cytoskeleton in coccolith morphogenesis: the effect of cytoskeleton inhibitors. J. Phycol. 46(6): 1252–1256.

Lanier, W., A. Moustafa, D. Bhattacharya and J.M. Comeron. 2008. EST analysis of *Ostreococcus lucimarinus*, the most compact eukaryotic genome, shows an excess of introns in highly expressed genes. PLoS ONE 3(5): e2171.

Lapina, T., V. Statinov, R. Puzanskiy and E. Ermilova. 2022. Arginine-dependent nitric oxide generation and S-nitrosation in the non-photosynthetic unicellular alga *Polytomella parva*. Antioxidants (Basel) 11(5): 949.

Lara, E., D. Vaque, E.L. Sà, J.A. Boras, A. Gomes, E. Borrull, C. Díez-Vives et al. 2017. Unveiling the role and life strategies of viruses from the surface to the dark ocean. Sci. Adv. 3(9): e1602565.

Larkum, A.W.D. and M. Kühl. 2005. Chlorophyll *d*: the puzzle resolved. Trends Plant Sci. 10(8): 355–357.

Larsen, S.H. and J. Beardall. 2021. The effect of CO_2 concentration on DMSP production in *Gephyrocapsa oceanica* (Isochrysidales, Coccolithophyceae). Phycologia 60: 439–448.

Larsson, J., J.A.A. Nylander and B. Bergman. 2011. Genome fluctuations in cyanobacteria reflect evolutionary, developmental and adaptive traits. BMC Ecol. Evol. 11: 187.

Larsson, J., N. Celepli, K. Ininbergs, C.L. Dupont, S. Yooseph, B. Bergma et al. 2014. Picocyanobacteria containing a novel pigment gene cluster dominate the brackish water Baltic Sea. ISME J. 8(9): 1892–1903.

Laudenback, D.E. and N.A. Staus. 1988. Characterization of a cyanobacterial iron stress-induced gene similar to psbC. J. Bacteriol. 170: 5018–5026.

Lauro, F.M., D. McDougald, T. Thomas, T.J. Williams, S. Egan, S. Rice et al. 2009. The genomic basis of trophic strategy in marine bacteria. Proc. Natl. Acad. Sci. USA 106: 15527–15533.

Lavaut, E., M.-L. Guillemin, S. Colin, A. Faure, J. Coudret, C. Destombe et al. 2022. Pollinators of the sea: A discovery of animal-mediated fertilization in seaweed. Science 377 (6605): 528–530.

Lavrentyev, P.J., G. Franzè, J.J. Pierson and D.K. Stoecker. 2015. The effect of dissolved polyunsaturated aldehydes on microzooplankton growth rates in the Chesapeake Bay and Atlantic coastal waters. Mar. Drugs 13(5): 2834–2856.

Lawson, C.A., J.-B. Raina, T. Kahlke, J.R. Seymour and D.J. Suggett. 2018. Defining the core microbiome of the symbiotic dinoflagellate, *Symbiodinium*. Environ. Microbiol. Rep. 10(1): 7–11.

Lay, F.T. and M.A. Anderson. 2005. Defensins–components of the innate immune system in plants. Curr. Protein Pept. Sci. 6: 85–101.

Le, J., E.C. Yang, L. Graf, J.H. Yang, H. Qiu, U. Zelzion et al. 2018. Analysis of the draft genome of the red seaweed *Gracilariopsis choda* provides insights into genome size evolution in Rhodophyta. Mol. Biol. Evol. 35(8): 1869–1886.

Le, S.Q. and O. Gascuel. 2008. An improved general amino acid replacement matrix. Mol. Biol. Evol. 25(7): 1307–1320.

Le, T.B.K., M.V. Imakev, L.A. Mirny and M.T. Laub. 2013. High resolution mapping of the spatial organization of a bacterial chromosome. Science 342: 731–734.

Le Chevanton, M., M. Garnier, E. Lukomska, N. Schreiber, J.-R. Cadoret, B. Saint-Jean et al. 2016. Effects of nitrogen limitation on *Dunaliella* sp.-*Alteromonas* sp. interactions: from mutualistic to competitive relationships. Front. Mar. Sci. 3: 123.

Lee, J.H., H. Lin, S. Joo and U. Goodenough. 2008. Early sexual origins of homeoprotein heterodimerization and evolution of the plant KNOX/BELL family. Cell 133: 829–840.

Lee, J.J. 2006. Algal symbiosis in larger foraminifera. Symbiosis 42: 63–75.

Lee, J.J., L.J. Crockett, J. Hagen and R. Stone. 1974. The taxonomic identity and physiological ecology of *Chlamydomonas hedleyi* sp. from the foraminifer *Archaias angulatus*. British Phycol. J. 9: 407–422.

Lee, J.J. and P. Hallock. 1987. Algal symbiosis as the driving force in the evolution of larger foraminifera. Annal. New York Acad. Sci. 503: 330–347.

Lee, J.M., K.M. Kim, E.C. Yang, K.A. Miller, S.M. Boo, D. Bhattacharya and H.S. Yoon. 2016. Reconstructing the complex evolutionary history of mobile plasmids in red algal genomes. Sci. Rep. 6: 23744.

Lee, J.J., M.E. McEnery and J.R. Garrison. 1980. Experimental studies of larger foraminifera and their symbionts from the Gulf of Elat on the Red Sea. J. Foraminiferal Res. 10: 31–47.

Lee, J.J., M.E. McEnery, E. Kahn and F. Schuster. 1979. Symbiosis and the evolution of larger foraminifera. Micropaleontology 25: 118–140.

Lee, J.J., J. Morales, S. Bacus, A. Diamant, P. Hallock, J. Pawlowski et al. 1997. Progress in characterizing the endosymbiotic dinoflagellates of soritid foraminifera and related studies on some stages of the life cycle of *Marginopora vertebralis*. J. Foraminiferal Res. 27: 254–263.

Lee, J.J., K. Sang, B. ter Kuile, E. Strauss, P.J. Lee and W.W. Faber. 1991. Nutritional and related experiments on laboratory maintenance of three species of symbiont-bearing foraminifera. Mar. Biol. 109: 417–425.

Lee, J.M., E.C. Yang, L. Graf, J.H. Yang, H. Qiu, U. Zelzion et al. 2018. Analysis of the draft genome of the red seaweed *Gracilariopsis chorda* provides insights into genome size evolution in Rhodophyta. Mol. Biol. Evol. 35(8): 1869–1886.

Lee, J.W., C.V. Tevault, T.G. Owens and E. Greenbaum. 1996. Oxygenic photoautotrophic growth without photosystem I. Science 273 (5273): 364–367.

Lee, R.E. 1999. Phycology. 3rd Edition. Cambridge University Press, Cambridge, UK. 614 pp.

Leedale, G.F. 1969. Observations on endonuclear bacteria in euglenoid flagellates. Österr. Bot. Z. 116: 279–294.

Leganés, F., F. Fernández-Piñas and C.P. Wolk. 1994. Two mutations that block heterocyst differentiation have different effects on akinete differentiation in *Nostoc ellipsosporum*. Mol. Microbiol. 12(4): 679–684.

Legris, M., Y.C. Ince and C. Fankhauser. 2019. Molecular mechanisms underlying phytochrome-controlled morphogenesis in plants. Nat. Commun. 10: 5219.

Lehmann, A., K. Myrberg, P. Post, I. Chubarenko, I. Dailidiene, H.-H. Hinrichsen et al. 2022. Salinity dynamics of the Baltic Sea. Earth Syst. Dynamics 13(1): 373–392.

Lemay, M.A., M.Y. Chen, F. Mazel, K.R. Hind, S. Starko, P.J. Keeling et al. 2020. Morphological complexity affects the diversity of marine microbiomes. ISME J. 15: 1372–1386.

Lehner, C, H.H. Kerschbaum and U. Lütz-Meindl. 2009. Nitric oxide suppresses growth and development in the unicellular green alga *Micrasterias denticulata*. J. Plant Physiol. 166(2): 117–127.

Lemieux, C., C. Otis and M. Turmel. 2014. Six newly sequenced chloroplast genomes from prasinophyte green algae provide insights into the relationships among prasinophyte lineages and the diversity of streamlined genome architecture in picoplanktonic species. BMC Genomics 15: 857.

Lepelletier, F., S.A. Karpov, E. Alacid, S. Le Panse, E. Bigeard, E. Garcés et al. 2014. *Dinomyces arenysensis* gen. et sp. nov. (Rhizophydiales, Dinomycetaceae fam. Nov.), a chytrid infecting marine dinoflagellates. Protist 165(2): 230–244.

Lesser, M.P. 2010. Coral bleaching: causes and mechanisms. pp. 405–419. *In:* Dubinsky, Z. and N. Stambler [eds.]. Coral Reefs: An Ecosystem in Transition. Springer Nature, Switzerland AG.

Letcher, P.M., S. Lopez, R. Schmieder, P.A. Lee, C. Behnke, M.J. Powell et al. 2013. Characterization of *Amoeboaphelidium protococcarum*, an algal parasite new to the *Cryptomycota* isolated from an outdoor algal pond used for the production of biofuel. PLoS ONE 8: e56232.

Leutenegger, S. 1984. Symbiosis in benthic foraminifera; specificity and host adaptations. J. Foraminifer. Res. 14: 16–35.

Leveson, A. and J.T.Y. Wong 1999. PCNA-like proteins in dinfoalgellates. J. Phycol. 35: 798–805

Levi-Setti, R., K.L. Gavrilov and P.J. Rizzo. 2008. Divalent cation distribution in dinoflagellate chromosomes imaged by high-resolution ion probe mass spectrometry. Eur. J. Cell Biol. 87: 963–976.

Levitan, O., J. Dinamarca, G. Hochman and P.G. Falkowski. 2014. Diatoms: a fossil fuel of the future. Trends Biotechnol. 32(3): 117–124.

Lewis, L.A. and R.M. McCourt. 2004. Green algae and the origin of land plants. Am. J. Bot. 91(10): 1535–1556.

Lewitus, A.J., H.B. Glagow and J.M. Burkholder. 1999. Kleptoplastidy in the toxic dinoflagellate *Pfiesteria piscicida* (Dinophyceae). J. Phycol. 35(2): 303–312.

Lhee, D., J. Ha, S. Kim, M.G. Park, D. Bhattacharya and H.S. Yoon. 2019. Evolutionary dynamics of the chromatophore genome in three photosynthetic *Paulinella* species. Sci. Rep. 9: 2560.

Li, C. and T. Todo. 2005. The cryptochromes. Genome Biol. 6(5): 220.

Li, C., Y. Li, J. Guo, Y. Wang, X. Shi, Y. Zhang et al. 2024. Abundant mRNA m^1A modification in dinoflagellates: a new layer of gene regulation. EMBO Rep. DOI: 10.1038/s44319-024-00234-2.

Li, C.Y., J.C. Crack, S. Newton-Payne, A.R.J. Murphy, X.L. Chen, B.J. Pinchbeck et al. 2022. Mechanistic insights into the key marine dimethylsulfoniopropionate synthesis enzyme DsyB/DSYB. mLife 1: 114–130.

Li, C.Y., X.J., Wang, X.L. Chen, Q. Sheng, S. Zhang, P. Wang et al. 2021. A novel ATP dependent dimethylsulfoniopropionate lyase in bacteria that releases dimethyl sulfide and acryloyl-CoA. eLife. 10: e64045.

Li, D.-W., J.-Z. Tan, Z.-F. Li and L.-J. Ou. 2024. Membrane lipid remodeling and autophagy to cope with phosphorus deficiency in the dinoflagellate *Prorocentrum shikokuense*. Chemosphere 349: 140844.

Li, H., J. Chen, L. Yu, G. Fan, T. Li, L. Li et al. 2024. In situ community transcriptomics illuminates CO_2-fixation potentials and supporting roles of phagotrophy and proton pump in plankton in a subtropical marginal sea. Microbiol. Spectr. 12(3): e02177–23.

Li, H., L. Li, L. Yu, X. Yang, X. Shi, J. Wang et al. 2020. Transcriptome profiling reveals versatile dissolved organic nitrogen utilization, mixotrophy, and N conservation in the dinoflagellate *Prorocentrum shikokuense* under N deficiency. Sci. Tot. Environ. 763: 143013.

Li, J., W. Li, Y. Huang, H. Bu, K. Zhang and S. Lin. 2024. Phosphorus limitation intensifies heat-stress effects on the potential mutualistic capacity in the coral-derived *Symbiodinium*. Sci. Tot. Environ. 945: 173912.

Li, J., K. Zhang, L. Li, Y. Wang and S. Lin. 2023. Phosphorus nutrition strategies in a symbiodiniacean species: implications in coral-alga symbiosis facing increasing phosphorus deficiency in future warmer oceans. Glob. Chang. Biol. 29(23): 6558–6571.

Li, J., S. Wu, K. Zhang, X. Sun, W. Lin, C. Wang et al. 2024. Clustered regularly interspaced short palindromic repeat-CRISPR-associated protein and its utility all at sea: status, challenges, and prospects. Microorganisms 12(1): 118.

Li, L., R. Hong and J.W. Hastings. 1997. Three functional luciferase domains in a single polypeptide chain. Proc. Natl. Acad. Sci. USA 94(17): 8954–8958.

Li, L., S. Wang, H. Wang, S.K. Sahu, B. Marin, H. Li et al. 2020. The genome of *Prasinoderma coloniale* unveils the existence of a third phylum within green plants. Nat. Ecol. Evol. 4: 1220–1231.

Li, L., H. Wang, S. Wang, Y. Xu, H. Liang, H. Liu and E.C. Sonnenschein. 2021. The draft genome of the centric diatom *Conticribra weissflogii* (Cosinodiscophyceae, Ochrophyta). Protist 172 (5-6): 125845.

Li, L., H. Peng, S. Tan, J. Zhou, Z. Fang, Z. Hu et al. 2020. Effects of early cold stress on gene expression in *Chlaymonas reinhardtii*. Genomics 112: 1128–1138.

Li, M., L. Li, X. Shi, L. Lin and S. Lin. 2015. Effects of phosphorus deficiency and adenosine 5'-triphosphate (ATP) on growth and cell cycle of the dinoflagellate *Prorocentrum donghaiense*. Harmful Algae 47: 35–41.

Li, M., X. Shi, C. Guo and S. Lin. 2016. Phosphorus deficiency inhibits cell division but not growth in the dinoflagellate Amphidinium carterae. Front. Microbiol. 7: 826.

Li, P., C.Y. Liu, H. Liu, Q. Zhang and L. Wang. 2013. Protective function of nitric oxide on marine phytoplankton under abiotic stresses. Nitric Oxide 33: 88–96.

Li, S., T. Nosenko, J.D. Hackett and D. Bhattacharya. 2006. Phylogenomic analysis identifies red algal genes of endosymbiotic origin in the chromalveolates. Mol. Biol. Evol. 23(3): 663–674.

Li, T., X. Lin, L. Yu, S. Lin, I.B. Rodriguez and T.Y. Ho. 2020. RNA-seq profiling of *Fugacium kawagutii* reveals strong responses in metabolic processes and symbiosis potential to deficiencies of iron and other trace metals. Sci. Tot. Environ. 705: 135767.

Li, T., J. Wu, H. Du, P. Pei, C. Yang, J. Huang et al. 2022. Environmental nitrogen and phosphorus nutrient variability triggers intracellular resource reallocation in *Gracilariopsis lemaneiformis* (Rhodophyta). Algal Res. 66: 102778. Doi: 10.1016/j.algal.2022.102778.

Li, X., J. Li, X. Hu, L. Huang, J. Xiao, J. Chan and K. Mi. 2015. Differential roles of the hemerythrin-like proteins of *Mycobacterium smegmatis* in hydrogen peroxide and erythromycin susceptibility. Sci. Rep. 5: 16130.

Li, X., T. Liang and H.T. Liu 2021. How plants coordinate their development in response to light and temperature signals. The Plant Cell 24: 955–966.

Li, X., T. Liu, K. Wang and T.D. Waite. 2015. Light-induced extracellular electron transport by the marine raphidophyte *Chattonella marina*. Environ. Sci. Technol. 49: 1392–1399.

Li, X.Y., Z.H. Li, F.F. Wang, S.H. Zhao, C.Z. Xu, Z.Y. Mao et al. 2024. Structures and organizations of PSI–AcpPCI supercomplexes from red tidal and coral symbiotic photosynthetic dinoflagellates. Proc. Natl. Acad. Sci. USA. 121(7): e2315476121.

Liang, H., S. Lin, Y. Chen, M. Aslam, J. Chen, J. Chen et al. 2024. Urea amidolyase as the enzyme for urea utilization in algae: functional display in *Chlamydomonas reinhardtii* and evolution in algae. bioRxiv DOI: 10.1101/2024.05.16.594469.

Liang, Y., J.A. Koester, J.D. Liefer, A.J. Irwin and Z.V. Finkel. 2019. Molecular mechanisms of temperature acclimation and adaptation in marine diatoms. ISME J. 13(10): 2415–2425.

Liao, X., W. Liu, H.-Q. Yang and G.I. Jenkins. 2020. A dynamic model of UVR8 photoreceptor signalling in UV-B acclimated Arabidopsis. New Phytol. 227(3): 857–866.

Lickwar, C.R., F. Mueller, S.E. Hanlon, J.G. McNally and J.D. Lieb. 2012. Genome-wide protein-DNA binding dynamcis suggest a molecular clutch for transcription factor function. Nature 484: 251–255.

Lidie, K.B., J.C. Ryan, M. Barbier and F.M. Van Dolah. 2005. Gene expression in Florida red tide dinoflagellate *Karenia brevis*: analysis of an expressed sequence tag library and development of DNA microarray. Mar. Biotechnol. 7: 481–493.

Lieberman-Aiden, E., N.L. van Berkum, L. Williams, M. Imakaev, T. Ragoczy, A. Telling et al. 2009. Comprehensive mapping of long-range interactions reveals folding priciples of the human genome. Science 326(5950): 289–293.

Liedvogel, M. and H. Mouritsen. 2010. Cryptochromes-a potential magnetoreceoptor: what do we know and what do we want to know? J. R. Soc. Interface 7(suppl. 2): S147–S162.

Lietz, G., A. Oxley, C. Boesch-Saadatmandi and D. Kobayashi. 2012. Importance of β, β-carotene 15, 15′-monooxygenase 1 (BCMO1) and β, β-carotene 9′, 10′-dioxygenase 2 (BCDO2) in nutrition and health. Mol. nutrition food res. 56(2): 241–250.

Lin, S. 2005. The smallest dinoflagellate genome is yet to be found: a comment on LaJeunesse et al. "Symbiodinium (Pyrrophyta) genome sizes (DNA content are smallest among dinoflagellates". J. Phycol. 42(3): 746–748.

Lin, S. 2011. Genomic understanding of dinoflagellates. Res. Microbiol. 162: 551–569.

Lin, S. 2023. Phosphate limitation and ocean acidification co-shape phytoplankton physiology and community structure. Nat. Commun. 14: 2699.

Lin, S. 2024. A decade of dinoflagellate genomics illuminating an enigmatic eukaryote cell. BMC Genomics 25: 932.

Lin, S. and E.J. Carpenter. 1995. Growth characteristics of phytoplankton determined by cell cycle proteins: The cell cycle of *Ethmodiscus rex* in the southwestern North Atlantic Ocean and Caribbean Sea. J. Phycol. 31: 778–785.

Lin, S. and E.J. Carpenter. 1998. Detection and preliminary characterization of the PCNA gene in marine phytoplankton. Mol. Mar. Biol. Biotech. 7: 62–71.

Lin, S. And E.J. Carpenter. 1999. A PSTTLRE-form of cdc2-like gene in the marine micro alga *Dunaliella tertiolecta.* Gene 239(1): 39–48.

Lin, S. and P.L.A.M. Costjiens. 2002. Molecular cloning and expression of the proliferating cell nuclear antigen gene from the coccolithophorid Pleurochrysis carterae (Haptophyceae). J. Phycol. 38: 164–173.

Lin, S. and H. Zhang. 2003. Mitogen-activated protein kinase in Pfiesteria piscicida and its growth rate-relatd expression. Appl. Environ. Microbiol. 69(1): 343–349.

Lin, S., J. Chang and E.J. Carpenter. 1994. Detection of proliferating cell nuclear antigen (PCNA) analog in four species of marine phytoplankton. J. Phycol. 30: 449–456.

Lin, S., J. Chang and E.J. Carpenter. 1995. Growth characteristics of phytoplankton determined by cell cycle proteins: PCNA immunostaining of *Dunaliella tertiolecta.* J. Phycol. 31: 388–395.

Lin, S., E.J. Carpenter and J. Chang. 1996. Detection of p34cdc2- and cyclin B-like proteins in *Dunaliella tertiolecta* (Chlorophyceae). Mar. Biol. 125: 603–610.

Lin, S., J. Chang and E.J. Carpenter. 1997. Can a non-terminal event of the cell cycle be used for phytoplankton species-specific growth rate estimation? Mar. Ecol. Prog. Ser. 151: 283–290.

Lin, S., S. Cheng, B. Song, X. Zhong, X. Lin, W. Li et al. 2015. The genome of *Symbiodinium kawagutii* illuminates dinoflagellate gene expression and coral symbiosis. Science 350: 691-694. DOI:10.1126/science.aad0408.

Lin, S., H. Zhang and M.W. Gray. 2008. RNA editing in dinoflagellates and its implications for the evolutionary history of the editing machinery. pp. 280–309. *In:* Smith, H. [ed.]. RNA and DNA editing: Molecular Mechanisms and Their Integration into Biological Systems. John Wiley & Sons, Inc.

Lin, S., S. Henze, P. Lundgren, B. Bergman and E.J. Carpenter. 1998. Whole-cell immunolocalization of nitrogenase in marine diazotrophic cyanobacteria, *Trichodesmium* spp. Appl. Environ. Microbiol. 64(8): 3052–3058.

Lin, S., Z. Hu, X. Song, C.J. Gobler, Y.Z. Tang. 2022. Vitamin B12-auxotrophy in dinoflagellates caused by incomplete or absent methionine synthase genes (metE). Fundamental Res. 2: 727–737.

Lin, S. and H. Zhang. 2010. Dinoflagellate meta-transcriptomics enabled by spliced leader. In 13th International Conference on Harmful Algae, Hong Kong, pp. 166–170.

Lin, S., R.W. Litaker and W.G. Sunda. 2016. Phosphorus physiological ecology and molecular mechanisms in marine phytoplankton. J. Phycol. 52: 10–36.

Lin, S., E. Magaletti and E.J. Carpenter. 2000. Molecular cloning and antiserum development of cyclin box in the brown tide alga *Aureococcus anophagefferens*. Mar. Biotechnol. 2: 577–586.

Lin, S., M.R. Mulholland, H. Zhang, T.N. Feinstein, F.J. Jochem and E.J. Carpenter. 2004. Intense grazing and prey-dependent growth of Pfiesteria piscicida (Dinophyceae). J. Phycol. 40(6): 1062–1073.

Lin, S., B. Song, D. Morse. 2021. Spatial organization of dinoflagellate genomes: novel insights and remaining critical question. J. Phycol. 57: 1674–1678.

Lin, S., L. Wang, L. Zheng, Y. Dong, S. Liu, S. Ding et al. 2014. Current status and future prospect of DNA barcoding in marine biology. Acta Oceanologica Sinica 36(12): 1–17 (In Chinese with English abstract).

Lin, S., L. Yu, Z. Wu, M. Li, Y. Zhang, H. Luo et al. 2022. Active meiosis during dinoflagellate blloms: A 'sex for proliferation' hypothesis. Harmful Algae 118: 102307.

Lin, S., H. Zhang, D. Spencer, J. Norman and M.W. Gray. 2002. Widespread and extensive editing of mitochondrial mRNAs in dinoflagellates. J. Mol. Biol. 320: 727–739.

Lin, S., H. Zhang, Y. Zhuang, B. Tran and J. Gill. 2010. Spliced leader-based metatranscriptomic analyses lead to recognition of hidden genomic features in dinoflagellates. Proc. Natl. Acad. Sci. USA 107(46): 20033–20038.

Lin, W.Y., T.K. Huang and T.J. Chiou. 2013. Nitrogen limitation adaptation, a target of microRNA827, mediates degradation of plasma membrane-localized phosphate transporters to maintain phosphate homeostasis in *Arabidopsis*. Plant Cell 25(10): 4061–4074.

Lin, X., C. Guo, L. Li, T. Li and S. Lin. 2019. Non-conventional metal ion cofactor requirement of dinoflagellate alkaline phosphatase and translational regulation by phosphorus limitation. Microogranisms 7: 232.

Lin, X., L. Wang, X. Shi and S. Lin. 2015. Rapidly diverging evolution of an atypical alkaline phosphatase ($PhoA^{aty}$) in marine phytoplankton: insights from dinoflagellate alkaline phosphatases. Front. Microbiol. 6: 868.

Lin, X., H. Zhang, Y. Cui and S. Lin. 2012. High sequence variability, diverse subcellular localizations, and ecological implications of alkaline phosphatase in dinoflagellates and other eukaryotic phytoplankton. Front. Microbiol. 3: 235.

Lin, Y.-L., C.-L. Chung, M.-H. Chen, C.-H. Chen and S.-C. Fang. 2020. SUMO protease SMT7 modulates ribosomal protein L30 and regulates cell-size checkpoint function. Plant Cell 32(4): 1285–1307.

Lin, Y.-T., T. Takeuchi, B. Youk, J. Umen, B.B. Sears and C. Benning. 2022. Chlamydomonas CHT7 is involved in repressing DNA replication and mitotitc genes during synchronous growth. G3 (Bethesda) 12(3): jkac023.

Linares-Maurizi, A., G. Reversat, R. Awad, V. Bultel-Poncé, C. Oger, J.-M. Galano et al. 2023. Bioactive oxylipins profile in marine microalgae. Mar. Drugs 21: 136.

Liu, D., Q. Ma, I. Valiela, D.M. Anderson, J.K. Keesing, K. Gao et al. 2000. Roles of C4 carbon fixation in Ulva prolifera, the macroalgal responsible for the world's largest green tides. Comm. Biol. 3: 494.

Liu, H., S. Aris-Brosou, I. Probert and C. de Vargas. 2010. A time line of the environmental genetics of the haptophytes. Mol. Biol. Evol. 27(1): 161–176.

Liu, H., I. Probert, J. Uitz, H. Claustre, S. Aris-Brosou, M. Frada et al. 2009. Extreme diversity in noncalcifying haptophytes explains a major pigment paradox in open oceans. Proc. Natl. Acad. Sci. USA 106: 12803–12808.

Liu, H., T.G. Stephens, R.A. González-Pech, V.H. Beltran, B. Lapeyre, P. Bongaerts et al. 2018. *Symbiodinium* genomes reveal adaptive evolution of functions related to coral-dinoflagellate symbiosis. Commun. Biol. 1: 95.

Liu, H., Y. Zhou, W. Xiao, L. Ji, X. Cao and C. Song. 2012. Shifting nutrient-mediated interactions between algae and bacteria in a microcosm: evidence from alkaline phosphatase assay. Microbiol. Res. 167: 292–298.

Liu, J., N. Jiao, H. Hong, T. Luo and H. Cai. 2005. Proliferating cell nuclear antigen (PCNA) as a marker of cell proliferation in the marine dinoflagellate *Prorocentrum donghaiense* Lu and the green alga *Dunaliella salina* Teodoresco. J. Apple. Phycol. 17: 323–330.

Liu, J., L. Yang, M. Luan, Y. Wang, C. Zhang, B. Zhang et al. 2015. A vacuolar phosphate transporter essential for phosphate homeostasis in *Arabidopsis*. Proc. Natl. Acad. Sci. USA 112(47): E6571–6578.

Liu, K.-H. and Y.-F. Tsay. 2003. Switching between the two action modes of the dual-affinity nitrate transporter CHL1 by phosphorylation. EMBO J. 22: 1005–1013.

Liu, N. and T. Pan. 2015. RNA Epigenetics. Transl. Res. 165: 28–35.

Liu, Q.Y. and M.E. Reith. 1994. Isolation of a gametophyte-specific cDNA encoding a lipoxygenase from the red alga Porphyra purpurea. Mol. Mar. Biol. Biotechnol. 3: 206–209.

Liu, X., K. Bogaert, A.H. Engelen, F. Leliaert, M.Y. Roleda and O. De Clerck. 2017. Seaweed reproductive biology: environmental and genetic controls. Bot. Mar. 60(2): 89–108.

Liu, X., Y.-Z. Hu, Y.-B. Pan, J. Liu, Y.-B. Jiang, Y.-A. Zhang and X.-J. Zhang. 2023. Comparative study on antibacterial characteristics of the multiple liver expressed antimicrobial peptides (LEAPs) in the teleost fish. Front. Immunol. 14: 1128138.

Liu, X., Z. Zuo, X. Xie, S. Gao, S. Wu, W. Gu et al. 2024. SLC24A-mediated calcium exchange as an indispensable component of the diatom cell density-driven signaling pathway. ISME J. 18(1): wrae039.

Liu, Z. and C. Wu. 2012. Response of alkaline phosphatases in the cyanobacterium *Anabaena* sp. FACHB709 to inorganic phosphate starvation. Curr. Biol. 64: 524–529.

Liu, Z., A.C. Jones, V. Campbell, K.D. Hambright, K.B. Heidelberg and D.A. Caron. 2015. Gene expression in the mixotrophic prymnesiophyte, *Prymnesium parvum*, responds to prey availability. Front. Microbiol. 6: 319.

Llaveria, G., Figueroa, R.I., Garcés and E. Berdalet. 2009. Cell cycle and cell mortality of *Alexandrium minutum* (dinophyceae) under small-scale turbulence conditions. J. Phycol. 45: 1106–1115.

Lo, R., K.E. Dougan, Y. Chen, S. Shah, D. Bhattacharya and C.X. Chan. 2022. Alignment-free analysis of Symbiodiniaceae whole-genome sequences reveals differential phylogenetic signals in distinct regions. Front. Plant Sci. 13: 815714.

Lodeyro, A.F., R.D. Ceccoli, J.J.P. Karlusich and N. Carrillo. 2012. The importance of flavodoxin for environmental stress tolerance in photosynthetic microorganisms and transgenic plants: Mechanism, evolution and biotechnological potential. FEBS Lett. 586(18): 2917–2924.

Logares, R., K. Shalchian-Tabrizi, A. Boltovskoy and K. Rengefors. 2007. Extensive dinoflagellate phylogenies indicate infrequent marine-freshwater transitions. Mol. Phylogenet. Evol. 45(3): 887–903.

Lohr, M. and C. Wilhelm. 1999. Algae displaying the diadinoxanthin cycle also possess the violaxanthin cycle. Proc. Natl. Acad. Sci. USA 96: 8784–8789.

Lomas, M.W., J.A. Bonachela, S.A. Levin and A.C. Martiny. 2014. Impact of ocean phytoplankton diversity on phosphate uptake. Proc. Natl. Acad. Sci. USA 111(49): 17540–17545.

Lommer, M., M. Specht, A.-S. Roy, L. Kraemer, R. Andreson, M.A. Gutowska et al. 2012. Genome and low-iron response of an oceanic diatom adapted to chronic iron limitation. Genome Biol. 13: R66.

Long, M., D. Marie, J. Szymczak, J. Toullec, E. Bigeard, M. Sourisseau et al. 2021. Dinophyceae can use exudates as weapons against the parasite *Amoebophrya* sp. (Syndiniales). ISME Commun. 1: 34.

Longworth, J., D. Wu, M. Huete-Ortego, P.C. Wright and S. Caidyanathan. Proteome response of *Phaeodactylum tricornutum*, during lipid accumulation induced by nitrogen depletion. Algal Res. 18: 213–224.

Lopez, E. 1979. Algal chloroplasts in the protoplasm of three species of benthic foraminifera: taxonomic affinity, viability and persistence. Mar. Biol. 53: 201–211.

López, M.D., J.J.M. Guerra and T. Samuelson. 2010. Analysis of gene order conservation in eukaryotes identifies transcriptionally and functionally linked genes. PLoS ONE 5(5): e10654.

López-Legentil, S., B. Song, M. DeTure and D.G. Baden. 2010. Characterization and localization of a hybrid non-ribosomal peptide synthetase and polyketide synthase gene from the toxic dinoflagellate *Karenia brevis*. Mar. Biotechnol. 12: 32–41.

Lovecchio, S., E. Climent, R. Stocker and W. Durham 2019. Chain formation can enhance the vertical migration of phytoplankton through turbulence. Sci. Adv. 5(10): eaaw7879.

Lu, C., Y. Tian, S. Wang, Y. Su, T. Mao, T. Huang et al. 2017. Phosphorylation of SPT5 by CDKD;2 is required for VIP5 recruitment and normal flowering in *Arabidopsis thaliana*. Plant Cell 29: 277–291.

Lu, D-C., F.-Q. Wang, R.I. Amann, H. Teeling and Z.-J. Du. 2023. Epiphytic common core bacteria in the microbiomes of co-located green (*Ulva*), brown (*Saccharina*) and red (*Grateloupia*, *Gelidium*) macroalgae. Microbiome 11: 126.

Lu, P., D. Ma, Y. Chen, Y. Guo, G.Q. Chen, H. Deng et al. 2013. L-glutamine provides acid resistance for *Escherichia coli* through enzymatic release of ammonia. Cell Res. 23: 635–644.

Lu, Y., S. Wohlrab, G. Glöckner, L. Guillou and U. John 2014. Genomic insights into processes driving the infection of *Alexandrium tamarense* by the parasitoid *Amoebophrya* sp. Euk. Cell 13(11): 1439–1449.

Lu, Z., Z. Chen, L. Xiao, L. Lei, B.-P. Han and H.W. Paerl. 2022. Cyanophycin accumulated under nitrogen-fluctuating and high-nitrogen conditions facilitates the persistent dominance and blooms of *Raphidiopsis raciborskii* in tropical waters. Water Res. 214: 118215.

Lucas, I.A.N. 2012. Symbionts of the tropical dinophysiales (Dinophyceae). Ophelia 33: 213–224.

Lukeš, J., B.S. Leander and P.J. Keeling. 2009. Cascades of convergent evolution: the corresponding evolutionary histories of euglenozoans and dinoflagellates. Proc. Natl. Acad. Sci. USA 106(Suppl 1): 9963–9970.

Lund, P., A. Tramonti and D. De Biase. 2014. Coping with low pH: molecular strategies in neutralophilic bacteria. FEMS Microbiol. Rev. 38(6): 1091–1125.

Luo, H., X. Lin, L. Li, L. Lin, C. Zhang and S. Lin. 2017. Transcriptomic and physiological analyses of the dinoflagellate *Karenia mikimotoi* reveal non-alkaline phosphatase-based molecular machinery of ATP utilization. Environ. Microbiol. 19(11): 4506–4518.

Luo, H., J. Wang, J.I. Goes, H.d.R. Gomes, K. Al-Hashmi, C. Tobias et al. 2022. A grazing-driven positive nutrient feedback loop and active sexual reproduction underpin widespread *Noctiluca* green tides. ISME Comm. 2: 103.

Luo, R., B. Liu, Y. Xie, Z. Li, W. Huang, J. Yuan et al. 2012. SOAPdenovo2: An empirically improved memory-efficient short-read *de novo* assembler. GigaScience 1: 18.

Luo, T., K. Mohan, P.A. Iglesias and D.N. Robinson 2013. Molecular mechanisms of cellular mechanosensing. Nat. Mater. 12: 1064–1071.

Luo, W., K. Inomura, H. Zhang and Y.-W. Luo. 2022. N_2 fixation in *Trichodesmium* does not require spatial segregation from photosynthesis. mSystems 7(4): e00538-22.

Lürling, M. 2021. Grazing resistance in phytoplankton. Hydrobiologia 848: 237–249.

Luxem, K.E., M.J. Ellwood and R.F. Strzepek. 2017. Intraspecific variability in *Phaeocystis antarctica*'s response to iron and light stress. PLoS ONE 12: e0179751.

Lynch, M. and J.S. Conery. 2003. The origins of genome complexity. Science 302: 1401–1404.

Lyons, T. W., C.T. Reinhard and N.J. Planavsky. 2014. The rise of oxygen in Earth's early ocean and atmosphere. Nature 506: 307–315.

Ma, F., X. Zhang, X. Zhu, T. Li, J. Zhan, H. Chen et al. 2017. Dynamic changes of isiA-containing complexes during long-term iron deficiency in *Synechocystis* sp. PCC 6803. Mol. Plant 10(1): 143–154.

Ma, M., X. Shi and S. Lin. 2020. Heterologous expression and cell membrane localization of dinoflagellate opsins (rhodopsin proteins) in mammalian cells. Mar. Life Sci. Technol. 2: 302–308.

Ma, M., H. Li, C. Wang, T. Li, J. Wang, H. Yuan, L. Yu et al. 2023. A comparative study reveals the relative importance of prokaryotic and eukaryotic proton pump rhodopsins in a subtropical marginal sea. ISME Commun. 3(1): 79.

Ma, M., Y. You, Y. Huang, X. Sun, J. Li, L. Li et al. 2024. Proton-pumping rhodopsin of the coral symbiont *Breviolum minutum* and its potential role in coping with phosphorus deficiency in future warmer ocean. Coral Reefs 43(4): 867–879.

Mackiewicz, P., A. Body and K. Moszczyński. 2013. The case of horizontal gene transfer from bacteria to the peculiar dinoflagellate plastid genome. Mob. Genet. Elements 3(4): e25845.

Macorano, L., T.M. Binny, T. Spiegi, V. Klimenko, A. Singer, L. Oberleitner et al. 2023. DNA-binding and protein structure of nuclear factors likely acting in genetic information processing in the *Paulinella* chromatophore. Proc. Natl. Acad. Sci. USA 120(27): e2221595120.

Maeda, H.A. and A.R. Fernie. 2021. Evolutionary history of plant metabolism. Annu. Rev. Plant Biol. 72: 185–216.

Maeda, Y., R. Kobayashi, K. Watanabe, T. Yoshino, C. Bowler, M. Matsumoto et al. 2022. Chromosome-scale genome assembly of the marine oleaginous diatom *Fistulifera solaris*. Mar. Biotechnol. 24: 788–800.

Maere, S., S. De Bodt, J. Raes, T. Casneuf, M. Van Montagu, M. Kuiper et al. 2005. Modeling gene and genome duplications in eukaryotes. Proc. Natl. Acad. Sci. USA 102(15): 5454–5459.

Maire, J. S.K. Girvan, S.E. Barkla, A. Perez-Gonzalez, D.J. Suggett, L.L. Blackall et al. 2021. Intracellular bacteria are common and taxonomically diverse in cultured and *in hospite* algal endosymbionts of coral reefs. ISME J. 15: 2028–2042.

Makita, Y., S. Shigekatsu, K. Fushimi, S. Shimada, A. Suehisa, M. Hirata et al. 2021. Identification of a dual orange/far-red and blue light photoreceptor from an oceanic green picoplankton. Nat. Commun. 12: 3593.

Malcolm, H. and H. April. 2012. The magnesium inhibition and arrested phagosome hypotheses: new perspectives on the evolution and ecology of Symbiodinium symbioses. Biol. Rev. Camb. Philos. Soc. 87(4): 804–821.

Maldonado, E.M. and M.I. Latz. 2007. Shear-stress dependence of dinoflagellate bioluminescence. Biol. Bull. 212: 242–249.

Maliet, O., D.E. Shelton and R.E. Michod. 2015. A model for the origin of group reproduction during the evolutionary transition to multicellularity. Biol. Lett. 11(6): 20150157.

Malik, S.B., A.W. Pightling, L.M. Stefaniak, A.M. Schurko and J.M. Logsdon. 2008. An expanded inventory of conserved meiotic genes provides evidence for sex in *Trichomonas vaginalis*. PLoS ONE 3: 2879.

Malitsky, S., C. Ziv, S. Rosenwasser, S. Zheng, D. Schatz, Z. Porat et al. 2016. Viral infection of the marine alga *Emiliania huxleyi* triggers lipidome remodeling and induces the production of highly saturated triacylglycerol. New Phytol. 210: 88–96.

Mallén-Ponce, M.J., M.E. Pérez-Pérez and J.L. Crespo. 2022. Deciphering the function and evolution of the target of rapamycin signaling pathway in microalgae. J. Exp. Bot. 73: 6993–7005.

Mallick, N., L.C. Rai, F.H. Mohn and C.J. Soeder. 1999. Studies on nitric oxide (NO) formation by the green alga *Scenedesmus obliquus* and the diazotrophic cyanobacterium *Anabaena doliolum*. Chemosphere 39(10): 1601–1610.

Malviya, S., E. Scalco, S. Audic and C. Bowler. 2016. Insights into global diatom distribution and diversity in the world's ocean. Proc. Natl. Acad. Sci. USA 113(11): E1516–E1525.

Mann, D.G. and S.J.M. Droop. 1996. Biodiversity, biogeography and conservation of diatoms. Hydrobiologia 336: 19–32.

Mann, D.G. and P. Vanormelingen. 2013. An inordinate fondness? The number distributions, and origins of diatom species. J. Euk. Microbiol. 60: 414–420.

Mann, M., M. Serif, T. Jakob, P.G. Kroth and C. Wilhelm. 2017. PtAUREO1a and PtAUREO1b knockout mutants of the diatom *Phaeodactylum tricornutum* are blocked in photoacclimation to blue light. J. Plant Physiol. 217: 44–48.

Mann, M., M. Serif, T. Wrobel, M. Eisenhut, S. Madhuri, S. Flachbart et al. 2020. The aureochrome photoreceptor PtAUREO1a is a highly effective blue light switch in diatoms. iScience 23: 101730.

Mano, Y. and K. Nemoto. 2012. The pathway of auxin biosynthesis in plants. J. Exp. Bot. 63: 2853–2872.

Mao, X., J. Chen, C. van Oosterhout, H. Zhang, G. Liu, Y. Zhuang, et al. 2022. Diversity, prevalence, and expression of cyanase genes (cynS) in planktonic marine microorganisms. ISME J. 16(2): 602–605.

Maor-Landaw, K., M.J.H. van Oppen and G.I. McFadden. 2020. Symbiotic lifestyle triggers drastic changes in the gene expression of the algal endosymbiont *Breviolum minutum* (Symbiodiaceae). Ecol. Evol. 10(1): 451–466.

Marahiel, M.A., T. Stachelhaus and H.D. Mootz. 1997. Modular peptide synthetases involved in nonribosomal peptide synthesis. Chem. Rev. 97(7): 2651–2674. DOI:10.1021/cr960029e.

Marchenkov, A.M., D.P. Petrova, A.A. Morozov, Y.R. Zakharova, M.A. Grachev and A.A. Bondar. 2018. A family of silicon transporter structural genes in a pennate diatom *Synedra ulna* subsp. *danica* (Kütz.) Skabitsch. PLoS ONE 13(8): e0203161.

Marchetti, A., D. Catlett, B.M. Hopkinson, E. Kelsey and N. Cassar. 2015. Marine diatom proteorhodopsins and their potential role in coping with low iron availability. ISME J. 9(12): 2745–2748.

Marchetti, A., M.S. Parker, L.P. Moccia, E.O. Lin, A.L. Arrieta, F. Ribalet et al. 2009. Ferritin is used for iron storage in bloom-forming marine pennate diatoms. Nature 457: 467–470.

Marchetti, A., D.M. Schruth, C.A. Durkin, M.S. Parker, R.B. Kodner, C.T. Berthiaume et al. 2012. Comparative metatranscriptomics identifies molecular bases for the physiological responses of phytoplankton to varying iron availability. Proc. Natl. Acad. Sci. USA 109(6): E317–E325.

Mardones, J.I., J.J. Dorantes-Aranda, P.D. Nichols and G.M. Hallegraeff. 2015. Fish gill damage by the dinoflagellates *Alexandrium catenella* from Chilean fjords: synergistic action of ROS and PUFA. Harmful Algae 49: 40–49.

Margoliash, E. 1963. Primary structure and evolution of cytochrome C. Proc. Natl. Acad. Sci. USA 50(4): 672–679.

Margulies, E.H., A.M. White and S. Sherry. 1980. Sulfinpyrazone: a review of its pharmacological properties and therapeutic use. Drugs 20(3): 179–197.

Margulis, L. 1970. Origin of Eukaryotic Cells. Yale University Press. pp. 349.

Marin, B., E.C.M. Nowack and M. Melkonian. 2005. A plastid in the making: evidence for a second primary endosymbiosis. Protist 156: 425–432.

Marin, B., E.C.M. Nowack, G. Glöckner and M. Melkonian. 2007. The ancestor of the *Paulinella* chromatophore obtained a carboxysomal operon by horizontal gene transfer from a *Nitrococcus*-like γ-proteobacterium. BMC Evol. Biol. 7: 85.

Marinov, G.K. and M. Lynch. 2016. Diversity and divergence of dinoflagellate histone proteins. G3 Genes Genomes Genetics 6: 397–422.

Marinov, G.K., X. Chen, M.P. Swaffer, T. Xiang, A.R. Grossman and W.J. Greenleaf. 2024. Genome-wide distribution of 5-hydroxymethyluracil and chromatin accessibility in the *Breviolum minutum* genome. Genome Biol. 25(1): 115.

Marinov, G.K., A.E. Trevino, T. Xiang, A. Kundaje, A.R. Grossman and W.J. Greenleaf. 2021. Transcription-dependent domain-scale three-dimensional genome organization in the dinoflagellate *Breviolum minutum*. Nat. Genet. 53: 613–617.

Markson, J.S., J.R. Piechura, A.M. Puszynska and E.K. O'Shea. 2013. Circadian control of global gene expression by the cyanobacterial master regulator RpaA. Cells 155: 1396–1408.

Martin, P., S.T. Dyhrman, M.W. Lomas, N. J. Poulton and A.S. van Mooy. 2014. Accumulation and enhanced cycling of polyphosphate by Sargasso Sea plankton in response to low phosphorus. Proc. Natl. Acad. Sci. USA 111: 8089–8094.

Martin, W., T. Rujan, E. Richly, A. Hansen, S. Cornelsen, T. Lins et al. 2002. Evolutionary analysis of *Arabidopsis*, cyanobacterial, and chloroplast genomes reveals plastid phylogeny and thousands of cyanobacterial genes in the nucleus. Proc. Natl. Acad. Sci. USA 99: 12246–12251.

Martin, W.F., D.A. Bryant and J.T. Beatty. 2018. A physiological perspective on the origin and evolution of photosynthesis. FEMS Microbiol. Rev. 42: 205–231.

Martinac, B. and C.D. Cox. 2017. Mechanosensory transduction: focus on ion channels. Reference Module in Life Sciences 1–47. DOI: 10.1016/B978-0-12-809633-8.08094-8.

Martinez-Gutierrez, C.A. and F.O. Aylward. 2022. Evolutionary genomics of marine bacteria and archaea. pp. 327–354. *In*: Stal, L.J. and M.S. Cretoiu [eds.]. The Marine Microbiome (2nd edition). Springer.

Martínez-Núñez, M.A. and V.E.L.y. López. 2016. Nonribosomal peptides synthetases and their applications in industry. Sustain. Chem. Proc. 4: 13.

Maruyama, S., T. Suzaki, A.PM. Weber, J.M. Archibald and H. Nozaki. 2011. Eukaryote-to-eukaryote gene transfer gives rise to genome mosaicism in euglenids. BMC Evol. Biol. 11: 105.

Masclaux-Daubresse, C., Q. Chen and M. Havé. 2017. Regulation of nutrient recycling via autophagy. Curr. Opin. Plant Biol. 39: 8–17.

Maselli, M., K. Anestis, K. Klemm, P.J. Hansen and U. John. 2021. Retention of prey genetic material by the kleptoplastidic ciliate *Strombidium* cf. *basimorphum*. Front. Microbiol. 12: 694508.

Mathieu-Rivet, E., N. Mati-Baouche, M.-L. Walet-Balieu, P. Lerouge and M. Bardor. 2020. *N*- and *O*-glycosylation pathways in the microalgae polyphyletic group. Front. Plant Sci. 11: 609993.

Mather, R.L., S.E. Reynolds, G.A. Wolff, R.G. Williams, S. Torres-Valdes, E.M.S. Woodward et al. 2008. Phosphorus cycling in the North and South Atlantic Ocean subtropical gyres. Nat. Geosci. 1: 439–443.

Matsui, H., B.M. Hopkinson, K. Nakajima and Y. Matsuda. 2018. Plasma membrane-type aquaporins from marine diatoms function as CO_2/NH_3 channels and provide photoprotection. Plant Physiol. 178(1): 345–357.

Matsuo, E., K. Morita, T. Nakayama, E. Yazaki, C. Sarai, K. Takahashi et al. 2022. Comparative plastid genomics of green-colored dinoflagellates unveils parallel genome compaction and RNA editing. Front. Plant Sci. 13: 918543. doi: 10.3389/fpls.2022.918543.

Matsuzaki, M., O. Misumi, T. Shin-I, S. Maruyama, M. Takahara, S.-y. Miyagishima et al. 2004. Genome sequence of the ultrasmall unicellular red alga *Cyanidioschyzon merolae* 10D. Nature 428: 653–657.

Matthews, J.L., M. Ueland, N. Bartels, C.A. Lauson, T.E. Lockwood, Y. Wu et al. 2024. Multi-chemical omics analysis of the Symbiodiniaceae *Durusdinium trenchii* under heat stress. Microorganisms 12(2): 317.

Matthews, J.L., A. Khalil, M. Siboni, J. Bougoure, P. Guagliardo, U. Kuzhiumparambil et al. 2023. Coral endosymbiont growth is enhanced by metabolic interactions with bacteria. Nat. Commun. 14: 6864.

Maurel, C., L. Verdoucq, D.-T. Luu and V. Santoni. 2008. Plant aquaporins: membrane channels with multiple integrated functions. Annu. Rev. Plant Biol. 59: 595–624.

Mazéas, L., R. Yonamine, T. Barbeyron, B. Henrissat, E. Drula, N. Terrapon et al. 2022. Assembly and synthesis of the extracellular matrix in brown algae. Semin. Cell Dev. Biol. 134:112–124.

McCormick, A.J., P. Bombelli, A.M. Scott, A.J. Phillips, A.G. Smith, A.C. Fisher et al. 2011. Photosynthetic biofilms in pure culture harness solar energy in a mediatorless bio-photovoltaic cell (BPV) system. Energy Environ. Sci. 4: 4699–4709.

McDowell, R.E., C.D. Amsler, D.A. Dickson, J.B. McClintock and B.J. Baker. 2014a. Reactive oxygen species and the Antarctic macroalgal wound response. J. Phycol. 50(1): 71–80.

McDowell, R.E., C.D. Amsler, J.B. McClintock and B.J. Baker. 2014b. Reactive oxygen species as a marine grazing defense: H_2O_2 and wounded *Ascoseira mirabilis* both inhibit feeding by an amphipod grazer. J. Exp. Mar. Biol. Ecol. 458: 34–38.

McGaughey, S.A., S.D. Tyerman and C.S. Byrt. 2021. An algal PIP-like aquaporin facilitates water transport and ionic conductance. Biochim. Biophys. Acta Biomembr. 1863(10): 183661.

McGinn, P.J. and F.M.M. Morel. 2008. Expression and inhibition of the carboxylating and decarboxylating enzymes in the photosynthetic C_4 pathway of marine diatoms. Plant Physiol. 146: 300–309.

McKie-Krisberg, Z.M., R.W. Sanders and R.J. Gast. 2018. Evaluation of mixotrophy-associated gene expression in two species of polar marine algae. Front. Mar. Sci. 5: 273. DOI:10.3389/fmars.2018.00273.

McKinnie, L.J., S.F. Cummins and M. Zhao. 2023. Identification of incomplete annotations of biosynthesis pathways in rhodophytes using a multi-omics approach. Mar. Drugs 22(1): 3.

McLachlan, J. and J.S. Craigie. 1966. Chitan fibres in *Cyclotella cryptica* and growth of *C. cryptica* and *Thalassiosira fluviatilis*. pp. 511–517. *In:* Barnes, H. [ed.], Some Contemporary Studies in Marine Science. George Allen and Unwin Ltd., London, UK.

McManus, G.B., W. Liu, R.A. Cole, D. Biemesderfer and J.L. Mydosh. 2018. *Strombidium rassoulzadegani*: a model species for chloroplast retention in oligotrich ciliates. Front. Mar. Sci. 5: 205.

McParland, E.L., M.D. Lee, E.A. Webb, H. Alexander and N.M. Levine. 2021. DMSP synthesis genes distinguish two types of DMSP producer phenotypes. Environ. Microbiol. 23: 1656–1669.

McQuaid, J.B., A.B. Kustka, M. Obornik, L. Horák, J.P. McCrow, B.J. Karas et al. 2018. Carbonate-sensitive phytotransferrin controls high-affinity iron uptake in diatoms. Nature 555(7697): 534–537.

Medlin, L.K. 2016. Evolution of the diatoms: major steps in their evolution and a review of the supporting molecular and morphological evidence. Phycologia 55(1): 79–103.

Medlin, L.K., W.C.H.F. Kooistra, P. Sims and U. Wellbrock. 1997. Is the origin of the diatoms related to the end-Permian mass extinction? Nova Hedwigia 65(1-4): 1–11.

Meeks, J.C. and R.W. Castenholz. 1971. Growth and photosynthesis in an extreme thermophile, *Synechococcus lividus* (Cyanophyta). Arch. Mikrobil. 78(1): 25–41.

Mei, Q. and V. Dvornyk. 2015. Evolutionary history of the photolyase/cryptochrome superfamily in eukaryotes. PLoS ONE 10(9): e0135940.

Mendez, G.S., C.F. Delwiche, K.E. Apt and J.C. Lippmeier. 2015. Dinoflagellate gene structure and intron splice sites in a genomic tandem array. J. Euk. Microbiol. 62(5): 679–87.

Meng, R., L. Zhang, C. Zhou, K. Liao, P. Xiao, Q. Luo et al. 2022. Genome sequence of *Chryotila roscoffensis*, a coccolithophore contributed to global biogechemical cycles. Genes 13(1): 40.

Merchant, S.S., M.D. Allen, J. Kropat, J.L. Moseley, J.C. Long, S. Tottey et al. 2006. Between a rock and a hard place: trace element nutrition in *Chlamydomonas*. Biochim. Biophys. Acta 1763: 578–594.

Merchant, S.S., S.E. Prochnik, O. Vallon, E.H. Harris, S.J. Karpowicz, G.B. Witman et al. 2007. The Chlamydomonas genome reveals the evolution of key animal and plant functions. Science 318(5848): 245–250.

Merchant, S.S., S. Schmollinger, D. Strenkert, J.L. Moseley and C.E. Blady-Haas. 2020. From economy to luxury: copper homeostasis in Chlamydomonas and other algae. Biochim. Biophys. Acta-Mol. Cell Res. 1867: 118822.

Merk, M., R.A. Mitchell, S. Endres and R. Bucala. 2012. D-Dopachrome tautomerase (D-DT or MIF-2): doubling the MIF cytokine family. Cytokine 59(1): 10–17.

Michaeli, S. 2011. Trans-splicing in trypansomes: machinery and its impact on the parasite transcriptome. Future Microbiol. 6(4): 459–474.

Michelet, C., E.G.J. Danchin, M. Jaouannet, J. Bernhagen, R. Panstruga, K.H. Kogel et al. 2019. Cross-kingdom analysis of diversity, evolutionary history, and site selection within the eukaryotic macrophage migration inhibitory factor superfamily. Genes 10(10): 740. DOI:10.3390/genes10100740.

Middlebrooks, M., S.S. Bell, N.E. Curtis and S.K. Pierce. 2014. Atypical plant-herbivore association of algal food and a kleptoplastic sea slug (*Elysia clarki*) revealed by DNA barcoding and field surveys. Mar. Biol. 161: 1429–1440.

Miller, J.L., A. Harupa, S.H.I. Kappe and S.A. Mikolajczak. 2012. *Plasmodium yoelii* macrophage migration inhibitory factor is necessary for efficient liver-stage development. Infect. Immun. 80: 05861-11. DOI:10.1128/iai.05861-11.

Minge, M.A., K. Shalchian-Tabrizi, O.K. Tφrresen, K. Takishita, I. Probert, Y. Inagaki et al. 2010. A phylogenetic mosaic plastid proteome and unusual plastid-targeting signals in the green-colored dinoflagellate *Lepidodinium chlorophorum*. BMC Ecol. Evol. 10: 191.

Miralto, A., G. Barone, G. Romano, S.A. Poulet, A. Ianora, G.L. Russo et al. 1999. The insidious effect of diatoms on copepod reproduction. Nature 402: 173–176.

Miranda, L.N., K. Hutchison, A.R. Grossman and S.H. Brawley. 2013. Diversity and abundance of the bacterial community of the red macroalga *Porphyra umbilicalis*: Did bacterial farmers produce macroalgae? PLoS ONE 8: e58269.

Mitra, A., K.J. Flynn, U. Tillmann, J.A. Raven, D. Caron, D.K. Stoecker et al. 2016. Defining planktonic protist functional groups on mechanisms for energy and nutrient acquisition: incorporation of diverse mixotrophic strategies. Protist 167(2): 106–120.

Miyachi, K., M.J. Fritzler and E.M. Tan. 1978. Autoantibody to a nuclear antigen in proliferating cells. J. Immunol. 121(6): 2228–2234.

Miyashita, H., H. Ikemoto, N. Kurano, K. Adachi, M. Chihara and S. Miyachi. 1996. Chlorophyll d as a major pigment. Nature 383: 402.

Mizuno, K. and R.D. Sloboda. 2017. Protein arginine methyltransferases interact with intraflagellar transport particles and change location during flagellar growth and resorption. Mol. Biol. Cell 28(9): 1208–1222.

Mobley, H.L.T. and R.P. Hausinger 1989. Microbial ureases: Significance, regulation, and molecular characterization. Microbiol. Rev. 53: 85–108.

Mock, T., R.P. Otillar, J. Strauss, M. McMullan, P. Paajanen, J. Schnutz et al. 2017. Evolutionary genomics of the cold-adapted diatom *Fragilariopsis cylindrus*. Nature 541: 536–540.

Mock, T., M.P. Samanta, V. Iverson, C. Berthiaume, M. Robison, K. Holtermann et al. 2008. Whole-genome expression profiling of the marine diatom *Thalassiosira pseudonana* identifies genes involved in silicon bioprocesses. Proc. Natl. Acad. Sci. USA 105(5): 1579–1584.

Moeys, S., J. Frenkel, C. Lembke, J.T.F. Gillard, V. Devos, K.V. den Berge et al. 2016. A sex-inducing pheromone triggers cell cycle arrest and mate attraction in the diatom *Seminavis robusta*. Sci. Rep. 6: 19252.

Möglich, A., R.A. Ayers and K. Moffat. 2009. Structure and signaling mechanism of Per-srnt-Sim domains. Structure 17(10): 1282–1294.

Mohamed, A.R., N. Andrade, A. Moya, C.X. Chan, A.P. Negri, D.G. Bourne et al. 2019. Transcriptomic insights into the establishment of coral-algal symbioses from the symbiont perspective. https://doi.org/10.1101/652131.

Mohanty, A., K. Satpathy, G. Sahu, K.J. Hussain, M. Prasad and S. Sarkar. 2010. Bloom of *Trichodesmium erythraeum* (Ehr.) and its impact on water quality and plankton community structure in the coastal waters of southeast coast of India. Ind. J. Mar. Sci. 39(3): 323–333.

Mohibbullah, M., M.M.H. Bhuiyan, M.A. Hannan, P. Getachew, Y.-K. Hong, I.S. Choi et al. 2016. The edible red algal *Porphyra yezoensis* promotes neuronal survival and cytoarchitecture in primary Hippocampal neurons. Cell Mol. Neurobiol. 36(5): 669–682.

Moison, M., J.M. Pacheco, L. Lucero, C. Fonouni-Farde, J. Rodrígue-Melo, N. Mansilla et al. 2021. The lncRNA APOLO interacts with the transcription factor WRKY42 to trigger root hair cell expansion in response to cold. Mol. Plant 14: 937.

Mojzeš, P., L. Gao, T. Ismagulova, J. Pilátová, Š. Moudříková, O. Gorelova et al. 2020. Guanine, a high-capacity and rapid-turnover nitrogen reserve in microalgal cells. Proc. Natl. Acad. Sci. USA 117(51): 32722–32730.

Monier, A., A. Chambouver, D.S. Miner, V. Attah, R. Terrado, C. Lovejoy et al. 2017. Host-derived viral transporter protein for nitrogen uptake in infected marine phytoplankton. Proc. Natl. Acad. Sci. USA 114(36): E7489–7498.

Monier, A., A. Pagarete, C. de Vargas, M.J. Allen, B. Read, J.-M. Claverie et al. 2009. Horizontal gene transfer of an entire metabolic pathway between a eukaryotic alga and its DNA virus. Genome Res. 19: 1441–1449

Monod, J. 1950. La technique de culture continue: Theorie et applications. Annales de l'Institut Pasteur Lille 79: 390–410.

Monroe, E.A. and F.M. Van Dolah. 2008. The toxic dinoflagellate *Karenia brevis* encodes novel type I-like polyketide synthases containing discrete catalytic domains. Protist 159(3): 471–482.

Monteiro, F.M., L.T. Bach, C. Brownlee, P. Brown, R.E.M. Rickaby, A.J. Poulton et al. 2016. Why marine phytoplankton calcify. Sci. Adv. 2(7): e1501822.

Montressor, M., L. Vitale, D. D'Alelio and M.I. Ferrante. 2016. Sex in marine planktonic diatoms: insights and challenges. Perspect. Phycol. 3(2): 61–75.

Moon, J.Y., K. Lee, W.A. Lim, E. Lee, M. Dai, Y.H. Choi et al. 2021. Anthropogenic nitrogen is changing the East China and Yellow Seas from being N deficient to being P deficient. Limnol. Oceanogr. 66(3): 1–11.

Moore, L.R., R. Goericke and S.W. Chisholm. 1995. Comparative physiology of *Synechococcus* and *Prochlorococcus*: influence of light and temperature on growth, pigments, fluorescence and absorptive properties. Mar. Ecol. Prog. Ser. 116(1-3): 259–275.

Moore, C.M. M.M. Mills, K.R. Arrigo, I. Berman-Frank, L. Bopp, P.W. Boyd et al. 2013. Processes and patterns of oceanic nutrient limitation. Nat. Geosci. 6: 701–710.

Moore, J.K., S.C. Doney, D.M. Glober and I.Y. Fung. 2001. Iron cycling and nutrient-limitation patterns in surface waters of the World Ocean. Deep Sea Res. Part I Oceanogr. Res. Pap. 49: 463–507.

Moran, X.A., A. Lopez-Urrutia, A. Calvo-Diaz and W.K.W. Li. 2010. Increasing importance of small phytoplankton in a warmer ocean. Glob. Change Biol. 16: 1137–1144.

Mordecai, G.J., F. Verret, A. Highfield and D.C. Shroeder. 2017. Schrödinger's Cheshire Cat: Are haploid *Emiliania huxleyi* cells resistant to viral infection or not? Viruses 9(3): 51.

Moreau, H., B. Verhelst, A. Couloux, E. Derelle, S. Rombauts, N. Grimsley et al. 2012. Gene functionalities and genome structure in *Bathycoccus prasinos* reflect cellular specializations at the base of the green lineage. Genome Biol. 13(8): R74.

Moreira, D., R. Tavera, K. Benzerara, F. Skouri-Panet, E. Couradeau, E. Gérard et al. 2017. Description of *Gloeomargarita lithophorra* gen. nov. sp. nov. a thylakoid-bearing basal branching cyanobacteria with carbonates, and proposal for Gloemargaritales ord. nov. Int J. Syst. Evol. Microbiol. 67(3): 653–658.

Morel, F.M.M., P.J. Lam and M.A. Saito. 2020. Trace metal substitution in marine phytoplankton. Ann. Rev. Earth Planet. Sci. 48: 491–517.

Morey, J.S., E.A. Monroe, A.L. Kinney, M. Beal, J.G. Johnson, G.L. Hitchcock et al. 2011. Transcriptomic response of the red tide dinoflagellate, *Karenia brevis*, to nitrogen and phosphorus depletion and addition. BMC Genom. 12: 346.

Mori, T., B. Binder and C.H. Johnson. 1996. Circadian gating of cell division in cyanobacteria growing with average doubling times of less than 24 hours. Proc. Natl. Acad. Sci. USA 93: 10183–10188.

Moriano-Gutierrez, S., E.J. Koch, H. Bussan, K. Romano, M. Belcaid and F.E. Rey. 2019. Critical symbiont signals drive both local and systemic changes in diel and developmental host gene expression. Proc. Natl. Acad. Sci. 116: 7990–7999.

Moriano-Gutierrez, S., C. Bongrand, T. Essock-Burns, L. Wu, M.J. McFall-Ngai and E.G. Ruby. 2020. The noncoding small RNA SsrA is released by *Vibrio fischeri* and modulates critical host responses. PLoS Biol. 18(11): e3000934.

Morozov, A.A. and Y.P. Galachyants. 2019. Diatom genes originating from red and green algae: implications for the secondary endosymbiosis models. Mar. Benomics 45: 72–78.

Morrissey, J., R. Sutak, J. Paz-Yepes, A. Tanaka, A. Moustafa, A. Veluchamy et al. 2015. A novel protein, ubiquitous in marine phytoplankton, concentrates iron at the cell surface and facilitates uptake. Curr. Biol. 25: 364–371.

Morse, D. 2019. A transcriptome-based perspective of meiosis in dinoflagellates. Protist 170: 397–403.

Morse, D., P.M. Milos, E. Roux and J.W. Hastings. 1989. Circadian regulation of bioluminescence in Gonyaulax involves translational control. Proc. Natl. Acad. Sci. USA 86: 172–176.

Morse, D., P. Salois, P. Markovic and J.W. Hastings. 1995. A nuclear-encoded form II Rubisco in dinoflagellates. Science 268(5217): 1622–1624.

Moustafa, A., B. Beszteri, U.G. Maier, C. Bowler, K. Valentin and D. Bhattacharya. 2009. Genomic footprints of a cryptic plastid endosymbiosis in diatoms. Science 324(5935): 1724–1726.

Moustafa, A., A.N. Evans, D.M. Kulis, J.D. Hackett, D.L. Erdner, D.M. Anderson et al. 2010. Transcriptome profiling of a toxic dinoflagellate reveals a gene-rich protist and a potential impact on gene expression due to bacterial presence. PLoS ONE 5(3): e9688.

Movassagh, T., K.H. Bui, H. Sakakibara, K. Oiwa and T. Ishikawa. 2010. Nucleotide-induced global conformational changes of flagellar dynein arms revealed by in situ analysis? Nat. Struct. Mol. Biol. 17: 761–767.

Mujer, C.V., D.L. Andrews, J.R. Manhart, S.K. Pierce and M.E. Rumpho. 1996. Chloroplast genes are expressed during intracellular symbiotic association of *Vaucheria litorea* plastids with the sea slug *Elysia chlorotica*. Proc. Natl. Acad. Sci. USA 93: 12333–12338.

Mukherjee, K., B. Futcher and J. Leatherwood. 2018. Mmi and rep2 mRNA are novel RNA targets of the Mei2 RNA-binding protein during early meiosis in *Schizosaccharomyces pombe*. Open Biol. 8(9): 180110.

Mulkidjanian, A. Y., E.V. Koonin, K.S. Makarova, S.L. Mekhedov, A. Sorokin, Y.I. Wolf et al. 2006. The cyanobacterial genome core and the origin of photosynthesis. Proc. Natl. Acad. Sci. USA 103(35): 13126–13131.

Müller, H., J. Lukas, A. Schneider, P. Warthoe, J. Bartek, M. Eilers et al. 1994. Cyclin D1 expression in regulated by the retinoblastoma protein. Proc. Natl. Acad. Sci. USA 91(8): 2945–2949.

Mullis, K., F. Faloona, S. Scharf, R. Saiki, G. Horn and H. Erlich. 1986. Specific enzymatic amplification of DNA in vitro: The polymerase chain reaction. Cold Spring Harb. Symp. Quant. Biol. 51: 263–273.

Mungpakdee, S., C. Shinzato, T. Takeuchi, T. Kawashima, R. Koyanagi, K. Hisata et al. 2014. Massive gene transfer and extensive RNA editing of a symbiotic dinoflagellate plastid genome. Genome Biol. Evol. 6: 1408–1422. doi: 10.1093/gbe/evu109.

Muñoz-García, J. and S. Ares. 2016. Formation and maintenance of nitrogen-fixing cell patterns in filamentous cyanobacteria. Proc. Natl. Acad. Sci. USA 113: 6218–6223.

Muñoz-Gómez, S.A., F.G. Mejia-Franco, K. Durnin, M. Colp, C.J. Grisdale, J.M. Archibald et al. 2019. The new red algal subphylum Proteorhodophytina comprises the largest and most divergent plastid genomes known. Curr. Biol. 27: 1677–1684.

Muñoz-Gómez, S.A., M. Kreutz and S. Hess. 2021. A microbial eukaryote with a unique combination of purple bacteria and green algae as endosymbionts. Sci. Adv. 7(24): eabg4102.

Murray, A. and T. Hunt. 1993. The Cell Cyle: An Introduction. Oxford University Press. 264 pp.

Murray, S.A., M. Wiese, A. Stuken, S. Brett, R. Kellmann, G. Hallegraeff et al. 2011. sxtA-based quantitative molecular assay to identify saxitoxin-producing harmful algal blooms in marine waters. Appl. Environ. Microbiol. 77: 7050–7057.

Murray, S.A., R. Ruvindy, G.S. Kohli, D.M. Anderson and M.L. Brosnahan. 2019. Evaluation of sxtA and rDNA qPCR assays through monitoring of an inshore bloom of *Alexandrium catenella* Group 1. Sci. Rep. 9: 14532.

Murata, N. and D.A. Los. 2006. Histidine kinase Hik33 is an important participant in cold-signal transduction in cyanobacteria. Physiol. Plant 126: 17–27.

Muscatine, L. 1967. Glycerol excretion by symbiotic algae from corals and tridacna and its control by the host. Science 156(3774): 516–519.

Muscatine, L. 1990. The role of symbiotic algae in carbon and energy flux in reef corals. pp. 75–87. *In*: Dubinsky, Z. [ed.]. Ecosystems of the world, Vol. 25, Coral reefs. Elsevier, Amsterdam.

Mutalipassi, M., G. Riccio, V. Mazzella, C. Galasso, E. Somma, A. Chiarore et al. 2021. Symbioses of cyanobacteria in marine environments: ecological insights and biotechnological perspectives. Mar. Drugs 19(4): 227.

Naessens, E., G. Dubreuil, P. Giordanengo, O.L. Baron, N. Minet-Kebdani, H. Keller et al. 2015. A secreted MIF cytokine enables aphid feeding and represses plant immune responses. Curr. Biol. 25(14): 1898–1903.

Nagao, R., M. Yokono, Y. Ueno, T. Suzuki, K. Kato, K.-H. Kato et al. 2021. Molecular organizations and function of iron-stress-induced-A protein family in *Anabaena* sp. PCC7120. Biochim. Biophys. Acta-Bioenergetics 1862(1): 148327.

Nagasaki, K. 2008. Dinoflagellates, diatoms, and their viruses. J. Microbiol. 46(3): 235–243.

Nagata, N., R. Tanaka, S. Satoh and A. Tanaka. 2005. Identification of a vinyl reductase gene for chlorophyll synthesis in *Arabidopsis thaliana* and implications for the evolution of *Prochlorococcus* species. Plant Cell. 17: 233–240.

Nagel, G., D. Ollig, M. Fuhrmann, S. Kateriya, A.M. Musti, E. Bamberg et al. 2002. Channelrhodopsin-1: a light-gated proton channel in green algae. Science 296(5577): 2395–2398.

Nagel, G., T. Szellas, W. Huhn, S. Kateriya, N. Adeishvili, P. Berthold et al. 2003. Channelrhodopsin-2, a directly light-gated cation-selective membrane channel. Proc. Natl. Acad. Sci. USA 100(24): 13940–13945.

Nakajima, K., A. Tanaka and Y. Matsuda. 2013. SLC4 family transporters in a marine diatom directly pump bicarbonate from seawater. Proc. Natl. Acad. Sci. USA 110: 1767–1772.

Nakajima, Y., K. Kojima, Y. Kashiyama, S. Doi, R. Nakai, Y. Sudo et al. 2020. Bacterium lacking a known gene for retinal biosynthesis constructs functional rhodopsins. Microb. Environ. 35(4): ME20085.

Nakamura, Y., N. Sasaki, M. Kobayashi, N. Ojima, M. Yasuike, Y. Shigenobu et al. 2013. The first symbiont-free genome sequence of marine red alga, Susabi-nori (*Pyropia yezoensis*). PLoS ONE 8(3): e57122.

Nakatogawa, H., K. Suzuki, Y. Kamada and Y. Ohsumi. 2009. Dynamics and diversity in autophagy mechanisms: Lessons from yeast. Nat. Rev. Mol. Cell Biol. 10: 458–467.

Nakayama, T., M. Nomura, Y. Takano, G. Tanifuji, K. Shiba, K. Inaba et al. 2019. Single-cell genomics unveiled a cryptic cyanobacterial lineage with a worldwide distribution hidden by a dinoflagellate host. Proc. Natl. Acad. Sci. USA 116(32): 15973–15978.

Nakov, T., J.M. Beaulieu and A.J. Alverson. 2019. Diatoms diversify and turn over faster in freshwater than marine environments. Evolution 73(12): 2497–2511.

Nambiar, M. and G.R. Smith. 2016. Repression of harmful meiotic recombination in centromeric regions. Semin. Cell Dev. Biol. 54: 188–197.

Nand, A., Y. Zhan, O.R. Salazar, M. Aranda, C.R. Voolstra and J. Dekker. 2021. Genetic and spatial organization of the unusual chromosomes of the dinoflagellate *Symbiodinium microadriaticum*. Nat. Genet. 53: 618–629.

Nanjappa, D., R. Sanges, M.I. Ferrante and A. Zingone. 2017. Diatom flagellar genes and their expression during sexual reproduction in *Leptocylindrus danicus*. BMC Genomics 18: 813.

Narayanan, N.N., U. Ihemere, W.T. Chiu, D. Siritunga, S. Rajamani, S. Singh et al. 2011. The iron assimilatory protein, FEA1, from *Chlamydomonas reinhardtii* facilitates iron-specific metal uptake in yeast and plants. Front. Plant Sci. 2: 67.

Nash, E.A., A.C. Barbrook, R.K. Edwards-Stuart, K. Bernhardt, C.J. Howe and R.E.R. Bisbet. 2007. Organization of the mitochondrial genome in the dinoflagellate *Amphidinium carterae*. Mol. Biol. Evol. 24(7): 1528–1536.

Nash, E.A., R.E.R. Nisbet, A.C. Barbrook and C.J. Howe. 2008. Dinoflagellates: a mitochondrial genome all at sea. Trends Genet. 24(7): 328–335.

Naumova, N., M. Imakaev, g. Fudenberg, Y. Zhan, B.R. Lajoie, L.A. Mirny et al. 2013. Organization of the mitotic chromosome. Science 342(6161): 948–953.

Navarathna, D.H.M.L.P., A. Das, J. Morschhäuser, K.W. Nickerson and D.D. Roberts. 2011. Dur3 is the major urea transporter in *Candida albicans* and is co-regulated with the urea amidolyase Dur1,2. Microbiology 157: 270–279.

Nelson, D.R. 2018. Cytochrome P450 diversity in the tree of life. Biochim. Biophys. Acta-Proteins Proteom. 1866: 141–154.

Nelson, D.R., B. Khraiwesh, W. Fu, S. Alseekh, A. Jaiswal, A. Chaiboonchoe, K.M. Hazzouri et al. 2017. The genome and phenome of the green alga *Chloroidium* sp. UTEX 3007 reveal adaptive traits for desert acclimatization. Elife 6: e25783.

Nelson, D.R., A. Mystikou, A. Jaiswal, C. Rad-Menendez, M.J. Preston, F. De Boever et al. 2024. Macroalgal deep genomics illuminate multiple paths to aquatic, photosynthetic multicellularity. Mol. Plant 17: 747–771.

Neupane, S., S.E. Schweitzer, A. Neupane, E.J. Andersen, A. Fennell, R. Zhou et al. 2019. Identification and characterization of mitogen-activated protein kinase (MAPK) genes in sunflower. Plants 8: 28.

Needham, D.M., S. Yoshizawa, T. Hosaka, C. Poirier, C.J. Choi, E. Hehenberger et al. 2019. A distinct lineage of giant viruses brings a rhodopsin photosystem to unicellular marine predators. Proc. Natl. Acad. Sci. USA 116(41): 20574–20585.

Nica, A.C. and E.T. Dermitzakis. 2013. Expression quantitative trait loci: present and future. Philos. Trans. R. Soc. Lond. B. Biol. Sci. 368(1620): 20120362. doi: 10.1098/rstb.2012.0362.

Nieves-Morión, M., E. Flores and R.A. Foster. 2020. Predicting substrate exchange in marine diatom-heterocystous cyanobacteria symbioses. Environ. Microbiol. 22: 2027–2052.

Niklas, K.J., R. Wayne, M. Benitez and S.A. Newman. 2019. Polarity, planes of cell division, and the evolution of plant multicellularity. Protoplasma 256(3): 585–599.

Nikolov, D. and S. Burley. 1997. RNA polymerase II transcription initiation: a structural view. Proc. Natl. Acad. Sci. 94(1): 15–22.

Nimmo, I.C., A.C. Barbrook, I. Lassadi, J.E. Chen, K. Geisler, A.G. Smith et al. 2019. Genetic transformation of the dinoflagellate chloroplast. eLife 8: e45292.

Nirody, J.A., I. Budin and P. Rangamani. 2020. ATP synthase: evolution, energetics, and membrane interactions. J. Gen. Physiol. 152(11): e201912475.

Nishitani, G., S. Nagai, S. Hayakawa, Y. Kosaka, K. Sakurada, T. Kamiyama et al. 2012. Multiple plastids collected by the dinoflagellate Dinophysis mitra through kelptoplastidy. App. Environ. Microbiol. 78(3): 813–821.

Nomura, M., K. Atsuji, K. Hirose, K. Shiba, R. Yanase, T. Nakayama et al. 2019. Microtubule stabilizer reveals requirement of Ca^{2+}-dependent conformational changes of microbutubles for rapid coiling of haptonema in haptophyte algae. Biol. Open. 8(2): bio036590.

Norén, F., Ø. Moestrup and A.-S. Rehnstam-Holm. 1999. *Parvilucifera infectans* Norén et Moestrup gen. et sp. nov (Perkinsozoa phylum nov.): a parasitic flagellate capable of killing toxic microalgae. Eur. J. Protistol. 35(3): 233–254.

Nosenko, T. and D. Bhattacharya. 2007. Horizontal gene transfer in chromalveolates. BMC Evol. Biol. 7: 173.

Nosenko, T., K.L. Lidie, F.M. van Dolah, E. Lindquist, J.-F. Cheng and D. Bhattacharya. 2006. Chimeric plastid proteome in the Florida "red tide" dinoflagellate *Karenia brevis*. Mol. Biol. Evol. 23(11): 2026–2038.

Novosel, N., T. Mišić Radić, J. Zemla, M. Lekka, A. Čačković, D. Kasum et al. 2022. Temperature-induced response in algal cell surface properties and behaviour: an experimental approach. J. Appl. Phycol. 34: 243–259.

Norman, J.E. and M.W. Gray. 2001. A complex organization of the gene encoding cytochrome oxidase subunit 1 in the mitochondrial genome of the dinoflagellate, *Crypthecodinium cohnii*: homologous recombination generates two different cox1 open reading frames. J. Mol. Evol. 53(4-5): 351–363.

Nowack, E.C.M., M. Melkonian and G. Glockner. 2008. Chromatophore genome sequence of *Paulinella* sheds light on acquisition of photosynthesis by eukaryotes. Curr. Biol. 18: 410–418.

Nowack, E.C.M., H. Vogel, M. Growth, A.R. Grossman, M. Melkonian and G. Glockner. 2011. Endosymbiotic gene transfer and transcriptional regulation of transferred genes in *Paulinella chromatophora*. Mol. Biol. Evol. 28(1): 407–422.

Nowack, E.C.M., D.C. Price, D. Bhattacharya, A. Singer, M. Melkonian and A.R. Grossman. 2016. Gene transfers from diverse bacteria compensate for reductive genome evolution in the chromatophore of *Paulinella chromatophora*. Proc. Natl. Acad. Sci. USA 113(43): 12214–12219.

Nowack, E.C.M. and A.P.M. Weber. 2018. Genomics-informed insights into endosymbiotic organelle evolution in photosynthetic eukaryotes. Annu. Rev. Plant Biol. 69(1): 51–84.

Nurse, P. 1990. Universal control mechanism regulating onset of M-phase. Nature 344: 503–508.

Ochiai, K.K., D. Hanawa, H.A. Ogawa, H. Tanaka, K. Uesaka, T. Edzuka et al. 2024. Genome sequence and cell biological toolbox of the highly regenerative, coencytic green algal Bryopsis. Plant J. doi: 10.1111/tpj.167642.

Ochsenkühn, M.A., T. Röthig, C. D'Angelo, J. Wiedenmann and C.R. Voolstra. 2017. The role of floridoside in osmoadaptation of coral-associated algal endosymbionts to high-salinity conditions. Sci. Adv. 3(8): e1602047.

Oesterhelt, D. and W. Stoeckenius. 1971. Rhodopsin-like protein from the purple membrane of *Halobacterium halobium*. Nat. New Biol. 233: 149–152.

Ogura, A., Y. Akizuki, H. Imoda, K. Mineta, T. Gojobori and S. Nagai. 2018. Comparative genome and transcriptome analysis of diatom, *Skeletonema costatum*, reveals evolution of genes for harmful algal bloom. BMC Genomics 19: 765.

Okamoto, O.K. and J.W. Hastings. 2003. Novel dinoflagellate clock-related genes identified through microarray analysis. J. Phycol. 39(3): 519–526.

Okie, J.G., A.T. Poret-Peterson, Z.M.P. Lee, A. Richter, L.D. Alcaraz, L.E. Eguiarte et al. 2020. Genomic adaptations in information processing underpin trophic strategy in a whole-ecosystem nutrient enrichment experiment. eLife 9: e49816.

Oliver, M.J., O. Schofield and K. Bidle. 2010. Density dependent expression of a diatom retrotransposon. Mar. Genomics 3: 145–150.

Oliver, T., T.D. Kim, J.P. Trinugroho, V. Cordón-Preciado, N. Wijayatilake, A. Bhatia et al. 2023. The evolution and evolvability of photosystem II. Annu. Rev. Plant Biol. 74: 225–257.

Olson, R.J. and S.W. Chisholm. 1986. Effects of light and nitrogen limitation on the cell cycle of the dinoflagellate *Amphidinium carteri*. J. Plankton Res. 8(4): 785–793.

Onyshchenko, A., W.R. Roberts, E.C. Ruck, J.A. Lewis and A.J. Alverson. 2021. The genome of a nonphotosynthetic diatom provides insights into the metabolic shift to heterotrophy and constraints on the lass of photosynthesis. New Phytol. 232: 1750–1764.

Oravecz, A., A. Baumann, Z. Máté, A. Brzezinska, J. Molinier, E.J. Oakeley et al. 2006. Constitutively Photomorphogenic1 is required for the UV-B response in *Arabidopsis*. Plant Cell 18(8): 1975–1990.

Orr, J.C., V.J. Fabry, O. Aumont, L. Bopp, S.C. Doney, R.A. Feely et al. 2005. Anthropogenic ocean acidification over the twenty-first century and its impact on calcifying organisms. Nature 437(7059): 681–686.

Osman, M.B., S.B. Das, L.D. Trusel, M.J. Evans, H. Fischer, M.M. Grieman et al. 2019. Industrial-era decline in subarctic Atlantic productivity. Nature 569: 551–555.

Osorio, H., P. Tapia-Reyes, D. Espinoza, D. Laporte, A. Gonzalez, E. Castro-Nallar et al. 2022. The genome of the marine alga *Ulva compressa* (Chlorophyta) reveals protein-coding genes with similarity to plants and green microalgae, but also to animal, bacterial, and fungal genes. Int. J. Mol. Sci. 23: 7279.

Ou, L., X. Huang, B. Huang, Y. Qi and S. Lv. 2015. Growth and competition for different forms of organic phosphorus by the dinoflagellate *Prorocentrum donghaiense* with the dinoflagellate *Alexandrium catenella* and the diatom *Skeletonema costatum* s.l. Hydrobiol. 754: 29–41.

Oviatt, C.A. 2004. The changing ecology of temperate coastal waters during a warming trend. Estuaries 27(6): 895–904.

Paasche, E. 2001. A review of the coccolithophorid *Emiliania huxleyi* (Prymnesiophyceae), with particular reference to growth, coccolith formation, and calcification-photosynthesis interactions. Phycologia 40(6): 503–529.

Padfield, D., G. Yvon-Durocher, A. Buckling, S. Jennings and G. Yvon-Durocher. 2016. Rapid evolution of metabolic traits explains thermal adaptation in phytoplankton. Ecol. Lett. 19(2): 133–142.

Page, M.D., M.D. Allen, J. Kropat, E.I. Urzica, S.J. Karpowicz, S.I. Hseih et al. 2012. Fe sparing and Fe recycling contribute to increased superoxide dismutase capacity in iron-starved *Chlamydomonas reinhardtii*. Plant Cell 24(6): 2649–2665.

Pakos-Zebrucka, K., I. Koryga, K. Mnich, M. Ljujic, A. Samali and A.M. Gorman. 2016. The integrated stress response. EMBO Rep. 17(10): 1374–1395.

Palacios, O.A., G. Gomez-Anduro, Y. Bashan and L.E. de-Bashan. 2016. Tryptophan, thiamine and indole-3-acetic acid exchange between *Chlorella sorokiniana* and the plant growth-promoting bacterium *Azospirillum brasilense*. FEMS Microbiol. Ecol. 92(6): fiw077.

Palenik, B. 2015. Molecular mechanisms by which marine phytoplankton respond to their dynamic chemical environment. Ann. Rev. Mar. Sci. 7: 325–340.

Palenik, B., B. Brahamsha, F.W. Larimer, M. Land, L. Hauser, P. Chain et al. 2003. The genome of a motile marine *Synechococcus*. Nature 424: 1037–1042.

Palenik, B., J. Grimwood, A. Aerts, P. Rouzé, A. Salamov, N. Putnam et al. 2007. The tiny eukaryote *Ostreococcus* provides genomic insights into the paradox of plankton speciation. Proc. Natl. Acad. Sci. USA 104(18): 7705–7710.

Pančić, M. and T. Kiorboe. 2018. Phytoplankton defence mechanisms: traits and trade-offs. Biol. Rev. 93(2): 1269–1303.

Pang, C., J. Chai, P. Zhu, J. Shanklin and Q. Liu. 2023. Structural mechanism of intracellular autoregulation of zinc uptake in ZIP transporters. Nat. Commun. 14: 3404.

Pansch, C., O. Cerda, M. Lenz, M. Wahl and M. Thiel. 2009. Consequences of light reduction for anti-herbivore defense and bioactivity against mussels in four seaweed species from northern-central Chile. Mar. Ecol. Progr. Ser. 381: 83–97.

Panstruga, R., K. Baumgarten and J. Bernhagen. 2015. Phylogeny and evolution of plant macrophage migration inhibitory factor/D-dopachrome tautomerase-like proteins. BMC Evol. Biol. 15: 64.

Pant, B.D., A. Burgos, P. Pant, A. Cuadros-Inostroza, L. Willmitzer and W.R. Scheible. 2015. The transcription factor PHR1 regulates lipid remodeling and triacylglycerol accumulation in *Arabidopsis thaliana* during phosphorus starvation. J. Exp. Bot. 66: 1907–1918.

Parages, M.L., F.L. Figueroa, R.M. Conde-Álvarez and C. Jiménez. 2014. Phosphorylation of MAPK-like proteins in three intertidal macroalgae under stress conditions. Aquat. Biol. 22: 213–226.

Parfrey, L.W. and L.A. Katz. 2010. Dynamic genomes of eukaryotes and the maintenance of genomic integrity. Microbe 5(4): 156–163.

Parfrey, L.W., D.J.G. Lahr, A.H. Knoll and L.A. Katz. 2011. Estimating the timing of early eukaryotic diversification with multigene molecular clocks. Proc. Natl. Acad. Sci. USA 108(33): 13624–13629.

Park, G. and H.G. Dam. 2021. Cell-growth gene expression reveals a direct fitness cost of grazer-induced toxin production in red tide dinoflagellate prey. Proc. Biol. Sci. 288(1944): 20202480.

Parker, D.J., J. Bast, K. Jalvingh, Z. Bumas, M. Robinson-Rechavi and T. Schwander. 2019. Repeated evolution of asexuality involves convergent gene expression changes. Mol. Biol. Evol. 36(2): 350–364.

Parra, G., K. Bradnam and I. Korf. 2007. CEGMA: a pipeline to accurately annotate core genes in eukaryotic genomes. Bioinformatics 23(9): 1061–1067.

Parrow, M.W. and J.M. Burkholder. 2003. Reproduction and sexuality in *Pfiesteria shumwayae* (Dinophyceae). J. Phycol. 39(4): 697–711.

Partensky, F. and L. Garczarek. 2010. Prochlorococcus: advantages and limits of minimalism. Annu. Rev. Mar. Sci. 2: 305–331.

Pavlinska, Z., D. Chorvat, Jr., A. Mateasik, M. Jerigova, D. Velic, N.I. DeNardis et al. 2020. Fluorescence responsiveness of unicellular marine algae Dunaliella to stressors under laboratory conditions. J. Biotechnol. 324S: 100018.

Paytan, A. and K. McLaughlin. 2007. The oceanic phosphorus cycle. Chem. Rev. 107(2): 563–576.

Paz, Y., E. Shimoni, M. Weiss and U. Pick. 2007. Effects of iron deficiency on iron binding and internalization into acidic vacuoles in *Dunaliella salina*. Plant Physiol. 144(3): 1407–1415.

Pearson, G.A., A. Lago-Leston, F. Cànovas, C.J. Cox, F. Verret, S. Lasternas et al. 2015. Metatranscriptomes reveal functional variation in diatom communities from the Antarctic Peninsula. ISME J. 9: 2275–2289.

Pecoraro, V., K. Zerulla, C. Lange and J. Soppa. 2011. Quantification of ploidy in proteobacteria revealed the existence of monoploid, (mero-)oligoploid and polyploid species. PLoS ONE 6: e16392.

Peers, G. and N.M. Price. 2006. Copper-containing plastocyanin used for electron transport by an oceanic diatom. Nature 441: 341–344.

Peers, G., S.-A. Quesnel and N.M. Price. 2005. Copper requirements for iron acquisition and growth of coastal and oceanic diatoms. Limnol. Oceanogr. 50: 1149–1158.

Pelusi, A., F. Margiotta, A. Passarelli, M.I. Ferrante, M.R. d'Alcala and M. Montresor. 2020. Density-dependent mechanisms regulate spore formation in the diatom *Chaetoceros socialis*. Limnol. Oceanogr. 5: 371–378.

Penberthy, W.T., M. Sadari and J. Zempleni. 2020. Biotin. pp. 289–304. *In*: Marriott, B.P., D.F. Birt, V.A. Stallings and A.A. Yates [eds.]. Present Knowledge in Nutrition, Eleventh Edition. London, United Kingdom: Academic Press (Elsevier).

Peng, M., C. Hannam, H. Gu, Y.M. Bi and S.J. Rothstein. 2007. A mutation in NLA, which encodes a RING-type ubiquitin ligase, disrupts the adaptability of *Arabidopsis* to nitrogen limitation. Plant J. 50(2): 320–337.

Peng, M., C.Y. Li, X.L. Chen, B.T. Williams, K. Li, Y.N. Gao et al. 2022. Insights into methionine S-methylation in diverse organisms. Nat. Commun. 13: 2947.

Peñuelas, J., B. Poulter, J. Sardans, P. Vialis, M. van der Velde, L. Bopp et al. 2013. Human-induced nitrogen-phosphorus imbalances alter natural and managed ecosystems across the globe. Nat. Commun. 4: 2934.

Peredo, E.L. and Z.G. Cardon. 2020. Shared up-regulation and contrasting down-regulation of gene expression distinguish desiccation-tolerant from intolerant green algae. Proc. Natl. Acad. Sci. USA 117: 17438–17445.

Pereira, D.A. and A. Giani. Cell density-dependent oligopeptide production in cyanobacterial strains. FEMS Microbiol. Ecol. 88: 175–183.

Perez, R., K. Forchhammer, G. Salerno and I. Maldener. 2016. Clear differences in metabolic and morphological adaptations of akinetes of two Nostocales living in different habitats. Microbiology (Reading) 162(2): 214–223.

Perez, R., L. Wörmer, P. Sass and I. Maldener. 2018. A highly asynchronous developmental program triggered during germination of dormant akinetes of filamentous diazotrophic cyanobacteria. FEMS Microbiol. Ecol. 94(1): 1–11.

Pérez-Pérez, M.E., M. Esther and J.L. Crespo. 2014. Autophagy in algae. Perspect. Phycol. 1(2): 93–101.

Pérez-Pérez, M.E., C. Inmaculada and J.L. Crespo. 2017. The TOR signaling network in the model unicellular green alga *Chlamydomonas reinhardtii*. Biomolecules 7(3): 54.

Peschansky, V.J. and C. Wahlestedt. 2014. Non-coding RNAs as direct and indirect modulators of epigenetic regulation. Epigenetics. 9: 3–12.

Petrov, D.A. 2002. Mutational equilibrium model of genome size evolution. Theor. Popul. Biol. 61: 531–544.

Petroll, R., M. Schreiber, H. Finke, J.M. Cock, S.B. Gould and S.A. Rensing. 2021. Signatures of transcription factor evolution and the secondary gain of red algae complexity. Genes (Basel) 12(7): 1055.

Picciano, A.L. and B.R. Crane. 2019. A nitric oxide synthase-like protein from *Synechococcus* produces NO/NO3- from L-arginine and NAPDH in a tetrahydrobiopterin- and Ca^{2+}-dependent manner. J. Biol. Chem. 294(27): 10708–10719.

Pierella Karlusich, J.J., E. Pelletier, F. Lombard, M. Carsique, E. Dvorak, S. Colin et al. 2021. Global distribution patterns of marine nitrogen-fixers by imaging and molecular methods. Nat. Commun. 12: 4160.

Pińko, D., S. Abramovich, E. Rahav, N. Belkin, M. Rubin-Blum, M. Kucera et al. 2023. Shared ancestry of algal symbiosis and chloroplast sequestration in foraminifera. Sci. Adv. 9(41): eadi3401.

Pisciotta, J.M., Y.J. Zou and I.V. Baskakov. 2011. Role of the photosynthetic electron transfer chain in electrogenic activity of cyanobacteria. Appl. Microbial. Biotechnol. 91: 377–385.

Pleissner, D. and N.T. Eriksen. 2012. Effects of phosphorus, nitrogen, and carbon limitation on biomass composition in batch and continuous flow cultures of the heterotrophic dinoflagellate *Crypthecodinium cohnii*. Biotechnol. Bioeng. 109(8): 2005–2016.

Plotkin, J.B. and G. Kudla. 2011. Synonymous but not the same: the causes and consequences of Codon bias. Nat. Rev. Genet. 12: 32–42.

Podar, D. and F.J.M. Maathuis. 2022. Primary nutrient sensors in plants. iScience 25(4): 104029.

Ponmani, T., R. Guo and J.S. Ki. 2015. A novel cyclophilin gene from the dinoflagellate *Prorocentrum minimum* and its possible role in the environmental stress response. Chemosphere 139: 260–267. DOI: 10.1016/j.chemosphere.2015.06.036.

Port, J.A., M.S. Parker, R.B. Kodner, J.C. Wallace, E.V. Armbrust and E.M. Faustman. 2013. Identification of G protein-coupled receptor signaling pathway proteins in marine diatoms using comparative genomics. BMC Genomics 14: 503.

Poulet, S.A., M. Laabir, A. Ianora and A. Miralto. 1995. Reproductive response of *Calanus helgolandicus*. I. Abnormal embryonic and naupliar development. Mar. Ecol. Prog. Ser. 129: 85–95.

Prahl, F.G. and S.G. Wakeham. 1987. Calibration of unsaturation patterns in long-chain ketone compositions for palaeotemperature assessment. Nature 330: 367–369.

Prechtl, J., C. Kneil, P. Lockhart, K. Wenderoth and U.-G. Maier. 2004. Intracellular spheroid bodies of *Rhopalodia gibba* have nitrogen-fixing apparatus of cyanobacterial origin. Mol. Biol. Evol. 21(8): 1477–1481.

Prevett, A., J. Lindström, J. Xu, B. Karlson and E. Selander. 2019. Grazer-induced bioluminescence gives dinoflagellates a competitive edge. Curr. Biol. 29(12): R564–565.

Price, G.D. 2011. Inorganic carbon transporters of the cyanobacterial CO_2 concentrating mechanism. Photosynth. Res. 109(1-3): 47–57.

Price, D.C., N. Farinholt, C. Gates, A. Shumaker, N.E. Wagner, P. Bienfang et al. 2016. Analysis of *Gambierdiscus* transcriptome data supports ancient origins of mixotrophic pathways in dinoflagellates. Environ. Microbiol. 18(12): 4501–4510.

Prigent, M., H. Jean-Jacques, D. Naquin, S. Chédin, M.-H. Cuif, R. Legouis et al. 2024. Sulfur starvation-induced autophagy in *Saccharomyces cerevisiae* involves SAM-dependent signaling and transcription activator Met4. Nat. Commun. 15: 6927.

Prochnik, S.E., J. Umen, A.M. Nedelcu, A. Hallmann, S.M. Miller, I. Nishii et al. 2010. Genomic analysis of organismal complexity in the multicellular green alga *Volvox carteri*. Science 329: 223–226.

Prokopowich, C.D., T.R. Gregory and T.J. Crease. 2003. The correlation between rDNA copy number and genome size in eukaryotes. Genome 46: 48–50.

Proteau, P.J. and W.H. Gerwick. 1992. Cymathere ethers A and B: bicyclic oxylipins from the marine brown alga *Cymathere triplicata*. Tetrahedron Lett 33: 4393–4396.

Puga, M.I., I. Mateos, R. Charukesi, Z. Wang, J.M. Franco-Zorrilla, L. de Lorenzo et al. 2014. SPX1 is a phosphate-dependent inhibitor of Phosphate Starvation Response 1 in *Arabidopsis*. Proc. Natl. Acad. Sci. USA 111(41): 14947–14952.

Pulsford, S.B., M.A. Outram, B. Förster, T. Rhodes, S.J. Williams, M.R. Badger et al. 2024. Cyanobacterial α-carboxysome carbonic anhydrase is allosterically regulated by the Rubisco substrate RuBP. Sci. Adv. 10: eadk7283.

Pushkarev, A., K. Inoue, S. Larom, J. Flores-Uribe, M. Singh, M. Konno et al. 2018. A distinct abundant group of microbial rhodopsins discovered using functional metagenomics. Nature 558(7711): 595–599.

Puxty, R.J., A.D. Millard, D.J. Evans and D.J. Scanlan. 2016. Viruses inhibit CO2 fixation in the most abundant phototrophs on Earth. Curr. Biol. 26: 1585–1589.

Qi, D., Z. Ouyang, L. Chen, Y. Wu, R. Lei, B. Chen et al. 2022. Climate change drives rapid decadal acidification in the Arctic Ocean from 1994 to 2020. Science 377(6614): 1544–1550.

Qian, H., W. Chen, J. Li, J. Wang, Z. Zhou and W. Liu et al. 2009. The effect of exogenous nitric oxide on alleviating herbicide damage in *Chlorella vulgaris*. Aquat. Toxicol. 92(4): 250–257.

Qin, N., D. Xu, J. Li and X.W. Deng. 2020. COP9 signalosome: Discovery, conservation, activity, and function. J. Integr. Plant Biol. 62(1): 90–103.

Qiu, D., L. Huang and S. Lin 2016. Cryptophyte farming by symbiotic ciliate host detected in situ. Proc. Natl. Acad. Sci. USA 113(43): 12208–12213.

Qiu, D., L. Huang, S. Liu and S. Lin. 2011. Nuclear, mitochondrial and plastid gene phylogenies of *Dinophysis miles* (Dinophyceae): evidence of variable types of chloroplasts. PLoS ONE 6(12): e29398.

Qiu, H., D.C. Price, E.C. Yang, H.S. Yoon and D. Bhattacharya. 2015. Evidence of ancient genome reduction in red algae (Rhodophyta). J. Phycol. 51: 624–636.

Qu, P.-P., F.-X. Fu, X.-W. Wang, J.D. Kling, M. Elghazzawy, M. Huh et al. 2022. Two co-dominant nitrogen-fixing cyanobacteria demonstrate distinct acclimation and adaptation responses to cope with ocean warming. Environ. Microbiol. Rep. 14(2): 203–217.

Quigg, A., Z.V. Finkel, A.J. Irwin, Y. Rosenthal, T.-Y. Ho, J.R. Reinfelder et al. 2003. The evolutionary inheritance of elemental stoichiometry in marine phytoplankton. Nature 425: 291–294.

Raghukumar, C.1986. Thraustochytrid fungi associated with marine algae. Indian J. Mar. Sci. 15: 121–122.

Raghukumar, C. 1987. Fungal parasites of marine algae from Mandapam (South India). Dis. Aquat. Org. 3: 137–145.

Rai, A.N., E. Söderbäck and B. Bergmann. 2000. Tansley Review No. 116. Cyanobacterium-plant symbioses. New Phytol. 147(3): 449–481.

Rai, L.C. and J.P. Gaur 2001. Algal Adaptation to Environmental Stresses: Physiological, Biochemical and Molecular Mechanisms. Springer-Verlag Berlin Heidelberg New York. 421 pp.

Raitsos, D.E., X. Yi, T. Platt, M.-F. Racault, R.J.W. Brewin, Y. Pradhan et al. 2015. Monsoon oscillations regulate fertility of the Red Sea. Geophys. Res. Lett. 42(3): 855–862.

Raj, S., A.M. Kuniyil, A. Sreenikethanam, P. Gugulothu, R.B. Jeyakumar and A.K. Baihaiya. 2021. Microalgae as a source of mycosporine-like amino acids (MAAs): Advances and future prospects. Int. J. Environ. Res. Public Health 18(23): 12402.

Ramírez-Gamboa, D., A.L. Díaz-Zamorano, E.R. Meléndez-Sánchez, H. Reyes-Pardo, K.R. Villaseñor-Zepeda and M.E. López-Arellanes. 2022. Photolyase production and current applicatios: A review. Molecules 27(18): 5998.

Rantanen, M., A.Y. Karpechko, A. Lipponen, K. Nordling, O. Hyvärinen, K. Ruosteenoja et al. 2022. The Arctic has warmed nearly four times faster than the globe since 1979. Comm. Earth & Environ. 3: 168.

Ratcliffe, S., E.M. Meyer, C.E. Walker, M. Knight, H.M. McNair, P.G. Matson et al. 2023. Characterization of the molecular mechanisms of silicon uptake in coccolithophores. Environ. Microbiol. 25(2): 315–330.

Raven, J.A. 2013. The evolution of autotrophy in relation to phosphorus requirement. J. Exp. Bot. 64(13): 4023–4046.

Raven, J.A., J. Beardall, K.J. Flynn and S.C. Maberly. 2009. Phagotrophy in the origins of photosynthesis in eukaryotes and as a complementary mode of nutrition in phototrophs: relation to Darwin's insectivorous plants. J. Exp. Bot. 60(14): 3975–3987.

Raven, J.A., J. Beardall and P. Sánchez-Baracaldo. 2017. The possible evolution and future of CO_2-concentrating mechanisms. J. Exp. Bot. 68(14): 3701–3716.

Rawat, D., U. Sharma, P. Poria, A. Finlan, B. Parker, R.A. Sharma et al. 2022. Iron-dependent mutualism between *Chlorella sorokiniana* and *Ralstonia pickettii* forms the basis for a sustainable bioremediation system. ISME Comm. 2: 83.

Rayko, E., F. Maumus, U. Maheswari, K. Jabbari and C. Bowler. 2010. Transcription factor families inferred from genome sequences of photosynthetic stramenopiles. New Phytol. 188: 52–66.

Raymond, J., O. Zhaxybayeva, J.P. Gogarten, S.Y. Gerdes and R.E. Blankenship. 2002. Whole-genome analysis of photosynthetic prokaryotes. Science 298: 1616–1620.

Razeghi, J. and A. Kianianmomeni. 2019. UV-B response is modulated by cell-type specific signaling pathway in multicellular green algae *Volvox carteri*. Plant Growth Regulation 87: 303–315.

Read, B.A., J. Kegel, M.J. Klute, A. Kuo, S.C. Lefebvre, F. Maumus et al. 2013. Pan genome of the phytoplankton *Emiliania* underpins its global distribution. Nature 499: 209–213.

Redfield, A.C. 1934. On the proportions of organic derivatives in sea water and their relation to the composition of plankton. University Press of Liverpool. pp. 17.

Redfield, A.C. 1958. The biological control of chemical factors in the environment. Amer. Sci. 46(3): 205–221.

Reedy, J.L., A.M. Floyd and J. Heitman. 2009. Mechanistic plasticity of sexual reproduction and meiosis in the *Candida* pathogenic species complex. Curr. Biol. 19: 891–899.

Rees, D.A., I.W. Steele and F.B. Williamson. 1969. Conformational analysis of polysaccharides. III. The relation between stereochemistry and properties of some natural polysaccharide sufates. J. Polyler Sci. Part C: Polymer Symposia 28(1): 261–276.

Rees, T.A.V. and P.J. Syrett. 1979. The uptake of urea by the diatom, *Phaeodactylum tricornutum*. New Phytol. 82: 169–178.

Reinfelder, J.R. 2011. Carbon concentrating mechanisms in eukaryotic marine phytoplankton. Ann. Rev. Mar. Sci. 3: 291–315.

Reinfelder, J.R., A.M.L. Kraepiel and F.M.M. Morel. 2000. Unicellular C_4 photosynthesis in a marine diatom. Nature 407: 996–999.

Reinfelder, J.R., A.J. Milligan, F.M.M. Morel. 2004. The role of the C_4 pathway in carbon accumulation and fixation in a marine diatom. Plant Physiol. 135(4): 2106–2111.

Reiskind, J.B. and G. Bowes. 1991. The role of phosphoenolpyruvate carboxykinase in a marine macroalga with C4-like photosynthetic characteristics. Proc. Natl. Acad. Sci. USA 88(7): 2883–2887.

Rempt, M., F. Weinberger, K. Grosser and G. Pohnert. 2012. Conserved and species-specific oxylipin pathways in the wound-activated chemical defense of the noninvasive red alga *Gracilaria chilensis* and the invasive *Gracilaria vermiculophylla*. Beilstein J. Org. Chem. 8: 283–289.

Reñé, A., E. Alacid, A.E. Vishnyakov, K. Seto, C.S. Tcvetkova, J. Gordi et al. 2022. The new chytridiomycete *Paradinomyces triforaminorum* gen. et sp. nov. co-occurs with other parasitoids during a *Kryptoperidinium foliaceum* (Dinophyceae) bloom in the Baltic Sea. Harmful Algae 120: 102352.

Repetti, S.I., C. Iha, K. Uthanumallian, C.J. Jackson, Y. Chen, C.X. Chan et al. 2021. Nuclear genome of a pedinophyte pinpoints genomic innovation and streamlining in the green algae. New Phytol. 233(5): 2144–2154.

Rhiel, E., M. Westermann, F. Steiniger and C. Hoischen. 2020. The proteorhodopsins of the dinoflagellate *Oxyrrhis marina*: ultrastructure and localization by immunofluorescence light microscopy and immunoelectron microscopy. Protoplasma 257(6): 1531–1541.

Rhiel, E., C. Hoischen and M. Westermann. 2022. Rhodopsins build up the birefringent bodies of the dinoflagellate *Oxyrrhis marina*. Protoplasma 259(4): 1047–1060.

Riebesell, U., I. Zondervan, B. Rost, P.D. Tortell, R.E. Zeebe and F.M. Morel. 2000. Reduced calcification of marine plankton in response to increased atmospheric CO_2. Nature 407(6802): 364–367.

Riediger, M., Y. Hihara and W.R. Hess. 2018. From cyanobacteria and algae to land plants: the RpaB/Ycf27 regulatory network in transition. Perspect. Phycol. 5(1): 13–25.

Riediger, M., T. Kadowaki, R. Nagayama, J. Georg, Y. Hihara and W.R. Hess. 2019. Biocomputational analyses and experimental validation identify regulon controlled by the redox-responsive transcription factor RpaB. iScience 15: 316–331.

Rieseberg, T.P., A. Dadras, J.M.R. Fürst-Jansesn, A.D. Ashok, T. Darienko, S. de Vries et al. 2023. Crossroads in the evolution of plant specialized metabolism. Semin. Cell Dev. Biol. 134: 37–58.

Risser, D.D. and S.M. Callahan. 2009. Genetic and cytological evidence that heterocyst patterning is regulated by inhibitor gradients that promote activator decay. Proc. Natl. Acad. Sci. USA 106: 19884–19888.

Ritter, A., L. Cabioch, L. Brillet-Guéguen, E. Corre, A. Cosse, L. Dartevelle et al. 2017. Herbivore-induced chemical and molecular responses of the kelps *Laminaria digitata* and *Lessonia spicata*. PLoS ONE 12(3): e0173315.

Ritter, A., S. Goulitquer, J.P. Salaün, T. Tonon, J.A. Correa and P. Potin. 2008. Copper stress induces biosynthesis of octadecanoid and eicosanoid oxygenated derivatives in the brown algal kelp *Laminaria digitata*. New Phytologist 180(4): 809–821.

Rius, M., J.S. Rest, G.V. Filloramo, A.M.G.N. Vancloá, J.M. Archibald and J.L. Collier. 2023. Horizontal gene transfer and fusion spread carotenogenesis among diverse heterotrophic protists. Genome Biol. Evol. 15(3): evad029.

Rizzo, P.J. 1991. Those amazing dinoflagellate chromosomes. Cell Res. 13: 215–217.

Rodriguez, M., J.W. Cho, H.W. Sauer and P.J. Rizzo. 1993. Evidence of the presence of a Cdc2-like protein kinase in the dinoflagellate *Crypthecodinium cohnii*. J. Eukaryot. Microbiol. 40(1): 91–96.

Roberts, K., E. Granum, R.C. Leegood and J.A. Raven. 2007. C3 and C4 pathways of photosynthetic carbon assimilation in marine diatoms are under genetic, not environmental, control. Plant Physiol. 145(1): 230–235.

Rocha, E.P. and A. Danchin. 2002. Base composition Bias might result from competition for metabolic resources. Trends Genet. 18: 291–294.

Rocap, G., F.W. Larimer, J. Lamerdin, S. Malfatti, P. Chain, N.A. Ahlgren et al. 2003. Genome divergence in two *Prochlorococcus* ecotypes reflects oceanic niche differentiation. Nature 424: 1042–1047.

Rocha, E.P.C. 2004. Codon usage bias from tRNA's point of view: redundancy, specialization, and efficient decoding for translation optimization. Genome Res. 12(11): 2279–2286.

Rocha, G.S., C.C. Parrish and E.L.G. Espíndola. 2021. Effects of copper on photosynthetic and physiological parameters of a freshwater microalga (Chlorophyceae). Algal Res. 54: 102223.

Rockwell, N.C., J.C. Lagarias and D. Bhattacharya. 2014. Primary endosymbiosis and the evolution of light and oxygen sensing in photosynthetic eukaryotes. Front. Ecol. Evol. 2: 66.

Rodriguez, F., J. Lillington, S. Johnson, C.R. Timmel, S.M. Lea and B.C. Berks. 2014. Crystal structure of the *Bacillus subtilis* phosphodiesterase PhoD reveals an iron and calcium-containing active site. J. Biol. Chem. 289(45): 30889–30899.

Rodríguez, J.J.G., A.S. Mirón, F.G. Camacho, M.C.C. García, E.H. Belarbi, Y. Chisti et al. 2009. Causes of shear sensitivity of the toxic dinoflagellate *Protoceratium reticulatum*. Biotechnol. Prog. 25(3): 792–800.

Rodríguez-Ezpeleta, N., H. Brinkmann, S.C. Burey, B. Roure, G. Burger, W. Löffelhardt et al. 2005. Monophyly of primary photosynthetic eukaryotes: Green plants, red algae, and glaucophytes. Curr. Biol. 15: 1325–1330.

Rohde, J.R. and M.E. Cardenas. 2003. The TOR pathway regulates gene expression by linking nutrient sensing to histone acetylation. Mol. Cell Biol. 23(2): 629–635.

Roleda, M.Y. and C.L. Hurd. 2019. Seaweed nutrient physiology: application of concepts to aquaculture and bioremediation. Phycologia 58(5): 552–562.

Rose, S.L., J.M. Fulton, C.M. Brown, F. Natale, B.A.S. Van Mooy and K.D. Bidle. 2014. Isolation and characterization of lipid rafts in *Emiliania huxleyi*: a role for membrane microdomains in host-virus interactions. Environ. Microbiol. 16: 1150–1166.

Rosic, N.N., M. Pernice, S. Dunn, S. Dove and O. Hoegh-Guldberg. 2010. Differential regulation by heat stress of novel cytochrome P450 genes from the dinoflagellate symbionts of reef building corals. Appl. Environ. Microbiol. 76(9): 2823–2829.

Rosenberg, G. and H.W. Paerl. 1981. Nitrogen fixation by blue-green algae associated with the siphonous green seaweed *Codium decorticatum*: Effects on ammonium uptake. Mar. Biol. 61: 151–158.

Rosenwasser, S., M.A. Mausz, D. Schatz, U. Sheyn, S. Malitsky, A. Aharoni et al. 2014. Rewiring host lipid metabolism by large viruses determines the fate of *Emiliania huxleyi*, a bloom-forming algal in the ocean. Plant Cell 26(6): 2689–2707.

Rosenwasser, S., C. Ziv, S.G. van Creveld and A. Vardi. 2016. Virocell metabolism: metabolic innovations during host-virus interactions in the ocean. Trends Microbiol. 24(10): 821–832.

Rossoni, A.W. and A.P.M. Weber. 2019. Systems biology of cold adaptation in the polyextremophilic red alga *Galdieria sulphuraria*. Front. Microbiol. 10: 927.

Roth, M.S., S.J. Cokus, S.D. Gallaher, A. Walter, D. Lopez, E. Erickson et al. 2017. Chromosome-level genome assembly and transcriptome of the green alga *Chromochloris zofingiensis* illuminates astaxanthin production. Proc. Natl. Acad. Sci. USA 114(21): E4296–E4305.

Roy, S. and D. Morse. 2012. A full suite of histone and histone modifying genes are transcribed in the dinoflagellate *Lingulodinium*. PLoS One 7(4): e34340.

Roy, S., R. Jagus and D. Morse. 2018. Translatoin and translational control in dinoflagellates. Microorganisms 6: 30.

Roy, S., M. Beauchemin, S. Dagenais-Bellefeuille, L. Letourneau, M. Cappadocia and D. Morse. 2014. The *Lingulodinium circadian* system lacks rhythmic changes in transcript abundance. BMC Biol. 12: 107.

Roy, S.W., L. Gozashti, B.A. Bowser, B.N. Weinstein, G.E. Larue and R. Corbett-Detig. 2023. Intron-rich dinoflagellate genomes driven by Introner transposable elements of unprecedented diversity. Curr. Biol. 33(1): 189–196. e4.

Rozenberg, A., I. Kaczmarczyk, D. Matzov, J. Vierock, T. Nagata, M. Sugiura et al. 2022. Rhodopsin-bestrophin fusion proteins from unicellular algae form gigantic pentameric ion channels. Nature Struct. Mol. Biol. 29(6): 592–603.

Ruban, A.V. 2016. Nonphotochemical chlorophyll fluorescence quenching: Mechanism and effectiveness in protecting plants from photodamage. Plant Physiol. 170: 1903–1916.

Rue, E. and K.W. Bruland. 2001. Domoic acid binds iron and copper: a possible role for the toxin produced by the marine diatom *Pseudo-nitzschia*. Mar. Chem. 76: 127–134.

Rumpho, M.E., K.N. Pelletreau, A. Moustafa and D. Bhattacharya. 2011. The making of a photosynthetic animal. J. Exp. Biol. 214: 303–311.

Rumpho, M.E., E.J. Summer and J.R. Manhard. 2000. Solar-powered sea slugs: Mollusc/algal chloroplast symbiosis. Plant Physiol. 123(1): 29–38.

Rumpho, M.E., J.M. Worful, J. Lee, K. Kannan, M.S. Tyler, D. Bhattacharya et al. 2008. Horizontal gene transfer of the algal nuclear gene psbO to the photosynthetic sea slug *Elysia chlorotica*. Proc. Natl. Acad. Sci. USA 105(46): 17867–17871.

Rungruangsak-Torrissen, R. Moss, L.H. Andresen, A. Berg and R. Waagbø. 2006. Different expressions of trypsin and chymotrypsin in relation to growth in Atlantic salmon (*Salmo salar* L.). Fish Physiol. Biochem. 32(1): 7–23.

Russo, E., G. d'lppolito, A. Fontana, D. Sarno, D. D'Alelio, G. Busseni et al. 2020. Density-dependent oxylipin production in natural diatom communities: possible implications for plankton dynamics. ISME J. 14: 164–177.

Ruvindy, R., A. Barua, C.J.S. Bolch, C. Sarowar, H. Savela and S.A. Murray. 2023. Genomic copy number variability at the genus, species, and population levels impact in situ ecological analyses of dinoflagellates and harmful algal blooms ISME Commun. 3(1): 70.

Sabehi, G., A. Loy, K.-H. Jung, R. Partha, J.L. Spudich, T. Isaacson et al. 2005. New insights into metabolic properties of marine bacteria encoding proteorhodopsins. PLoS Biol. 3(8): e273.

Sabine, C.L., R.A. Feely, N. Gruber, R.M. Key, K. Lee, J.L. Bullister et al. 2004. The ocean sink for anthropogenic CO2. Science 305(5682): 367–371.

Sach, J.L. and T.P. Wilcox. 2006. A shift to parasitism in the jellyfish symbiont *Symbiodinium microadriaticum*. Proc. Biol. Sci. 273(1585): 425–429.

Sagan, L. 1967. On the origin of mitosing cells. J. Theor. Biol. 14(3): 255–274.

Sagi, M., O. Davydov, S. Orazova, Z. Yesbergenova, R. Ophir, J.W. Stratmann et al. 2004. Plant respiratory burst oxidase homologs impinge on wound responsiveness and development in *Lycopersicon esculentum*. Plant Cell 16(3): 616–628.

Saidi, Y., A. Finka and F. Goloubinoff. 2010. Heat perception and signaling in plants: a tortuous path thermotolerance. New Phytol. 190(3): 556–565.

Saint, N., H. Cadiou, Y. Bessin and G. Molle. 2002. Antibacterial peptide pleurocidin forms ion channels in planar lipid bilayers. Biochim. Biophys. Acta Biomembr. 1564(2): 359–364.

Saita, E.A. and D. de Mendoza. 2015. Thermosensing via transmembrane protein–lipid interactions. Biochim. Biophys. Acta Biomembr. 1848: 1757–1764.

Sakakibara, K., S. Ando, H.K. Yip, Y. Tamada, Y. Hiwatashi, T. Murata et al. 2013. KNOX2 genes regulate the haploid-to-diploid morphological transition in land plants. Science 339: 1067–1070.

Sakihama, Y., S. Nakamura and H. Yamasaki. 2002. Nitric oxide production mediated by nitrate reductase in the green alga *Chlamydomonas reinhardtii*: an alternative NO production pathway in photosynthetic organisms. Plant Cell Physiol. 43(3): 290–297.

Saldivia, P., M. Hernandez, A. Isla, R. Fritz, D. Varela, J. M-Gonzalez-Jartin et al. 2023. Proteomic and toxicological analysis of the response of dinoflagellate *Alexandrium catenella* to changes in $NaNO_3$ concentration. Harmful Algae 125: 102428.

Samanta A., T.E.T. Hughes and V.Y. Moiseenkova-Bell. 2018. Transient receptor potential (TRP) channels. pp. 141–165. *In:* Harris J. and Boekema E. [eds.]. Membrane Protein Complexes: Structure and Function. Springer, Singapore.

Sancar, A. 2003. Structure and function of DNA photolyase and cryptochrome blue-light photoreceptors. Chem Rev. 103: 2203–2237.

Sanchez, N., E.A. Brown, Y. Olsen, O. Vadstein, J.L. Iriarte, H.E. Gonzalez et al. 2018. Effect of siderophore on iron availability in a diatom and a dinoflagellate species: contrasting response in associated bacteria. Front. Mar. Sci. 5: 118.

Sánchez-Baracaldo, P. 2015. Origin of marine planktonic cyanobacteria. Sci. Rep. 5: 17418.

Sánchez-Baracaldo, P., A. Ridgwell and J.A. Raven. 2014. A Neoproterozoic transition in the marine nitrogen cycle. Current Biology 24: 652–657.

Sánchez-Baracaldo, P., J.A. Raven, D. Pisani and A.H. Knoll. 2017. Early photosynthetic eukaryotes inhabited low-salinity habitats. Proc. Natl. Acad. Sci. USA 114(37): E7737–E7745.

Sánchez-Baracaldo, P. and T. Cardona. 2019. On the origin of oxygenic photosynthesis and cyanobacteria. New Phytol. 225(4): 1440–1446.

Sandh, G., L. Xu and B. Bergman. 2012. Diazocyte development in the marine diazotrophic cyanobacterium *Trichodesmium*. Microbiology 158: 345–352.

Sandkvist, M. 2001. Type II secretion and pathogenesis. Infect Immun. 69(6): 3523–3535.

Sañudo-Wilhelmy, S.A., L.S. Cutter, R. Durazo, E.A. Smail, L. Gómez-Consarnau, E.A. Webb et al. 2012. Multiple B-vitamin depletion in large areas of the coastal ocean. 109(35): 14041–14045.

Sanz-Luque, E., S. Saroussi, W. Huang, N. Akkawl and A.R. Grossman. 2020. Metabolic control of acclimation to nutrient deprivation dependent on polyphosphate synthesis. Sci. Adv. 6: eabb5351.

Sapp, M., A.S. Schwaderer, K.H. Wiltshire, H.-G. Hoppe, G. Gerdts and A. Wichels. 2007. Species-specific bacterial communities in the phycosphere of microalgae? Microb. Ecol. 53: 683–699.

Sarai, C., G. Tanifuji, T. Nakayama, R. Kamikawa, K. Takahashi, E. Yazaki et al. 2020. Dinoflagellates with relic endosymbiont nuclei as models for elucidating organellogenesis. Proc. Natl. Acad. Sci. U S A 117: 5364–5375.

Sarich, V.M. and A.C. Wilson. 1967. Rates of albumin evolution in primates. Proc. Natl. Acad. Sci. USA. 58(1): 142–148.

Sasaki, K. and R. Imai. 2012. Pleiotropic roles of cold shock domain proteins in plants. Front. Plant Sci. 2: 116.

Sato, M., M. Seki, Y. Suzuki and S. Ueki. 2023. The dataset of de novo assembly and inferred functional annotation of the transcriptome of *Heterosigma akashiwo*, a bloom-forming, cosmopolitan raphidophyte. Data in Brief 48: 109071.

Sato, S., G. Beakes, M. Idei, T. Nagumo and D.G. Mann. 2011. Novel sex cells and evidence for sex pheromones in diatoms. PLoS ONE 6(10): e26923.

Saunders, G.W. and D.C. McDevit. 2012. Methods for DNA barcoding photosynthetic protists emphasizing the macroalgae and diatoms. *In*: Kress, W. and D. Erickson [eds.]. DNA Barcodes: Methods in Molecular Biology 858: 207–222. Humana Press, Totowa, New Jersey, USA.

Savela, H., K. Harju, L. Spoof, E. Lindehoff, J. Meriluoto, M. Vehniainen et al. 2016. Quantity of the dinoflagellate sxtA4 gene and cell density correlates with paralytic shellfish toxin production in *Alexandrium ostenfeldii* blooms. Harmful Algae 52: 1–10.

Scalco, E., A. Amato, M.I. Ferrante and M. Montresor. 2016. The sexual phase of the diatom *Pseudo-nitzschia multistriata*: cytological and time-lapse cinematography characterization. Protoplasma 253: 1421–1431.

Schirrmeister, B.E., A. Antonelli and H.C. Bagheri. 2011. The origin of multicellularity in cyanobacteria. BMC Evol. Biol. 11: 348.

Schlüter, L., K.T. Lohbeck, J.P. Gröger, U. Riebesell and T.B.H. Reusch. 2016. Long-term dynamics of adaptive evolution in a globally important phytoplankton species to ocean acidification. Sci. Adv. 2: e1501660.

Schmelling, N.M., N. Scheurer, C. Köbler, A. Wilde and I.M. Axmann. 2021. Diversity of timing systems in cyanobacteria and beyond. pp. 179–202. *In*: Johnson, C.H. and M.J. Rust [eds.]. Circadian Rhythms in Bacteria and Microbiomes. Springer Nature Switzerland AG.

Schmitt, M., A. Telusma, E. Bigeard, L. Guillou and C. Alves-de-Souza. 2022. Temperature affects the biological control of dinoflagellates by the generalist parasitoid *Parvilucifer rostrata*. Microorganisms 10(2): 385.

Schneider, M.J., M. Ulland and D. Sloboda. 2008. A protein methylation pathway in Chlamydomonas flagella is active during flagellar resorption. Mol. Biol. Cell 19(10): 4319–4327.

Schoberer, J., Y.-J. Shin, U. Vavra, C. Veit and R. Strasser. 2018. Protein glycosylation in the ER. Methods Mol. Biol. 1691: 205–222.

Schurko, A.M. and J.M. Jr. Logsdon. 2008. Using a meiosis detection toolkit to investigate ancient asexual "scandals" and the evolution of sex. Bioessays. 30: 579–589.

Schvarcz, C.R., S.T. Wilson, M. Caffin, R. Stancheva, Q. Li, K.A. Turk-Kubo et al. 2022. Overlooked and widespread pennate diatom-diazotroph symbioses in the sea. Nat. Comm. 13: 799.

Schwarzenberger, A., M. Hasselmann and E.V. Elert. 2020. Positive selection of digestive proteases in Daphnia: A mechanism for local adaptation to cyanobacterial protease inhibitors. Mol. Ecol. 29: 912–919.

Schwechheimer, C., G. Serino and X.W. Deng. 2002. Multiple ubiquitin ligase-mediated processes require COP9 signalosome and AXR1 function. Plant Cell 14: 2553–2563.

Schwertmann, U. 1991. Solubility and dissolution of iron oxides. Plant Soil 130: 1–25.

Scott, C., T.W. Lyons, A. Bedder, Y. Shen, S.W. Poulton, X. Chu et al. 2008. Tracing the stepwise oxygenation of the Proterozoic ocean. Nature 452: 456–459.

Secco, D., C. Wang, B.A. Arpat, Z. Wang, Y. Poirier, S.D. Tyerman et al. 2012. The emerging importance of the SPX domain-containing proteins in phosphate homeostasis. New Phytol. 193(4): 842–851.

Selander, E., E.C. Berglund, P. Engstrom, F. Berggren, J. Eklund, S. Harðardóttir et al. 2019. Copepods drive large-scale trait-mediated effects in marine plankton. Sci. Adv. 5: eaat5096.

Selander, E., H.H. Jakobsen, F. Lombard and T. Kiørboe. 2011. Grazer cues induce stealth behavior in marine dinoflagellates. Proc. Natl. Acad. Sci. USA 108: 4030–4034.

Selander, E., J. Kubanek, M. Hamberg, M.X. Andersson, G. Cervin and H. Pavia. 2015. Predator lipids induce paralytic shellfish toxins in bloom-forming algae. Proc. Natl. Acad. Sci. USA 112: 6395–6400.

Selander, E., P. Thor, G.B. Toth and H. Pavia. 2006. Copepods induce paralytic shellfish toxin production in marine dinoflagellates. Proc. R. Soc. Lond. Ser. B-Biol. Sci. 273(1594): 1673–1680.

Seo, K.S. and L. Fritz. 2006. Karyology of a marine non-motile dinoflagellate, *Pyrocystis lunula*. Hydrobiologia 563: 289–296.

Seppala, J., T. Tamminen and S. Kaitala. 1999. Experimental evaluation of nutrient limitation of phytoplankton communities in the Gulf of Riga. J. Mar. Syst. 23: 107–126.

Seymour, J.R., S.A. Amin, S.-B. Raina and R. Stocker. 2017. Zooming in on the phycosphere: the ecological interface for phytoplankton-bacteria relationships. Nat. Microbiol. 2: 17065.

Shah, S., Y. Chen, D. Bhattacharya and C.X. Chan. 2020. Sex in Symbiodiniaceae dinoflagellates: genomic evidence for independent loss of the canonical synaptonemal complex. Sci. Rep. 10: 9792.

Shah, S., K.E. Dougan, Y. Chen, R. Lo, G. Laird, M.D. Fortuin, et al. 2024. Massive genome reduction predates the divergence of Symbiodiniaceae dinoflagellates. ISME J. 18(1): wrae059.

Shao, H., B. Gontero, S.C. Maberly, H. Jiang, Y. Cao, W. Li et al. 2017. Responses of *Ottelia alismoides*, an aquatic plant with three CCMs, to variable CO_2 and light. J. Exp. Bot. 68: 3985–3995.

Shao, X., N. Lv, J. Liao, J. Long, R. Xue, N. Ai et al. 2019. Copy number variation is highly correlated with differential gene expression: a pan-cancer study. BMC Med. Genet. 20: 175.

Sharma, P. 2007. Mining microsatellites in eukaryotic genomes. Trends Biotechnol. 25: 490.

Sharples, F.P., P.M. Wrench, K. Ou and R.G. Hiller. 1996. Two distinct forms of the peridinin-chlorophyll *a*-protein from *Amphidinium carterae*. Biochim. Biophys. Acta 1276: 117–123.

Shemi, A., S. Ben-Dor, R. Rotkopf, O. Dym and A. Vardi. 2023. Phylogeny and biogeography of the algal DMS-releasing enzyme in the global ocean. ISME Commun. 3: 72.

Shemi, A., D. Schatz, H.F. Fredricks, B.A.S. van Mooy, Z. Porat and A. Vardi. 2016. Phosphorus starvation induces membrane remodeling and recycling in *Emiliania huxleyi*. New Phytol. 211: 765–766.

Shen, G., F. Gan and D.A. Bryant. 2016. The siderophilic cyanobacterium *Leptolyngbya* sp. strain JSC-1 acclimates to iron starvation by expressing multiple isiA-family genes. Photosynth. Res. 128: 325–340.

Shen, S., Y. Jian, Z. Cai, F. Li, M. Lv, Y. Liu et al. 2022. Structural insights reveal the specific recognition of meiRNA by the Mei2 protein. J. Mol. Cell Biol. 4(5): mjac029.

Shetty, P., M.M. Gitau and G. Maróti. 2019. Salinity stress responses and adaptation mechanisms in eukaryotic green microalgae. Cells 8(12): 1657.

Shi, D., Y. Xu, B.M. Hopkinson and F.M.M. Morel. 2010. Effects of ocean acidification on iron availability to marine phytoplankton. Science 327: 676–679.

Shi, Q. 2019. Expression profiling of gene coding for abundant proteins in the alkenone body of marine haptophyte alga *Tisochrysis lutea*. BMC Microbiol. 19: 56.

Shi, Q., H. Araie, R.K. Bakku, Y. Fukao, R. Rakwal, I. Suzuki et al. 2015. Proteomic analysis of lipid body from the alkenone-producing marine haptophyte alga *Tisochrysis lutea*. Proteomics 15(23-24): 4145–4158.

Shi, X., M. Ma and S. Lin. 2017a. Cell cycle-dependent expression dynamics of G1/S specific cyclin, cellulose synthase and cellulase in the dinoflagellate *Prorocentrum donghaiense*. Front. Microbiol. 8: 1118.

Shi, X., H. Zhang and S. Lin. 2013. Tandem repeats, high copy number and remarkable diel rhythm in transcription of form II Rubisco in dinoflagellate *Prorocentrum donghaiense*. PLoSONE 8(8): e71232.

Shi, X., L. Li, C. Guo, X. Lin, M. Li and S. Lin. 2015. Rhodopsin gene expression regulated by the light dark cycle, light spectrum and light intensity in the dinoflagellate *Prorocentrum*. Front. Microbiol. 6: 55.

Shi, X., X. Lin, L. Li, M. Li, B. Palenik and S. Lin. 2017b. Transcriptomic and microRNAomic profiling reveals multi-faceted mechanisms to cope with phosphate stress in a dinoflagellate. ISME J. 11: 2209–2218.

Shi, X., L. Liu, Y. Li, Y. Xiao, G. Ding, S. Lin et al. 2018. Isolation of an algicidal bacterium and its effects against the harmful-algal-bloom dinoflagellate *Prorocentrum donghaiense* (Dinophyceae). Harmful Algae 80: 72–79.

Shi, X., Y. Xiao, L. Liu, Y. Xie, R. Ma and J. Chen. 2021. Transcriptome responses of the dinoflagellate *Karenia mikimotoi* driven by nitrogen deficiency. Harmful Algae 103: 101977.

Shi, X., Y. Zou, Y. Zhang, G. Ding, Y. Xiao, S. Lin et al. 2024. Salinity decline promotes growth and harmful blooms of a toxic alga by diverting carbon flow. Glob. Chang. Biol. 30: 17348.

Shikata, T., F. Takahashi, H. Nishide, S. Shigenobu, Y. Kamei, S. Sakamoto et al. 2019. RNA-seq analysis reveals genes related to photoreception, nutrient uptake, and toxicity in a noxious red-time raphidophyte *Chattonella antiqua*. Front. Microbiol. 10: 1764.

Shimasaki, Y., K. Mukai, Y. Takai, X. Qiu and Y. Oshima. 2021. Recent progress in the study of peroxiredoxin in the harmful algal bloom species *Chattonella marina*. Antioxidants 10(2): 162.

Shimoda, C., M. Uehira, M. Kishida, H. Fujioka, Y. Lino, Y. Watanabe et al. 1987. Cloning and analysis of transcription of the mei2 gene responsible for initiation of meiosis in the fission yeast *Schizosaccharomyces pombe*. J. Bacteriol. 169(1): 93–96.

Shin, W., S.M. Boo and J. Longcore. 2001. *Entophlyctis apiculata*, a chytrid parasite of *Chalmydomonas* sp. (Chlorophyceae). Can. J. Bot. 79: 1083–1089.

Shintani, M., Z.K. Sanchez and K. Kimbara. 2015. Genomics of microbial plasmids: classification and identification based on replication and transfer systems and host taxonomy. Front. Microbiol. 6: 242.

Shiu, C.T. and T.M. Lee. 2005. Ultraviolet-b-induced oxidative stress and responses of the ascorbate-glutathione cycle in a marine macroalgae *Ulva fasciata*. J. Exp. Bot. 56(421): 2851–2865.

Shoguchi, E. 2022. Gene clusters for biosynthesis of mycosporine-like amino acids in dinoflagellate nuclear genomes: possible recent horizontal gene transfer between species of Symbiodiniaceae (Dinophyceae). J. Phycol. 58: 1–11.

Shoguchi, E., G. Beedessee, K. Hisata, I. Tada, H. Narisoko, N. Satoh et al. 2021. A new dinoflagellate genome illuminates a conserved gene cluster involved in sunscreen biosynthesis. Genome Biol. Evol. 13(2): evaa235.

Shoguchi, E., G. Beedessee, I. Tada, K. Hisata, T. Kawashima, T. Takeuchi et al. 2018. Two divergent *Symbiodinium* genomes reveal conservation of a gene cluster for sunscreen biosynthesis and recently lost genes. BMC Genomics 19: 458.

Shoguchi, E., C. Shinzato, T. Kawashima, F. Gyoja, S. Mungpakdee, R. Koyanagi et al. 2013. Draft assembly of the *Symbiodinium minutum* nuclear genome reveals dinoflagellate gene structure. Curr. Biol. 23(15): 1399–1408.

Shoguchi, E., Y. Yoshioka, C. Shinzato, A. Arimoto, D. Bhattacharya and N. Satoh. 2020. Correlation between organelle genetic variation and RNA editing in dinoflagellates associated with the coral *Acropora digitifera*. Genome Biol. Evol. 12(3): 203–209.

ShuZon, W. and M. Bezanilla. 2014. Myosin VIII associates with microtubule ends and together with actin plays a role in guiding plant cell division. eLife 3: e03498.

Sibbald, S.J., J.F. Hopkins, G.V. Filloramo and J.M. Archibald. 2019. Ubiquitin fusion proteins in algae: implications for cell biology and the spread of photosynthesis. BMC Genomics 20: 38.

Sibly, R.M. and P. Calow. 1989. A life-cycle theory of responses to stress. Biol. J. 37(1-2): 101–116.

Siegenthaler, U., T.F. Stocker, E. Monnin, D. Lüthi, J. Schwander, B. Stauffer et al. 2005. Stable carbon cycle-climate relationship during the late Pleistocene. Science 310: 1313–1317.

Sieracki, M.E., N.J. Poulton, O. Jaillon, P. Wincker, C. de Vargas, L. Rubinat-Ripoll et al. 2019. Single cell genomics yields a wide diversity of small planktonic protists across major ocean ecosystems. Sci. Rep. 9(1): 6025.

Sigee, D.C. 1986. The dinoflagellate chromosome. Adv. Bot. Res. 12: 205–264.

Sims, P.S., D.G. Mann and L.K. Medlin. 2006. Evolution of the diatoms: insights from fossil, biological and molecular data. Phycologia 45: 361–402.

Singer, A., G. Poschmann, C. Mühlich, C. Valadez-Cano, S. Hänsch and V. Hüren et al. 2017. Massive protein import into the early evolutionary-stage photosynthetic organelle of the amoeba *Paulinella chromatophore*. Curr. Biol. 27: 2763–2773.

Singh, H., K. Kaur, M. Singh, G. Kaur and P. Singh. 2020. Plant cyclophilins: Multifaceted proteins with versatile roles. Front. Plant Sci. 11: 585212.

Singh, K. and C. Argáez. 2018. Cyclosporine for moderate to severe plaque psoriasis in adults: A review of clinical effectiveness and safety. Can. Agency Drugs Technol. Health, Ottawa (ON). PMID: 30329243.

Šírová, J., M. Sedlářová, J. Piterková, L. Luhová and M. Petřivalský. 2011. The role of nitric oxide in the germination of plant seeds and pollen. Plant Sci. 181: 560–572.

Six, C., J.-C. Thomas, L. Garczarek, M. Ostrowski, A. Dufresne, N. Blot et al. 2007. Diversity and evolution of phycobilisomes in marine *Synechococcus* spp.: a comparative genomics study. Genome Biol. 8: R259.

Skovgaard, A., I. Meneses and M.M. Angelico. 2009. Identifying the lethal fish egg parasite *Ichthyodinium chabelardi* as a member of Marine Alveolate Group I. Environ. Microbiol. 11(8): 2030–2041.

Skrobonja, A., Z. Gojkovic, A.L. Soerensen, P.-O. Westlund, C. Funk and E. Björn. 2019. Uptake kinetics of methylmercury in a freshwater alga exposed to methylmercury complexes with environmentally relevant thiols. Environ. Sci. Technol. 53(23): 13757–13766.

Slamovits, C.H. and P.J. Keeling. 2008. Widespread recycling of processed cDNAs in dinoflagellates. Curr. Biol. 18(13): R550–R552.

Slamovits, C.H., N. Okamoto, L. Burri, E.R. James and P.J. Keeling. 2011. A bacterial proteorhodopsin proton pump in marine eukaryotes. Nat. Commun. 2: 183.

Slatko, B.E., A.F. Gardner and F.M. Ausubel. 2018. Overview of next generation sequencing technologies. Curr. Protoc. Mol. Biol. 122(1): e59.

Slavov, C., V. Schrameyer, M. Reus, P.J. Ralph, R. Hill, C. Büchel et al. 2016. "Super-quenching" state protects *Symbiodinium* from thermal stress-implications for coral bleaching. Biochim. Biophys. Acta-Bioenergetics 1857(6): 840Björn847.

Smertenko, A., F. Assaad, F. Baluška, M. Bezanilla, H. Buschmann et al. 2017. Plant cytokinesis: terminology for structures and processes. Trend. Cell Biol. 27(12): 885Björn894.

Smith, F.A. and J.A. Raven. 1979. Intracellular pH and its regulation. Annu. Rev. Plant Physiol. 30: 289Björn311.

Smith, K.M., D.A. Goff, K.M. Fajer and K.M. Barkigia. 1982. Chirality and structures of bacteriochlorophylls *d*. J. Am. Chem. Soc. 104: 3747Björn3749.

Snuggs, J.W., S. Tessier, R.A.B. Bunning, I.M. Shapiro, M.V. Risbud and C.L. Le Maitre. 2021. TonEBP regulates the hyperosmotic expression of aquaporin 1 and 5 in the intervertebral disc. Sci. Rep. 11: 3164.

Sobotka, R. 2014. Making proteins green: biosynthesis of chlorophyll-binding proteins in cyanobacteria. Photosynth. Res. 119(1-2): 223–232.

Socha, A.L. and M.L. Guerinot. 2014. Mn-euvering manganese: the role of transporter gene family members in manganese uptake and mobilization in plants. Front. Plant Sci. 5: 106.

Solórzano, L. and J.D.H. Strickland. 1968. Polyphosphate in seawater. Limnol. Oceanogr. 13(3): 515Björn518.

Söllner, R., G. Glässer, G. Wanner, C.R. Somerville, G. Jürgens and F.F. Assaad. 2002. Cytokinesis-defective mutants of *Arabidopsis*. Plant Physiol. 129(2): 678Björn690.

Sommer, U. 1991. A comparison of the Droop and the Monod models of nutrient limited growth applied to natural populations of phytoplankton. Func. Ecol. 5(4): 535Björn544.

Song, B., D. Morse, Y. Song, Y. Fu, X. Lin, W. Wang et al. 2017. Comparative genomics reveals two major bouts of gene retroposition coinciding with crucial periods of *Symbiodinium* evolution. Genome Biol. Evol. 9(8): 2037–2047.

Soo, R., H. Hemp, D.H. Parks, W.W. Fischer and P. Hugenholtz. 2017. On the origins of oxygenic photosynthesis and aerobic respiration in cyanobacteria. Science 355(6332): 1436–1440.

Sournia, A., M.J. Chrdtiennot-Diner and M. Ricard. 1991. Marine phytoplankton: How many species in the world ocean? J. Plankton Res. 13: 1093–1099.

Soyer M.-O. and O.K. Haapala. 1974. Electron microscopy of RNA in dinoflagellate chromosomes. Histochemistry 42: 239–246.

Soyer, M.-O. And O.K. Haapala. 1974a. Structural changes of dinoflagellate chromosome by pronate and ribonuclease. Chromosoma 47: 179–192.

Spaepen, S. and J. Vanderleyden. 2011. Auxin and plant—microbe interactions. Cold Spring Harb. Perspect. Biol. 3: a001438.

Speijer, D., J. Lukes and M. Elias. 2015. Sex is a ubiquitous, ancient, and inherent attribute of eukaryotic life. Proc. Natl. Acad. Sci. USA. 112: 8827–8834.

Spielmeyer, A. and G. Pohnert. 2012. Influence of temperature and elevated carbon dioxide on the production of dimethylsulfoniopropionate and glycine betaine by marine phytoplankton. Mar. Environ. Res. 73: 62–69.

Spitz, F. and E.E.M. Furlong. 2012. Transcription factors: from enhancer binding to developmental control. Nat. Rev. Genet. 13: 613–626.

Spoerner, M., T. Wilchard, T. Bachhuber, J. Stratmann and W. Oertel. 2012. Growth and thallus morphogenesis of *Ulva mutabilis* (Chlorophyta) depends on a combination of two bacterial species excreting regulatory factors. J. Phycol. 48: 1433–1447.

Sprecher, B.N., H. Zhang and S. Lin. 2020. Nuclear gene transformation in a dinoflagellate. Microorganisms 18(1): 126.

Spudich, J.L. and R. Sager. 1980. Regulation of the *Chlamydomonas* cell cycle by light and dark. J. Cell Biol. 85: 136–145.

Stanley, G.D. and S.D. Cairns. 1988. Constructional azooxanthellate coral communities: an overview with implications for the fossil record. Palaios 3: 233–242.

Stat, M., E. Morris and R.D. Gates. 2008. Functional diversity in coral-dinoflagellate symbiosis. Proc. Natl. Acad. Sci. USA 105(27): 9256–9261.

Statello, L., C.-J. Guo, L.-L. Chen and M. Huarte. 2021. Gene regulation by long non-coding RNAs and its biological functions. Nat. Rev. Mol. Cell Biol. 22: 96–118.

Steinacher, M., F. Joos, T.L. Frölicher, G.-K. Plattner and S.C. Doney. 2009. Imminent ocean acidification in the Arctic projected with the NCAR global coupled carbon cycle-climated model. Biogeosciences 6: 515–533.

Stephens, T.G., R.A. Gonzalez-Pech, Y. Cheng, A.R. Mohamed, D.W. Burt, D. Bhattacharya et al. 2020. Genomes of the dinoflagellate *Polarella glacialis* encode tandemly repeated single-exon genes with adaptive functions. BMC Biol. 18: 56.

Stephens, T.G., M.A. Ragan, D. Bhattacharya and C.X. Chan. 2018. Core genes in diverse dinoflagellate lineages include a wealth of conserved dark genes with unknown functions. Sci. Rep. 8(1): 17175.

Stief, P., C. Schauberger, M.B. Lund, A. Greve, R.M.M. Abed, M.A. Al-Najjar et al. 2022. Intracellular nitrate storage by diatoms can be an important nitrogen pool in freshwater and marine ecosystems. Comm. Earth. Environ. 3: 154.

Stiller, J.W., J. Schreiber, Y. Yue, H. Guo, Q. Ding and J. Huang. 2014. The evolution of photosynthesis in chromist algae through serial endosymbiosis. Nat. Comm. 5: 5764.

Stoecker, D.K. 1999. Mixotrophy among dinoflagellates. J. Eukaryot. Microbiol. 46(4): 397–401.

Strassert, J.F.H., A. Karnkowska, E. Hehenberger, J. del Campo, M. Kolisko, N. Okamoto et al. 2018. Single cell genomics of uncultured marine alveolates shows paraphyly of basal dinoflagellates. ISME J. 12: 304–308.

Strzalka, W. and A. Ziemienowicz. 2011. Proliferating cell nuclear antigen (PCNA): a key factor in DNA replication and cell cycle regulation. Ann. Bot. 107(7): 1127–1140.

Stuart, K.D., A. Schnaufer, N.L. Ernst and A.K. Panigraphi. 2005. Complex management: RNA editing in trypanosomes. Trends Biochem. Sci. 30: 97–105.

Stuken, A., P. Riobo, J. Franco, K.S. Jakobsen, L. Guillou and R.I. Figueroa. 2015. Paralytic shellfish toxin content is related to genomic sxtA4 copy number in *Alexandrium minutum* strains. Front Microbiol. 6: 404.

Sturm, S., J. Engelken, A. Gruber, S. Vugrinec, P.G. Kroth, I. Adamska et al. 2013. A novel type of light-harvesting antenna protein red algal origin in algae with secondary plastids. BMC Evol. Biol. 13: 159.

Strychar, K.B., M. Coates and P.W. Sammarco. 2004. Loss of *Symbiodinium* from bleached Australian scleractinian corals (*Acropora hyacinthus*, *Favites complanata* and *Porites solida*). Mar. Fresh. Res. 55: 135–144.

Subrahmanyam, D.A. 1996. Heavy metals induce changes in chlorophyll pattern in tomato and *Chlorella pyrenoidosa*. Annu. Rev. Plant Physiol. 10(1): 17.

Suikkanen, S., H. Kaartokallio, S. Hällfors, M. Huttunen and M. Laamanen. 2010. Life cycle strategies of bloom-forming, filamentous cyanobacteria in the Baltic Sea. Deep. Sea Res. Part II Top. Stud. Oceanogr. 57: 199–209.

Sukenik, A., R.N. Kaplan-Levy, J.M. Welch and A.F. Post. 2012. Massive multiplication of genome and ribosomes in dormant cells (akinetes) of *Aphanizomenon ovalisporum* (Cyanobacteria). ISME J. 6(3): 670–679.

Sullivan, M.B., D. Lindell, J.A. Lee, L.E. Thompson, J.P. Bielawski and S.W. Chisholm. 2006. Prevalence and evolution of core photosystem II genes in marine cyanobacterial viruses and their hosts. PLoS Biol. 4(8): e234.

Sumiya, N., T. Fujiwara, A. Era and S.-Y. Miyagishima. 2016. Chloroplast division checkpoint in eukaryotic algae. Proc. Natl. Acad. Sci. USA 113(47): E7629–E7638.

Sumper, M. and N. Kröger. 2004. Silica formation in diatoms: the function of long-chain polyamines and silaffins. J. Mater. Chem. 14: 2059–2065.

Sun, X., Y. Zhong, Z. Huang and Y. Yang. 2014. Selenium accumulation in unicellular green alga *Chlorella vulgaris* and its effects on antioxidant enzymes and content of photosynthetic pigments. PLoS ONE 9(11): 2112270.

Sun, Y., T. Fukamachi, H. Saito and H. Kobayashi. 2012. Adenosine deamination increases the survival under acidic conditions in *Escherichia coli*. J. Appl. Microbiol. 112: 775–781.

Sun, Y., N. Zhang, J. Zhou, S. Dong, X. Zhang, L. Guo et al. 2020. Distribution, contents, and types of mycosporine-like amino acids (MAAs) in marine macroalgae and a database for MAAs based on these characteristics. Mar. Drugs 18(1): 43.

Sunda, W.G. and S.A. Huntsman. 1983. Effects of competitive interactions between manganese and copper on cellular manganese and growth in estuarine and oceanic species of the diatom *Thalassiosira*. Limnol. Oceanogr. 28(5): 924–934.

Sunda, W.G. and S.A. Huntsman. 2000. Effect of Zn, Mn, and Fe on Cd accumulation in phytoplankton: implications for oceanic Cd cycling. Limnol. Oceanogr. 45: 1501–1516.

Sunda, W.G. and S.A. Huntsman. 1998a. Control of Cd concentrations in a coastal diatom by interactions among free ionic Cd, Zn, and Mn in seawater. Environ. Sci. Technol. 32: 2961–2968.

Sunda, W.G. and S.A. Huntsman. 1998b. Interactions among Cu^{2+}, Zn^{2+}, and Mn^{2+} in controlling cellular Mn, Zn, and growth rate in the coastal alga *Chlamydomonas*. Limnol. Oceanogr. 43: 1055–1064.

Surek, B. and M. Melkonian. 1983. Intracellular bacteria in the Euglenophyceae: prolonged axenic culture of an alga-bacterial system. *In:* Schenk, H.E.A. and W. Schwemmler [eds.]. Endocytobiology 2: 475–486. De Gruyter, Berlin, Germany.

Sutak, R., J.-M. Camadro and E. Lesuisse. 2020. Iron uptake mechanisms in marine phytoplankton. Front. Microbiol. 11: 566691.

Suttle, C.A. 2007. Marine viruses—Major players in the global ecosystem. Nat. Rev. Microbiol. 5: 801–812.

Sutton, D.C. and O. Hoegh-Guldberg. 1990. Host-zooxanthella interactions in four temperate marine invertebrate symbioses: Assessment of effect of host extracts on symbionts. Biol. Bull. 178(2): 175–186.

Sverdrup, H.U. 1953. On conditions for the vernal blooming of phytoplankton. J. du Conseil International pour l'Exporation de la Mer 18: 287–295.

Swan, B.K., B. Tupper, A. Sczyrba, F.M. Lauro, M. Martinez-Garcia, J.M. Gonzalez et al. 2013. Prevalent genome streamlining and latitudinal divergence of planktonic bacteria in the surface ocean. Proc. Natl. Acad. Sci. USA 110: 11463–11468.

Swatek, K.N. and D. Komander. 2016. Ubiquitin modifications. Cell Res. 26(4): 399–422.

Swindell, W.R., M. Huebner and A.P. Weber. 2007. Transcriptional profiling of *Arabidopsis* heat shock proteins and transcription factors reveals extensive overlap between heat and non-heat stress response pathways. BMC Genomics 8: 125.

Tachibana, M., A.E. Allen, S. Kikutani, Y. Endo, C. Bowler and Y. Matsuda. 2011. Localization of putative carbonix anhydrases in two marine diatoms, *Phaeodactylum tricornutum* and *Thalassiosira pseudonana*. Photosynth. Res. 109(1-3): 205–221.

Taft, R.J., M. Pheasant and J.S. Mattick. 2007. The relationship between non-protein-coding DNA and eukaryotic complexity. BioEssays 29(3): 288–299.

Takai, N., M. Nakajima, T. Oyama, R. Kito, C. Sugita, M. Sugita et al. 2006. A KaiC-associating SasA-RpaA two-component regulatory system as a major circadian timing mediator in cyanobacteria. Proc. Natl. Acad. Sci. USA 103(32): 12109–12114.

Takahashi, K., G. Benico, W.M. Lum and M. Iwataki. 2019. *Gertia stigmatica* gen. et sp. nov. (Kareniaceae, Dinophyceae), a new marine unarmored dinoflagellate possessing the peridinin-type chloroplast with an eyespot. Protist 170(5): 125680.

Takahashi, F., D. Yamagata, M. Ishikawa, Y. Fukamatsu, Y. Ogura, M. Kasahara et al. 2007. AUREOCHROME, a photoreceptor required for photomorphogenesis in stramenopiles. Proc. Natl. Acad. Sci. USA 104(49): 19625–19630.

Takahashi, N., N. Ogita, T. Takahashi, S. Taniguchi, M. Tanaka, M. Seki et al. 2019. A regulatory module controlling stress-induced cell cycle arrest in *Arabidopsis*. eLife 8: e43944.

Takeuchi, T. and C. Benning. 2019. Nitrogen-dependent coordination of cell cycle, quiescence and TAG accumulation in Chlamydomonas. Biotechnol. Biofuels 12: 292.

Takishita, K., K. Koike, T. Maruyama and T. Ogata. 2002. Molecular evidence for plastid robbery (kleptoplastidy) in *Dinophysis*, a dinoflagellate causing diarrhetic shellfish poisoning. Protist 153: 293–302.

Tamaki, S., K. Mochida and K. Suzuki. 2021. Diverse biosynthetic pathways and protective functions against environmental stress of antioxidants in microalgae. Plants 10(6): 1250.

Tamogami, J. and T. Kikukawa. 2021. Functional mechanism of proton pump-type rhodopsins found in various microorganisms as a potential effective tool in optogenetics. *In*: Anwar, M., Z. Farooq, R. Rather, M. Tauseef and T. Heinbockel [eds.]. Epigenetics to Optogenetics-A new Paradigm in the Study of Biology. IntechOpen. Pp. 176. DOI: 10.5772/intechopen.97589.

Tamura, M., S. Shimada and T. Horiguchi. 2005. *Galeidinium rugatum* gen. et sp. nov (Dinophyceae), a new coccoid dinoflagellate with a diatom endosymbiont. J. Phycol. 41: 658–671.

Tan, I.H., J. Blomster, G. Hansen, E. Leskinen, C.A. Maggs, D.G. Mann et al. 1999. Molecular phylogenetic evidence for a reversible morphogenetic switch controlling the gross morphology of two common genera of green seaweeds, *Ulva* and *Enteromorpha*. Mol. Biol. Evol. 16(8): 1011–1018.

Tan, Y.-J., X.-Y. Zhang, G.-L. Ding, R. Li, L. Wang, L. Jin et al. 2015. Aquaporin 7 plays a crucial role in tolerance to hyperosmotic stress and in the survival of oocytes during cryopreservation. Sci. Rep. 5: 17741.

Tanaka, R., S. Kikutani, A. Mahardika and Y. Matsuda. 2014. Localization of enzymes relating to C4 organic acid metabolisms in the marine diatom, *Thalassiosira pseudonana*. Photosynth. Res. 121(2-3): 251–263.

Tanaka, T., Y. Maeda, A. Veluchamy, M. Tanaka, H. Abida, E. Marechal et al. 2015. Oil accumulation by the oleaginous diatom *Fistulifera solaris* as revealed by the genome and transcriptome. Plant Cell 27: 162–176.

Tang, E.P.Y. 1996. Why do dinoflagellates have lower growth rates. J. Phycol. 32: 80–84.

Tang, J., D. Yao, H. Zhou, M. Wang and M. Daroch. 2023. Distinct molecular patterns of two-component signal transduction systems in thermophilic cyanobacteria as revealed by genomic identification. Biology 12(2): 271.

Tang, J., Z. Hu, J. Zhang and M. Daroch. 2024. Genome-wide identification and comparative analysis of transcription factors in thermophilic cyanobacteria. BMC Genomics 25: 44.

Tang, K.W. 2003. Grazing and colony size development in *Phaeocystis globosa* (Prymnesiophyceae): the role of a chemical signal. J. Plankton Res. 25(7): 831–842.

Tang, X., J. Chen, W.-H. Wang, T.-W. Liu, J. Zhang and Y.-H. Gao. 2011. The changes of nitric oxide production during the growth of *Microcystis aerugrinosa*. Environ. Pollut. 159(12): 3784–3792.

Tang, Y., C. Xie, X. Jin, Z. Wang and R. Hu. 2022. Inhibitory effect of isolated bacteria from the phycosphere of *Levanderina fissa* on the growth of different microalgae. Front. Mar. Sci. 9: 908813.

Tang, Y.Z. and F.C. Dobbs. 2007. Green autofluorescence in dinoflagellates, diatoms, and other microalgae and its implications for vital staining and morphological studies. Appl. Environ. Microbiol. 73(7): 2306–2313.

Tang, Y.Z., F. Koch and C.J. Gobler. 2010. Most harmful algal bloom species are vitamin B1 and B12 auxotrophs. Proc. Natl. Acad. Sci. USA 107(48): 20756–20761.

Taschner, M. and E. Lorentzen. 2016. The intraflagellar transport machinery. Cold Spring Harbor Perspect Biol. 8: a028092.

Taylor, F.J.R. 1982. Symbioses in marine microplankton. Ann. Inst. Oceanogr. (Paris) 58: 61–90.

Taylor-Kearney, L.J., S. Madden, J. Wilson, W.K. Myers, D.M. Gunawardana, E. Pires et al. 2022. Plant cysteine oxidase oxygen-sensing function is conserved in early land plants and algae. ACS Bio. Med. Chem. Au. 2: 521–528.

Tengs, T., O.J. Dahlberg, K. Shalchian-Tabrizi, D. Klaveness, K. Rudi, C.F. Delwiche et al. 2000. Phylogenetic analyses indicate that the 19′hexanoyloxy-fucoxanthin-containing dinoflagellates have tertiary plastids of haptophyte origin. Mol. Biol. Evol. 17: 718–29.

Thamatrakoln, K. and A.B. Kustka. 2009. When to say when: can excessive drinking explain silicon uptake in diatoms? Bioessays 31: 322–327.

Thamatrakoln, K., B. Bailleul, C.M. Brown, M.Y. Gorbunov, A.B. Kustka, M. Frada et al. 2013. Death-specific protein in a marine diatom regulates photosynthetic responses to iron and light availability. Proc. Natl. Acad. Sci. USA 110(50): 20123–20128.

Thangaraj, S. and J. Sun. 2023. Ocean warming and acidification affect the transitional C:N:P ratio and macromolecular accumulation in the harmful raphidophyte *Heterosigma akashiwo*. Commun. Biol. 6: 151.

Thomas, M.K., C.T. Kremer, C.A. Klausmeier and E. Litchman. 2012. A global pattern of thermal adaptation in marine phytoplankton. Science 338(6110): 1085–1088.

Thompson, A.W., R.A. Foster, A. Krupke, B.J. Carter, N. Musat, D. Vaulot et al. 2012. Unicellular cyanobacterium symbiotic with a single-celled eukaryotic alga. Science 337(6101): 1546–1550.

Thompson, L.R., Q. Zeng, L. Kelly, K.H. Huang, A.U. Singer, J. Stubbe et al. 2011. Phage auxiliary metabolic genes and the redirection of cyanobacterial host carbon metabolism. Proc. Natl. Acad. Sci. USA 108(39): E757–764.

Tilbrook, K., A.B. Arongaus, M. Binkert, M. Heijde, R. Yin and R. Ulm. 2013. The UVR8 UV-B photoreceptor: perception, signaling and response. Arabidopsis Book 11: e0164.

Tillich, U.M., N. Wolter, P. Franke, U. Dühring and M. Frohme. 2014. Screening and genetic characterization of thermo-tolerant *Synechocystis* sp. PCC6803 strains created by adaptive evolution. BMC Biotechnol. 14: 66.

Ting, C.S., G. Rocap, J. King and S.W. Chisholm. 2002. Cyanobacterial photosynthesis in the oceans: The origins and significance of divergent light-harvesting strategies. Trends Microbiol. 10: 134–142.

Tischner, R., E. Planchet and W.M. Kaiser. 2004. Mitochondrial electron transport as a source for nitric oxide in the unicellular green alga *Chlorella sorokiniana*. FEBS Letters 576(1): 151–155.

Tkacz, A., M. Hortala and P.S. Poole. 2018. Absolute quantitation of microbiota abundance in environmental samples. Microbiome 6: 110.

Tomaru, Y., K. Toyoda and K. Kimura. 2015. Marine diatom viruses and their hosts: resistance mechanisms and population dynamics. Perspect. Phycol. 2(2): 69–81.

Tomo, T., T. Okubo, S. Akimoto, M. Yokono, H. Miyashita, T. Tsuchiya et al. 2007. Identification of the special pair of photosystem II in a chlorophyll *d*-dominated cyanobacterium. Proc. Natl. Acad. Sci. USA 104(17): 7283–7288.

Tortorelli, G., C. Rautengarten, A. Bacic, G. Segal, B. Ebert, S. Davy et al. 2022. Cell surface carbohydrates of symbiotic dinoflagellates and their role in the establishment of cnidarian-dinoflagellate symbiosis. ISME J. 16(1): 190–199.

Toseland, A., S.J. Daines, J.R. Clark, A. Kirkham, J. Strauss, C. Uhlig et al. 2013. The impact of temperature on marine phytoplankton resource allocation and metabolism. Nat. Clim. Change 3: 979–984.

Tourlousse, D.M., S. Yoshiike, A. Ohashi, S. Matsukura, M. Noda and Y. Sekiguchi. 2017. Synthetic spike-in standards for high-throughput 16S rRNA gene amplicon sequencing. Nuc. Acid. Res. 45(4): e23.

Trench, R. 1971. The physiology and biochemistry of zooxanthellae symbiotic with marine coelenterates I. The assimilation of photosynthetic products of zooxanthellae by two marine coelenterates. Proc.Roy. Soc. Lond. B. Biol. Sci. 177(1047): 225–235.

Trench, R., D. Wethey and J. Porter. 1981. Observations on the symbiosis with zooxanthellae among the *Tridacnidae* (Mollusca, Bivalvia). Biol. Bull. 161(1): 180–198.

Tripathy, B.C. and R. Oelmüller. 2012. Reactive oxygen species generation and signaling in plants. Plant Signal Behav. 7(12): 1621–1633.

Trottmann, F., K. Ishida, J. Franke, A. Stanišić, M. Ishida-Ito, H. Kries et al. 2020. Sulfonium acids loaded onto an unusual thiotemplate assembly line construct the cyclopropanol warhead of a *Burkholderia* virulence factor. Angew. Chemie - Int. Ed. 59: 13511–13515.

Tsai, C.-H., J. Warakanont, T. Takeuchi, B.B. Sears, E.R. Moellering and C. Benning. 2014. The protein Compromised Hydrolysis of Triacylglycerols 7 (CHT7) acts as a repressor of cellular quiescence in *Chlamydomonas*. Proc. Natl. Acad. Sci. USA 111(44): 15833–15838.

Tschitschko, B., M. Esti, M. Philippi, A.T. Kidane, S. Littmann, K. Kitzinger et al. 2024. Rhizobia-diatom symbiosis fixes missing nitrogen in the ocean. Nature 630: 899–904.

Tsuchiya, M., S. Miyawaki, K. Oguri, T. Toyofuku, A. Tame, K. Uematsu, et al. 2020. Acquisition, maintenance, and ecological roles of kleptoplastis in *Planoglabratella opercularis* (Foraminifera, Rhizaria). Front. Mar. Sci. 7: 585.

Tsujimoto, R., N. Kamiya and Y. Fujita. 2014. Transcriptional regulators ChlR and cnFR are essential for diazotrophic growth in nonheterocystous cyanobacteria. Proc. Natl. Acad. Sci. USA 111(18): 6762–6767.

Tujula, N.A., G.R. Crocetti, C. Burke, T. Thomas, C. Holmström and S. Kjelleberg. 2010. Variability and abundance of the epiphytic bacterial community associated with a green marine Ulvacean alga. ISME J. 4: 301–311.

Tun, N.N., C. Santa-Catarina, T. Begum, V. Silveira, W. Handro, E.I. Floh et al. 2006. Polyamines induce rapid biosynthesis of nitric oxide (NO) in *Arabidopsis thaliana* seedlings. Plant Cell Physiol. 47: 346–354.

Turkina, M.V., J. Kargul, A. Blanco-Rivero, A. Villarejo, J. Barber and A.V. Venter. 2006. Environmentally modulated phosphoproteome of photosynthetic membranes in the green alga *Chlamydomonas reinhardtii*. Mol. Cell. Proteomics 5(8): 1412–1425.

Turner, J. T. and P.A. Tester. 1997. Toxic marine phytoplankton, zooplankton grazers, and pelagic food webs. Limnol. Oceanogr. 42(5, part 2): 1203–1214.

Turnšek, J., J.K. Brunson, M.d.P.M. Viedma, T.J. Feerinck, A. Horák, M. Obornik et al. 2021. Proximity proteomics in a marine diatom reveals a putative cell surface-to-chloroplast iron trafficking pathway. eLife 10: e52770.

Twining, B.S. and S.B. Baines. 2013. The trace metal composition of marine phytoplankton. Ann. Re. Mar. Sci. 5: 191–215.

Tyrrell, T. 1999. The relative influences of nitrogen and phosphorus on oceanic primary production. Nature 400: 525–531.

Umen, J.G. 2014. Green algae and the origins of multicellularity in the plant kingdom. Cold Spring Harb. Perspect. Biol. 6(11): a016170.

Unal, D. and F.O. Cekic. 2019. Cold acclimation of SnRK2.2 kinases mutant *Chlamydomonas reinhardtii*. Phycol. Res. 67: 202–207.

Underhill, D.M. and A. Ozinsky. 2002. Toll-like receptors: key mediators of microbe detection. Curr. Opin. Immunol. 14(1): 103–110.

Ungerer, M.C., L.C. Johnson and M.A. Herman. 2007. Ecological genomics: understanding gene and genome function in the natural environment. Heredity 100: 178–183.

Upadhyaya, S., S. Agrawal, A. Gorakshakar and B.J. Rao. 2020. TOR kinase activity in *Chlamydomonas reinhardtii* is modulated by cellular metabolic states. FEBS Letters 594: 3122–3141.

Urica, E.I., D. Casero, H. Yamasaki, S.I. Hsieh, L.N. Adler, S.J. Karpowicz et al. 2012. Systems and trans-system level analysis identifies conserved iron deficiency responses in the plant lineage. Plant Cell 24(10): 3921–3948.

Uribe-Querol, E. and C. Rosales. 2020. Phagocytosis: our current understanding of a universal biological process. Front. Immunol. 11: 1066.

Urra, M., J. Buezo, B. Royo, A. Cornejo, P. López-Gómez, D. Cerdán et al. 2022. The importance of the urea cycle and its relationships to polyamine metabolism during ammonium stress in *Medicago truncatula*. J. Exp. Bot. 73(16): 5581–5595.

Ustick, L.J., A.A. Larkin, C.A. Garcia, N.S. Garcia, M.L. Brock, J.A. Lee et al. 2021. Metagenomic analysis reveals global-scale patterns of ocean nutrient limitation. Science 372(6539): 287–291.

Valenzuela, J.J., A.L.G. de Lomana, A. Lee, E.V. Armbrust, M.V. Orellana and N.S. Baliga. 2018. Ocean acidification conditions increase resiliences of marine diatoms. Nat. Commun. 9: 2328.

Valiadi, M. and D. Iglesias-Rodriguez. 2013. Understanding bioluminescence in dinoflagellates-How far have we come? Microorganisms 1(1): 3–25.

Vallon, O. 2005. Chlamydomonas immunophilins and parvulins: Survey and critical assessment of gene models. Eukaryot. Cell 4: 230–241.

van Creveld, S.G., S.N. Coesel, S. Blaskowski, R.D. Groussman, M.J. Schatz and V. Armbrust. 2023. Divergent functions of two clades of flavodoxin in diatoms mitigate oxidative stress and iron limitation. eLife 12: e84392.

van Dam, T.J.P., M.J. Townsend, M. Turk, A. Schlessinger, A. Sali, M.C. Field et al. 2013. Evolution of modular intraflagellar transpfort from a coatomer-like progenitor. Proc. Natl. Acad. Sci. 110(17): 6943–6948.

van Dam, B.R., M.A. Zeller, C. Lopes, A.R. Smyth, M.E. Bottcher, C.L. Osburn et al. 2021. Calcification-driven CO_2 emissions exceed "Blue Carbon" sequestration in a carbonate seagrass meadow. Sci. Adv. 7(51): eabj1372.

van de Heuvel, J., C. Ashiono, L.C. Gillet, K. Dörner, E. Wyler, I. Zemp et al. 2021. Processing of the ribosomal ubiquitin-like fusion protein FUBI-eS30/FAU is required for 40S maturation and depends on USP36. eLife 10: e70560.

Van Dolah, F.M., G.S. Kohli, J.S. Morey and S.A. Murray. 2017. Both modular and single-domain type I polyketide synthases are expressed in the brevetoxin-producing dinoflagellate, *Karenia brevis* (Dinophyceae). J. Phycol. 53(6): 1325–1339.

Van Dolah, F.M., T.A. Leighfield, H.D. Sandel and C.K. Hsu. 1995. Cell division in the dinoflagellate *Gambierdiscus toxicus* is phased to the diurnal cycle and accompanied by activation of the cell cycle regulatory protein, CDC2 kinase. J. Phycol. 31(3): 395-400.

Van Dolah, F.M., J.S. Morey, S. Milne, A. Ung, P.E. Anderson and M. Chinain. 2020. Transcriptomic analysis of polyketide synthases in a highly ciguatoxic dinoflagellate, *Gambierdiscus polynesiensis* and low toxicity *Gambierdiscus pacificus*, from French polynesia. PLoS ONE 15(4): e0231400.

van Etten, J. and D. Bhattacharya. 2020. Horizontal gene transfer in eukaryotes: not if, but how much? Trend. Genetc. 36(12): 915–925.

van Mooy, B.A.S., H.F. Fredricks, B.E. Pedler, S.T. Dyhrman, D.M. Karl, M. Kolízek et al. 2009. Phytoplankton in the ocean use non-phosphorus lipids in response to phosphorus scarcity. Nature 458: 69–72.

van Straalen, N.M. and D. Roelofs. 2012. An Introduction to Ecological Genomics. 2nd edition. Oxford University Press, New York, NY, USA. 363 pp.

Vardi, A. 2008. Cell signaling in marine diatoms. Commun. Integra. Biol. 1: 134–136.

Vardi, A., F. Formiggini, R. Casotti, A. De Martino, F. Ribalet, A. Miralto et al. 2006. A stress surveillance system based on calcium and nitric oxide in marine diatoms. PLoS Biol. 4(3): e60.

Vardi, A., B.A. Van Mooy, H.F. Fredricks, K.J. Popendorf, J.E. Ossolinski, L. Haramaty et al. 2009. Viral glycosphingolipids induce lytic infection and cell death in marine phytoplankton. Science 326: 861–865.

Vaulot, D., R.J. Olson and S.W. Chisholm. 1986. Light and dark control of the cell cycle in two marine phytoplankton species. Exp. Cell Res. 167(1): 38–52.

Vaulot, D. and S.W. Chisholm. 1987. Flow cytometric analysis of spermatogenesis in the diatom *Thalassiosira weissflogii* (Bacillariophyceae). J. Phycol. 23(1): 132–137.

Vega, J., G. Schneider, B.R. Moreira, C. Herrera, J. Bonomi-Barufi and F.L. Figueroa. 2021. Mycosporine-like amino acids from red macroalgae: UV-photoprotectors with potential cosmeceutical applications. Appl. Sci. 11(11): 5112.

Veglia, A.J., K.S. Bistolas, C.R. Voolstra, B.C. Hume, H.-J. Ruscheweyh, S. Planes et al. 2023. Endogenous viral elements reveal associations between a non-retroviral RNA virus and symbiotic dinoflagellate genomes. Commun. Biol. 6(1): 566.

Veldhuis, M.W., T.L. Cucci and M.E. Sieracki. 1997. Cellular DNA content of marine phytoplankton using two new fluorochromes: taxonomic and ecological implications. J. Phycol. 33: 527–541.

Verma, A., G.S. Kohli, D.T. Harwood, P.J. Ralph and S.A. Murray. 2019. Transcriptomic investigation into polyketide toxin synthesis in *Ostreopsis* (Dinophyceae) species. Environ Microbiol. 21(11): 4196–4211.

Vesteg, M. and J. Krajcovic. 2011. On the origin of meiosis and sex. Curr. Genet. 100: 147–161.

Vieira-Silva. S. and E.P. Rocha. 2010. The systemic imprint of growth and its uses in ecological (meta)genomics. PLOS Genet. 6: e1000808.

Vila, E., D. Hornero-Méndez, G. Azziz, G. Azziz, C. Lareo and V. Saravia. 2019. Carotenoids from heterotrophic bacteria isolated from Fildes Peninsula, King George Island, Antarctica. Biotechnol. Rep. 21: e00306.

Villanova, V., and C. Spetea. 2021. Mixotrophy in diatoms: Molecular mechanism and industrial potential. Physiol. Plant. 173(2): 603–611.

Villareal, T.A. 1992. Marine nitrogen-fixing diatom-cyanobacteria symbioses. pp. 163–175. *In*: Carpenter, E.J., D.G. Capone and J.G. Reuter [eds.]. Marine pelagic cyanobacteria: *Trichodesmium* and other diazotrophs. Kluwer, Dordecht, Netherlands.

Villareal, T.A., L. Joseph, M.A. Brzezinski, R.F. Shipe. F. Lipschultz and M.A. Altabet. 1999. Biological and chemical characteristics of the giant diatom Ethmodiscus (Bacillariophyceae) in the central North Pacific gyre. J. Phycol. 35(5): 896–902.

Vincent, F. and A. Vardi. 2023. Viral infection in the ocean-A journey across scales. PLoS Biol. 21(1): e3001966.

Vincent, F., M. Gralka, G. Schleyer, D. Schatz, M. Cabrera-Brufau, C. Kuhlisch et al. 2023. Viral infection switches the balance between bacterial and eukaryotic recyclers of organic matter during coccolithophore blooms. Nat. Commun. 14: 510.

Vincent, F., U. Sheyn, Z. Porat, D. Schatz and A. Vardi. 2021. Visualizing active viral infection reveals diverse cell fates in synchronized algal bloom demise. Proc. Natl. Acad. Sci. USA 118(11): e2021586118.

Vinga, S. and J. Almeida. 2003. Alignment-free sequence comparison-a review. Bioinformatics 19(4): 513–523.

Volkman, J.K., G. Eglinton, E.D.S. Corner and T.E.V. Forsberg. 1980. Long-chain alkenes and alkenones in the marine coccolithophorid *Emiliania huxleyi*. Photochemistry 19(12): 2619–2622.

Volkman, J.K., S.M. Barrett, S.I. Blackburn and E.L. Sikes. 1995. Alkenones in *Gephyrocapsa aceanica*: implications for studies of paleoclimate. Geochim. Cosmochim. Acta 59(3): 513–520.

Von Lintig, J. 2010. Colors with functions: elucidating the biochemical and molecular basis of carotenoid metabolism. Ann. Rev. Nutrition 30: 35–56.

Vogt, T. 2010. Phenylpropanoid biosynthesis. Mol. Plant 3: 2–20.

Vreede, J., M.A. van der Horst, K.J. Hellingwerf, W. Crielaard and D.M.F. van Aalten. 2003. PAS domains: common structure and common flexibility. J. Biol. Chem. 278(20): 18434–18439.

Wada, T., K. Shimono, T. Kikukawa, M. Hato, N. Shinya, S.Y. Kim, et al. 2011. Crystal structure of the eukaryotic light-driven proton-pumping rhodopsin, *Acetabularia* rhodopsin II, from marine alga. J. Mol. Biol. 411(5): 986–998.

Wägele, H., O. Deusch, K. Händeler, R. Martin, C. Schmitt, G. Christa et al. 2011. Transcriptomic evidence that longevity of acquired plastids in the photosynthetic slugs *Elysia timida* and *Plakobranchus ocellatus* does not entail lateral transfer of algal nuclear genes. Mol. Biol. Evol. 28(1): 699–706.

Wagner, D., X. Pochon, L. Irwin, R.J. Toonen and R.D. Gates. 2011. Azooxanthellate? Most Hawaiian black corals contain *Symbiodinium*. Proc. R. Soc. B 278: 1323–1328.

Wagstaff, B.A., E.S. Hems, M. Rejzek, J. Pratscher, W. Brook, S. Kuhaudomlarp et al. 2018a. Insights into toxic *Prymnesium parvum* blooms: the role of sugars and algal viruses. Biochem. Soc. Trans. 46(2): 413–421.

Wagstaff, B.A., M. Rejzek and R.A. Field. 2018. Identification of a Kdn biosynthesis pathway in the haptophyte *Prymnesium parvum* suggests widespread sialic acid biosynthesis among microalgae. J. Biol. Chem. 293(42): 16277–16290.

Waheed, S., M. Anwar, M.A. Saleem, J. Wu, M. Tayyab and Z. Hu. 2021. The critical role of small RNAs in regulating plant innate immunity. Biomolecules 11(2): 184.

Waller, R.F. and C.J. Jackson. 2009. Dinoflagellate mitochondrial genomes: stretching the rules of molecular biology. Bioessays 31(2): 237–245.

Wang, C., X. Lin, L. Li and S. Lin. 2016. Differential growth responses of marine phytoplankton to herbicide glyphosate. PLoS ONE 11(3): e0151633.

Wang, C., H. Zhang, J. Wang, B. Sprecher and S. Lin. 2024. Glyphosate (Roundup) as phosphorus nutrient enhances carbon and nitrogen accumulation and up-regulates phosphorus metabolism in the haptophyte *Isochrysis galbana*. Sci. Tot. Environ. 913: 169715.

Wang, H. and G. Oster. 1998. Energy transduction in the F1 motor of ATP synthase. Nature 396(6708): 279–282.

Wang, D., X. Yu, K. Xu, G. Bi, M. Cao, E. Zelzion et al. 2020. *Pyropia yezoensis* genome reveals diverse mechanisms of carbon acquisition in the intertidal environment. 11: 4028.

Wang, H., H. Kim and J.-S. Ki. 2021. Transcriptomic identification and expression analysis of cold shock domain protein (CSP) genes in the marine dinoflagellate *Prorocentrum minimum*. J. Appl. Phycol. 33: 843–854.

Wang, H., X. Niu, X. Feng, R.J. Gonçalves and W. Guan. 2019. Effects of ocean acidification and phosphate limitation on physiology and toxicity of the dinoflagellate *Karenia mikimotoi*. Harmful Algae 87: 101261.

Wang H. and J. Siemens. 2015. TRP ion channels in thermosensation, thermoregulation and metabolism. Temp. Multidiscip. Biomed. J. 2: 178–187.

Wang, J., L. Li and S. Lin. 2023. Active viral infection during blooms of a dinoflagellate indicates dinoflagellate-viral co-adaptation. Appl. Environ. Microbiol. 89(11): e0115623.

Wang, J., H. Li, L. Li, Y. Wang and S. Lin. 2024. Comparative genomics illuminates adaptive evolution of DVNP with lifestyle and with loss of histone H1 in dinoflagellates. bioRxiv DOI: 10.1101/2024.02.09.579734.

Wang, L., X. Jia, Y. Zhang, L. Xu, B. Menand, H. Zhao et al. 2021. Loss of two families of SPX domain-containing proteins required for vacuolar polyphosphate accumulation coincides with the transition to phosphate storage in green plants. Mol. Plant 14(5): 838–846.

Wang, S., L. Lin, Y. Shi, W. Qian, N. Lin, X. Yan et al. 2020. First draft genome assembly of the seaweed *Sargassum fusiforme*. Front. Genet. 11: 590065.

Wang, S., F. Zhao, X. Wei, B. Lu, D. Duan and G. Wang. 2013. Preliminary study on flg22-induced defense responses in female gametophytes of *Saccharina japonica* (Phaeophyta). J. Appl. Phycol. 25: 1215–1223.

Wang, S., L. Lin, Y. Shi, W. Qian, N. Li, X. Yan et al. 2020. First draft genome assembly of the seaweed *Sargassum fusiforme*. Front. Genet. 11: 590065.

Wang, S. and M. Wu. 2023. The draft genome of the "Golden Tide" seaweed, *Sargassum horneri*: characterization and comparative analysis. Genes 14: 1969.

Wang, W.-H., B. Köhler, F.-Q. Cao and L.-H. Liu. 2008. Molecular and physiological aspects of urea transport in higher plants. Plant Sci. 175(4): 467–477.

Wang, X. and L. Wang. 2016. GMATA: An integrated software package for genome-scale SSR mining, marker development and viewing. Front. Plant Sci. 7: 1350.

Wang, X., Z. Lu, A. Gomez, G.C. Hon, Y. Yue, D. Han et al. 2014. N^6-Methyladenosine-dependent regulation of messenger RNA stability. Nature 505: 117–120.

Wang, X., Z. Wu, Y. Wu, M. An, Z. Zhou and S. Lin. 2021. Differential affinities of a *Pocillopora damicornis* galectin to five genera of *Symbiodiniaceae* at different temperatures. Front. Mar. Sci. 8: 754808.

Wang, Y., J. Liu and J.E. Segall. 1998. MAP kinase function in amoeboid chemotaxis. J. Cell Sci. 111: 373–383.

Wang, Z., W. Ruan, J. Shi, L. Zhang, D. Xiang, C. Yang et al. 2014. Rice SPX1 and SPX2 inhibit phosphate starvation responses through interacting with PHR2 in a phosphate-dependent manner. Proc. Natl. Acad. Sci. USA 111(41): 14953–14958.

Ward, L.M. and P.M. Shih. 2021. Granick revisited: synthesizing evolutionary and ecological evidence for the late origin of bacteriochlorophyll via ghost lineages and horizontal gene transfer. PLoS ONE 16(1): e0239248.

Wasiel, A.A., H.J. Rozeboom, D. Hauke, B.J. Baas, E. Zandvoort, W.J. Quax et al. 2010. Structural and functional characterization of a macrophage migration inhibitory factor homologue from the marine cyanobacterium *Prochlorococcus marinus*. Biochemistry 49(35): 7572–7581.

Watanabe, Y. and M. Yamamoto. 1994. S. pomb mei2+ encodes an RNA-binding protein essential for premeiotic DNA synthesis and meiosis I, which cooperates with a novel RNA species meiRNA. Cell 78: 487–498.

Waters, E.R. and I. Rioflorido 2007. Evolutionary analysis of the small heat shock proteins in five complete algal genomes. J. Mol. Evol. 65(2): 162–174.

Waters, E.R. and E. Vierling. 2020. Plant small heat shock proteins-evolutionary and functional diversity. New Phytol. 227: 24–37.

Wawrzynska, A., G. Moniuszko and A. Sirko. 2015. Links between ethylene and sulfur nutrition-A regulatory interplay or just metabolite association? Front. Plant Sci. 6: 1053.

Weake, V.M. and J.L. Workman. 2010. Inducible gene expression: diverse regulatory mechanisms. Nat. Rev. Genet. 11: 426–437.

Weber, A.P., M. Linka and D. Bhattacharya. 2006. Single, ancient origin of a plastid metabolite translocator family in Plantae from an endomembrane-derived ancestor. Eukaryotic Cell 5: 609–612.

Wege, S., G.A. Khan, J.Y. Jung, E. Vogiatzaki, S. Pradervand, I. Aller et al. 2016. The EXS domain of PHO1 participates in the response of shoots to phosphate deficiency via a root-to-shoot signal. Plant Physiol. 170(1): 385–400.

Wei, N. and X.W. Deng. 1992. COP9: A new genetic locus involved in light-regulated development and gene expression in *Arabidopsis*. Plant Cell 4: 1507–1518.

Wei, N. and X.W. Deng. 2003. The COP9 signalosome. Annu. Rev. Cell Dev. Biol. 19: 261–286.

Weiberg, A., M. Wang, F.-M. Lin, H. Zhao, Z. Zhang, I. Kaloshian et al. 2013. Fungal small RNAs suppress plant immunity by hijacking host RNA interference pathways. Science 342(6154): 118–123.

Weider, L.J., J.J. Elser, T.J. Crease, M. Mateos, J.B. Cotner and T.A. Markow. 2005. The functional significance of ribosomal (r)DNA variation: impacts on the evolutionary ecology of organisms. Annu. Rev. Ecol. Evol. Systemat. 36: 219–242.

Weis, V.M., S.K. Davy, O. Hoegh-Gulderg, M. Rodriguez-Lanetty and J.R. Pringle. 2008. Cell biology in model systems as the key to understanding corals. Trends Ecol. Evol. 23: 369–376.

Weinberger, F. 2007. Pathogen-induced defense and innate immunity in macroalgae. Biol. Bull. 213(3): 290–302.

Weits, D.A., B. Giuntoli, M. Kosmacz, S. Parlanti, H.M. Hubberten, H. Riegler et al. 2014. Plant cysteine oxidases control the oxygen-dependent branch of the N-end-rule pathway. Nat. Commun. 5: 3425.

Wells, M.L., C.G. Trick, W.P. Cochlan, M.P. Hughes and V.L. Trainer. 2005. Domoic acid: the synergy of iron, copper, and the toxicity of diatoms. Limnol. Oceanogr. 50: 1908–1917.

Wendel, J.F. 2000. Genome evolution in polyploids. Plant Mol. Biol. 42: 225–249.

Werthén, M. and T. Lundgren. 2001 Intracellular Ca^{2+} mobilization and kinase activity during acylated homoserine lactone-dependent quorum sensing in *Serratia liquefaciens*. J. Biol. Chem. 276: 6468–6472.

West, A.H. and A.M. Stock. 2001. Histidine kinases and response regulator proteins in two-component signaling systems. Trends Biochem. Sci. 26: 369–376.

Westermann, M., C. Hoischen, L. Wohlbrand, R. Rabus and E. Rhiel. 2023. Light and prey influence the abundances of two rhodopsins in the dinoflagellate *Oxyrrhis marina*. Protoplasma 260: 529–544.

Wetherbee, R., T.T. Bringloe, J.F. Costa, A. van de Meene, R.A. Andersen and H. Verbruggen. 2021. New pelagophytes show a novel mode of algal colony development and reveal a perforated theca that may define the class. J. Phycol. 57(2): 396–411.

Wheeler, G.L., D. Sturm and G. Langer. 2023. *Gephyrocapsa huxleyi* (*Emiliania huxleyi*) as a model system for coccolithophore biology. J. Phycol. 59: 1123–1129.

Whelan, S. and N. Goldman. 2001. A general empirical model of protein evolution derived from multiple protein families using a maximum-likelihood approach. Mol. Biol. Evol. 18: 691–699.

White, M.D., J. Kamps, S. East, L.J. Taylor-Kearney and E. Flashman. 2018. The plant cysteine oxidases from *Arabidopsis thaliana* are kinetically tailored to act as oxygen sensors. J. Biol. Chem. 293(30): 11786–11795.

Wichard, T. 2015. Exploring bacteria-induced growth and morphogenesis in the green macroalga order Ulvales (Chlorophyta). Front. Plant Sci. 6: 86.

Wiener, D. and S. Schwartz. 2021. The epitranscriptome beyond m^6A. Nat. Rev. Genet. 22(2): 119–131.

Wilson, W.H., D.C. Schroeder, M.J. Allen, M.T.G. Holden, J. Parkhill, B.G. Barrell et al. 2005. Complete genome sequence and lytic phase transcription profile of a Coccolithovirus. Science 309: 1090–1092.

Wisecaver, J.E. and J.D. Hackett. 2010. Transcriptome analysis reveals nuclear-encoded proteins for the maintenance of temporary plastids in the dinoflagellate *Dinophysis acuminata*. BMC Genomics 11: 366.

Wisecaver, J.E., R.P. Auber, A.L. Pendleton, N.F. Watervoort, T.R. Fallon, O.L. Riedling et al. 2023. Extreme genome diversity and cryptic speciation in a harmful algal-bloom-forming eukaryote. Curr. Biol. 33: 2246–2259.

Wilkinson, K.A. and J.M. Henley. 2010. Mechanisms, regulation and consequences of protein SUMOylation. Biochem. J. 428(2): 133–145.

Wilks, J.C. and J.L. Slonczewski. 2007. pH of the cytoplasm and periplasm of *Escherichia coli*: Rapid measurement by green fluorescent protein fluorimetry. J. Bacteriol. 189: 5601–5607.

Williams, A., E.N. Chiles, D. Conetta, J.S. Pathmanathan, P.A. Cleves, H.M. Putnam et al. 2021. Metabolomic shifts associated with heat stress in coral holobionts. Sci. Adv. 7(1): eabd4210.

Wisecaver, J.H. and J.D. Hackett. 2010. Transcriptome analysis reveals nuclear-encoded proteins for the maintenance of temporary plastids in the dinoflagellate *Dinophysis acuminata*. BMC Genomics 11: 366.

Wisecaver, J.H. and J.D. Hackett. 2011. Dinoflagellate genome evolution. Ann. Rev. Microbiol. 65: 369–387.

Wisecaver, J.H., M.L. Brosnahan and J.D. Hackett. 2013. Horizontal gene transfer is a significant driver of gene innovation in dinoflagellates. Genome Biol. Evol. 5: 2368–2381.

Wohlrab, S., M.H. Iversen and U. John. 2010. A molecular and co-evolutionary context for grazer induced toxin production in *Alexandrium tamarense*. PLoS ONE 5(11): e15039.

Wolf, Y.I. and E.V. Koonin. 2013. Genome reduction as the dominant mode of evolution. Bioessays 35(9): 829–837.

Wolfe, K.H. and D.C. Shields. 1997. Molecular evidence for an ancient duplication of the entire yeast genome. Nature 387: 708.

Wong, J.T.Y. 2019. Architectural organization of dinoflagellate liquid crystalline chromosomes. Microorganisms 7(2): 27.

Wong, J.L. and M.A. Johnson. 2010. Is HAP2-GCS1 an ancestral gamete fusogen? Trends Cell Biol. 20(3): 134–141.

Wong, F.C.Y. and J.C. Meeks. 2002. Establishment of a functional symbiosis between the cyanobacterium *Nostoc punctiforme* and the brophyte *Anthoceros punctatus* requires genes involved in nitrogen control and initiation of heterocyst differentiation. Microbiology 148: 315–323.

Wood-Charlson, E.M., L.L. Hollingsworth, D.A. Krupp and V.M. Weis. 2006. Lectin-glycan interactions play a role in recognition in a coral/dinoflagellate symbiosis. Cell. Microbiol. 8(12): 1985–1993.

Wooldridge, S.A. 2020. Excess seawater nutrients, enlarged algal symbiont densities and bleaching sensitive reef locations: 1. Identifying thresholds of concern of the Great Barrier Reef, Australia. Mar. Pollut. Bull. 152: 107667.

Wooldridge, S.A. 2013. Breakdown of the coral-algae symbiosis: towards formalising a linkage between warm-water bleaching thresholds and the growth rate of the intracellular zooxanthellae. Biogeoscience 10: 1647–1658.

Wong, F.C.Y. and J.C. Meeks. 2002. Establishment of a functional symbiosis between the cyanobacterium *Nostoc punctiforme* and the bryophyte *Anthoceros punctatus* requires genes involved in nitrogen control and initiation of heterocyst differentiation. Microbiology (Reading). 148(Pt 1): 315–323.

Wong, J.L. and M.A. Johnson. 2010. Is HAP2-GCS1 an ancestral gamete fusogen? Trends Cell Biol. 20: 134–141.

Wong, J.T.Y., D.C. New, J.C.W. Wong and V.K.L. Hung. 2003. Histone-like proteins of the dinoflagellate *Crythecodinium cohnii* have homologies to bacterial DNA-binding proteins. Euk. Cell 2(3): 646–650.

Worden, A.Z., J.-H. Lee, T. Mock, P., Rouzé, M.P. Simmons, A.L. Aerts et al. 2009. Green evolution and dynamic adaptations revealed by genomes of the marine picoeukaryotes Micromonas. Science 324(5924): 268–272.

Worden, A.Z., J. Janouskovec, D. McRose, A. Engman, R.M. Welsh and S. Malfatti. 2012. Global distribution of a wild alga revealed by targeted metagenomics. Current Biol. 22(7): R675–R677.

Wu, G. and E.P. Spalding. 2007. Separate functions for nuclear and cytoplasmic cryptochrome 1 during photomorphogenesis of *Arabidopsis* seedlings. Proc. Natl. Acad. Sci. USA 104(47): 18813–18818.

Wu, J.-T., H.-C. Lin, Y.-C. Hu and C.-T. Chien. 2005. Neddylation and deneddylation regulate Cul1 and Cul3 protein accumulation. Nat. Cell Biol. 7(10): 1014–1020.

Wu, X., D.W. Lee, R.A. Mella and J.W. Golden. 2007. The *Anabaena* sp. strain PCC7120 asr1734 gene encodes a negative regulator of heterocyst development. Mol. Microbiol. 64(3): 782–794.

Wu, T.M., Y.T. Hsu, M.S. Sung, Y.T. Hsu and T.M. Lee. 2009. Expression of genes involved in redox homeostasis and antioxidant defense in a marine macroalga *Ulva fasciata* by excess copper. Aquat. Toxicol. 94(4): 275–285.

Wu, Y., P. Zhang, Z. Liang, Y. Yuan, M. Duan, Y. Liu et al. 2022. Genome-wide identification and analysis of the aureochrome gene family in *Saccharina japoinica* and a comparative analysis with six other algae. Plants (Basel) 11(16): 2088.

Wu, Z., X. Yang, S. Lin, W.H. Lee and P.K.S. Lam. 2022. A Rhizobium bacterium and its population dynamics under different culture conditions of its associated toxic dinoflagellate *Gambierdiscus balechii*. Mar. Life Sci. Technol. 3: 2570–2580.

Xiang, T., R.E. Jinkerson, S. Clowez, C. Tran, C.J. Krediet, M. Onishi et al. 2018. Glucose-induced trophic shift in an endosymbiont dinoflagellate with physiological and molecular consequences. Plant Physiol. 176: 1793–1807.

Xiang, T., E. Lehnert, R.E. Jinkerson, S. Clowez, R.G. Kim, J.C. DeNofrio et al. 2020. Symbiont population control by host-symbiont metabolic interaction in Symbiodiniaceae-cnidarian associations. Nat. Commun. 11: 108.

Xiao, Y., L. Gao and Z. Li. 2023. Unique high-temperature tolerance mechanisms of zoochlorellae *Symbiochlorum hainanensis* derived from scleractinian coral *Porites lutea*. mBio DOI: 10.1128/mbio.02780-23.

Xie, B., S. Bishop, D. Stessman, D. Wright, M.H. Spalding and L.J. Halverson. 2013. *Chlamydomonas reinhardtii* thermal tolerance enhancement mediated by a mutualistic interaction with vitamin B_{12}-producing bacteria. ISME J. 7: 1544–1555.

Xing, C., J. Li, H. Yuan and J. Yang. 2022. Physiological and transcription level responses of microalgal *Auxenochlorella protothecoides* to cold and heat induced oxidative stress. Environ. Res. 211: 113023.

Xiong, J., W.M. Fischer, K. Inoue, M. Nakahara and C.E. Bauer. 2000. Molecular evidence for the early evolution of photosynthesis. Science 289: 1724–1730.

Xue, S., Y. Zhang, J. Chen, S. Shang, L. Gao and X. Tang. 2022. Ultraviolet-B radiation stress triggers reactive oxygen species and regulates the antioxidant defense and photosynthesis systems of intertidal red algae *Neoporphyra haitanensis*. Front. Mar. Sci. 9: 1043462.

Yamaguchi, A., N. Yubuki and B.S. Leander. 2012. Morphostasis in a novel eukaryote illuminates the evolutionary transition from phagotrophy to phototrophy: description of *Rapaza viridis* n. gen. et n. sp. (Euglenozoa, Euglenida). BMC Evol. Biol. 12: 29.

Yamamoto, M. 1996. The molecular control mechanisms of meiosis in fission yeast. Trends Biol. Sci. 21(1): 18–22.

Yamamoto, H., K. Uesaka, Y. Tsuzuki, H. Yamakawa, S. Itoh and Y. Fujita. 2022. Comparative genomic analysis of the marine cyanobacterium *Acaryochloris marina* MBIC10699 reveals the impact of phycobiliprotein reacquisition and the diversity of *Acaryochloris* plasmids. Microorganisms 10(7): 1374.

Yamano, H., C. Tsurumi, J. Gannon and T. Hunt. 1998. The role of the destruction box and its neighbouring lysine residues in cyclin B for anaphase ubiquitin-dependent proteolysis in fission yeast: defining the D-box receptor. EMBO J. 17 (19): 5670–5678.

Yang, J., Y. Pan, C. Bowler, L. Zhang and H. Hu. 2016. Knockdown of phosphoenolpyruvate carboxykinase increases carbon flux to lipid synthesis in *Phaeodactylum tricornutum*. Algal Res. 15: 50–58.

Yang, A., W.I. Suh, N.K. Kang, B. Lee and Y.K. Chang. 2018. MAPK/ERK and JNK pathways regulate lipid synthesis and cell growth of *Chlamydomonas reinhardtii* under osmotic stress, respectively. Sci. Rep. 8: 13857.

Yang, W., L. Zhou, J. Wang, L. Wang, S. Gao and G. Wang. 2022. Knockout of a diatom cryptochrome by CRISPR/Cas9 causes an increase in light-harvesting protein levels and accumulation of fucoxanthin. Algal Res. 66: 102822.

Yang, Y., Y. Li, Q. Chen, Y. Sun and Z. Lu. 2019. WGDdetector: a pipeline for detecting whole genome duplication events using the genome or transcriptome annotations. BMC Bioinformatics 20: 75.

Yasuda, A., D. Inami and M. Hanaoka. 2020. RpaB, an essential response regulator for high-light stress, is extensively involved in transcriptional regulation under light-intensity upshift conditions in *Synechococcus elongatus* PCC7942. J. Gen. Appl. Microbiol. 66(2): 73–79.

Yau, S., C. Hemon, E. Derelle, H. Moreau, G. Piganeau and N. Grimsley, 2016. A viral immunity chromosome in the marine picoeukaryote, *Ostreococcus tauri*. PLoS Pathog. 12(10): e1005965.

Ye, N., X. Zhang, M. Miao, X. Fan, Y. Zheng, D. Xu et al. 2015. *Saccharina* genomes provide novel insight into kelp biology. Nat. Commun 6: 6986.

Yee, D.P., T.J. Samo, R.M. Abbriano, B. Shimasaki, M. Vernet, X. Mayali et al. 2023. The V-type ATPase enhances photosynthesis in marine phytoplankton and further links phagocytosis to symbiogenesis. Curr. Biol. 33(12): 2541–2547.

Yin, K., P.-Y. Qian, J.C. Chen, D.P.H. Hsieh and P.J. Harrison. 2000. Dynamics of nutrients and phytoplankton biomass in the Pearl River estuary and adjacent water of Hong Kong during summer: preliminary evidence for phosphorus and silicon limitation. Mar. Ecol. Prog. Ser. 194: 295–305.

Yin, L., W. Li, T.V. Madsen, S.C. Maberly and G. Bowes. 2017. Photosynthesis inorganic carbon acquisition in 30 freshwater macrophytes. Aquat. Bot. 140: 48–54.

Yoch, D.C. 2002. Dimethylsulfoniopropionate: its sources, role in the marine food web, and biological degradation to dimethylsulfide. Appl. Environ. Microbiol. 68(12): 5804–5815.

Yong, S.C., P. Roversi, J. Lillington, F. Rodriguez, M. Krehenbrink, O.B. Zeldin et al. 2014. A complex iron-calcium cofactor catalyzing phosphotransfer chemistry. Science 345(6201): 1170–1173.

Yoon, H.S. and J.W. Golden. 1998. Heterocyst pattern formation controlled by a diffusible peptide. Science 282: 935–938.

Yoon, H.S., J.D. Hackett, C. Ciniglia, G. Pinto and D. Bhattacharya. 2004. A molecular timeline for the origin of photosynthetic eukaryotes. Mol. Biol. Evol. 21: 809–818.

Yoon, H.S., A. Reyes-Prieto, M. Melkonian and D. Bhattacharya. 2006. Minimal plastid genome evolution in the *Paulinella* endosymbiont. Curr. Biol. 16: R670–R672.

Yordanova, Z.P., E.T. Iakimova, S.M. Cristescu, F.J.M. Harren, V.M. Kapchina-Toteva and E.J. Woltering. 2010. Involvement of ethylene and nitric oxide in cell death in mastoparan-treated unicellular alga *Chlamydomonas reinhardtii.* Cell Biol. International 34(3): 301–308.

Yoshiie, T., M. Maeda, M. Kimura, Y. Hama, M. Uchida and Y. Kimura. 2012. Structural features of N-glycans of seaweed glycoproteins: predominant occurrence of high-mannose type N-glycans in marine plants. Biosci. Biotechnol. Biochem. 76: 1996–1998.

Yoshimura, K. 1998. Mechanosensitive channels in the cell body of *Chlamydomonas*. J. Membr. Biol. 166: 149–155.

Yoshizawa, S., T. Azuma, K. Kojima, K. Inomura, M. Hasegawa, Y. Nishimura et al. 2023. Light-driven proton pumps as a potential regulator for carbon fixation in marine diatoms. Microbes Environ. 38(2): ME23015.

You, X., X. Zhang, J. Cheng, Y. Xiao, J. Ma, S. Sun et al. 2023. *In situ* structure of the red algal phycobilisome-PSII-PSI-LHC mega complex. Nature 616: 199–206.

You, Y., X. Sun, M. Ma, J. He, L. Li, F. Porto et al. 2022. Trypsin is a coordinate regulator of N and P nutrients in marine phytoplankton. Nat. Commun. 13: 4022.

Young, K.H. 1998. Yeast two-hybrid: so many interactions, (in) so little time. Biol. Reprod. 58(2): 302–311.

Young, J.N., J.A.L. Goldman, S.A. Kranz, P.D. Tortell and F.M.M. Morel. 2015. Slow carboxylation of Rubisco constrains the rate of carbon fixation during Antarctic phytoplankton blooms. New Phytol. 205(1): 172–181.

Yu, G., k. Nakajima, A. Gruber, C. Rio Bartulos, A.F. Schober, B. Lepetit et al. 2022. Mitochondrial phosphoenolpyruvate carboxylase contributes to carbon fixation in the diatom *Phaeodactylum tricornutum* at low inorganic carbon concentrations. New Phytol. 235(4): 1379–1393.

Yu, L., T. Li, H. Li, M. Ma, L. Li and S. Lin. 2023. In situ molecular ecological analyses illuminate distinct factors regulating formation and demise of a harmful dinoflagellate bloom. Microbiol. Spectr. 11(3): e05157-22.

Yu, L., Y. Zhang, M. Li, C. Wang, X. Li, L. Li et al. 2020. Comparative metatranscriptomic profiling and microRNA sequencing to reveal active metabolic pathways associated with a dinoflagellate bloom. Sci. Tot. Environ. 699: 134323.

Yudkovsky, Y., M. Shteinberg, T. Listovsky, M. Brandeis and A. Hershko. 2000. Phosphorylation of Cdc20/Fizzy negatively regulates the mammalian cyclosome/APC in the mitotic checkpoint. Biochem. Biophys. Res. Commun. 271: 299–304.

Yue, C., Z. Chai, Z. Hu, L. Shang, Y. Deng and Y.Z. Tang. 2022. Deficiency of nitrogen but not phosphorus triggers the life cycle transition of the dinoflagellate *Scrippsiella acuminata* from vegetative growth to resting cyst formation. Harmful Algae 118: 102312.

Yvon-Durocher, gz., C.-E. Schaum and M. Trimmer. 2017. The temperature dependence of phytoplankton stoichiometry: investigating the roles of species sorting and local adaptation. Front. Microbiol. 8: 2003.

Zaccaria, S. and B.J. Raphael. 2020. Accurate quantification of copy-number aberrations and whole-genome duplications in multi-sample tumor sequencing data. Nat. Commun. 11: 4301.

Zaccara, S., R.J. Ries and S.R. Jaffrey. 2019. Reading, writing and erasing mRNA methylation. Nat. Rev. Mol. Cell Biol. 20: 608–624.

Zachar, I. and G. Boza. 2022. The evolution of microbial facilitation: sociogenesis, symbiogenesis, and transition in individuality. Front. Ecol. Evol. 10: 798045.

Zachleder, V., I. Ivanov, M. Vitová and K. Bišová. 2019. Cell cycle arrest by supraoptimal temperature in the alga *Chlamydomonas reinhardtii*. Cells 8(10): 1237.

Zachleder, V., K. Bišová and M. Vitová. 2016. The cycle of microalgae. pp. 3–46. *In*: Borowitzka, M.A., J. Beardall and J.A. Raven [eds.]. The Physiology of Microalgae. Springer, New York.

Zaheri, B. and D. Morse. 2022. An overview of transcription in dinoflagellates. Gene. 829: 146505.

Záhonová, K., Z. Füssy, E. Birčák, A.M.G. Novák Vaclova, V. Klimeš, M. Vesteg et al. 2018. Peculiar features of the plastids of the colorless alga *Euglena longa* and the photosynthetic euglenophytes unveiled by transcriptome analyses. Sci. Rep. 8: 17012.

Zakryś, B., R. Milanowski and A. Karnkowska. 2017. Evolutionary origin of *Euglena*. Adv. Exp. Med. Biol. 979: 3–17.

Zalutskaya, Z., U.S. Skryabina and E.V. Ermilova. 2019. Generation of hydrogen peroxide and transcriptional regulation of antioxidant enzyme expression in *Chlmydomonas reinhardtii* under hypothermia. Russ. J. Plant Physiol. 66: 223–230.

Zan, J., A. Li, M.D. Tianero, J. Davis, R.T. Hill and M.S. Donia. 2019. A microbial factory for defensive kahalalides in a tripartite marine symbiosis. Science 364: eaaw6732.

Zauner, S., M. Fraunholz, J. Wastl, S. Penny, M. Beaton, T. Cavalier-Smith et al. 2000. Chloroplast protein and centrosomal genes, a tRNA intron, and odd telomeres in an unusually compact eukarotic genome, the cryptomonad nucleomorph. Proc. Natl. Acad. Sci, USA 97: 200–205.

Zavala, J.A., A.G. Patankar, K. Gase and I.T. Baldwin. 2004. Constitutive and inducible trypsin proteinase inhibitor production incurs large fitness costs in *Nicotiana attenuate*. Proc. Natl. Acad. Sci. USA 101(6): 1607–1612.

Zehr, J.P., J.B. Waterbury, P.J. Turner, J.P. Montoya, E. Omoregie, G.F. Steward et al. 2001. Unicellular cyanobacteria fix N_2 in the subtropical North Pacific Ocean Nature 412(6847): 635–638.

Zehr, J.P., S.R. Bench, B.J. Carter, I. Hewson, F. Niazi, T. Shi et al. 2008. Globally distributed uncultivated oceanic N_2-fixing cyanobacteria lack oxygenic photosystem. Science 322(5904): 1110–1112.

Zehr, J.P., I. Shilova, H.M. Farnelid, M.d.C. Muñoz-Marín and K.A. Turk-Kubo. 2017. Unusual marine unicellular symbiosis with the nitrogen-fixing cyanobacterium UCYN-A. Nat. Microbiol. 2: 16214.

Zhang, B., X. Xie, X. Liu, L. He, Y. Sun and G. Wang. 2020. The carbonate concentration mechanism of *Pyropia yezoensis* (Rhodophyta): evidence from transcriptomics and biochemical data. BMC Plant Biol. 20: 424.

Zhang, C., H. Luo, L. Huang and S. Lin. 2017. Molecular mechanism of glucose-6-phosphate utilization in the dinoflagellate *Karenia mikimotoi*. Harmful Algae 67: 74–84.

Zhang, C. and S. Lin. 2019. Initial evidence of functional siRNA machinery in dinoflagellates. Harmful Algae 81: 53–58.

Zhang, C., S. Lin, L. Huang, W. Lu, M Li and S. Liu. 2014. Suppression substraction hybridization analysis revealed regulation of some cell cycle and toxin genes in *Alexandrium cetenella* by phosphate limitation. Harmful Algae 39: 26–39.

Zhang, C.C., S. Laurent, S. Sakr, L. Peng and S. Bédu. 2006. Heterocyst differentiation and pattern formation in cyanobacteria: a chorus of signals. Mol. Microbiol. 59(2): 367–375.

Zhang, F., Y. Fan, D. Zhang, S. Chen, X. Bai, X. Ma et al. 2020. Effect and mechanism of the algicidal bacterium Sulfitobacter porphyrae ZFX1 on the mitigation of harmful algal blooms caused by *Prorocentrum donghaiense*. Environ. Poll. 263: 114475.

Zhang, F., W. Zuo, S. Wang, W. Tang, Y.-W. Luo, S.A. Kranz et al. 2022. Phosphate limitation intensifies negative effects of ocean acidification on globally important nitrogen fixing cyanobacterium. Nat. Commun. 13: 6730.

Zhang, F., W. Gu, M.E. Hurles and J.R. Lupski. 2009. Copy number variation in human health, disease, and evolution. Ann. Rev. Genomics Hum. Genet. 10: 451–481.

Zhang, H. and S. Lin. 2003. Complex gene structure of the form II Rubisco in the dinoflagellate *Prorocentrum minimum* (Dinophyceae). J. Phycol. 39: 1160–1171.

Zhang, H., Y. Hou, L. Miranda, D.A. Campbell, N.R. Sturm, T. Gaasterland et al. 2007. Spliced leader RNA *trans*-splicing in dinoflagellates. Proc. Natl. Acad. Sci. USA 104: 4618–4623.

Zhang, H., B. Gu, Y. Zhou, X. Ma, T. Liu, H. Xu et al. 2022. Multi-omics profiling reveals resource allocation and acclimation strategies to temperature changes in a marine dinoflagellate. Appl. Environ. Microbiol. 88(17): e0121322.

Zhang, H., X. Wang, M. Qu, H. Yu, J. Yin, X. Liu et al. 2024. Genome of *Halimeda opuntia* reveals differentiation of subgenomes and molecular bases of multinucleation and calcification in algae. Proc. Natl. Acad. Sci. USA 121(39): e2403222121.

Zhang, H.Y., G. Lei, H.W. Zhou, C. He, J.L. Liao and Y.J. Huang. 2017. Quantitative ITRAQ-based proteomic analysis of rice grains to assess high night temperature stress. Proteomics 17: 1600365.

Zhang, H., Y. Zhou, T.-Q. Liu, X.-J. Yin, L. Lin, Q. Lin et al. 2021. Initiation of efficient C4 pathway in response to low ambient CO_2 during the bloom period of a marine dinoflagellate. Environ. Microbiol. 23(6): 3196–3211.

Zhang, J. 2003. Evolution by gene duplication: an update. Trends Ecol. Evol. 18(6): 292–298.

Zhang, K., J. Li, J. Chen and S. Lin 2023. Alkaline phosphatase PhoD mutation induces fatty acid and long chain polyunsaturated fatty acid (LCPUFA)-bound phospholipid production in the diatom *Phaeodactylum tricornutum*. Mar. Drugs 21(11): 560.

Zhang, K., J. Li, J. Wang, X. Lin, L. Li, Y. You et al. 2022. Functional differentiation and complementation of alkaline phosphatases and choreography of DOP scavenging in a marine diatom. Mol. Ecol. 31: 3389–3399.

Zhang, K., J. Li, Z. Zhou, R. Huang and S. Lin. 2021. Roles of alkaline phosphatase PhoA in algal metabolic regulation under phosphorus-replete conditions. J. Phycol. 57(3): 703–707.

Zhang, K., A.C. Martiny, N.B. Reppas, K.W. Barry, J. Malek, S.W. Chisholm et al. 2006. Sequencing genomes from single cells by polymerase cloning. Nat. Biotechnol. 24(6): 680–686.

Zhang, L., D. Hou, X. Chen, D. Li, L. Zhu, Y. Zhang et al. 2012. Exogenous plant MIR168a specifically targets mammalian LSLRAP1: evidence of cross-kingdom regulation by microRNA. Cell Res. 22: 107–126.

Zhang, L., W. Yu, T. He, J. Yu, R.E. Caffrey, E.A. Dalmasso et al. 2002. Contribution of human alpha-defensin 1, 2, and 3 to the anti-HIV activity of CD8 antiviral factor. Science 298: 995–1000.

Zhang, M., L. Wang and D. Zhong. 2016. Photolyase: dynamics and mechanisms of repoair of sun-induced DNA damage. Photochem. Photobiol. 93(1): 78–92.

Zhang, Q., E.M. Bendif, Y. Zhou, B. Nevado, R. Shafiee and R.E.M. Rickaby. 2022. Declining metal availability in the Mesozoic seawater reflected in phytoplankton succession. Nat. Geosci. 15: 932–941.

Zhang, R., E.C.M. Nowack, D.C. Price, D. Bhattacharya and A.R. Grossman. 2017. Impact of light intensity and quality on chromatophore and nuclear gene expression in *Paulinella chromatophora*, an amoeba with nascent photosynthetic organelles. Plant J. 90: 221–234.

Zhang, S.-F., K. Zhang, H.-M. Cheng, L. Lin and D.-Z. Wang. 2020. Comparative transcriptomics reveals colony formation mechanism of a harmful algal bloom species *Phaeocystis globosa*. Sci. Tot. Environ. 719: 137454.

Zhang X., W. Han, X. Fan, Y. Wang, D. Xu, K. Sun et al. 2023. Gene duplication and functional divergence of new genes contributed to the polar acclimation of Antarctic green algae. Mar. Life Sci. Technol. 5: 511–524.

Zhang, X., S. Lin and D. Liu. 2020. Transcriptomic and physiological responses of *Skeletonema costatum* to ATP utilization. Environ. Microbiol. 22(5): 1861–1869.

Zhang, X., S. Cheng, Z. Gao, Y. Cui, Q. Yao, J. Qin et al. 2023. Transcriptomic and physiological analyses unveil the distinct mechanisms of ATP and glucose-6-phosphate utilization in *Phaeodactylum tricornutum*. Front. Mar. Sci. 10: 1163189.

Zhang, Y., X. Lin, X. Shi, L. Lin, H. Luo, L. Li et al. 2019. Metatranscriptomic signatures associated with phytoplankton regime shift from diatom dominance to a dinoflagellate bloom. Front. Microbiol. 10: 590.

Zhang, Y., L. Yin, H. Jiang, W. Li, B. Gontero and S.C. Maberly. 2014. Biochemical and biophysical CO_2 concentrating mechanisms in two species of freshwater macrophyte within the genus *Ottelia* (Hydrocharitaceae). Photosynth. Res. 121: 285–297.

Zhang, Y., A. Pfeiffer, J.M. Tepperman, J. Dalton-Roesler, P. Leivar, E.G. Grandio et al. 2020. Central clock components modulate plant shade avoidance by directly repressing transcriptional activation activity of PIF proteins. Proc. Natl. Acad. Sci. USA 117(6): 3261–3269.

Zhang, Y.-J., S.-F. Zhang, Z.-P. He, L. Lin and D.-Z. Wang. 2015. Proteomic analysis provides new insights into the adaptive response of a dinoflagellate *Prorocentrum donghaiense* to changing ambient nitrogen. Plant. Cell Environ. 38(10): 2128–2142.

Zhang, Z., B.R. Green and T. Cavalier-Smith. 1999. Single gene circles in dinoflagellate chloroplast genomes. Nature 400: 155–159.

Zhang, Z., D. Li, R. Xie, R. Guo, S. Nair, H. Han et al. 2023. Plastoquinone synthesis inhibition by tetrabromo biphenyldiol as a widespread algicidal mechanism of marine bacteria. ISME J. 17(11): 1979–1992.

Zhang, Z., C. Qu, R. Yao, Y. Nie, C. Xu, C. Xu et al. 2019. Genome Biol. Evol. 11(7): 1897–1908.

Zhao, J., L. Zhu, C. Fan, Y. Wu and S. Xiang. 2018. Structure and function of urea amidolyase. Biosci. Rep. 38(1): BSR20171617.

Zhao, L.-S., N. Wang, K. Li, C.-Y. Li, J.-P. Guo, F.-Y. He et al. 2024. Architecture of symbiotic dinoflagellate photosystem I-light-harvesting supercomplex in *Symbiodinium*. Nat. Commun. 15: 2392.

Zhao, Z., C. Dent, H. Liang, J. Lv, G. Shang and Y. Liu. 2022. CRY2 interacts with CIS1 to regulate thermosensory flowering via FLM alternative splicing. Nat. Commun. 13(1): 7045.

Zheng, F., H. Liu, M. Jiang, Z. Xu, Z. Wang, C. Wang et al. 2018. The complete mitochondrial genome of the *Caulerpa lentillifera* (Ulvophyceae, Chlorophyta): sequence, genome content, organization structure and phylogenetic consideration. Gene 673: 225–238.

Zheng, S., G. Wang and S. Lin. 2012. Heat shock effects and population survival in the polar dinoflagellate Polarella glacialis. J. Exp. Mar. Biol. Ecol. 438: 100–108.

Zheng, S., J. Guo, F. Cheng, Z. Gao, L. Du, C. Meng et al. 2022. Cytochrome P450s in algae: bioactive natural product biosynthesis and light-driven bioproduction. Acta Pharm. Sin. B. 12(6): 2832–2844.

Zheng, Y., D.F. Harris, Z. Yu, Y. Fu, S. Poudel, R.N. Ledbetter et al. 2018. A pathway for biological methane production using bacterial iron-only nitrogenase. Nat. Microbiol. 3: 281–286.

Zhou, L., S. Gao, W. Yang, S. Wu, L. Huan and X. Xie. 2022. Transcriptomic and metabolic signatures of diatom plasticity to light fluctuations. Plant Physiol. 190(4): 2295–2314.

Zhou, W., G. Liang, P.L. Molloy and P.S. Jones. 2020. DNA methylation enables transposable element-driven genome expansion. Proc. Natl. Acad. Sci. USA 117(32): 19359–19366.

Zhu, Y., X. Liao, T. Han, J.-Y. Chen, C. He and Z. Lu. 2022. Symbiodiniaceae microRNAs and their targeting sites in coral holobionts: A transcriptomics-based exploration. Genomics 114(4): 110404.

Zhu, Z.-M., F.-C. Huo and D.-S. Pei. 2020. Function and Evolution of RNA N^6-Methyladenosine Modification. Int. J. Biol. Sci. 16: 1929–1940.

Zhu, Z., F. Qian, R. Yang, J. Chen, Q. Luo, H. Chen and X. Yan. 2015. A lipoxygenase from red alga *Pyropia haitanensis*, a unique enzyme catalyzing the free radical reactions of polyunsaturated fatty acids with triple ethylenic bonds. PLoS ONE 10: 23.

Zhuang, Y., H. Zhang and S. Lin. 2013. Cyclin B gene and its cell cycle-dependent differential expression in the toxic dinoflagellate *Alexandrium fundyense* Atama Group I/Clade I. Harmful Algae 26: 71–79.

Zhuang, Y., H. Zhang, L. Hannick and S. Lin 2015. Metatranscriptome profiling reveals versatile N-nutrient utilization, CO_2 limitation, oxidative stress, and active toxin production in an *Alexandrium fundyense* bloom. Harmful Algae 42: 60–70.

Ziegler, K., A. Diener, C. Herpin, R. Richter, R. Deutzmann and W. Lockau. 1998. Molecular characterization of cyanophycin synthetase, the enzyme catalyzing the biosynthesis of the cyanobacterial reserve material multi-L-arginyl-poly-L-aspartate (cyanophycin). Eur. J. Biochem. 254 (1): 154–159.

Zielezinski, A., S. Vinga, J. Almeida and W.M. Karlowski. 2017. Alignment-free sequence comparison: benefits, applications, and tools. Genome Biol. 18: 186.

Zirbel, M.J., F. Veron and M.I. Latz. 2000. The reversible effect of flow on the morphology of *Ceratocorys horrida* (Peridiniales, Dinophyta). J. Phycol. 36: 46–58.

Ziv, C., S. Malitsky, A. Othman, S. Ben-Dor, Y. Wei, S. Zheng et al. 2016. Viral serine palmitoyltransferase induces metabolic switch in sphingolipid biosynthesis and is required for infection of a marine alga. Proc. Natl. Acad. Sci. USA 113: E1907–E1916.

Zones, J.M., I.K. Blaby, S.S. Merchant and J.G. Umen. 2015. High-resolution profiling of a synchronized diurnal transcriptome from *Chlamydomonas reinhardtii* reveals continuous cell and metabolic differentiation. Plant Cell. 27(10): 2743–2769.

Zuckerkandl, E. and L.B. Pauling 1962. Molecular disease, evolution, and genetic heterogeneity. pp. 189–225. *In*: Kasha, M. and B. Pullman [eds.]. Horizons in Biochemistry. Academic Press, New York, USA.

Zurbriggen, M.D., V.B. Tognetti and N. Carrillo. 2007. Stress-inducible flavodoxin from photosynthetic microorganisms: The mystery of flavodoxin loss from the plant genome. IUBMB Life 59: 355–360.

Index

A

abiotic stress 86, 176, 191, 220, 316
Adaptation 1, 7, 18, 46, 56, 62, 64, 65, 72, 80, 81, 84, 86, 94, 102–105, 107, 109, 114, 126, 142, 143, 149, 165, 169, 175, 184, 192, 205, 206, 216, 217, 242, 247, 248, 250, 251–253, 267, 270, 297, 313, 315, 358, 363, 373, 382, 384
adaptive evolution 80, 143, 176, 178, 255, 312, 313, 317
akinete development 69, 72, 73, 179, 329
Algae 1, 2, 4, 6, 7, 9, 11–13, 21, 22, 27, 28, 30,–32, 34–40, 43–48, 50–52, 56–58, 60, 67, 75–82, 84–90, 92–97, 99, 103, 104, 106, 108–110, 112, 138, 142, 144, 145, 147, 151, 152, 157–160, 162, 164, 166, 167, 169, 172, 173, 175–184, 186–200, 203, 205–215, 218, 220–229, 231–242, 244, 248–251, 255–261, 263, 265, 269–275, 278–288, 292, 293, 295–299, 301, 303, 304, 306–308, 310, 312–318, 320, 321, 323, 324, 326–330, 332, 334–347, 349–356, 358, 360, 362–367, 369–371, 373, 374, 376–382, 383, 384
Algae-Microbes interactions 377
alkaline phosphatase 71, 72, 94, 96, 134, 179, 265, 266, 270, 286, 295, 300, 301, 305–307
alkenoates 159
alkenones 159, 160
allelopathic interactions 241

B

Bacteria 4, 12, 27, 35, 36, 40, 46, 47, 50, 51, 56, 57, 59, 60, 70, 72, 84, 86, 101, 102, 124, 127, 128, 143, 149, 151, 159, 167, 169, 171, 175, 176, 183, 184, 189, 194, 196, 203, 206, 213, 226, 228, 240, 257–259, 267–273, 276–280, 282, 283, 287, 292, 294, 296, 297, 305, 306, 308, 317, 321–324, 343, 355, 365, 366, 369, 371, 375, 377, 380–383
blue-green alga 40, 51, 60, 259, 287, 296, 297, 303, 307

C

C4 pathway 38, 102, 106, 107, 263
calcification 82, 89, 150, 153, 155, 261, 266
Carbon Concentrating Mechanisms (CCMs) 37, 263
Cell division cycle 57, 118, 192, 332, 343, 354
charophytes 74, 107, 179, 215, 227, 303, 378
Chemical defense 315, 317, 318, 326
chlorophyll-based 273, 275
Chlorophyta 2, 28, 39, 45, 50, 75, 87–89, 172, 174, 180, 215, 223, 226, 228, 256, 292, 313, 320, 334, 344
Chlorophytes 28, 33, 36–38, 44, 48, 74, 75, 79, 80, 82, 87, 88, 95, 172, 190–192, 196, 205, 209, 215, 224, 227, 231, 232, 234, 235, 238, 251, 254, 275, 278, 292, 310, 317, 318, 331, 341, 350–352, 354, 362, 372, 378
Co-adaptation 384
Coccolithphore 151, 152, 154, 157, 240, 257, 259, 265, 301, 331, 353
cold tolerance 81
Concepts 1
coral bleaching 373–375
Cryptophyte 31–33, 37, 38, 44, 50, 67, 110, 147, 151, 162, 163, 171, 188, 190, 196, 215, 227, 238, 278, 288, 305, 350, 352, 363, 371, 376
Cyanobacteria 4, 27–31, 34–37, 40, 50–52, 56, 60–73, 89, 92, 114, 144, 167, 170, 171, 176, 187, 189, 190, 192, 194, 196, 203–206, 223, 224, 227, 228, 231, 233, 235, 237–239, 248, 251–253, 255, 260, 263, 272, 273, 275, 278, 283, 286, 293, 301, 303, 313, 315–317, 320, 325, 329, 336, 344, 354–360, 362, 365, 371, 378, 380, 383

D

Diatoms 9, 22, 30–33, 35, 37–39, 43–46, 50, 52, 65, 66, 84, 95, 98–107, 109, 110, 112–115, 127, 139, 151–153, 159, 164, 166, 167, 172, 178, 187, 188, 190–192, 194–196, 199–201, 206, 209, 213, 220–224, 226–228, 231–233, 235, 237, 238, 248, 249, 251–253, 257–259, 263, 267, 274, 278–280, 282, 286–289, 293, 298, 299, 301–303, 306, 310, 314–317, 320, 321, 325, 328, 329, 331, 333, 336, 341, 344, 350, 352, 355–358, 362, 363, 369, 371, 375, 376, 378, 380, 382
Dinoflagellates 3, 6–8, 12, 13, 17, 19, 22, 30–34, 37, 38, 43–52, 55, 56, 58, 65, 67, 95, 110, 113, 116–118, 120, 122–124, 126–130, 132, 134, 135, 137–149, 151, 153, 159, 162, 169, 188, 190, 192–196, 199, 202, 203, 205, 208, 209, 213, 214, 219, 220, 222–224, 227, 228, 230–233, 235, 237–239, 248, 251–253, 255, 257–259, 263, 265, 270, 271, 274, 277–280, 286, 288, 289, 293–295, 298, 299, 301–303, 305, 306, 310, 314–317, 320, 321, 324, 327–329, 331, 334, 336, 338, 341, 343, 346, 347, 349, 351, 353, 356, 360, 362, 364, 366, 369, 370, 373, 375, 376, 378, 379, 382–384
DMS 151–153, 159, 160, 172, 173, 175, 239, 257–260, 272, 315, 322
DMSP 151–153, 159, 160, 167, 172, 173, 175, 257–259, 322

E

ecological genomics 1, 2, 11
Elevated CO_2 261, 263, 265, 266, 269, 271, 297
energy niche 280
Environmental factors 2, 179, 298, 331
environmental stimuli 7, 9, 147, 188, 189, 203, 207
environmental stress 70, 101, 102, 147, 216, 219, 252, 312, 315, 329, 336, 351, 352, 373–375
Eulgenophyte 6, 7, 31, 32, 34–36, 38, 39, 162, 163, 169, 365

F

feedback 151, 152, 205, 219, 233, 257, 259, 260, 266, 310, 314
Fungi 4, 12, 40, 80, 84, 107, 145, 186, 196, 209, 259, 287, 292, 323, 350, 377, 379, 380, 382

G

gene copy number variation (CNV) 58
Gene family expansion and contraction 49
gene regulation 4, 7, 12, 26, 137, 149, 200, 217, 257, 269
Gene transfer 19, 27, 31, 32, 37, 48, 50, 51, 60, 64, 65, 84, 86, 92, 102, 103, 105, 109, 126, 129, 136, 153, 164, 169, 170, 171, 173, 176, 229, 240, 270, 365, 366, 376, 382, 383
Genome evolution 18, 46, 48, 122
genome size variation 46, 47, 49, 51, 64, 124, 126, 177
genome structure 2, 5, 8, 48, 87, 102
genomic mechanism 255, 267
global warming 206, 242, 244, 245, 248, 259–261, 267
grazing 108, 150, 155, 184, 266, 267, 310–318, 320–322, 326, 328
green algae 22, 28, 30, 31, 34, 35, 38, 40, 51, 60, 74, 75, 77–82, 85–89, 110, 167, 169, 173, 179, 187, 205, 206, 209, 225, 227–229, 231, 240, 259, 270, 278, 287, 293, 296, 302, 303, 307, 350, 352, 360, 378, 380

H

Haptophytes 28, 31–33, 37, 38, 44, 65, 66, 110, 138, 140, 150–152, 154, 155, 157–160, 188, 196, 205, 232, 233, 238, 257–259, 263, 265, 275, 278, 288, 301, 303, 310, 320, 350, 352, 356, 358, 359, 363, 371, 376
homeostasis 101, 107, 112, 149, 155, 166, 211, 217, 219, 222, 239, 247, 253, 261, 263, 265, 268, 269, 296–298, 307, 342
Horizontal gene transfer 19, 31, 37, 48, 50, 51, 60, 64, 65, 84, 86, 92, 102, 103, 105, 109, 126, 129, 136, 171, 173, 229, 240, 270, 365, 366, 376, 383

I

iron stress 112, 114, 302, 303

L

Life cycle 72, 81, 90, 144, 148, 150, 157, 158, 175, 177, 178, 182, 196, 329–336, 341, 353, 354, 378
light 8, 9, 22, 28, 34–36, 43, 51, 56, 61, 62, 64–68, 72, 73, 78, 80, 82, 85, 87, 89, 92, 99, 101, 102, 104, 107, 109, 114, 116, 117, 127, 135, 137, 139, 144, 149, 151, 153, 155, 158, 164–169, 171–173, 176, 177, 179, 184–186, 188, 190, 194–197, 199, 200, 203–206, 213, 216, 220–222, 225, 231–235, 237, 240, 243, 245, 247, 250, 253, 255, 259, 267, 270, 273, 275, 276, 279, 280, 288, 298, 303, 314, 323, 331, 335, 353, 354, 359, 360, 362, 364, 369, 376

M

mechanosensitive ion channel 214
metabolites 34, 40, 94, 114, 145, 154, 220, 239, 240, 272, 315, 316, 365, 374, 375, 377, 379, 380, 382
microbial attack 321
molecular mechanism 138, 204, 297, 352, 353, 372
mortality 164, 242
multicellularity 88, 172, 173, 175, 177, 179–184
multinucleate 4, 48, 75, 77, 81, 82, 88, 179, 180, 183
mutualism 142, 143, 247, 355, 356, 358, 363, 364, 367, 369, 373–375, 379, 380

N

nitrogen fixation 64, 69–72, 194, 248, 265, 301
nutrient 2, 3, 7, 12, 43, 56–59, 62, 64, 65, 69, 70, 72, 73, 75, 80, 84, 94–99, 101, 102, 105, 107, 109, 112, 127, 134, 135, 138, 142, 144, 145, 149, 155, 160, 165–169, 177, 179, 184, 186, 194, 208, 211, 217, 241–244, 248, 265, 267, 269, 272, 276, 280–289, 292, 293, 295, 296, 298, 300, 305–308, 310–312, 315, 332, 333, 335, 336, 347, 352–354, 362, 363, 369–371, 373–375, 381–383
nutrient limitation 101, 134, 265, 292, 333, 335, 352, 353
Nutrient uptake 56, 84, 96, 99, 102, 109, 135, 142, 167, 168, 186, 208, 217, 242, 272, 282–284, 287–289, 298, 373, 383

O

ocean acidification (OA) 37, 261
ornithine-urea cycle (OUC) 99, 105, 286, 288

P

Parasitoid 377, 378
Paulinella 30, 162, 163, 170, 171
Pelagophyte 31, 44, 163–167, 188, 200, 350, 352, 354
Phaeophyta 2, 167, 172, 174, 180, 183, 192, 223, 256, 292, 320, 334, 344
Photoenergy harvesting and conversion 2, 43, 66, 200, 201, 273, 275, 276, 278–280
photosynthetic apparatus 36, 37, 165, 224, 254, 263, 265, 275
plasticity 1, 11, 46, 89, 101, 103, 165, 216, 217, 239, 242, 246–248, 259, 263
Prasinophyte 33, 36, 74, 77, 79, 82, 83, 85–87, 139, 143, 147, 163, 165, 169, 190–192, 195, 205, 209, 220, 227, 230, 232, 238, 257, 282, 308, 325, 333, 347, 350, 352, 373, 383, 384

prasinophytes 36, 74, 79, 82, 83, 85–87, 139, 163, 190–192, 205, 209, 220, 227, 230, 232, 238, 257, 282, 308, 350, 352
Primary endosymbiosis 28–30, 34, 89, 162, 170, 175, 300, 355, 371
proton pump rhodopsin 2, 99, 107, 127, 153, 200–202, 265, 273, 278–280, 383

R

Raphidophyte 31, 163, 167, 200, 205, 224, 235, 238, 265, 350, 353
red algae 4, 30, 35, 89, 92–94, 175–177, 184, 186, 187, 235, 321, 355, 369
Redfield ratio 282
regulator 8, 62, 69, 70, 82, 92, 104, 175, 189–191, 195, 205, 240, 250, 297, 337, 352–354, 371
Regulatory molecular mechanisms 332
(Retro)Transposition 124
Rhodophyta 2, 39, 45, 50, 89, 93, 97, 172, 174, 177, 180, 183, 226, 231, 292, 317, 320, 344, 362

S

Secondary endosymbiosis 7, 28, 30–34, 109, 274, 299, 303, 355, 371
Sensing 73, 102, 147, 166, 188, 189, 193–195, 199, 203, 206–208, 213, 237, 249, 250, 314, 327
Serial secondary endosymbiosis 7, 30, 32, 33
Sexual reproduction 90, 92, 148–150, 182, 194, 315, 329–331, 334, 336, 338, 341, 362
signal transduction pathway 8, 188, 326, 379
single nucleotide polymorphism 92, 384
Stoichiometry 67, 96, 245, 282, 310
stress response 1, 56, 86, 101, 104, 107, 134, 142, 147, 148, 169, 186, 192, 207, 216–219, 234, 240, 241, 247, 250, 253, 255, 268, 299, 321, 373–375, 379
symbiosis 56, 57, 62, 124, 135, 162, 247, 274, 355, 357, 358, 362–375, 377, 384

T

techniques 1, 12, 15, 104, 110
temperature 7, 26, 51, 57–59, 65, 73, 81, 99, 107, 109, 122, 149, 155, 159, 168, 179, 184, 185, 196, 197, 199, 206, 207, 219, 241–255, 257, 267, 296, 298, 299, 311, 315, 321, 366, 374, 375, 378
Temperature variability 149, 253
Tertiary replacement 7, 33, 355
transcription factor 9–11, 25, 81, 86, 87, 101, 134, 135, 168, 173, 175, 179, 194, 196, 199, 205–209, 213, 216, 217, 219, 233, 234, 240, 249, 250, 252, 253, 255, 257, 333, 341, 342, 344, 352, 368
transporter 37, 44, 56, 64, 65, 71, 72, 78, 84, 93, 94, 96, 101, 107, 109–112, 135, 144, 155, 166, 168, 186, 187, 203, 208, 209, 217, 234, 240, 263, 270, 283, 284, 286–289, 292–299, 301, 302, 304–306, 363, 365, 369–373, 383

V

Viruses 6, 23, 46, 47, 50, 51, 54, 80, 110, 132, 155, 158, 184, 321, 323–326, 377, 382–384